Compendium of thermophysical property measurement methods

Compendium of Thermophysical Property Measurement Methods

Volume 1

Survey of Measurement Techniques

Compendium of Thermophysical Property Measurement Methods

Volume 1

Survey of Measurement Techniques

Edited by

K. D. Maglić
Boris Kidrič Institute of Nuclear Sciences
Belgrade, Yugoslavia

A. Cezairliyan
National Bureau of Standards
Washington, D.C., USA

and

V. E. Peletsky
Institute for High Temperatures
Moscow, USSR

Plenum Press • New York and London

Library of Congress Cataloging in Publication Data

Main entry under title:

Compendium of thermophysical property measurement methods.

Includes bibliographical references and index.
Contents: v. 1. Survey of measurement techniques.
1. Solids—Thermal properties—Measurement—Handbooks, manuals, etc. I. Maglić, Kosta D. II. Cezairliyan, A. III. Peletsky, V. E.
QC176.8.T4C66 1984 530.4′1 84-3268
ISBN 0-306-41424-4

A Division of Plenum Publishing Corporation
233 Spring Street, New York, N.Y. 10013

Printed in the United States of America

Contributors

JEAN FRANÇOIS BAUMARD, Centre de Recherches sur la Physique des Hautes Temperatures CNRS, Orleans, France

ARED CEZAIRLIYAN, Thermophysics Division, National Bureau of Standards, Washington, D.C.

VITALIY YA. CHEKHOVSKOI, Institute of High Temperatures, USSR Academy of Sciences, Moscow, USSR

W.R. DAVIS, British Ceramic Research Association Ltd., Penkhull, Stoke-on-Trent, England

R. DE CONINCK, Materials Science Department, SCK/CEN, Mol, Belgium

DAVID A. DITMARS, National Bureau of Standards, Washington, D.C.

D.N. KAGAN, Institute of High Temperatures, USSR Academy of Sciences, Moscow, USSR

A.S. KASPEROVICH, Department of Physics, Belorussian State University, Minsk, USSR

R.K. KIRBY, National Bureau of Standards, Washington, D.C.

S. KLARSFELD, Isover Saint-Gobain, Centre de Recherches Industrielles de Rantigny, Rantigny, France

YA. A. KRAFTMAKHER, Institute of Inorganic Chemistry, USSR Academy of Sciences, Novosibirsk, USSR

M.J. LAUBITZ, Division of Physics, National Research Council of Canada, Ottawa, Ontario

K.D. MAGLIĆ, Boris Kidrič Institute of Nuclear Sciences, Institute for Thermal Engineering and Energy Research, Belgrade, Yugoslavia

J.P. MOORE, Metals and Ceramics Division, Oak Ridge National Laboratory, Oak Ridge, Tennessee

V.E. PELETSKY, Institute of High Temperatures, USSR Academy of Sciences, Moscow, USSR

L.P. PHYLIPPOV, Moscow State University, Moscow, USSR

M.J. RICHARDSON, National Physical Laboratory, Teddington, England

J.C. RICHMOND, National Bureau of Standards, Washington, D.C.

G. RUFFINO, Microtecnica, Turin, Italy

R.E. TAYLOR, Thermophysical Properties Research Laboratory and CINDAS, Purdue University, West Lafayette, Indiana

G.M. VOLOKHOV, Department of Physics, Belorussian State University, Minsk, USSR

Preface

The need for reliable data on thermophysical and thermal optical properties of solid materials grows continually and increasingly. Existing property data, except for selected pure elements and for some simple alloys and compounds, are often not reliable, so in many cases the need for correct and acceptably accurate information can only be met through the measurement of a given property. The measurement—that is, the selection of the measurement method, building or purchase of the apparatus, and the measurement procedure itself—carries many hidden hazards because methods and their variants are numerous and not appropriate for all materials and temperature ranges, and have many subtle sources of systematic errors, known only to those who have thoroughly studied them.

The need for a concise yet complete reference work describing thermophysical and thermal optical property measurement techniques, and ultimately, reliable and detailed directions for property measurement discussed at the Sixth European Thermophysical Properties Conference in Dubrovnik, Yugoslavia in 1978, led its International Organizing Committee to launch an international cooperative project with these objectives. This reference work, the *Compendium of Thermophysical Property Measurement Methods*, is the result of the first phase of work on this program. It is a summary of the state-of-the-art methods for the measurement of thermal and electrical conductivity, thermal diffusivity, specific heat, thermal expansion, and thermal radiative properties of solid materials, from room temperature to very high temperatures.

The primary objective of this volume has been to systematically record the critical knowledge accumulated, particularly during the last three decades of rapid development of thermophysical measurement methods. This volume is also the basis for the second phase of work, which will appear subsequently in Volume 2: the description of standard recommended practices for the determination of thermophysical and thermal optical properties of selected groups of materials.

Together, this two-volume reference work will serve as a guide in selecting the best technique to be employed in measuring the required property of a substance. Several of the more common measurement techniques are described in sufficient detail that an experimental apparatus could be designed and built. Thus, the reference work is sufficiently complete that it should not be necessary, in most cases, to consult the primary literature, which could be time consuming and may not be readily available in technologically less-developed countries. Exhaustive information in it makes it valuable and of practical utility to newcomers in the field of thermophysical and thermal optical property measurement. The broad international composition of the authors, specialists in their particular fields of thermophysics, hopefully brings new information and perspectives for the reader.

Since the book has not been written by one person in the process of developing a continuous review but by a number of researchers, it is more of an organized composite summary than a textbook. Depending on the topic and the scope of coverage, chapters differ in the details of treatment, some giving particulars of equipment and measurement procedures, while others concentrate on systematic variants of a given measurement technique. In some cases details are left for the second volume, which will contain recommended practices for the measurement of thermophysical properties of selected groups of materials.

The material in the text is presented in seven major sections comprising twenty chapters. The first section presents six chapters on six of the most important methods for the measurement of thermal conductivity. Methods appropriate for metals are covered along with methods for insulating materials and refractory ceramics. The second section is devoted to the measurement of electrical resistivity. The third section has four chapters dealing with methods for the measurement of thermal diffusivity in transient, periodic, and monotonic heating regimes. Six methods for the measurement of specific heat comprise the fourth section, with six chapters, covering the classical method of mixtures, levitation calorimetry, adiabatic calorimetry, as well as pulse and modulation techniques. A chapter on the application of differential scanning calorimetry to the measurement of specific heat is also included. Methods for the measurement of thermal expansion are limited to a review of interferometric dilatometry, which comprises the sixth section. A critical review of pushrod and capacitance dilatometry, for which instruments are commercially available, is left for the second volume. The last section contains a summary of thermophysical property standard reference materials.

EDITORIAL BOARD

Acknowledgments

The editors are indebted to many people who helped in the preparation of this book. Special thanks, however, must go to the individuals who contributed most, the authors of chapters, and to M. L. Minges, Chairman of the CODATA Task Group on Thermophysical Properties.

In addition, thanks must be also extended to the referees of individual chapters, whose reviews contributed to the final form and quality of the manuscripts in this book. These are, in the order of appearance of particular chapters in the main section of the book:

For the section on thermal conductivity:

D.L. McElroy, *Oak Ridge National Laboratory, Oak Ridge, Tennessee, U.S.A.*

M.L. Minges, *Wright-Patterson AFB, Ohio, U.S.A.*

C.D. Pears, *Southern Research Institute, Birmingham, Alabama, U.S.A.*

K.-H. Bode, *Physikalisch-Technische Bundesanstalt, Braunschweig, Federal Republic of Germany*

F. De Ponte, *University of Padova, Padova, Italy*

A.E. Wechsler, *Arthur D. Little Inc., Cambridge, Massachusetts, U.S.A.,* (two chapters)

For the section on electrical resistivity:

F. Cabannes, *Centre de Recherches sur la Physique des Hautes Températures, Orléans, France*

For the section on thermal diffusivity:

R.U. Acton, *Sandia National Laboratories, Albuquerque, New Mexico, U.S.A.*

H.-E. Schmidt, *European Institute for Transuranian Elements, Karlsruhe, Federal Republic of Germany*

G. Neuer, *Institut für Kernenergetik and Energiesysteme der Universität Stuttgart, Stuttgart, Federal Republic of Germany*

A.E. Beck and J.H. Blackwell, *University of Western Ontario, London, Ontario, Canada*

For the section on specific heat:

C.R. Brooks, *University of Tennessee, Knoxville, Tennessee, U.S.A.*

V.Ya. Chekhovskoi, *Institute of High Temperatures, Moscow, U.S.S.R.*

J. Margrave, *Rice University, Houston, Texas, U.S.A.*

R. Palinsky, *Ispra Establishment, Ispra, Italy*

F. Righini, *Istituto di Metrologia "G. Colonnetti," Turin, Italy*

R.E. Taylor, *Thermophysical Properties Research Laboratory, West Lafayette, Indiana, U.S.A.*

For the section on thermal expansion:

R.K. Kirby, *National Bureau of Standards, Washington, D.C., U.S.A.*

For the section on thermal radiative properties:

V.A. Petrov, *Institute of High Temperatures, Moscow, U.S.S.R.*

F. Levadou, *European Space Research and Technology Centre, Noordwijk, Holland*

The editors wish to express grateful appreciation to referees: C. R. Brooks, J. Margrave, and H.-E. Schmidt for voluntary lectorial editing of the refereed manuscripts of non-English origin.

Thanks are extended to all organizations represented by the authors and referees throughout the world. The contributions of the U.S.-Yugoslav Joint Board for Scientific and Technological Cooperation and the Soviet Organizing Committee for International Thermophysical Properties Conferences to the coordination of work on this volume are also gratefully acknowledged.

Contents

II. ELECTRICAL CONDUCTIVITY MEASUREMENT METHODS

III. THERMAL DIFFUSIVITY MEASUREMENT METHODS

V. THERMAL EXPANSION MEASUREMENT METHODS

CHAPTER 18. THERMAL EXPANSION MEASUREMENT BY INTERFEROMETRY

G. Ruffino

VI. THERMAL RADIATIVE PROPERTY MEASUREMENT

CHAPTER 19. MEASUREMENT OF THERMAL RADIATION PROPERTIES OF MATERIALS

J.C. Richmond

VII. THERMOPHYSICAL PROPERTY STANDARD REFERENCE MATERIALS

CHAPTER 20. CERTIFIED REFERENCE MATERIALS FOR THERMOPHYSICAL PROPERTIES

R.K. Kirby

I

THERMAL CONDUCTIVITY MEASUREMENT METHODS

Introduction

In general, the measurement of transport properties presents some intrinsic difficulties compared to the measurement of other important fundamental properties such as thermodynamic quantities (e.g., heat capacity), because the determination of a flux is involved. To calculate the proportionality constant relating heat flux and temperature gradient, that is, the thermal conductivity, defining and controlling the heat flow path and then measuring the heat flux are the most difficult tasks. One-dimensional heat flow is usually the goal so that the corresponding temperature gradient is defined unambiguously and is readily measurable, and so that heat losses can be identified and appropriate corrections made.

The six chapters devoted to thermal conductivity determination cover the principal methods of measuring this property from cryogenic temperatures up to several thousand degrees on solid materials exhibiting a very wide range of conductivity. These techniques are axial flow, radial flow, direct electrical heating, guarded hot plate, hot-wire method, and panel test method. The first three methods are classical ones used generally in measuring rigid materials having moderate to high conductivity. The other three methods are generally used in measurements on materials with moderate to low conductivity where the specimens are often nonrigid. The number of different measurement system configurations is especially larger and varied in this low-conductivity range (e.g., thermal insulations). The latter three chapters in this section describe the more important approaches used here.

The configuration of a given measurement system and of the specimen itself is influenced most prominently by the magnitude of the thermal conductivity. When the thermal conductivity is high, the specimens are usually "long" (for example, in the form of cylinders). When the conductivity is low, the specimens are usually "flat" (for example, in the form of plates or disks). Simple thermal considerations indicate why this is so. When the specimen conductivity is high the heat flux is usually fairly high so that, relatively speaking, heat losses from the large lateral surface area of the specimen are small; a long specimen in the direction of flow helps establish a reasonably high temperature gradient which can then be accurately measured. When the specimen conductivity is low and the heat flux correspondingly low, only a relatively small thickness is required to generate a large, accurately measurable gradient. With this low specimen heat flux, lateral losses are of concern, thus a plate-type specimen itself tends to minimize these spurious flows since the lateral surface area is small. As a matter of fact, in some cases the lateral surfaces of the specimen are surrounded by pieces of the same specimen material to provide self-guarding.

The other independent parameter of fundamental importance is the magnitude of specimen conductivity relative to the surroundings. It is generally desired that the specimen effective conductance be as high as possible relative to that of the surrounding insulation. This generally becomes more of a problem as the temperature of the measurement system rises. With some measurement techniques used at very high temperatures, which will be discussed in this section, the lateral losses are allowed to be high; they are accounted for quantitatively in the conductivity measurement. Further, in these apparatus, the specimen geometry actually maximizes the lateral surface area; it is a very high aspect ratio cylinder, that is, a wire or a tube.

The electrical conductivity of the specimen is a parameter of great utility in thermal conductivity measurements. If the test material is a conductor then two beneficial factors arise: (1) the specimen may be self-heated electrically to develop the thermal flux and (2) there may be sufficient thermal transport via the electrons that a quantitative correlation exists between the electrical and thermal conductivity. Thus in some cases the latter may be determined based on experimental measurement of the electrical conductivity, generally an easier quantity to measure. In other cases at least the consistency of measured thermal conductivity can be monitored via the electrical transport properties. In some measurement systems described in this section, the electrical and thermal conductivities can be determined simultaneously, a very important feature. The fact that the ratio of specimen to surrounding insulation conductivity in the electrical case can be orders of magnitude higher than in the thermal case is an important reason the electrical measurements are more easily made.

In addition, contemporary theory can be of great practical value in estimation. For engineering applications, where conductivity values with ± 10% are required, estimation rather than direct measurement will likely be successful in many cases. Variations in the thermal conductivity with temperature can often be estimated from theoretical considerations, from literature data on similar materials, or from electrical/thermal resistivity relationships. Thus in some cases, where measurement is found to be necessary, only a limited amount of data need be taken, the temperature dependence being established through known relationships from theory or from available data bases.

Turning to the specific measurement techniques described in this section, a summary comparing the important features of all the techniques is given in the table on page 6. The ranges of temperatures, thermal conductivity, accuracy, and materials for which the techniques are well suited include some judgments based on experience. That is, they represent practical conditions and expected results when the measurements are conducted by experienced experimentalists exercising prudent, but not necessarily extraordinary, precautions.

Axial flow methods have been long established and have produced some of the most consistent, highest accuracy results reported in the literature. It is the method of choice at temperatures below 100 K. Key measurement issues center mainly on reduction of radial heat losses in the axial heat flow developed through the specimen from the electrical heater mounted at one end (the power dissipation of this heater is used in calculating column heat flux). These losses are minimal at low temperatures thus the special attraction of this technique in the cryogenic range. As the specimen temperature moves above room temperature moves above room temperature, control of heat losses becomes more and more difficult. Thus a great deal of attention centers on important experimental parameters such as the ratio of effective specimen conductance to lateral insulation conductance (the higher the better) and to the quality of guarding (that is the match of the axial gradient in the specimen to that of the surrounding insulation).

This chapter is built upon cylindrical symmetry heat transfer analyses of typical experimental arrangements. Two basic configurations are treated: guarded and unguarded. In the unguarded or Forbe's bar approaches, explicit quantitative methods are developed to determine radial heat loss corrections. Temperature field plots are presented in terms of key independent variables from which an excellent intuitive understanding can be developed on how practical measurement systems will perform over a wide range of conditions.

The section reviewing specific measurement systems is divided logically into three different temperature ranges: (a) $T < 40$ K; (b) $40 < T < 400$ K; (c) $300 < T < 1500$ K. These are natural groupings based on heat loss considerations: across the first range, where one specific measurement system is illustrated, losses are very small; across the second range, where three systems are illustrated, losses are finite but readily controlled; across the third range, where three systems are also illustrated, losses can be quite high and control/measurement is more difficult.

Radial flow techniques, described in the second order chapter of this section, are utilized over a wide temperature range, generally with moderate to low thermal conductivity specimens in either solid or powder form and readily produce results with engineering accuracy. Most measurements have been made with cylindrical specimens although other geometries have been

TABLE 1
Thermal Conductivity Measurement Techniques

Measurement technique	Key features		Temperature range (K)	Principal specimen materials	Conductivity range ($W\,m^{-1}\,K^{-1}$)	Uncertainty (%)
	Advantages	Disadvantages/limitations				
Axial heat flow	Simultaneous electrical resistivity measurements Temperature coverage (below and above RT) High accuracy	Control of radial heat losses above ~ 500 K	90–1300	Metals and alloys in cylindrical form	10–500	0.5–2.0
Radial heat flow	Accuracy	Large specimen size	RT–2600	Solids and powders in cylindrical form	0.01–200	3–15
	Flexibility at high temperatures	Long measurement time				
Direct electrical heating	Very high temperature measurements Multiple property determination including electrical properties	Sophisticated, often costly equipment Specimen must be an electrical conductor	400–3000	Wires, rods, tubes of electrical conductors	10–200	2–5
	Rapid measurements	Stable optical properties				
Guarded hot plate	Adaptable to wide range of commercial materials	Low conductivity materials	80–1500	Thermal insulations	< 1.0	2–5
	High accuracy	Long test times (3–12 h)		"Building" materials		
Hot wire method	Small temperature drop in specimen	Low conductivity materials	RT–1800	Refractory materials including powder and nonrigid forms	0.02–2	5–15
Panel test	Simple apparatus	High-temperature gradient in specimen Long test times	600–1600	Refractory materials in block form	0.05–15	15

tried. In addition, some successful measurements on materials with high conductivity have been reported. The method has been employed to very high temperatures since guarding to control heat losses is facilitated with the cylindrically shaped specimen configuration. Often relatively large specimens are required compared with those used, for example, with the axial flow techniques, and long test times may be required to produce data over the full temperature range. The main guarding task, of course, is control of axial (i.e., "nonradial") heat flow. End guarding issues are treated in some detail. The two basic modes of operation are described: (1) outward flow (from an electrical heater mounted on the center line; (2) inward heat flow (to a water calorimeter located on the center line).

Very comprehensive analytical treatment of the common measurement system configurations are presented. The results of this analysis are particularly valuable and practical since they build an intuitive foundation and understanding which shows just how this experimental configuration performs over a wide range of conditions. Temperature fields for several configurations as a function of important independent variables (e.g., specimen/insulation conductivity ratio) are very helpful in this regard. The analysis has been applied to five different specific measurement systems using solid specimens and two using powders which have been reported in the literature. Conclusions and recommendations provide a quick but very complete summary of the factors important in radial heat flow measurement techniques.

Direct electrical heating methods come into their strongest region of performance as the specimen temperature increases and measurements become problematic with the axial flow or radial flow techniques. These techniques are especially well-suited for measurements above 1200 K. The governing differential equations describe a long cylindrical specimen heated resistively by direct passage of an electrical current and dissipating heat by conduction radially or axially along the specimen (since a significant axial gradient is usually present) and by radiation from the surface. Depending on the relative magnitude of these terms and the boundary conditions, different specific measurement techniques result. This chapter reviews all of the important variants dividing them into three basic categories: (1) radial flow measurement at the center of the specimen where the axial gradient (because of symmetry) is zero; (2) the case where surface heat losses are small and the conductivity is determined based on axial gradients in the specimen – this is often referred to as the Kolrausch method and yields the best results below about 1000 K; and (3) the very high temperature situation when both axial conduction and radiation dissipation are significant. The technique also allows the direct measurement of a number of other thermophysical properties and is often referred to as the multiproperty technique.

The very high temperature system receives detailed treatment in the chapter because significant developments have occurred in recent years both in the analytical treatments and in sophisticated experimental measurement improvements. Much attention is devoted to the determination of temperature and its spatial derivatives since these quantities are critical measured quantities entering directly into the conductivity equations. With significant radiation boundary conditions specimen emittance is of great importance and its measurement is crucial. Once measured the stability of the emittance in subsequent conductivity experiments is essential since it enters as a constant in the conductivity equations.

Guarded hot plate is the most widely used and most versatile method for measuring the thermal conductivity of insulations. Although the specimens are often rather large, up to a meter square, this usually presents no difficulty since the costs of commercial building material samples or other insulations typically measured are modest. A flat, electrically heated metering section surrounded on all lateral sides by a guard heater section controlled through differential thermocouples supplies the planar heat source introduced over the hot face of the specimens. There are two basic measurement configurations: (1) the conventional symmetrically arranged guarded hot plate where the heater assembly is sandwiched between two specimens, and (2) the biguarded arrangement when the heat flow is through one specimen, the back of the main heat/guard plane rendered adiabatic by using another planar guard heater.

Three different categories of measurement systems can be distinguished: (1) apparatus working around room temperatures (up to 375 K), (2) apparatus working below room temperatures (down to around 80 K), and (3) apparatus working at high temperature (to 1500 K). A given apparatus is most often best adopted for measurement in one of these temperature ranges. Quantitative best practice guidance is given in this chapter describing the influence of measurement system variables on measured conductivity and permissible ranges of important parameters such as gap width between metering and guard sections, temperature unbalances between meter and guard, thermal contact resistance effects and edge loss effects.

A discussion is included on reference materials and interlaboratory measurement comparisons ("round-robin" tests). Measurement results are given on three of the more important reference materials: glass fiber, alumina fiber, and resin bonded fiber board.

Multiple modes of heat transfer (solid/gas phase conduction, convection, and radiation) are always possible in thermal insulations and under some circumstances the effective conductivity can become temperature gradient dependent, thus losing precise meaning. The chapter includes discussion on these issues important in establishing proper measurement conditions and in interpreting the results on porous materials which may also exhibit diathermanous behavior.

Hot wire methods are most commonly used to measure the thermal conductivity of "refractories" such as insulating bricks and powder or fibrous materials. Because it is basically a transient radial flow technique, isotropic specimens are most readily measured. The technique has been used in a more limited way to measure properties of liquids and electrically conducting materials of relatively low thermal conductivity.

Relatively recent interesting modifications of this long-established technique are described. The "heat probe" configuration is of particular practical utility where the specimen conductivity is determined from the response of a "hypodermic needle" heat flux/temperature probe inserted in the test specimen. Thus the method is conveniently applied to low-conductivity materials in powder or other semirigid form. A probe device was used during the Apollo missions to measure the thermal properties of the lunar surface.

Panel test method is a rather specialized technique for conductivity measurements and is detailed in the final chapter of this section. The method, first employed in the 1920s and 1930s, is especially suited for refractory thermal insulations in either rigid form (i.e., brick) or nonrigid form. Functionally, the measurement system is a guarded water calorimetric device. That is, the cold side of a specimen is fixed by a water flow calorimeter while the hot side and in turn the mean specimen temperature is established by a planar electrical heater. The cold side temperature could be raised by interposing a thin insulating layer between the specimen and the water calorimeter although this clearly cannot be adjusted during a given test run and could lead to additional heat loss sources.

The method is analogous to the inward flow approach in the cylindrical specimen radial methods discussed in the second chapter of this section, except that flat panel specimen specifications are used here. Since the cold-side temperature is fixed at a relatively low level, two disadvantages may arise (1) even moderately high mean specimen temperatures can be attained only by driving the hot-side temperatures to a very high level, (2) since the measured thermal conductivity represents an integrated value across the full temperature drop experienced by the specimen, any nonlinearities in the true conductivity–temperature curve will be obscured. Nevertheless, the method is a classical one, highly adaptable and relatively straightforward to use when data of engineering accuracy on refractory materials are sought.

1

Axial Heat Flow Methods of Measuring Thermal Conductivity

M. J. LAUBITZ

1. INTRODUCTION

The science of the thermal conductivity of solids is an old one, and can be seen, in retrospect, to have developed in a series of jumps. There was one flurry of activity after the turn of the century, when scientists became aware of and interested in the variation of the thermal conductivity with temperature. There was another major burst of activity in the late fifties and sixties when advances in technology demanded knowledge of the properties of many new substances or of old substances over broader temperature ranges, and developments of the theory of the solid state raised interesting scientific questions which one hoped to answer experimentally.

The periods of strong interest in thermal conductivity (λ) have led to advances and developments in the techniques of its measurement. Thus, the period at the turn of the century established the principles of virtually every variant of the techniques of measuring λ. The latter period, the fifties and sixties, served to refine and develop these techniques, to delineate more clearly the appropriate ranges for the various approaches, and to determine and minimize the various experimental errors. In consequence, the major techniques involving a steady state matured some time ago; there has been little significant development in this area in the last ten years. Thus little new exists to add to what has already been well summarized in existing tests and review articles – the two volumes on *Thermal Conductivity*, edited by Tye, for instance.[1]

A proper review, therefore, of existing experimental techniques of measuring λ, one describing *established* methods and procedures, would have a dated appearance and contain few references to modern measuring equipment,

M.J. LAUBITZ • Division of Physics, National Research Council of Canada, Ottawa K1A OR6, Canada.

in particular to such relatively recent developments as microprocessors. Yet there is no question that automation of the equipment, now so readily achievable, will greatly modify the details of the various techniques which have been developed for measuring λ. For example, the determination of λ has been traditionally a time-consuming and tedious process. It involved precise measurement of temperature, which could only be done manually with one of the variety of potentiometers then available. With the time required to make a single measurement on the order of minutes, and that required to complete the investigation of one specimen on the order of weeks, the emphasis of the techniques was on a "do it once and do it right" philosophy. A statistical approach to the problem simply did not make much sense – every experimental point was made to count. Furthermore, because of the relatively limited computational capacity available to the experimentalists, particularly in the dedicated mode, the experimental arrangements were specifically aimed at analytical simplicity, requiring little "real-time" computation during the progress of the experiment.

The advent and proliferation of micro- and minicomputers changes completely the old philosophy of measurement. Unfortunately, the new philosophy has not had, as yet, a large enough effect on the measurement of λ of solids, due to the relatively low level of activity in this field in the last few years. A low level of activity does not justify extensive development of hardware, and thus the development has not occurred even though, technically, it is perfectly feasible.

In view of this situation, rather than presenting descriptions of the current state of the art which is behind the times, this chapter will deal with the principles underlying the techniques, which remain unchanged. Given an adequate understanding of these principles, the intelligent reader can readily design the apparatus himself using the components of the eighties, rather than waste his time with solutions that were adequate in the fifties and sixties.

To this end, some general points regarding the measurement of λ are first considered (Section 2), followed by an outline and discussion of the detailed problems which affect this measurement by the axial heat flow technique (Section 3). Then some useful operating procedures are reviewed (Section 4), and the chapter ends with a very brief description of some existing experimental systems (Section 5), to serve as useful illustrations of possible solutions to existing problems, not as blueprints for new design.

2. GENERAL CONSIDERATIONS

Two properties determine the response of a solid to applied temperature gradients: the thermal conductivity, λ, which relates the heat flux to the temperature gradient, and the thermal diffusivity, α, which determines the temporal response of the solid to temperature changes. For homogeneous, isotropic solids these properties are trivially related: $\alpha = \lambda/c'$, where c' is the heat capacity per unit volume, and can be readily calculated from each other. It may be useful, therefore, to discuss briefly under what conditions one requires the knowledge of one or the other of these properties, and under what circumstances one should measure the other, rather than the required, property and resort to calculation. This is, of course, always assuming that a measurement is indeed required.

At this point, a brief degression on the necessity of direct measurement is in order. Experience dictates that, unless one has ready access to functioning and tested equipment, direct measurement of either λ or α should be avoided at all cost. When confronted with the requirement for either property, first recourse should always be had to the excellent data reference books readily available,[2] which, when coupled with some experience, can frequently fill the need. Failing this, one may wish to calculate the required property from other known parameters and established empirical or theoretical relations. For example, λ for metals may frequently be estimated from its electrical conductivity, σ, through an empirically modified Wiedemann–Franz law;[3] σ is invariably easier to measure than λ, and consequently has been studied more prolifically and precisely in the past. If only moderate accuracy is required for λ, say $\pm 10\%$, estimates of it from σ are quicker, cheaper, and probably just as reliable as measurements on new and untried apparatus. As a third choice, one should try to get somebody else to do the measurement, either a "commercial" laboratory, or an indebted friend. Only as a last resort should one set out to perform the measurements oneself.

Having concluded, by a process of elimination, that experimental measurements have indeed to be made, which property should one set out to measure? The answer depends on the specific circumstances, and can only be unequivocally given for some limiting conditions. For purely scientific purposes, such as fundamental investigations and detailed comparisons with theory, it is λ which is invariably required, and the inherent loss of accuracy in the indirect determination of λ from α, due to the unavoidable uncertainty in c', strongly favors the direct measurement of λ itself. For practical, engineering purposes, the answer is not nearly as clear: immediate questions arise whether in the real world true "steady state" ever exists, and, if so, whether the customary requirement of somewhat lower accuracy cannot be satisfactorily met from an "indirect" determination of λ.

The problem of steady state is in itself interesting for it turns out that just because temperatures change with time, and change continuously, this does not necessarily imply that the determining parameter is α. Usually, there is a characteristic transient time during which the temperature distribution is determined by α; for times much longer than the characteristic time, however,

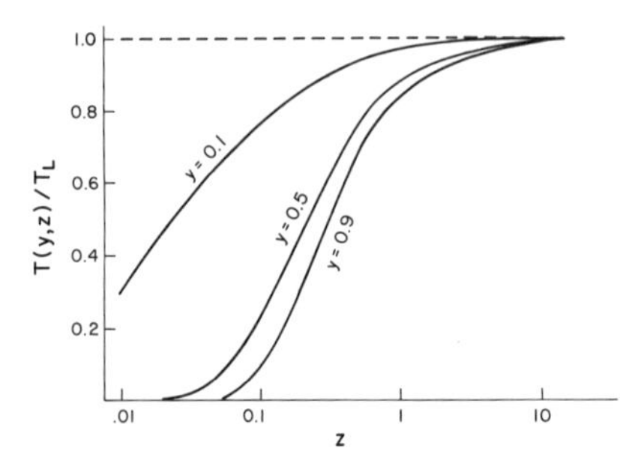

FIGURE 1. The ratio of the correct transient temperature of a slab $T(y, z)$, to the linear temperature, T_L. The slab is of thickness l, with the end at $x = l$ kept fixed at zero temperature, and the end at $x = 0$ heated at constant rate, β. $y = x/l$, and $z = t/t_0$, where t is the time, and t_0 the "characteristic time" of the system. $T_L = \beta t(1 - y)$.

the distribution is, for practical purposes, determined by λ. Take as an example one-dimensional heat flow in a slab of length l, one end of which is kept at constant temperature $T = 0$, and the other at a changing temperature $T_0(t)$, where t is the time. Only if $T_0(t) \propto e^{\gamma t}$ will the temperature distribution in the state be forever determined by α. For weaker boundary conditions, $T_0(t) \propto t^m$, there is a characteristic time $t_0 = ml^2/\alpha$ such that for $t \leqslant t_0$ the temperature distribution is determined by α, but for $t \gg t_0$ it is for practical purposes given by λ, i.e., as if a "steady state" existed in the slab; this is illustrated in Fig. 1. For example, the wall of a boiler made from steel ($\alpha = 0.1$ cm^2/s) of thickness 1 cm will be in a quasi "steady state" for any constant heating rate of its outside surface 10 s after the start of the warm-up. Thus, even in practical problems dealing with changing temperatures it is not necessarily α which is the required parameter. On the other hand, steady state invariably requires λ.

Let us now deal with the problem of which property to measure regardless of which may be the one required. The simplistic notion of measuring directly the property that is desired is confused by two aspects, one universally recognized and the other rather contentious, namely, that on the one hand it is much quicker to measure α than λ, but on the other hand the measurement of λ is intrinsically capable of greater accuracy. The question of the relative speed of the two determinations not being in contention, no more time need be wasted on it. The problem of relative accuracy, however, being a "sensitive issue," needs some justification; the following brief but general argument is an attempt at this. It relates to the basic defining equations for the two properties. For λ we have

$$\lambda = \bar{q}/\overline{\nabla_n T} \tag{1}$$

where $\bar{q}$ is the heat flux and $\overline{\nabla_n T}$ is the (normal) gradient of temperature. For α we have

$$\alpha = \nabla^2 T/(\partial T/\partial t) \tag{2}$$

where ∇^2 is the Laplacian operator. To determine α with any degree of accuracy,

$\nabla_n T$ cannot be very small, and under this condition the practical determination of the transient temperature, $\partial T/\partial t$, is not a trivial matter if one allows for contact resistances between the specimen and the temperature sensor and the heat capacity of the latter. Certainly this cannot be done any more accurately than the measurement of $\bar{q}$ in the steady-state measurement of λ. However, the measurement of $\nabla^2 T$ is always about an order of magnitude more uncertain than the determination of ∇T and, broadly speaking, this is the relationship between the accuracies of the two techniques. This conclusion is modified only by the obvious observation that α, unlike λ, depends only on relative, not absolute, temperature measurements, in the sense that the thermometers have to be linear devices but need not be calibrated; this is only a minor complication for the steady-state technique, of significance in extreme temperature ranges.

It is, of course, perfectly true that equation (2), on which the above argument is based, figures rarely if ever in modern diffusivity techniques. This, however, does not destroy the logic of the argument for, stated or not, this *is* the defining equation for α, and no amount of transformation or transposition of the $\nabla^2 T$ into measured or assumed boundary conditions changes the sensitivity of α to its explicit or implicit presence; it merely transfers that sensitivity to some other parameter which is determined by the details of the specific technique employed.

From the foregoing we may conclude that knowledge of λ is necessary for most scientific endeavors, all steady-state applications, and some time-dependent phenomena of long duration. The determination of λ, however, is only necessary if the highest experimental accuracy is required (currently $\leqslant 1\%$ near room temperature). For less precise applications, λ may be derived from the somewhat rougher but much quicker determinations of α.

The restriction of the measurement of λ to cases requiring high accuracy limits the choice of the suitable steady-state methods to the "absolute" ones; "comparative" methods, which relate the λ of the specimen to that of a "standard" (if there be such a thing) suffer in consequence a degradation of accuracy which can no longer successfully compete with diffusivity techniques.

Traditional classification distinguishes four absolute steady-state techniques for measuring λ: *linear (axial) heat flow, radial heat flow, guarded hot plate,* and *direct electrical heating.* A more sophisticated approach might distinguish only two: *direct electrical heating,* suitable for electrical conductors, and "*generalized nearly unidirectional heat flow,*" which encompasses all the rest, and is suitable for all solids. The latter clearly depends on the measurement of the temperature gradient due to a known heat flux traversing the specimen, usually in a geometrically simple way, and it is a spurious concentration on this geometry of the heat flow which has led to the traditional division of this technique into separate "methods." This is mildly regrettable, for the principles of these methods are identical, and the distinctions drawn more illusory than real.

An unbiased observer, for instance, may well argue, and correctly, that the guarded hot-plate technique is no more than axial heat flow with a short and fat specimen, and radial heat flow no more than guarded heat flow pushed to its logical conclusion: a short and fat specimen bent onto itself so as to be self-guarding in one dimension.

The choice between these geometries is essentially the choice between the procedures employed to ensure that the heat flux is mainly unidirectional and the need to generate measurable temperature differences by practical heat fluxes. It thus depends largely on the ratio of the λ of the specimen to that of its environment and, in consequence, on the temperature range over which the determination of λ is required. This article is devoted largely to the critical examination of the principles of operation underlying the axial (linear) heat flow method, an examination which will explicitly designate the conditions under which this technique, and its related variants, can be successfully employed. Implicitly, it will delineate the conditions more suitable for the other geometries. For those who have no patience with verbosity, however, a brief summary is here attached: the axial heat flow technique is most suitable for small specimens, for specimens of relatively large λ ($\geqslant 0.01$ W/cm K), and for investigations where simultaneous measurement of other transport properties, the electrical conductivity and thermoelectric power for instance, is required. It is *the* method for use at low temperatures ($T < 100$ K), but can be successfully used up to about 1500 K. it is definitely *not* recommended for temperatures in excess of that. For comparison the guarded hot plate is suitable only for relatively large specimens of low λ, and can be employed between 100 and 1000 K. The radial method is the most satisfactory for high-temperature work, can be successfully employed for temperatures in excess of 300 K, but requires substantial specimens for satisfactory operation. However, as mentioned before, these techniques have much in common and useful knowledge of general applicability can be derived from the detailed examination of each of these techniques.

3. BASIC PROBLEMS IN AXIAL HEAT FLOW

The general defining equation for λ of an isotropic, homogeneous body is equation (1), given above. For purely linear heat flow, the equation can be put into the form

$$\lambda = \frac{\dot{Q}}{A} \cdot \frac{\Delta z}{\Delta T} \tag{3}$$

where $\dot{Q}$ is the power traversing the specimen of cross-sectional area A and $\Delta T/\Delta z = [T(z_2) - T(z_1)]/(z_2 - z_1)$ is the constant temperature gradient along the specimen, usually determined from the temperatures T_2 and T_1 of sensors located at z_2 and z_1. The determination of λ therefore involves the following:

(1) The determination of the geometrical parameters A and $\Delta z = z_2 - z_1$;

(2) Assurance that the heat flow pattern is indeed unidirectional. This usually involves both careful design of experimental configurations, and active "temperature matching" of various components of the apparatus during the experiment;

(3) Compliance with the requirements of steady-state conditions;

(4) Measurement of $\dot{Q}$, usually generated in some heater attached to, or embedded in, the specimen. This measurement in itself is normally a trivial operation, but difficulties are encountered in ensuring that all the measured power, and only that power, is responsible for the measured temperature gradient $\Delta T/\Delta z$;

(5) The measurement of the temperature of the two (or more) sensors, T_2 and T_1. As with the power measurement, this, by itself, is again straightforward; however, ensuring that the measured sensor temperature corresponds to the specimen temperature at the presumed locations z_2 and z_1 is not trivial.

These five points are common to all "unidirectional" steady-state methods, although their relative importance may vary from technique to technique. In general, (1) and (3) are straightforward, the other three troublesome. For the axial heat flow technique, (5) is the most important factor at low temperatures, and (2) at high temperatures; all of them will be discussed in detail in the following sections of this chapter.

3.1. Geometry

Problems relating to the geometry are, for axial heat flow, relatively self-evident, and will be dealt with here only briefly.

For the idealized case of perfect axial heat flow, only two geometrical parameters are important – the cross-sectional area of the specimen, and the effective separation of the temperature sensors, Δz, which is used to determine the temperature gradient [cf. equation (3)]. The determination of the cross-sectional area is almost trivial for cylindrical specimens, although some obvious precautions may have to be exercised with other profiles. For most purposes, mechanical devices such as micrometers or calipers are perfectly adequate. Where the specimen is soft and easily deformed, when made of a well-annealed pure metal for instance, optical comparators can be used to refer the specimen transverse dimensions to a standard. Problems only arise with very soft and thin wires, used for very low-temperature studies, and recourse must then be had to more exotic techniques, such as calculations from known length, weight, and density.

The determination of Δz is more complicated, not so much because of technical difficulties as due to the uncertainty that the geometrical location of the sensor coincides with its effective thermal position. The geometrical Δz can always be measured by optical means, a traveling microscope for example, but it is then uncertain by about $\sqrt{2} \times \delta z$, where δz is the width of the contact between the thermometer and the specimen in the axial direction; for a thermocouple, this is at least of the order of the wire diameter. The imprecision can be substantially reduced for metallic specimens if the separation is measured electrically with reference to standard knife edges, the thermometers being used as potential contacts – all normal procedures associated with the measurement of the electrical resistivity must then be followed. If the electrical current flow matches exactly the heat flow in the test section of the specimen, then the effective electrical separation gives precisely the effective thermal separation. Unfortunately, the currents match only rarely, for it is neither possible to inject the electrical current precisely analogously to the power entering the specimen from the specimen heater, nor is it possible to isolate thermally the periphery of the specimen. Both these factors affect the correspondence between the effective thermal and electrical separations, and while the effect of the former can be minimized by locating the sensors at least one specimen diameter away from the ends of the specimen where power and current are injected and withdrawn,[4] the latter is an unavoidable difficulty. Its effect can only be guessed at, or estimated from repetitive measurements with various thermometer separations on a single specimen. Experience indicates that the electrical separation is less uncertain than the estimate given above for the optical separation by about a factor of 4 to 10.

Where the heat flow is not purely axial, and corrections for peripheral losses have to be made, the absolute location of the thermometers on the specimen with respect to the specimen heater must also be known. Usually, however, the required precision of this absolute distance is less than that required for Δz, for it is used only to determine corrections of a few percent, and under these circumstances optical measurements are perfectly satisfactory.

3.2. Heat Flow Patterns

When one considers potential experimental methods for the determination of λ, the axial heat flow technique comes readily to mind because of the similarity that it bears to the standard method of measuring electrical resistivity. In the latter case one passes a steady, known electrical current through an electrically insulated specimen, and determines the resistance from the voltage drop along the specimen. The thermal analog of this technique is illustrated schematically in Fig. 2. Here, a cylindrical specimen of length l_0 and radius r_1 is placed on top of a heat sink, inside a thermally insulating container. The speci-

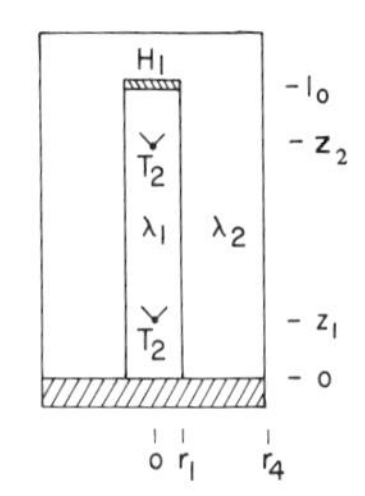

FIGURE 2. Schematic illustration of the simplest axial heat flow system: a specimen of conductivity λ_1, length l_0, and radius r_1, mounted on a base plate, and surrounded by a container, of radius r_4, containing insulation of conductivity λ_2. The specimen has one heater, H_1, and two temperature sensors, T_1 and T_2, mounted on it.

men has a heater attached to it, H_1, which generates a steady power $\dot{Q}$. At steady state, the measurement of the temperatures along the specimen at positions z_2 and z_1 permits the determination of λ through equation (3).

The identity of the principle of these two measurements is, unfortunately, misleading, chiefly due to the fact that while there are many near-perfect electrical insulators, the ratio of best to worst electrical conductivity at room temperature being about 10^{30}, thermal insulation is impossible except at very low temperatures; at room temperature, the ratio of the highest thermal conductivity to the lowest available, without recourse to vacuum, is barely 10^4. In practical terms, this means that neither the specimen heater H_1, nor the specimen itself can be effectively isolated without elaborate precautions. Consequently, one cannot equate the power entering the specimen with that developed in the heater, nor be certain that all the power entering the specimen traverses its whole length. Of these two difficulties, the first is easier to deal with, either through heater shielding or the utilization of a "split" specimen, and its discussion will be left to a later section, Section 3.4. The second difficulty, which is the chief reason why λ measurements are so inaccurate at moderate to high temperatures ($T > 100$ K), will be examined in detail in this section.

Let us first give a rough estimate of the magnitude of the effect that heat leakage can have on the measured λ if one were to employ as naive a system as that depicted in Fig. 2. Assume for simplicity that the container surrounding the specimen is filled with insulation of conductivity λ_2, and that its outside wall is maintained at the temperature of the heat sink. In very crude terms, then, the error in the measured conductivity, $\Delta\lambda$, will be given by the ratio of the heat lost radially by the specimen to the heat conducted along it. For small radial losses the temperature gradient along the specimen is approximately constant, and

$$\frac{\Delta\lambda_1}{\lambda_1} \sim \frac{\displaystyle\int_0^{l_0} \frac{2\pi\lambda_2}{\ln r_4/r_1} \cdot \frac{\Delta T}{\Delta z}\, z\, dz}{\displaystyle\pi r_1^2 \lambda_1 \frac{\Delta T}{\Delta z}} \sim \frac{(l_0/r_1)^2}{\rho_1 \ln r_4/r_1} \tag{4}$$

where $\rho_1 = \lambda_1/\lambda_2$.

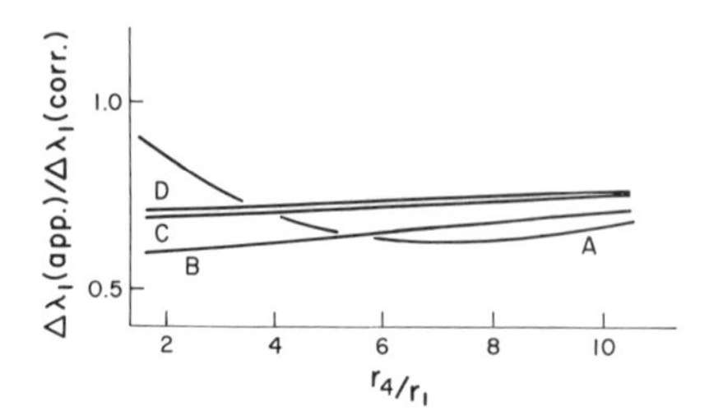

FIGURE 3. Comparison of approximate errors due to radial losses, $\Delta\lambda_1$ (app.), with exact ones, $\Delta\lambda_1$ (corr.), as a function of r_4/r_1. $l_0/r_1 = 20$. Curves A to D are for conductivity ratios, λ_1/λ_2, equal to 10, 10^2, 10^3, and 10^4, respectively.

For a specimen made of steel ($\lambda_1 \sim 0.1$ W/cm K), of dimensions 1 cm diam by 10 cm long, placed in a 2-cm-diam container filled with a good grade of powdered thermal insulation in air ($\lambda_2 \sim 10^{-3}$ W/cm K), the error in λ_1 at room temperature amounts to some 600%, an error so large as to be well beyond the range of validity of our simple-minded analysis. If the specimen were evacuated, thereby decreasing λ_2 by about a factor of 10, then $\Delta\lambda_1/\lambda_1$ would be reduced to some 60% at room temperature, and drop to an acceptable level of about 2% below 100 K.

The approximate equation (4) gives only the order of magnitude of the errors (a comparison of its predictions with detailed calculations[5] is shown in Figs. 3 and 4), nevertheless it is sufficient to underline the severity of the problems caused by heat losses from the specimen. Except for temperatures below about 100 K, these losses cannot be ignored, but must be treated in one of three possible ways:

(1) Application of theoretically calculated corrections to account for these losses; this approach is satisfactory only when applied to good conductors in vacuum for limited temperature ranges above 100 K;

(2) Experimental determination of corrections for the radial losses; several techniques have been developed employing this approach, which may be labeled the "Generalized Forbes' Bar" approach;[5]

(3) "Elimination" of the radial losses through "guarding." In this approach guards with heaters on them are placed around the specimen and "matched" in temperature to it. This approach, as will be shown below, is difficult and often misleading, but, with suitable precautions, may be usefully employed in certain circumstances.

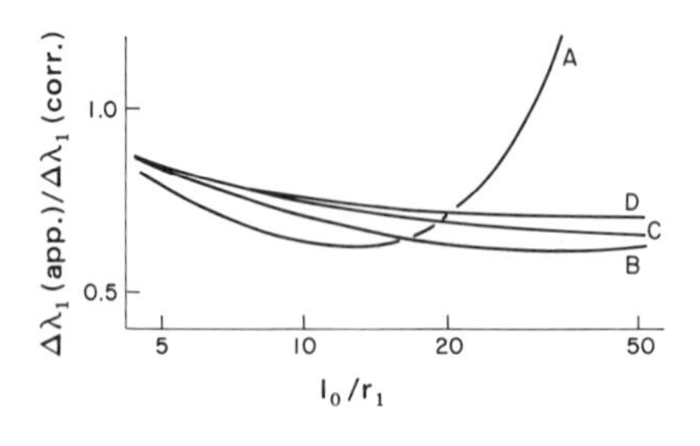

FIGURE 4. Comparison of approximate errors due to radial losses, $\Delta\lambda_1$ (app.), with exact ones, $\Delta\lambda_1$ (corr.), as a function of l_0/r_1. $r_4/r_1 = 4$. Legend as in Fig. 3.

The three approaches outlined above have one feature in common, in that they require a sound mathematical analysis for their application. This analysis is required, in the first place, to design the experimental system, optimize such of its parameters as are controllable, and yield an apparatus in principle capable of satisfactory operation – this is particularly pertinent to "guarded" systems. In the second place, it is required for the operation of the experimental system: the determination, be it experimentally or theoretically, of the peripheral losses; the derivation of suitable procedures and formulas for the calculation of λ, and the determination of the uncertainties and possible errors in the results. This applies equally to all three approaches. One example of such a mathematical analysis is outlined below.

3.2.1. *Mathematical Analysis*

The general mathematical treatment appropriate to the analysis of axial heat flow techniques has been described in detail previously.[6] Here, we will restate it only briefly, to provide the necessary definitions and nomenclature for subsequent discussion, and we summarize the resultant mathematical formulas in the Appendix. For detailed discussion, the reader is referred to the original article.

The model analyzed mathematically has cylindrical symmetry, and consists of four components: the specimen, a guard, and thermal insulation both between the specimen and the guard, and outside of the guard. The specimen is equipped with both wafer-thin transverse heaters and cylindrical surface heaters, and the guard with cylindrical heaters wound on its outer surface. The conductivities of the specimen and guard can differ, but are assumed uniform for each. The system is illustrated schematically in Fig. 5.

The specimen (1) has radius r_1, and length l and conductivity λ_1, and has radial heaters A_1 to A_l wound on its surface r_1. Each of these heaters is defined by its length $2\theta_l(A)$, and the location of its midpoint, $z_l(A)$ and is assumed to dissipate uniformly a total power of $\dot{Q}_l(A)$. The specimen also contains wafer-thin transverse heaters C_1 to C_j located at $z_j(C)$ and extending over the whole specimen cross-section; these are assumed to dissipate uniformly a power of $\dot{Q}_j(C)$.

The guard (3) has an outer radius r_3, inner radius r_2, conductivity λ_3, and extends over the whole length of the specimen. It has radial heaters along r_3, B_1 to B_m. These heaters have lengths $2\theta_m(B)$ and midpoints located at $z_m(B)$, and they dissipate uniformly a total power of $\dot{Q}_m(B)$.

The spaces between the specimen and the guard, (2), and the guard and the outer surface at r_4, (4), are filled with insulation having conductivities λ_2 and λ_4, respectively. The temperatures along the transverse end $z = 0$ and $z = l$, are assumed zero. Under these conditions, and with the further assumption that

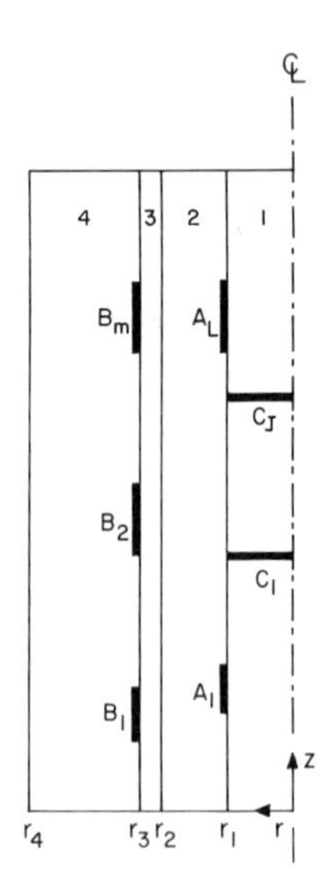

FIGURE 5. The four-component system used for the mathematical analysis: A_1 to A_l, radial specimen heaters; B_1 to B_m, radial guard heaters; C_1 to C_j, transverse specimen heaters.

all the conductivities, λ_1 to λ_4, are independent of temperature, we can write an analytical expression for the temperature at any point (r, z), $T(r, z)$, which is a linear function of the power dissipated in all the heaters; this analytical expression is given in the Appendix.

The full analysis, as given in the Appendex, is complex and involved, although well within the capabilities of modern minicomputers. It has to be employed in its entirety for the study of *guarded* axial heat flow systems. For "unguarded" systems, however, those that consist essentially of a specimen in an insulating medium (cf. Fig. 2), a simplified analysis is frequently perfectly acceptable. This approach, inappropriately called the "thin-rod approximation",[7, 8] assumes that all the isotherms in the specimen are planes perpendicular to the direction of heat flow, and all losses from the convex surface of it are determined by the purely radial temperature difference between the specimen and some arbitrary surface located at r_0. The distribution is then a solution of the equation

$$\frac{d^2 T_{\text{app}}}{dz^2} = \frac{\beta^2}{{r_1}^2} [T_{\text{app}} - T(r_0)] \tag{5}$$

where

$$\beta^2 = \frac{2}{\rho_1 \ln (r_0/r_1)} \tag{6}$$

and $\rho_1 = \lambda_1/\lambda_2$. T_{app} is the specimen temperature, a function of z only; $T(r_0)$ is the temperature along the surface at r_0, and may be a function of z. Solutions to this equation have been produced for various boundary conditions, and many of them are summarized by Carslaw and Jaeger.[8]

For the system that we have been discussing above (Fig. 2), the solution of equation (5) can be easily obtained. For instance, for a specimen of length l_0 with a transverse heater (C-type), enclosed in a container with zero temperature at $r_0 = r_4$ we have

$$T_{\text{app}} = \frac{\sinh \beta z/r_1}{\beta \cosh \beta l_0/r_1} \frac{\dot{Q}(C)}{\pi r_1 \lambda_1} \tag{7}$$

while for the same system with a radial (A-type) heater of length $2\theta(\mathrm{A})$ we have

$$T_{\text{app}} = f(z) \frac{\dot{Q}(A) \ln (r_4/r_1)}{4\pi\theta(A)\lambda_2} \tag{8}$$

where

$$f(z) = \frac{\sinh [2\beta\theta(\mathrm{A})/r_1] \sinh \beta z/r_1}{\cosh \beta l_0/r_1}, \qquad 0 \leqslant z \leqslant l_0 - 2\theta(\mathrm{A}) \tag{9}$$

$$= 1 - \frac{\cosh \beta[l_0 - 2\theta(\mathrm{A})]/r_1 \cosh \beta(l_0 - z)/r_1}{\cosh \beta l_0/r_1}, \qquad l_0 - 2\theta(\mathrm{A}) \leqslant z \leqslant l_0 \tag{10}$$

Solutions for other boundary conditions at r_4 can readily be obtained.

Figures 6 and 7 show representative comparisons of the approximate specimen temperature distributions, T_{app}, with those obtained from the exact analysis, $T(r, z)$, for a variety of geometrical and thermal parameters; more examples are given in Ref. 7. From these it is clearly seen that T_{app} is accurate to 1 or 2% for $\rho_1 \geqslant 10^2$, provided that the specimen aspect ratio, l_0/r_1, does not exceed 20.

To illustrate the applicability of the mathematical analysis, two examples will be briefly discussed: first, the use of the thin-rod approximation for an unguarded system, and, subsequently, some results of the full analysis performed on guarded systems.

3.2.2. Unguarded System – Thin-Rod Approximation

To begin, let us note that the employment of an unguarded technique does not demand the application of the approximate thin-rod analysis. The exact analysis may be used, but its complexity makes such an approach cumbersome, thus favoring the approximate approach. Proceeding on that basis, the first step is to ensure that the system satisfies the requirements for the validity of this

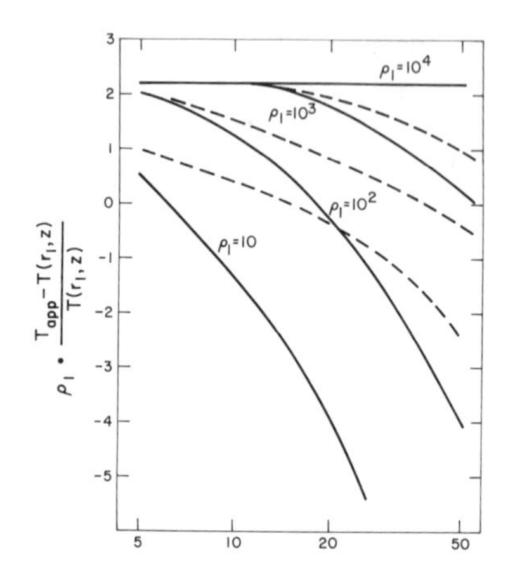

FIGURE 6. Departures of approximate specimen temperature distribution from exact as a function of l_0/r_1, for $r_4/r_1 = 4$, $T(r_4, z) = 0$, and transverse specimen heater. Solid lines, $z = 0.2l_0$; dashed lines, $z = 0.8l_0$.

approximation. In essence, this involves the careful choice of the geometrical parameters of the system, given that ρ_1 is usually beyond one's control: the (expected) specimen conductivity, λ_1, is of course fixed, and the choice of insulation, in any case not large, may well be further restricted by such extraneous factors as chemical compatibility with the specimen, or mechanical and physical integrity at the temperatures of interest. Having thus determined the geometrical parameters, l_0, r_1, and r_0, the design of the rest of the system is largely governed by the principles of this technique, which involve the determination of the specimen temperature distributions for various easily determined boundary conditions. To this end, the specimen is mounted inside the container, with the space between them filled with insulation. The container supports heaters and thermocouples, and is made sufficiently thermally massive

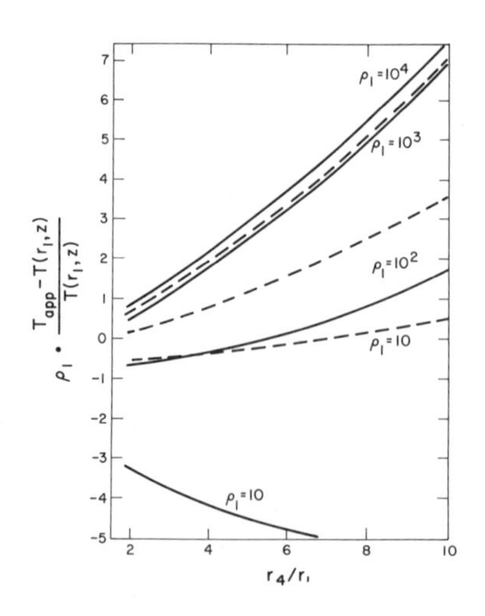

FIGURE 7. Departures of approximate specimen temperature distribution from exact as a function of r_4/r_1, for $l_0/r_1 = 20$, $T(r_4, z) = 0$ and transverse specimen heater. Solid lines, $z = 0.2l_0$; dashed lines, $z = 0.8l_0$.

so as to maintain an essentially uniform temperature – this is not absolutely necessary, but it minimizes the number of thermocouples necessary to prescribe its temperature, and simplifies the analysis. The specimen itself has at least two heaters attached to it, one to raise it to the required temperature, and the other, which shall be called the "specimen heater," to establish the gradient required to determine λ. The specimen heater must have provision for the measurement of the power that it dissipates, and precautions must be taken that all the power that it generates flows into the specimen, at least in the first instance. Furthermore, several thermocouples must be attached to the specimen to determine its temperature distribution, two being the minimum number. Note, however, that employment of more than two thermocouples is strongly recommended, to gain consistency checks on the operation of the system.

The experimental procedure requires the determination of at least three separate temperature distributions on both the specimen and the container, under the following conditions:

1. With the specimen and the container raised to the same mean temperature of the test, T_0, and the specimen heater off.
2. With all the heaters as in condition 1, but with the specimen heater dissipating a carefully measured power, $\dot{Q}$.
3. With the central heater off again, the the container and specimen purposefully mismatched by a few degrees, with the mean temperature of the system again approximately T_0.

The subsequent analysis depends on the superposition of heat flows, for which a sufficient condition is the effective temperature independence of the conductivities of the system. The easiest way to achieve this is to ensure that, once T_0 is established in condition 1, all subsequent temperature excursions in conditions 2 and 3 be small. The differences between the sets of temperature readings in conditions 2 and 1 then isolate the effects due to the central heater, and equation (7), or its equivalent, can be solved for λ_1 provided that β is known. β, on the other hand, can be readily determined from an easily obtained solution of equation (5) applicable to conditions 1 and 3, wherein the difference of the specimen temperature distribution is related to that of the container. The three sets of measurements thus determine both λ_1 and ρ_1 at the mean temperature T_0; corrections can be applied to λ_1 both for the inaccuracies of the thin-rod approximation, and the effects of the temperature variations of the λ's. Examples of systems employing this technique will be briefly discussed in Section 5.

3.2.3. Guarded Systems

By definition, the *ideal* axial heat flow technique requires that the temperature distribution of the specimen be perfectly linear, and hence only a function of z. This ideal distribution, the goal of all guarded axial heat flow techniques, is given without any loss of generality by

$$T(z) = \frac{\dot{Q}z}{\pi r_1^2 \lambda_1} \tag{11}$$

$\dot{Q}$ is the power flowing along the specimen which, in the simplest of models, illustrated in Fig. 2, is dissipated by a shielded transverse (C-type) heater. Departures of the "actual" specimen temperature distribution, $T(r, z)$, from the idealized one, $T(z)$, thus give a measure of the quality of the system, of the errors to be expected from it in the determination of λ_1, or of the corrections that may have to be applied to the observed results. Determination of these departures is the central problem for the mathematical analysis, be it applied in hindsight, to evaluate existing systems, or at the start, to design a new one. The conventional prescription for achieving this idealized distribution is as follows: if one surrounds the specimen with a heated guard, the temperature of which matches that of the specimen at every axial location z, then the isotherms across the guard-specimen are all planar and no heat is exchanged between the specimen and the guard. Consequently, the heat flow in the specimen is purely axial, following the ideal relation.

In practice, this conventional wisdom contains two fallacies: firstly, it is impossible to match the temperature of the guard to that of the specimen at every point; secondly, even if the guard were matched to the specimen perfectly, equation (11) would not be satisfied, since part of the heat flowing down the insulation between the specimen and the guard must come from the specimen heater, and thus the observed linear specimen temperature distribution is due to a flux less than $\dot{Q}/\pi r_1^2 \lambda_1$. The second difficulty is minor, and can be made small by suitable design. The first, however, is serious, and constitutes the major difficulty in the design and evaluation of guarded axial systems: the rest of this section will be devoted to it.

The difficulties involved in perfect guarding can perhaps be better appreciated from a more realistic restatement of the problem. In a system with many heaters, such as our model one of Fig. 5, the heat flow patterns due to any one heater are independent of the power dissipation of the other heaters.* Thus, the radial component of the heat flow from the specimen heaters, although affected by the presence of the guard, is not affected by the power dissipated by the guard heaters. Effective "guarding" can only be achieved if the guard and its heaters are so designed and operated that the radial component of their heat

*This is true only if all conductivities are independent of temperature. For weak temperature dependences, it is an acceptable approximation (cf. Section 3.2.4). For strong temperature dependences guarding is impossible.

flow pattern compensates precisely that of the specimen heater at every point of the specimen surface.

To achieve such precise cancellation in practice is simply impossible. One can neither design a perfect system, nor operate it. Experimentally, temperature matching can only be contemplated at a few locations, two or three at most, and that temperature matching is itself inaccurate. Practical systems of *reasonable* accuracy and ease of operation can be designed, however, and the approach to such designs, based on a full-scale analysis of the problem, has been given before.[6] Here we shall limit ourselves to a brief discussion of a simple system with a guard matched to the specimen at two points only: the end point, $z = 0$, and one other "matching point," z_m. In our mathematical modeling of the system, the matching at $z = 0$ is automatic and perfect, and hence will only be concerned with a single temperature match at z_m, accomplished through adjustments of a single guard heater. In practice, other heaters are necessary in the system, to achieve both the test temperature and the match at $z = 0$; these are ignored in the following discussion, but can be readily treated by suitable manipulation of the model.[6]

Two important design questions will be considered: the error in the specimen temperature distribution, and hence the λ_1 derived from it, at perfect match; and the sensitivity of the system to the inevitable departure from a perfect match encountered in practice.

In general, perfect matching of the guard to the specimen at a small number of points (two in our case) ensures neither linearity of the specimen temperature distribution at all z, nor, when linearity occurs, its correspondence to the ideal distribution of equation (11). The latter point is important, for it shows that linearity of the observed distribution is, by itself, no guarantee of acceptable operation of the system; errors may still exist caused by heat exchanges between the specimen and its surroundings in the nonlinear, mismatched but unobserved region of the specimen, close to its heaters. The departure of the observed temperature distribution from the ideal one can be viewed as its error, and is a function of the geometry of the system, the boundary conditions imposed upon it, the type and lengths of the specimen and guard heaters, and the ratios of the conductivities of the components. It is illustrated in Figs. 8–11 for some representative values of the various relevant parameters.

Figure 8 illustrates two separate aspects for a typical system: the variation of the specimen error with the position of the matching point, z_m, and the effect of the external boundary conditions, at r_4. From it can be seen that if the errors in the specimen temperature are to be kept within the acceptable limit of $\pm 1\%$, then the matching of the guard to specimen cannot be made in the immediate vicinity of the heaters ($z_m \sim l_0$), but must be made further down the length ($z_m \sim 0.8 l_0$). Even then unacceptable errors result unless the guard itself be roughly thermally shielded. In practice, the latter means that the guard must

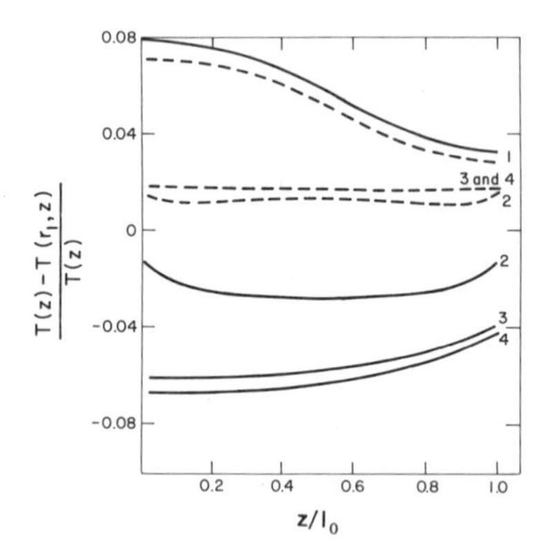

FIGURE 8. Errors in a guarded specimen temperature distribution for a variety of conditions. 1, $T(r_4, z) = 0$; 2, $T(r_4, z) = T(r_1, l_0/2)$; 3, no radial losses at r_4; 4, matched furnace. Solid lines, guard matched to specimen at $z_m = l_0$; dashed lines, guard matched at $z_m = 0.8 l_0$. The system parameters are $l_0/r_1 = 10$, $r_4/r_1 = 4$, $r_3/r_1 = 2$, $(r_3 - r_2)/r_1 = 0.1$. The guard has one longitudinal heater on it, $\theta(B)/r_1 = 0.6$, $z(B)/l_0 = 0.94$. The conductivity ratios are $\rho_1 = \rho_2 = \rho_3 = 100$.

be surrounded by a furnace matched roughly in temperature to it, a condition subsequently referred to as "matched furnace."

The dependence of the specimen temperature error on the matching point, z_m, is further illustrated in Fig. 9. Clearly, the exact position of the matching point is immaterial, as long as it is not very close to the guard, or specimen, heater. This figure also shows the strong dependence of this error on the length of the guard heater: the shorter the heater, the smaller the error. The limit, a guard heater of zero axial dimension, equivalent to a "transverse" specimen heater, would result in perfect matching. This, however, is impractical, and one has to be satisfied with as short a heater as possible. Both the illustrated features are the consequence of the strong variation of the temperature in the immediate heater region, which settles down quickly away from it. The shorter this region, the less the effect of this perturbation, while any attempt to match temperatures within the region is bound to lead to unsatisfactory results.

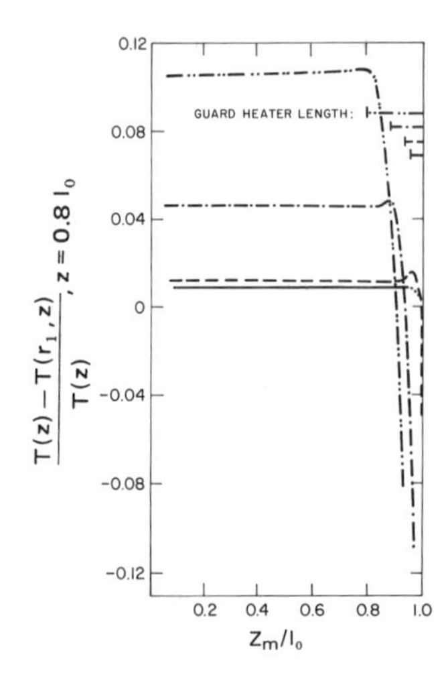

FIGURE 9. Errors in the specimen temperature distribution as a function of the guard to specimen match point, z_m, for a variety of guard heater lengths. $l_0/r_1 = 20$, $r_3/r_1 = 2$, $(r_3 - r_2)/r_1 = 0.4$, $\rho_1 = \rho_2 = 100$.

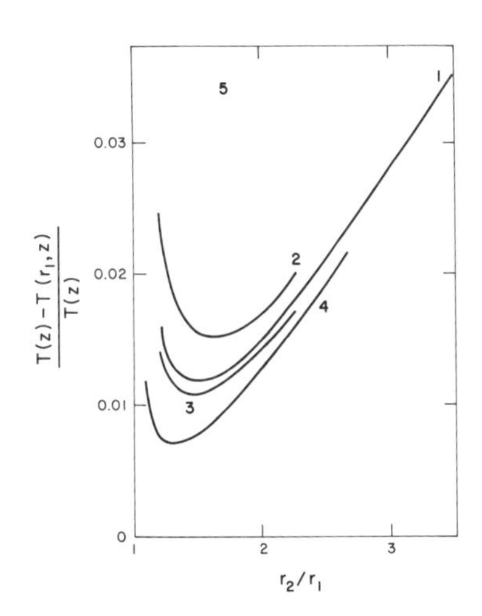

FIGURE 10. Errors in the specimen temperature distribution for a variety of systems as a function of r_2/r_1 for $\rho_1 = \rho_2 = 100$. Dimensions of these systems, in terms of the specimen radius, r_1, are as follows. 1, $l_0 = 20, r_3 - r_2 = 0.4; \theta(B) = 0.4$; 2, $l_0 = 20$, $r_3 - r_2 = 0.8$, $\theta(B) = 0.4$; 3, $l_0 = 20$, $r_3 - r_2 = 0.2$, $\theta(B) = 0.4$; 4, $l_0 = 20$, $r_3 - r_2 = 0.4$, $\theta(B) = 0.2$ and $l_0 = 10$, $r_3 - r_2 = 0.4$, $\theta(B) = 0.2$; 5, $l_0 = 20, r_3 - r_2 = 0.4, \theta(B) = 0.5$ and $l_0 = 50, r_3 - r_2 = 0.4, \theta(B) = 0.5$.

Figure 10 shows the variation of the specimen temperature error with details of the guard geometry. Three features are here illustrated. Firstly, that for any particular specimen geometry, (r_1, l_0), there is an optimum guard diameter, r_2. Secondly, that the errors are independent of l_0 for an otherwise fixed geometry. And thirdly, that the errors are strong functions of the guard thickness, $r_3 - r_2$, and, as already shown in Fig. 9, of the guard heater length. The existence of a minimum in the error when viewed as a function of the guard diameter is a unique feature, in the sense that all other parametric variations produce optimum results only in the limit. It is a consequence of the competition of two sources of error. One is the heat conducted axially in the insulation between the specimen and the guard, part of which must come from the specimen heater. This obviously decreases with decreasing r_2/r_1. The other is the heat exchanged between the specimen and the guard in the immediate vicinity of the heaters, a region where temperature matching cannot be accomplished. This error increases with decreasing r_2/r_1, and the sum of the two results in the minimum. Note that the second effect is absent in specimens insulated by vacuum, and hence one would not expect a minimum in the error in this case; this is indeed confirmed by calculation.[5]

The dependence of the specimen temperature error on the specimen length and on the guard heater length is further illustrated in Fig. 11. There it is seen that if the heater length is kept constant (dashed line), then the error is independent of l_0. Clearly, however, it is the absolute length of the guard heater that is important; scaling it with respect to l_0 results in a serious increase in the error as l_0/r_1 is increased (solid line). Note that all these illustrations refer to $\rho_1 \equiv \rho_2 = 100$, and that the errors shown are roughly inversely proportional to ρ_1.

The performance of a system at perfect match, is however, only one

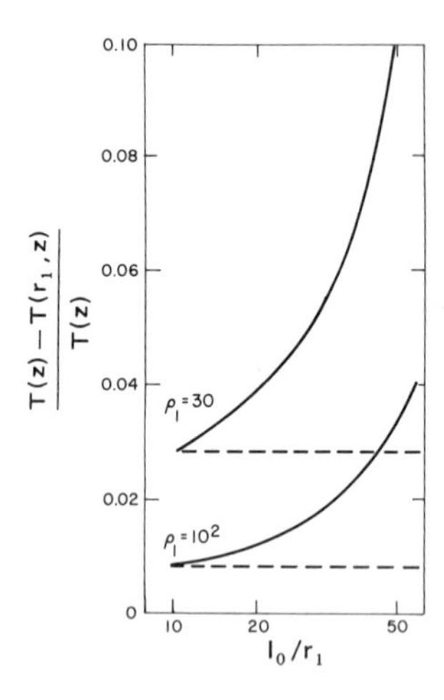

FIGURE 11. Error in the specimen temperature distribution as a function of specimen length, l_0, $r_2/r_1 = 1.6$, $(r_3 - r_2)/r_1 = 0.4$. Solid lines $\theta(B) = 0.02 l_0$; dashed lines $\theta(B) = 0.2 r_1$.

aspect of a guarded system; another, equally important one, is its sensitivity to departures from perfection, unavoidable in practice. Two factors conspire against perfect matching: the intrinsic errors of the temperature sensors, and one's inability to locate these sensors on the specimen and the guard at precisely corresponding positions. The second difficulty is further compounded by unequal thermal expansions in cases where the guard and specimen are made of different materials, the usual situation in systems designed for general utility.

A convenient way to study the effects of imperfect matching is through the "mismatch parameter" (MP), describing the fractional change in the specimen temperature at (r, z), $\Delta T(r, z)/T(r, z)$ caused by a (fractional) temperature mismatch at z_m of magnitude δT, where $\delta T = 1 - T(r_3, z_m)/T(r_1, z_m)$; the MP is defined as $[\Delta T(r, z)/T(r, z)]/\delta T$. The MP is a function of (r, z), and of all the geometrical and thermal parameters of the system, and, like the error at perfect match, can be systematically investigated through the mathematical analysis.[6] Figures 12 and 13 illustrate typical results for the MP, one as a function of ρ_1 for one specific geometry, the other as a function of r_2/r_1 for a variety of geometrical parameters for $\rho_1 \equiv \rho_2 = 100$. Noteworthy is the tremendous "amplification" of the matching error in the specimen temperature for $\rho_1 < 500$. For instance, for $\rho_1 = 100$, a 1% error in matching* affects the specimen temperature by 5%, and, since the effect depends on z, can lead to serious errors in λ. This amplification of the matching error results from the complex interaction between the specimen and the guard, which cannot be predicted by approximate analyses. For instance, the thin-rod approximation with a guard temperature

*Note that the mismatch δT is taken relative to the specimen temperature measured with reference to the specimen bottom ($z = 0$) and not its "absolute" value. Therefore, a 1 K matching error in a specimen at 1000 K but supporting a longitudinal gradient yielding 10 K over the length z_m, amounts to a 10%, not a 0.1%, matching error.

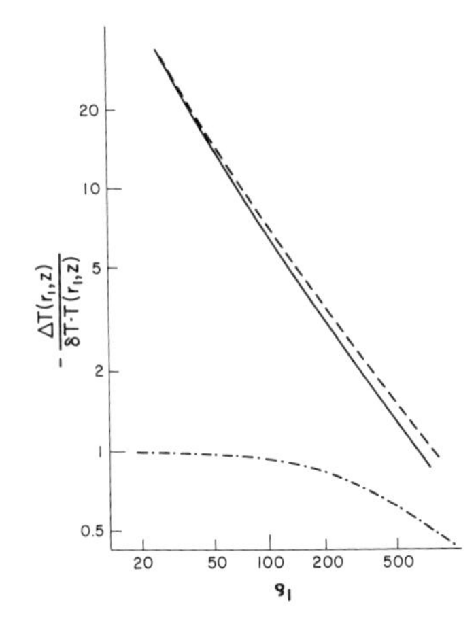

FIGURE 12. Mismatch parameter as a function of ρ_1. $l_0/r_1 = 20$, $r_3/r_1 = 2$, $(r_3 - r_2)/r_1 = \theta(B)/r_1 = 0.4$, $z_m = 0.8 l_0$. Solid line, $z = 0.8 l_0$; dashed line, $z = 0.2 l_0$; dash-dot line, predicted from approximate calculation.

assumed to be linear but mismatched by δT at z_m totally fails to predict the MP (Fig. 12); the simple assumption of linearity of the guard temperature, which makes it unresponsive to specimen temperature changes, wipes out completely the amplification of the mismatch.

As can be seen from the illustrations, both the error at perfect match and the MP are functions of the geometrical and thermal parameters of the system, and a good design must attempt to minimize them both. For the simple system that we have here chosen for illustration, one with a single explicit matching point and a matched furnace, some general conclusions can be drawn relatively easily from the mathematical analysis. As could be expected, minimization of the two factors makes different demands on the geometry of the system. Some of these demands can be reconciled; others conflict, and must be resolved by compromise. For instance, one has a free hand in the choice of the specimen

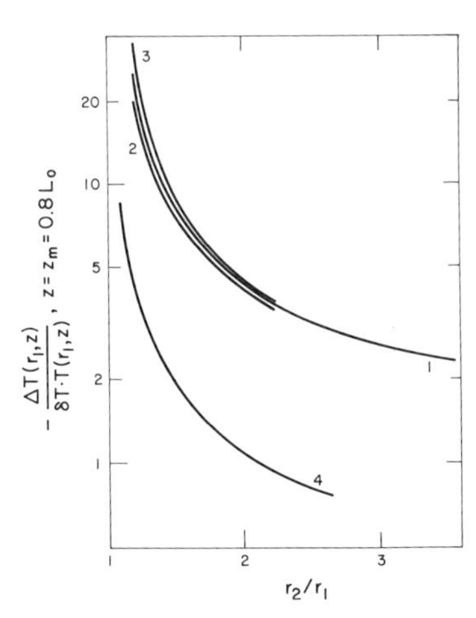

FIGURE 13. Mismatch parameter as a function of r_2/r_1, for a variety of systems. The system parameters in terms of the specimen radius, r_1, are as follows: 1, $l_0 = 20$, $r_3 - r_2 = 0.4$, $\theta(B) = 0.4$; 2, $l_0 = 20$, $r_3 - r_2 = 0.8$, $\theta(B) = 0.4$; 3, $l_0 = 20$, $r_3 - r_2 = 0.2$, $\theta(B) = 0.4$; 4, $l_0 = 10$, $r_3 - r_2 = 0.4$, $\theta(B) = 0.2$. $\rho_1 = 100$.

dimensions: the smaller the ratio of l_0/r_1 the lower the mismatch parameter, while, for other parameters fixed in relation to r_1, the variation of l_0/r_1 has no significant effect on the errors in the temperature distribution. Equally, one can select as low a value of $\theta(\mathrm{B})/r_1$, as is physically possible, minimizing in this way the error in the temperature distribution while leaving the mismatch parameter essentially unchanged. The thickness of the guard has also little effect on the mismatch parameter, and can be chosen on the basis of the temperature distribution once the guard diameter is fixed.

Conflicting demands are met in the choice of the matching point, z_m, and in the diameter of the guard. The former can be resolved relatively easily: for a short guard heater one would select the match point about one heater length [$2\theta(\mathrm{B})$] below its end, which is sufficiently removed from it not to affect the temperature distribution yet close enough to have little adverse effect on the mismatch error. No such effective resolution is possible in the conflict over the guard diameter; the choice here has to be made on the basis of how much accuracy one is willing to sacrifice in order to compensate for one's inability to measure precisely the temperature as a definite location. This problem has no ready answer, however, as is clear from Fig. 13, its importance decreases with decreasing l_0/r_1. In general, we may conclude that guarded systems can be designed with errors due to heat flow problems of about 1%, provided that the specimen conductivity is not too small ($\rho_1 > 100$) and that careful attention is paid in the design to the optimization of the geometrical parameters.

Guarded specimen systems more complex than the one here discussed can also be treated mathematically, and some specific examples have been reported.[6] However, general conclusions or systematic trends cannot be readily discerned, as the number of possibilities and permutations is prohibitive; such problems have to be dealt with on an individual basis, as required.

3.2.4. Temperature-Dependent Conductivities

The only serious limitation of the mathematical analysis discussed above is the assumption that the conductivities are independent of temperature. In certain problems, notably those dealing with single, isotropic bodies where the prescribed boundary conditions involve *either* the temperature or the heat flux, $-\lambda\bar{\nabla}T$, the temperature variation of λ can be treated exactly analytically through the substitution

$$\theta = \frac{1}{\lambda_0}\int_0^T \lambda(T)\,dT \qquad (12)$$

For steady state, this changes the pertinent differential equation from $\bar{\nabla}\cdot(\lambda\bar{\nabla}T) = 0$ to the usual Laplace equation $\nabla^2\theta = 0$. Hence, known solutions for constant conductivity can be taken over directly for θ, and the inverse transformation of (12) then yields T in terms of $\lambda(T)$.

Unfortunately, this approach cannot be extended to a multibody system, where at each two-body interface *both* T and $\lambda\overline{\nabla T}$ must match; the substitution of equation (12) will preserve the heat flux across the boundary, but is incapable of coping with the temperature matching, except in the unlikely event that both temperature-dependent conductivities are proportional to each other for all T. In general, this problem involves the solution of nonlinear equations, and is analytically intractable; recourse must be had to numerical techniques for cases of specific interest.

However, some general conclusions can be drawn from approximate treatments applicable for large ρ_1. As pointed out in the Appendix, if the λ's are not functions of temperature, then the specimen temperature can always be written in the form

$$T(r, z) = \sum_j c_j(\rho_1)\dot{Q}_j(\mathrm{C}) + \sum_l a_l(\rho_1)\dot{Q}_l(\mathrm{A}) + \sum_m b_m(\rho_1)\dot{Q}_m(\mathrm{B}) \quad (13)$$

where the coefficients a_l, b_m, and c_j are not only functions of ρ_1, as here indicated, but of the other geometrical and thermal parameters as well. For large ρ_1, c_j, and a_l reduce to functions of the geometry only, while b_m decreases as $1/\rho_1$. As is intuitively obvious, the dominant contributions to T come from the specimen heaters ("C" and "A") and these are largely independent of ρ_1. Furthermore, in the limit $\rho_1 \to \infty$, the solution

$$\frac{1}{\lambda_1(0)} \int_0^{T(r,z)} \lambda_1(T)dT = \theta_1(r, z) = \sum_j c_j(\infty)\dot{Q}_j(\mathrm{C}) + \sum_l a_l(\infty)\dot{Q}_l(\mathrm{A}) \quad (14)$$

obtained from the standard substitution, is exact. It is therefore reasonable to assume that the solution

$$\frac{1}{\lambda_1(0)} \int_0^{T(r,z)} \lambda_1(T)dT = \sum_j c_j(\rho_1)\dot{Q}_c(\mathrm{C}) + \sum_l a_l(\rho_1)\dot{Q}_l(\mathrm{A}) + \sum_m b_m(\rho_1)\dot{Q}_m(\mathrm{B}) \quad (15)$$

is a good approximation to the correct solution for the temperature dependent conductivities for large ρ_1. One would expect the relative accuracy of T given by this relation to be of order

$$\left[1 - \frac{1}{\lambda_1(0)T} \int_0^T \lambda_1(T)\,dT\right] \Big/ \rho_1$$

The approximation described above states the essentially obvious fact that for a reasonably well insulated specimen, the only temperature dependence that matters is that of the specimen itself. This applies to both guarded and unguarded techniques. However, there is an important distinction between the role that this temperature dependence plays in the two situations. For the guarded technique, it produces second-order effects, i.e., it involves $d^2\lambda_1/dT^2$. For the unguarded technique, the effects are of first order, $d\lambda_1/dT$. This difference is important and easily explained. Qualitatively, in the unguarded system, the lateral heat losses are determined from the curvature of the specimen temperature distribution. Since $d\lambda_1/dT$ itself introduces a curvature as well, it effectively mimics a lateral heat loss, and, as these losses introduce first-order corrections to λ_1, so does $d\lambda_1/dT$. In the guarded heat flow method, on the other hand, the measured temperature difference is associated with a λ_1 for a mean specimen temperature, T_M, and while the position of T_M along the specimen will depend on $d\lambda_1/dT$, its magnitude will not; hence, the resultant $\lambda_1(T_M)$ is independent of the first derivative.

The whole problem of temperature-dependent conductivities may be stated as follows. The mathematical analysis will yield analytical expressions only under the assumption that all conductivities are temperature independent. As this obviously is not so, one expects large errors to result from such an analysis, due to the gross conductivity changes produced by the large temperature variations encountered in real systems operating well away from ambient temperature. The effects of these errors can be minimized by formulating the problem in terms of relatively small temperature differences due to changes in the power dissipated in some selected heaters, specifically the specimen heater in the unguarded technique, or the specimen heater and one matching guard heater in the guarded method (cf. Section 4). In this approach, the changes in the temperature-dependent conductivities are small as the changes in temperature are small, and they introduce relatively minor errors into the analysis, errors which turn out to be of first order in the unguarded specimen method, and of second order in the guarded one. Corrections for these errors may be obtained from an approximate analysis, valid for large ρ_1, which considers only the temperature dependence of the specimen conductivity, the dominant term, and ignores that of the other components.

3.3. Departures from Steady State

The thermal conductivity is calculated under the assumption of steady state. Unfortunately, this ideal condition cannot normally be attained experimentally. The best that one can hope for is a slow, steady drift, and it is necessary to correct the calculated results for the effects of it. Given a detailed time–temperature history of the whole system, exact corrections to the conductivity can be computed. Such a complex approach is, however, seldom necessary. In properly operated systems the drift corrections are small, and approximate treatments suffice.

In axial heat flow techniques, the important source of temperature instability is the specimen "base" heater, which raises the specimen to its operating temperature T_o. The specimen heater, H_1 in our schematic illustration in Fig. 2, provides only the small gradient along the specimen, ΔT, necessary for the determination of λ. ΔT is usually anywhere from one to two orders of magnitude smaller than T_o and hence, for equal power regulation to the heaters, so is its contribution to the temperature fluctuations of the specimen. The guard heaters are also not as important as the base heater, for their power fluctuations are smoothed by the thermal resistance between guard and specimen, and are thus reduced by a factor on order of ρ_1.

The effects of the drifts of the base heater can be calculated approximately as follows: if we assume that the heat flow is one dimensional, and that the temperature of the specimen varies linearly with time at a (constant) rate β_0 throughout, then the measured temperature difference between the specimen thermocouples at z_1 and z_2, ΔT_m, is given by

$$\Delta T_m = \Delta T_c + \frac{\beta_0}{\alpha}(z_2 - z_1)\left(\frac{z_2 + z_1}{2} - l_0\right) \tag{16}$$

where ΔT_c is the correct temperature difference due to H_1 and α is the thermal diffusivity of the specimen. The measured λ, λ_m, is then related to the correct λ_c through

$$\lambda_c = \lambda_m \left(1 - \frac{\beta_0 t_0}{\Delta T_c} \cdot G_0\right)^{-1} \tag{17}$$

where $G_0 = (z_2 - z_1)/2l_0$ and $t_0 = l_0{}^2/\alpha$; t_0 is the "characteristic time" of the specimen. The necessity for such a characteristic time becomes apparent from a more detailed analysis,[8] wherein one supposes that the base heater starts drifting at time $t = 0$, and drifts so that the base temperature (at $z = 0$) varies linearly with time: $T(0, t) = T_0 + \beta_0 t$. The resultant measured λ_m is then a

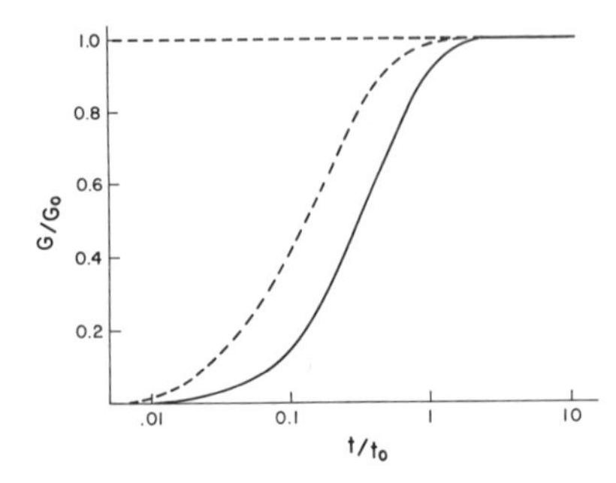

FIGURE 14. The ratio of the exact parameter G, to the approximate one, G_0, as a function of t/t_0, where t_0 is the characteristic time of the system. The dashed line shows the same ratio if G_0 is calculated from the specimen thermocouples, rather than the true temperature drift of the heater (see text).

function of t/t_0, and is related to λ_c through the same equation as above (17), except that G_0 is replaced by $G(t/t_0)$. The ratio $G(t/t_0)/G_0$ is shown in Fig. 14 for one specific geometry, $z_2/l_0 = 0.8$ and $z_1/l_0 = 0.2$.

A casual glance at this figure may lead one to conclude that the approximate relation seriously underestimates the true drift correction for $t < t_0$. This, however, would only be the case were one to use for the β_0 therein the value derived from the time variation of the base temperature, at $z = 0$. If, instead, one derives an apparent β_0' from the observed time variation of the same specimen temperature sensors as are used to measure ΔT, a natural procedure, then this β_0' is itself a function of t/t_0 even though β_0 is constant. Insertion of β_0' into equation (17) removes some of the discrepancy between $G(t/t_0)$ and G_0, as shown in the figure. The simple approximate equation can therefore be used satisfactorily at all times, provided that the drift is small and uniform, and that its magnitude is determined from the specimen temperature sensors.

As an example, let us consider a specimen 10 cm long, made from an alloy with an α of 0.1 cm^2/sec; t_0 is then 10^3 sec. For $(z_2 - z_1) \sim 0.6 l_0$, $\Delta T_c \sim 10$ K, and a drift of 1 K/h, the drift correction to λ is approximately 0.8% for $t \geqslant t_0$. For $t \sim t_0/10$, it is 0.3%, and the fact that equation (17) predicts it with an error of about a factor of 2 is not very important. Note that for the same specimen with a drift of 0.1 K/min the correction for $t \geqslant t_0$ becomes 5%, and the error at $t \sim t_0/10$ due to the inaccuracies of equation (16) is no longer negligible if the required overall inaccuracy of λ is on the order of 1%. This is not to say that such high drift rates must be avoided, only that when they do occur, more sophisticated corrections than the one above must be used in the analysis of results. This may well be the right way to proceed in the future in view of the availability of fast data acquisition equipment, as such an approach would remove the chief drawback of so-called steady-state techniques, their intrinsic slowness.

3.4. *Power Input*

The determination of λ involves the measurement of the temperature distribution in the specimen due to a given heat flux within it. Usually, this heat flux is produced by a specimen heater powered from a stabilized electrical dc source, and, since four-terminal dc power measurements can be performed with great precision, the power *generated* in the heater can be determined with high accuracy. Unfortunately, that measured power need not correspond to the heat flux in the specimen. In the first place, not all the power generated in the specimen heater may enter the specimen in the first instance, and in the second place, heat generated in other, auxiliary heaters may be also traversing the specimen. The auxiliary heaters which may contribute spurious heat flow fall into two categories: essential ones, such as those used for matching in guarded systems, and secondary ones, which merely serve to maintain the specimen at the ambient temperature. The effects of the former are an essential problem of heat flow, and have already been discussed in Section 3.2.3; the effects of the latter may be readily minimized by a "double-difference" technique, as explained in Section 4. In this section, we limit ourselves specifically to problems connected with the specimen heater itself.

A typical heater used for axial heat flow techniques is illustrated schematically in Fig. 15. The actual heat source is a wire of resistance R embedded in the body of the heater, having current and potential leads for the measurement of the power, $\dot{Q}$. The heater itself, H, is either attached firmly to the specimen, or may be part of it if the specimen is electrically insulating. It is surrounded by a thermal shield, S. In general, there are six important considerations entering the design and construction of such a heater:

(1) The heat exchange between the specimen and the shield must be sufficiently small relative to $\dot{Q}$ not to affect the accuracy of λ. This calls for as large a thermal resistance between specimen and shield as possible, and implies a large physical separation.

(2) The shield, while matching the temperature of the heater, must not contribute any heat flux to the specimen proper, via the exposed specimen periphery, for instance. This calls for as close a physical proximity of the shield to the heater as possible, in direct contradiction to (1). It also implies that both heater and shield must be physically small, as large heaters and shields tend to affect severely the specimen temperature distribution, and, being "essential"

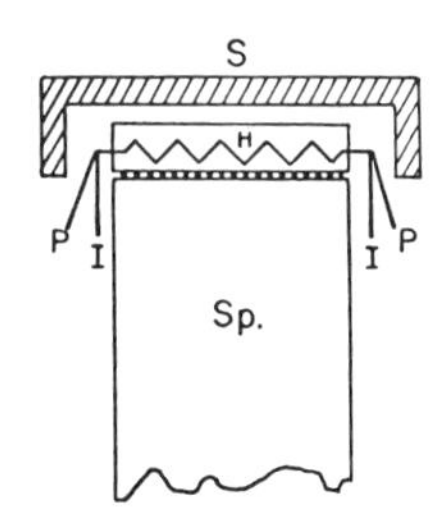

FIGURE 15. Schematic diagram of a specimen heater. Sp., specimen; H, heater, with current (I) and potential (P) leads, attached to the specimen; S, shield.

heaters, their effect cannot be minimized by the difference technique mentioned above.

(3) The electrical resistance of the current leads to the heater must be much less than the resistance of the heater itself, otherwise the power generated in those leads outside the potential probes causes complications. Even when thermally tied down to the shield, guard, or special "binding post," there is considerable uncertainty in the fraction of that unmeasured power which flows into the specimen directly or indirectly.

(4) The thermal conductance of all the leads into the heater must be small, so that they do not cause significant heat exchanges due to experimentally unavoidable temperature mismatches. This is simple to achieve for the potential leads, but, since λ and σ are roughly related for all metals, runs contrary to (3) for the current leads. Thermal tie-points for these leads at temperatures matched to those of the heater are essential.

(5) The thermal resistance between the specimen heater and the specimen must be as low as possible, ideally zero. Large contact resistances result both in augmented heat leaks from the heater, and in heat exchanges between the specimen and the heater shield through the surrounding insulation. Hence the frequent employment of heaters that are embedded in the specimen.

(6) The heater resistance wire must be thermally well bonded to the heater body, otherwise its temperature will rise substantially during operation, especially in vacuum. This increases the heat leaks mentioned in (4), and effectively precludes remedial measures through attempted temperature matching, unless that matching be done on the heater wire itself, a tricky proposition at best.

The implementation of these design criteria, and the resolution of their conflicting requirements, is very much a function of the operating temperature of the envisaged system. It is easiest to accomplish at low temperatures ($T < 100$ K) where the high thermal insulation value of the usual environment, vacuum, and the availability of all kinds of suitable glues, epoxies, electrically insulating varnishes, and physically small wire-wound resistors, make the design and fabrication of the heater a straightforward proposition; here, almost anything goes. For intermediate temperatures, $T \leqslant 600$ K, reasonably satisfactory design can still be achieved with small resistors "plotted" into the specimen, and surrounded by a close-fitting matched shield; one such approach[9] is shown

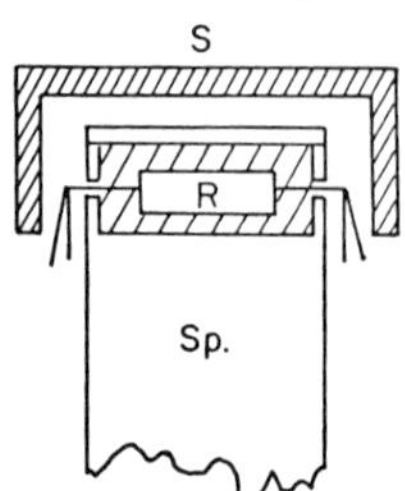

FIGURE 16. Schematic diagram of a specimen heater consisting of a resistor, R, "potted" into a cavity in the specimen, Sp. S is the shield.

schematically in Fig. 16. For the high-temperature region ($T > 600$ K), the design of the specimen heater is very important and very difficult; indeed, the question arises whether shielded heaters can be employed at all. Recourse then has to be taken to double-ended, symmetrical specimens with the heater embedded in the middle.[10] Even then substantial corrections have to be applied to the results to allow for unavoidable heat losses from the heater windings.

3.5. *Temperature Measurement*

In this section we will presuppose familiarity with temperature measurement techniques in general, for which many excellent review articles exist,[11] and discuss only specific aspects of it pertinent to the measurement of λ. Steady-state techniques have three distinct requirements related to temperature measurement:

1. the determination of the mean temperature, T_0, with which to associate the measured λ;
2. the determination of the temperature gradient on the specimen; and
3. the establishment of the various temperature matches required for guarding or thermal shielding.

The determination of the mean temperature involves an absolute temperature measurement, one, however, of low accuracy by thermometric standards. If the experimental uncertainty in the measurement of λ is $\delta\lambda$, then there is no necessity for knowing the mean temperature of the test with an uncertainty much less than δT given by $(d \ln \lambda/dT)\delta T \leqslant \delta\lambda/\lambda$. As $\delta\lambda/\lambda$ is at best approximately 1% and, for $T > 100$ K, $|d \ln \lambda/dT|$ is seldom much larger than 10^{-3} K^{-1}, the permissible uncertainty in T_0 is on order of 10 K, a very modest requirement which can be satisfied by a variety of temperature sensors. Only at very low temperatures, where $(d \ln \lambda/dT)$ may become quite large ($\gtrsim 1$ K^{-1}), does the required precision increase to a point where recourse may have to be had to precisely calibrated thermometers.

The determination of the gradient along the specimen presents a totally different requirement. Here, it is necessary to measure precisely a small temperature difference, ΔT, between two points of precisely known separation, Δz, the required precision being directly related to the accuracy required for λ. By metrological standards, such precision for ΔT is not difficult to obtain. However the necessity of knowing precisely the spatial separation to which it pertains severely restricts the size of the thermometers that can be employed for all but the lowest temperature range. Only for $T < 40$ K does there exist sufficiently good thermal insulation to allow the thermometers to be physically separated from the specimen while maintaining effective thermal contact through high-conductivity leads. Such an arrangement permits the thermometers to be of substantial size without compromising the precision of separation, determined only by the lead size. At all other temperatures the thermometers must be in intimate contact with the specimen to ensure temperature equality, and thus they must be of dimensions less than $\Delta z \cdot (\delta\lambda/\lambda)$, typically 0.5 mm. It is this requirement which is responsible for the almost universal choice of thermocouples for the temperature sensors for all temperatures above about 10 K, in spite of their somewhat questionable thermometric performance.

While thermocouples may present problems in the determination of the gradient on the specimen, the resolution of which we will discuss below, they are almost ideal sensors for the null detectors required for temperature matching. They are almost universally used in such applications, in a differential configuration. The only requirements there are adequate sensitivity, which need not be accurately known, relative stability, and small physical size. Thermocouples suitable for this purpose can be readily found for the whole temperature range where axial heat flow techniques are successfully used and temperature matching is required.

Thus, from the point of view of thermometry, the temperature range spanned by axial heat flow techniques divides into two overlapping domains. The first, below about 40 K, where size is no restriction, a number of different sensors can be employed, such as gas and vapor pressure thermometers, electrical resistance and magnetic susceptibility thermometers, and certain thermocouples. Since I have had no personal experience in this range, I prefer to refer the reader to other review articles dealing with the techniques of temperature measurement[12] and their application to the measurement of λ.[13] In the other domain, above about 10 K, where heat leak problems become progressively more and more important, thermocouples are the conventional sensors.

In utilizing thermocouples, the following precautions must be exercised:

(1) The thermocouple junction must be in intimate thermal contact with the specimen. In metallic specimens, one may wish to have the junction, and only the junction, in electrical contact with it. The thermocouple separation can then be measured electrically, and the wires of the thermocouple can be used to measure the electrical resistivity of the specimen concurrently with λ. Utilizing the same probes for both measurements presents a number of advantages, as will be discussed in the next section.

(2) The thermocouple leads near the junction must not be exposed to any temperature gradients, for two reasons: firstly, heat flowing along the thermocouple will invariably produce a temperature drop between the specimen and the junction due to contact thermal resistance, and thus cause the thermocouple to read incorrectly; secondly, the thermal emf of the thermocouple is poorly defined in the region of the junction, where alloying may occur – any gradient along the region yields erroneous emfs. This means that either the thermocouple must be thermally tempered to isothermal regions of the specimen,[14] or run

over and be thermally tempered to an auxiliary post or to the guard matched in temperature to the specimen.[15] It is important to remember that any thermal tempering must not upset the knowledge of the location of the temperature that the thermocouple records; for instance, good thermal tempering to the specimen along significantly nonisothermal regions wipes out any possibility of determining the effective Δz precisely.

(3) Adequate allowance must be made for the well-known drawbacks of thermocouples as temperature sensors.[16] Basically, thermocouples are poor absolute sensors, and must normally be calibrated for successful absolute operation. However, since the thermocouples generate their signal not in the isothermal regions, which the test environment usually tends to be, but in regions where temperature gradients exist, calibration is largely meaningless unless the gradients and their locations can be duplicated between the test and calibration runs. This is only possible for *in situ* calibrations in the λ apparatus, and where the absolute temperature must be known accurately ($T \leqslant 200$ K), or the thermocouples used are particularly variable and irreproducible, provision for such calibration must be built into the system.[15] In more normal circumstances, one obviates the inadequacies of the thermocouples by modifying the procedure so that the small temperature difference, ΔT, necessary for the determination of λ, is not simply a small difference between two large readings obtained on different thermocouples, but a difference of differences, $\Delta\,\Delta T = \Delta T_2 - \Delta T_1$, where ΔT_2 and ΔT_1 are changes in the output of the thermocouples located at z_2 and z_1 due to changes in the power dissipation of the specimen heater, $\Delta Q = \dot{Q}_2 - \dot{Q}_1$. This procedure effectively calibrates the thermocouples relative to each other, and eliminates individual differences between them.[17] The result then depends on the thermopower of the thermocouple, $S(T)$, and not differences of emfs, $\Delta E(T)$, where $S(T) = dE(T)/dT$. $S(T)$ is more reproducible from thermocouple to thermocouple than $E(T)$, and less sensitive to differences in cold work on installation and the contamination that may occur in the experiment, particularly at high temperature. The utilization of the double-difference technique is absolutely essential for high-accuracy work when employing thermocouples without *in situ* calibration, and is valuable even in the case of the latter, for it facilitates such a calibration.

4. OPERATING PROCEDURES

In the previous sections we have discussed design principles from the point of view of heat flow patterns, and practical difficulties connected with power generation and temperature measurement. Once the design of the system has been optimized for the expected specimen properties and the temperature range required, there remains the question of optimum operating procedures, which will realize the potential accuracy of the experimental equipment. The most important of the procedures is the one which permits the substitution of temperature differences obtained on individual sensors for their absolute outputs, variously called "double-difference,", "base-line," or "internal calibration" techniques. This not only improves the accuracy of λ for thermometric reasons, as pointed out in the section above, but isolates the effects due to a specific heater, and thus minimizes error due to unavoidable stray heat flows. As will become clear further on, however, the requirement to achieve both these effects need not be identical, and the best practical procedure to be adopted in any specific case may well be a compromise that has to be determined by detailed calculations.

The principles of this procedure are most clearly demonstrated in the case of the "unmatched" technique in the ideal situation where all the components have temperature independent conductivities. As shown in the Appendix, the specimen (surface) temperature at z, $T(z)$, can then be written as

$$T(z) = \sum_n f_n(z)\dot{Q}_n/\lambda_1 \tag{18}$$

where the sum over n extends over all heaters in the system, and $f_n(z)$ depends implicitly on its geometry and thermal parameters. If we now measure $T(z)$ for two values of the power dissipated in some selected specimen heater, s, denoted by $\dot{Q}_s(1)$ and $\dot{Q}_s(2)$, then

$$\Delta T(z) = T_2(z) - T_1(z) = \frac{f_s(z)}{\lambda_1}[\dot{Q}_s(2) - \dot{Q}_s(1)]$$

$$= \frac{f_s(z)}{\lambda_1}\Delta\dot{Q}_s \tag{19}$$

For two specimen thermometers at z_1 and z_2

$$\Delta\Delta T = \Delta T(z_2) - \Delta T(z_1) = \frac{\Delta\dot{Q}_s}{\lambda_1}[f_s(z_2) - f_s(z_1)]$$

$$= \frac{\Delta\dot{Q}_s}{\lambda_1}\Delta f_s$$

and

$$\lambda_1 = \frac{\Delta\dot{Q}_s \cdot \Delta f_s}{\Delta\Delta T} \tag{20}$$

becomes the equation determining the specimen thermal conductivity. Δf_s is determined from the principles of the specific technique employed and the measured geometry. $\Delta \dot{Q}_s$ is measured directly; frequently $\dot{Q}_s(1) = 0$, and test 1 is then called the "base line." $\Delta \dot{Q}_s$ then equals $\dot{Q}_s$, the power dissipated in the specimen heater during test 2, the "gradient" test. Note that

$$\begin{aligned}\Delta \Delta T &= [T_2(z_2) - T_1(z_2)] - [T_2(z_1) - T_1(z_1)) \\ &= [T(z_2) - T(z_1)]_{\dot{Q}_s(2)} - [T(z_2) - T(z_1)]_{\dot{Q}_s(1)} \\ &= \Delta T(2) - \Delta T(1)\end{aligned}$$

In many applications, ΔT can be measured directly along the specimen for each test, either by differential thermocouples on electrically insulating specimens, or, on metallic specimens, by a differential measurement incorporating the specimen. This procedure yields additionally the thermopower of the specimen, and, furthermore, removes the usual difficulties encountered in determining precisely a small difference of two large numbers. Note, however, that this technique requires careful "alignment" of the thermocouples, to ensure that both "legs" of every sensor touch the specimen at the same isotherm. Otherwise, corrections may have to be applied.

As mentioned before, the double-difference technique accomplishes two functions: It isolates the effects of one specimen heater, and it minimizes the undesirable intrinsic differences between thermometers. The procedure required to isolate the effect of a specific specimen heater is, in our idealized case, absolutely clear: only that one heater must be changed between the two sets of readings, while all the other heaters must remain at the same power level. Although essential to accomplish the separation of heat flows, this is not the ideal procedure from the point of view of thermometry. There, the ideal operation is such that the mean temperature of each test is approximately the same, i.e., $T_2(z_2) + T_2(z_1) \sim T_1(z_2) + T_1(z_1)$. Obviously, this equality cannot be achieved when only one heater is adjusted, and the temptation exists to adjust some auxiliary heaters to maintain equality of mean temperatures. Such adjustments are permissible only if they can be shown to have identical effects on both specimen sensors, that is, if they contribute precisely zero to $\Delta T(2)$, an improbable occurrence. On balance, the thermometric errors being usually much less important than stray heat flows, the requirement for precise heat flow separation given above must prevail in the ideal case.

The situation becomes less definite in the most realistic case of temperature dependent conductivities. As discussed before, the specimen temperature is then no longer linearly dependent on the heater powers, and precise heat flow separation is impossible. One cannot therefore unequivocally maintain that only the specimen heater may be changed between the two tests; conceivably, more precise results could be obtained if several heaters were adjusted. Such a procedure, however, cannot be established by any general arguments, but must rely on detailed calculations for specific cases. In their absence, and knowing that the limit of large ρ_1 additivity of $\theta = (1/\lambda_0) \int_0^T \lambda(T)dT$ exists, the technique of single-heater adjustment is the best ***general*** procedure that can be employed. Of course, λ must now be calculated from a modified form of equation (20), where $\Delta\Delta T$ is replaced by an analogously defined $\Delta\Delta\theta$.

The technique for single-heater adjustment cannot, of course, be applied to guarded systems. Here, at least two heaters will have to be adjusted to maintain the specimen-guard temperature match, and in practice usually more than two, for matching must be achieved at several locations. Hence, the role of the double-difference technique in heat-flow separation is neither clear nor required in principle, and its thermometric benefit may well be the more important of the two. This being the case, one should obviously strive for equality of mean temperatures of the individual tests, and determine the effects that such a procedure may have on the principle of guarding through a detailed analysis of the specific system.

Having thus dealt with the method of operation, I should now like to conclude the section by a very brief discussion of the possible tests that can be performed to establish that the experimental system is indeed functioning satisfactorily. By far the best test of all is the measurement of some "standard" specimen of conductivity similar to that of the materials in which one is interested. Although, to the best of my knowledge, there are no "official," internationally accepted thermal conductivity standards, a number of materials have been measured by more than one laboratory with satisfactory agreement of results[17] – any one of these could form acceptable test specimens. Where this is impossible, the system may be checked out for internally consistent operation. For instance, the performance of the various heaters and effect of temperature mismatches can be predicted from model calculations and compared with experiment. Equally, the system may be used to measure λ in more than one way: as a test, a guarded linear system can be run in the unmatched mode, and consistency in the resultant λ tested. Furthermore, in a new system, more than one specimen of the same material should be measured, to determine errors due to differences in individual instrumentation. Finally, for metallic specimens, the two other "normal" transport properties, the electrical conductivity σ, and the thermopower S, should always be measured concurrently with λ. As mentioned above, the determination of S comes out automatically from the measurement of $\Delta\Delta T$ if the latter is measured differentially through the specimen. The measurement of σ involves only little extra instrumentation and very little extra time. The measurement of these additional properties allows one to characterize properly the specimen and make useful comparisons among

TABLE 1
Malfunctions Which Affect the Experimental Determination of the Various Transport Properties[a]

Property	Specimen contamination	Changes in geometry	Thermocouple contamination	Heater or heat flow problem
Electrical conductivity	√	√	—	—
Thermal conductivity	√	√	√	√
Thermopower	√	—	√	—

[a]"√" Indicates that the effect is strong, "—" that it is weak or null.

various experiments. Furthermore, it helps immensely in determining the possible malfunctioning of the apparatus. This is true particularly at high temperature where frequently, on cycling, the original results are no longer reproduced, without any clear-cut indication of the cause of the fault. Since the three transport properties are affected differently by the various possibilities, as shown in Table 1, inspection of which of them do or do not reproduce can usually pinpoint the trouble.

5. EXPERIMENTAL SYSTEMS

This section is a very brief summary of the material discussed heretofore. In it, we consider each temperature range in turn, restate the important problems in that range, and illustrate and briefly describe some successful systems that have been designed to overcome these problems. These so-called "successful" systems were selected on the basis of their experimental performance in relation to other apparatus. Except for the lowest temperature range, the selection is largely based on the already published[17] comparison of eight experimental systems with but minor updating. As mentioned in the Introduction, they serve to illustrate the variety of approaches that are possible, but are not meant to serve as blueprints for new work.

$T < 40\,K$. In this temperature range, the thermal insulation provided by vacuum is sufficiently good that purely axial heat flow is obtained for any reasonable specimen geometry without external guarding. The major problems in this area are connected with thermometry[12] and with heat leaks along the various leads that may be joined to the specimen,[18] and the specimen configuration may well be selected on these grounds and not, as is the case at higher temperatures, to satisfy the demands of the heat flow patterns. In consequence, the experimental arrangement may range from the fairly obvious rectilinear rod with heat sink, specimen heater, and two thermometers in between[19] to the interesting arrangement of Zaitlin and Anderson[20] with two small specimen heaters located between the single thermometer and the heat sink, and illustrated in Fig. 17. In the latter case, λ of the specimen may be computed either from

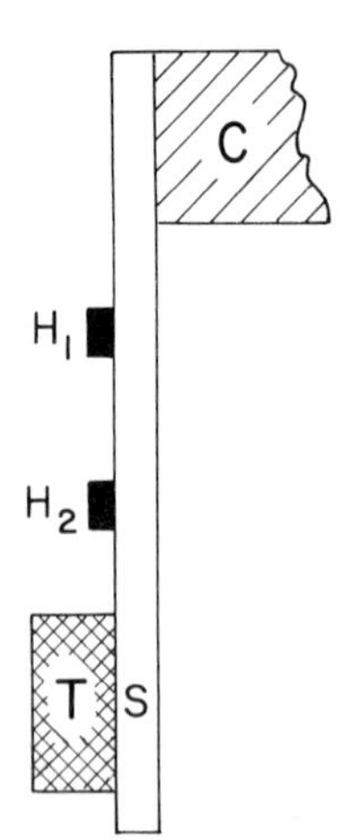

FIGURE 17. Experimental arrangement of Zaitlin and Anderson,[20] with one thermometer (T) and two heaters (H_1 and H_2) on the specimen (S); the specimen is attached to a cold finger (C).

the difference of the specimen temperature when equal power is dissipated by either specimen heater in succession, or from the difference in power required by either heater for the same specimen temperature, the choice depending on whether one trusts more the calibration of the thermometer or the quality of the heat sink.

$40 \leqslant T \leqslant 400$. This is the temperature range wherein the heat leaks, although no longer negligible, are still comparatively small, and their elimination either by guarding or the Forbes' bar approach is a relatively simple matter. The other problems normally encountered in measurements of λ are also manageable. Although thermal tempering of leads is important, it can be readily accomplished by attaching the wires to equithermal parts of the apparatus using the numerous

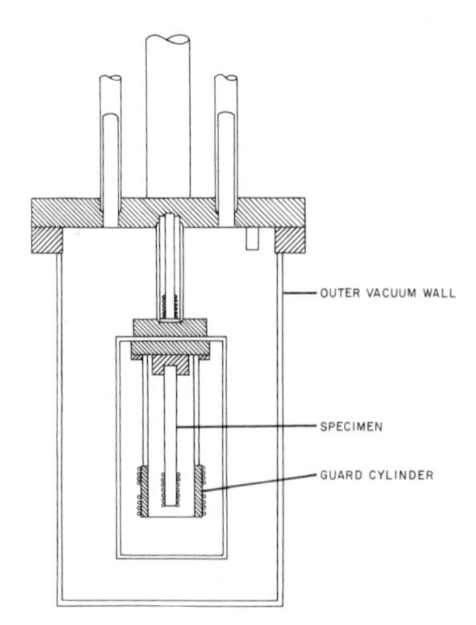

FIGURE 18. Guarded longitudinal heat-flow apparatus of Moore *et al.*[21]

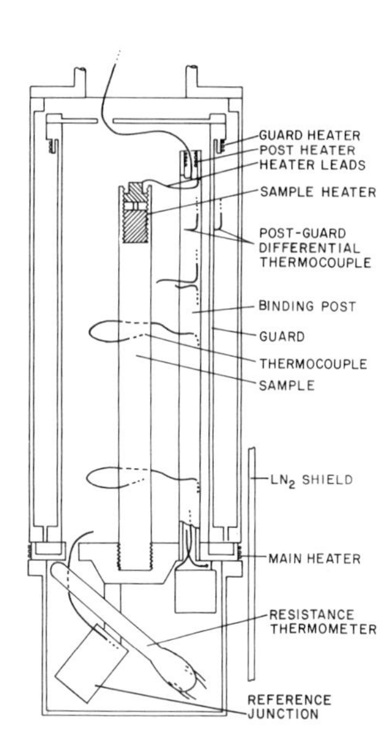

FIGURE 19. The guarded longitudinal heat-flow apparatus of Cook *et al.*[15]

varnishes or epoxies available. Temperature measurement in this range is probably at its most precise, and contamination, either of specimen or of the thermometers, usually nonexistent. In consequence, the measurement of λ is here at its best, and uncertainties can be reduced below the level of 1%.

Three axial heat flow systems are here illustrated, to show the variety of approaches. The first[21] (Fig. 18) and the second[15] (Fig. 19) are guarded systems. Both use thermocouples to measure the specimen temperature, but incorporate platinum resistance thermometers for their *in situ* calibration. Their specific differences lie in the construction of the specimen heater, one being a wire winding directly on the surface of the specimen while the other has a heater embedded inside it, and in the thermal tempering of the leads, one utilizing the guard for that purpose while the other employed a "binding post" matched in temperature to the specimen. The last facilitates the assembly of the apparatus which, due to the reactivity of the specimens for which it was designed, had to be accomplished under oil.

The third system[22] (Fig. 20) is of the Forbes' bar type. Here, the guard is thermally decoupled from the rest of the system, and has a uniform heater wound on it; thus, its temperature is essentially uniform, and can be adjusted to any value independently of the specimen. For systems in this category several sets of readings are taken at each temperature: with the specimen heater off and the guard temperature close to that of the specimen; with the specimen heater on, and the guard unchanged; and finally with the specimen heater off again, and the guard deliberately mismatched. As described before, these readings

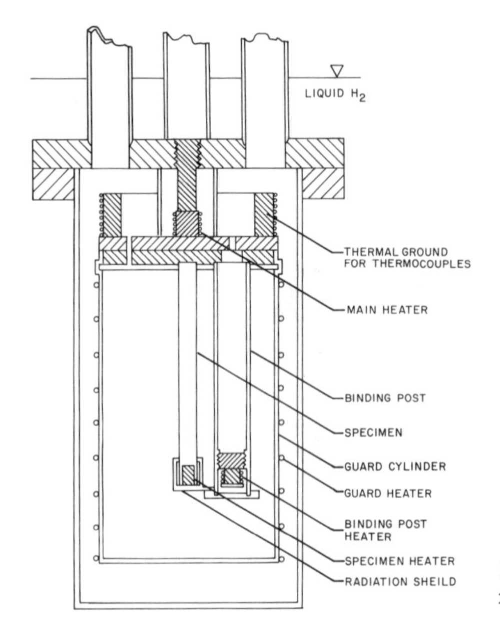

FIGURE 20. The "Forbes' bar" apparatus of Matsumura and Laubitz.[22]

are then used to calculate both the specimen conductivity and the ratio of conductivities, ρ_1. Because of the deliberate temperature mismatches between the specimen and the guard, thermal tempering of the leads going to the specimen was accomplished via a "binding post." Although no resistance thermometer was employed in the actual system, one could be readily incorporated if necessary.

$300 \leqslant T \leqslant 1500$. For temperatures above 300 K the radial heat losses from the specimen become very important, and guarding progressively more and more difficult. Indeed, for accurate work in this temperature range, the Forbes bar is the preferred approach. It can be employed successfully at least to the practical limit of thermocouples,[23] about 1500 K.* The limit for guarded axial flow is about 1000 K for the very best designed systems. Even in that restrcited range the guarded system can be recommended only if special circumstances demanded such an approach – for instance, the requirement for flat isotherms in liquid specimens to minimize effects of convection.

Of the three systems illustrated here, two fall into the guarded class. The first, illustrated in Fig. 21, was developed by Moore *et al.*[24] for small (1 cm diam × 7 cm long) specimens which were radioactive, so that speed and ease of assembly of the system were a paramount design consideration. This prompted the use of threaded connections between the specimen and the sink and heater,

*This is not a limit on thermocouples per se, but on their application in the measurement of λ, in environments likely to lead to their contamination.

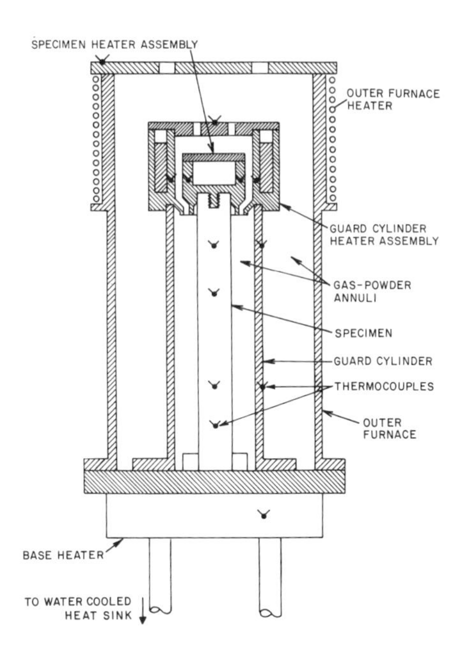

FIGURE 21. The high-temperature guarded longitudinal heat-flow apparatus of Moore *et al.*[24]

as illustrated, which in practice gave results markedly inferior to those obtained with brazing of these joints; with brazing, the experimental errors could be kept at or below 2% in the range of 300 to 1000 K. It is worth noting, however, that even though the authors label this a "guarded" system, they do make allowances in the computation of λ for small radial exchanges between specimen and guard due to imperfect thermal matching.

The second system, illustrated in Fig. 22, is a pure guarded system developed by Cook[25] for the study of liquid metals. As mentioned above, guarding in this case is essential to eliminate convective heat transfer. The range of operation of this system is up to about 700 K, and its experimental inaccuracy about ± 1.5%. This system has several well-conceived features, which could be easily duplicated in other guarded systems. These are the double-walled guard, which improves the temperature distribution next to the specimen; the encapsulation of the specimen heater within the specimen, to minimize heat leaks; and the carefully designed specimen-heater shield, which extends between specimen heater and the top guard heater. The last fulfills two additional important functions besides shielding the specimen heater itself: it shields the specimen from the inevitable specimen-guard temperature mismatch in the region of the guard heater; and it provides the heat flowing down the insulation between the specimen and the guard, part of which could otherwise come from the specimen heater and thus result in errors in λ even at perfect matching.

It is worth noting that both these guarded systems are placed inside

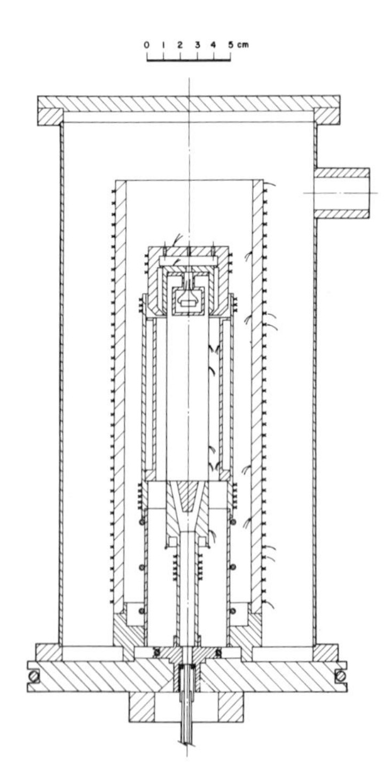

FIGURE 22. The high-temperature guarded longitudinal heat flow apparatus of Cook,[25] designed for liquid metal specimens.

another heated container, one that is roughly matched in temperature to the guard. This double guarding, as pointed out in the section dealing with heat flow, is a necessary requirement for satisfactory operation at high temperatures.

The third system,[26] illustrated in Fig. 23, belongs to the Forbes' bar class, the preferred approach in this temperature range. It contains an axially symmetrical specimen with a thin transverse specimen heater, CH, in the centre, and supported, on end heaters, inside a heated furnace. The space between the furnace and specimen is filled with finely powdered alumina insulation. As with all Forbes' bar systems, several sets of readings are taken at the same tempera-

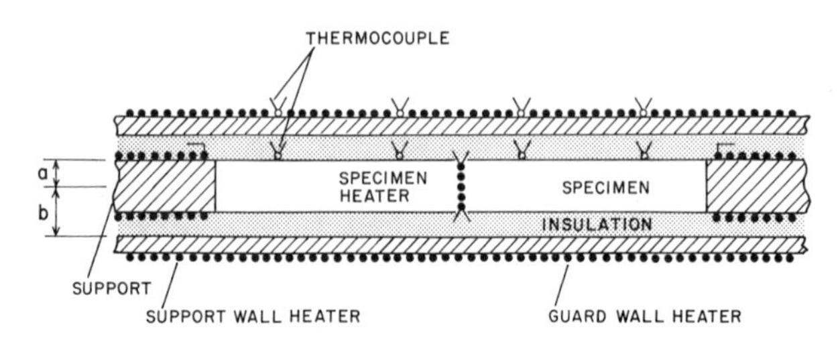

FIGURE 23. Schematic drawing of the high-temperature "Forbes bar" apparatus of Laubitz.[10]

ture, four in this case: (1) a "base line," with CH off and the furnace roughly at the specimen temperature; (2) a "gradient" set, with CH on and all other heaters as in (1); this is followed by two sets with CH off, and furnace and specimen mismatched by + and -10 K, but with the mean temperature close to (1). The last two readings serve to calculate ρ_1 at the mean test temperature, which is then used in conjunction with readings (1) and (2) to determine λ of the specimen. All calculations are done in the "thin-rod" approximation: however, the further simplification in which the hyperbolic functions were replaced by the first few terms of their expansions, employed in the first experiments described in the earlier reference, were dropped in the subsequent work, the solution of the exact transcendental equation being obtained numerically. The range of this system extends from 300 to 1300 K,[26] with an experimental inaccuracy of less than $\pm 1.5\%$.

Most of the systems here illustrated have an established experimental provenance. Perhaps the clearest indication of this comes from what appears to be a unique attempt[17] at a systematic comparison of a number of experimental systems to establish their reliability, using data accumulated over a number of years. This involved the intercomparison of the results obtained on a variety of materials by eight different experimental systems, six of which were of the axial heat flow type, either guarded or unmatched (Forbes' bar); four* of the systems chosen for illustration in this chapter were taken from that group of tested systems, some of them in a modified, improved version. A brief summary of the materials which served to test the systems, and the temperature ranges of both the measurements and intercomparisons; are given in Table 2. The intercomparison showed that for the materials selected, which ranged in conductivity by a factor of 35, the eight systems agreed within their combined estimated experimental inaccuracies, which in all cases were equal to or less than $\pm 2\%$; This provides impressive evidence of the accuracy that can be consistently achieved with well designed systems when coupled with careful experimental procedure. That the majority of these tested systems fall into the axial heat flow class amply attests to the importance of this type of apparatus for precise measurement of λ.

APPENDIX

For the system described in Section 3.2.1, and illustrated in Fig. 5, the temperature at a point (r, z), $T(r, z)$, located in region X, $X = 1$ to 4, is given by

$$T(r,z) = \sum_n [F_n(X)I_0(\alpha_n r) + G_n(X)K_0(\alpha_n r)] \sin \alpha_n z + \delta(1, X)T(z) \qquad \text{(A.1)}$$

*These are the systems of Moore *et al.*,[21] Cook *et al.*,[15] Matsumura and Laubitz,[22] and Laubitz.[26]

TABLE 2
Summary of the Materials Used for the Intercomparison of Eight Experimental Systems for the Measurement of λ^a

Material	λ (273.15 K) (W/cm K)	ρ (273.15 K) ($\mu\Omega$ cm)	No. of systems compared	Temperature range (K)		Accuracy of individual systems (±, %)	Observed difference in overlap region (%)
				of intercomparison	of complete measurement		
Cu	4.02_0	1.545	2	300–400	90–1250	1–1.2	<1
Ag	4.35_5	1.468	2	290–360	80–1100	1	<1
Au	3.17_8	2.052	3	90–400	90–1200	0.8–1.2	<1
Pd	0.71_7	9.806	2	300–360	80–1300	1	<1
Pt	0.71_1	9.847	4	120–1200	90–1200	0.6–2	3.5
W	1.70_0	5.002	2	300–500	80–1300	1.2–1.5	<1
Cr	0.96_8	11.66	2	90–300	90–350	1–1.2	<1
Fe (Armco)	0.74_2	9.9	3	100–1000	100–1300	1.2–2	<3
Nb–10 W	0.45_5	15.02	2	300–400	90–1100	1.2–1.5	1.5
Inconel 702:							
Solution annealed	0.114	121.3	3	400–1400	120–1400	1–2	2
Age hardened	0.119	123.0	3	400–1000	120–1000	1–2	2

[a] Reference 17.

where

$$\delta(1, 1) = 1 \tag{A.2}$$

$$\delta(1, X) = 0, \qquad X \neq 1 \tag{A.3}$$

$$\alpha_n = \frac{n\pi}{l} \tag{A.4}$$

$$T(z) = T(z_i(\mathrm{C})) + G_{i+1}[z - z_i(\mathrm{C})], \qquad z_i(\mathrm{C}) \leqslant z \leqslant z_{i+1}(\mathrm{C}) \tag{A.5}$$

$$\sum_{i=1}^{j+1} G_i[z_i(\mathrm{C}) - z_{i-1}(\mathrm{C})] = 0 \tag{A.6}$$

$$z_0(\mathrm{C}) = 0 \tag{A.7}$$

$$z_{j+1}(\mathrm{C}) = l \tag{A.8}$$

$$G_i - G_{i+1} = \frac{\dot{Q}_j(\mathrm{C})}{\pi r_1{}^2 \lambda_1} \tag{A.9}$$

I and K are the Bessel functions of purely imaginary argument. $T(z)$ is the linear temperature distributions due to the "C" heaters which would obtain if the specimen were perfectly insulated. The coefficients $F_n(X)$ and $G_n(X)$ are determined by the internal continuity requirements and boundary conditions along $r = r_4$.

The internal continuity conditions are fixed: the temperature must be continuous across each boundary, and so must be the heat flow, subject, of course, to the power generated in the heaters. These conditions we can therefore write explicitly, if for convenience we define a number of otherwise rather meaningless parameters

$$\eta_n = \frac{I_0(\alpha_n r_1)\beta_n}{n} \sum_i \frac{\dot{Q}_l(\mathrm{A})}{\pi^2 \theta_l(\mathrm{A})\lambda_2} \sin \alpha_n \theta_l(\mathrm{A}) \sin \alpha_n z_l(\mathrm{A}) \tag{A.10}$$

$$\sigma_n = \frac{1}{n} \sum_m \frac{\dot{Q}_m(\mathrm{B})}{\pi^2 \theta_m(\mathrm{B})\lambda_4} \sin \alpha_n \theta_m(\mathrm{B}) \sin \alpha_n z_m(\mathrm{B}) \tag{A.11}$$

$$\tau_n = \frac{I_1(\alpha_n r_1)\beta_n}{n} \sum_j \frac{2\dot{Q}_j(\mathrm{C})}{\pi^2 r_1 \lambda_1} \sin \alpha_n z_j(\mathrm{C}) \tag{A.12}$$

$$\rho_1 = \lambda_1/\lambda_2, \qquad \rho_2 = \lambda_3/\lambda_2, \qquad \rho_3 = \lambda_3/\lambda_4 \tag{A.13}$$

$$\beta_n = \frac{1}{1 + (\alpha_n r_1)(\rho_1 - 1) I_1(\alpha_n r_1) K_0(\alpha_n r_1)} \tag{A.14}$$

$$\gamma_n = (\alpha_n r_1)(\rho_1 - 1) I_1(\alpha_n r_1) I_0(\alpha_n r_1) \beta_n \tag{A.15}$$

$$\phi_n = \frac{(\rho_2 - 1)}{\rho_2} (\alpha_n r_2) K_0(\alpha_n r_2) K_1(\alpha_n r_2) \tag{A.16}$$

$$\xi_n = \frac{1}{\rho_2} [1 + (\alpha_n r_2)(\rho_2 - 1) I_1(\alpha_n r_2) K_0(\alpha_n r_2)] \tag{A.17}$$

$$\chi_n = \xi_n \left[\gamma_n + \frac{I_1(\alpha_n r_2)}{K_1(\alpha_n r_2)}\right] - \frac{I_1(\alpha_n r_2)}{K_1(\alpha_n r_2)} \tag{A.18}$$

$$\theta_n = 1 - \phi_n \left[\gamma_n + \frac{I_1(\alpha_n r_2)}{K_1(\alpha_n r_2)}\right] \tag{A.19}$$

We can now write the following relations describing the continuity conditions.

Continuity at $r = 0$:

$$G_n(1) = 0 \tag{A.20}$$

Continuity at $r = r_1$:

$$F_n(1) = \beta_n F_n(2) + \frac{K_0(\alpha_n r_1)}{I_0(\alpha_n r_1)} \eta_n - \frac{K_1(\alpha_n r_1)}{I_1(\alpha_n r_1)} \tau_n \tag{A.21}$$

$$G_n(2) = -\gamma_n F_n(2) + \eta_n + \rho_1 \tau_n \tag{A.22}$$

Continuity at $r = r_2$:

$$F_n(3) = \theta_n F_n(2) + \phi_n [\eta_n + \rho_1 \tau_n] \tag{A.23}$$

$$G_n(3) = -\chi_n F_n(2) + \xi_n [\eta_n + \rho_1 \tau_n] \tag{A.24}$$

There remain the two conditions due to continuity across $r = r_3$, and the boundary condition along $r = r_4$, which provide three equations to solve for $F(2)$, $F(4)$, and $G(4)$.

For example, we may want to describe a system that is water cooled along $r = r_4$, so that $T(r_4, z) = 0$. Then

$$K_0(\alpha_n r_4)G_n(4) = -I_0(\alpha_n r_4)F_n(4) \tag{A.25}$$

and continuity at $r = r_3$ gives

$$F_n(2) = \frac{\sigma_n}{\mu_n} - v_n[\eta_n + \rho_1 \tau_n] \tag{A.26}$$

where

$$\mu_n = \frac{\epsilon_n \theta_n}{K_0(\alpha_n r_3)} - \frac{\epsilon_n - \rho_3}{I_0(\alpha_n r_3)} \chi_n \tag{A.27}$$

$$v_n = \frac{\epsilon_n \phi_n I_0(\alpha_n r_3) + (\epsilon_n - \rho_3)\xi_n K_0(\alpha_n r_3)}{\epsilon_n \theta_n I_0(\alpha_n r_3) - (\epsilon_n - \rho_3)\chi_n K_0(\alpha_n r_3)} \tag{A.28}$$

$$\epsilon_n = \frac{I_0(\alpha_n r_4)K_0(\alpha_n r_3)}{I_0(\alpha_n r_4)K_0(\alpha_n r_3) - K_0(\alpha_n r_4)I_0(\alpha_n r_3)} + (\rho_3 - 1)(\alpha_n r_3)K_0(\alpha_n r_3)I_1(\alpha_n r_3) \tag{A.29}$$

From which relations all coefficients but $F_n(4)$, which is of little importance, can be deduced explicitly.

Equally, we could have assumed that there are no radial heat losses at $r = r_4$. This would correspond to a system placed in an evacuated vessel at fairly low temperatures, or a system of fairly large external diameter sitting in still air. Under these conditions

$$K_1(\alpha_n r_4)G_n(4) = I_1(\alpha_n r_4)F_n(4) \tag{A.30}$$

and, as it turns out, $F_n(2)$ is given by the same formulas as before, except that in all parameters $K_1(\alpha_n r_4)$ replaces $K_0(\alpha_n r_4)$ and $-I_1(\alpha_n r_4)$ replaces $I_0(\alpha_n r_4)$.

Other conditions could of course be assumed for the boundary at r_4. In the general case, one would obtain a relation between $F_n(4)$ and $G_n(4)$ by a Fourier analysis of this assumed condition and, as before, proceed to obtain $F_n(2)$ through the equations describing continuity at r_3.

The solution obtained above is not the only one that can be derived for a multicomponent system. We can produce an analogous set of equations using the Bessel functions of real argument, J_0 and Y_0, with the circular functions replaced by hyperbolic ones. A solution of this type allows Fourier analysis along transverse boundaries, and would be preferable to the solution developed above if nonuniform transverse heaters were considered. Indeed, one can combine the two types of solutions; the result allows of an analysis, in principle at least, of a very general system containing not only nonuniform transverse heaters and cylindrical heaters, but also a composite specimen, the various parts of which have differing conductivities. The solution for such a system can readily be written down formally: the difficulties lie in obtaining the coefficients of (at least) two series, the coefficients of each being represented by an infinite series of those of the other.

Although the solution here developed is not the most general one, it does describe precisely the fairly versatile model of Fig. 5. As it stands, it represents a system that has a constant temperature at each of its transverse ends, $z = 0$ and $z = l$, and, in its simplest form, the same constant temperature or zero heat loss along its outer surface $r = r_4$. We can, however, make the plane $z = l/2$ a mirror plane, simply by restricting the summation over n to odd terms only, and thus obtain a very close representation of a system of length $l_0 = l/2$ with one of its ends "free" and shielded, either by a heated specimen cap, or thick layers of insulation. We can further treat the plane $z = 0$ as a mirror plane, as our solution contains only odd functions of z, and the model thus represents a system within a furnace the uniform temperature of which is matched to the specimen mid-point. Alternatively, we can set λ_4 to zero, and obtain a fairly good representation of a system in a close-fitting furnace, the temperature of which is matched ideally to that of the guard. Thus even the simplest form of the model can be made to simulate exactly or approximately most of the situations encountered in practice.

The only serious limitation of the mathematical analysis is the assumption that the conductivities are independent of temperature. As discussed in the text, in some problems the temperature variation of the conductivity can be treated exactly by employing the following substitution for the temperature, T

$$\lambda_0\theta = \int\lambda(T)dT \qquad \text{(A.31)}$$

where θ is the new variable, $\lambda(T)$ the temperature-dependent conductivity, and λ_0 is a (convenient) constant conductivity. θ satisfies the ordinary Laplace equation, from the solution of which the temperature T is obtained by the above relation. Although this approach can be used for several boundary conditions in a homogeneous body, it cannot be applied to the multicomponent system. In that case, one has to resign oneself to the fact that the solution will only describe systems where the temperature variation over the important region of the system is small enough so that the temperature dependence of λ can be largely ignored, and such small effects as persist be treated approximately (see text). This, in any event, is the best way of operating the equipment experimentally if the temperature variation of λ is to be obtained accurately. Systems in which large gradients are used and thus yield what is more properly described as the conductance of the specimen rather than its conductivity have to be analysed by approximate solutions, into which (A.31) can be incorporated.

The restriction of the allowable temperature variation over the length of the specimen for which our solution yields acceptable results is probably most severe for those cases wherein the model is supposed to represent systems using vacuum, rather than insulation, between the specimen and guard and outside of the guard. We can only deal with these systems if the temperature variation is small enough so that the radiant-heat transfer can be adequately represented by the "$4T^3\Delta T$" approximation to the normal fourth-power law. If this condition is satisfied, then the radiant-heat transfer can be approximated by a conductive transfer with an effective conductivity, λ_e, which, for the gap between two concentric cylindrical surfaces having radii r_1 and r_2 ($r_2 > r_1$) and emissivities ϵ_1 and ϵ_2, is given by

$$\lambda_e = \frac{\epsilon_1 \epsilon_2 r_1 (r_2/r_1) \ln (r_2/r_1)}{\epsilon_1(1-\epsilon_2) + \epsilon_2(r_2/r_1)} 4\sigma T^3 \tag{A.32}$$

where σ is the Stefan–Boltzmann constant. Fortunately, problems concerned with the selection of a suitable value for ϵ in any realistic system are outside the scope of this article.

The formulas for the temperature distribution given above can be rearranged so that they appear as a sum over all the heaters of the system, each term of the sum being the product of two factors, an absolute one and a relative one. The absolute factor is a function of the power input into the particular heater, a basic length of the system, say l, and the conductivity of any one of the system components; it determines the magnitude of the temperature in the system. The relative factor depends only on the ratios of the geometrical and thermal parameters of the system, and describes the relative temperature distribution in it; this is the factor which determines the performance of the system.

The possibility of such a rearrangement of the formulas is a direct consequence of our assumption of temperature independence of the conductivities, and reflects the well-known superposition theorem of heat flow. The reason that they are not represented here in such a fashion is that there are many possible arrangements, none of them of universal preference. This detail is left, therefore, to the prospective user, to choose the arrangement best suited to his individual predictions, computer requirements and available subroutines.

NOTATION

Symbol	*Definition*
A	Cross-sectional area
A_l	Designation of lth radial heater on specimen
B_m	Designation of mth radial heater on guard

C_j	Designation of jth transverse heater on specimen
c'	Heat capacity per unit volume
E	Thermocouple emf
f_n	Function of geometrical and thermal parameters
G, G_0	Temperature drift parameters
l	Length of one-dimensional slab, or of total guarded specimen
l_0	Length of specimen up to its heater
$\bar{q}$	Heat flow density (vector)
$\dot{Q}$	Power flowing through specimen
$\dot{Q}_l(A), \dot{Q}_m(B), \dot{Q}_j(C)$	Power dissipated by heaters A_l, B_m, C_j
r	Radial position
r_0	Radial position of prescribed temperature boundary
r_1	Specimen radius
r_2	Inner guard radius
r_3	Outer guard radius
r_4	External system boundary
S	Thermocouple thermoelectric power
t	Time
t_0	Characteristic time of response
T	Temperature
T_o	Operating temperature
T_c	Correct temperature
T_M	Mean temperature
T_{app}	Temperature calculated from thin-rod approximation
z	Axial position
z_i	Location of ith temperature sensor
$z_l(A), z_m(B), z_j(C)$	Location of heater A_l, B_m, C_j
z_m	Axial location at which the specimen and guard are matched in temperature
α	Thermal diffusivity
β	Loss parameter in thin-rod approximation
β_0	Correct constant drift rate
β_0'	Apparent constant drift rate
γ	Rate constant
δ	Uncertainty (of some quantity)
Δ	Difference (of some quantity)
θ	Transformed temperature
$\theta_l(A), \theta_m(B), \theta_j(C)$	Half-length of heaters A_l, B_m, C_j
λ	Thermal conductivity
λ_i	Thermal conductivity of ith component
λ_c	Correct thermal conductivity
λ_m	Measured thermal conductivity
ρ_1	Ratio of thermal conductivities, λ_1/λ_2
ρ_2	Ratio of thermal conductivities, λ_3/λ_2
∇	Gradient operator
∇_n	Normal gradient operator
∇^2	Laplacian operator
∂	Partial derivative operator

REFERENCES

1. *Thermal Conductivity,* Vols. 1 and 2 (R.P. Tye, ed.), Academic Press, New York (1969).
2. (i) *Thermophysical Properties of Matter,* The TPRC Data Series, Vols. 1 and 2 (Y.S. Touloukian, R.W. Powell, C.Y. Ho, and P.G. Klemens, eds.), IFI/Plenum, New York (1970). (ii) R.W. Powell, C.Y. Ho, and P.E. Liley, *Thermal Conductivity of Selected Materials,* NSRDS-NBS 8 (1966) and NSRDS-NBS 16 (1968). (iii) C.Y. Ho, R.W. Powell, and P.E. Liley, *Thermal Conductivity of the Elements, J. Phys. Chem. Ref. Data* **1** (2) (1972).
3. (i) J.G. Hust and L.L. Sparks, Lorenz Ratios of Technically Important Metals and Alloys, NBS Technical Note 634 (1973). (ii) R.W. Powell, Correlation of the Thermal and Electrical Conductivity of Metals, Alloys and Compounds, Proc. of 3rd Conf. on Thermal Conductivity, Oak Ridge National Laboratory, Oak Ridge, Tennessee, Vol. 1, p. 79 (1963).
4. S.H. Jury, D. Arnurius, T.G. Godfrey, D.L. McElroy, and J.P. Moore, *J. Franklin Inst.* **298,** 151 (1974).
5. Reference 1, Vol. 1, Chap. III.
6. Reference 1, Vol. 1, p. 111.
7. Reference 1, Vol. 1, p. 128.
8. H.S. Carslaw and J.C. Jaeger, *Conduction of Heat in Solids,* Oxford University Press, London (1959).
9. J.G. Cook, *Can. J. Phys.* **59,** 25 (1981).
10. M.J. Laubitz, *Can. J. Phys.* **45,** 3677 (1967).
11. (i) C.A. Swenson, *CRC Crit. Rev. Solid State Sci.* **1,** 99 (1970). (ii) R.E. Bedford, T.M. Dauphinee, and H. Preston-Thomas, *Tools and Techniques in Physical Metallurgy* (F. Weinberg, ed.), **1,** 1 (1970). (iii) R.L. Anderson and T.G. Kollie, *CRC Crit. Rev. Anal. Chem.* **6,** 171 (1976). (iv) L.A. Guildner and G.W. Burns, *High Temp. High Pressure* **11,** 173 (1979).
12. A.C. Anderson, J.H. Anderson, and M.P. Zaitlin, *Rev. Sci. Instrum.* **47,** 407 (see also Ref. 11i) (1976).
13. (i) G.K. White, Ref. 1, Chap. 2. (ii) A.C. Anderson, *Rev. Sci. Instrum.* **51,** 1603 (1980).
14. D.R. Flynn and M.E. O'Hagan, *J. Res. NBS* **71C,** 255 (1967).
15. J.G. Cook, M.P. van der Meer, and M.J. Laubitz, *Can. J. Phys.* **50,** 1386 (1972).
16. Reference 11(iv).
17. M.J. Laubitz and D.L. McElroy, *Metrologia* **7,** 1 (1971).
18. (i) J.G. Hust, *Rev. Sci. Instrum.* **41,** 622 (1970). (ii) A.C. Anderson, *Rev. Sci. Instrum.* **40,** 1502 (1969). (ii) A.C. Anderson, *Rev. Sci. Instrum.* **51,** 1603 (1980).
19. J.L. Vorhaus and A.C. Anderson, *Phys. Rev.* **B 14,** 3526 (1976).
20. M.P. Zaitlin and A.C. Anderson, *Phys. Rev. Lett.* **33,** 1158 (1974). See also R.B. Roberts and R.S. Crisp, *Phil. Mag.* **36,** 81 (1977).
21. J.P. Moore, R.K. Williams, and R.S. Graves, *Rev. Sci. Instrum.* **45,** 87 (1974).
22. T. Matsumura and M.J. Laubitz, *Can. J. Phys.* **48,** 1499 (1970).
23. M.J. Laubitz and T. Matsumura, *Can. J. Phys.* **50,** 196 (1972).
24. J.P. Moore, D.L. McElroy, and R.S. Graves, Oak Ridge National Laboratory Report ORNL-4986 (1974).
25. J.G. Cook, *Can. J. Phys.* **59,** 25 (1981).
26. Reference 10; M.J. Laubitz, *Can. J. Phys.* **47,** 2633 (1969); Ref. 23; M.J. Laubitz and T. Matsumura, *Can. J. Phys.* **51,** 1247 (1973); M.J. Laubitz, T. Matsumura, and P.J. Kelly, *Can. J. Phys.* **54,** 92 (1976).

2

Analysis of Apparatus with Radial Symmetry for Steady-State Measurements of Thermal Conductivity

J. P. MOORE

1. INTRODUCTION

Fourier's law states that heat flow per unit area, q', between two isotherms in an opaque and homogeneous material is given by

$$q' = -\lambda \nabla T \tag{1}$$

where ∇T is the temperature gradient and λ is a proportionality constant defined as the thermal conductivity.[1] There are many different techniques for measuring this material property, and each technique has specific advantages and disadvantages that become apparent when error sources are examined, when the apparatus is tested with standards or materials with known λ, and occasionally when theoretically inconsistent results are obtained. Techniques where heat flows radially away from a central heater or into a central heat sink have proved very successful in the measurement of λ. In these systems ∇T is the gradient in the radial direction, $\nabla_r T$. The general group includes[2]

1. systems with right circular symmetry where all components are cylindrical and the isotherms are assumed to be cylindrical;
2. ellipsoidal envelope systems;
3. systems where the heater, specimen, and heat sink are spherical. This is actually a special case of 2 where the major and minor axes are equal;

J.P. MOORE • Metals and Ceramics Division, Oak Ridge National Laboratory, Oak Ridge, Tennessee 37830. Research sponsored by the Division of Materials Sciences, U.S. Department of Energy, under contract W-7405-Eng-26 with the Union Carbide Corporation.

4. comparative systems where a right-circular cylinder of a material with a known λ is concentric with a material of unknown λ. The gradients across the two concentric cylinders are compared to determine the λ of the unknown.

In the first three systems the outward heat flow through the specimen is usually determined by measurements of the electrical power dissipated in the central heater. There are, however, systems with right-circular symmetry where heat flow is inward to a fluid-cooled heat sink on the central axis and the temperature rise and flow rate of this coolant are measured and used to calculate the power flow through the specimen. Heat flow through the specimen in comparative sytems is calculated by using the measured temperature gradient in the material with a known thermal conductivity. The comparative systems have limited use because of their inaccuracy and size, but they have been especially useful with fluid specimens because the material with an unknown λ can be poured into an annulus cut into the standard material. This technique has been especially useful for measurements of the λ of corrosive fluids[3] and for measurements at high pressures.[4]

Before 1954, apparatus with ellipsoidal and spherical symmetry were used occasionally for measuring thermal conductivity.[5-9] In 1954, three papers appeared that described experimental results obtained with ellipsoidal or spherical systems[10-12]; however, many of these data were later shown to be suspect. Because these two systems have been used infrequently, they will be ignored in this chapter. Rather, the more important systems with right circular symmetry will be described in more detail.

Radial apparatus with right-circular symmetry have been used successfully on all types of solids including pure metals, metal alloys, semiconductors, ceramics, and many materials of engineering interest such as concretes and rocks.[2] Nonsolid materials such as soils and powders have been studied in this type apparatus. The latter requires the addition of a chamber for specimen containment and special precautions concerning temperature sensor positioning. The upper measurement temperature of these apparatus is usually restricted by problems such as heater shorting. The lowest measurement temperature is limited only by the coolant temperature. Longitudinal techniques, which require specimens much smaller than those required for radial heat flow apparatus (RHFA), are excellent for measuring the thermal conductivity of solids at low temperatures. This has usually led to a practical lower limit of room temperature for RHFA with solid specimens. Thermocouples can be used as temperature sensors up to about 1400 K, and pyrometers have been used for measuring temperatures up to 2770 K.

This chapter will include a description of a fundamental RHFA with right circular symmetry. Advantages, disadvantages, determinate errors, and indeterminate errors will be discussed. One source of indeterminate error is nonideal heat flow at the specimen midplane, and the magnitude of this error is calculated for the RHFA. The calculations demonstrate the influence of system parameters and guarding procedures on this error for specimens with high thermal conductivity, such as metals, and ones with low thermal conductivity, such as metal oxides. The influence of a second thermal configuration, normally referred to as an "isothermal correction," will be examined for several cases of the fundamental RHFA.

Heat flow in selected experimental RHFA was examined in an attempt to determine how closely the RHFA met the requirements for ideal radial heat flow at the specimen midplane. These apparatus were selected for analysis because they illustrated many of the problems and features mentioned during the discussion of the fundamental RHFA. In addition, these apparatus were used to obtain data that are still interesting and sometimes controversial, and an analysis of heat flow within the individual apparatus is of direct importance in assessments of the validity of the experimental results.

Thermometry is a problem that RHFA have in common with all techniques for measuring thermophysical properties and is a subject within itself. Nevertheless, a discussion of thermometry problems relevant to the RHFA is included.

2. FUNDAMENTAL RADIAL HEAT FLOW APPARATUS

2.1. *Operation*

Although thermal conductivity measurements with RHFA are usually precise, the simplifying assumptions required for calculations of thermal conductivity values are often invalid, and measurements of temperatures, voltages, and thermometer positions may be subject to error. It is nearly impossible to quantitatively define all potential errors since they are strong functions of many parameters. Before examining any specific apparatus that has been used to measure λ, it would be helpful to examine a fundamental three-component RHFA for definition of the general problem areas. Figure 1 shows a simple RHFA where the specimen, central core heater, and muffle heater are shown suspended in a water-cooled chamber. The specimen has the shape of a right-circular cylinder, and the hole on the central axis contains a core heater. In some cases this central heater would be replaced with a water-cooled calorimeter so that heat would flow primarily inward toward the central specimen axis. The core heater is assumed to protrude from each end of the specimen, as would be observed in actual practice. The specimen is surrounded by a furnace or "muffle heater" for varying the specimen temperature.† The three annuli and other

†Conceptually, the simplest RHFA would not include a muffle heater. However, the elimination of the muffle heater would be impractical if any flexibility concerning temperature and specimen λ range is desired.

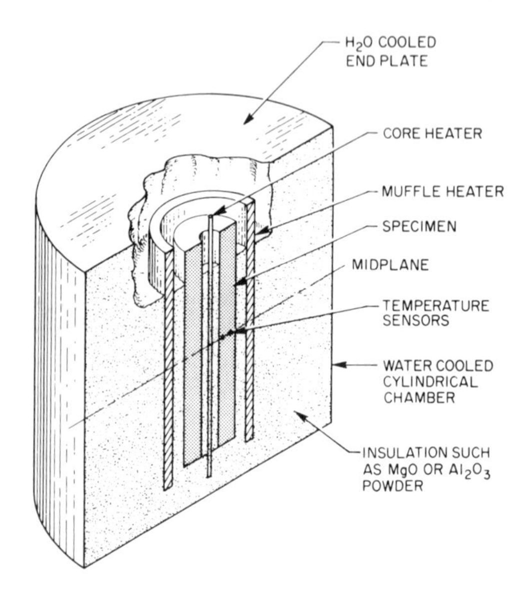

FIGURE 1. Schematic of a fundamental RHFA without end guards on the core heater or specimen. The temperature sensor positions are denoted by •. Current input leads to the core heater and to the muffle heater; supports for the system are omitted for simplicity.

spaces between system components are assumed to be filled with a powder of Al_2O_3 or MgO particles and a gas such as argon or helium. Supports necessary for holding the components in place have been omitted for simplicity, but these supports will be considered in the actual RHFA that have been used in the past. Thermometers (usually thermocouples) are mounted in the specimen at a minimum of two radii near the specimen midplane.

Determination of the specimen's thermal conductivity would consist of the follow steps:

1. A stable electrical current is passed through the core heater to generate heat that is assumed to flow radially outward, at least at the system midplane. This establishes a temperature difference between the thermometers.
2. The power to the muffle heater is adjusted to raise the system temperature to any desired value.
3. After the system reaches steady state, the thermometer outputs are measured and converted to temperatures.
4. The current through the core heater and the voltage drop across a known core heater length, which spans the specimen midplane, are measured.
5. The thermal conductivity is calculated by using

$$\lambda = \left(\frac{P}{l}\right)\frac{\ln(r_2/r_1)}{2\pi(T_1 - T_2)} \tag{2}$$

where P/l is the power per unit length from the core heater over the specimen midplane and T_n is the temperature at radius r_n. It can be shown that thermal conductivity data obtained with a RHFA do not require a correction for the thermal expansion of the specimen, but a correction must be applied for core heater expansion so that the length in equation (2) is the distance between potential taps at the core heater temperature.

This configuration will yield accurate data if the ratio of specimen length, l, to diameter, d, is sufficiently large to ensure that all heat flow at the specimen midplane is radial. Thermal conductivity has been measured many times with the simplifying assumption that the specimen was "infinitely long." A summary of these measurements obtained from the previous review by McElroy and Moore[2] is given in the Appendix.

This fundamental configuration has been modified by using end guard heaters and/or stacked-disk specimens to improve heat flow. These are the most accurate RHFA with some authors reporting errors approaching 1%. A complete summary of these systems is also given in the Appendix.

2.2. *Advantages and Disadvantages of a Radial Heat Flow Apparatus*

The greatest advantage of a properly designed RHFA is its high accuracy at elevated temperatures. A secondary advantage accrues from the fact that these systems can be operated with relatively simple instrumentation although sophisticated control systems for maintaining heater temperatures reduce operation times. An additional advantage of the RHFA is its wide range of applicability on specimens with very high and very low thermal conductivities. They have been used above room temperature on solid materials with λ values as high as 200 W/m K and on loose fill materials with λ approaching 10^{-2} W/m K. However, the specimens with extremely low or high λ values do require different guarding schemes.

The greatest disadvantage of a RHFA is the large specimen size, which can lead to prohibitive specimen cost, assembly difficulties, and slow operation. The large size can be advantageous when the specimen is heterogeneous by averaging the effects of local inhomogeneities. This is especially true for multiphase materials such as composites. At very high temperatures the slow operation, which can sometimes amount to a few hours, can lead to specimen and/or temperature sensor contamination.

One additional disadvantage of a RHFA is the difficulty of measuring the electrical resistivity, ρ, and Seebeck coefficient, S, of metallic specimens simultaneously with the thermal conductivity. These two properties are often essential for understanding the thermal conductivity of a specimen. In addition, simultaneous measurements of these two properties can indicate specimen degradation due to reaction with the environment. Three ways have been described for measuring one or both of these properties, and these are discussed in a later section on experimental apparatus.

2.3. *Determinate Errors*

There are two specific types of errors in λ determinations from using this approach:[2] those that include uncertainties in the measurements of the various parameters in equation (2) and those that are due to deviation of the system from assumed boundary conditions. The absolute magnitude of the fractional determinate error, $\Delta\lambda/\lambda$, in a thermal conductivity measurement is the sum of the fractional uncertainties in the parameters[2] of equation (2) or

$$\left|\frac{\Delta\lambda}{\lambda}\right| = \left|\frac{\Delta P}{P}\right| + \left|\frac{\Delta l}{l}\right| + \left|\frac{\Delta r_1}{r_1 \ln (r_2/r_1)}\right| + \left|\frac{\Delta r_2}{r_2 \ln (r_2/r_1)}\right| + \left|\frac{\Delta(T_1 - T_2)}{T_1 - T_2}\right| \tag{3}$$

The fractional uncertainty in the power, $\Delta P/P$, includes uncertainty in the voltage measurement across a standard resistor to determine current flow in the core heater and uncertainty in the voltage measurement between the potential taps. Since these voltages can be measured with high accuracy and since the uncertainty in the value of the standard resistor is negligible in this case, the fractional uncertainty in $\Delta P/P$ is usually less than 0.001 or 0.1%. The fractional uncertainty in the distance between taps, $\Delta l/l$, can usually be restricted to 0.005 or 0.5%.

The fractional uncertainties in the radial location of the temperature sensors, $\Delta r_1/r_1 \ln(r_2/r_1)$ and $\Delta r_2/r_2 \ln(r_2/r_1)$, are usually large. Since Δr_1 is usually the same as Δr_2, and $r_1 \ll r_2$, the fractional term involving the inside radius is usually dominant. These errors include the uncertainties in thermometry well location in addition to uncertainties in the temperature sensor position within the well. For example, Godfrey *et al.*[13] reported thermal conductivity measurements where r_1 and r_2 had values of 12.7 and 33.3 mm, respectively, and it was assumed that Δr_1 and Δr_2 were each equal to 0.25 mm. For this case $\Delta r_1/r_1 \ln(r_2/r_1)$ and $\Delta r_2/r_2 \ln(r_2/r_1)$ were equal to 0.024 and 0.0078 or 2.4% and 0.78%, respectively. If there are n thermocouples located at each radius and if the positional errors are random, the above errors can be divided by $\sqrt{n}$, which reduces the error significantly. In the above case six thermocouples were at each radius so that the errors in r_1 and r_2 would be closer to 0.98% and 0.31%, respectively.

The errors from the last term in equation (3) are dependent on the temperature sensor, the instrument used to measure the sensor output, and the accuracy of the conversion of the sensor output to temperature. This term is usually between 0.01 (1.0%) and 0.05 (5%). Therefore the total determinate error in λ would be between 0.0289 (2.89%) and 0.0689 (6.89%). Most experimentalists claim measurement errors based on this general approach.

2.4. Indeterminate Errors

When λ is calculated by using equation (2), it is tacitly assumed that the radial specimen is infinitely long so that all heat generated in the central metered section of the core heater flows radially outward. Although the measurements of thermal conductivity using RHFA contain many small determinate errors, the invalidity of this assumption for all except very long systems can lead to errors that dwarf all others. One of the worse aspects of this error is that it is reproducible and does not decrease data precision, and the high precision has often been misinterpreted for high accuracy. Since many of the details of radial techniques have been discussed elsewhere, most of this chapter will be used to describe the deviation of actual systems from boundary conditions necessary for the use of equation (2) and the determination of experimental errors due to this deviation. This will be done for various configurations of the fundamental RHFA in Fig. 1 and then for several selected experimental techniques that have been used for measuring the thermal conductivities of solids and powders.

3. THERMAL ANALYSIS – HEATING5

The thermal performance of the fundamental radial system and the performances of several specific apparatus were examined by using a finite difference heat conduction code[14] known as HEATING5. This code is multiregional so that complex heat transfer problems containing many materials, boundary conditions, heat generation rates, and all modes of heat transfer such as radiation, conduction, and convection can be readily solved with a high-speed computer. The code was used to numerically model systems of interest in the following manner. A thermal model was formulated to approximate the experimental apparatus of interest. Geometrically and dimensionally this model was the same as the experimental apparatus except that some unimportant simplifications were made to decrease the complexity of the calculations, which usually decreased the computational time and expense for a given problem. Thermal conductivities were assigned to each system component, and some of these were assumed to be temperature dependent when they varied over the temperature range of interest. Values for the thermal conductivities were obtained from several sources but especially from tabulations by Touloukian[15] and Ho *et al.*[16] Specific boundary conditions were imposed. One of the most common was an isothermal surface, which is usually observed experimentally when water-cooled surfaces are used for system heat sinks. Power generation rates were assigned to any regions within the model, such as core heater, muffle heater, and end guards, that were used to establish temperature gradients or to raise the system temperatures above those of the boundaries. The geometrical space covered by the problem was then subdivided into a two-dimensional grid of nodal points. Since it was assumed that all configurations had right-circular symmetry, the two coordinates were always z and r. The heat conduction code was then used to obtain solutions to the finite difference equations, which approximated the partial differential equations that describe steady-state heat flow. The calculated temperatures at all the system nodes were printed and then used to determine the shapes of isotherms within the systems. Since the thermal model was representative of the actual system, the temperatures calculated for the nodes within the specimen permit calculation of the gradient in the specimen. The power density to the core heater could also be used with equation (2) to calculate the gradient if nonradial heat flow was negligible. The difference in these two gradients at the system midplane is a direct indication of the error due to nonideal radial heat flow.

The heating code has been routinely used to obtain numerical solutions to the temperature distributions in systems too complex for analytical treatment, and solutions from it agree excellently with results from simple experiments and with results from exact solutions, where they exist.[17] Although equations have been solved exactly for heat losses from the ends of a radial system with right-circular symmetry, this finite difference code can provide much more information about large multiregional systems and can be used for parameter sensitivity calculations. One should remember, however, that simplifying assumptions must be made in formulating any model, and this can lead to calculated predictions which may be only qualitative for a given apparatus.

4. HEAT FLOW IN A FUNDAMENTAL RADIAL HEAT FLOW APPARATUS WITH RIGHT-CIRCULAR SYMMETRY

4.1. *Fundamental Radial Heat Flow Apparatus without End Guards*

The temperature distribution in the fundamental three-component RHFA shown in Fig. 1 was analyzed to show the effects of various parameters on thermal conductivity measurements. The thermal model for this system is shown in Fig. 2, where all dimensions are reduced by the specimen diameter. For example, the lower end of the core heater is at $z = 0.4$, which would be 0.4 times the specimen diameter. If the specimen diameter were 80 mm, the distance

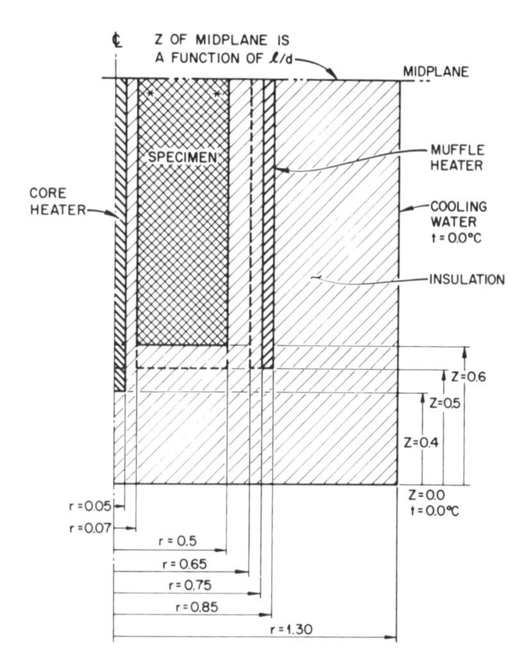

FIGURE 2. Thermal model for analysis of the thermal errors of the RHFA shown in Fig. 1. The dashed lines indicate regions that are assumed to be filled with powder except for some calculations pertaining to end guarding.

from the chamber base to the core heater would be 32 mm. The outside chamber radius at $r = 1.30$ would be at 104 mm for the same size specimen. The specimen midplane was assumed to be a plane of symmetry so that this plane and the central axis of the specimen could be treated as adiabatic boundaries, and only one-quarter of the actual problem need be modeled. The core heater protrusion has been divided into two regions, as shown by dashed lines. These two regions, the region at the lower end of the specimen defined by dashed lines and the region adjacent to the muffle heater defined by dashed lines, will be discussed later with regard to end guarding. For the initial calculations without end guards the last two regions were assumed to be filled with powder. The following assumptions were common to all calculations for the model RHFA:

1. The outside chamber wall at $r = 1.30$ and the lower end of the chamber at $z = 0.0$ were assumed to be fixed temperature boundaries at 0.0°C.
2. The power density (W/m^3) in the muffle heater was selected so that the nominal specimen temperature would be about 1000°C.
3. The power density in the core heater was selected so that the temperature difference between the outside and inside specimen radii would be 15°C in the absence of nonradial heat flow at the specimen midplane.
4. The λ of the powder was assumed to be constant at 0.6 W/m K although the λ of a powder actually increases with increasing temperature. This

assumption does not influence the calculated results but does decrease computation time and, hence expense.

5. The λ of the core heater was assumed to be constant at 15 W/m K.

These assumptions were made without loss of generality for these illustrative calculations. Parameters of interest for unguarded systems included the specimen length-to-diameter ratio, l/d; the specimen thermal conductivity; and the magnitude of the heat losses from the core heater and specimen ends.

Figure 3 shows the calculated temperature at a given position along the core heater minus the temperature on the core surface at the system midplane as a function of distance from the midplane for a specimen with a λ of 100 W/m K. The four curves shown are for systems with l/d values of 2.0, 4.0, 6.0, and 8.0. The ends of these four systems are 1.6, 2.6, 3.6, and 4.6 from the midplane, respectively. For an experimental system to have a negligible error due to heat losses from the end of the system, the curve for a particular system would have to have a constant value of zero close to the system midplane. The inset in Fig. 3 shows that this condition would not be met even when the l/d is as large as 8.0. Similar curves are shown in Fig. 4 for a specimen with a λ of 5 W/m K, which is a factor of 20 lower than the specimen of Fig. 3. In each of the four cases the negative deviation of $T\text{-}T_{\text{midplane}}$ is much greater than that calculated for the

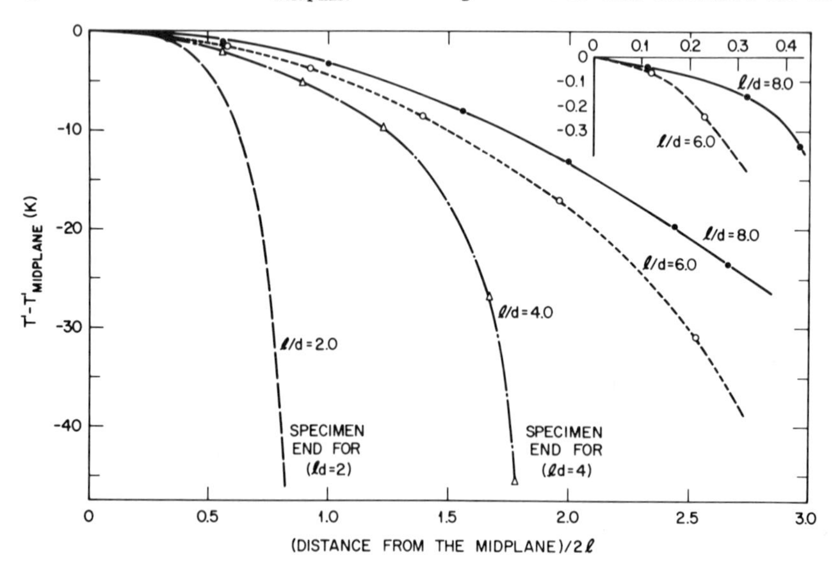

FIGURE 3. Calculated temperature of the core heater surface minus the surface temperature at the midplane of the core vs. the distance from the midplane for a specimen with a thermal conductivity of 100 W/m K and l/d values of 2.0, 4.0, 6.0, and 8.0. The positions of the ends of systems with l/d values of 2.0 and 4.0 are noted on the horizontal axis. The region near the specimen midplane is shown on an expanded scale in the upper corner.

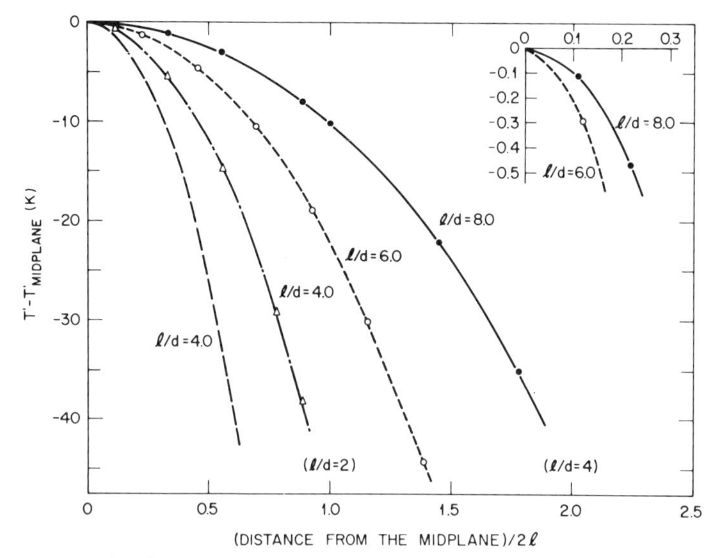

FIGURE 4. Calculated temperature of the core heater surface minus the surface temperature at the midplane for a specimen with a thermal conductivity of 5 W/m K and l/d values of 2.0, 4.0, 6.0, and 8.0. The region near the specimen midplane is shown on an expanded scale in the upper right-hand corner.

higher thermal conductivity specimen. Therefore, this indicates that the error for a system of a specific size would be less for specimens with high thermal conductivities.

Percentage errors in λ as calculated for the model RHFA are shown in Fig. 5 as a function of the l/d of the system. The percentage error for each case was calculated by using equation (2) with calculated values for ΔT from the HEATING5 analysis of the system. The values of the predicted temperatures at the inner and outer specimen radii were used to determine ΔT. The P/l was obtained from the input power density in the core heater. The top curve shows that the greatest errors would occur for the specimen with a λ of 50 W/m K. A system with an l/d of 2.0 would be in error by nearly 90%. The lower curve shows that the system errors would be much lower for the specimen with the high thermal conductivity.

The errors are caused by axial heat flow, which means that equation (2) cannot be used to calculate an accurate thermal conductivity value because the necessary boundary conditions are being satisfied. The errors are not simple, as can be seen by an examination of the system isotherms shown in Fig. 6 for a specimen with an l/d of 4.0 and a thermal conductivity of 5.0 W/m K.† The

† The thermal configurations for the model radial were calculated for a specimen temperature near 1300 K and a boundary temperature of 273 K. The isotherms on subsequent figures were labeled in terms of the temperature difference or °C.

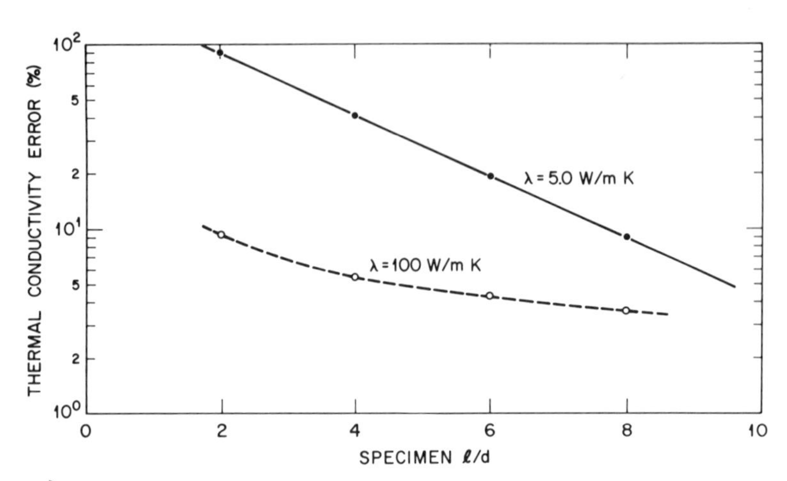

FIGURE 5. Percentage error for an unguarded RHFA as a function of the specimen l/d for specimen thermal conductivities of 5 and 100 W/m K.

abscissa of the graph is the z or axial direction given in Fig. 6, and the ordinate is the radial direction on a logarithmic scale. Ideally, the heat flow in the specimen should only be radial, so the isotherms would all be straight lines parallel to the core heater. Since the isotherms differ drastically from this ideal situation even near the specimen midplane, only part of the heat generated in the core heater is flowing in the radial direction through the specimen. A large fraction of the heat flows out the end of the core heater or into the specimen and then out the end of the specimen. This axial heat flow is so great that the isotherms toward the end of the specimen are nearly horizontal. These nonideal isotherms are caused partly by heat flow from the muffle heater into the specimen and, thence, out the specimen end. This inward flow of heat from the muffle heater contributes as much of a problem as does the axial heat flow in the core heater.

Since the shapes of the isotherms change as one passes from the specimen midplane near 980°C to the lower end near 800°C, it becomes obvious that the error in a thermal conductivity measurement using this unguarded system would be a strong function of temperature sensor position. If one had inside and outside temperature sensors at radii of 1.0 and 4.0 mm and axial positions at 14.8 and 15.4 mm, respectively, both sensors would be near the 920°C isotherm, as shown in Fig. 6, and this would yield an absurdly high value for the thermal conductivity.

Thus, it is extremely difficult to obtain a specimen that is long enough to ensure ideal radial heat flow at the specimen midplane. Admittedly, the actual errors calculated are sensitive to every dimension and thermal conductivity value used in the fundamental system, and numerous combinations of these parameters could be found to reduce the system error. However, this is

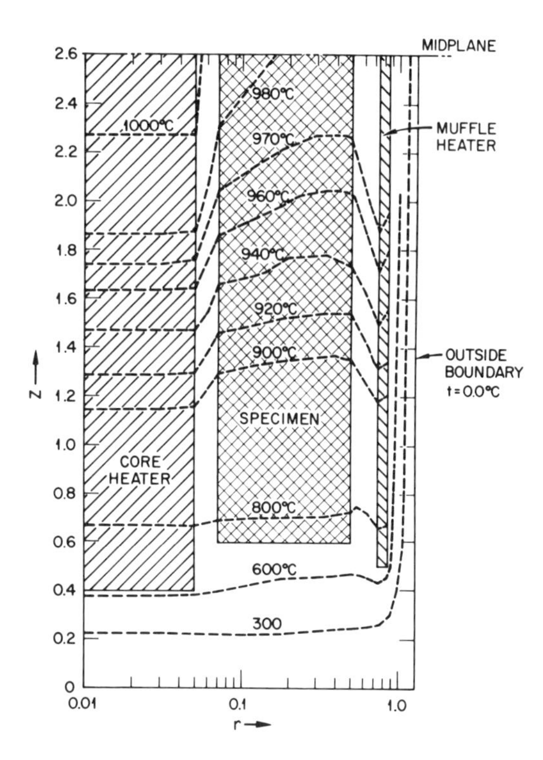

FIGURE 6. Isotherms without right circular cylindrical symmetry in the specimen of a RHFA with an $l/d = 4.0$ and specimen $\lambda = 5.0$ W/m K. Percentage error calculated for this system was +41%.

impractical, and some additional concepts are more promising. Several of these will be discussed in the next section. All these concepts were analyzed for a specimen with an l/d of 4.0 and a thermal conductivity of 5 W/m K. These values were selected because the l/d is close to that generally used in experimental RHFA, and the system errors are generally more apparent with low thermal conductivity specimens.

4.2. Techniques to Simulate Infinitely Long Systems

4.2.1. Interfaces

One way that has been advocated for the reduction of axial heat flow is the introduction of high thermal resistance interfaces or gaps in the specimen. The influence of the gaps could probably be adequately simulated by assuming that the specimen had an anisotropic thermal conductivity with λ in the z direction less than λ in the radial direction, but it was decided that the best approach was to model the system by inserting three gas-filled regions into the specimen, as shown in Fig. 7, for a RHFA with an l/d of 4.0 and a specimen λ

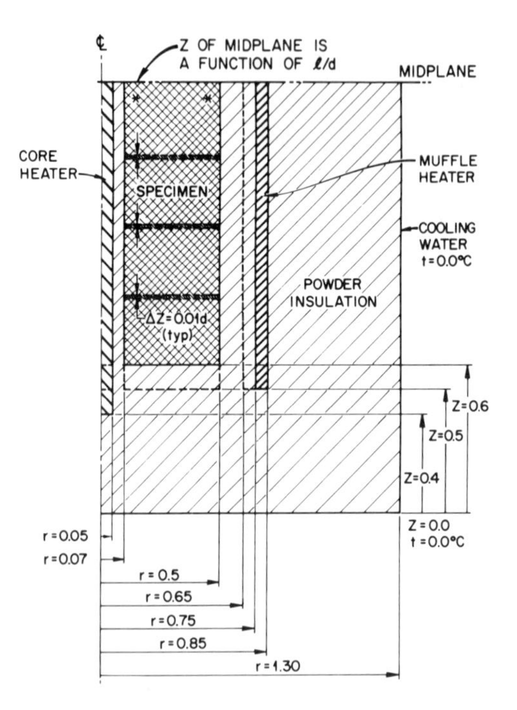

FIGURE 7. Thermal model for examining the influence of interfaces on a RHFA with an l/d of 4.0 and a specimen thermal conductivity of 5.0 W/m K.

of 5.0 W/m K. Heat transfer across gaps is generally the sum of point-to-point contact through high regions on the opposing solid specimen surfaces, radiation, gaseous conduction, and gaseous convection. For narrow gaps in the absence of high contact pressure directed normally to the gap, heat transfer by gaseous conduction and radiation dominate. The remaining two terms have been combined into one term, and the gap material is assumed to have a thermal conductivity equal to that of the powder within the system. The gap widths were chosen to be representative of expectations with mating machined surfaces. The isotherms for this configuration are shown in Fig. 8, where the radial direction is on a logarithmic scale. Comparison of this figure with Fig. 6 for a solid specimen shows that the isotherms have not been improved by the presence of the interfaces, and in fact they are even farther from the desired right-circular symmetry. The model predicts that the gaps in this system would reduce the axial heat flow by approximately 26%, but the system with the interfaces would have an error of + 61%, which is much worse than the + 41% encountered without the interfaces. This surprising result is caused by most of the axial heat flow in the specimen coming from the muffle heater. When the interfaces are inserted, the heat flow from the muffle tends to raise the temperatures at the outer radius of the specimen. Thus the temperature drop across the specimen in the radial direction would be lower than that observed in Fig. 6, where no inter-

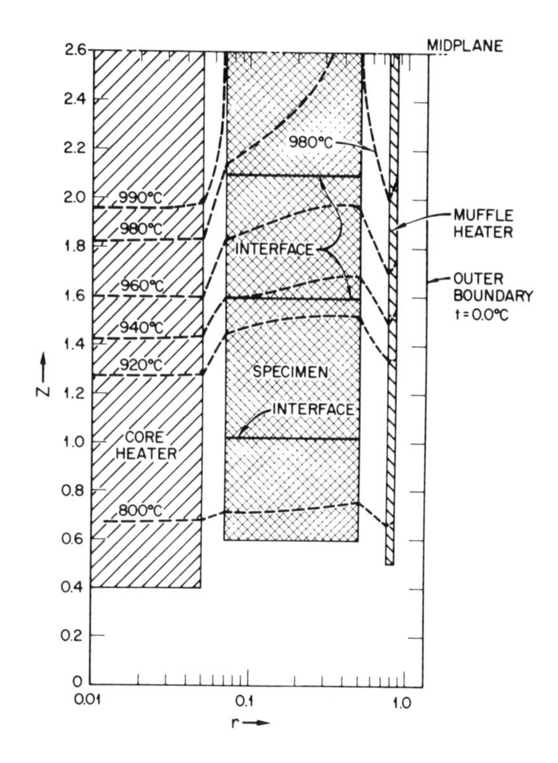

FIGURE 8. Isotherms in the specimen of a RHFA when the specimen contains interfaces or gaps to block axial heat flow. Specimen $\lambda = 5.0$ W/m K and $l/d = 4.0$. Percentage error calculated for system was + 61%.

faces were present. This lower temperature would lead to an increased positive error in λ.

However, there are situations where the presence of interfaces might be helpful in a small way. One case can be seen by reducing the muffle heater output to zero so that the nominal specimen temperature is only 53°C above the boundary temperature. The presence of interfaces would then reduce the error from 2.7% to 2.4%.

4.2.2. End Guard Heaters

The most positive way to reduce axial heat flow is the employment of guard heaters on the end of the core heater, the specimen, and/or both. Although guard heaters do reduce axial heat flow, their presence causes some surprising and disheartening effects on the isotherms within the specimen. The relative effectiveness of the two types of end guard heaters is a function of system dimensions and the thermal conductivities of system components. These dependences will be ignored, and the effect of end guards will be examined for a system with an l/d of 4.0 and a specimen with a thermal conductivity of

5.0 W/m K. A core heater end guard was simulated by assuming that heat was generated in the region of the core heater immediately below the specimen shown in Fig. 2. In all previous calculations it was assumed that no heat was generated in the core heater below the bottom plane of the specimen.

Calculations were made with several different heat generation rates in the end guard to determine the value required to raise the temperature of the end of the core heater to that of the midplane. The calculations showed that measurements of λ would be in error by + 38% even when the core heater end was hotter than the middle. Since the core heater guard reduced the error for this size system by only 3%, the largest part of the error cannot be caused by simple axial heat flow out the end of the core heater. Figure 9 shows that heat flow in this system becomes complex as a result of the introduction of a relatively simple guard scheme. The power input to the end of the core heater causes a local hot spot on the end of the core heater and in the surrounding powder. The presence of this hot spot does not significantly affect the core heater temperature near the specimen midplane. The core heater temperature decreases from a midplane value of 970°C to a minimum value of 923°C above the local hot spot. The gradient along the central region of the core heater indicates that heat would flow down the core heater from the midplane, into the specimen, and

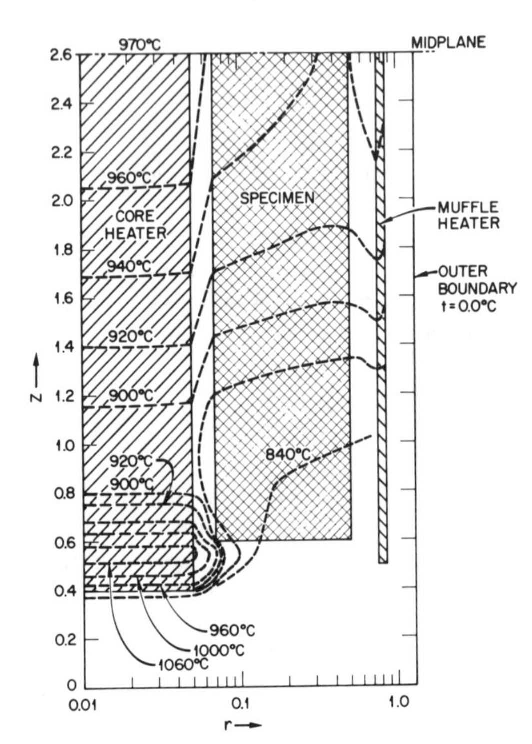

FIGURE 9. Isotherms in the model RHFA with an end guard heater on the core heater showing the local hot spot that would develop at the lower end of the core heater. Calculated system error was + 38%.

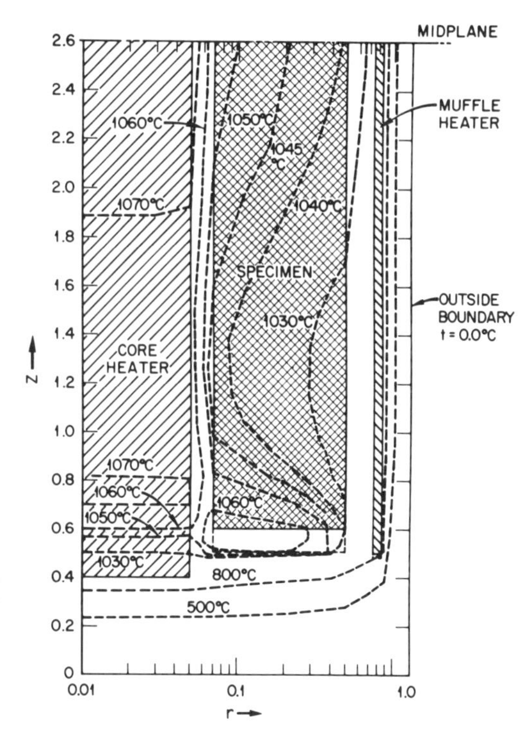

FIGURE 10. Isotherms in the model RHFA with a flat disk guard heater at the bottom of the specimen with an l/d of 4.0 and a λ of 5.0 W/m K. Calculated error for this system was + 12%.

thence out the end of the specimen. The system isotherms are very similar to those of Fig. 6 with the exception of the high temperatures around the core heater end.

Thus, for this particular combination of geometry, size, and component conductivities, the addition of a core heater end guard is a useless complication. One of the greatest complications would be the difficulty of selecting the optimum position along the core heater for measuring the temperature for matching purposes.

Axial heat flow in the fundamental RHFA could also be reduced by using a flat disk heater at the bottom of the specimen, as shown in Fig. 2. This disk was assumed to have the same thermal conductivity as porous Al_2O_3.† Calculations were made for several heat generation rates in this region until the proper rate was found to ensure that the average temperatures at the specimen midplane and end were equal. The isotherms from the calculation for this system are shown in Fig. 10. The most encouraging things to note are that the isotherms

†During all previous calculations this region was assumed to contain powder without any heat generation. Merely inserting the higher thermal conductivity material reduced the system error from + 41% to + 40%.

within the specimen are now approaching cylindrical symmetry near the specimen midplane and that the temperature gradient within the core is not nearly as large as was observed for the unguarded system shown in Fig. 6. The calculated error for this system is only 12%. The bow shape of the specimen isotherms is largely caused by a combination of gradient along the muffle heater and heat generation in the guard heater. The muffle heater temperature decreases with increasing axial distance from the midplane. This causes a similar decrease in the temperature of the specimen. The decrease in the specimen is arrested in the lower end of the specimen by the action of the guard heater which has been used to raise the average temperature at the end of the specimen to that at the midplane. Thus the minima in the isotherms at the lower end of the specimen are caused by a large heat flow radially outward to the cold end of the muffle heater. This problem was first observed by Godfrey,[13] who showed that the addition of a liner with a high thermal conductivity on the inside circumference of the muffle heater would reduce the bow in the isotherms. The influence of such a liner was examined by assuming that the region adjacent to the muffle heater shown in Fig. 2 contained a material with a thermal conductivity of 400 W/m K. The power to the muffle heater was adjusted to maintain the calculated specimen temperatures near 1000°C, and the heat generated to the end guard was again adjusted to maintain the specimen end and midplane at the same average temperature. Figure 11 shows that the large bow in the isotherms obtained when the liner region was powder was markedly reduced by the presence of the liner. The metal liner modifies the system so that the isotherms within the specimen now approach right-circular symmetry. The error for this system is + 2.6% compared with the 12% error predicted without the liner.

4.2.3. *Isothermal Corrections*

The influences of l/d, specimen λ, interfaces, guarding the end of the specimen, and guarding the core heater have been illustrated for the fundamental RHFA. One procedure that is very useful for improving the accuracy of thermal conductivity measurements and can be used for all of the above configurations is to obtain temperature sensor readings when the core heater power is zero and the power density in the muffle heater (and any guard heaters) has been adjusted so that the specimen has the same average temperature that it had when the core heater was energized.† This additional thermal configuration has been used

†In some longitudinal heat flow apparatus[18] the "isothermal correction" has been obtained by reducing the power to the specimen heater to zero and by leaving the power to every other heater constant. However, when a specimen with a high λ is in a RHFA, the core heater power is a significant fraction of the total; de-energizing the core heater causes a reduction in the specimen temperature and, probably, a significant shift in the thermal gradients along the temperature sensor wire. In addition, operation of some radial systems requires adjustments of end guard heaters to compensate for power generated in the lead wire to the specimen core heater.

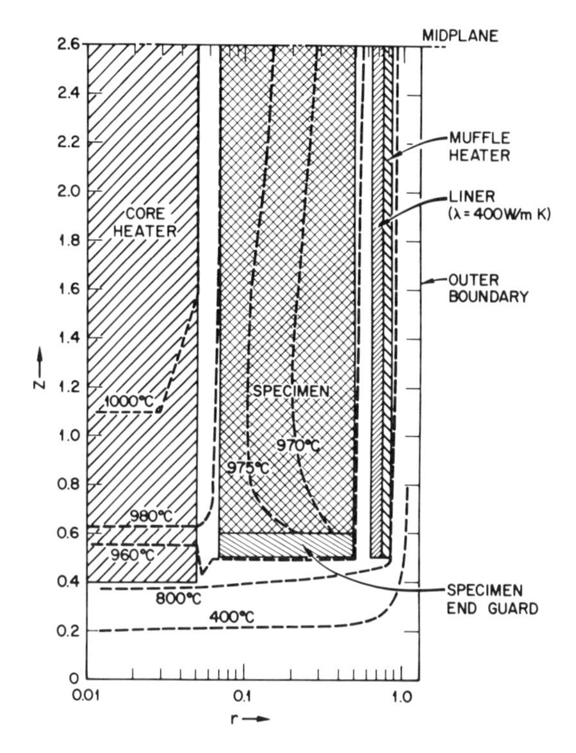

FIGURE 11. Isotherms in the model RHFA with a flat disk guard heater and a liner material with a high thermal conductivity inserted between the specimen and muffle. Thermal conductivities of the specimen and liner were assumed to be 5.0 and 400 W/m K, respectively, and assumed specimen l/d was 4.0. The calculated system error was + 2.6%.

successfully with both longitudinal and radial techniques[19] and is usually called the "isothermal correction." The original idea was to create boundary conditions around the specimen so that the specimen would be isothermal and all the temperature sensors within the specimen could be intercompared. Sometimes a working temperature standard has been included in the specimen so that this could serve as a sensor calibration. This approach generally improves the data accuracy, but it is truly a misnomer since it is impossible to guarantee that a specimen is isothermal. At the very least, errors in temperature sensors used to monitor the boundary conditions and errors in temperature controllers used for maintaining the temperatures cause unavoidable temperature mismatches that lead to undesired heat flow. In practice, the "isothermal correction" usually corrects for some inconsistencies in thermometry and for spurious heat flow *as long as* the cause of the latter is the same for both thermal conditions. The

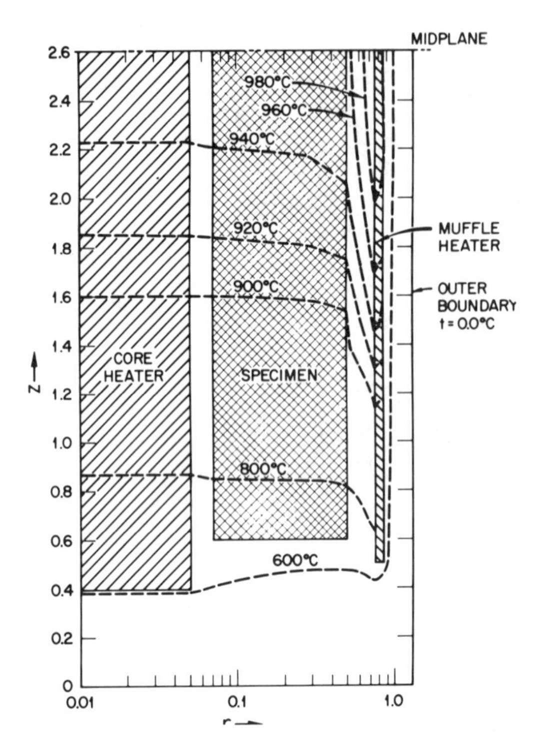

FIGURE 12. Isotherms for the model RHFA during the "isothermal correction" configuration for the unguarded specimen with an l/d of 4.0 and a λ of 5.0 W/m K.

"isothermal correction" is used in a modified form of equation (2). Now

$$\lambda = \frac{P}{l} \frac{\ln (r_2/r_1)}{2\pi(\Delta T_1 - \Delta T_2)} \tag{4}$$

where ΔT_1 is the difference in the temperatures at positions r_1 and r_2 while the core heater is energized, and ΔT_2 is the difference when the core heater power has been reduced to zero, the muffle heater has been adjusted to give the same specimen temperature, and the end guards, if any, have been adjusted to give the same temperature difference between the specimen midplane and the specimen end as that encountered while the core heater was energized.

An "isothermal correction" was calculated for the unguarded system shown in Fig. 2 for an l/d of 4.0 and a specimen λ of 5 W/m K. The isotherms generated by using HEATING5 are shown in Fig. 12, and these should be compared with the isotherms shown in Fig. 6, which were calculated for the same system with power to the core heater. Figure 12 shows that the specimen is far from isothermal when the core heater is off. Heat flows inward from the muffle heater into the specimen and thence out the end, as in the previous

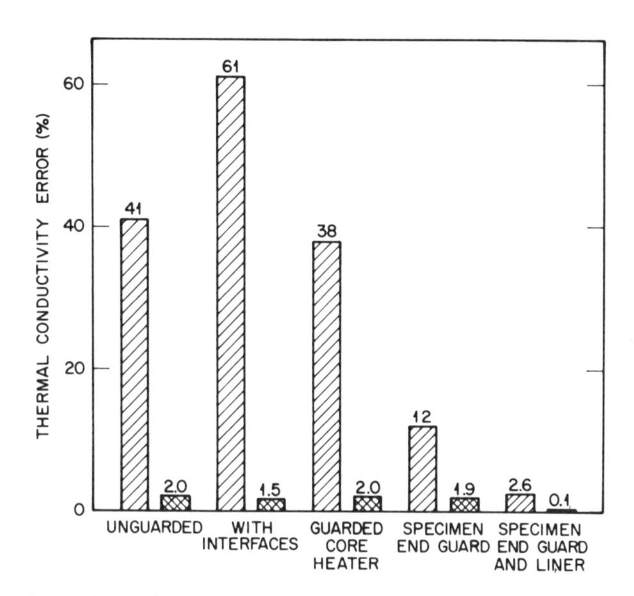

FIGURE 13. Comparison of predicted thermal errors for various systems showing the large reduction expected by using an "isothermal correction" appropriate for each system.

configuration. In this case, ΔT_2 would be negative since heat flows radially inward. Neglect of the "isothermal correction" led to an error of $+41\%$, whereas use of the correction reduces the error for this particular system to only 2%.

Application of the above concept beneficially affects all the configurations examined, as shown in Fig. 13. In all cases the "isothermal correction" reduces the measurement error due to nonradial heat flow to less than 2%, and in the case of the specimen end guard plus liner the error is reduced to 0.1%. Therefore, although many things can be done to a system with a specimen of finite length to simulate an infinitely long specimen, the "isothermal correction" concept appears to be universally beneficial for reducing the thermal error in RHFA. McElroy and Moore[2] give numerical values showing the importance of this correction for an experimental RHFA.

5. HEAT FLOW IN EXPERIMENTAL APPARATUS FOR MEASUREMENTS ON SOLIDS

5.1. Introduction

This section describes analysis results for five RHFA that have been used extensively for measuring the thermal conductivity of solids. This selection is not meant to imply that these are necessarily the "best" apparatus but rather that they represent the good and bad features of RHFA. This group of five includes RHFA for high and low temperatures, with and without end guards, and with heat flowing radially outward from a central heater and radially inward toward a heat sink at the specimen center. Some suggested changes are included to improve heat flow so that the boundary conditions required for use of equation (2) are approached. Since an earlier report[2] was published, a RHFA was developed and used by Ondracek *et al.*[20] for measuring the thermal conductivity of cermets. However, this apparatus will not be described here.

5.2. Systems without End Guards

Thermal conductivity has been measured many times by using RHFA without end guards,[2] and their success has depended on many factors already discussed for the model system. Two of these have been selected for thermal analysis.

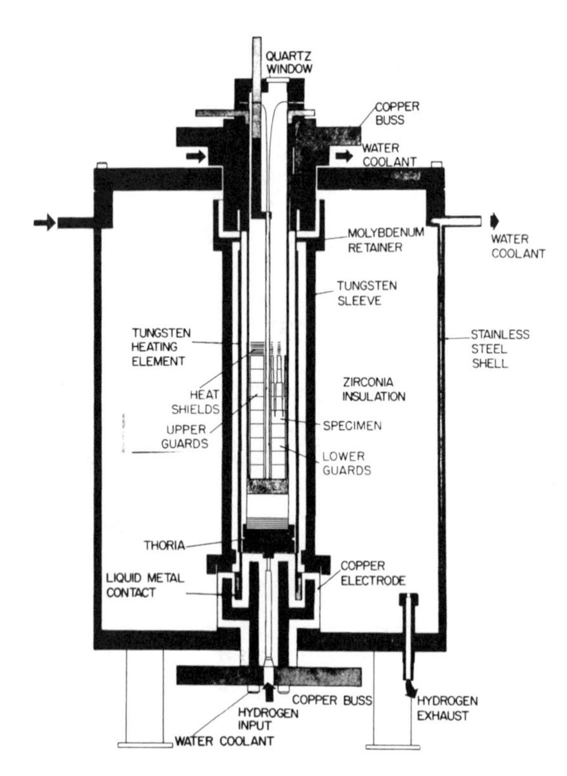

FIGURE 14. A cross-sectional view of the RHFA used by Feith[21] for measuring thermal conductivity at high temperatures.

5.2.1. High-Temperature Radial Heat Flow Apparatus Used by Feith

One example of a RHFA that did not employ end guard heaters was used by Feith[21] for measuring the thermal conductivity of refractory metals and of UO_2 at high temperatures. A cross-sectional view of this apparatus is shown in Fig. 14. The specimen consisted of a stack of disks resting at the bottom of a 520-mm-deep molybdenum crucible with 4.8-mm-thick walls. The central heater was a 6.35-mm-O.D. by 5.33-mm-I.D. molybdenum tube. Four axial holes at two radii on the specimen served as blackbodies so that an optical pyrometer could be used for measuring the temperatures within the specimen by sighting into the bottoms of the holes from the top of the apparatus. Zirconia insulation was contained in the annulus between a thick-walled (13-mm) tungsten sleeve and the outer water-cooled walls of the chamber. Stacks of radiation shields rested on the bottom of the chamber and on the top of the specimen to reduce heat exchange by radiation. Feith's estimated uncertainty for this apparatus was $+6\%$.

Several aspects of this apparatus complicated the thermal modeling. First, the midplane of the specimen stack was not a plane of symmetry because the long core heater protruded from the top end of the specimen stack and terminated at the lower end of the stack by threading into a tapped hole in a 25-mm-thick molybdenum disk that formed the bottom of the crucible. This asymmetry dictated that the entire system – not just one half as in the previous sections – be analyzed. The second difficulty involved the tungsten tube, which was used as a muffle heater by passage of an electrical current. Each end of this heater extended to a water-cooled surface so that the temperature of each heater end was normally near that of the cooling water. Thus, the variation of temperature along this heater could be expected to be much greater than that encountered with the fundamental radial because the muffle heater in the latter stops far short of the water-cooled end plates. This steeper temperature profile is made even worse because the temperature dependence of the electrical resistivity of tungsten, $d\rho/dt$, is positive, which causes a higher heat generation rate near the specimen midplane than at either end.

A cross-sectional view of the thermal model of this RHFA is shown in Fig. 15. The core heater centerline was assumed to be an adiabatic boundary, and the outer chamber wall at $r = 240$ mm was assumed to be an isothermal boundary. Since the upper and lower electrodes were water cooled, the upper ($z = 450$ mm) and lower ($z = 0.0$ mm) ends of the thermal model were terminated at the electrode surfaces nearest to the specimen. These two boundaries were extended to $r = 240$ mm, which is farther than the electrodes actually went, and means that the thermal model ignored part of the apparatus. This simplification would not influence the calculated results since the temperature in the omitted spaces is near that of the outer boundaries. Heat transfer in the regions

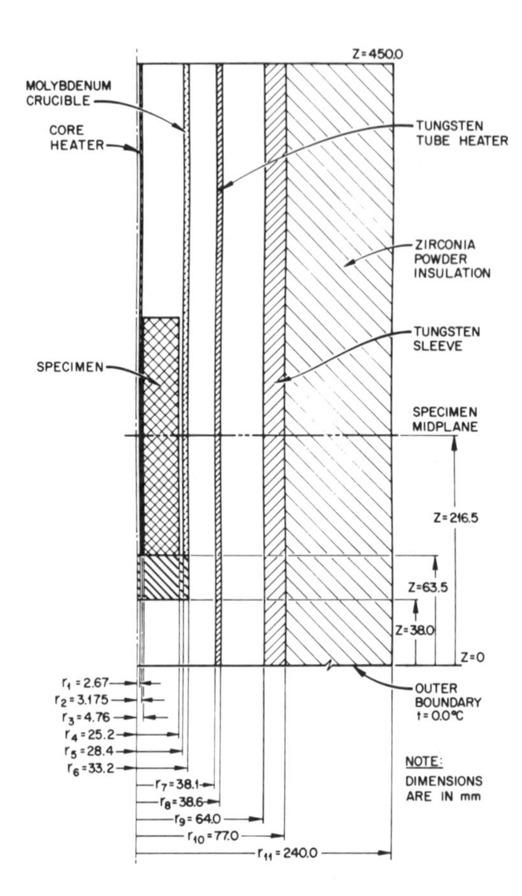

FIGURE 15. The thermal model used with HEATING5 to study Feith's RHFA.

containing argon gas was due to a combination of conduction, convection, and radiation. Heat transfer coefficients were selected to describe heat transfer in the numerous gaps with different sizes and geometries.

The calculated thermal errors are given in Table 1 for several conditions of the muffle heater. In the first two cases it was assumed that the power per unit volume, P/V, in the muffle heater was a constant, but the magnitude of P/V was different so that the resultant specimen temperature would differ for the two cases. Calculations indicated thermal errors of + 7.9% and + 8.4% at temperatures of 1100 and 1322°C, respectively. The calculated thermal error increased to + 13.3% at 1213°C for case 3 when the muffle heater power was assumed to be a function of temperature to reflect the $d\rho/dt$ of tungsten. Based on calculations not shown this error would be higher at higher temperatures.

The asymmetry of this apparatus is apparent in Fig. 16, which shows the calculated isotherms for case 3 of Table 1. The temperatures in the top half of

TABLE 1
Summary of the Calculations for Feith's Apparatus with a Specimen with a Thermal Conductivity of 2.5 W/m K

Case	Muffle heater	Specimen temperature (°C)	Thermal error (%)
1	$P/V = \text{const}$	1100	+ 7.9
2	$P/V = \text{const}$	1322	+ 8.4
3	$P/V = f(t)$	1213	+ 13.3

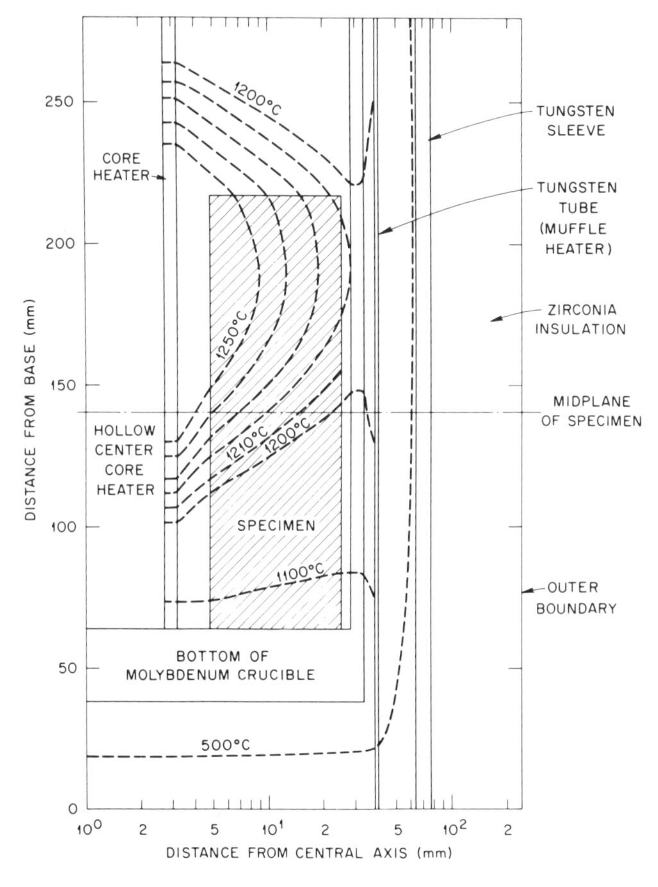

FIGURE 16. Calculated isotherms for Feith's RHFA containing a UO_2 specimen showing the hot spot on the top of the core heater.

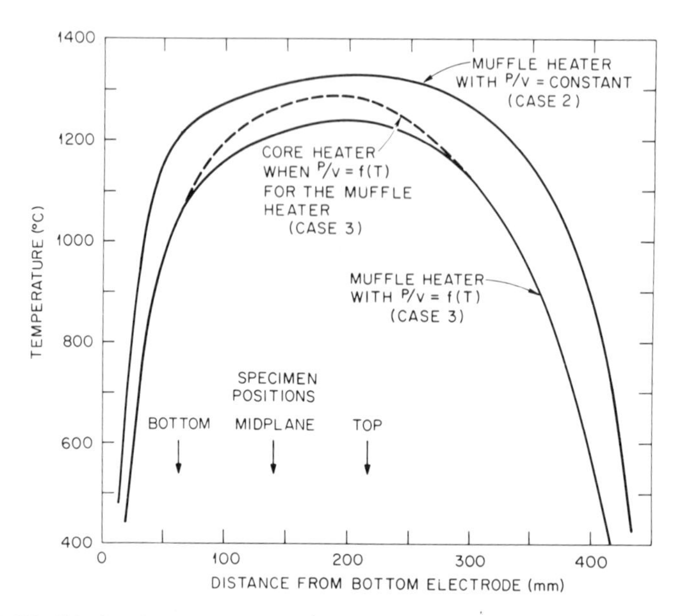

FIGURE 17. Calculated temperature profiles along the muffle heater in Feith's apparatus when the power per unit volume in the heater was assumed to be a constant and a function of temperature.

the specimen are much higher than those in the bottom half. The isotherms are not ideal at the specimen midplane where radial temperatures were measured in the experimental apparatus, and there is no plane within the specimen where the radial temperatures could give an exact value of the thermal conductivity.† There are two reasons for the asymmetry in this apparatus. One is the protrusion of the core heater from the top of the specimen but not at the bottom. The second is simply that the specimen is not at the best axial position within the apparatus. This can be seen in Fig. 17, which compares the temperature profiles along the muffle heater for cases 2 and 3 with respect to the position of the specimen. The broad maximum in the profile occurs at 200 mm, which is close to the top of the specimen at 216.5 mm. The temperature difference between the maximum temperature of the muffle heater and its temperature at the top plane of the specimen is only 3.9°C, but the same difference with respect to the bottom plane of the specimen is over 200°C. The profile for case 3 is not as broad as that for case 2 because of the lower amount of power generated in the lower ends of the muffle heater in case 3. The temperature profile of the muffle

†Feith measured the temperatures in the pyrometer sight holes with the specimen at 1300°C, where thermocouples could be used. He reported only a few degrees variation from specimen top to midplane. This agrees with the calculated results. If he had been able to measure to the bottom of the specimen, the asymmetry would have been apparent.

heater is mirrored by that of the core heater as shown by the dashed line in Fig. 17.

Feith's apparatus was chosen as an example of an unguarded system because of the interesting features such as the tungsten tube muffle heater and the high temperature capabilities. However, based on the above analysis, this apparatus could be improved by

1. increasing the specimen l/d even if it were necessary to use dummy disks at each end of the specimen stack,
2. improving the symmetry by locating the specimen closer to the midplane of the apparatus and by allowing the core heater to protrude from the bottom of the specimen,
3. decreasing the end losses – especially at the bottom – by the addition of end guards.

5.2.2. A Simple Radial Heat Flow Apparatus for Pipe Insulations

The analyses of the fundamental RHFA showed that nonradial heat flow and, hence experimental errors would be greatest for materials with low thermal conductivity. Therefore, it is difficult to accurately measure the thermal conductivity of materials used for pipe insulations that have very low thermal conductivities. A novel RHFA that was developed by Jury *et al.*[22] for measuring the λ pipe insulations is shown in Fig. 18. Calcium silicate pipe insulation was installed between two sets of brass flanges and around a cylindrical heater fabricated from type 316 stainless steel screen (40 × 40 mesh per 25.4 mm). This

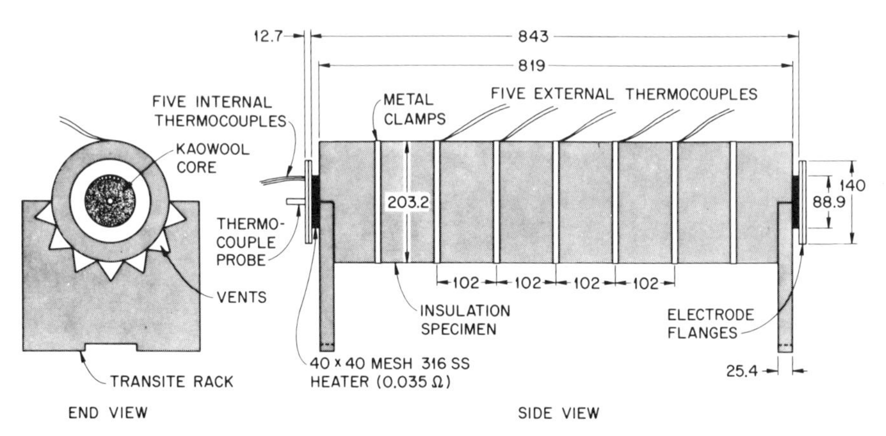

FIGURE 18. A cross-sectional view of a simple RHFA for measuring the thermal conductivity of pipe insulations taken from Jury *et al.*[17]

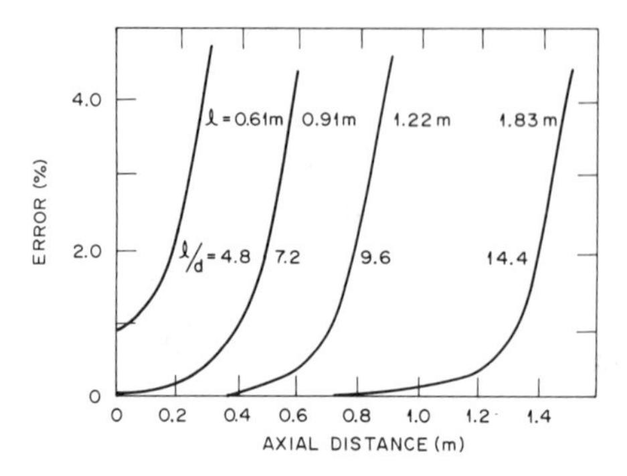

FIGURE 19. Percentage errors as a function of the axial distance between the specimen midplane and the system thermocouples for a pipe insulation with l/d values of 4.8, 7.2, 9.6, and 14.4.

screen had five platinum vs. $Pt_{90}Rh_{10}$ thermocouples attached at various intervals to measure the screen temperature. The temperatures on the outside radius of the specimen were determined by measuring the temperatures of five metal clamps around the specimen at the same axial positions as the thermocouples attached to the screen. The center of the screen core heater contained fibrous insulation and a quartz thermocouple well. Direct current was passed through the central heater to establish a temperature gradient in the specimen, and the values of the current flow in the screen heater and the voltage drop per unit length were measured so that the heat generated per unit length in the screen heater could be calculated. The thermocouple outputs were measured at steady state and converted to temperatures. The thermal conductivity of the specimen was calculated by using equation (2). Although the authors do not state an error limit based on the uncertainties of the parameters used in equation (2), a fair estimate would be about ± 3%.

Before apparatus operation the authors examined the influence of l/d and l/s, where s is the specimen wall thickness, on system accuracy using HEATING5. Figure 19 shows the calculated percentage thermal errors for four l/d ratios as a function of the axial position at which the inner and outer radii temperatures were measured. The errors are all less than 1% for all l/d when the temperatures of the inside and outside radii are measured at the specimen midplane. Calculated thermal errors for a specimen with an l/d of 14.4 are less than 1% for temperature measurement positions up to 1.3 m from the midplane. If the temperature measurements were made at the midplane of the pipe insulation tester with an l/d of 4.8, the thermal error would only be + 1%, which is much lower than the + 41% calculated for the model RHFA with a specimen l/d of 4.0 and a λ of 5.0 W/m K. This is especially surprising because the assumed thermal conductivity of the pipe insulation was 0.035 W/m K, which is more than two orders of magnitude less.† The accuracy of this RHFA has been

†The thermal conductivity of the calcium silicate was actually closer to 0.06 W/m K than the 0.035 W/m K used in the calculations, but this only means that the apparatus would be even better than shown by the calculated results of Fig. 19.

confirmed by comparison of results with other techniques. The causes of these low errors are the low thermal conductivity of the screen wire heater in the axial direction and the direct contact of the central heater with the specimen.

5.3. Apparatus with End Guard Heaters

5.3.1. The Stacked-Disk Radial Heat Flow Apparatus of the National Physical Laboratory

One of the best known RHFA that used specimen end guards was developed by R.W. Powell[23] and was used to measure the thermal conductivities of iron and several steels with a claimed uncertainty less than ± 2%. The specimen of this apparatus, which is shown in Fig. 20, consisted of a stack of right-circular disks. These disks were aligned so that a movable thermocouple could be

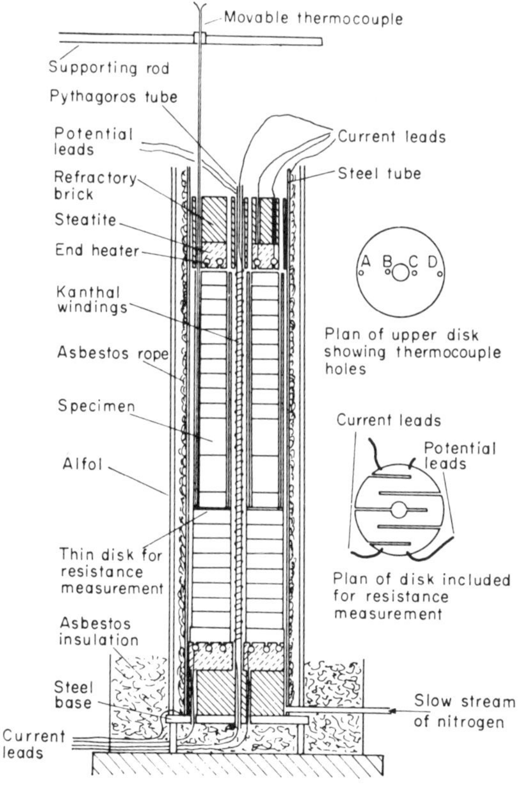

FIGURE 20. The stacked-disk RHFA of the National Physical Laboratory showing the movable thermocouple used for all temperature measurements in the specimen.

inserted into longitudinal holes at radii of 7.1 and 29.6 mm. This arrangement permitted temperature measurements at any axial position within the long holes. This movable thermocouple could also be used to measure the temperature along the central axis of the hollow core heater. Figure 21 shows the cross section of the thermal model used with the heat conduction code to study heat flow in this apparatus. The primary approximations were as follows:

1. All interfaces in the specimen stack were ignored, and the specimen was treated as a solid.
2. The support rods of the lower steel plate were ignored, and this plate was assumed to be thermally attached to the base plate only by the asbestos insulation.
3. The thermal conductivity of the asbestos rope and insulation were assumed to be the same.
4. The "Pythagoras" tube used for the central heater wire support was assumed to be a solid rod with a thermal conductivity equal to one-half that of lavite.(24)
5. The specimen midplane was assumed to be adiabatic.

All calculations were for a specimen of iron with a temperature-independent thermal conductivity of 28.9 W/m K, and the power density in the central

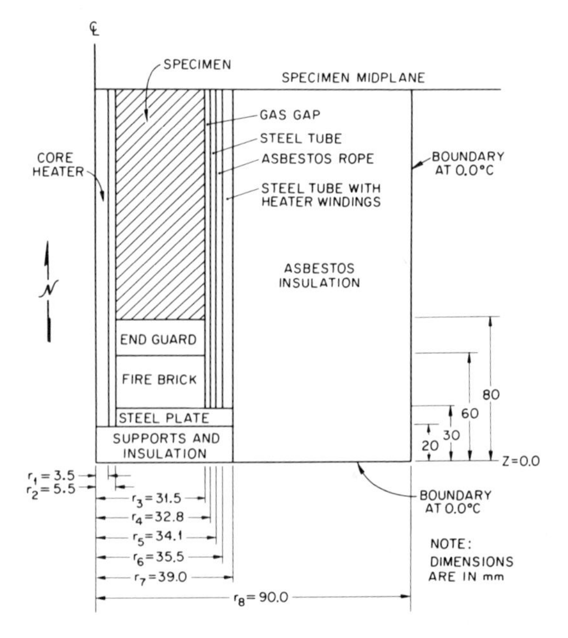

FIGURE 21. The thermal model of the National Physical Laboratory RHFA.

core heater was adjusted to give a temperature difference of 11.0°C between the specimen inner and outer radii. The muffle heater power density was selected to yield a specimen temperature near 750°C, and the end guard power was adjusted to ensure a temperature difference less than 0.5°C between the specimen midplane at 205 mm and specimen end at 80 mm.

The temperature distribution calculated with HEATING5 has some interesting features, as shown in Fig. 22. The lower part of the core heater in this system has a distinct hot spot with a maximum temperature nearly 300°C hotter than the central region of the core heater. Since the power generation density within the core heater is uniform, this hot spot could only be caused by the very low thermal conductivity of the refractory brick and – to a lesser extent – by the low thermal conductivity of the end guard heater compared with that of the specimen. The two isotherms at 800 and 900°C in the gap between core heater and specimen are nearly cylindrical between axial positions of 110 to 205 mm. Although the isotherms within the specimen indicate that axial heat flow near the specimen midplane would be small, these isotherms have slight curvatures that are smaller than, but similar to, those previously observed for the fundamental RHFA when a specimen end guard was used to block axial heat flow (Fig. 10). This curvature of the isotherms was verified experimentally

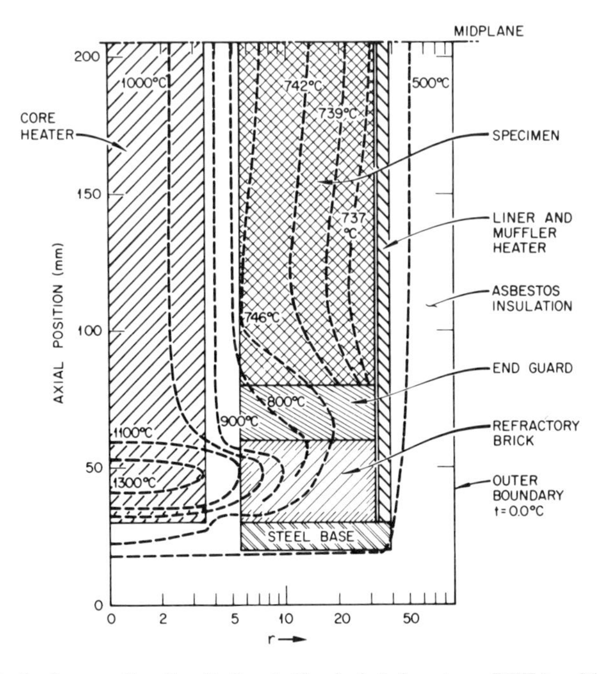

FIGURE 22. Isotherms for the National Physical Laboratory RHFA with a specimen composed of iron disks. System error is about 1%.

by Powell,[23] who reported decreasing temperatures at both inside and outside radii as the movable thermocouple was moved away from the specimen midplane. The temperature decreases observed by Powell are in good quantitative agreement with the calculated curvature for these isotherms. The thermal map for this system would indicate that it would be in error by about 1%, but this small error could probably be eliminated by the application of an "isothermal correction." However, Powell did not obtain such a correction because he felt that the single thermocouple concept would negate any potential benefit. These calculations ignored the potential influence of the multiple interfaces caused by use of the stacked-disk specimen. Introduction of these interfaces into the analysis, however, would not influence the thermal error significantly. Although this RHFA would have an error less than 1% for materials with thermal conductivities near that of iron, the errors would probably be greater for specimens with lower thermal conductivities.

The electrical resistivity, ρ, and Seebeck coefficient, S, of a metallic material are important properties that assist in understanding the thermal conductivity of the material. Simultaneous measurements of these properties can indicate specimen degradation caused by reaction with the environment of the RHFA. Unfortunately, these properties are more difficult to measure on a RHFA specimen than on a long rod normally used in longitudinal heat flow apparatus. Merely positioning a small rod within the hot zone of a RHFA is not adequate because the rod would perturb heat flow.

Powell[23] used a thin disk with an outer and inner diameter equal to that of the specimen as shown in Fig. 20 to measure resistance vs. temperature, which was then converted to ρ vs. T by using the measured room temperature ρ of a rod cut from the same material. The disk had narrow slots machined from top to bottom to increase the electrical resistance between the potential taps. The disk was inserted into the specimen column with a thin layer of electrically insulating material on the top and bottom surfaces to prevent shorting. A standard four-probe dc technique was used to measure the specimen resistance.

5.3.2. The Stacked-Disk Radial Heat Flow Apparatus of the Oak Ridge National Laboratory

A stacked-disk RHFA, which employed fixed thermocouples and a three-piece muffle heater, was developed at the Oak Ridge National Laboratory and was used to measure the thermal conductivities of materials with conductivities between 0.1 and 200 W/m K at temperatures as high as 1570 K.[25-27] A comparison of results from this apparatus with results from others[19] has substantiated the general accuracy claims of ± 1.5%. The apparatus flexibility was achieved by small but important changes, which were required to extend its use to materials with very high or very low thermal conductivity. The first form of

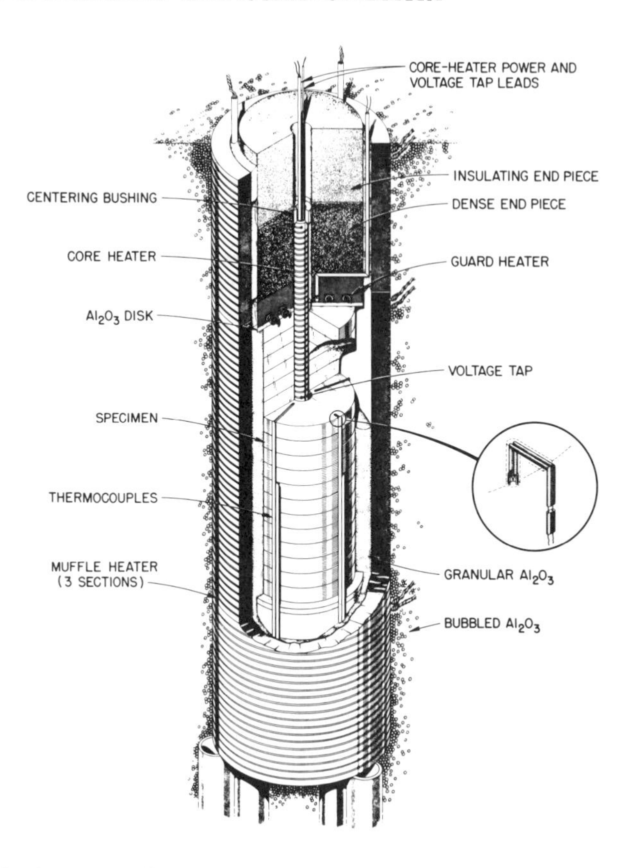

FIGURE 23. A cutaway drawing of the stacked-disk RHFA at the Oak Ridge National Laboratory showing the wire-wound alumina tube core heater that was used for thermal conductivity measurements in UO_2 and iron. Obtained from Godfrey *et al.*[13]

this apparatus that saw widespread use is shown in Fig. 23 as described in detail by Godfrey *et al.*[25] The muffle heater was composed of three independently controlled sections, which were fabricated by winding molybdenum wire into the grooves of 127-mm-O.D. by 101-mm-I.D. alumina tubing. The thermocouples in the specimen were insulated with round two-hole alumina insulation, and the thermocouple assemblies were pushed down into the bottom of thermocouple wells that had been drilled at two radii. Each insulated thermocouple was positioned in a radial slot that extended from the outside radius of the specimen to the radius of the thermocouple well. Normally two instrumented disks were at the specimen midplane, and each disk contained three thermocouples at an inner radius and three thermocouples at an outer radius. Two thermocouples were installed in each end disk to monitor the temperatures at the ends of the specimen. The core heater consisted of Pt–10% Rh wire which was bifilar wound into grooves cut into the surface of a 380-mm-long alumina tube. This core heater was so long that it extended beyond the end guard heaters on each end of the specimen. Potential taps consisted of fine wire welded onto the heater wire at known positions over the central 76-mm length of the core heater. This core heater (and the two others yet to be described) was powered by passing a stable direct current through the heater. The specimen for this RHFA normally consisted of a stack of disks making it a right circular cylinder with height and diameter of 228 and 76 mm, respectively, for an l/d of 3.0. The two small annuli contained fine granular alumina, whereas the large gap between muffle heater and outside system boundary contained large ($\sim$ 6 mm in diameter) bubbled alumina. The bottom and top muffles were always controlled at a lower temperature than the center muffle so that a nonzero power would always be required in the end guards. If T was the datum temperature with respect to that of the outer boundary, the middle temperature of the end muffles would be controlled approximately 0.1 T below that of the center muffle at the mid-plane. An "isothermal correction" was always obtained for each datum. The apparatus was used in this configuration to obtain data on Inor-8, UO_2,[25] iron,[26] and Inconel 702.

The power output of the wire-wound alumina core heater was insufficient to establish accurately measurable temperature gradients in specimens with high thermal conductivity. Therefore, another core heater was prepared by threading each end of a 305-mm-long, 6.1-mm-diam spectrographic carbon rod into graphite electrodes. Potential taps were attached either in shallow grooves on the carbon rod or in 0.25-mm-diam holes that had been drilled radially through the rod. This core heater was shorter than the wire-wound one and barely extended through the end guard heaters. This type core heater was used for measurements on W, Mo, Nb, several graphites, and several refractory metal alloys over a nominal temperature range from 330 to 1400 K with an uncertainty less than $\pm$ 1.5%.

The thermal model used for analysis of this configuration with HEATING5 is shown in Fig. 24. The inside and outside radii of the specimen shown are not as great as those of the end guard heater and dense brick because these components had originally been designed for a larger specimen. The change to a specimen with a smaller inside radius permitted location of the inside thermocouples in the specimen closer to the system central axis; this gave, for a fixed power, a larger temperature difference between the inside and outside thermocouples since the temperature gradient is steeper near the central axis. The "low-density brick" is shown covering a region that was actually composed of a mixture of low-density fire brick, bubbled Al_2O_3, and ceramic support tubes. Calculations with this system showed that the thermal errors would be + 1% at 1000°C for a specimen with a thermal conductivity of 65 W/m K both with and without a metal liner ($\lambda = 120$ W/m K) inserted on the inside of the muffle

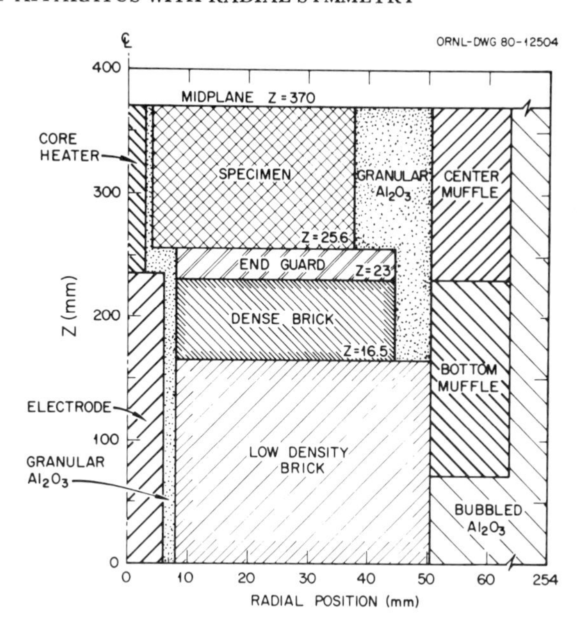

FIGURE 24. The thermal model used with HEATING5 to analyze the Oak Ridge National Laboratory RHFA with a carbon rod core heater.

heaters. For a specimen with a thermal conductivity of 4 W/m K, the calculated errors at the same temperature would be + 2.5% and + 1.6% with and without the metal liner, respectively. All four of these values are exclusive of a calculated "isothermal correction," which when applied would reduce the thermal errors below 0.1% in every case.

The good accuracy of this RHFA is not too surprising when one examines the isotherms for one of the above combinations. Figure 25 shows the calculated results for the specimen with the lower thermal conductivity and without a metal liner. The isotherms within the specimen are parallel to the central axis to a depth of about 30 mm below the midplane. Below that depth the isotherms on the inside of the specimen (1015, 1010, and 1000°C) remain approximately parallel another 50 mm, whereas those on the outside, such as the one at 990°C, show the distinct bows similar to those in other RHFA. The isotherms in this system are better than those in any RHFA analyzed for this section with the possible exception of the apparatus of the National Physical Laboratory.

In most RHFA the addition of a liner with a high thermal conductivity reduces the bow in the outer specimen isotherms, but in this case the bow was increased, as shown by the one isotherm at 975°C for the same system with the metal liner. Calculations showed that the bow increase was accompanied by a thermal error increase from 1.6 to 2.5%.

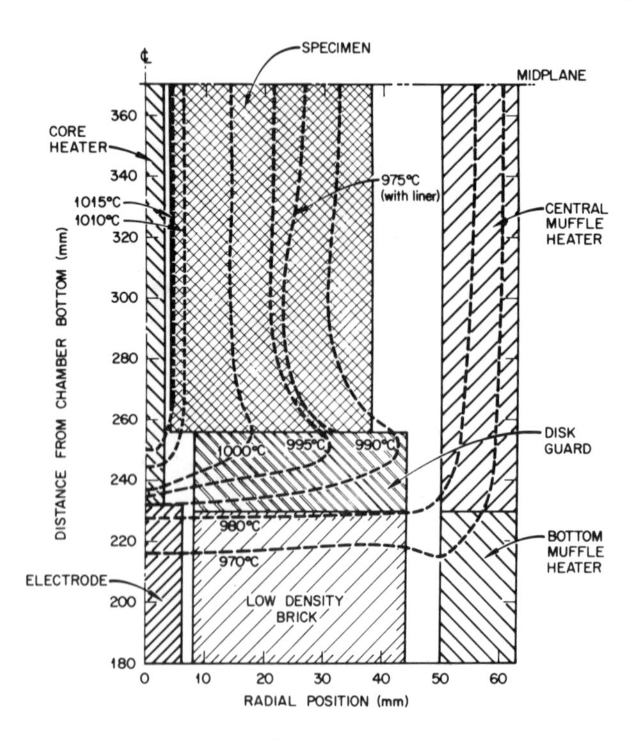

FIGURE 25. Calculated isotherms for the Oak Ridge National Laboratory RHFA with a specimen with a thermal conductivity of 4 W/m K and without a liner on the inside of the muffle heater.

This RHFA was also used for materials with thermal conductivities as low as 0.1 W/m K, but this required a third type core heater.[27] This heater, which had lower axial conductance than either of the two previously described, was fabricated from a 6.35-mm-diam stainless steel tube with a wall thickness of 0.5 mm. Radially wound heaters were strapped to the ends of the tube immediately outside the specimen stack, and five thermocouples were spaced inside the tube along its length and welded to the tube walls. These thermocouples provided information about the axial temperature distribution in the core heater, which was used to adjust the core heater end guards until the axial temperature was nearly constant. The voltage drop between adjacent pairs of thermocouples could be used with the measured current to calculate the power in the core heater. This type core heater was used to measure the thermal conductivities of low-density fibrous carbon solids[27] and powders encapsulated in a cannister. This cannister was simply placed in between the end guards in the position normally occupied by the solid specimen. The powder measurements were based on a concept by Flynn[28] that will be described in a later section.

The Seebeck coefficient of a radial specimen was measured in the apparatus of Fig. 23 by winding platinum wire around the hot junction of Pt–10% Rh thermocouples and then forcing the assemblies into thermocouple wells. Normally, electrically insulating (Al_2O_3) alignment bushings were in the holes, but these were removed from one well each at an inner and outer radii for insertion of these wire-wound hot junctions. Since these thermocouples were in electrical contact with parts of the specimen that were at different temperatures, the thermocouple outputs and the emf from the two platinum wires could be used to calculate the Seebeck coefficient. This method tends to give imprecise data because of wire movements during thermal cycling.

A specimen for a technique that was used successfully for measuring both ρ and S is shown in Fig. 26. This specimen was machined from a 6.3-mm-thick disk with an inner and outer diameter the same as the RHFA specimen. Two slots were machined through the disk to leave a rod with a rectangular cross section of 4.5 by 3.0 mm. The top and bottom surfaces of this rod were milled so that neither surface would contact the metal disk that would be on each side of the rod during measurements. Two Pt vs. Pt–10% Rh thermocouples were welded to the top of the rod at known positions. These thermocouples were electrically insulated with Al_2O_3 tubing and passed through the slots to the outside of the specimen and then downward toward electrical junctions at the bottom of the apparatus. Current leads were attached to the free end of the specimen and the disk so that a standard four-probe dc measurement could be used to determine ρ. While this thin disk was in the stack of thicker disks used to form the 230-mm-high stack, its temperature was always near that of the λ specimen, and ρ and S were measured in the normal fashion.

5.2.3. A High-Temperature Radial Inflow Apparatus

The RHFA shown in Fig. 27 was developed by Rasor and McClelland[29] and Taylor[30] and was used to measure the thermal conductivities of graphite, molybdenum, tungsten, and TiC at temperatures up to 2770 K. Part of the heat generated in the graphite heater surrounding the specimen flowed inward through the specimen to a water-cooled calorimeter mounted at the center of the specimen. The heat adsorbed by the calorimeter and the temperature gradient in the specimen were measured and used to calculate the specimen thermal conductivity with equation (2). Moeller[31] has reported on a similar apparatus with a greater l/d.

The helix heater shown in Fig. 27 was fabricated by milling slots through the wall of a 76-mm-long graphite sleeve with inside and outside diameters of 63.5 and 69.9 mm, respectively. Each end of this helix was screwed into a graphite ring from which two graphite electrodes extended. The temperatures of the ring–electrode assemblies could be varied independently by adjusting

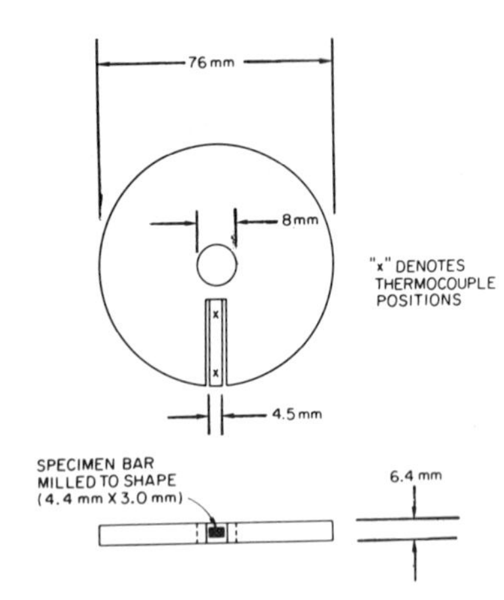

FIGURE 26. A rod specimen machined from a thin disk for measurements of ρ and S during thermal conductivity measurements. Two thermocouples were welded to the rod, and a current input lead was attached to the free end.

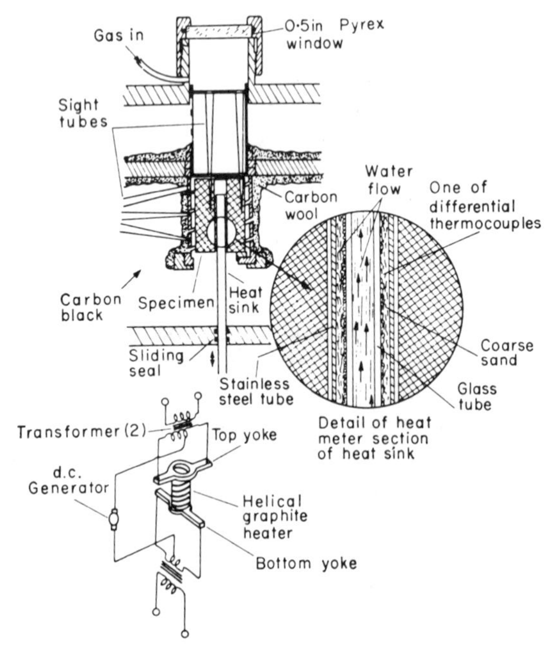

FIGURE 27. The high-temperature RHFA with a water-cooled calorimeter on the central axis as developed by Rasor and McClelland[29] and Taylor.[30] Taken from Rasor and McClelland.[29]

the ac current between the two electrodes on each ring. This selective heating in the two end assemblies provided the capability of adjusting the temperatures at the ends of the specimen until they were at the same temperatures as the middle. Direct current generators were used as a power source for the helix.

The block on the top of the specimen contained six tapered sight holes so that the midplane temperatures could be measured at two radii with a pyrometer. The block was either lamp black, which had been packed around removable mandrels to form the holes, or was machined from porous carbon.

The retractable water-cooled calorimeter extended from the bottom of the chamber to the top of the specimen. This calorimeter consisted of an outer stainless steel tube concentric with an inner glass tube. Water flowed upward in the inside glass tube and then downward in the annulus. The temperature change of this water over a 25-mm span at the specimen midplane and the water flow rate were measured and used to calculate the heat absorbed by the calorimeter over this span. Rasor and McClelland gave an estimated uncertainty of ± 5% for this apparatus.

A cross section of the model used for thermal analysis is shown in Fig. 28.

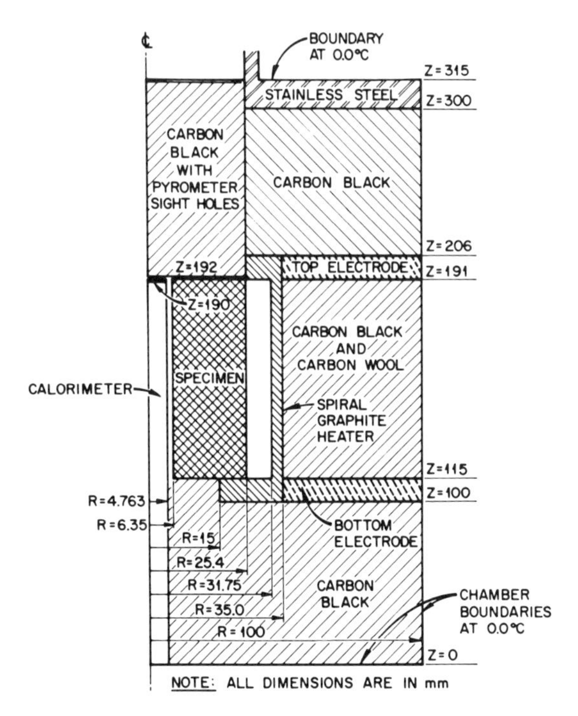

FIGURE 28. The thermal model used for analysis of a high-temperature RHFA with a water-cooled calorimeter.

The analysis assumed the following:

1. The sight holes in the top block could be ignored.
2. The electrodes at the top and bottom of the muffle heater were thin disks rather than the actual rods used in this experiment. The conductances (product of conduction area for heat flowing radially and thermal conductivity) of these disks were selected to approximate that of the actual rods. This was necessary to reduce the problem to two dimensions.
3. The helix heater was assumed to be solid with a thermal conductivity equal to one-half that of graphite to simulate the milled slots.
4. The outer surface of the calorimeter was treated as an isothermal boundary at a temperature of 25°C. This assumption will not adversely affect the analysis because Taylor[32] reports that the maximum temperature rise of the water was 10°C, which is negligible compared with the large temperature differences in the region surrounding the calorimeter.
5. The specimen was graphite with a λ of 40 W/m K.
6. Heat transfer between the specimen and calorimeter was by a gas with a λ of 0.3 W/m K.

Calculations were made for two cases. First, it was assumed that heat transfer between the specimen and muffle heater was by a combination of conduction through the argon gas plus radiation. The second case was for conduction only. The calculated isotherms are shown in Fig. 29 for the first case for a specimen near 2430°C and with the heat generation rates in the electrodes adjusted so that the temperatures at the top and bottom of the specimen are -0.8 and -6.1°C, respectively, relative to the temperature at the specimen midplane at 126 mm.

Some of the heat dissipated by the muffle heater flows radially inward through the specimen and into the calorimeter. Heat flow into the calorimeter in the span denoted by the shaded area in Fig. 28 was calculated by using

$$\frac{P}{l} = \frac{2\pi\lambda_{\text{gas}}[T_{r_1} - T_{r_2}]}{\ln(r_2/r_1)} \tag{5}$$

where λ_{gas} is the thermal conductivity of the gas between specimen and calorimeter, T_{r_2} is the average temperature over the span at a radius of 4.763 mm as calculated by HEATING5, $r_2 = 4.763$ mm, and $r_1 = 4.590$ mm. The temperature gradient in the specimen was used with the P/l from equation (5) to calculate a predicted specimen thermal conductivity of 44.1 W/m K, which is in error by + 10.3%. Since the isotherms in the specimen are not ideal the error should depend on the exact axial position of the temperature measurement; this poses a

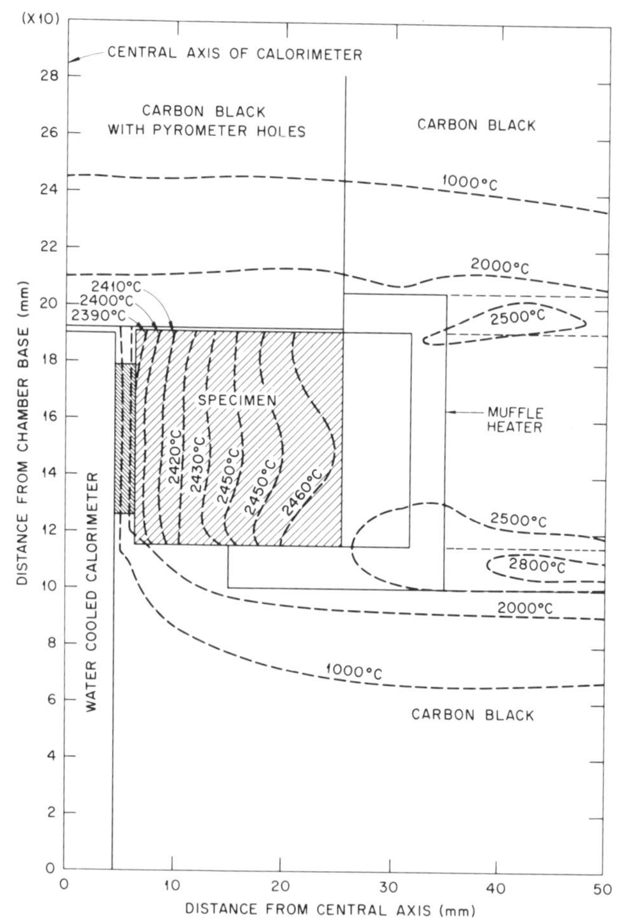

FIGURE 29. Calculated isotherms for the high-temperature RHFA with a water-cooled calorimeter on the central axis.

special problem when a pyrometer is sighting into a long hole. The calculated temperature distribution indicates that the thermal error would remain near + 10.3% for temperatures measured 10 mm above the midplane, but the thermal error would decrease to + 3.2% for temperatures measured 10 mm below the midplane.

Calculations for case 2, where heat transfer between the muffle heater and the specimen was due solely to conduction, indicated a thermal error of + 4.0% at a specimen temperature of 2328°C. The heat generation rate in the end electrodes was adjusted so that the temperatures at the bottom and top of the specimen were + 11.8 and − 3.4°C relative to the temperature at the specimen midplane.

Although the first case is more realistic, both cases indicate that the error due to axial heat flow in this apparatus is small – especially for operation at such a high temperature. This does not mean, however, that this apparatus is easy to construct and operate because Taylor[30] has reported many frustrating problems, such as fogging of the sight port and sagging of the muffle heater at high temperatures.

6. THERMAL CONDUCTIVITY MEASUREMENTS ON POWDERS USING RADIAL HEAT FLOW APPARATUS

6.1. Introduction

The thermal conductivities of powders have been measured numerous times by using RHFA with cylindrical symmetry, and the general effects discussed for the fundamental radial are observed with powder specimens as with solid specimens. However, three differences make the measurements on the powder more difficult:

1. The powder must be enclosed, and the container walls can cause extraneous heat flow. The enclosure can lead to data hysteresis if the thermal expansions of the powder and enclosure material are not the same.
2. The thermal conductivities of powders are much lower than those of solids, and it has already been shown that errors due to nonideal heat flow will increase with decreasing specimen thermal conductivity.
3. The temperature gradient inside a powder is difficult to measure without the temperature sensors causing perturbation of the heat flow.

Two diverse apparatus were selected for thermal analysis, but this selection does not imply that they are necessarily the best of the many that have been used. These apparatus employ different concepts for determining the temperature gradient with a powder specimen and are markedly different in size. One requires a large sample volume, whereas the other is amenable to use with small sample volumes.

6.2. A Radial Heat Flow Apparatus for Powders Developed by Godbee

A RHFA for thermal conductivity measurements on large powder specimens has been developed and used by Godbee[33] on MgO and Al_2O_3. A cutaway drawing of the sample container is shown in Fig. 30. A 610-mm-long Inconel pipe formed the outside container wall, and a tubing entered this container at the bottom for gas insertion. The top of this pipe was welded to a flange. The

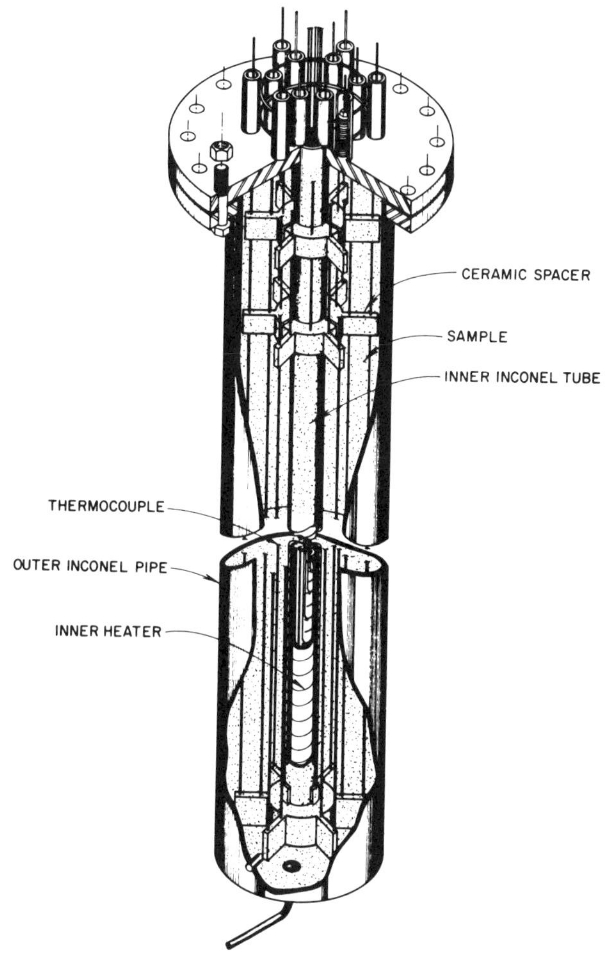

FIGURE 30. A cutaway drawing of the specimen chamber used by Godbee for measuring the thermal conductivity of powders. Taken from Godbee.[33]

powder specimen filled the annulus between this pipe and an inner Inconel tube that was welded to the center of the top flange and centered on the bottom by a spacer. The thermocouples for measuring the temperature gradients were inside the powder and were spring loaded on the top plate to minimize bowing because of thermal expansion. The thermocouple positions were also assured by the presence of ceramic spacers. The core heater for this apparatus consisted of a wire that had been wound onto a ceramic tube. Additional wires had been attached to the heater wire so that the potential drop across the central 25.4-mm section of the core heater could be measured. This central heater was positioned inside the Inconel tube. High temperatures were obtained by lowering this chamber into the bore of a large furnace, which had four independently controlled heater sections so that isothermal furnace walls could be obtained. Godbee[33] estimated the errors from parameter measurement uncertainty and stated that, "the square root of the sum of the squares of the above error estimates is approximately ± 10%." About 9% of this was due to uncertainty in the thermocouple positions.

The thermal model for this apparatus is shown in Fig. 31. Since the container bottom had nearly the same temperature as the furnace wall, whereas the container top lost heat by radiation, conduction, and convection into the room, the system did not have a plane of symmetry, and the entire system was modeled. There were several large gas regions in this model, and heat exchange through these was assumed to be a combination of conduction, convection, and radiation. The three ceremic spacers in the powder were assumed to extend all the way to the outside Inconel pipe, but the thermal conductivity was divided by 5 to reflect the fact that these were spokes and not solid beyond a radius of 25.4 mm. The outer furnace wall and bottom were assumed to be isothermal surfaces at 700°C, and the core heater power density was selected to give a temperature difference of 40°C across the powder annulus. The isotherms calculated for this system are shown in Fig. 32.† There is not a single position within the powder specimen that would be free of some axial heat flow. Above an axial position of 180 mm much of the heat flowing from the core heater goes into the specimen and then flows axially through the specimen and into the room via heat transfer from the top plate. The calculated temperature difference across the annulus at the 329.5-mm specimen midplane was 37.46°C instead of the 40°C that one would obtain by using equation (2). Therefore, the apparatus would have a predicted error of + 6.4%.¶ An "isothermal correction" of −2.54°C was calculated for this apparatus at the same average temperature. This "isothermal correction" would reduce the error to 0.0%.

Although Godbee[33] did not employ an "isothermal correction," the effect of the correction on his thermal conductivity results was studied experimentally in a later attempt to understand the differences between powder results from different techniques.[34] Experimental measurements showed that this RHFA would have an isothermal correction of −3°C at a temperature near 740°C, which is close to the calculated value of −2.54°C. This agreement between the predicted and experimentally determined value of the "isothermal correction" is extremely good, and we have another example of the correction concept improving the accuracy of a RHFA.

†The dimensions on the two axes differ which accentuates the deviation of the isotherms from right circular cylinders.

¶The calculated (and experimental) error is actually a function of the thermocouple immersion depth.

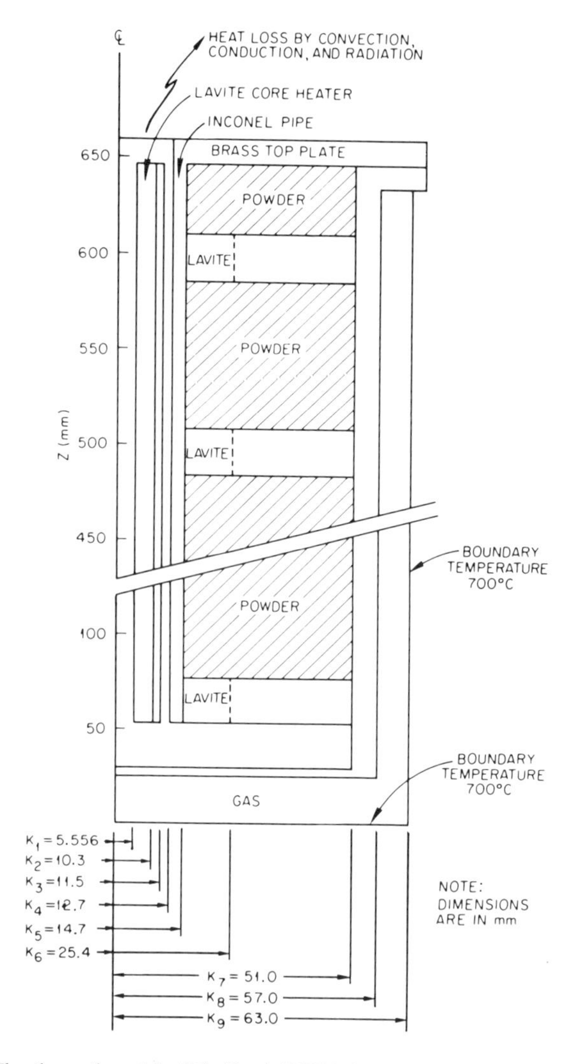

FIGURE 31. The thermal model of Godbee's RHFA for thermal conductivity measurements on powders.

Since the axial heat flow and, therefore, the large negative "isothermal correction" are caused by the large heat loss out the top plate, the best way to improve the system would be to add a large heater above the top of the chamber and control its temperature a few degrees below that of the furnace. It should be noted that the thermal model ignored the presence of the long metal thermocouples suspended in the powder. Since there is so much axial heat flow in this apparatus, this assumption may be an oversimplification although there is good agreement between the calculated isothermal and the one obtained experimentally.

6.3. The Radial Heat Flow Apparatus for Powders Developed by Flynn

A RHFA was developed by Flynn[28] for measuring the thermal conductivity of loose-fill insulations, such as powder, over a temperature range from 400 to 1350 K with an accuracy better than 2% if any potential error due to uncertainties in core heater expansion corrections are ignored. The best features

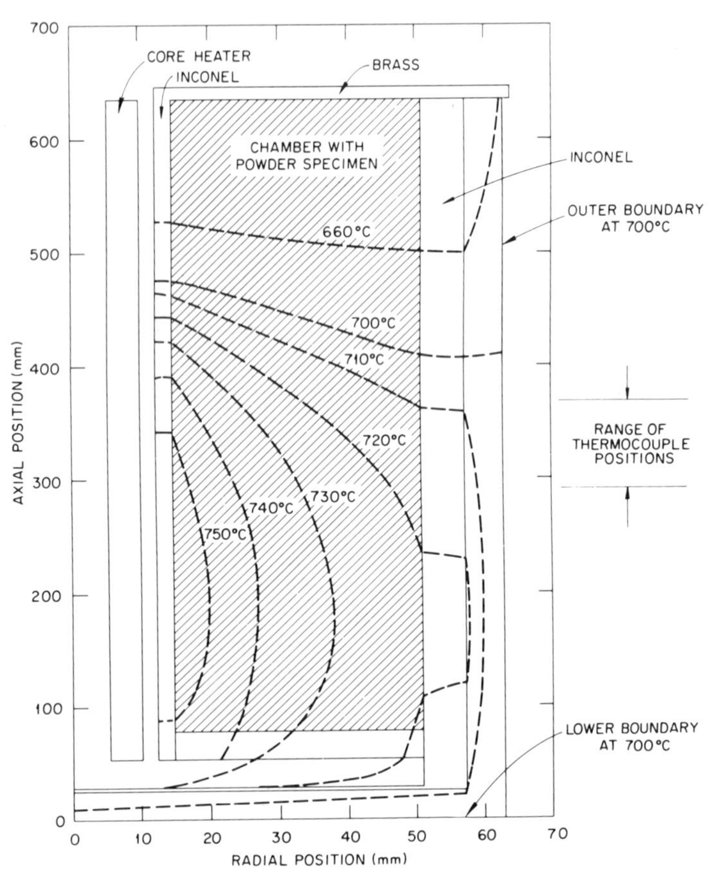

FIGURE 32. Isotherms calculated for the powder apparatus of Godbee showing the marked asymmetry caused by heat flow out the top.

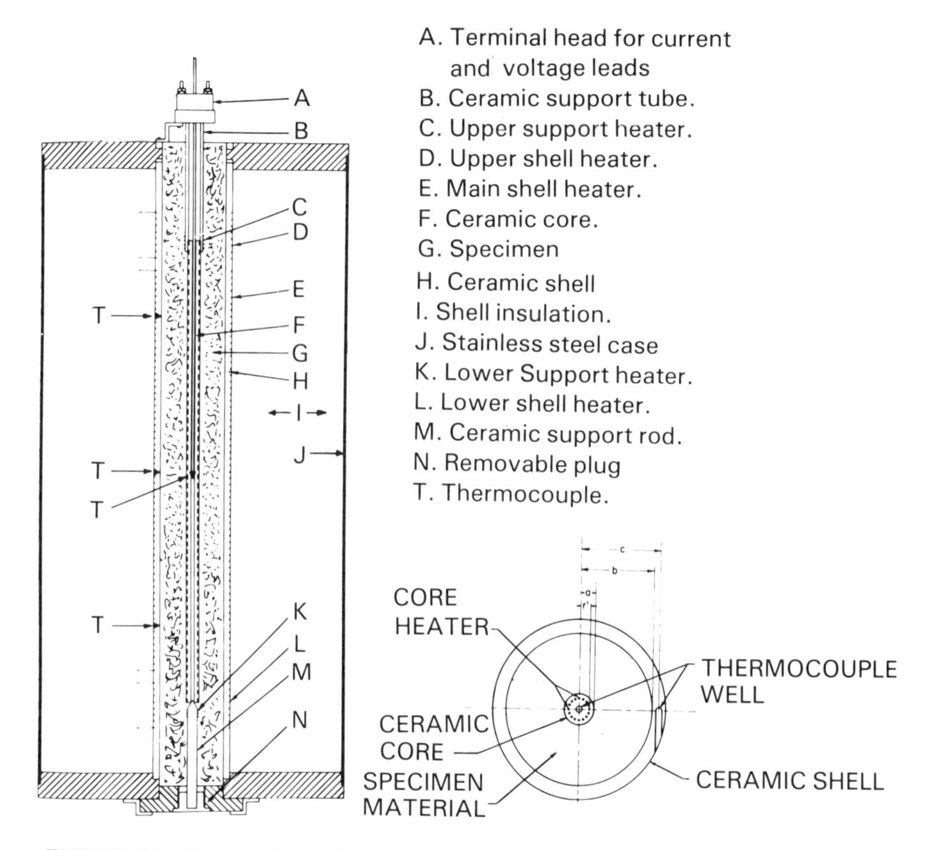

FIGURE 33. Cross-sectional views of the apparatus used by Flynn for measuring the thermal conductivity of loose-fill insulation. Taken from Flynn.[28]

of this apparatus are the small amount of material required and the operation without thermocouples immersed in the specimen. A horizontal cross section and a vertical cross section of the lower half of the apparatus are shown in Fig. 33. The ceramic core heater contained numerous axial holes on a concentric ring, and a heater wire was passed through each hole. The center of the ceramic rod was hollow so that a thermocouple could be inserted to any axial position. The bottom of this ceramic core heater rested on a rod, which contained an independently controlled radially wound heater for blocking axial heat flow. The top of the core heater was positioned inside a ceramic tube, which also contained a heater for blocking axial heat flow. The specimen was contained in the annulus between this ceramic rod and a ceramic muffle heater, which contained a long central heating section and a shorter heater section on each end to block axial heat flow. The temperatures of the muffle heater were measured by using three thermocouples at the axial positions shown. The l/d of the specimen was approximately 9, which is much greater than that for any previously considered apparatus.

The muffle heater temperature controllers were set at the desired value, and the controllers for the end muffle heaters were adjusted until the temperatures of the muffle at the outside thermocouple positions agreed with the midplane temperature to within 1 or 2°C. The power to the core heater was adjusted to give the desired temperature gradient across the specimen annulus, and the core heater end guard heaters were adjusted so that there would not be a significant gradient down the core heater. After the system achieved a steady-state condition, the thermocouple outputs were read and converted to temperatures, and the power input to the core heater was measured. The specimen thermal conductivity was calculated by using a form of equation (2) that corrected for the temperature drop both in the ceramic core heater between the ring of wire heaters and the heater–specimen interfaces and between the outside specimen–muffle heater interface.

This apparatus was the subject of extensive analyses by Flynn[(28)] and Peavy,[(35)] who showed that the thermal error due to nonideal radial heat flow at the midplane would be negligible. Flynn[(28)] intentionally mismatched the temperature at the ends of the core heater to substantiate some of the calculations. Although further analysis would appear to be unnecessary, this system was modeled for calculations with HEATING5, and the calculated isotherms in the apparatus are shown in Fig. 34. The isotherms in the specimen have ideal symmetry from the midplane to about 100 mm on each side of the midplane, and the experimental error would be less than 1% when the end heaters for the muffle heater and the core heaters were used to minimize the axial gradients in the two heaters. The present calculated influence of temperature depression on the ends of the core heater agree with those obtained experimentally by Flynn.[(28)]

This concept has been used extensively for measuring the λ of a variety of loose fill materials including fibrous insulations, liquid, soils,[(36)] and powders where the sample was limited in quantity. For example, the λ of nuclear fuel microspheres in various gases were measured by replacing the stacked-disk specimen shown in Fig. 23 with a suitably instrumented chamber containing the microsphere powder.[(37)]

There are two hidden traps with this approach that must be considered. The first is the influence of temperature jump at the interfaces between the specimen and the core heater and between the specimen and the outer chamber wall.[(38)] This temperature jump can lead to negative errors for measured thermal conductivity values and must be considered for each apparatus. The width of the specimen can usually be made large enough that this error in determination of the appropriate temperature gradient in the specimen is much less than the error encountered when thermocouples are suspended in the specimen. The high error for the latter is caused by uncertainties in the positions of the thermocouples[33] which can sometimes move on temperature cycling.

A second problem can be encountered with a slight variation of this apparatus. Sometimes an apparatus similar to the one developed by Flynn is used with a heavy wall metal tube inserted between the central core heater and the specimen.[36] This variation leads to errors due to axial conduction down the metal tube even when axial end guards are used on the core heater.

7. THERMOMETRY IN RADIAL HEAT FLOW APPARATUS

Thermal conductivity measurements with RHFA require precise measurements of temperatures at accurately known positions with a minimum of heat flow perturbation. This is extremely important because the temperature drops across solid specimens (especially pure metals) are only a few degrees. The primary problems, the sensor types most influenced by the problems, and the possible solutions are given in Table 2. The first three items deal with the agreement of the temperature sensor wells.

Figure 35 shows three possible sensor wells. The one in Fig. 35a is not recommended because the wires into the sensor would be in a temperature gradient that would cause an error due to heat flow out of the sensor. If this configuration were used for high-temperature measurement, the walls of the hole would be in a thermal gradient, which would make it impossible to measure the temperature of the hole bottom with an optical pyrometer.

Figure 35b shows temperature sensor wells in the form of axial holes extending the entire length of the specimen. The temperature sensors could be positioned permanently at the specimen midplane or a movable sensor could be used as demonstrated by Powell.[23]

In the configuration shown in Fig. 35c the temperature sensor resides in the bottom of an axial hole, and there is no gradient on the sensor wires as long as the axial gradient in the specimen is zero. This configuration is much superior for temperature measurements with all three sensor types mentioned in Table 2. A variation of this configuration as used by Godfrey *et al.*[25] and Fulkerson *et al.*[26] is shown in Fig. 36 for disks with two (left) and three (right) thermocouples at each radius. These disks were mounted near the midplane of the apparatus shown in Fig. 23. The thermocouple assemblies were passed through radial slots and then were turned downward into holes so that the wires leaving the thermocouple hot junctions were in isothermal zones. The thermocouples were insulated with round two-hole ceramic (Al_2O_3) insulation. Single-hole ceramic tubes were placed in the bottom of each hole to center the thermocouple hot junction and to prevent electrical contact with the specimen.

TABLE 2
Temperature Measurement Problems in RHFA and Possible Solutions

No. Problem	Type of sensor affected	Error magnitude (%)	Solution
1. Location uncertainty	All types	5	Accurate machining
2. Sensor orientation	All types	6	Ensure that sensors are not in a gradient near the measurement position
3. Heat flow perturbation	All types	< 3	Position sensor and select sensor to minimize error
4. Relation between sensor output and temperature	All types	0.1	Careful calibration
5. Sensor instability due to			
a. Cold working	TC[a]	*b*	Anneal before use
b. Contamination	TC, R[a]	*b*	Select sensors suitable for environment and use an additional sensor type
c. Insulator shorting	TC, R[a]	*b*	Careful selection of pure insulators and limit the upper temperature.
d. Window fogging	P[a]	*b*	Blow gas across window and/or cover window with a baffle when not in use
e. Instrument failure	All types	*b*	Routine calibration

[a] TC, R and P stand for thermocouple, resistance thermometer, and pyrometer, respectively.
[b] Cannot be stated in general.

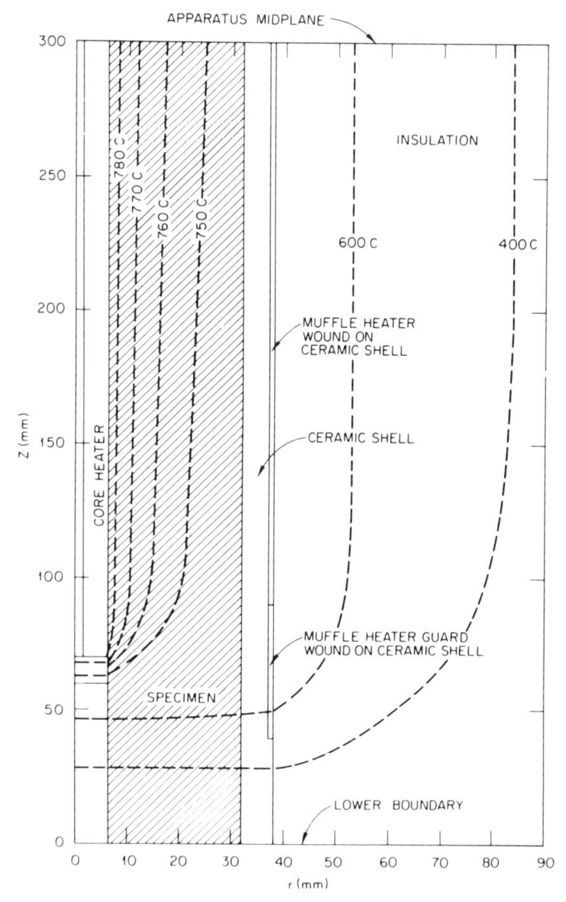

FIGURE 34. Isotherms with right circular symmetry calculated for the apparatus developed by Flynn[28] for measurements of the λ of loose fill material.

Although the configuration shown in Fig. 35c is the best, it still can lead to errors by perturbing the heat flow through the disk. The inside sensor wells have a greater effect than do the outside ones because they subtend a larger area normal to the heat flow. The magnitude of this effect cannot be described in detail it is a function of so many variables, including the thermal conductivities of the material between the core heater and the specimen, the sensor well, and specimen in addition to all the dimensions of the disk and its temperature sensor wells.

Assumed parameters and calculated errors are given in Table 3 for the configuration shown in Fig. 35b. The calculated error was -0.8% and -0.6% when the ratio of specimen λ to assumed sensor well λ was 10 and 3, respectively.

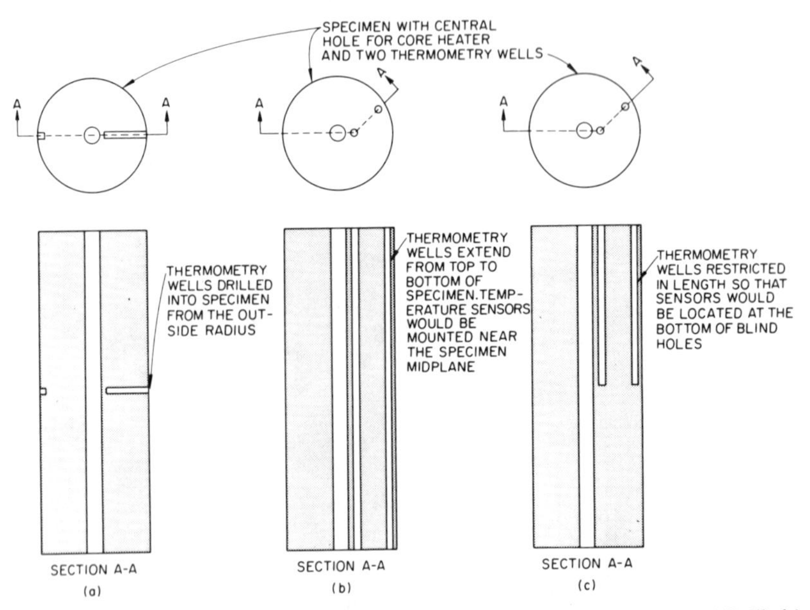

FIGURE 35. Three different types of temperature sensor wells for a RHFA. (a) Drilled into the side of the specimen. (b) Axial holes drilled through the specimen at different radii. (c) Axial holes drilled partway through the specimen.

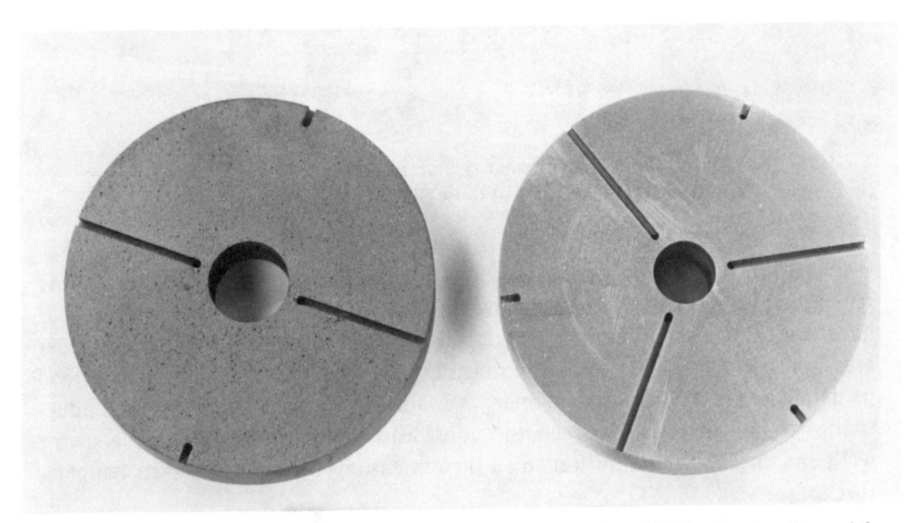

FIGURE 36. Disks taken from the midplane of a stacked-disk RHFA showing the axial thermocouple wells and the radial slots for passages of the insulated thermocouples. Taken from McElroy and Moore.[2]

TABLE 3
Calculated Errors Due to Heat Flow Perturbation by Three Temperature Wells Near the Inside Radius of a RHFA Specimen

Sensor well diameter (mm)	Average λ of sensor well and sensor (W/m K)	P/l[a] (W/m)	Error[b] indicated for λ measurement (%)
1.59	10	1.3860×10^4	−0.8
1.59	30	1.3857	−0.6
3.18	10	1.3808	−3.0

[a] P/l = power per unit length.
[b] Calculated for a specimen with an O.D. of 76.2 mm, and I.D. of 8.0 mm, and a λ of 100 W/m K. The distance between the center of the sensor wells and the specimen inner radius was assumed to be 3.2 mm.

When the ratio was 10 and the sensor diameter was doubled, calculated error for a measurement was −3.0%. These calculations show that the sensor well should be as small as practical and that the average λ of the well should be as close as possible to that of the specimen.

Extension of these calculations to three dimensions showed that the errors for the configuration shown in Fig. 35c would be about one-half those given in Table 3 for the case with through holes.

Temperatures measurement problem No. 4 listed in Table 2 is relatively minor. At low temperatures ($T < 800$ K) items 5a, 5b, 5c, and 5d are all minor, but these problems can cause dominant errors at high temperatures. Problems with sensor contamination and insulator shunting can best be avoided by careful selection of sensors† and insulators, by using sheathed thermocouples, or by improving the environment. One way to detect these errors is to include at least one temperature sensor of another type in the apparatus on the assumption that the effects of contamination and insulator shorting will not be the same on the different types of temperature sensors. Godfrey *et al.*[25] used a central core heater wound with a Pt–10% Rh wire, and the temperature calculated from the measured resistance of this wire was routinely compared with the temperature indicated by the thermocouples during the "isothermal correction" thermal configuration.

8. CONCLUSIONS

Radial heat flow apparatus have several advantages for measuring the thermal conductivities of solid and loose-fill materials such as powders. Their

† Many papers are available that describe the use and selection of temperature sensors.[39-41]

high accuracy over a thermal conductivity range from 1×10^{-2} to 200 W/m K and a temperature range from 300 to 2600 K is the primary advantage. The primary disadvantage is the usual requirement for a relatively large specimen although this can become an advantage for heterogeneous specimens.

One of the primary sources of indeterminate error with a RHFA is nonradial heat flow near the specimen midplane. Calculations of the errors due to nonideal heat flow in a fundamental radial apparatus with right circular symmetry show the following:

1. For an unguarded system, a specimen l/d up to 8.0 may be insufficient to ensure an error less than 4% due to nonradial heat flow.
2. Although multiple interfaces do reduce axial heat flow, they are ineffective for reducing errors due to axial heat flow.
3. End guard heaters offer the best solution for the problem of axial heat flow, but these heaters cause local hot spots which raise some question about the optimum location of thermometers on the guards.
4. The application of an "isothermal correction" reduces the thermal error in every system. In the three worse cases the errors were reduced from + 61 to + 41%, and + 38% to less than 2%.
5. Much of the heat that flows axially through the specimen in an unguarded system comes from the muffle heater. The temperature configuration of this muffle heater is extremely important to system behavior and the heater should be maintained as isothermal as possible by adding a metal liner inside the muffle heater and/or by increasing the power density at the ends of the heater. This is especially important for specimens with a low thermal conductivity.

The thermal errors of several experimental RHFA that have been used for measuring the thermal conductivities of solids and powders were calculated and the results are summarized in Table 4. Table 4 and information obtained during the calculations lead to the following additional conclusions:

6. The stacked-disk RHFA described by Godfrey *et al.*[13] and that described by Powell[23] have thermal errors approaching 1% even without isothermal corrections.
7. The high-temperature unguarded apparatus of Feith[21] has a hot spot at the top of the core heater, which improves the isotherms in the top of the specimen. The thermal error, however, appears to be positive in all cases studied because of axial heat losses from the bottom of the specimen.
8. The thermal error of a tester described by Jury *et al.*[17] for measuring λ of pipe insulations is less than 1%.
9. At 740°C, the powder apparatus of Godbee[33] has an error of + 6.4% due to nonradial heat flow at the midplane. A calculated "isothermal

TABLE 4
Calculated Percentage Errors from Nonideal Heat Flow in Experimental RHFA

Apparatus	Reference	Specimen[a] l/d	Specimen λ (W/m K)	Error claimed by experimentalist (%)	Calculated error[b] (%)	Calculated error with isothermal correction (%)	Comments
Feith	21	6.1	2.5	± 6	+ 13.3		$P/V = f(T)$[c]
Jury *et al.*	17	4.8	0.035	± 3	+ 1.0		
Powell	23	4.0	29.8	± 2	+ 1.0	< 0.1	
Godfrey *et al.*	13, 25	3.0	65	± 1.5	+ 1.0 + 1.6	< 0.1	
Taylor	30	1.5	4	± 5	+ 10.3		Heat transfer in voids by gaseous conduction and convection
Godbee	33	12	0.77	± 10	+ 6.4	0.0	
Flynn	28	9	*d*	± 3	< 1.0		

[a] l/d = length to diameter ratio.
[b] Since the apparatus error claimed by the experimentalist normally excludes the one calculated from nonideal heat flow, the actual experimental error would be closer to the sum of the two.
[c] P/V is the power per unit volume dissipated in the tungsten muffle heater.
[d] λ of the specimen assumed to be $\lambda = 0.27 + 1.0 \times 10^{-4}\ T$ where λ is in W/m K and T is in K.

correction" value would reduce this error to 0.0%, and this agrees with experimental measurements of the "isothermal correction." The error obtained without use of an "isothermal correction" is probably caused by the heat loss from the specimen chamber at the top.

10. The RHFA developed by Flynn[28] for measuring the thermal conductivity of powders without immersing thermocouples directly into the powder appears to have a negligible thermal error.

9. RECOMMENDATIONS

Previous comparisons,[14] careful analysis of determinate errors, and, now, analysis of the thermal errors have indicated that the RHFA of Godfrey *et al.*[25] is one of the best for measurements of the thermal conductivity of solids over temperature ranges suitable for thermocouple use ($T < 1400$ K). The three-section muffle should be operated at near isothermal conditions for measurements with low thermal conductivities. Modern thermocouple technology should permit use of smaller diameter thermocouples, which would reduce the uncertainty in thermocouple position and, hence, determinate error.

The RHFA developed by Rasor and McClelland[29] and Taylor[30] and a duplicate of their apparatus with a greater specimen l/d as developed by Moeller[31] appear to be the best for high-temperature measurements requiring an optical pyrometer.

Variations of the apparatus developed by Flynn[28] are probably the best for measuring the thermal conductivity of loose fill material such as powder. This superiority is based on the lack of temperature sensors within the loose-fill specimen, which could move during thermal cycling and, thus, lead to large measurement errors. Additionally, Flynn's[28] concept can be easily used to measure the thermal conductivity of loose fill material with different gases at different pressures in the specimen chamber and in the remainder of the system.

When possible a RHFA should be analyzed for thermal error before construction and operation. However, this is sometimes impossible, and one should then consider system construction based on the following:

1. The ends of the system should be symmetrical.
2. The specimen should have an l/d of at least 3.0. This specimen can either be a solid or a series of stacked disks. The only considerations would be specimen raw material shape and method of thermocouple insertion.
3. Thermocouples should be installed so that the thermocouple wire adjacent to the hot junction is isothermal or, if optical pyrometry is used, blackbody holes can be sighted from the ends of the specimen.

4. Whenever possible a long hole should be included from end to end of the specimen on an outer radius. A traversing thermocouple should then be used in this hole to detect nonideal isotherms, which lead to thermal error.
5. The I.D. and O.D. of the specimen guard heaters should be the same as those of the specimen.
6. The muffle heater should be constructed so that it can be operated at a nearly isothermal temperature with the ends only cold enough to allow operation of the specimen guard heaters.
7. An "isothermal correction" should always be used.

ACKNOWLEDGMENTS. The author gratefully acknowledges Sharon Phillips for her patience and typing skill, and D.W. Yarbrough, G.H. Llewellyn, and D.L. McElroy for their text reviews and suggestions. The general support and encouragement of the latter is also greatly appreciated. Lastly, T.G. Godfrey is acknowledged for his past research with the local radial heat flow apparatus and for many conversations that have added to my understanding of that apparatus. Research sponsored by the Division of Materials Sciences, U.S. Department of Energy, under contract W-7405-Eng-26 with the Union Carbide Corporation.

APPENDIX

The λ of solids has been measured many times by using RHFA that were assumed to be so long that axial heat flow at the specimen was negligible. These systems do not employ electrically heated end guards to control the temperatures at each end of the specimen stack to the midplane temperature. Operationally, these are the simplest apparatuses since the only two heaters are the core heater and the muffle heater, and only the latter requires control circuitry. Table A.1 lists several techniques of this general type. The number and type of temperature sensors used in each, the temperature ranges, and, most importantly, the specimen length-to-diameter ratios, l/d, are given for each technique. This ratio ranged from a low value of 1.7 for the RHFA of Poole[2] to a high value of 30 for that of Stephens.[3] Furthermore, the number of temperature sensors employed varies widely. The authors' overall inaccuracies are suspect in some cases in light of the potential errors described in the text for unguarded systems.

Table A.2 lists techniques that employed stacked-disk specimens. Some of these specimens were fabricated from stacked disks in hopes that the interfaces would reduce axial heat flow and thereby reduce experimental error. Some

TABLE A.1
Comparison of Experimental RHFA That Do Not Contain Electrically Powered End Heaters to Minimize End Losses

Author and date	References	Temperature sensor	Sensor configuration No. inside	No. outside	Temperature range (K)	Specimen[a] l/d	Stated overall inaccuracy (%)
Niven (1905)	1	Pt wire resistance, Fe vs. German silver thermocouple	1	1	273	1.67	± 5
Poole (1912)	2	Pt vs. Pt–Ir thermocouples	4	4	313–813	1.73	*c*
Stephens (1932)	3	Pt wire resistance	1	1	93–488	30.0	*c*
Kingery (1954)	4	Pt–10% Rh vs. Pt	2	2	473–1573	12.3	*c*
Wray and Connolly (1959)	5	W wire resistivity	1	0	200–2000	14.0	*c*
Slack and Glassbrenner (1960)	6	Pt–10% Rh vs. Pt	1	1	300–1020	4.8	± 5
Neel *et al.* (1962)	7	Pt–10% Rh vs. Pt	3	3	< 1373	1.95	± 10
		Pyrometer	3	3	1373–3023		
Pears (1962)	8	Pyrometer	3	3	1373–3023	3	± 10
Glassbrenner and Slack (1964)	9	Pt–10% Rh vs. Pt			300–1580	5.0	± 5

[a] l/d = length to diameter ratio.
[b] Three thermocouples, all at different radii.
[c] Not stated.

TABLE A.2
Features of Experimental Unguarded RHFA That Used Stacked-Disk Specimens

Author	Reference	Temperature sensor	No. of sensor configuration		Temperature range (K)	Specimen[a] l/d	No. of disks	Stated overall inaccuracy (%)
			Inside	Outside				
Flinta	10	Pt–10% Rh vs. Pt and W–Mo	*b*	*b*	323–2073	6.7	5	± 5 to ± 10
Paine *et al.*	11	Pt–10% Rh vs. Pt	2	2	973–1698	3.0	3	± 5 to ± 10
Lowrance	12	Thermocouples and optical pyrometer			623–2323	3.0	3	
Feith	13	Pt–10% Rh vs. Pt and optical pyrometer	2	2	1473–2523	2.8	11	± 6.1
Banev and Chekhovskoi	14	Chromel vs. Alumel	1	1	423–1113	6	16	± 3.0
Feith	15	W–25% Re vs. W	2	2	1233–2273	2.8	11	*c*

[a] l/d = length to diameter ratio.
[b] Three thermocouples mounted on a spiral.
[c] Not stated.

TABLE A.3
Comparison of Several Features of RHFA That Used Stacked-Disk Specimens and End Guard Heaters

			Sensor configuration					Stated overall
Author and date	Reference	Temperature sensor	No. inside	No. outside	Temperature range (K)	Specimen[a] l/d	Disks	inaccuracies (%)
Powell	16	Pt–10% Rh vs. Pt	2	2	368–1273	4.0	10	*b*
Burr	17	Chromel vs. Alumel	1	1	373–1073	1.7	5	± 5
Mikol	18	Pt–10% Rh vs. Pt	2	3	608–1423	1.5	15	± 5
Rasor and McClelland	19	Pt–10% Rh vs. Pt	3	3	1273–2973	1.5	1 or 3	± 5
	20	Optical pyrometer						
Fieldhouse *et al.*	21	Pt–10% Rh vs. Pt	2	2	323–1873	3.0	15	± 5
Brophy and Sinnott	22	Chromel vs. Alumel	2	2	423–873	4.4	Many	± 3
Howard and Gulvin	23	Pt–13% Rh vs. Pt–1% Rh	2	2	373–1773	1.4	7	± 6
Godfrey *et al.*	24	Pt–10% Rh vs. Pt	2	2	216–1373	3.0	9	± 1.5
Hayes and DeCrescente	25	Optical pyrometer	3	3	1173–1873	1.5	3	± 6 to ± 16
Fulkerson *et al.*	26	Pt–10% Rh vs. Pt	3	3	323–1273	3.0	9	± 1.5

[a] l/d = length-to-diameter ratio.
[b] Not stated.

were fabricated because of the convenience of cutting disks from flat plates of the specimen material. The apparatus by Feith had a stated overall inaccuracy of ± 6.1%. However, calculations in this text show that the thermal error alone for a UO_2 specimen would be about + 13.3% at a temperature of 1486 K.

Table A.3 lists several RHFA that had stacked-disk specimens in addition to end guards on the specimen to minimize axial heat flow. The apparatus of Powell[16] and Godfrey *et al.*[24] were analyzed in the text and found to be relatively free of thermal errors due to axial heat flow. The apparatus of Rasor and McClelland[19] had a stated overall inaccuracy of ± 5%. Calculations described in the text show that the thermal error alone would be about + 10% at 2700 K with a graphite specimen ($\lambda = 40.0$ W/m K). Although this exceeds the stated inaccuracy, it is still low when one considers the fact that the measurements were made at such a high temperature.

REFERENCES TO APPENDIX

1. C. Niven *Proc. R. Soc.* **76**, 34 (1905).
2. H.H. Poole, *Phil. Mag.* **24**, 45 (1912).
3. R.W.B. Stephens, *Phil. Mag.* **14**, 897 (1932).
4. W.D. Kingery, *J. Am. Ceram. Soc.* **37**, 88 (1954).
5. K.L. Wray and T.J. Connolly, *J. Appl. Phys.* **30**, 1702, 1702 (1959).
6. G.A. Slack and C. Glassbrenner, *Phys. Rev.* **120**, 782 (1960); L.L. Sparks and R.L. Powell, NBS Report 8750 (1965); R.P. Stein and M.U. Gutstein, in *Heat Transfer* (Sterrs, ed.), Chemical Engineering Progress Symposium Series 56 (30), p. 167, American Institute of Chemical Engineers.
7. D.S. Neel, C.D. Pears, and S. Oglesby, Southern Research Inst., WADD-60-924, 58-201 (1962).
8. C.D. Pears, Southern Research Inst. Technical Documentary Report ASD-TDR-62-765, 20-402 (1962).
9. C.J. Glassbrenner and G.A. Slack, *Phys. Rev.* **134A**, 1058 (1964).
10. J.E. Flinta, USAEC Report TID-7546, Book 2, 516–525 (1957).
11. R.M. Paine, A.J. Stonehouse, and W.W. Beaver, WADC TR-59-29 (1959).
12. D.T. Lowrance, Chance Vought Corp., Report No. 2-253420/2R375, 1–95 (1962); D.L. McElroy, T.G. Godfrey, and T.G. Kollie, *Trans. Am. Soc. Metals* **55**, 749 (1962).
13. A.D. Feith, GETM-296 (1964).
14. A.M. Banacy and V.Y. Chekhovskoi, *Teplofiz. Vysokikh Temp.* **3**, 47 (1965).
15. A.D. Feith, GETM-65-10-1 (1965b).
16. R.W. Powell, *Proc. Phys. Soc. Lond.* **51**, 407 (1939).
17. A.C. Burr, *Can. J. Tech.* **29**, 451 (1951).
18. E.P. Mikol, USAEC Report ORNL-1131, 1–7 (1952).
19. N.S. Rasor and J.D. McClelland, *J. Phys. Chem. Solids* **15**, 17 (1960a).
20. N.S. Rasor and J.D. McClelland, *Rev. Scient. Instrum.* **31**, 595 (1960b).
21. I. B. Fieldhouse, J.C. Hedge, and J.I. Lang, WADC Technical Report, 58-274 (1958).
22. J.H. Brophy and M.J. Sinnott, *Trans. Am. Soc. Metals* **52**, 567 (1960).
23. V.C. Howard and T.F. Gulvin, UKAEA IG Rept. 51 (RD/C), 1–23 (1961).
24. T.G. Godfrey, W. Fulkerson, T.G. Kollie, J.P. Moore, and D.L. McElroy, ORNL-

3556 (1964); T.G. Godfrey, W. Fulkerson, T.G. Kollie, J.P. Moore, and D.L. McElroy, *J. Am. Ceram. Soc.* **48**, 297 (1965).

25. B.A. Hayes and M.A. DeCrescente, PWAC-480 (1965).
26. W. Fulkerson, J.P. Moore, and D.L. McElroy, *J. Appl. Phys.* **37**, 2639 (1966); C.J. Glassbrenner, *Rev. Sci. Instrum.* **36**, 984 (1965).

REFERENCES

1. H.S. Carslaw and J.C. Jaeger, *Conduction of Heat in Solids,* Oxford Univ. Press, Oxford, 1959.
2. D.L. McElroy and J.P. Moore, Chap. 4 in *Thermal Conductivity,* Vol. 1 (R.P. Tye, ed.), Academic Press, London (1969).
3. R.K. Williams and W.O. Philbrook, *Rev. Sci. Instrum.* **39**, 1104–1114 (1968).
4. P. Andersson and G. Backstrom, *High Temp. High Pressures* **4**, 101 (1972).
5. F.A. Laws, F.L. Bishop, and P. McJunkin, *Proc. Am. Acad. Arts Sci.* **41**, 454 (1905).
6. F.L. Bishop, *Proc. Am. Acad. Arts Sci.* **41**, 671 (1906).
7. A. Eucken and H. Laube, *Tonindustriezentung* **53**, 1599 (1929).
8. S.E. Green, *Proc. Phys. Soc.* **44**, 295 (1932).
9. J.R. Winckler, *J. Am. Ceram. Soc.* **26**, 339 (1943).
10. M. Adams and A. L. Loeb, *J. Am. Ceram. Soc.* **37**, 73 (1954).
11. M. Adams, *J. Am. Ceram. Soc.* **37**, 74 (1954).
12. M. McQuarrie, *J. Am. Ceram. Soc.* **37**, 84 (1954).
13. T.G. Godfrey, W. Fulkerson, T.G. Kollie, J.P. Moore, and D.L. McElroy, ORNL-3556, Oak Ridge National Laboratory, Oak Ridge, Tennessee, June 1964.
14. W.P. Turner, P.C. Elrod, and I.I. Siman-Tov, ORNL/CSD/TM-15, Oak Ridge National Laboratory, Oak Ridge, Tennessee, March 1977.
15. Y.S. Touloukian, *Thermophysical Properties of High Temperature Solid Materials,* Vols. 1–6, The MacMillan Company, New York (1967).
16. C.Y. Ho, R.W. Powell, and P.E. Liley, *Thermal Conductivity of the Elements* (Category 5, Thermodynamic and Transport Properties) NSRDS-NBS, Washington, D.C. 20402.
17. S.H. Jury, D. Arnurius, T.G. Godfrey, D.L. McElroy, and J.P. Moore, *J. Franklin Inst.* **298** (3), 151 (1974).
18. M.J. Laubitz, Chap. 3 in *Thermal Conductivity,* Vol. 1 (R.P. Tye, ed.), Academic Press, New York. (1969).
19. M.J. Laubitz and D.L. McElroy, *Int. J. Sci. Metrology* **7** (1), 1 (1971).
20. G. Ondracek, B. Schulz, and F. Thummler, *High Temp. High Pressures* **1**, 439 (1969).
21. A.D. Feith, GETM-296 (1965).
22. S.H. Jury, D.L. McElroy, and J.P. Moore, *Thermal Transmission Measurements of Insulation,* ASTM STP 660 (R.P. Tye, ed.), American Society for Testing and Materials, Philadelphia (1978), p. 310.
23. R.W. Powell, *Proc. Phys. Soc. Lond.* **51**, 407 (1939).
24. R.D. Redin and G.A. MacGregor, *Proceedings of the 5th Conference on Thermal Conductivity,* University of Denver, Denver, Colorado, October 20–22 (1965).
25. T.G. Godfrey, W. Fulkerson, T.G. Kollie, J.P. Moore, and D.L. McElroy, *J. Am. Ceram. Soc.* **48** (6), (1965).
26. W. Fulkerson, J.P. Moore, and D.L. McElroy, *J. Appl. Phys.* **37**, 2639 (1966).
27. T.G. Godfrey and D.L. McElroy, *Nucl. Tech.* **22**, 94 (April 1974).
28. D.R. Flynn, *J. Res. Natl. Bur. Stds.-C. Engineering and Instrumentation* **67C** (2), 129 (April–June 1969).
29. N.D. Rasor and J.D. McClelland, *Rev. Sci. Instrum.* **31**, 595 (1960).

30. R.E. Taylor, *J. Am. Ceram. Soc.* **44**, 525 (1961).
31. C.E. Moeller, *Proceedings Europaische Konferenz uber Thermophysikalische Eigenschaften Bonfestoffen bei Hohen Temperaturen BMBW-FB*, K70-01, February (1970), p. 125.
32. R.E. Taylor, private communication, April (1980).
33. H.W. Godbee, ORNL-3510, Oak Ridge National Laboratory, April (1966).
34. J.P. Moore and D.L. McElroy, unpublished work.
35. B.A. Peavy, *J. Res. Natl. Bur. Stds.-C. Engineering and Instrumentation* **67C** (2), 119 (April–June 1963).
36. M.S. Kersten, "Thermal Properties of Soils," Bulletin No. 28, U. of Minn. Inst. of Tech. 52 (21), June 1 (1949).
37. J.P. Moore, D.L. McElroy, and R.S. Graves, unpublished work on the thermal conductivity of MgO powder in nitrogen (1967).
38. E.H. Kennard, *Kinetic Theory of Gases*, McGraw-Hill, New York (1938).
39. D.L. McElroy and W. Fulkerson, Chap. 2 in *Techniques of Metals Research*, Vol. 1 (1) (R.F. Bunshah, ed.), Interscience Publishers, New York (1968), p. 105.
40. F.R. Caldwell, *Temperature Its Measurement and Control in Science and Industry*, Vol. 3 (A.I. Dahl, ed.), Reinhold Publishing Corporation, New York, p. 81.
41. R.L. Anderson and T.G. Kollie, *CRC Crit. Rev. Anal. Chem.* **July**, 171 (1976).

3

Thermophysical Property Determinations Using Direct Heating Methods

R. E. TAYLOR

1. INTRODUCTION

There is a general class of thermophysical determinations which utilize Joulean heating generated by electrical current flowing through the sample to control the sample temperature gradients during the property measurements. This general class we designate as "direct electrical heating methods." They are distinguished by the lack of external furnaces to control the sample temperature. This in turn greatly shortens the time to attain an equilibrium temperature and easily permits transient measurements. These methods are limited to reasonable electrical conductors. Samples are in the form of rods, tubes, or wires. Direct electrical heating methods have been used by a number of investigators since Kohlrausch[1] used the method before the turn of the century (see Table 1).

Direct heating methods have been used to measure thermal conductivity, electrical resistivity, specific heat, hemispherical total emissivity, normal spectral emissivity, thermal expansion, other electrical properties (Thomson, Seebeck, Peltier, and Richardson coefficients), and properties derived from the primary properties such as enthalpy, thermal diffusivity, coefficient of linear expansion, and Wiedemann–Franz–Lorenz ratio. Usually the technique yields more than one property determination and, in fact, all the properties listed above can be measured simultaneously/consecutively on the same specimen. Measurements have been made from cryogenic temperatures to temperatures in excess of 2500 K. Materials whose thermal conductivity values range from 600 to 1 $\mathrm{W\,m^{-1}\,K^{-1}}$ have been measured routinely using these techniques, and measurements have been made on materials whose values fall outside this range.

R.E. TAYLOR • Properties Research Laboratory and CINDAS, Purdue University, 2595 Yeager Road, West Lafayette, Indiana, 47906. Work supported by the National Science Foundation.

TABLE 1
Summary of Events (Development in Theory, Experiments, and Error Analysis) in Direct-Heating Methods

No.	Investigator	Ref.	Year	Contribution
1.	Kohlrausch	39	1874	Originator
2.	Jaeger and Diesselhorst	29	1900	Measurements
3.	Angell	20	1911	Radial heat flow measurements
4.	Worthing	3	1914	LTRA[a] measurements
5.	Holm and Störmer	40	1930	Necked-down configuration
6.	Osborn	10	1941	LTRA measurements
7.	Krishnan and Jain	24	1954	LTRA variation
8.	Mikryukov	22	1959	Modify Kohlrausch method
9	Lebedev	25	1960	LTRA variation
10.	Bode	27	1961	LTRA variation
11.	Gumenyuk and Lebedev	11	1961	LTRA variation
12.	Rudkin, Parker, and Jenkins	9	1962	LTRA variation
13.	Filippov and Simonova	41	1964	LTRA variation
14.	Platunov	42	1964	LTRA variation
15.	Flynn and O'Hagan	37	1967	Exact comparison of direct heating to other techniques
16.	Flynn	19	1969	Comprehensive review
17.	Taylor, Davis, and Powell	17	1969	LTRA, spline and shooting technique
18.	Taylor and Hartge	35	1970	LTRA error evaluation
19.	Taylor	43	1978	Measurement of SRM
20	James	28	1979	LTRA
21.	Taylor	44	1981	Multiproperties

[a]Long thin rod approximation.

The mathematics describing directly heated specimens often involves second-order nonlinear differential equations with no known closed solutions. Historically, simplifying assumptions were made in order to yield simple mathematical expressions relating measured parameters (current, voltage, temperature, and temperature gradients) to thermophysical properties. This severely limited the accuracy and applicability of the methods. As a consequence, direct heating methods have not been very popular for thermal conductivity determinations at low and moderate temperatures (below 1000°C). However, direct electrical heating methods afford a means of obtaining thermal conductivity values at higher temperatures that was so much simpler experimentally than other techniques that they have been used extensively. This simplicity in experimental setup and easily fabricable sample geometries combined with the possibility of simultaneous thermophysical property determinations and advances in technology (particularly in optical pyrometry) led to an upsurge in usage of direct electrical heating methods. This upsurge has been greatly aided by recent advances in sophisticated data handling procedures such as SPLINE and numerical techniques, and the proliferation of easily accessible computers and these data-handling techniques has negated the disadvantage of complex mathematics that had heretofore severely limited the accuracy and use of direct heating techniques. Finally, the use of rapid data acquisition systems that are compatible with the inherently short times to steady state or the use of transient conditions has brought direct heating methods into a position where they have become very attractive. We might divide measurement techniques into two broad classes – one of tedious experimentation and readily calculable results and the other of relatively simple experimentation and complex mathematics. An example of the first class would be the radial inflow method for measuring thermal conductivity, where we need extensive heaters, temperature measurement holes at radii r_2 and r_1 in large samples, long times to steady states, etc., but the applicable equation

$$\lambda = [Q \ln (r_2/r_1)]/2\pi L \Delta T$$

can readily be used with a calculator, slide rule, or even by hand. Direct heating methods are an example of the second class, where a sample is simply connected between two electrodes, heated by passing current through it, and measuring I, voltage drops, and temperature gradients. In general no holes are required in the sample. Steady-state conditions can be achieved in a matter of minutes. However, the data may have to be entered into a computer in order to yield accurate values of thermal conductivity or other thermophysical properties. These descriptions are admittedly oversimplifications, but they should illustrate the general situation.

We might further clarify the advantages and disadvantages of direct electrical heating methods with Table 2.

2. BACKGROUND

This chapter is primarily aimed at thermal conductivity determinations. However, we must always keep in mind when discussing direct electrical heating methods that they are equally applicable to a number of other property determinations including specific heat (covered in a separate chapter by Cezairliyan), electrical resistivity, thermal expansion, and emissivity. We will briefly describe measurements of these other properties in conjunction with thermal conductivity determinations since the measurements are often intertwined.

Analyses of the data obtained by various researchers have revealed that the results obtained using the various direct heating techniques have not yielded consistent, reliable data. For example, the thermal conductivity data obtained at high temperatures for tungsten by various researchers using direct electrical heating methods are shown in Fig. 1 along with a recommended curve (Ho

TABLE 2
Advantages and Disadvantages of Direct Electrical Heating

1. Advantages
   a. No external furnaces required; therefore short times involved (no thermal inertia)
   b. Simple sample geometry with few or no holes required (rods, tubes, wires)
   c. Multiple property determinations on same sample
2. Disadvantages
   a. Limited electrical conductors (metals, alloys, graphitic materials)
   b. Usually complex mathematics involved; therefore access to computer highly desirable
   c. Control of heat losses essential – requires high-vacuum enclosure for many materials to keep emissivity reproducible.

et al.[2]) which is based on analysis of experimental data, coupled with theoretical considerations.

Determinations by Worthing,[3] Zwikker,[4] Jenkins *et al.*,[5] Allen *et al.*,[6] Simonova and Filippov,[7] and of Forsythe and Worthing[8] above 2000 K lie well above the recommended curve, and those of Rudkin *et al.*[9] lie below it, the spread being of the order of 50%, 80%, and 70% at 1800, 2200, and 2600 K, respectively. Determinations by Forsythe and Worthing below 2000 K, Osborn,[10] Gumenyuk and Lebedev,[11] Gumenyuk *et al.*,[12] Platunov and Fedorov,[13] and Chekhovskoi and Vertogradskii[15] lie within about 10% of the recommended curve.

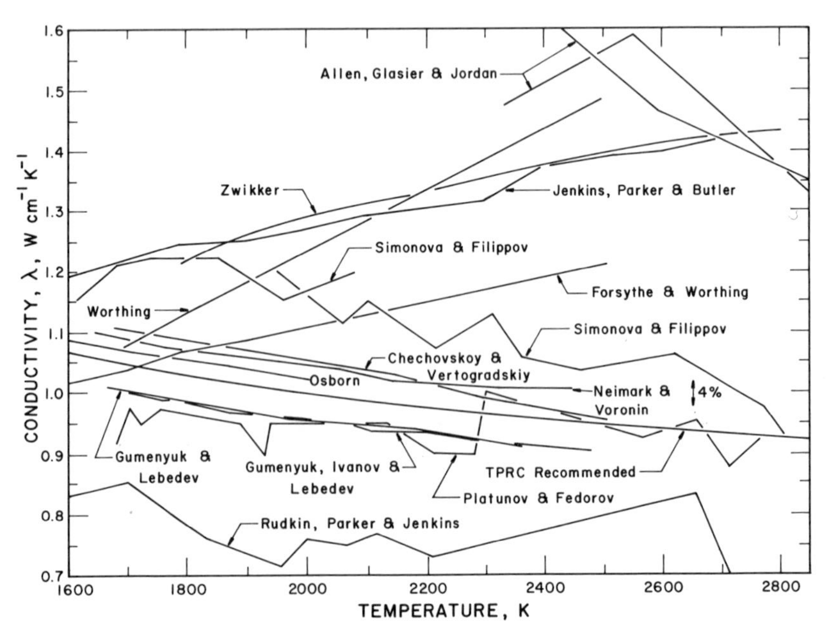

FIGURE 1. Thermal conductivity of tungsten (direct heating method).

Other examples of discrepancies among conductivity results obtained by direct-heating methods could be given for other high melting metals, such as molybdenum, stainless steel, and platinum (Powell *et al.*;[16] Taylor *et al.*[17]). There are several reasons for these discrepancies and they are discussed in a separate section. However, similar examples could be given for thermal conductivity values obtained at high temperatures using nonelectrical methods (Powell *et al.*;[18] Ho *et al.*[2]).

The equation governing energy transport in a conductor at steady state, heated by the passage of current, is (Flynn[19])

$$\nabla \cdot \lambda \nabla T + J \cdot (\rho J - \mu \nabla T) = 0 \tag{1}$$

where λ is the thermal conductivity, J the current density, ρ the electrical resistivity, T temperature, and μ the Thomson coefficient. Equation (1) assumes no external sources of heat (including sample–electrode junctions), homogeneous samples, and only electrons involved in mass transport. For the case of isotropic cylindrical samples (i.e., no angular dependence) and with the cylindrical surface an electrical insulator, equation (1) becomes

$$\frac{\partial}{\partial Z}\left(\lambda \frac{\partial T}{\partial Z}\right) + \frac{1}{r}\left(r\lambda \frac{\partial T}{\partial r}\right) + \frac{\rho I^2}{A^2} - \mu \frac{I}{A}\frac{\partial T}{\partial Z} = 0 \tag{2}$$

where I is the current, r the radius, Z an axial distance, and A the cross-sectional area.

There are three important cases:

a. Long rods or hollow cylinders whose surface near the center is at a uniform temperature, i.e., $\partial T/\partial Z = 0$ near the center. The thermal conductivity is determined from radial temperature gradients. Examples are the Angell,[20] and the Powell and Schofield[21] methods.
b. Rods or hollow cylinders in which heat loss from the outer surface is small. Examples are the Kohlrausch,[1] Mikryukov,[22] and Callendar[23] methods.
c. Long thin rods in which the heat loss from the surface is appreciable. Usually this heat loss is by radiation and equation (2) becomes

$$\lambda \frac{d^2T}{dZ^2} + \frac{d\lambda}{dT}\left(\frac{dT}{dZ}\right)^2 + \frac{I^2\rho}{A^2} - \frac{P\epsilon_H \sigma (T^4 - T_0^4)}{A} - \mu \frac{I}{A}\frac{dT}{dZ} = 0 \tag{3}$$

where P is the perimeter, ϵ_H is the total hemispherical emittance, σ is the Stefan–Boltzmann constant, and T_0 is the temperature of the sur-

> rounding vacuum enclosure. Equation (3) may be called the long thin rod approximation and is of particular importance for high temperature experimental determinations of λ. Examples are the Krishnan and Jain[24] and Lebedev[25] methods.

The experimental involves measuring $T(Z)$, whereas equation (3) contains dT/dZ and d^2T/dZ^2 in addition to T. No closed solution of equation (3) is known even for the simpler case of λ, ρ, and ϵ_H independent of temperature and $\mu = 0$. Therefore, it is necessary to employ very restrictive mathematical approximations in order to linearize equation (3), use numerical integration techniques, or differentiate an appropriate expression for $T(Z)$ to obtain dT/dZ and d^2T/dZ^2 in order to obtain λ from the experimental data. Because of the loss of accuracy in differentiation, the latter procedure had not been used before Taylor *et al.*[17] applied SPLINE techniques. Several variants of the long thin rod case involving mathematical approximations and boundary conditions have been developed over the years prior to the application of sophisticated numerical integration techniques or SPLINE programs. The relations of most of these variants to the general solution are summarized in Fig. 2.

The various methods are described in some detail and the appropriate equations are derived elsewhere (Flynn[19] and Powell *et al.*[16]). When we

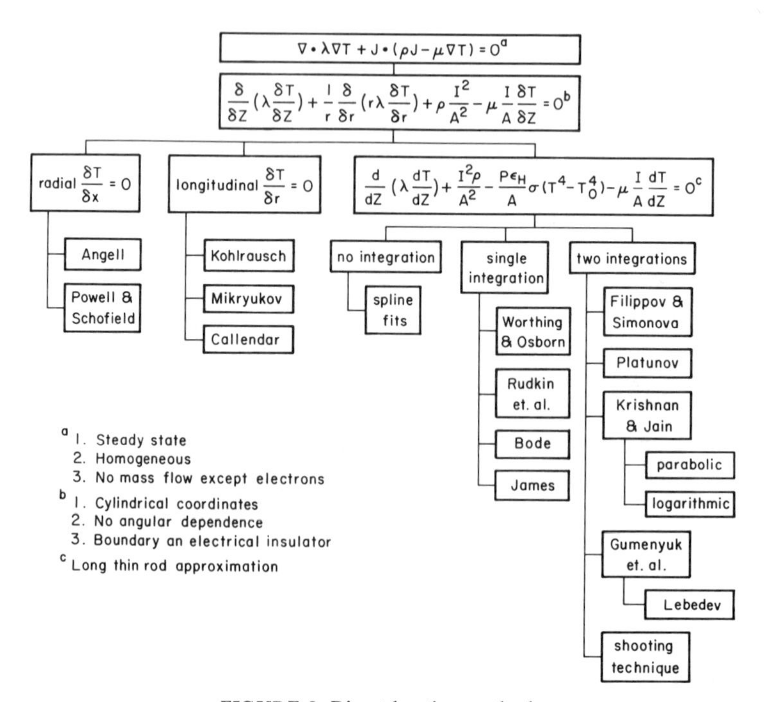

FIGURE 2. Direct heating methods.

employ the long thin rod approximation, we reduce the number of independent variables from two to one. The two-independent-variable problem is elliptic and a solution may be determined using boundary values (Sokolnikoff and Redheffer[26]). In this case one does not need to know $\partial T/\partial r$ in the interior of the sample. The presence of heat sources and sinks located near the electrode–sample interfaces requires that additional terms be incorporated in equation (3). These sources and sinks include the Peltier heating and cooling of the electrodes. These effects complicate equation (3) and are an impetus for making the long thin rod approximation. Although apparently no one has included the appropriate terms and solved equation (2), it is not obvious that this procedure could not be successful.

Various methods which employ the long thin rod approximation are arbitrarily divided into three categories in Fig. 2. These categories are based upon the number of mathematical integrations used in the derivation of the working equation, i.e., whether the working equation for λ contains terms involving d^2T/dZ^2, dT/dZ, or only T. For example, the equation derived by Rudkin *et al.*[9] is

$$\lambda = \frac{2}{A}\left(\frac{dT}{dZ}\right)^{-2} \int_{T_c}^{T} \left(\frac{I^2\rho}{A} - P\epsilon_H \sigma T^4\right) dT \tag{4}$$

where a mathematical integration results in the $(dT/dZ)^{-2}$ term while the integral is determined graphically from the experimental data. Therefore this method is classified as a single (mathematical) integration with dT/dZ terms present. Other examples are the Bode[27] and James[28] methods. On the other hand, the Lebedev[24] derivation yields

$$\lambda = \frac{1}{2}\rho\,\frac{I_s^2 - I_\infty^2}{A^2}\,\frac{Z^2}{\Delta T} \tag{5}$$

where I_s and I_∞ are the currents through a short and long sample, respectively. Since equation (5) contains no dT/dZ or d^2T/dZ^2 terms, the Lebedov method is classified with the double integration methods.

In order to perform the mathematical integrations, various assumptions and restrictive conditions have been employed. These include neglecting the $\mu(I/A)(dT/dZ)$ term, neglecting the temperature dependencies of λ, ρ, ϵ_H, and μ, and various assumptions to linearize the differential equation, such as substituting $4T_\infty^3(T_\infty - T)$ for $T_\infty^4 - T^4$, where T_∞ is the temperature at the center of an infinitely long rod.

As a result of the sophisticated mathematical techniques developed in recent times, it is no longer necessary to make these assumptions. In essence the various long thin rod approximation methods in which the assumptions are made, such as those of Worthing and Osborn, Rudkin *et al.*, Krishnan and Jain, and others, are now outdated.

3. DESCRIPTION OF SEVERAL DIRECT HEATING APPARATUS

Before discussing direct heating methods further, it would be instructive to describe certain apparatuses. A Kohlrausch-type apparatus will be considered first as it is the basis for most of the lower-temperature ($T < 1000$ K) longitudinal direct-heating techniques. The Thermophysical Properties Research Laboratories (TPRL) Kohlrausch apparatus may be considered to be a modified Kohlrausch apparatus since an external heater is added to minimize heat losses, whereas the original analysis assumed that the surface was a thermal barrier.

The Kohlrausch method involves the determination of the product of the thermal conductivity "λ" and the electrical resistivity "ρ." Since the electrical resistivity is also measured at the same time, λ can be calculated. The method involves passing constant direct current through the specimen to heat the sample while the ends are kept at constant temperature (Fig. 3). Radial heat losses are minimized by an external heater maintained at the sample's midpoint temperature. With these provisions, at steady state a parabolalike axial temperature profile is obtained. Thermocouples are placed at the center and at an equal distance apart on each side of the center. The thermocouples may also act as voltage probes. Numbering the center thermocouple as the "2" position and the other positions as "1" and "3" (Fig. 3), it is possible to get the products of λ

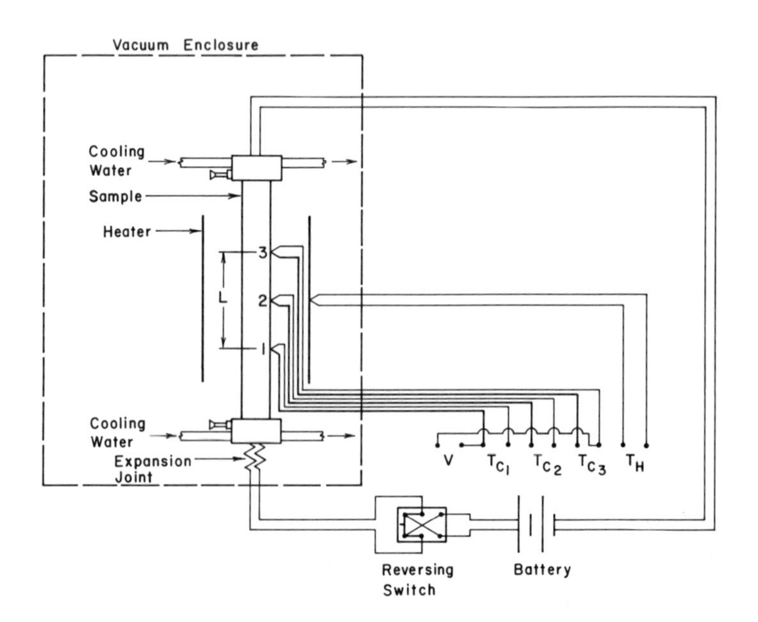

FIGURE 3. Kohlrausch apparatus (schematic).

and ρ:

$$\lambda\rho = \frac{(V_3 - V_1)^2}{4[2T_2 - (T_1 + T_3)]} \tag{6}$$

where $V_3 - V_1$ is voltage drop between the third and first thermocouple, $T_1 + T_3$ is the sum of the temperatures at the outside thermocouples, and T_2 is the center temperature. Since ρ is also measured simultaneously [$\rho = (V_3 - V_1)A/IL$, where A is the cross-sectional area, I is the current, and L is the distance between positions 1 and 3], λ can be calculated. The data collection (T_1, T_2, T_3, $V_3 - V_1$, I) may be computerized. The results are calculated for a set of measurements performed while the sample is under vacuum and the heater temperature matched to that of T_2. Then additional current is used, a new set of equilibrium conditions is obtained, and the process repeated.

Voltage outputs are measured with an integrating digital voltmeter (IDVM) which is part of a digital data acquisition system (DDAS) described later as part of the TPRL multiproperty apparatus. A computer printout of a typical experiment is given in Table 3. The emf outputs of chromel–alumel thermocouples

TABLE 3
Computer Output for Kohlrausch Experiment; Experiment Date: 8/17/78

TC1	TC2	TC3	V	I/1000
0.00537690	0.00597460	0.00537150	0.03964802	0.07965803
0.00537920	0.00597550	0.00537260	0.03964902	0.07965803
0.00537970	0.00597600	0.00537270	0.03965002	0.07966003
0.00538010	0.00597640	0.00537280	0.03965002	0.07965904
0.00538030	0.00597650	0.00537320	0.03965002	0.07966003
Average of IDVM reading in volts for point 1				
0.00537924	0.00597580	0.00537256	0.03964942	0.07965904
0.00534600	0.00599700	0.00537720	−0.03964602	−0.07965803
0.00534600	0.00599710	0.00537740	−0.03964402	−0.07965703
0.00534600	0.00599720	0.00537750	−0.03964402	−0.07965601
0.00534610	0.00599730	0.00537760	−0.03964502	−0.07965703
0.00534620	0.00599730	0.00537810	−0.03964502	−0.07965803
Average of IDVM readings in volts for point 2				
0.00534606	0.00599718	0.00537756	−0.03964481	−0.07965723

Distance between voltage probes: 1.95360 cm
Area of sample: 0.09379 cm²

$T1$	$T2$	$T3$	V	I
404.061	419.622	404.369	0.039647	79.6581

Thermal conductivity = 0.5337 watts per cm K at 414.5 K
Thermal conductivity = 370.1 BTU in. h^{-1} ft^{-2} F^{-1} at 286.4°F
Resistivity = 0.238948E–04 Ω cm at 414.5 K
Electronic thermal conductivity = 0.4233 W/cm K

located at positions 1, 2, and 3 are given under the headings "TC1," "TC2," and "TC3." The voltage drop $V_3 - V_1$ is given under the heading V and the voltage drop across a 0.001-Ω standard shunt is given under the heading $I/1000$. Five readings of each of these quantities were taken in sequence by an integrating digital voltmeter (IDVM). The results for each of these five sets of readings are averaged for point 1. The current flow is then reversed and the process repeated to obtain average values for point 2. Then the averaged values are themselves averaged, the emf outputs converted to temperatures, and the voltage drop across the shunt converted to current. These results are outputed as $T1$, $T2$, $T3$, V, and I. The geometrical constants L and A are also printed out. From these data, the electrical resistivity is calculated:

$$\rho = \frac{A}{L}\frac{V}{I} \tag{7}$$

The thermal conductivity is then calculated from equation (6). Also the electronic component to the thermal conductivity, λ_e, is calculated from the Weidemann–Franz–Lorenz relationship,

$$\lambda_e = \frac{LT}{\rho} \tag{8}$$

where T is in kelvin and L is assumed to have the classical value of $2.443 \times 10^{-8}\ \Omega\,\mathrm{K}^{-2}$. Should the calculated electronic component exceed the measured λ value, a warning is printed out. Such an occurrence may signal experimental difficulties or it may be legitimate. The latter case occurs, for example, for pure metals below room temperature and also for some composite materials.

Some of the significant research efforts involving Kohlrausch's method or modifications of it include those of Jaeger and Diesselhorst,[29] O'Day,[30] and Mikryukov.[22]

The radial heat flow method consists of passing electric current in the axial direction of a long, but thick rod or cylinder. As in the Kohlrausch method, it is assumed that over a small temperature range the thermal conductivitity and electrical resistivity are independent of temperature. Referring to Fig. 4, the heat flow across dr is $-2\pi rL\lambda\, dT/dr$.

At steady state

$$\pi r^2 LE\bar{i} = -2\pi rL\lambda \frac{dT}{dr} \tag{9}$$

where E is the electrical potential drop over a length L at the central region of the long specimen and $\bar{i}$ is

$$\bar{i} = \frac{I}{\pi r_1^2 L} \tag{10}$$

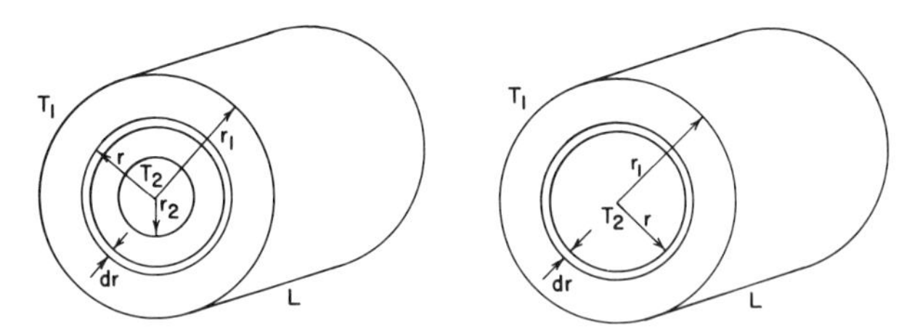

FIGURE 4. Radial heat flow specimen geometry.

Integrating both sides of equation (10) from the center to the circumference leads to

$$\frac{E\bar{i}}{2}\int_0^{r_1} r\,dr = -\int_{T_2}^{T_1} dT \tag{11}$$

Solving for λ we get

$$\lambda = \frac{E\bar{i}r_1^2}{4(T_2 - T_1)}$$

Substituting equation (10) into the above equation yields

$$\lambda = \frac{EI}{4\pi L(T_2 - T_1)} \tag{12}$$

For a hollow cylindrical sample at steady state

$$\pi(r^2 - r_2^2)LE\bar{i} = 2\pi L\lambda\frac{dT}{dr} \tag{13}$$

Solving for λ, we obtain

$$\lambda = \frac{E\bar{i}}{2(T_2 - T_1)}\left|\frac{r_1^2 - r_2^2}{2} - r_2^2 \ln\frac{r_1}{r_2}\right| \tag{14}$$

Substituting for $\bar{i}$ into the above equation yields

$$\lambda = \frac{EI}{2(T_2 - T_1)\pi L}\left|\frac{1}{2} - \frac{r_2^2}{(r_1^2 - r_2^2)}\ln\frac{r_2}{r_2}\right| \tag{15}$$

This method was used by Scott[31] to determine the thermal conductivity of uranium dioxide and also by Powell and Schofield[21] to measure the thermal and electrical conductivities of carbon and graphite for temperatures up to 2400°C. Johnson and Watt[32] later determined the thermal conductivity of pyrolitic graphite in the temperature range 1100–1960°C by this method. One difficulty of the method involves measuring the temperature at two different radii, one of which is on the surface.

The Thermophysical Properties Research Laboratory Multiproperty Apparatus has been used in an extensive investigation of high-temperature direct heating methods. This apparatus has been duplicated at the Oak Ridge National Laboratory and has been used as a model for the apparatus constructed at the Boris Kidrič Institute (Belgrade). The TPRL multiproperty apparatus consists of a high-vacuum system (10^{-7} Torr), large bell jar equipped with two long windows, interior piping, and sample holders with provisions for sample expansion and contraction, and regulated dc power supplies. An automatic optical pyrometer and elevating stand are required, and twin telemicroscopes and stand are used for thermal expansion. Samples in the form of thin rods, tubes, or wires are supported vertically between water-cooled movable electrodes. The electrode separation distance is adjustable between 0 and 35.6 cm. Sample expansion and contraction is maintained stress-free through a spring network mounted on a movable "c-cell" equipped with strain gauges. The bell jar which covers the sample support system is raised and lowered by a hoist. The bell jar rests on a feed-through collar which contains rotary feed-throughs for protecting the window and for moving the c-cell, instrumentation leads, electrical connections, and water lines. Usually instrumentation readout is accomplished using a mini-computer based digital data acquisition system. The bank of regulated power supplies is equipped with remote controls, reversing switches, and calibrated current shunts. Temperature measurements are made using the automatic optical pyrometer mounted on a positioning stand external to the vacuum system and viewing the sample through an optical window. The system is shown schematically in Fig. 5 and a picture of the apparatus is shown in Fig. 6. Linear thermal

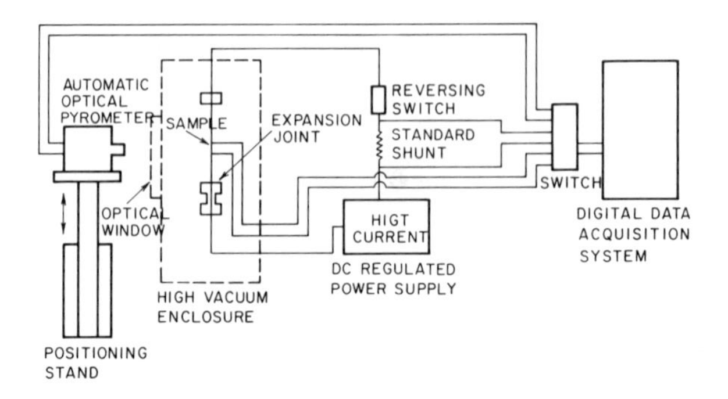

FIGURE 5. Schematic of multiproperty apparatus.

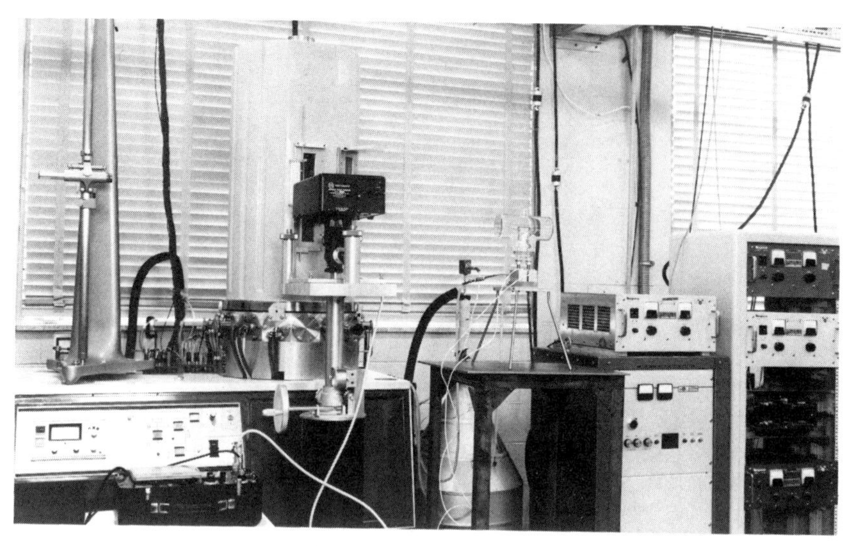

FIGURE 6. Multiproperty apparatus.

expansion measurements use twin telemicroscopes mounted on a second stand for viewing the sample through a second window.

Figure 7 shows a close-up view of the test enclosure. A sample (A) is mounted eccentrically to reduce back-scattered radiation between two copper electrodes (B and C). The lower is joined to a thermal expansion take-up assembly (D) and the upper fixed to a movable stainless steel plate (E). The samples can extend through each electrode. The plates supporting the electrodes are clamped to hollow stainless-steel water-cooled columns (F) with ceramic rings as electrical insulation. Regulated dc power flows to the electrodes through water-cooled copper lines. Water lines to the electrodes are of copper and those to the stainless-steel columns are of stainless steel. All copper parts, except the movable parts of the lower electrode are nickel plated to facilitate cleaning and attainment of good vacuum. Also, all bolts are slotted.

The test cell is enclosed by a water-cooled steel bell jar (C) (raised in Fig. 7), 45.7 cm in diameter and 91.4 cm high and fitted with two 33 x 5.1 cm vertical windows. It is operated by a hoist, and, in the closed position, rests on a O-ring seal fitted to the universal feed-through collar (H). A shutter (I) protects the window when not in use for optical pyrometer measurements.

All control mechanisms, leads, and water feeds operate through seals in the collar (H). The unit of Fig. 3 is mounted on an automatic high-vacuum facility (Veeco), capable of maintaining the bell-jar enclosure in the mid-10^{-7} Torr range. The multiproperty apparatus is connected to a digital data acquisition system which controls experiments, collects data, calculates results, and outputs the results in tabular and graphical form.

FIGURE 7. Closeup of test cell enclosure.

The Digital Data Acquisition System (DDAS), shown in Fig. 8, serves the multiproperty apparatus. The system contains a PDP-8/E minicomputer (second cabinet from the right) with 32 000 words of core connected to an omnibus. Connected to the omnibus through a "data break" is a controller and disk drive for a disk system (right-hand cabinet) which is subdivided into four parts, System (SYS), disk 1, disk 2, and disk 3. The SYS system with 1×10^6 word capacity and DSK1 with 6×10^5 word capacity reside on a removable cartridge with a floating head. The other two parts (DSK2 and DSK3 with 1×10^6 and 6×10^5 word capacity, respectively) reside on a fixed head cartridge. Information can be swapped readily among the various subdivisions. The capacity of the DDAS is sufficiently large (3.3×10^6 words) that programs and experimental data may be stored and used without input–output to any other device such as magnetic tape, punched tape, punched cards, etc., although a high-speed paper tape reader/punch and magnetic tape units (both housed in second cabinet from right) are included. Other disk cartridges may be readily substituted for the

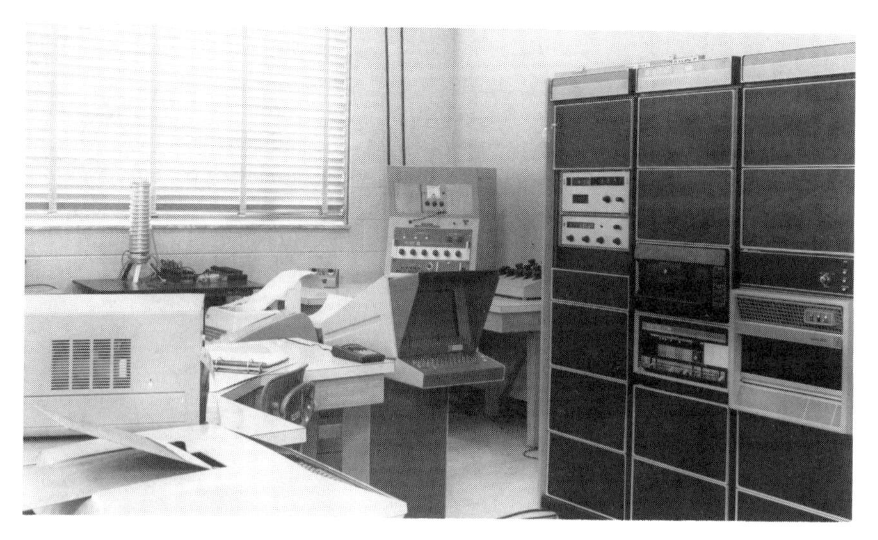

FIGURE 8. Digital data acquisition system.

removable cartridge, providing additional storage and back-up capabilities. A floating point processor (left-hand cabinet) is used in the processing of experimental data so that calculations can be performed in double precision and FORTRAN IV programs can be readily handled.

Instructions and programs may also be entered through the Tektronix Display Terminal (model 4010-1), which is the master input–output device or through the Decwriter (high-speed teletype) or standard teletype units. The Decwriter is essentially used as a line printer to give printed output. In addition, hard copies of whatever is displayed on the Display Terminal can be made by the Tektronix Hard Copy Unit (model 4610).

Experimental data are directly entered into the system through an analog-to-digital converter (ADC) or through a VIDAR model 521-01 integrating digital voltmeter (IDVM). The IDVM (third cabinet from right) is controlled via a master Scanner (Vidar 610) and is equipped for 100 input channels. The Vidar DVM has full scale ranges from 10 MV to 1000 V in steps of 10 and has three integration periods (166.7, 16.7, and 1.67 ms) so chosen as to essentially provide infinite rejection of 60-Hz signals. The resolution of the IDVM depends upon the integration period and for the 10-MV full scale range is $\pm 0.1\ \mu V$ for the longest integration period. Operating in conjunction with the IDVM is a special crystal-controlled timer which accurately records the time that data are taken without resort to software timing.

In addition, the DDAS is equipped with a real time clock and with programmable switches and digital-to-analog converters (DAC) for controlling experiments.

In practice the sample diameter (and bulk density) are determined. Then voltage probes are attached over a central portion of the sample. The electrical resistivity at room temperature, the effective voltage probe separation distance, and the appropriate factors (A/L and $PL\sigma$) are determined prior to insertion of the sample into the multiproperty apparatus. This is done with the aid of a knife blade holder, which holds two knife blades a precisely known distance apart. The sample is connected to a regulated dc power supply in series with a reversing switch and a precision shunt. The voltage drops across the shunt, the voltage probes, and the knife blades are measured when the knife blades are located over the same general region of the sample as the voltage probes. The current flow is reversed and the procedure repeated. The measured voltage drops are averaged to eliminate stray emf effects. The current is determined from the voltage drop across the standard shunt. The resistivity is determined from equation (7) and the effective distance between the voltage probe wires is calculated from the relation

$$L_{\text{voltage probes}} = L_{\text{knife blades}} \frac{V_{\text{voltage probes}}}{V_{\text{knife blades}}} \tag{16}$$

A computer program collects the data and averages the measured voltages and outputs A, L, A/L, $PL\sigma$, and ρ. The system is evacuated, and the sample heated to a temperature above or equal to the highest test temperature desired. After appropriate aging, the voltage drop, current flow, and temperature are measured, the current flow reversed, and the process repeated. The sample temperature is lowered to a new level and measurements repeated. This procedure is followed until the desired temperature range has been covered.

A computer program collects the E, I, and T data; averages the measurements; corrects the temperature for spectral emissivity and window absorption; calculates ρ and ϵ_H at each temperature; fits ρ versus T and ϵ_H versus T data to least-squares polynomials; compares the fitted data to the experimental results giving calculated values, differences, maximum, average, and standard deviations; plots the results and stores the coefficients for thermal conductivity determinations and for comparison with repeat runs.

Typical data acquisitions for resistivity and total hemispherical emissivity of a graphite sample are shown in Table 4. The first column is voltage output from an automatic optical pyrometer, the second column is voltage drop across a 0.1 Ω shunt, and the third column is voltage drop across the voltage probes. Successive readings of these three quantities were taken five times and the results averaged for point 1. The current flow was reversed and the process repeated for point 2. Then the current was decreased and the above procedure repeated for points 3 and 4. For this particular experiment 24 data sets (points) were obtained. Next the pyrometer voltage data were converted to true temperatures. This procedure, all done by the same DDAS that took the data, involves con-

TABLE 4
Long Sample Data (Partial Output); Experiment Date: 12/7/79; Sample ID: 115E01PA1

Pyro	I/1000	V
0.37012010	−0.14021110	−1.44461989
0.37012010	−0.14021009	−1.44461989
0.37012010	−0.14020913	−1.44456028
0.37013012	−0.14021009	−1.44458055
0.37013012	−0.14020913	−1.44448041
Average of IDVM readings in volts for point 1		
0.37012410	−0.14020991	−1.44457221
0.37040013	0.14021509	1.44507050
0.37041014	0.14021509	1.44508063
0.37041014	0.14021611	1.44506096
0.37041014	0.14021909	1.44510030
0.37041014	0.14021909	1.44502997
Average of IDVM readings in volts for point 2		
0.37040811	0.14021688	1.44506871
0.35248017	0.13002508	1.31375074
0.35248017	0.13002210	1.31379008
0.35247015	0.13002413	1.31390035
0.35247015	0.13002312	1.31390035
0.35248017	0.13002413	1.31395041
Average of IDVM readings in volts for point 3		
0.35247623	0.13002371	1.31385862
0.35229009	−0.13004213	−1.31313085
0.35230010	−0.13004410	−1.31335079
0.35230010	−0.13004314	−1.31299078
0.35230010	−0.13004314	−1.31340026
0.35229009	−0.13003814	−1.31321072
Average of IDVM readings in volts for point 4		
0.35229611	−0.13004213	−1.31321609

TABLE 5
Temperature Conversion; E-Coefs: E1 = 0.9400005E+00, E2 = −0.4750005E−04, E3 = 0.0000000E+00, E4 = 0000000E+00

Point No.	True temp.	Brightness temp.	Point No.	True temp.	Brightness temp.
1	2234.2348	2172.4277	7	1707.4412	1672.3757
2	2145.0537	2087.2515	8	1646.1855	1614.0371
3	2066.5913	2012.6688	9	1564.1868	1535.7786
4	1954.5453	1906.7400	10	1485.2148	1460.2386
5	1869.3858	1826.6814	11	1382.8867	1362.1169
6	1778.2992	1741.4802	12	1313.1159	1295.0652

verting the voltage readings to brightness temperatures, correcting for window absorption, and converting brightness temperatures to true temperatures (TT), i.e., blackbody temperatures. The conversion of pyrometer voltages to brightness temperatures is based on calibration data using standard optical lamps calibrated by the National Bureau of Standards. The window absorption corrections were determined as a function of brightness temperature by comparing pyrometer outputs with the window inserted between the standard lamp and the pyrometer to pyrometer outputs at the same brightness temperatures with the window removed. Both the pyrometer calibration and window absorption corrections are in data files which reside in the DDAS.

The spectral emissivity at the wavelength of the automatic optical pyrometer (0.65 μm) has been entered into the computer for each sample as a set of coefficients $E1$, $E2$, $E3$, and $E4$, where $\epsilon_{0.65} = E_1 + E_2 T + E_3 T^2 + E_4 T^3$. The results of the temperature conversion data of Table 3 plus window corrections are given in Table 5. In this particular case the spectral emissivity was linear, i.e., $E_3 = E_4 = 0$. The spectral emissivity coefficients have to be determined as a separate experiment (see Section 4). Using the values for $\epsilon_{0.65}$ and an iterative procedure based on the Planck's relation

$$\mathrm{TT} = \frac{1.438/(0.65 \times 10^{-4})}{\ln\left[\epsilon_{0.65}\left(e^{1.4381/(0.65 \times 10^{-4}\,BT)} - 1\right) + 1\right]} \tag{17}$$

the true temperature "TT" values corresponding to the experimentally measured brightness temperatures "BT" are calculated and tabulated both in degrees kelvin (Table 5).

The true temperature data of Table 4 are combined with the corresponding average voltage and current values in the DDAS to calculate resistivity (ρ) and hemispherical emissivity (EH) values in Table 6. First the A/L (AL) and $PL\sigma$ (PLS) values are entered into the computer and then the DDAS produces Tables 6 where TT is the true temperature in K, V is the average voltage in volts, I is the average current in amperes, and FT is the temperature in Fahrenheit. The output of Table 6 is used to generate least-squares fit data for resistivity (Table 7) and emissivity (Table 8). These fits (within one standard deviation) represent the data to about 0.1%. The results are plotted in Fig. 9.

After the electrical resistivity and total emissivity have been determined as functions of temperature (long sample data), the effective length of the sample is shortened. The upper portion of the sample slides through the upper electrode clamp and the lower portion of the sample slides through the lower electrode clamp.

The test cell is evacuated, and the sample heated so that the maximum temperature (near the sample center) reaches the desired test temperature. A

TABLE 6
Calculation of Resistivity and Emissivity, AL = 0.97810018E−01, PLS = 0.92095994E+01

Rho	EH	TT	V	I	FT
1007.86	0.88306	2234.23	1.4448255	140.21402	3561.9
988.03	0.87633	2145.05	1.3135427	130.03349	3401.4
970.61	0.87376	2066.59	1.2066018	121.59049	3260.1
945.26	0.87205	1954.54	1.0640192	110.09818	3058.5
925.37	0.87027	1869.38	0.9619898	101.67998	2905.2
903.78	0.86816	1778.30	0.8592016	91.98543	2741.2
886.74	0.86771	1707.44	0.7843358	86.51411	2613.7
871.82	0.86681	1646.18	0.7224799	81.05480	2503.4
851.62	0.86379	1564.18	0.6434907	73.90527	2355.8
832.23	0.86032	1485.21	0.5722668	67.25710	2213.7
807.71	0.85491	1382.88	0.4870908	58.98418	2029.5
791.55	0.84944	1313.11	0.4332636	53.53688	1093.9

temperature profile, consisting of about 20 temperatures measured as a function of the z coordinate, is obtained by moving the pyrometer from an arbitrary zero reference point near the lower end of the sample, to a roughly corresponding position near the upper electrode. The current flow is reversed and second temperature profile obtained. A computer program is used to collect the temperature versus position data. With each temperature versus position point, the current flow is also measured. A portion of the output is shown in Table 9. The IDVM readings for the pyrometer and the current flow as measured by the voltage drop across the 0.001-Ω standard shunt are taken five times and the results are averaged. The first point is taken at position zero which is an arbitrary position on the sample. The data in the first column are the current readings divided by 1000. Then the pyrometer is raised 1 mm and the process repeated.

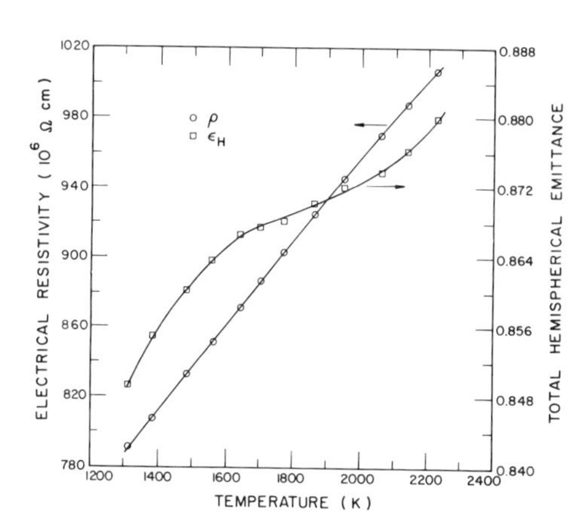

FIGURE 9. Typical plots of resistivity and emissivity.

TABLE 7
Least-Squares Fit of Resistivity Data, $Y = A + B \cdot X + C \cdot X \cdot X$

A	B	C	
0.4391943812E–03	0.2860011458E–06	–0.1400445103E–10	
X	Y	Calculated Y	Deviation
0.2234200835E+04	0.1107900238E–02	0.1008272767E–02	0.3725290894E–06
0.2145100235E+04	0.9880399703E–03	0.9882543087E–03	0.2144370675E–06
0.2066599726E+04	0.9706197977E–03	0.9704334735E–03	0.1862645149E–06
0.1954500079E+04	0.9452600479E–03	0.9446856975E–03	0.5745096206E–06
0.1869400143E+04	0.9253801107E–03	0.9249041080E–03	0.4761387705E–06
0.1778300404E+04	0.9037799835E–03	0.9035030603E–03	0.2769520878E–06
0.1707400083E+04	0.8867399692E–03	0.8866869211E–03	0.5331821441E–07
0.1646199822E+04	0.8718303442E–03	0.8720576763E–03	0.2273591160E–06
0.1564199924E+04	0.8516299724E–03	0.8522922992E–03	0.6622867584E–06
0.1485199928E+04	0.8322299718E–03	0.8330719470E–03	0.8417993783E–06
0.1382899880E+04	0.8077100515E–03	0.8079229593E–03	0.2129236459E–06
0.1313100457E+04	0.7915599346E–03	0.7905954122E–03	0.9645010232E–06

Avg. dev. = 0.4219183325E–06
Std. dev. = 0.5025196075E–06
Max. dev. = 0.9645010232E–06

TABLE 8
Least-Squares Fit of Emissivity Data, $Y = A + B \cdot X + C \cdot X \cdot X$

A	B	C	
0.7701818943E+00	0.8133971691E–04	–0.1452378034E–07	
X	Y	Calculated Y	Deviation
0.2234200835E+04	0.8830600976E+00	0.8794133663E+00	0.3646612763E–02
0.2145100235E+04	0.8763401508E+00	0.8778331279E+00	0.1493096351E–02
0.2066599726E+04	0.8737697601E+00	0.8762499094E+00	0.2480149269E–02
0.1954500079E+04	0.8720600605E+00	0.8736785650E+00	0.1618504524E–02
0.1869400143E+04	0.8072700138E+00	0.8714826107E+00	0.1212716102E–02
0.1778300404E+04	0.8681598901E+00	0.8688989877E+00	0.7390975952E–03
0.1707400083E+04	0.8677200078E+00	0.8667213916E+00	0.9986162185E–03
0.1646199822E+04	0.8668199777E+00	0.8647243976E+00	0.2095699310E–02
0.1564199924E+04	0.8637899160E+00	0.8618777990E+00	0.1912117004E–02
0.1485199928E+04	0.8603198528E+00	0.8589508533E+00	0.1368999481E–02
0.1382899880E+04	0.8549098968E+00	0.8548911809E+00	0.1871585845E–04
0.1313100457E+04	0.8494498729E+00	0.8519465923E+00	0.2496838569E–02

Avg. dev. = 0.1673430204E–02
Std. dev. = 0.1903261542E–02
Max. dev. = 0.3646612763E–02

TABLE 9
Partial Output of Profile Data Collection; Experiment Date: 12/7/79; Sample ID: 115E01PA2

Pryo	$I/1000$	Pyro	$I/1000$
0.26060110	−0.08421502	0.27748012	−0.08420600
0.26061916	−0.08421401	0.27746415	−0.08420504
0.26057618	−0.08421401	0.27746218	−0.08420504
0.26058107	−0.08421303	0.27745705	−0.08420600
0.26057314	−0.08421201	0.27745705	−0.08420504
Average of IDVM readings in volts for point 1 at 0.00 mm		Average of IDVM readings in volts for point 3 at 2.00 mm	
0.26059019	−0.08421362	0.27746415	−0.08420542
0.26914310	−0.08420403	0.28522014	−0.08420301
0.26916408	−0.08420403	0.28522914	−0.08420301
0.26916711	−0.08420301	0.28523510	−0.08420301
0.26915115	−0.08420301	0.28522014	−0.08420103
0.26914310	−0.08420202	0.28521609	−0.08420103
Average of IDVM readings in volts for point 2 at 1.00 mm		Average of IDVM readings in volts for point 4 at 3.00 mm	
0.26915174	−0.08420323	0.28522419	−0.08420219

Usually between 18 and 25 sets of readings are taken per profile. Two extra data points are taken on the second profile. For these two points, the current is varied but the position is maintained constant. Thus the temperature change caused by a slight change in current is determined. At the conclusion of the profile determinations, the average current is calculated and the individual temperature data are corrected to what the temperature would be if the average current had been flowing when the temperature measurement was taken. Typical

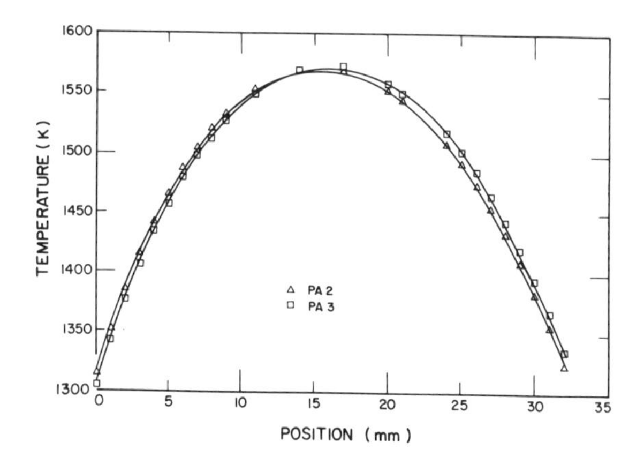

FIGURE 10. Typical profile data.

TABLE 10
Typical Profile Data Results

Position (mm)	Run 2 (K)	Run 3 (K)	Position (mm)	Run 2 (K)	Run 3 (K)
0	1316.4	1305.4	20	1553.8	1558.3
1	1352.5	1342.7	21	1545.8	1550.2
2	1385.9	1376.3	24	1509.9	1518.1
3	1415.8	1406.5	25	1493.1	1502.1
4	1442.3	1433.7	26	1475.8	1484.9
5	1466.2	1457.9	27	1456.2	1465.8
6	1487.9	1480.1	28	1434.6	1444.0
7	1505.1	1498.3	29	1410.4	1420.5
8	1520.6	1512.3	30	1384.0	1394.3
9	1534.4	1527.9	31	1356.3	1366.6
11	1553.1	1549.3	32	1324.9	1335.1
14	1569.4	1568.6	33		1300.7
17	1570.5	1572.4			
Average current:	84.2163	−84.2635			

profile data are plotted in Fig. 10 (these plots are from the visual display screen of the DDAS) and are tabulated in Table 10. The two profiles are offset from each other due to the Thomson term [equation (3)].

The temperature profiles are analyzed using advanced mathematical techniques involving SPLINE functions to yield values of $T, dT/dZ$, and d^2T/dZ^2 at any location Z. The outputs of the SPLINE program for one of the profiles of Table 9 are given in Tables 11–13. Table 10 gives the SPLINE equations for $T(Z)$. Note that these equations represent the observed temperatures within an average difference of 0.4 even though the temperatures increased from 1305.9 to 1572.4 and then decreased to 1300.7. Each SPLINE equation is valid between the particular points (knots) indicated. These equations have the property that the values for $T, dT/dZ$, and d^2T/dZ^2 calculated at a knot are the same for the SPLINE equation above and below the knot. For example, the values corresponding to $Z = 0.471429$ cm for $T, dT/dZ$, and d^2T/dZ^2 for the equation

$$T = 0.130612 \times 10^4 + 0.37839 \times 10^3 Z - 0.13989 \times 10^3 Z^2 - 0.16302 \times 10^2 Z^3$$

are the same as those for

$$T = 0.14517 \times 10^4 + 0.23563 \times 10^3 (Z - 0.471429) - 0.16293 \times 10^3 (Z - 0.471429)^2 + 0.569772 \times 10^2 (Z - 0.471429)^3$$

TABLE 11
Spline Output for $T(Z)$

Knots	Cubic coefficients	Knots	Cubic coefficients
XI(1) = 0.000000		XI(5) = 1.885719	
	C(1) = 0.130612E+04		C(1) = 0.156528E+04
	C(2) = 0.378396E+03		C(2) = −0.491486E+02
	C(3) = −0.139890E+03		C(3) = −0.778632E+02
	C(4) = −0.163020E+02		C(4) = −0.152533E+02
XI(2) = 0.471429		XI(6) = 2.357148	
	C(1) = 0.145171E+04		C(1) = 0.152321E+04
	C(2) = 0.235630E+03		C(2) = −0.132733E+03
	C(3) = −0.162937E+03		C(3) = −0.994348E+02
	C(4) = 0.569772E+02		C(4) = −0.754159E+01
XI(3) = 0.942858		XI(7) = 2.828577	
	C(1) = 0.153255E+04		C(1) = 0.143775E+04
	C(2) = 0.119992E+03		C(2) = −0.231514E+03
	C(3) = −0.823549E+02		C(3) = −0.110101E+03
	C(4) = −0.119691E+02		C(4) = −0.305768E+02
XI(4) = 1.414290		XI(8) = 3.300006	
	C(1) = 0.156956E+04	Least-squares error = 0.521941E+00	
	C(2) = 0.343626E+02	Average error = 0.383716E+00	
	C(3) = −0.992838E+02	Maximum error = 0.126188E+01	
	C(4) = 0.151474E+02	at 0.800001	

TABLE 12
Spline Output for dT/dZ

Knots	Cubic coefficients	Knots	Cubic coefficients
XI(1) = 0.000000		XI(4) = 1.980006	
	C(1) = 0.378920E+03		C(1) = −0.656508E+02
	C(2) = −0.275747E+03		C(2) = −0.178017E+03
	C(3) = −0.123756E+03		C(3) = 0.640961E+01
	C(4) = 0.137063E+03		C(4) = −0.313802E+02
XI(2) = 0.660001		XI(5) = 2.640007	
	C(1) = 0.182424E+03		C(1) = −0.189372E+03
	C(2) = −0.259990E+03		C(2) = −0.210564E+03
	C(3) = 0.147630E+03		C(3) = −0.557244E+02
	C(4) = −0.820948E+02		C(4) = −0.756365E+01
XI(3) = 1.320005		XI(6) = 3.300008	
	C(1) = 0.515354E+02	Least-squares error = 0.174956E+01	
	C(2) = −0.172401E+03	Average error = 0.115725E+01	
	C(3) = −0.149188E+02	Maximum error = 0.360666E+01	
	C(4) = 0.107719E+02	at 1.100006	

TABLE 13
Array for Calculating Conductivity Values

Z	T	dT/dZ	d^2T/dZ^2	ρ	ϵ_H
0.0000E+00	0.1306E+04	0.3789E+03	−0.2757E+03	0.7889E−03	0.8516E+00
0.1737E+00	0.1368E+04	0.3280E+03	−0.3063E+03	0.8041E−03	0.8543E+00
0.3474E+00	0.1420E+04	0.2739E+03	−0.3121E+03	0.8171E−03	0.8564E+00
0.5211E+00	0.1463E+04	0.2210E+03	−0.2931E+03	0.8276E−03	0.8581E+00
0.6947E+00	0.1497E+04	0.1736E+03	−0.2500E+03	0.8359E−03	0.8594E+00
0.8684E+00	0.1523E+04	0.1339E+03	−0.2092E+03	0.8423E−03	0.8604E+00
0.1042E+01	0.1544E+04	0.1001E+03	−0.1831E+03	0.8473E−03	0.8611E+00
0.1216E+01	0.1559E+04	0.6943E+02	−0.1720E+03	0.8510E−03	0.8617E+00
0.1389E+01	0.1569E+04	0.3949E+02	−0.1743E+03	0.8534E−03	0.8620E+00
0.1563E+01	0.1573E+04	0.8888E+01	−0.1777E+03	0.8543E−03	0.8622E+00
0.1737E+01	0.1571E+04	−0.2214E+02	−0.1792E+03	0.8539E−03	0.8621E+00
0.1911E+01	0.1564E+04	−0.5326E+02	−0.1788E+03	0.8522E−03	0.8619E+00
0.2084E+01	0.1552E+04	−0.8417E+02	−0.1777E+03	0.8494E−03	0.8615E+00
0.2258E+01	0.1535E+04	−0.1153E+03	−0.1817E+03	0.8453E−03	0.8608E+00
0.2432E+01	0.1513E+04	−0.1476E+03	−0.1914E+03	0.8398E−03	0.8600E+00
0.2605E+01	0.1484E+04	−0.1821E+03	−0.2068E+03	0.8328E−03	0.8589E+00
0.2779E+01	0.1449E+04	−0.2197E+03	−0.2265E+03	0.8242E−03	0.8575E+00
0.2953E+01	0.1407E+04	−0.2609E+03	−0.2476E+03	0.8139E−03	0.8559E+00
0.3126E+01	0.1358E+04	−0.3058E+03	−0.2701E+03	0.8018E−03	0.8539E+00
0.3300E+01	0.1301E+04	−0.3548E+03	−0.2940E+03	0.7876E−03	0.8514E+00

λ_0 = 0.891106E+00
λ_1 = −0.157509E−03
λ_2 = −0.596657E−07
μ = 0.703611E−04

T (K)	λ (W/cm K)	μ (V/K)	T (°F)	BTU (in/hr ft² F)
0.135000E+04	0.569727E+00	0.703611E−04	0.197033E+04	0.395018E+03
0.145000E+04	0.553647E+00	0.703611E−04	0.206033E+04	0.383870E+03
0.145000E+04	0.537269E+00	0.703611E−04	0.215033E+04	0.372514E+03
0.150000E+04	0.520593E+00	0.703611E−04	0.224033E+04	0.360952E+03
0.155000E+04	0.503619E+00	0.703611E−04	0.233033E+04	0.349182E+03

Next these SPLINE equations are differentiated once and the resulting values for dT/dZ are fit to SPLINE equations with two less knots (Table 12). In turn these equations are differentiated to yield d^2T/dZ^2 values.

Next, values of T, dT/dZ, d^2T/dZ^2, ρ, and ϵ_H are calculated for selected values of Z. These values are calculated from the SPLINE programs and the least-squares coefficients for resistivity and emissivity (Tables 7 and 8). Thus a table of about 20 sets of values required for equation (3) is set up as shown in Table 13. Then this array of equations is solved for the first values, in a least-

TABLE 14
Comparison of Conductivity Values Calculated for Two Profiles

Temp. (K)	$\lambda(1)$ ($W\,cm^{-1}\,K^{-1}$)	$\lambda(2)$ ($W\,cm^{-1}\,K^{-1}$)
1350	0.5607	0.5765
1400	0.5536	0.5531
1450	0.5373	0.5333
1500	0.5206	0.5172
1550	0.5036	0.5048

squares sense, of the three unknowns, λ, $d\lambda/dT$, and u. Since λ and $d\lambda T/dT$ are related, the actual solution involves λ_0, λ_1, and λ_2 where $\lambda = \lambda_0 + \lambda_1 T + \lambda_2 T^2$. Because the temperature range over which the profile is limited to several hundred degrees, $\lambda = \lambda_0 + \lambda_1 T$ is usually an adequate representation. The thermal conductivity results for the two profiles of Fig. 10 are compared in Table 14. The maximum difference is 2% and the average difference is less than 1%.

After the conductivity values have been determined over one temperature interval, profile data at higher temperatures are obtained and the process repeated. Additional long sample data are taken in between some profile determinations to determine any changes in the emissivity and resistivity. To obtain optimum thermal conductivity data, the short sample length could be decreased as one goes to higher temperatures so that the profile data $T(Z)$, approximate a reasonably "sharp" parabola and not a "shallow" one. Typical averaged thermal conductivity results are given in Fig. 11. In this figure, the results of the Kohlrausch method up to 1200 K are combined with the multiproperty results above 1200 K.

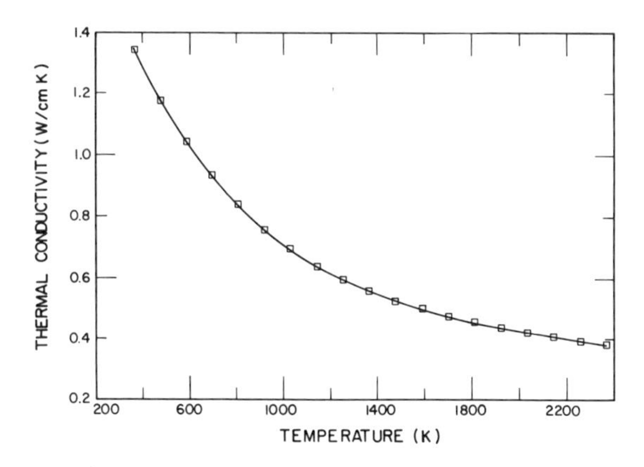

FIGURE 11. Typical thermal conductivity results (averaged values).

4. EXPERIMENTAL DIFFICULTIES AND CRITICAL EVALUATION OF DIRECT-HEATING METHODS

Direct-heating methods require good experimental techniques in regard to vacuum equipment, temperature measurements, etc., just as do other thermophysical property testing. However, there are several experimental difficulties which are either unique or at least unusually important in the case of direct-heating methods. As will be demonstrated in this section, it is very possible that the effects of errors will be to greatly magnify the uncertainty in the resulting thermal conductivity values. For example, a 1% error in emissivity may easily cause a 10% error in thermal conductivity values. In fact, in reexamining Fig. 1, the observation can be made that the agreement between the various authors was really quite good considering the possibility of large uncertainties in their results caused by the inadequate data reduction techniques which were available to them. Fortunately this problem can be overcome readily today and accuracy levels of 1%–2% can be achieved.

In this section we will divide experimental considerations into the following categories (not necessarily in order of importance):

1. Temperature measurements
2. Heat flux measurements and heat losses and emissivity control
3. Electrical resistivity measurements
4. End effects
5. Sample geometry
6. Strain control and expansion corrections
7. Data reduction
8. Thermomigration and thermionic emission

If one has a good understanding of these experimental considerations and has adequate experimental facilities, then one would be assured of success in determining the thermal conductivity of electrical conducting materials by direct heating techniques.

4.1. *Temperature Measurements*

One of the major sources of error in all thermal conductivity determinations is the inaccuracies in the measurements of temperatures and temperature gradients. Therefore the limitations associated with different types of temperature measurements along thin current-carrying samples has been investigated by Taylor *et al.*[33] Three general methods of temperature measurements applicable for direct-heating methods are (1) thermocouples, (2) optical pyrometry, and (3) the use of a physical property whose temperature dependence is known.

In order to establish the accuracy of reproducibility of each of these types of measurements, several measurements were performed concurrently. Since there was no independent way of establishing the accuracy of the various techniques for these particular applications, it was necessary to intercompare the results of different types of simultaneous measurements made under controlled conditions in which one parameter was varied at a time.

As an example, consider the determination of the accuracy achievable using thermocouples spot-welded to the surface of thin samples. The thermal conductivity determinations will be influenced by the relative rate of the heat loss to the wires compared to the rate of heat generation in the vicinity of the hot junctions, while the error in temperature measurement will depend upon the rate of heat loss down the thermocouple legs. The rate of heat loss by the thermocouple is a function of the following parameters: (1) diameter of the wire; (2) the thermal conductivities and total hemispherical emittance of the thermocouple materials; (3) the position of the thermocouple wire relative to the sample; (4) the temperature of the hot junction; and (5) the manner in which the hot junction is fabricated (contact conductance).

Other factors which influence the accuracy achievable by spot-welded thermocouples are: the Seebeck coefficients of the thermocouple materials relative to each other; to the sample material and to the emf gradient; and also the accuracy of the thermocouple calibration; the effect of strains in the thermocouple wires; the sensitivity of the measuring circuits; noise level; stability of the power supply used to heat the samples, etc.

In order to determine the effects of these parameters, one must be able to ascertain the temperature of the sample at the hot junction by an independent means. For absolute accuracies the temperature of the junction must be known precisely, but the relative effects can be determined if the reproducibility of the independent temperature-measuring device is satisfactory.

The reproducibility of temperature measurements obtained with thermocouples spot-welded to the surface and the magnitude of the error obtainable with this type of thermocouple were explored by Taylor *et al.*[33] For comparison purposes, an internal thermocouple was used in small bore thick tubing samples along with electrical resistivity and optical pyrometer. The legs of the spot-welded thermocouple were located perpendicular to the sample surface for 0.3 cm, and then were bent to run parallel to the surface for 1.3 cm to minimize the gradient in the thermocouple near the hot junction. The reproducibilities of the test thermocouple were determined by locating thermocouples along a constant temperature region of a long sample and also by replacing thermocouples at a fixed location by new thermocouples.

Two types of thermocouples were tested: (1) thermocouples in which the beads were formed and then spot-welded to the sample, and (2) thermocouples formed by spot-welding the legs individually to the sample at adjacent locations. The alignment of the legs relative to each other is very important, particularly for the second type of thermocouple. Typical voltage gradients were 0.1 V per 2.5 cm. Since this is 40 μV per 0.001 cm, a misalignment of 0.002 cm generates 80 μV which, dependent upon the polarity, adds to or subtracts from the thermocouple output. This corresponds to about 8°C for Pt–Pt/10 Rh thermocouples. Thus proper reversing of the polarity and averaging of the results is much more important than in the case where current is not flowing through the specimen. It was possible to align the individual legs so that this effect for the "nonbeaded" thermocouple approached that of the "beaded" thermocouple provided the welding was done under controlled conditions at a bench but not while the sample was mounted in the apparatus. From the standpoint of ease of operation, the beaded thermocouples were superior. The reproducibility was worse for the nonbeaded thermocouple than for the beaded thermocouple (± 1°C compared to ± 0.5°C), but the accuracy was usually higher. Figure 12 shows typical results for 0.005 and 0.010 cm Pt–Pt/10 Rh nonbeaded thermocouples spot-welded to type 304 stainless steel specimens. The presence of a temperature gradient in the sample at the hot junction did not significantly influence the accuracy of the spot-welded thermocouples. Since the absolute accuracy of the internal thermocouple was within 1°C during these experiments, the data of Fig. 12 essentially represent the absolute accuracies achieved with the spot-welded thermocouples. Thus, at a given location, if the sample temperature were raised from 1000 to 1100 K, a 0.010-cm spot-welded thermocouple would record this as a change from 995 to 1093 K. For this example, the spot-welded thermocouple would be 20% off in measuring the change of temperature. The magnitude of this error is a function of the absolute temperature and the wire diameter. Of course if data similar to those shown in Fig. 12 were available for the particular sample–thermocouple combination to be used in an experiment, corrections could be made for most of this error.

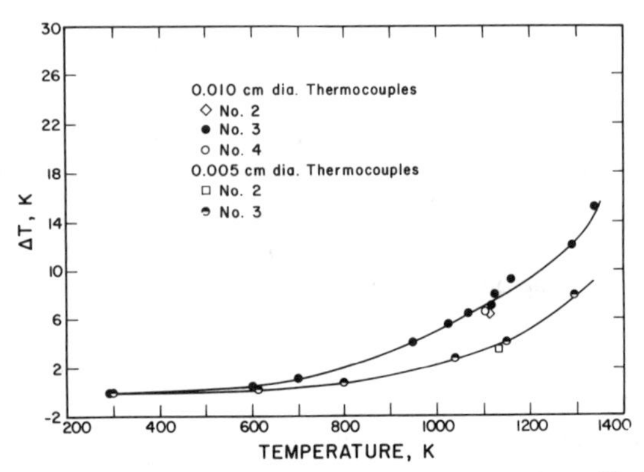

FIGURE 12. Temperature measurement error of thermocouple welded to surface of a directly heated rod.

A knowledge of temperature gradients is required for thermal conductivity determinations. In order to determine temperature gradients using spot-welded thermocouples, the reproducibility obtained from thermocouple to thermocouple must be considered. While this error depends upon the techniques employed in spot-welding thermocouples to the surface, it may be of the same order as that indicated in the preceding example (if small temperature differences are involved, the error caused by nonreproducibility will be greater than 2%). The nonreproducibility also limits the wire diameter. Judging from Fig. 12, the use of 0.002- or 0.001-cm wires would significantly reduce the absolute error obtainable with spot-welded thermocouples. However, it was found (with the techniques employed) that the additional nonreproducibilities obtained with the smaller diameter thermocouples negated the increased accuracies obtained.

The internal traversing thermocouple proved to be very accurate for measuring temperatures along the central two-thirds of stainless steel samples whose i.d. (0.12 cm) was only slightly larger than the o.d. of the alumina thermocouple tubing (0.10 cm) employed. It is believed that an accuracy within $\pm 0.2^\circ$C was achieved. This belief is based upon a number of tests. One of the more important tests was a demonstration that the measured temperature profiles were not affected by the presence or absence of alumina tubing "riding" on top of the bead. The test showed that heat conduction to or from the bead via the ceramic tubing was negligible. Other tests involved comparison of results obtained by other measuring techniques including optical pyrometry and electrical resistivity. However, in the case of a larger-bore (0.23 cm) molybdenum sample, the internal traversing thermocouple gave results which were too low below a point near the midplane and too high above this point unless the temperature gradient in the sample was small. The pyrometer results were obtained from the brightness temperatures using a spectral emissivity of 0.44 (determined experimentally). Temperatures measured by the thermocouples were less than those measured by the optical pyrometer below a position near the center, but were greater than the pyrometer results above this position. At a position on the left side of the profile, the thermocouple result was 18°C less than the optical pyrometer while at a similar position on the right side of the profile, the thermocouple was 9°C higher. When power was supplied to end heaters so that the temperature gradients in this part of the sample were considerably reduced, the thermocouple and optical pyrometer results were within 1°C of each other at these locations. Thus, the internal traversing thermocouple is very reliable when the temperature gradients are small in the central regions (samples with low thermal conductivities), but not when large thermal gradients extend to the central portion (samples with high conductivities). When using internal thermocouples traversing through large thermal gradients, the accuracy can be significantly improved by causing thermocouple tubing to "ride" on top of the bead. The presence of the additional tubing tends to counteract the heat loss or gain from the tubing supporting the thermocouple.

External traversing thermocouples which were independently heated to match the sample temperature were also tested. Two types of heated thermocouples were tried. In one type the thermocouple wire was heated by passing an ac current through it while the dc potential was measured. The other type consists of a thermocouple heated by a resistance coil embedded in ceramic on the thermocouple tubing. Radiation shields were located around the heater to minimize heat exchange with the sample.

For both types of heated thermocouples, the procedure was to raise the thermocouple temperature to about the sample temperature. This was done while the thermocouple bead was located very near the sample. Otherwise the change in temperature of the thermocouple upon being brought closer to the sample was too large to be allowed for conveniently. The thermocouple position could be observed through the cathetometer and its motion could be closely and smoothly controlled by means of the rotary feed-through and gear train. The thermocouple was held within one bead diameter (about 0.02 cm) from the sample surface during its temperature adjustments. Contact with the sample surface was ascertained by observing the thermocouple wire deflect. The change in thermocouple output was observed while contact was being made. The thermocouple was then withdrawn one bead diameter from the surface and its temperature adjusted in accordance with the direction and magnitude of the output change. This procedure was repeated until the thermocouple output change upon contact was negligible.

Because of the small mass of the thermocouple wire, the thermal inertia of the ac heated thermocouple was very small. Relatively little power was required to heat the thermocouple and the power input had to be controlled quite finely. The frequency of the ac current used to heat the thermocouple was varied to match impedances with the power amplifier used. The dc signal was filtered using several "π" networks. Although the dc resistance between the sample and thermocouple was several megohms when the two were not in contact, and ac short developed between the regulated dc power supply and the power amplifier causing ac current to flow from the thermocouple through the sample, and these tests were discontinued.

The second type of heated external thermocouple proved to be erratic. Part of the problem was traced to the changes which occurred in the heat conduction from the heated thermocouple tubing and the thermocouple wires as the wires were displayed during contact of the bead with the sample surface. As contact was made, the wires would shift within the tubing and this would affect the bead temperature. Cementing the wires in place reduced this effect considerably. However, the bead would still be pushed closer to the thermocouple heater during contact. Since the thermocouple heater was several hundred degrees hotter than the thermocouple bead, the bead temperature was affected. In addition, the presence of the heated thermocouple was found to raise the local temperature of the sample by about 3°C at 1000°C. Agreement between this type of heated external traversing thermocouple and the internal thermocouple was sometimes very close – within 0.3°C.

The development of the heated external thermocouples was not carried very far during the course of this program. They were not very convenient and the thermocouples tended to weld to the sample at higher temperatures.

Temperature measurements were also made using several types of optical pyrometers. The Automatic Optical Pyrometers have proven to be very stable. In fact, the short-term stability of the pyrometer has not been accurately determined since it is better than that of the power supplies (± 0.02% regulation) used to heat the samples and strip lamps. The automatic pyrometer has the capability of measuring brightness temperatures (or blackbody temperatures) to a sufficient accuracy and reproducibility to permit accurate thermal conductivity determinations. The limitation lies in the uncertainties involved in reducing brightness temperatures to absolute temperatures. This conversion, which involves a knowledge of the spectral emittance at the effective wavelength of this instrument, is important for two reasons: (1) it determines the effective temperature of the property (thermal conductivity, electrical resistivity, emittance, etc.) determination, and (2) the values for gradients (and therefore the value of the thermal conductivity) are different when determined from absolute temperatures rather than from brightness temperatures. This latter statement is true even when the spectral emittance is constant over the same surface. The dependence of the gradient on the spectral emittance is illustrated in Table 15.

The difference in brightness temperatures of 100 K (between 1400 and 1300 K) corresponds to an actual difference of 123 K for a spectral emittance of 0.2. This difference in actual temperatures depends upon the magnitude of the brightness temperature and becomes 129 K between brightness temperatures of 1700 and 1600 K. Thus brightness temperatures must be converted to absolute temperatures in order to obtain accurate thermal conductivity values.

TABLE 15
Effect of Spectral Emittance of the Determination of Temperature Gradient[a]

	ϵ_λ			
	0.2	0.3	0.4	1.0
Temperature (A)	1559	1515	1486	1400
Temperature (B)	1436	1399	1374	1300
ΔT^b	123	116	112	100

[a] Brightness temperature of A = 1400 K; of B = 1300 K.
[b] Temperature (A) minus temperature (B).

4.2. Heat Flux Measurements, Heat Losses, and Emissivity Control

Heat flux measurements are relatively easy to make in direct-heating experiments since the energy is simply VI. This energy is generated directly in the test section and the problem of transferring the measured heat flux to the sample does not exist. However, heat loss considerations at high temperatures are much more severe than for other types of high-temperature thermal conductivity determinations.

This situation arises because heat losses at high temperature are not a relatively minor correction but instead are very large and are an integral part of the governing equation [see equation (3)]. Thus the values of the hemispherical total emissivity play an extremely important role in the long thin rod cases (see section on data reduction). Unfortunately the total emissivity is a surface property and is subject to marked changes depending upon time–temperature–environment history.

The effect of a change in the total emittance was dramatically demonstrated during one experiment in which stainless steel was heated to 1170 K (Taylor *et al.*[(17)]). A small leak developed in the vacuum system during this experiment, causing the pressure in the bell jar to rise from 5×10^{-7} to 2×10^{-6} Torr. The experiment was terminated and the sample was observed to have become somewhat discolored except near the water-cooled electrodes. The leak was repaired and the sample was reheated. When a center temperature of 1320 K was obtained, the temperatures nearer the ends were about 1420 K; i.e., a large inverse temperature gradient was noted. An attempt to use data obtained under these conditions in equations derived on the basis of constant emittance would yield a negative value for the thermal conductivity. Upon continued heating above 127 K at pressures in the mid-10^{-7}-Torr range, the surface coating was removed and the temperature profile approached its usual shape. Since the total hemispherical emittance of the oxidized sample was more than twice its usual value, the power supplied to the oxidized sample exceeded that required to melt a nonoxidized sample. Stainless steel is known to perform badly when heated in vacuum. However, other materials exhibit these same tendencies and changes in spectral and total hemispherical emittances are a distinct possibility.

Several techniques are used to precondition the sample's surface so that reproducible emissivity (both total and spectral) are obtained. In the case of metallic samples which have volatile oxides (such as tantalum and molybdenum), the sample is preheated under high vacuum to a temperature at which the surface coating is removed before each determination. In the case of graphitic materials, it is often possible to obtain stable surfaces by slightly oxidizing the surface before starting the initial measurements. In any event, a stable and reproducible emissivity is a necessary condition for those methods based on equation (3).

4.3. Electrical Resistivity Measurements

Electrical resistivity measurements at low temperatures as a function of true temperature or at high temperatures as a function of brightness temperature are rather straightforward. However, it may be desirable to measure resistivity as a function of true temperature at high temperatures and this presents some challenges. Such determinations are used, for example, to convert brightness temperatures to true temperatures with the electrical resistivity serving as the transfer medium. The measurement of resistivity as a function of true temperature involves the use of an external heater surrounding the sample. The simplest technique is to control the power to the sample and heater separately until the sample "disappears." This is similar to the manual disappearing filament pyrometer in which the power to the filament is raised until the color of the filament matches that of the sample. Under these conditions the furnace enclosure acts as blackbody and the brightness temperature measured by the pyrometer (corrected for window absorption) becomes equal to the true temperature.

Special care is required for the voltage probes when the sample is enclosed by a heater. The wires from the probes must not touch the heater and in fact they must remain in relatively fixed positions in regard to the sample and heater even though all three expand as they are heated. It is possible to get large signal amplifications since the two voltage probes and the heater can act as a giant triode. The presence of this amplification becomes obvious upon sample current reversal. When it occurs, the voltage probe setup must be changed.

4.4. End Effects

Direct heating methods are subject to problems arising from end effects. These effects arise from two sources: (1) contact resistance and (2) Peltier heating and cooling. It is desired that the temperature profiles be nearly symmetrical and it is assumed that end effects are negligible as they are not included in the governing equations. If there is a poor electrical contact between the sample and one of the electrodes additional heat is generated and the temperature of this end may be increased substantially. While measurements are made in the central portion of the sample, there may still be end effects caused by Peltier heating of one end of the sample and cooling at the other end. This heating and cooling is reversed upon reversing the direction of current flow and thus is distinguishable from the poor contact case. The magnitude of the Peltier effect depends upon the magnitude of the current flow and upon the nature of the sample and the electrodes. It can be minimized by matching the Seebeck values for the electrode material and sample. For example one could use electrode clamps made of graphite for graphitic materials. However, if one uses a higher-resistivity material for the clamps, one has to make sure that the geometry of the clamps is sufficient to minimize Joulean heating.

4.5. Sample Geometry

The diameter and length-to-diameter ratio of the sample should be optimized for the particular technique and temperature range of interest. A too thick rod requires excessive power and may violate long thin rod assumptions. It is difficult to measure temperature gradients in very thin rods and the optimum sample length may be too short to be practical. It is important that temperature profiles be reasonably "sharp" (see section on data reduction), and the same sample geometry may not be suitable over an extended temperature range. No firm numbers for optimum diameter to length can be given as these depend upon the properties of the sample. For example for the Kohlrausch method, samples with larger thermal conductivity values should be longer and thinner than samples with lower conductivity values. Figure 13 shows how the temperature profile flattens out as the sample length increases. Excellent conductivity values could be calculated for the profiles marked 10, 12, and 16 cm but poor values

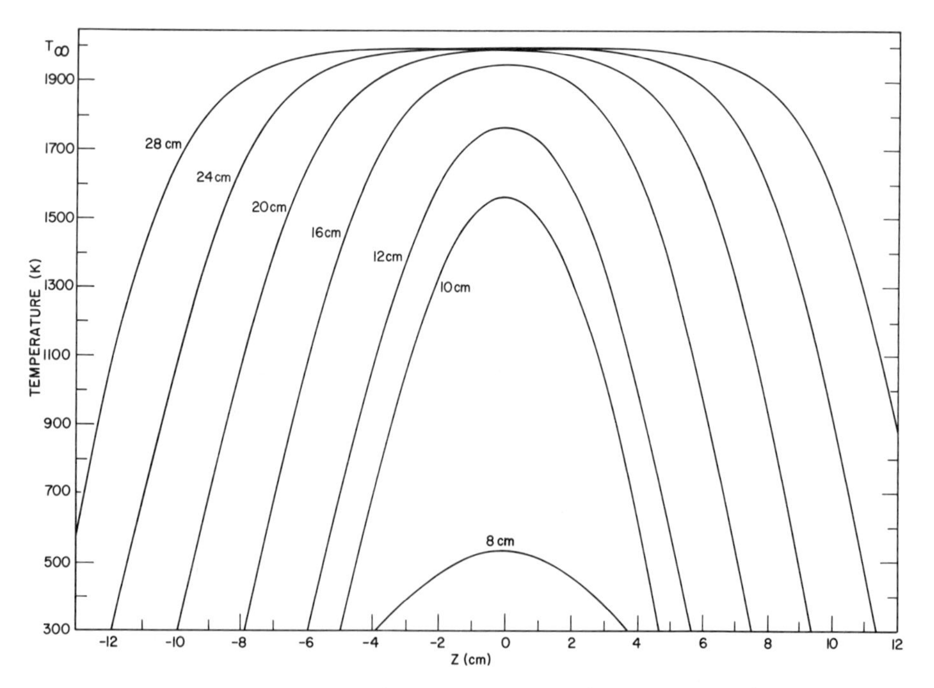

FIGURE 13. Temperature profiles for different length tungsten samples (0.3175 cm diam, 173.5 A).

would be obtained for the other profiles. Thus the effective sample length is changed for different temperature ranges in the TPRL apparatus by moving the electrode clamps.

Available sample lengths may not always be sufficient to satisfy the "infinitely" long configuration ($d^2T/dZ^2 = 0$ over test section). It has been found to be possible to join several sections together to obtain a long sample. Joining together sections without appreciable contact resistance requires good machining capabilities.

Hollow cylindrical samples have been measured successfully when the inside diameter is relatively small. Larger inside diameters can lead to an appreciable radiation heat transfer along the interior of the sample and this will lead to sizable errors.

The presence of perpendicular holes in the sample not only disturbs heat flow, they may also cause excessive localized heating due to increased power generation resulting from a higher resistance of the cross section. Thus thermocouple holes or sight holes for optical pyrometer measurements should be avoided.

4.6. *Strain Control and Expansion Corrections*

Direct heating samples are usually thin rods, tubes, or strips and have relatively heavy electrodes attached at each end. Provisions for thermal expansion and contraction of the samples as they are heated and cooled should be provided. Otherwise the samples will flex. If the sample is suspended vertically and the lower electrode is suspended freely from the sample, the electrode mass may cause the sample to extrude. Thus the electrodes should be free to move and be counterbalanced.

It should be noted that thermal expansion corrections to thermal conductivity values obtained for the long thin rod methods using noncontact temperature sensing (i.e., optical pyrometry) involves a $(\Delta L/L_0)^3$ term (Taylor[34]) even though the units for thermal conductivity are $\mathrm{W\,cm^{-1}\,K^{-1}}$. Since expansions of 2% are common at high temperatures, the expansion correction is often 8% at these temperatures. At the same time the expansion correction for total emissivity involves a $(\Delta L/L_0)^{-2}$ term, even though emissivity is a dimensionless quantity. However, the correction term for electrical resistivity involves $\Delta L/L_0$, which corresponds to the units Ω cm.

4.7. *Data Reduction*

Large discrepancies in thermal conductivity values obtained by various researchers using the techniques shown in Fig. 2 have often been noted. Extensive evaluation was conducted by Taylor and Hartge[35] and the reasons for the large errors were delineated. The major cause of the errors was found to lie in the fact that most researchers applied mathematical approximations to equation (3) which were valid only in the central portion of a temperature profile. In this region the magnitudes of the third and fourth terms of equation (3) are nearly equal and the second and fifth terms are very small, so that the magnitude of the term containing λ is essentially obtained as the difference between two large quantities. For example consider the profiles marked "12 cm" and "16 cm" in Fig. 13. At distances of 0.6, 1.2, and 2.4 cm from the center, the magnitudes of the various terms entering into equation (3) for these profiles are given in Table 16. Note that at 1.2 cm from the center of the 16-cm-long sample, the term containing thermal conductivity is less than 13% of the radiation term and less than 11% of the heat generation term. At smaller Z's, the percentage is even less and even at 2.4 cm, the magnitude of the $\lambda d^2T/dZ^2$ term is less than 30% of the radiation term and 21% of the heat generation term. The situation is considerably improved for the 12 cm profile, but even in this case, the conduction term is less than 40% of the heat generation term. Consequently a small error in ρ or ϵ_H causes large errors in the computed value for λ. When one considers that ϵ_H depends upon the immediate past history of the sample including environmental conditions and that its value can be changed by appreciable percentages, it becomes evident that precise knowledge and control of this quantity must be obtained. Thus Taylor *et al.*[36] and Taylor and Hartge[35] showed the necessity of measuring ρ and ϵ_H on the same sample in the same apparatus concurrently with λ determinations. Furthermore, the temperature dependencies of these quantities must be taken into account. Mathematical techniques such as the SPLINE and numerical techniques were developed which incorporated these temperature dependencies as well as utilizing a larger fraction of the temperature profile in order to include regions in which the conduction term is relatively large (Taylor *et al.*,[17] Taylor and Hartge[35]). The results yield the thermal conductivity as a function of temperature over intervals of several hundred degrees for each profile. Interestingly, these mathematical techniques for computing λ can be performed on modern computers at a cost of less than 50¢ per profile.

The shapes of T, dT/dZ, and dT^2/dZ^2 are shown in Fig. 14. Note that the values for dT/dZ go from large positive numbers, through zero, and to large negative numbers. The values for dT^2/dZ^2 are negative except at the very ends of the sample. It has been found to be important to include data beyond the minimums (Fig. 14) in d^2T/dZ^2. Otherwise the computed values of dT/dZ and d^2T/dZ^2 are subject to large uncertainties near the minimum in d^2T/dZ^2 and this adversely affects the computed conductivity values.

Although the various methods listed under the long thin rod approximation in Fig. 2 can be considered to be outmoded, a summary of the results of a critical evaluation (Taylor and Hartge[35]) of these methods is interesting

TABLE 16
Magnitude of Terms in Equation (3) for 12- and 16-cm Long Tungsten Sample (173.5 A, 0.3175 cm diam)

Length (cm)	Z (cm)	T (K)	dT/dZ ($K\ cm^{-1}$)	d^2T/dZ^2 ($K\ cm^{-2}$)	$\lambda d^2T/Z^2$ ($W\ cm^{-3}$)	$I^2\rho/A^2$ ($W\ cm^{-3}$)	Rad.[a] ($W\ cm^{-3}$)	Thom.[b] ($W\ cm^{-3}$)	$(T')^2 d\lambda/dT$[c] ($W\ cm^{-3}$)
12	0.6	1753.6	−50.03	−86.08	−90.533	230.383	141.075	−1.645	−0.400
12	1.2	1707.6	−104.29	−95.32	−100.974	222.976	123.654	−3.428	−1.777
12	2.4	1508.8	−230.46	−112.09	−122.561	191.488	66.981	−7.577	−9.523
16	0.6	1942.6	−15.30	−26.73	−23.314	261.626	234.364	−0.503	−0.070
16	1.2	1913.2	−32.83	−32.23	−28.396	259.373	218.925	−1.079	−0.323
16	2.4	1861.1	−85.01	−58.24	−52.122	248.202	189.490	−2.795	−2.168

[a] $\text{Rad} = P_{\epsilon H\sigma}(T^4 - T^4)/A$.
[b] $\text{Thom} = \mu[d\lambda/dT(T')^2]$.
[c] $T' = dT/dZ$.

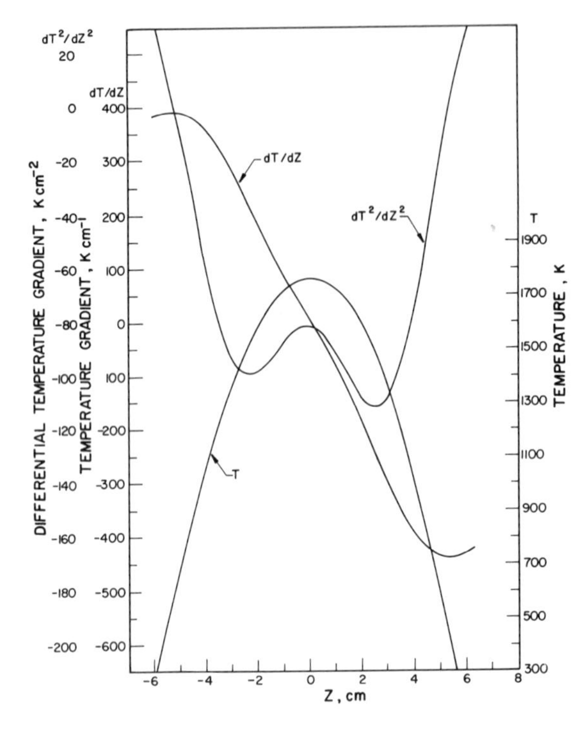

FIGURE 14. Plots of T, dT/dZ, and d^2T/dZ^2 for a typical profile.

because it reveals why data reported in the literature were often so discordant. In the case of the Lebedev and Krishnan and Jain methods, it was found that acceptable accuracy could be obtained in the case of ideal profiles and that measurement errors were not unduly magnified in this case. However, for non-optimum profiles, the errors became unacceptably large even if all quantities were measured precisely and, in addition, measurement errors were greatly magnified. The profile limitations were clearly demonstrated and it seems likely that these limitations were often exceeded.

The methods of Worthing[3] and of Rudkin, Parker, and Jenkins[9] are basically similar to the method employed in the shooting technique except that in the shooting technique the Thomson effect is included, and computers do the numerical integration. Thus the Worthing and the Rudkin, Parker, and Jenkins method should yield the same λ values (for the precise measurement case) that are obtained by the shooting technique using $d\lambda/dT = \mu = 0$. The errors introduced by ignoring the Thomson effect and the temperature dependency of λ depend strongly on the portion of the profile used in the calculations (see Table 15). Similar remarks can be made for the other methods. The results of the evaluation have shown that under the conditions of ideal profile shape and no measurement errors, the various direct heating methods are capable of yielding thermal conductivity values which are accurate within 5%.

Another significant result is the importance of self-consistency. In order to obtain accurate conductivity values, it is essential that all parameters that enter into the governing equation be measured on the same sample using the same measuring equipment. When this procedure is followed, very gross measurement errors cause relatively small errors in the computed conductivity values. This remarkable result is the result of the "Principle of Superposition," even though this is not obvious from equation (3). Experimentally, ρ and ϵ_H are obtained from the long-sample case in which the sample temperature is uniform. Then a temperature gradient is superimposed on the uniform temperature case and the thermal conductivity is computed from the differences in the responses of the measuring equipment between the long- and short-sample cases. Thus, when one makes a consistent error which applies to both the long and short-sample cases, the effects on the calculated thermal conductivity values are minimized.

One final point concerning data reduction should be made. Conductivity values are calculated over a few hundred degrees interval for each profile. Power input and effective sample lengths are changed and new profiles generated over an overlapping temperature range. The conductivity values calculated at the same temperatures for the various profiles are averaged together to yield a composite conductivity curve. This means that conductivity values calculated near the maximum of one profile are averaged with conductivity values calculated along the steep sides of other profiles. Since the values of dT/dZ and especially d^2T/dZ^2 are grossly different at the same temperatures, these quantities and the values for I are grossly different in equation (3), even though the values of temperature, resistivity, and emissivity are the same. Thus comparing conductivity values calculated from a region near the maximum of one profile with those calculated from the slope of another profile is equivalent to changing experimental conditions by almost two orders of magnitude – yet these values are usually in agreement within 2%. This may be considered to be a very stringent test of the validity of the experimental results.

4.8. *Thermomigration and Thermionic Effects*

There are several effects which could influence thermal conductivity determinations performed in the presence of a current flow. At high temperatures it is possible for atoms of the specimen material or impurity atoms to migrate under the influence of the electric potential gradient (electromigration) or under the influence of the temperature gradient (Soret effect). These processes involve energy flow and appropriate terms would have to be incorporated in equation (3) if these processes were significant. A discussion of these effects are given by

Flynn and O'Hagan.[37] To date, measurements of thermal conductivity in the presence of electrical flow and in its absence have yielded the same results.

Samples heated to very high temperatures may "boil off" ions or electrons. This process could result in a substantial energy loss from the samples. Also, it can certainly affect resistivity determinations (Loup *et al.*[38]), especially with high-resistance materials, but even with metals. However, significant heat losses due to thermionic emission have not been observed in direct heating methods. The reason for this may be the formation of a "space charge" in the vicinity of the sample surface and this space charge repels additional emission. If a charged collector were to be placed in close proximity to the sample's surface, then the ions or electrons could continue to flow and one would have an experiment to measure Richardson's coefficient.

5. SUMMARY

Direct heating methods can be used from cryogenic temperatures up to temperatures approaching or even exceeding the melting points of electrical conductors. The methods are capable of yielding state-of-the-art accuracy (2%–3%) at very high temperatures ($T > 2000$ K), but rather sophisticated data reduction procedures must be used. Much of the high-temperature work published in the literature did not satisfy certain limitations imposed by the associated mathematics; consequently, there are large errors associated with these papers. Data obtained at lower temperatures ($T < 1000$ K) have generally been at least as good as that obtained by other techniques and data reduction has not been a problem in this temperature range. The methods are quick, use small samples and are readily adaptable to modern data acquisition. In addition they often have the feature of permitting multiple property determinations on the same specimen. However, they are limited to reasonably good electrically conducting materials.

NOTATION

A	Cross-sectional area
E	Electrical potential
FT	Temperature in Fahrenheit
I	Current
I_∞	Current through a short sample
I_s	Current through a long sample
J	Current density
L	Distance between positions
P	Perimeter
r	Radius

T	Temperature
T_0	Temperature of the surrounding vacuum enclosure
TT	True temperature in K
T_∞	Temperature at the center of an infinitely long rod
V	Average voltage in V
Z	Axial distance
ϵ_H	Total hemispherical emissivity
ϵ_λ	Spectral normal emissivity
λ	Thermal conductivity
μ	Thomson coefficient
ρ	Electrical resistivity
σ	Stefan–Boltzmann constant

REFERENCES

1. F. Kohlrausch, *Sitz. Berlin. Akad.* **38**, 711–718 (1899).
2. C.Y. Ho, R.W. Powell, and P.E. Liley, "Thermal Conductivity of Selected Materials," Part 2, NSRDS-NBS-16, U.S. Government Printing Office, Washington, D.C. 20402 (1968).
3. A.G. Worthing, *Phys. Rev.* **4**, 535–543 (1914).
4. C. Zwikker, "Physical Properties of Tungsten at High Temperatures," Doctoral thesis, Amsterdam; *Arch. Nederland. Sci.* **9**, 207–339 (1925).
5. R.J. Jenkins, W.J. Parker, and C.P. Butler, "A Method of Determining the Thermal Properties of Metals above 1000 K Using Electrically Heated Wires," USNPDL-TR-348, 1–24 (1959) [AD 226 896].
6. R.D. Allen, L.F. Glasier, and P.D. Jordan, *J. Appl. Phys.* **31**, 1382–1387 (1960).
7. N. Simonova, and L.P. Filippov, *Zh. Prikl. Mekhan. i Tekhn. Fiz.*, (1), 112-12; English transl., NASA-TT-F-9553 (1965).
8. W.E. Forsythe, and A.G. Worthing, *Astrophys. J.* **61**, 146–185 (1925).
9. R.L. Rudkin, W.J. Parker, and R.J. Jenkins in *Temperature – Its Measurement and Control in Science and Industry* (C.M. Herzfeld and A.I. Dahl, eds.), Vol. 3(2), pp. 523–534, Reinhold, New York (1962).
10. R.H. Osborn, *J. Opt. Soc. Am.* **31**, 428–432 (1941).
11. V.S. Gumenyuk, and V.V. Lebedev, *Fiz. Metal. Metalloved.* **11**, 23–33; English transl., *Phys. Metals Metallog (USSR)* **11**, 30–35 (1961).
12. V.S. Gumenyuk, V.E. Ivanov, and V.V. Lebedev, *Pribory Tekhn. Eksperim.* **7**, 185–189; English transl., *Instr. Exptl. Tech. (USSR)* (1), 188–192 (1961).
13. E.S. Platunov, and V.B. Fedorov, *Teplofiz. Vys. Temp.* **2**, 628–633; English transl., *High Temp.* **2**, 568–572 (1964).
14. B.E. Neimark, and L.K. Voronin, *Teplofiz. Vys. Temp.* **6**, 1044–1051; English transl., *High Temp.* **6**, 999–1010 (1968).
15. V.Ya. Chekhovskoi, and V.A. Vertogradskii, in *Ninth Conference on Thermal Conductivity* (H.R. Shanks, ed.), CONF-691002-Physics (TID-4500), U.S. Atomic Energy Commission, pp. 300–306 (1970).
16. R.W. Powell, D.P. DeWitt, and M. Nalbantyan, "The Precise Determination of Thermal Conductivity and Electrical Resistivity of Solids at High Temperatures by Direct Electrical Heating Methods," AFML-TR-67-241, Air Force Systems Command, Wright-Patterson Air Force Base, Ohio (1967).

17. R.E. Taylor, F.E. Davis, and R.W. Powell, *High Temp. High Pressures* **1,** 663–673 (1969).
18. R.W. Powell, C.Y. Ho, and P.E. Liley, "Thermal Conductivity of Selected Materials," NSRDS-NBS-8, US Government Printing Office, Washington, DC 20402 (1966).
19. D.R. Flynn, in *Thermal Conductivity,* Vol. 1 (R.P. Tye, ed.), Academic Press, London, pp. 241–300 (1969).
20. M.F. Angell, *Phys. Rev.* **33,** 421–432 (1911).
21. R.W. Powell, and F.H. Schofield, *Proc. Phys. Soc. (London)* **51,** 152–172 (1939).
22. V.E. Mikryukov, in *Teploprovodnost i Elektroprovodnost Metallov i Splavov* (Thermal Conductivity and Electrical Conductivity of Metals and Alloys), Metallurgizdat, Moscow, pp. 93–117 (1959).
23. H.L. Callendar, "Conduction of Heat," in *Encyclopaedia Britannica,* 11th edition (1911).
24. K.S. Krishnan, and S.C. Jain, *Brit. J. Appl. Phys.* **5,** 426–430 (1954).
25. V.V. Lebedev, *Fiz. Metal. Metalloved.* **10,** 187–190; English transl., *Phys. Metals Metallog. (USSR)* 31–34 (1954).
26. I.S. Sokolnikoff, and R.M. Redheffer, *Mathematics of Physics and Modern Engineering,* McGraw-Hill, New York, p. 505 (1958).
27. K.H. Bode, *Allgem. Warmtech.* **10,** 110–119, 125–142 (1979).
28. H.M. James, *High Temp. High Pressures* **11,** 669–681 (1979).
29. W. Jaeger, and H. Diesselhorst, *Wiss. Abh. Phys.-Tech. Reichsanst.* **3,** 269 (1900).
30. M.D. O'Day, *Phys. Rev.* **23,** 245 (1924).
31. R. Scott, AERE M/R 2526, UKAEA Research Group (1958).
32. W. Johnson, and W. Watt, in *Special Ceramics 1962* (P. Popper, ed.), Academic Press, New York, pp. 237–259 (1963).
33. R.E. Taylor, R.W. Powell, M. Nalbantyan, and F. Davis, "Evaluation of Direct Electrical Heating Methods for the Determination of Thermal Conductivity at Elevated Temperatures," AFML-TR-68-227 (1968).
34. R.E. Taylor, "On Correcting Thermophysical Property Data for Thermal Expansion Effects," in *Thermal Conductivity, 15* (V.V. Mirkovich, ed.), Plenum Press, New York, pp. 177–186 (1978).
35. R.E. Taylor, and L.H. Hartge, Jr., *High Temp. High Pressures* **2,** 641–650 (1970).
36. R.E. Taylor, F.E. Davis, R.W. Powell, and W.D. Kimbrough, in *Ninth Conference on Thermal Conductivity* (H.R. Shanks, ed.), CONF-691002-Physics (TID-4500), US Atomic Energy Commission, pp. 601–610 (1970).
37. D.R. Flynn, and M.E. O'Hagan, *J. Res. Natl. Bur. Stand.* **71C,** 255 (1967).
38. J.-P. Loup, N. Jonkiere, and A.M. Anthony, *High Temp. High Pressures* **2,** 75–88 (1970).
39. F. Kohlrausch, "On Thermoelectricity, Heat and Electrical Conduction," *Göttingen Nachr.,* Feb. 7 (1874).
40. R. Holm, and R. Stormer, "Measurement of the Thermal Conductivity of a Platinum Probe in the Temperature Range 19–1020°C," *Wiss. Veroff. Siemens Konzern* **9**(2), 312–322 (1930).
41. L.P. Filippov, and Yu.N. Simonova, "Measurement of Thermal Conductivity of Metals at High Temperatures," *High Temp. (USSR)* **2**(2), 165–168 (1964).
42. E.S. Platunov, "Measurement of Heat Capacity and Heat Conductivity of Rod Subjected to Monotonic Heating and Cooling," *High Temp. (USSR)* **2**(3), 346–350 (1964).

43. R.E. Taylor, "Thermal Properties of Tungsten SRM's 730 and 799," *J. Heat Transfer* **100**(2), 330–333 (1978).
44. R.E. Taylor, "Determination of Thermophysical Properties by Direct Electric Heating," *High Temp. High Pressures* **13**, 9–22 (1981).

4

Guarded Hot Plate Method for Thermal Conductivity Measurements

S. KLARSFELD

1. INTRODUCTION

At the present time, the so-called "guarded hot plate" method is considered the more representative measurement method of thermal conductivity for the poorly conductive materials ($\lambda < 1$ W/m K). It is a reference measurement method, performed in thermal stationary conditions, standardized in certain countries for many years. The guarded hot plate apparatus are used in priority for construction materials and especially for thermal insulants in building applications (these materials are in most cases porous materials used at around ambient temperature).

The numerous theoretical and experimental research works undertaken in the last 25 years have made possible significant metrological improvements in the control of this measurement method. Its field of application has extended to various applications towards low temperatures (in particular Cyrogenics) as well as high (600°C) and very high (1200°C) temperatures. The energy crisis has contributed to a recrudescence of interest among research workers and users of insulating materials for the development of new and more sophisticated "guarded hot plate" installations. The decrease of the measuring time, the increase at the samples dimensions, the continuous improvement of measuring probes and controlling units, the automatic monitoring of the apparatus based on studies involving elaborated computations in order to predict and optimize the working conditions have been the object of many research efforts. Indeed, many studies have been devoted to new insulants, new applications and to a

S. KLARSFELD • Isover Saint-Gobain, Centre de Recherches Industrielles de Rantigny, BP. 19, 60290 Rantigny, France.

better understanding of the heat transfer mechanisms. A very detailed international standard is under elaboration at ISO/TC 163.[14]

Despite all this progress, the uncertainties in the measurement results of guarded hot plate apparatus remain high. Results lying within ± 2% at ambient temperature and within ± 5%–10% at high temperature can be considered stationary. One can therefore understand the present interest in reference materials and the different attempts to organize interlaboratory comparative measurements with these materials in order to explain and attempt to reduce the differences between the results obtained.

2. PRINCIPLE OF THE METHOD

The quantitative relationship on which this measurement method is based is Fourier's law:

$$\boldsymbol{\phi} = -\lambda \operatorname{grad} \mathbf{T} \tag{1}$$

It states, in each point $P(x, y, z)$ of an isotropic medium, the proportionality between the heat flux density $\boldsymbol{\phi}$ and the thermal gradient, grad $\mathbf{T}$. The proportionality constant is the thermal conductivity λ of the medium at the point P under consideration; it depends on the temperature of the medium at this point. If the medium is homogeneous (that is, the natural and physical properties are constant in each point), equation (1), relative to an elementary volume, can be extended – with given boundary conditions to be defined – to a finite volume allowing the effective measurement of $\boldsymbol{\phi}$ and grad $\mathbf{T}$ from which λ can be computed.

For example, we can consider the case of a medium limited by two isothermal plane, parallel, and infinite surfaces (the cases of the "infinite wall"). The components of the vector grad $\mathbf{T}$ will be:

$$\frac{\partial T}{\partial x}\mathbf{i} = \frac{\partial T}{\partial x}\mathbf{j} = 0, \qquad \frac{\partial T}{\partial z}\mathbf{k} \tag{2}$$

If we relate to this configuration the following boundary conditions (Fig. 1):

$$\begin{aligned} z_1 &= 0, & T(0) &= T_1 \\ z_2 &= e, & T(e) &= T_2 \neq T_1 \\ \frac{\partial T_1}{\partial t} &= \frac{\partial T_2}{\partial t} = 0 \end{aligned} \tag{3}$$

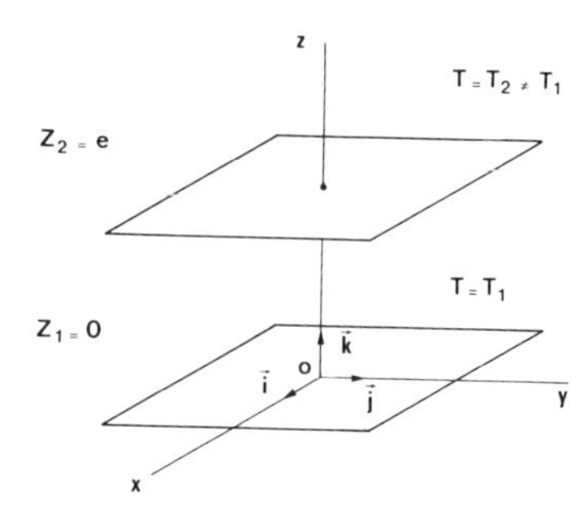

FIGURE 1. Infinite wall configuration.

assuming that $\Delta T = T_2 - T_1$ is small, which allows us to consider that $\lambda(T)$ is constant, we can write

$$\frac{\partial T}{\partial z} = \text{const} \tag{4}$$

The thermal field in the considered medium can be expressed as

$$T(z) = \frac{T_2 - T_1}{e} z + T_1 \tag{5}$$

This relation can be obtained very easily by integration of equation (4), the integration constants being determined from the boundary conditions in equation (3).

In the conditions stated by (2), equation (1) can be written in a scalar form:

$$\phi = \frac{Q}{St} = -\lambda \frac{dT}{dz} \tag{1'}$$

This gives, on differentiating (5),† the relation

$$\frac{Q}{tS} = -\lambda \frac{T_2 - T_1}{e} \tag{6}$$

which allows one to compute λ and gives simultaneously its physical meaning. Indeed, the thermal conductivity represents the quantity of heat flowing through

†A similar result can be obtained from the general heat flow equation $\nabla T = a\ \partial T/\partial t$ by making the same assumptions. We preferred to start from Fourier's law since it is the basis for the definition of the thermal conductivity.

a medium limited by two infinite parallel planes submitted to a thermal gradient (variation of temperature by unit of length) of 1 K/m, by unit of time and unit of area. The thermal conductivity expresses the capacity of the medium to conduct heat. In SI units, it is expressed in W/m K.

The computation formula (6) will be valid only if relation (4) is verified, i.e, if the thermal field is unidirectional, and if $\lambda(T)$ is constant. In this case the points of equal temperatures lie in planes parallel to each other and to the limit surfaces: consequently the heat flow lines are unidirectional.

The method of the "guarded hot plate" is nothing else but an ingenious means to effectively realize the necessary boundary conditions to ensure a unidirectional heat flow, through a material plate of finite size. The measurement error will be higher or lower depending on how much the measurement conditions depart from this basic assumption.

The same type of reasoning can be applied to anisotropic media, for which equation (1) becomes

$$\boldsymbol{\phi} = -\bar{\bar{\lambda}} \operatorname{grad} \mathbf{T} \tag{6'}$$

where the thermal conductivity is a second order tensor which can be written with nine elements λ_{ij} $(i, j = 1, 2, 3)$:

$$\bar{\bar{\lambda}} = \begin{pmatrix} \lambda_{xx} & \lambda_{xy} & \lambda_{xz} \\ \lambda_{yx} & \lambda_{yy} & \lambda_{yz} \\ \lambda_{zx} & \lambda_{zy} & \lambda_{zz} \end{pmatrix} \tag{7}$$

(6′) then becomes the system

$$\begin{aligned} \phi_x &= -\left(\lambda_{xx} \frac{\partial T}{\partial x} + \lambda_{xy} \frac{\partial T}{\partial y} + \lambda_{xz} \frac{\partial T}{\partial z}\right) \\ \phi_y &= -\left(\lambda_{yx} \frac{\partial T}{\partial x} + \lambda_{yy} \frac{\partial T}{\partial y} + \lambda_{yz} \frac{\partial T}{\partial z}\right) \\ \phi_z &= -\left(\lambda_{zx} \frac{\partial T}{\partial x} + \lambda_{zy} \frac{\partial T}{\partial y} + \lambda_{zz} \frac{\partial T}{\partial z}\right) \end{aligned} \tag{6''}$$

In the case of a material presenting a single anisotropy (stratified material; for instance, wood, asbestos, . . . , etc.) the matrix representing $\bar{\bar{\lambda}}$ will reduce to

$$\bar{\bar{\lambda}} = \begin{pmatrix} \lambda_x & 0 & 0 \\ 0 & \lambda_y = \lambda_x & 0 \\ 0 & 0 & \lambda_z \end{pmatrix} \tag{7'}$$

x, y being the stratification plane and z the direction perpendicular to this plane.

Of course, for such stratified materials, the thermal conductivity will be represented by its two components: λ_v along the direction perpendicular to the stratification planes and λ_H along the direction parallel to the stratification. These two components will be measured separately.

3. APPARATUS TYPES

Most of the realizations of installations of the "guarded hot plate" type are described in papers presented in seminars, symposiums, or ASTM, I.I.R., meetings, thermal conductivity conferences, etc. The research results and the successive improvements made to this method have widely contributed to the elaboration of national standard in Refs. 9–13. The ANSI/ASTM C 177 standard is by far the more complete and the one having had the more re-editions. The last edition, revised in 1976, mentions 23 bibliographical references describing research works undertaken in the United States as well as in Canada and Europe.

The common point of these different standards is the fact that they all attempt to standardize a method and a measurement principle and not a certain type of apparatus. Therefore they only indicate fundamental elements and details which have revealed themselves by experience to be essential for obtaining satisfactory results and which are requirements. Only ASTM C 177 and ISO Draft, in addition to these requirements, give a set of criteria of qualification of the apparatus allowing the designer to check the performance and the possible limitations due to the materials studied. The recommendations given in the standards mentioned are not always agreed upon.

3.1. *Two-Specimen Apparatus*

This is the most frequent implementation that one can find in laboratories. Two identical specimens constituting the sample to be measured are put symmetrically on each side of a square or sometimes circular plate comprising the hot source; the cold plates, of similar shape to the hot plate, are facing the

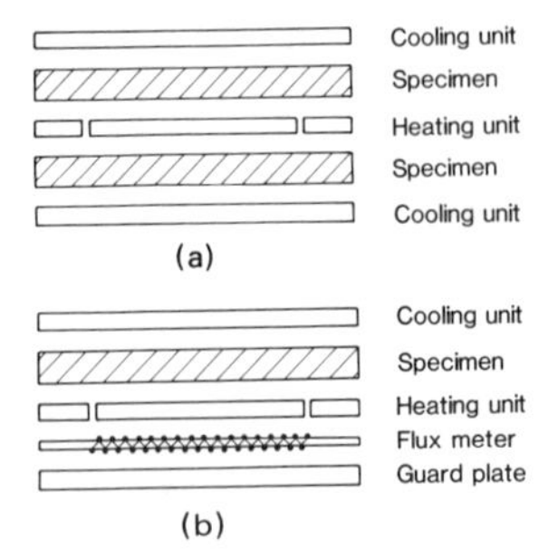

FIGURE 2. (a) Guarded hot plate apparatus. (b) Biguarded hot plate apparatus.

other face of the specimens (Fig. 2a). The hot plate is surrounded by a guard of similar shape from which it is completely or partially separated by a gap. The total area of the guarded hot plate, central metering area, and guard ring, is equal to the area of the cold plate. Heat, generated by the Joule effect from the central plate in which electrical current is dissipated, flows through the two specimens.

Electrical current is dissipated in the guard independently of the central plate; this guard is maintained at a temperature as close as possible to the temperature of the central plate. The guard is essential for a correct functioning of the method: its function consists in maintaining the heat flow lines parallel in the central measuring zone. Therefore, the density of heat flow is maintained constant and perpendicular to the hot surfaces and the isothermal planes are parallel in the whole thickness of the sample. In other words, the purpose of the guard is to reduce the lateral losses as much as possible, in the central plate as well as in the corresponding central zone inside the specimens. The conditions of thermal transfer specific to the case of an infinite wall are then realized for media of finite dimensions.

3.2. One-Specimen Apparatus (Biguarded Hot Plate Apparatus)

This implementation is a variant of the two-specimen method; its use has appeared more frequent in the last 10 years. The modification of the standard setup consists on the one hand, in suppressing one of the two cold plates and one of the two specimens, and on the other hand in replacing them by a second hot plate working as a back guard with respect to the central plate (Fig. 2b).

Electrical current is dissipated through this back-guard plate independently of the central plate; it is maintained, as the lateral guard at a temperature as close as possible or identical to the temperature of the central plate. The heat flow will be orientated in a single direction, towards the unique specimen in between the two hot and cold plates.

3.3. Computation Formula

The "guarded hot" plate imposes the boundary conditions on the "hot side" in order to ensure heat flux density and temperatures constant with time and uniform. The computation of the heat flux density is made from the electrical power dissipated and measured in the central part of the hot plate. The boundary conditions on the "cold side" are ensured by the cold plate. The surface temperature is constant with time and uniform.

In the case of the one-specimen implementation, the computation of the thermal conductivity will be made from the relation

$$\lambda = \frac{P}{S} \cdot \frac{e}{T_2 - T_1} \tag{8}$$

where P is the electrical power (W) dissipated in the center of the hot plate, S is the area (m^2) of the central zone, conventionally increased by the half area of the joint central plate/lateral guard, T_2 is the temperature (K) of the "hot surface" of the specimen,† T_1 is the temperature (K) of the "cold surface" of the specimen, and e is the specimen thickness.

In the case of the two-specimen implementation, it is assumed that the electrical power is dissipated symmetrically towards the two surfaces limiting the central plate in contact with the two specimens constituting the sample. The computation formula is the following:

$$\lambda = \frac{P}{2S} \cdot \frac{e_m}{\Delta T_m} \tag{9}$$

where P is the electrical power (W) dissipated in the center of the hot plate, S is the area (m^2) of the center of the hot plate conventionally increased by the half area of the joint central plate/lateral guard, ΔT_m is the mean temperature difference (K)¶ between the surfaces delimiting the two specimens, and e_m is the mean thickness of the two specimens.

4. FIELD OF APPLICATION

The field of application of the method is limited by the capability of the apparatus to maintain the unidirectional constant thermal flux density in the

†One can assume that the two specimens are not rigorously identical in thickness and thermal conductivity.

¶In certain cases, the "cold" and "hot" temperatures of the specimen on one side, and of the apparatus plates on the other side, are almost identical. The possible differences between these temperatures are due to thermal contact resistances.

specimen. From this condition results a limitation of the field of application which is therefore restricted, as we already mentioned, to the poorly conductive materials.

> The method is limited for purpose of certification to specimen with thermal resistance greater than 0.1 K m² W⁻¹ provided that thickness limits mentioned in Section 7 are not exceeded:
>
> $$R > 0.1\ \mathrm{K\,m^2\,W^{-1}}$$
>
> The limit for the thermal resistance may be as low as $0.02\ \mathrm{K\,m^2\,W^{-1}}$, but the accuracy stated in Section 5 may not be achieved over the full range[14]
>
> $$R > 0.02\ \mathrm{K\,m^2\,W^{-1}}$$

The thicknesses of the specimens will depend on the type of material and on the limitations imposed by its nature: homogeneity, anisotropy, opacity to infrared radiation and surface conditions (which affect the contact resistances), and on the dimensions of the apparatus plates. With the existing apparatus (dimensions – side or diameter – of plates between 200 and 1000 mm) the specimen thicknesses may vary approximately between 10 and 250 mm. (It is mainly the building materials, light and thick, which require large thicknesses of specimens and large sized apparatus). The range of temperature lies between −196 and 1200°C.

5. APPARATUS DESCRIPTION AND DESIGN REQUIREMENTS

The principle of the setup of the two- or one-specimen versions of the guarded hot plate method have already been described in Section 3. In this section, we now come back to the description of the apparatus, but this time giving some design requirements, especially the technological and metrological characteristics which are necessary to carry out the project and allow the realization of a performing installation.

All the data are, in most cases, in accordance with the ISO draft standard[14] which is the most recent synthesis of the results obtained in this field. In this standard, it is mentioned that "considerable latitude both in the temperature range and in the geometry of the apparatus is given to the designer of new equipment since various forms have been found to give comparable results . . . Continuing research and development is in progress to improve the apparatus and measurement techniques. Thus it is not practical to mandate a specific design or size of apparatus, especially as total requirements may vary quite widely."

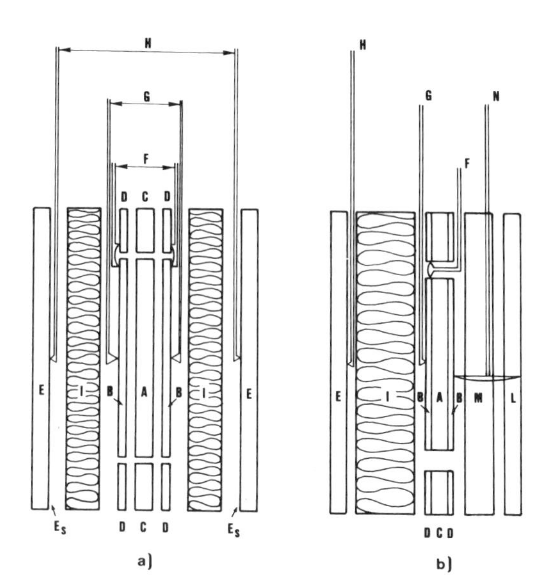

FIGURE 3. General features of two-specimen and single-specimen hot plate apparatus. (a) Two-specimen apparatus. (b) Single-specimen apparatus. KEY: A, metering area, heater; B, metering area surface plates; C, guard heater; D, guard surface plates; E, cooling units, E_s, cooling unit surface plates; F, differential thermocouples; G, heating unit surface thermocouples; H, cooling unit surface thermocouples; I, test specimens; L, guard plate; M, guard plate insulation; N, guard plate differential thermocouples.

5.1. Apparatus

5.1.1. Heating Unit

The heating unit consists of a central metering section and a guard section. The metering section consists of a metering section heater and metering section surface plates. The guard section consists of one or more guard heaters and the guard surface plates (Fig. 3a). For the single-specimen apparatus a back-guard plate realized in the same way as the guard section is necessary (Fig. 3b).

The surface plates are usually made of metal of high thermal conductivity. The materials used in the construction of the heating unit must be chosen carefully, giving adequate consideration to their performance at the temperatures at which the plate is to be operated. The working surfaces of the heating unit and cooling plates should not chemically react with the specimen and the environment.

The surfaces should be smoothly finished to conform to a true plane. The maximum departure of a surface from a plane should not exceed 0.025% (Fig. 4). The heating unit should be designed and constructed so that the two faces do not warp or depart from planarity at the operating temperatures.

Temperature uniformity: The heating unit (metering and guard section) should be designed and constructed so that when in operation deviations from temperature uniformity for each face are not greater than 2% of the temperature difference across the specimen. For the two-specimen apparatus the two faces of the metering section and of the guard section should be within 0.2 K of their

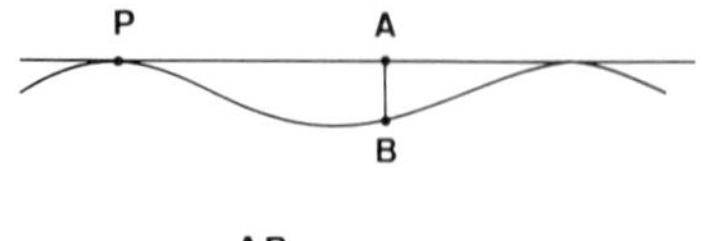

FIGURE 4. Surface departure from a true plane.

average temperature (at least for specimen having a thermal resistance greater than $0.1\,\mathrm{m^2\,K\,W^{-1}}$ and tested at a mean test temperature close to room temperature).

Emittance: The surfaces of all plates should have and maintain a total hemispherical emittance greater than 0.8 at the operating temperatures.

5.1.2. Gap and Metering Area

The heating unit should have a definite separation or gap between the surface plates of the metering and guard section. The area of the gap in the plane of the surface plate should be not more than 5% of the metering section area.

The separation between the heater windings and the gap between the metering section and the adjacent guard section should be designed so as to distribute heat to the surface plates uniformly according to the temperature uniformity criteria.

The dimension of the metering area should be determined by measurements to the center of the gap that surrounds this area, unless calculations or tests are used to define the area more precisely.

Imbalance through the gap: A suitable means such as a multijunction thermopile should be provided to detect the average temperature imbalance between the surface plates of the metering and guard sections (Fig. 5). When

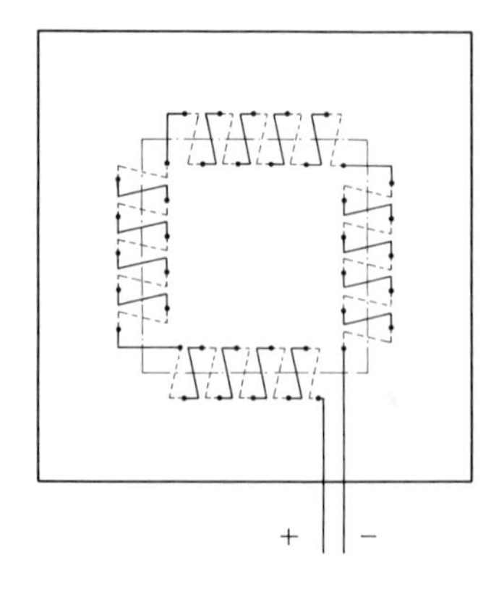

FIGURE 5. Multijunction thermopile to detect the average temperature imbalance.[12]

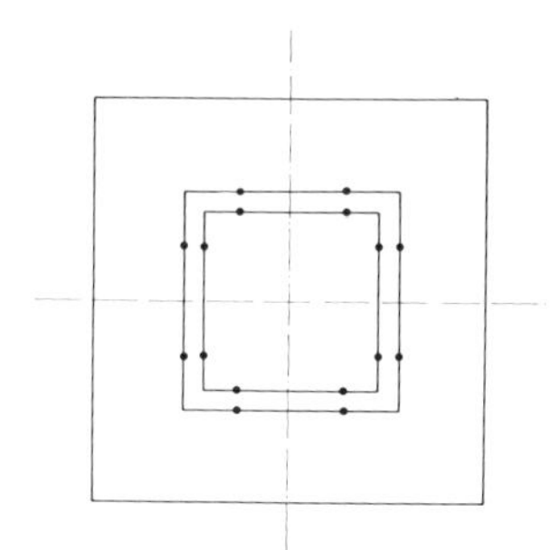

FIGURE 6. Limited number of sensing elements (recommended positions to detect gap balance).

only a limited number of sensing elements is used it is suggested that the most representative positions from which to detect the average balance will be those at a distance from the corner of the metering section equal to 1/4 of the side of the metering section along the gap; the corners and the axes should be avoided (Fig. 6).

When a temperature imbalance exists between the metering and guard sections, an amount of heat will flow between the two elements, partly through the specimen and partly through the gap itself. This heat transfer flow crossing the gap under thermal imbalance is the most severe limit when measuring high-resistance specimens. (See Section 6.)

5.1.3. Imbalance-Sensing Elements

If temperature-sensing elements are installed between the metal plates and the samples or within grooves in the metal plates on the side in contact with the samples, thermal resistances can exist between the sensing element and the metal plates, the surface of the specimen, and between the sensing elements mounted on the metering area and those mounted on the guard ring (see Fig. 7a). When the apparatus is in operation the temperature of the sensing elements is due to the combined effect of thermal balance between the metal plates of the metering section and of the guard ring, and the density of heat transfer rate flowing from the metal plates to the specimen. Correct balance is obtained only if the resistance between the metal plates and the sensing elements is made negligible with respect to the other resistances mentioned, or when the power flowing from the plates to the specimens does not cross the sensing elements, as in Fig. 7b or 7c.

The distance between the gap edge and the sensing elements should be smaller than 5% of the side or diameter of the metering section.

Since an uncertainty in the true temperature balance will always exist, the gap thermal resistance should be made as high as practically possible. Good general rules are: (i) any mechanical connections between metering section and

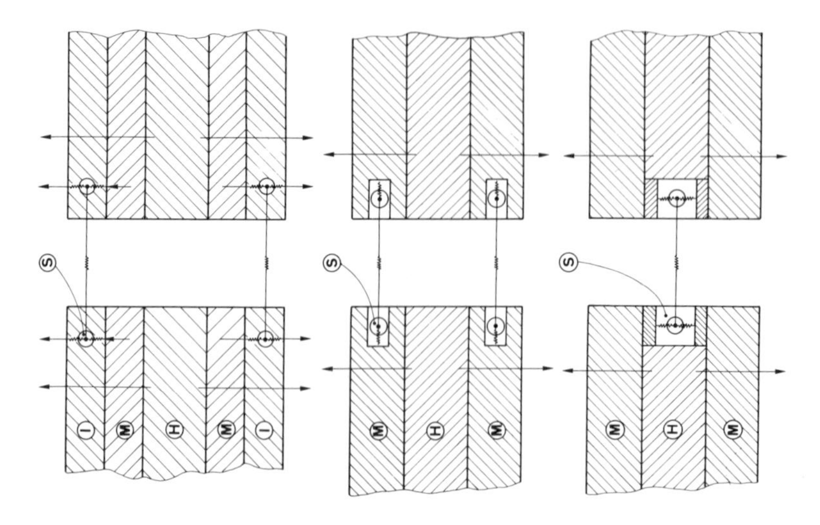

FIGURE 7. Sensing elements and related thermal resistance. KEY: H, heater; M, heating unit metal plate; I, insulation; S, sensing element.

guard ring should be made as small as possible, avoiding metal or continuous corrections where possible; (ii) electrical wire should cross the gap through oblique paths and should be of thin diameter and of metal with low thermal conductivity; and (iii) the use of copper should be limited to a minimum.

The reader will readily identify the requirements that apply to the single-specimen apparatus.

5.1.4. Cooling Unit

The cooling units should have surface dimensions at least as large as those of the heating unit, including the guard heater(s).

They should consist of metal plates maintained at constant and uniform temperature, within 2% of the temperature difference across the specimen. This can be accomplished by the use of a constant-temperature fluid, by the use of electrical heaters, by the use of thermal insulation of uniform thermal resistance applied between the outer most surfaces of the heating units, and by auxiliary cooling plates.

Fluid-cooled metal plates require particular care in their design in order to obtain temperature uniformity.[73,88] The best results will be obtained with a helical counterflow path for the fluid.

5.1.5. Edge Insulation and Edge Heat Losses

Deviation from one-dimensional heat flow in the specimen is due to nonadiabatic conditions at the edges of the heating unit and the specimen. Heat losses from the specimen edge introduce edge heat loss errors which can be computed only for homogeneous, isotropic, and opaque to IR radiation specimens under simplified boundary conditions (Section 6).

The heat losses from the outer edges of the guard section and the specimen can be accomplished by edge insulation, by controlling the surrounding ambient temperature, by an additional outer guard, by a linear gradient guard, or by a combination of these methods (Fig. 8). Edge heat loss from the specimen will be minimal if the surrounding temperature corresponds to the mean temperature of the specimen.

A very important heat flow path from the heating unit edges is along the wires of the heaters and temperature sensors. It is therefore necessary to provide an isothermal surface close to the hot plate and at the same temperature. All of the wires should be fastened securely to the surface. This isothermal surface may be an auxiliary guard or any other suitable surface. The level of thermal imbalance should be limited so that thermal flux exchanged through the wires will not exceed 10% of the flux crossing the samples in unidirectional ideal conditions.

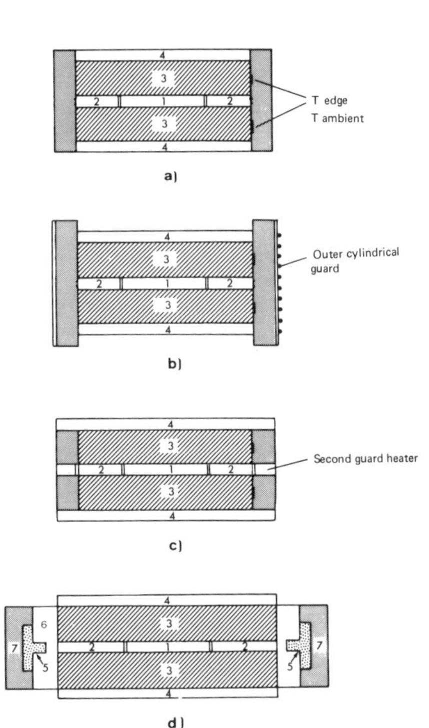

FIGURE 8. Possible configurations to restrict edge heat loss and gain. KEY: 1, hot plate central section; 2, hot plate guard ring; 3, specimen; 4, cold plate; 5, second T-shaped copper guard; 6, edge insulation; 7, second guard insulation.

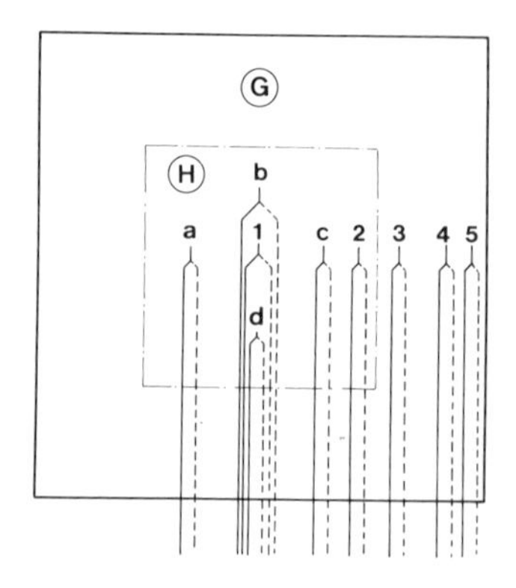

FIGURE 9. Permanent mounted temperature sensors.(12) KEY: H, Hot plate–central section; G, hot plate–guard ring.

5.2. Measuring Devices

5.2.1. Temperatures

5.2.1a. Imbalance Detection. The temperature-imbalance sensing elements may be read individually and the temperature difference calculated, or they may be connected differentially to indicate the temperature difference directly.

Small-diameter thermocouples, no greater than 0.3 mm, connected as a thermopile are often used for this purpose.

The detection system should be sufficiently sensitive to ensure that the error in the measured property due to gap temperature imbalance is restricted to 0.5% as determined experimentally or analytically.

For the one-sample-specimen apparatus the imbalance-sensing element used between the central metering section and the back-guard plate is a heat flowmeter.

5.2.1b. Temperature Differences in the Apparatus. The surface temperatures are often measured by means of permanently mounted temperature sensor, such as thermocouples, set in grooves in the plates or placed just under the surfaces in contact with the specimen (Fig. 9).

An accuracy of 1% may be used for the measurement of the temperatures in the apparatus.

Temperature sensors can be either completely insulated electrically from the plates or grounded to them only in a single point of the whole circuit (as a consequence, in differential connections only one junction of a thermocouple can be grounded. The amount of electrical insulation required depends on whether the sensors are shielded by grounded metal plates of the heating or cooling units or they are only insulated from other electrical circuits; in the latter case the insulation resistance should normally be larger than 100 MΩ. Computation and experimental verification should be made to be sure that other circuits do not affect the accuracy of measurements of thermal properties.

The number of temperature sensors on each side should be not less than $N\sqrt{A}$, or 2, whichever is greater, where $N = 10\,\text{m}^{-1}$ and A is the area in square meters of one face of the metering area plate. It is recommended that one sensing element should be placed in the center of the metering area section and that a similar number of sensors be permanently and similarly installed at corresponding positions in the facing cooling units.

5.2.1c. Temperature Difference Across the Specimen. For nonrigid specimens with flat uniform surfaces that conform well to flat surfaces of the plates, and with thermal resistance greater than $0.5\,\text{K}\,\text{m}^2\,\text{W}^{-1}$ the temperature difference across them is normally taken to be indicated by the thermocouples permanently mounted in the heating and cooling unit surfaces.

For rigid specimens, separate thermocouples will be mounted flush with or interior to the surfaces of the specimen. Thin layers of low-resistance material may be interposed between the specimen and the plates.

Another means to reduce the contact thermal resistance is to very carefully prepare the surfaces of the rigid specimens to ensure a good contact between the surfaces of the plates and the specimens.

5.2.1d. Type and Placement of Thermocouples. Thermocouples mounted in the surfaces of the plates should be made of wire not larger than 0.6 mm in diameter and preferably not larger than 0.2 mm in diameter for smaller apparatus.

Thermocouples placed against or set into the surfaces of the specimens should be made of wire not larger than 0.20 mm in diameter.

The thermocouples used to measure the temperatures of the hot and cold faces of the specimen should be fabricated from either calibrated thermocouple wire or from wire that has been certified by the supplier to be within the special limits of error given in Tables 1, 2, and 3.[9,14]

TABLE 1
Types of Thermocouples

Type B	Platinum–30% rhodium (+) versus platinum–6% rhodium (–).
Type E	Nickel–10% chromium (+) versus constantan (–).
Type J	Iron (+) versus constantan (–).
Type K	Nickel–10% chromium (+) versus Nickel–5% aluminum, silicon) (–) (Note 2).
Type R	Platinum–13% rhodium (+) versus platinum (–).
Type S	Platinum–10% rhodium (+) versus platinum (–).
Type T	Copper (+) versus constantan (–).

TABLE 2
Limits of Errors for Thermocouples[a]

Thermocouple type	Temperature range (°C)	Limits of error—reference junction 0°C: Standard [°C (whichever is greater)]	Special [°C (whichever is greater)]
T	0–350	± 1 or ± 0.75%	± 0.5 or 0.4%
J	0–750	± 2.2 or ± 0.75%	± 1.1 or 0.4%
E	0–900	± 1.7 or ± 0.5%	± 1 or ± 0.4%
K	0–1250	± 2.2 or ± 0.75%	± 1.1 or ± 0.4%
R or S	0–1450	± 1.5 or ± 0.25%	± 0.6 or ± 0.1%
B	800–1700	± 0.5%	
T[b]	– 200–0	± 1 or ± 1.5%	[c]
E[c]	– 200–0	± 1.7 or ± 1%	[c]
K[b]	– 200–0	± 2.2 or ± 2%	[c]

[a]Limits of error in this table apply to new thermocouple wire normally in the size range 0.25 to 3 mm in diameter (No. 30 to No. 8 Awg) and used at temperatures not exceeding the recommended limits of Table 2B. If used at higher temperatures these limits of error may not apply. Limits of error apply to new wire as delivered to the user *and do not allow for calibration drift during use.* The magnitude of such changes depends on such factors as wire size, temperature, time of exposure, and environment. Where limits of error are given in percent, the percentage applies to the temperature being measured when expressed in degrees Celsius.

[b]Thermocouples and thermocouple materials are normally supplied to meet the limits of error specified in the table for temperatures above 0°C. The same materials, however, may not fall within the subzero limits of error given in the second section of the table. If materials are required to meet the subzero limits, the purchase order must so state. Selection of materials usually will be required.

[c]Little information is available to justify establishing special limits of error for subzero temperatures. Limited experience suggests the following limits for types E and T thermocouples: type E – 200 to 0°C ± 1°C or ± 0.5% (whichever is greater); type T – 200 to 0°C ± 0.5°C or ± 0.8% (whichever is greater). These limits are given only as a guide for discussion between purchaser and supplier. Due to the characteristics of the materials, subzero limits of error for type J thermocouples and special subzero limits for type K thermocouples are not listed.

TABLE 3
Recommended Upper Temperature Limits for Protected Thermocouples[a]

Thermocouple type	Upper temperature limit for various wire sizes (°C): 3.25 mm	1.63 mm	0.81 mm	0.51 mm	0.33 mm	0.25 mm
T		370	260	200	200	150
J	760	590	480	370	370	320
E	870	650	540	430	430	370
K	1260	1090	980	870	870	760
R and S				1480		
B				1700		

[a]NOTE: This table gives the recommended upper temperature limits for the various thermocouples and wire sizes. These limits apply to protected thermocouples, that is, thermocouples in conventional closed-end protecting tubes. They do not apply to sheathed thermocouples having compacted mineral oxide insulation. Properly designed and applied sheathed thermocouples may be used at temperatures above those shown in the tables.

5.2.2. Thickness Measurement

Means should be provided for measuring the thickness of the specimen to within 0.5%. It is recommended that, when possible, specimen thickness be measured in the apparatus at the existing test temperature and compression conditions. Gaging points or measuring studs at the outer four corners of the cold plates or along the axis perpendicular to the plates at their corners, will serve for these measurements. For resilient materials a mechanism which permits a continuous displacement of the hot or cold plate can be used.

5.2.3. Clamping Force

Means should be provided either for imposing a reproducible constant clamping force upon the system, to promote good thermal contact, or for maintaining an accurate spacing between the plates.

5.2.4. Electrical Measurement System

A measuring system having a sensitivity and accuracy of at least 0.2% of the temperature difference across the specimen should be used for measurement of the output of all temperature and temperature difference detectors. Measurement of the power to the heating unit should be made to within 0.1% over the full operating range.

5.3. Encloser

The guarded hot plate apparatus should be placed in an enclosure equipped for maintaining the desired interior environmental gas temperature and dew or condensation point when the cooling unit temperature is below room temperature or when the mean temperature is substantially above room temperature. Means to control the environmental pressure and gas property should be provided if measurements are required in different gaseous environments and variable pressure.

5.4. Limitation Due to the Apparatus

5.4.1. Contact Resistance

The imperfect contact between the surfaces of the apparatus and of the specimens will introduce contact resistances not uniformly distributed. These will cause nonuniform heat transfer flow distribution and thermal field distortions. For specimens having a thermal resistance less than $0.1\ \mathrm{K\,m^2\,W^{-1}}$ special techniques for measuring surface temperatures will be required.

5.4.2. Upper Limits for the Thermal Resistance

These are limited by the stability of the power supplied to the heating unit; the ability of the instrumentation to measure the power level; the extent of the heat losses or gain due to temperature imbalance errors (see Section 6).

5.4.3. Limits to Temperature Differences

When measured differentially, temperature differences as low as 5 K can be used (provided that the uniformity and stability of the temperature of the hot and cold surfaces of the plates, the noise, resolution, and accuracy of the instrumentation, and the restrictions on temperature measurements can be maintained within the limits outlined in Sections 5.1 and 5.2).

If temperature measurements of each plate are made by means of independant thermocouples, temperature differences of at least 10 to 20 K are used in order to minimize temperature measurement errors. (The accuracy of the calibration of each thermocouple may be the limiting factor.)

Higher temperature differences are limited only by the capability of the apparatus to deliver enough power while maintaining required temperature uniformity.

5.4.4. Maximum Specimen Thickness

The maximum thickness of specimen will be limited by the boundary conditions of the edges of the specimen due to the effects of edge insulation, of auxiliary heaters, and of surrounding environment temperature.

5.4.5. Minimum Specimen Thickness

Minimum thickness is limited by contact resistance and by the accuracy of the instrumentation for measuring the thickness.

5.4.6. Maximum Operating Temperature

Maximum temperature may be limited by oxidation, thermal stress, or other factors which degrade the flatness and uniformity of the surface plate and by changes of electrical resistivity of electrical insulations which may affect the accuracy of all electrical measurements.

5.4.7. Apparatus Size

The overall size of a guarded hot plate will be governed by the specimen dimensions, which range normally within the limits of 0.2 to 1 m diameter or square. For ease of interlaboratory comparisons and for general improvement in collaborative measurements it is recommended that the design of future guarded hot plates be based upon one of the following suggested standard dimensions:

a. 0.3 m diameter, or square,
b. 0.5 m diameter, or square, and in addition,
c. 0.2 m diameter, or square, if only homogeneous materials are tested,
d. 1 m diameter or square if specimens are to be measured at a thickness that exceeds the limits permitted for a 0.5-m apparatus.

5.5. Design Parameters

When a guarded hot plate apparatus is to be designed, preliminary decisions must be taken on the following parameters:

- Minimum and maximum specimen thickness to be tested in the apparatus;
- Minimum and maximum thermal resistance;
- Minimum and maximum temperature difference across the specimen;
- Sensitivity of the balancing system for the guard plate section;
- Minimum cooling unit temperature;
- Maximum heating unit temperature;
- overall apparatus accuracy as maximum acceptable error in measured property in a defined worst-case condition (see Section 6);
- Surrounding environment.

6. ERROR EVALUATION. PERFORMANCES

From relations (8) and (9) in Section 3.3, which are used to compute the thermal conductivity, we can see that the uncertainty in the result obtained will be directly dependent upon the metrological quality of the means used to determine each parameter involved in the computation. To estimate this uncertainty, we must know the measurement errors

- on the power P, in the central part of the hot plate;
- on the temperatures, T_1 and T_2, or on the difference $\Delta T = T_2 - T_1$;
- on the thickness, e, of specimens;
- on the measurement area, S.

The evaluation of these errors does not raise particular questions and the different standards and recommendations give precise indications on this point.

The precision of the results obtained with a guarded hot plate apparatus depends also, but not directly and with no evidence from the computation formula, on how the apparatus, in normal working conditions, will be able to ensure the unidirectionality of the heat flow lines in the central measuring zone.

Two main causes of errors are involved in the possible distortions of the heat flow lines (Fig. 10): the unbalance of temperature between the center and the lateral guard of the hot plate ("unbalance error"), and the lateral losses through the specimens, considering the nonadiabatic boundary conditions ("edge heat losses"). Many theoretical and experimental works which cannot be analyzed here in detail, have been undertaken in this field (see Refs. 77–94). It must be noted that apparatus designers as wells as current standards or recommendations do not always take these results into account.

The criteria of elaboration of the projects of these apparatus are often empirical; nevertheless, because of the experience of the designers the performances can be excellent. It is also true that it is not always easy to take advantage of these research works: those that we took into consideration, to explain the treatment of this problem, have been obtained with simplifying assumption and two-specimen symmetrical apparatus. The materials are always considered as "ideal" material, that is to say homogeneous, isotropic, and opaque to radiation, which limit the field of application of the results.

Important work remains to be done in this field, concerning computation methods.

6.1. Measurement Errors Due to the Unbalance of Temperatures between the Center and the Guard of the Hot Plate (Unbalance Error)

At the present time, the results obtained by Woodside and Wilson[80] can still be considered as a reference, despite their age. Of relatively easy application, these results have been reexamined by many authors to be completed on a theoretical or experimental basis.[81,82,84,88] According to Woodside and Wilson, the heat flux $\Phi_{g'}'$ flowing through the gap between the center and the guard of the hot plate for a difference of temperature of 1 K (temperature unbalance) is expressed by

$$\Phi_g' = \Phi_0 + c\lambda \qquad \text{(W/K)} \tag{10}$$

where Φ_0 is the heat flux flowing directly through the gap between the center and the guard – this flux will mainly depend on mechanical and electrical connections as well as on convective and radiative transfers (Fig. 11); and where $c\lambda$ is the heat flux between the center and the guard flowing through the

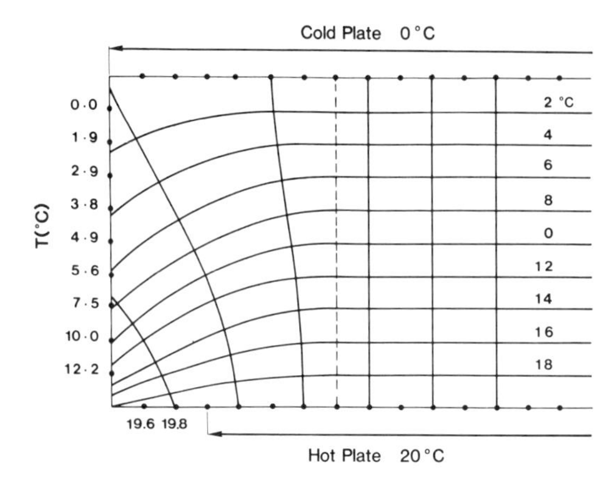

FIGURE 10. Distortion of heat flow lines and isothermals (from Pascal).[33]

two samples the thermal conductivity of which is λ; c is a parameter which depends on the apparatus, but also slightly on the measurement temperature and specimen thickness. This relationship was at first experimental: it has been very nicely and rigorously derived from correlations between the measured thermal conductivities, λ_m, and the temperature differences ΔT_g arbitrarily chosen between the center and the guard. The relation that can be established between λ_m and ΔT_g for different materials and apparatus designs is always linear; it is in fact an experimental approach for obtaining, by interpolation, a value of λ with no unbalance error (Figs. 12a and 12b):

$$\lim_{\Delta T_g \to 0} \lambda_m = \lambda \tag{11}$$

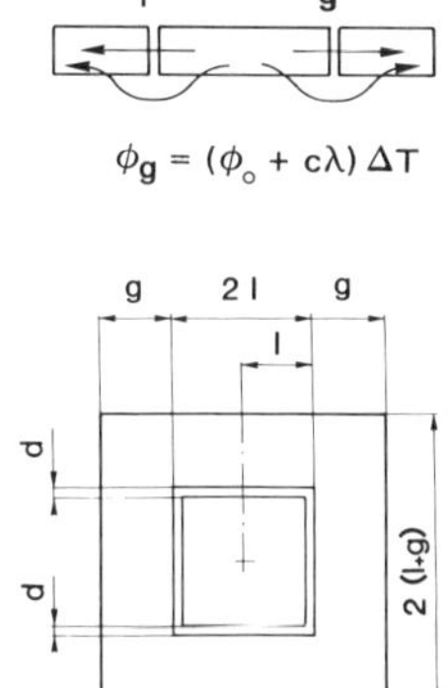

FIGURE 11. Unbalance error.

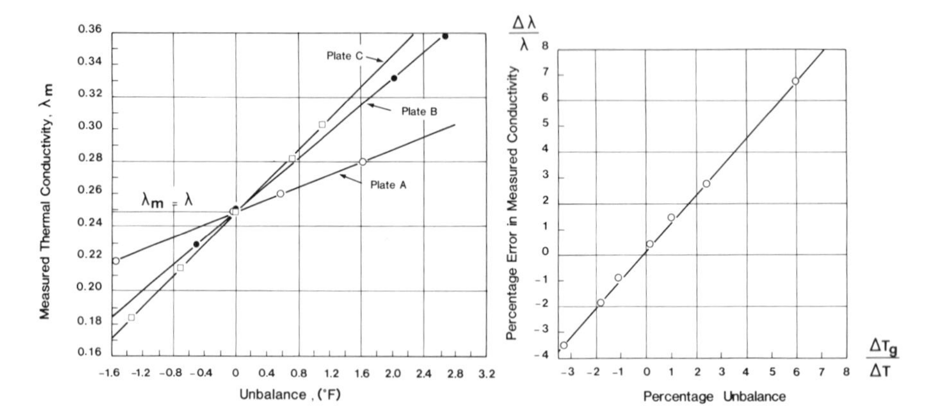

FIGURE 12(a). Effect of temperature unbalance on measured conductivity. (b) Percentage error in measured conductivity determined at several temperature differences between hot and cold plates versus the unbalance expressed as a percentage of the temperature difference.

Many experimental studies on errors use this type of representation to evaluate the measurement error as a function of ΔT_g.

From relation (10), it is possible to express the total lateral flux due to the unbalance error and corresponding to a difference of temperature ΔT_g by

$$\Phi_g = (\Phi_0 + c\lambda)\,\Delta T_g \tag{12}$$

and to introduce the measurement error, E_g, as equal to the ratio between this lateral thermal flux Φ_g and the unidirectional thermal flux Φ such that as it would be for a zero temperature difference between the center and the guard. One can write

$$E_g = \frac{\Phi g}{\Phi} = \frac{\lambda_m - \lambda}{\lambda} = \frac{\Delta\lambda}{\lambda} \tag{13}$$

Taking into account the relations expressing on one side Φ (Fourier's law) and on the other side Φ_g, the error due to the unbalance between the center and the guard can be written

$$E_g = \frac{\Delta\lambda}{\lambda} = \frac{e}{S} \cdot \frac{\Delta T_g}{\Delta T}\left(\frac{\Phi_0}{\lambda} + c\right) \tag{14}$$

where each symbol has the meaning already mentioned.

The parameters Φ_0 and c are specific for each apparatus and are slightly

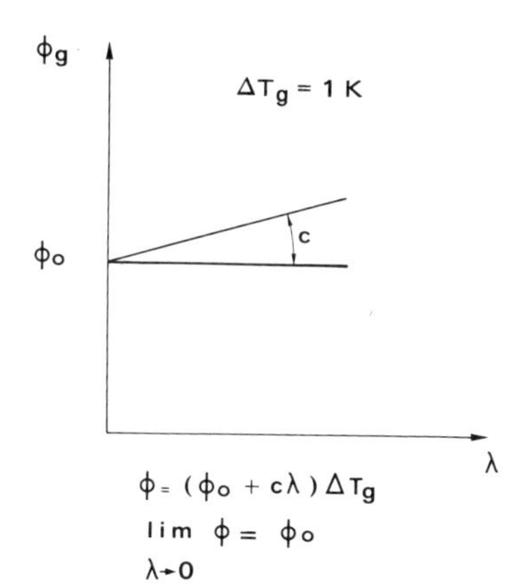

FIGURE 13. The parameters ϕ_0 and c characterizing a guarded hot plate.

dependent on the specimen and the test temperature; they can be determined experimentally according to the method described in Ref. 80, by choosing materials of different thermal conductivities and expressing λ_m for each of these materials as a function of ΔT_g (Fig. 13).

The interest of relation (14) is to put in evidence the part of the different parameters characterizing the apparatus and of the conditions of use; it allows us, for a given apparatus, to evaluate the error due to the unbalance between the guard and the center. From this evaluation, we can of course, depending on the quality of the temperature control of the center guard which is used, limit the value of ΔT_g in order to remain under a given limiting value of E_g and therefore optimize the operating conditions of the apparatus.

Experience has shown that for highly-performing apparatus, to obtain unbalance errors lower than 1%–2%, it is necessary to have temperature differences ΔT_g of the order of 10^{-2} K. In certain recommendations or studies, limit values of $\Delta T_g/\Delta T$ are often given, which represents an uncomplete use of relation (14).

The determination by computation of parameters Φ_0 and c is also possible following the theoretical studies of Woodside[81] and Donaldson[82] which lead to equivalent results, confirmed by experiments.

The parameter c can be expressed by the relation

$$c = \frac{16l}{\pi} \ln \frac{4}{1 - e^{-2\pi d/e}} \tag{15}$$

where $2l$ is the length of the central plate, d is the gap width, and e is the specimen thickness.

Figure 14 represents c/l as a function of d/e from Ref. 88. This correlation has been well verified by experimental results. We can see that the parameter

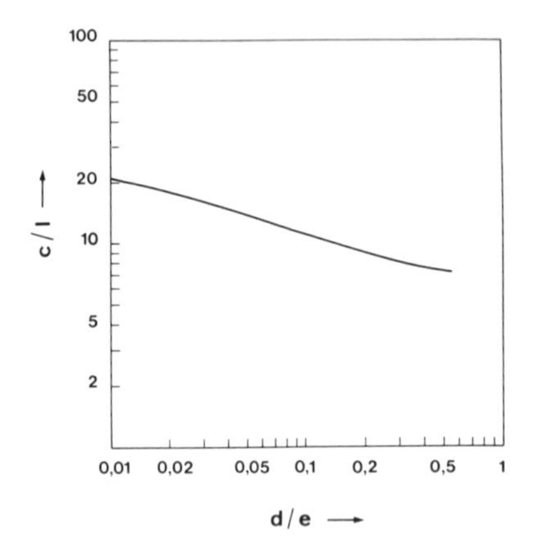

FIGURE 14. Plot of the unbalance parameter c/l versus the ratio between gap width and specimen thickness d/e.

c is proportional to the length of the gap, but that it is less influenced by the ratio d/e.

The parameter Φ_0 is mainly dependent on the plate design; it can be computed by taking into account the fixation points, electrical conductors, etc.(80,9,14)

Introducing expression (15) for c into relation (14) and writing e and l in dimensionless form: $e/2(l+g)$ (reduced thickness of the specimen), and $l/2(l+g)$ (reduced width of the center), the measurement error E_g due to the unbalance error center guard can be expressed from De Ponte(88) as

$$E_g = \frac{e}{2(l+g)}\left[\frac{2(l+g)}{l}\right]^2 \cdot \frac{1}{16(l+g)} \cdot \frac{\Delta T_g}{\Delta T}$$

$$\times \left[\frac{\Phi_0}{\lambda} + \frac{l}{2(l+g)}\,\frac{32(l+g)}{\pi}\ln\frac{4}{1-e^{-2\pi d/e}}\right] \tag{16}$$

We can conclude by noting that, according to relation (16),

1. For very effective insulating materials, the lateral heat flux resulting from a $\Delta T_g \neq 0$ is mainly due to the connection's center–guard, and the error E_g is inversly proportional to the square of the total length of the gap:

$$\text{If } \lambda \text{ low:} \quad \Phi_0 \gg c\lambda, \qquad E_g \sim 1/l^2$$

2. For less effective insulating materials, the lateral heat flux is mainly due to the transfer through the specimen; the error is inversely proportional

to the length of the gap:

$$\text{If } \lambda \text{ high:} \quad c\lambda \gg \Phi_0, \qquad E_g \sim 1/l.$$

In all these considerations, it is assumed that the temperature difference is uniformly distributed along the gap separating the guard and the center. Experience shows that the local differences of temperature between the center and the guard are not constant since the heat flux density through the guard is not uniform in each point because of lateral losses. In this case, the lateral heat flux is considered equal to zero on an average if[88]

$$\Delta T_g = \frac{1}{8l} \int_0^{8l} \Delta T_g l \, ds = 0 \tag{17}$$

6.2. *Measurement Error Due to Lateral Heat Losses through the Specimen (Edge Heat Losses)*

Even if the thermal balance between the center and the guard of the hot plate is perfect ($\Delta T_g = 0$) distortions of the heat flow lines can appear if the width of the guard is not large enough with respect to the thickness of the samples. These distortions show a lateral heat flow which will more or less affect the measurement precision depending on how large this lateral flow is compared to the main heat flow measured in the center of the plate.

The evaluation of the edge heat losses has been performed theoretically via an analytical solution; in particular by Somers and Cyphers[77] Woodside,[79] Donaldson,[82] and Pratt;[83] numerical and analogical solutions gave the same results. Experimental works must also be mentioned[85,86,92]; their conclusions are convergent. The analytical result obtained by Woodside with the conformal transformation method is remarkably simple and is often used despite the simplifying assumption taken into consideration.†

If we assume $\Delta T_g = 0$, the edge heat losses error, denoted E_l, will be defined as equal to the ratio of the lateral loss heat flow (from inside to outside or vice versa), Φ_l, to the unidirectional heat flow Φ such as it would be for no lateral losses:

$$E_l = \frac{\Phi_l}{\Phi} = \frac{\lambda_m - \lambda}{\lambda} \tag{18}$$

†The computations performed by Woodside suppose the following assumptions: isothermal cold and hot surfaces, ΔT_g being equal to zero; edges of the specimens maintained at a uniform temperature T between T_1 and T_2; $T = \epsilon T_2$, where $0 < \epsilon < 1$.

According to Woodside,[79] the ratio λ/λ_m appearing in the computation of E_l, is expressed by the relation

$$\left(\frac{\lambda}{\lambda_m}\right)^{1/2} = \frac{\pi\, l/e}{\epsilon \ln \dfrac{\cosh \pi\,[(g+l)/e] + 1}{\cosh \pi(g/e) + 1} + (1-\epsilon) \ln \dfrac{\cosh \pi[(g+l)/e] - 1}{\cosh \pi(g/e) - 1}} \tag{19}$$

with $0 < \epsilon < 1$.

From this relation, it results that E_l is function of the ratios $(g+l)/e$, g/e, l/e and of ϵ; this last parameter will depend on the temperature field at the specimen lateral surfaces and will be chosen so as to closely represent the true distribution.

In Fig. 15, the curves of equal error E_l (lying between 0.05 and 0.001) are represented as a function of the dimensionless ratios $e/2(l+g)$ (reduced thickness) and $g/2(l+g)$ (reduced guard width) for $\epsilon = 0.25$. These curves give us the correlation between the maximum specimen thickness and the guard width in order to obtain an edge heat losses error lower than a fixed value E_l. The value of ϵ can be chosen function of the true temperature distribution on the specimen edges.

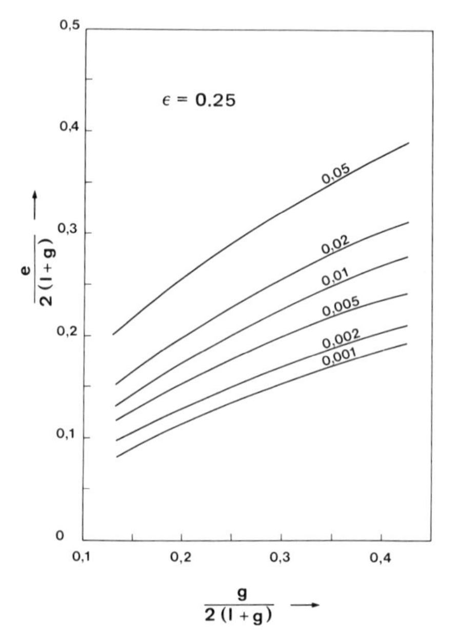

FIGURE 15. Edge heat losses error.

In the case where the specimen edges are not insulated, the temperature distribution and consequently E_l will depend on:

- the specimens thermal conductivity,
- the surface coefficient of heat transfer,
- the ambient air temperature,
- the temperature difference $\Delta T = T_2 - T_1$.

Recently, K.H. Bode[91] has reconsidered the evaluation of the edge heat losses error starting from a very general theoretical basis. The integration of the heat equation written for the three-dimensional case is performed analytically by using a power series development. The computations take into account two types of plate: square and circular shape.

As in previous cases the material specimen is considered homogeneous, isotropic, opaque to radiation, and with a constant thermal conductivity for the temperature differences studied. The hot and cold plate measurement surfaces are supposed to be isotherm ($\Delta T_g = 0$) and zero contact thermal resistances.

The ambient temperature of the air surrounding the lateral sides of the plate and the heat transfer between the specimen edge and the ambient air is described by an additional parameter ρ (function of several variables describing the boundary conditions) varying between 0 and the infinite (Fig. 16).

The edge heat losses error, E_l, is expressed by the relation

$$E_l = \frac{\lambda_m - \lambda}{\lambda} = F_1 + \frac{T_m - T_a}{\Delta T} F_2 \tag{20}$$

where ΔT is the temperature difference between hot and cold plate measurement surfaces, T_m is the mean test temperature, T_a is the ambient air temperature, and F_1 and F_2 are two error factors.

We can see from relation (20) that E_l will depend on the reduced temperature difference between the mean test temperature and the ambient temperature and also on the two error factors F_1 and F_2 characterizing the apparatus and the sample. F_1 and F_2 are very complicated functions expressing the reduced thickness of the specimen $g/2l$, of the guard width $g/2l$, and the parameter p under the form of a serie of powers of reduced variables.

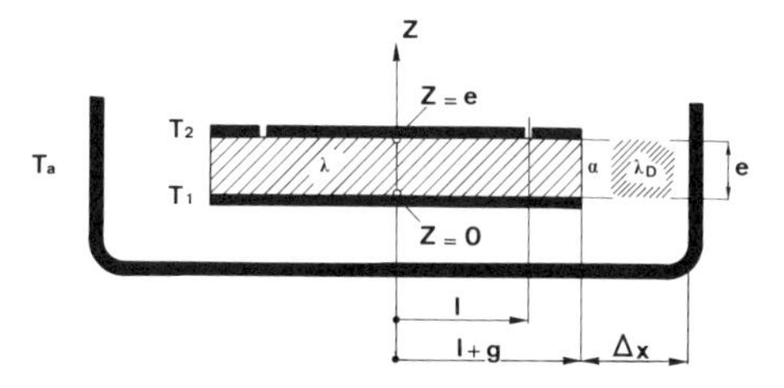

FIGURE 16. Guarded hot plate. Principle of operation; sketches from Ref. 91.

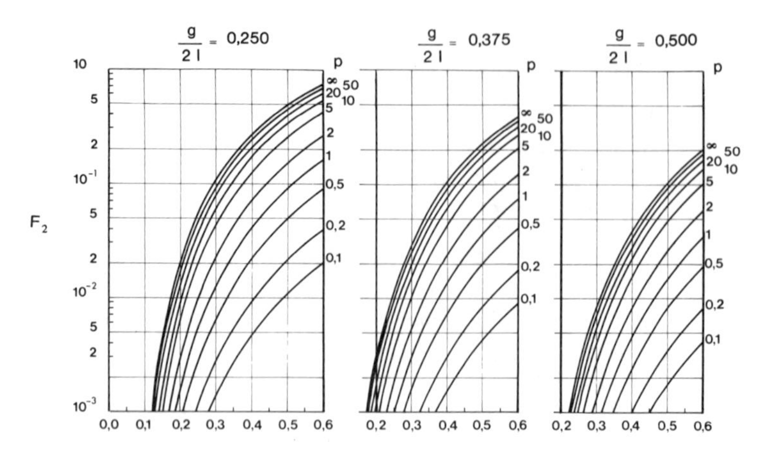

FIGURE 17. Curves allowing one to evaluate the error factor F_2 as a function of $e/2l$, $g/2l$, and p for a squared guarded hot plate.

The form of the functions F_1 and F_2 will depend on the shape of the plate (squared or circular). The computation of the error factors F_1 and F_2 requires powerful means of computation. These theoretical results clearly show the complexity of the thermal losses problem. Their advantage is to allow to have a set of curves, under a very general dimensionless form, permitting one in the limit of the assumptions considered to evaluate the error E_l, for any type of guarded hot plate apparatus (squared or circular). Such curves giving F_2 in the case of a squared guarded hot plate are shown in Fig. 17.

6.3. *Total Measurement Error*

The simultaneous treatment of edge heat losses and unbalance errors ($\Delta T_g \neq 0$) has led K.H. Bode[93] to determine theoretically the total measurement error $E_\Phi = E_l + E_g$. The relation expressing the total error E_Φ is of the form

$$E_\Phi = \frac{\lambda_m - \lambda}{\lambda} = Z_1 + \frac{T_m - T_a}{\Delta T} Z_2 + \frac{\Delta T_g}{\Delta T} Z_3 \tag{21}$$

where Z_1, Z_2, and Z_3 are the error factors in this very general case, function of the reduced dimensions of the sample, the lateral guard, the gap width between center and guard, of the sample conductivity, of the ambient air and of the factor characterizing the lateral heat transfer already mentioned.

The computation of the error factors Z_1, Z_2, and Z_3 as for the two factors F_1, F_2 requires powerful computational means but gives very general dimensionless curves. These results are not in contradiction with Woodside's.

Before concluding this section, we must also mention the use of the finite element method for the theoretical evaluation of the measurement errors. This numerical method has been recently applied to the study of the guarded hot plate by L.R. Troussart.[94] The whole set of results presented from Sections 6.1–6.3 allowing the evaluation of the unbalance error E_g, and the edge heat losses error, E_l, give us the possibility, either to determine experimentally the existing guarded hot plate characteristics by allowing us, on one hand, to correct the results obtained and, on the other, to optimize their use by limiting their field of application; or to evaluate, before construction, the characteristics required for $[g/2(l+g)]$ in order to limit the total error $E_\Phi = E_g + E_l$ under a fixed value, depending on the field of application (specimen thermal conductivity and maximum thickness).

6.4. *Accuracy and Precision. Performances*

The computation of the error of the guarded hot plate method is developed in Sections 6.1–6.3. The error is due not only to the apparatus itself, but also to the measurement devices (instrumentation) and to the specimens to be measured. Generally speaking the apparatus built and working according to the characteristics specified above are capable of measuring thermal conductivity accurately to withing ± 2% when the room temperature is near the mean temperature of the apparatus. The accuracy may be evaluated to ± 5% in the full operating range of an apparatus if all the adequate precautions in the design of the apparatus were taken.

It is necessary to adapt each type of apparatus to a particular range of temperature (for example, measurements at cryogenic and low temperatures, near the ambient temperature, at high and very high temperature, etc.). The precision (reproducibility) of the apparatus, when measurements are made on the same reference specimen removed and then mounted again after large time intervals is situated between ± 1% and ± 2% (depending on the range of temperature). For mean temperatures close to ambient temperature, the reproducibility of the method may be better than ± 1%. Differences in measurements of less than 2% may be considered insignificant.

7. LIMITATIONS DUE TO SPECIMEN

7.1. *Significance*

If the heat transfer through a specimen of a homogeneous material, at the mean temperature $\bar{T}$, is a purely conductive phenomenon (molecular agitation, vibration of the network, etc.), the density of heat flux and the temperature gradient are linearly dependent (Fourier's law), as we have shown in Section 1. In this case, and only in this case, using the guarded hot plate method and measuring the power dissipated in the central metering section, the "cold" and "hot" temperatures of the sample (equal in some cases to those of the plates), and the sample thickness, we can compute the thermal conductivity of this sample from equations (8) and (9) in Section 3.

In this case, the computed thermal conductivity is a specific property of the material, measured on a representative sample. Its value is independant of the measurement conditions except for the mean temperature $\bar{T}$.

Very often, the poorly conductive materials, and in particular insulants, are porous materials, that is to say they are heterogeneous materials consisting of two finely divided phases: a solid phase representing the matrix of the porous medium and a gaseous phase filling the voids of the solid matrix. In such materials, the mechanism of heat transfer is more complex with conduction in the solid and the gaseous phases, with radiation, and in some cases with natural convection due to movements of the interstitial gas under the influence of the temperature gradient.

If the total heat transfer is mainly conductive (nonmoving interstitial gas and low radiative component), the notion of thermal conductivity is still valid, as for homogeneous materials. It is commonly referred to as "equivalent" or "effective" or "apparent thermal conductivity." It is an extrapolation to porous media of a notion initially used only for homogeneous media.

In many cases, especially but not only for porous media of high porosity (semitransparent media), the heat transfer is represented by a coupling of condution and radiation. In such cases, we no longer have a linear relationship between the density of heat flux and the thermal gradient, and the notion of thermal conductivity is no longer valid. This fact shows itself experimentally in a dependence of the computed thermal conductivities and the measurement thicknesses, i.e., in a dependence of the results and the measurement conditions. But this is not the only parameter on which will depend the results of the measurements.

The guarded hot plate method can still be used for samples presenting such complex heat transfer mechanisms: in this case it is preferable to use the notion of thermal resistance R, at the mean temperature $\bar{T}$, in place of the thermal conductivity†

$$R = \Delta T/\phi$$

†It is only in the case of a purely conductive heat transfer that one can write

$$R = \frac{\Delta T}{\phi} = \frac{e}{\lambda}$$

where ΔT is the temperature differences between the "hot" and the "cold" faces of the sample, and ϕ is the density of heat flux. This thermal resistance characterizes the sample, the operator having to be precise with the measurement conditions: he must state, among other parameters, the sample thickness, the temperature difference, and the plate emissivity.

It is only after a critical analysis of the measurement results obtained from a guarded hot plate apparatus that one can determine if the thermal conductivity of the material may be computed or if only the thermal resistance of the specimen is relevant. (Experimentally precise criteria for such an analysis are given in Ref. 14).

7.2. *Factors Influencing Thermal Properties*

The thermal transmission properties of a specimen[14] of material may

1. vary due to the variability of the composition of the material or sample of it;
2. be affected by moisture or other factors;
3. change with time;
4. change with mean temperature;
5. depend upon the prior thermal history.

It must be recognized, therefore, that the selection of a typical value of thermal transmission properties (thermal conductivity or thermal resistance) representative of a material in a particular application should be based on a consideration of these factors and will not necessary apply without modification to all service conditions.

8. REALIZATIONS AND MEASUREMENT RESULTS

The principle of the guarded hot plate method, relying on a single physical basis consisting in the realization of a constant, unidirectional heat flow density quantitatively determined in the measurement zone, has been experimentally set up in different ways depending on the aim pursued and on the designer's imagination. The more representative versions will be indicated in Tables 4, 5, and 7 with references to the corresponding bibliography. The results obtained with these apparatus, despite the variety of the constructions, will often be comparable if the object of the measurement is the same and if the realization principles have been correctly applied.

In general, depending on the aims pursued, the apparatus differ by the size of the plates and consequently by the size of the specimens, and also by the range of temperatures which can be investigated. According to this last criterion, we will classify the apparatus in three categories:

a. apparatus working around the ambient temperature ($0 < T < 100^\circ C$);
b. apparatus working at very low and low temperatures (-200; $-80 < T < 0^\circ C$); and
c. apparatus working at high and very high temperatures ($100 < T < 400$, 700, 1000 $\cdots$ 1200°C).

We will say some words about the two first categories but will concentrate our attention on the third one, considering the objectives of this compendium. A more detailed description will be given for the apparatus realized or used in the CRIR Heat Transfer Laboratory.

8.1. Apparatus Working around the Ambient Temperature

8.1.1. Resolving Apparatus with a Biguarded Hot Plate for Measuring the Thermal Conductivity of Insulating Materials

This apparatus[39] was designed to allow the variation of the direction of the thermal gradient applied to the studied specimen, with respect to the gravitation field, in order to study the conditions of apparition of the natural convection movements inside an open cell porous medium. This aim implied realizing a one-specimen apparatus, with orientable plates. The version of the guarded hot plate method with a single specimen has been described in principle in Section 3.2. The construction details of the biguarded hot plate, as described in Ref. 39 are given in Fig. 18.

One can note the following characteristics:

- the limit temperature of use is 100°C;
- the lateral guard thermopile, made of a fine, 0.1-mm-diam wire, has 2 × 400 junctions points; the back-ground thermopile acting as a zero balance controller has 1000 junctions points;
- there is no metallic bridge between the central zone and the lateral guard; the gap of about $1\frac{1}{2}$ mm has four bridges in Epoxy resin;
- the plate in rectified "Duralumin," III.II, very carefully realized, ensures the planarity of the whole stack;

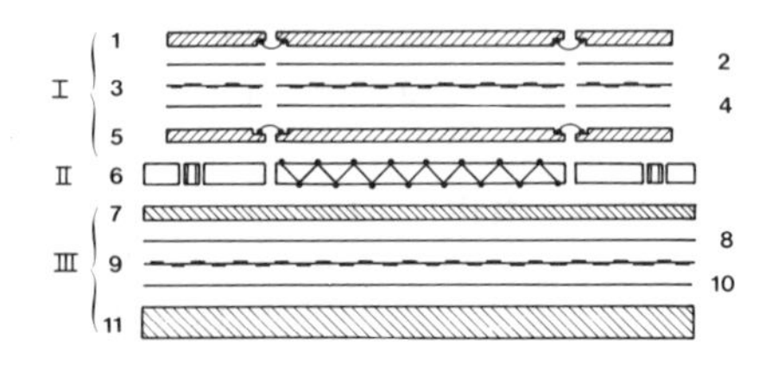

FIGURE 18. Structure of the biguarded hot plate.[39] (I) Guarded hot plate: I.1, I.5, copper plates; I.2, I.4, insulation plates (mica); I.3, heating resistances. (II) Zero balance controller. (III) Heating plate: III.7, III.11, copper plates; III.8, III.10, insulation plates; III.9, heating resistance.

- the structures of the heating elements are very similar to those described by Gilbo (ASTM, STP 217, 1957)[4];
- the exterior lateral dimensions are 40 × 40 cm, the central measuring zone being 23 × 23 cm;
- the heating plate of the back guard is thermally insulated on its free face.

The cold plate, realized in "Duralumin" represents a double spiral machined in the mass of the plate in which the liquid coming from a thermostat circulates at cross flow; its total thickness is 30 mm, the two faces being of 5 mm. The active face is rectified. A frame has been designed to support the whole stack consisting in the biguarded hot plate, the sample, and the cold plate, and also to allow its continuous rotation around a horizontal axis. The cold plate can be displaced continuously with respect to the hot plate while keeping a very good parallelism between the two measurement surfaces ("hot" and "cold").

The entire stack plates sample and frame is put inside an airtight enclosure (with respect to the ambient air of the laboratory). This enclosure allows one to maintain a temperature-controlled atmosphere around the structure and a sufficiently low degree of humidity in order to avoid any condensation on the cold plate or inside the specimen. This is obtained with a fan which makes the air of the enclosure flow through a cold radiator where water condenses and then through a heating element controlling the temperature at the desired level (mean temperature of the plates).

The zero balance controllers: center–lateral guard, and center–back guard are automatic; the heating of the central plate is realized with a stabilized direct current supply, at predetermined temperature set point.

For given conditions of temperature, the stationary regime is obtained within 3 to 12 h depending on the type of material under measurement. The reproductibility is better than ± 1% and the precision of about 2%. Figures 19 and 20 show comparative measurement results between this apparatus and two other apparatus of different design: a relative heat flowmeter apparatus working at a constant temperature of 24°C (I.I.F. Trondheim 1966,[6] and an apparatus of classic design: a guarded hot plate "Dynatech TCFG".[64] This last apparatus is described in Fig. 21. The results show a good agreement between the three different apparatus.

Figure 22 gives the variation of the thermal conductivity as a function of the temperature difference, for fiberglass samples saturated with liquid water and put in plastic bags (the area of the sealed samples is identical to the area of the center of the hot plate); mean temperature: 24°C; variable orientation of the thermal gradient: $\downarrow$ hot plate on top, no convection movement; $\uparrow$ hot

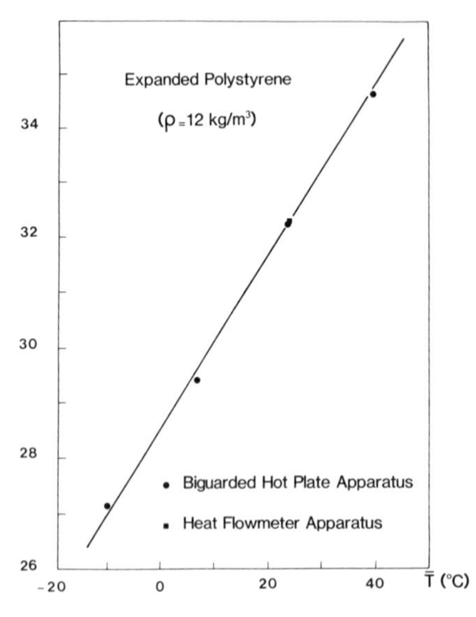

FIGURE 19. Comparison between a biguarded hot plate apparatus and a heat flowmeter apparatus.

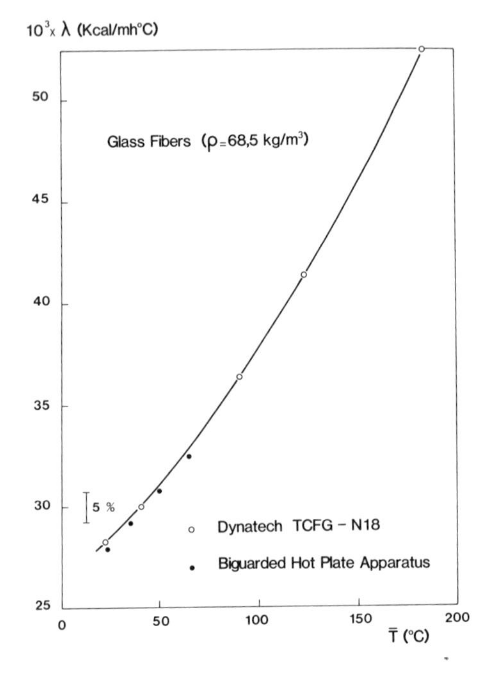

FIGURE 20. Comparison between results determined on a biguarded hot plate apparatus and on the Dynatech TCGF-N18.

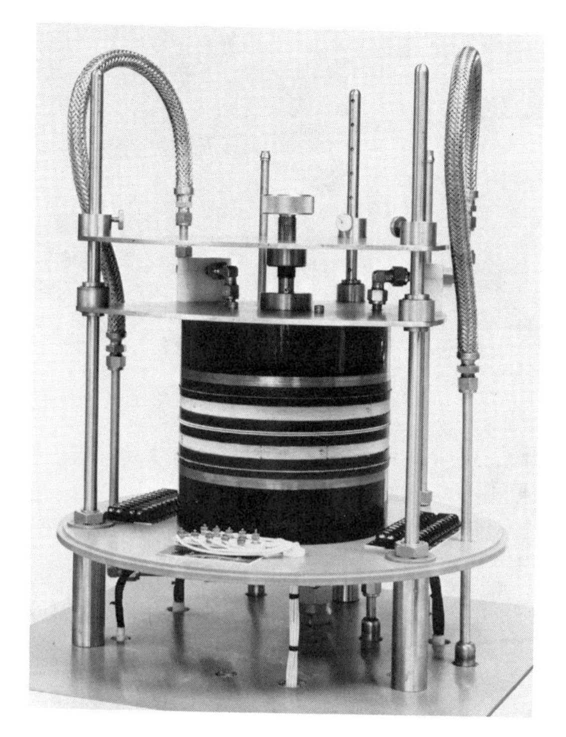

FIGURE 21. Model TCFGM guarded thermal conductivity apparatus.

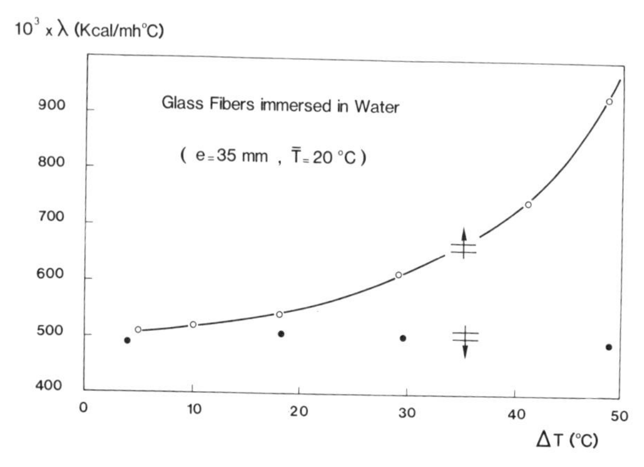

FIGURE 22. Variation of the thermal conductivity of a glass fiber specimen, immersed in water, as a function of the temperature difference and orientation of the thermal gradient with respect to the gravitational field.

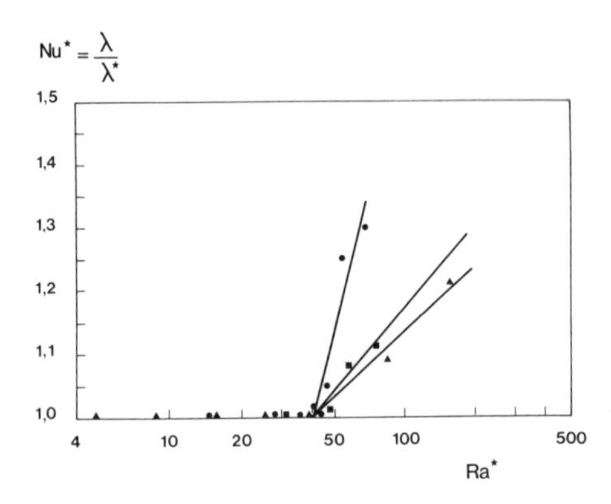

FIGURE 23. $Nu^* = f(Ra^*)$. Threshold of apparition of convective movements in a horizontal plane layer made of three types of glass fibers immersed in water.

plate at the bottom: convective movements starting for a critical difference of temperature ΔT and then function of ΔT.

The use of dimensionless numbers and of the similarity principle allows one to simulate the behavior of insulants with air or another gas as interstitial fluid, with open cell porous media (glass balls of fibers) saturated with water. Figure 23 expresses, in a more general way, with the filtration Nusselt, Nu^*, and Rayleigh, Ra^*, numbers associated to the fibrous media, the preceding result.

The filtration Rayleigh number, Ra^*:

$$Ra^* = g\,\frac{\beta(\rho c)_f}{\nu}\,\frac{4K_V K_H}{(\lambda_H^* K_V)^{1/2} + (\lambda_V^* K_H)^{1/2}} \cdot \Delta T \cdot d$$

Here

$$g\,\frac{\beta(\rho c)_f}{\nu} = f(\bar{T})$$

only depends on the nature of the interstitial gas and on the reference temperature; g is the gravity acceleration (m/s^2); β is the thermal expansion coefficient of the fluid (K^{-1}); ν is the kinematic viscosity of the fluid (m/s^2);

$$\frac{4K_V K_H}{(\lambda_H^* K_V)^{1/2} + (\lambda_V^* K_H)^{1/2}}$$

is the ratio characterizing the anisotropic porous medium; K_V, K_H are the vertical and horizontal permeabilities (m^2); λ_V^*, λ_H^* are the vertical and horizontal thermal conductivities with nonmoving interstitial fluid (W/mK); ΔT and

$\bar{T}$ characterize the thermal conditions; and d is the distance between hot and cold faces.

The filtration Nusselt number, Nu*:

$$\text{Nu}^* = \frac{\phi}{\phi^*} \quad \text{or} \quad \frac{\lambda}{\lambda^*}$$

Here ϕ is the density of heat flux in the presence of natural convection; ϕ^* is the density of heat flux with nonmoving interstitial fluid; λ is the thermal conductivity (equivalent or apparent) in the presence of natural convection; and λ^* is the thermal conductivity (equivalent or apparent) with nonmoving interstitial fluid. (In a $\underset{\downarrow}{=}$ configuration, the apparatus determines the λ^* values as a function of $\bar{T}$).

For a horizontal insulation, hot temperature at the bottom, the interstitial fluid will not move, therefore Nu = 1, if

$$\text{Ra}^* < \text{Ra}_c = 40 \qquad \text{for small } \Delta T$$

$$\text{Ra}^* < \text{Ra}_c = 40 - 6\,\Delta T/T \qquad \text{for large } \Delta T$$

In the case of Fig. 23, ΔT can be considered small and the criterion of nonappearance of convection is verified. Heat transfer with natural convection in horizontal, sloped, or vertical layers can be represented in a more general way by a function of the form

$$\text{Nu}^* = f(\text{Ra}^*, \mathscr{A}, \cos\alpha)$$

where $\mathscr{A}$ is the aspect ratio (ratio between the height l and d) (ratio between the height l and d), and α is the slope of the layer. For more detailed explanations concerning heat transfer by natural convection, see the articles given in Refs. 39, 59, and 7.

A more recent version of the biguarded hot plate working at temperatures near ambient temperature ($10 < T < 40°\text{C}$) has been designed for the measurement of the thermal resistance of thick low-density insulants (mineral fibers especially).[44] The new instrument is of the asymmetrical type, using one specimen with heat flowing from bottom to top. The plates dimensions are $500 \times 500\,\text{mm}^2$ and the heat flow density is measured within a central area of 250×250 mm.

The originality of this instrument stems from the fact that it offers to the operator the possibility of performing both relative and absolute measurements on full thickness specimen which can reach 120 mm. Relative measurements

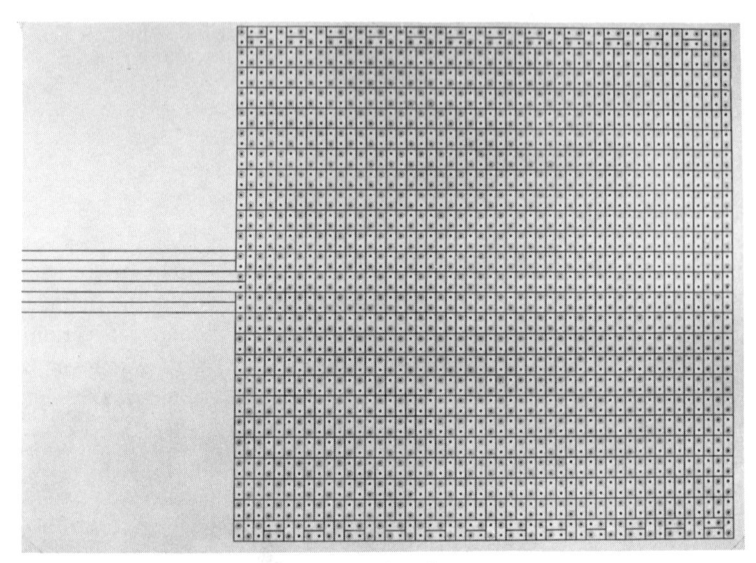

FIGURE 24. Heat flowmeter.

are made possible by a heat flowmeter located in contact with the cold plate. One face of the flowmeters makes up the isothermal cold surface. Both measurement and control heat flowmeters are built according to a new technique that wholly eliminates the use of thermocouples, wires, and welds. Cold and hot junctions are made with the aid of metallization and photoetching (Fig. 24). These techniques made possible the construction of heat flowmeters of large dimensions ($250 \times 250\,mm^2$ and more). The sensitivity of these flowmeters is high ($\approx 10^4\ W/m^2\ V$) due to the considerable increase in the number of junctions per unit area (1250 junctions for a metering area of $250 \times 250\,mm^2$). The aforementioned techniques allow the heat flowmeter to be produced with a high degree of uniformity and repeatibility.

The apparatus itself allows measurements of thermal resistance on thick specimens without slicing. It made possible the study of the correlation between dispersion of the measurement results and the specimen measurement area related to the maximum allowed thickness (which implies the slicing of the samples for small size apparatus[44]). This apparatus also allowed the evaluation of the "thickness effect," resulting in the coupling of the conductive and the radiative transfer in semitransparent light insulants at temperatures near ambient. A large number of studies have been devoted to this subject (see ASTM STP, No. 718).[4]

Comparisons between the measurement results obtained with this apparatus and the previous one on two reference materials are shown in Figs. 25 and 26.

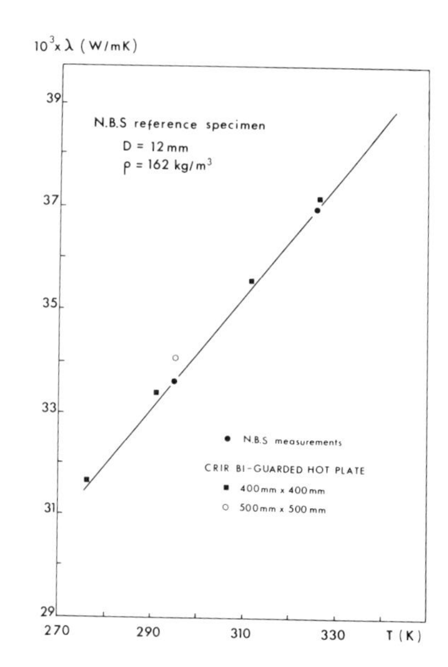

FIGURE 25. Thermal conductivity versus mean temperature for N.B.S. reference specimen.

In all cases, differences between results remain less than 2%. In Table 4, other designs of apparatus working at temperatures near ambient are mentioned. Among them, we must cite the "Robinson line heat source guarded hot plate apparatus," the principle of which, although very smart, departs in a certain way from the principle of a unidirectional density of heat flow.[41,45,46]

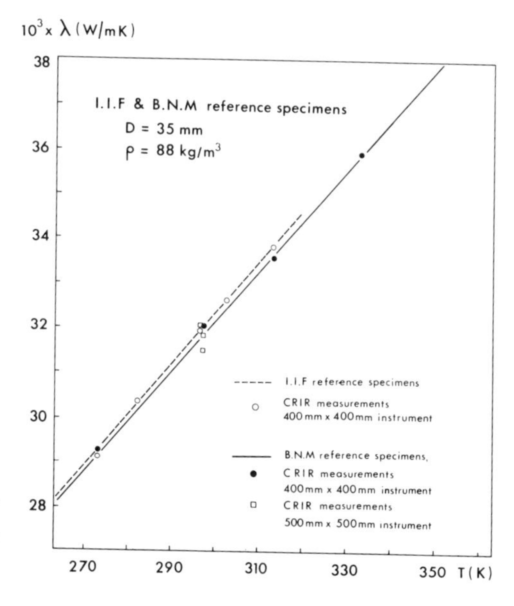

FIGURE 26. Thermal conductivity versus mean temperature for I.I.F. and B.N.M. reference specimens.

TABLE 4
Apparatus Working Around Ambient Temperature

Authors	Year	Range of temp. (°C)	Plates position and dimensions (mm)	Ambience Gas	Pressure	Materials studied	Scope	Particularities of the apparatus	Ref.
Verschoor, J.D., and Greebler, P.	1952	50...100	 ϕ 250	Air, He, CO_2, Freon	10^{-3}...760	Glass fibers 8–134 kg m^{-3}	Mechanism of heat transfer	Second lateral guard	32
Zabawsky, Z.	1957	12...30	 □ 200	Air	760	Cork, glass fibers		Automatic control	34
Marechal, J.-C.	1963	−10...100	 □ 500	Air	760	Building materials and insulants	Thermal parameters, $3 \cdot 10^{-2}$–2 W m^{-1} K^{-1}	Second lateral guard	96, 33
Fritz, W., and Bode, K.H.	1965	20...90	 80 ϕ 100 120	Air	760	Glass, quartzglass, Teflon, Ti alloy	Errors evaluation; reference materials	Control by fluid flow; uni-directional heat flow	35
Fournier, D., and Klarsfeld, S.	1969	−10...90	 400	Air	760	Glass fibers, reference materials	Natural convection	Ambient-controlled temperature; the thermal gradient can be revolved	39
Fritz, W., and Kuster, W.	1970	0...70	 ϕ 200	Air	760	Cellular materials, 16–150 kg/m^{-3}	Errors evaluation; thermal parameters; mechanism of heat transfer	Back and lateral guard plate controlled by fluid flow	38
Bankvall, C.	1972	0...50	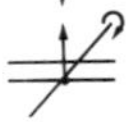 ϕ 350	Air	10^{-4}...760	Mineral fibers	Mechanism of heat transfer	Revolving thermal gradient; rapid steady state; fully automatic control	40
Hahn, M.H., Robinson, H.E., and Flynn, D.R.	1974	−40...260	 ϕ 300	Air	760		Theory of the method design; evaluation of errors	Homogeneous heat plate; linear heat source	41, 42
Degenne *et al.*	1978	10...40	 500 610	Air	760	Low-density and thick fibrous materials	Quality control; "thickness effect"	Rapid steady state; full thickness measurements	44
Hahn, M.H.; Hahn, M.H., and Ober, D.	1981	−20...200	 ϕ 1000	Air	Vacuum 760	Low-density and thick fibrous materials	Reference materials; 25, 75, 150 mm; "thickness effect"	Linear heat source; homogeneous heat plate; thickness of samples ⩽ 350 mm; controlled ambient temperature and humidity	45, 46
Bomberg, M., and Solvason, K.R.	1981	24	□ 450 600	Air	760	Low-density polysytrene	Errors evaluation and "thickness effect"		92

8.2. Apparatus Working at Low and Very Low Temperatures

The thermal conductivity measurements were made at mean temperatures ranging between 123 and 300 K by means of a modified guarded hot plate apparatus of Dynatech TCFG.[59] Since the thermal conductivity was to be measured without any natural convection effect, the normal apparatus was modified by introducing a heat flowmeter in place of the upper specimen of the guarded hot plate stack. The heat flowmeter was connected to a control device in order to have the lower surface of the upper "auxiliary heater unit" at the same temperature as the upper surface of the guarded hot plate by maintaining the corresponding heat flow at a mean null value. In this way the apparatus was modified to operate as a one-specimen apparatus, with heat flowing vertically from the top to the bottom. The thickness of 200-mm-diam specimens ranged from 24 to 28 mm (Fig. 27).

Measurements were taken at six mean temperatures in the range of 123 to 295 K. A cryostat was used as a cold basic supply to obtain the three upper values, and liquid-nitrogen cooling was used for the three lower values with the rate of flow of liquid nitrogen being controlled by means of an electrovalve and a thermocouple attached on the lower cold plate. All measurements were made with the test stack covered with a glass bell jar filled with dry air. The test stack and the interspace surrounding were packed with predried "Santocel" powder to reduce heat losses.

The temperature differences between the faces of the test specimen were maintained purposely at low values ranging between 20 and 70 K.

The results obtained for λ^* with this modified apparatus were compared with previous measurements made by means of the biguarded hot plate apparatus[39] at mean temperature ranging from 263 to 353 K and also with results obtained with a heat flowmeter apparatus (I.I.F, Trondheim 1966).[6] Values of λ^* provided by these three apparatus of different designs were in excellent agreement with each other in the temperature range common to them. The largest difference was 3% despite the measurements being made on specimens having different areas.

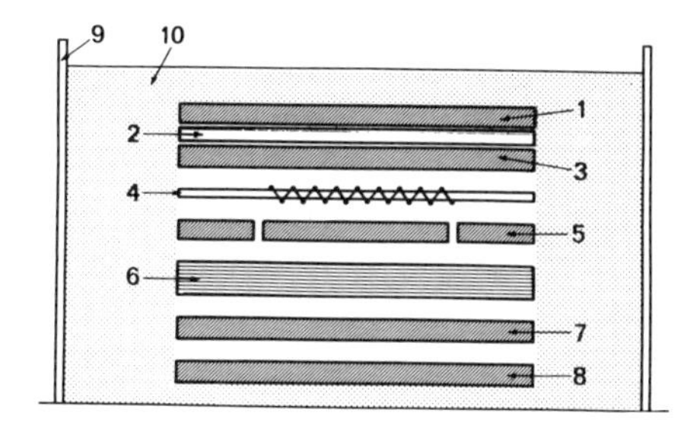

FIGURE 27. Schematic diagram of the modified Dynatech guarded hot plate apparatus. KEY: 1, top refrigerated heat sink; 2, insulation plate; 3, auxiliary heater unit (top "cold plate"); 4, heat flow transducer; 5, heating unit (central and guard sections); 6, test specimen; 7, auxiliary heater unit (bottom "cold plate"); 8, bottom refrigerated heat sink; 9, cooling shroud; 10, high-efficiency insulation.

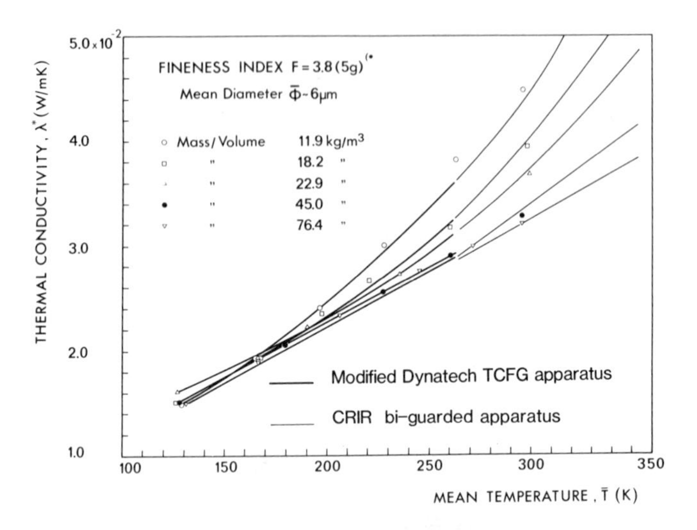

FIGURE 28. Thermal conductivity versus mean temperature for TEL glass fiber materials. Fineness index 3.8 (5 g). Five values are considered for mass/volume.

The results of the thermal conductivity measurements obtained on glass fiber insulants of different densities, ρ, are shown in Fig. 28. The "fineness" index F indicated in this figure is defined from a particular air-flow permeability measurement on a randomly compacted glass fiber plug of given mass (indicated in grams between parenthesis) introduced in a rigid cylindrical cell of fixed dimensions.

If we consider all of these results we can see λ^* values ranging at room temperature from 32 to 45 mW/m K for the whole set of glass-fiber materials tested. This interval becomes narrower as the temperature decreased and is especially small when $\bar{T}$ has decreased to below 198 K. When $\bar{T} = 123$ K all of the λ^* values are between 15 and 16 mW/m K. The consideration of these λ^* data leads to the conclusion that at very low temperatures the parameters ρ and F have very little influence on the thermal conductivities of the mineral fiber materials. This is due to the fact that in this low range of temperature the radiation heat transfer is not significant.

Below 198 K, the thermal conductivity values of mineral fiber insulants are very similar to those currently used in cold storage or cryogenic insulation, for example, cellular plastics (Fig. 29) or granulated insulations, but this also supposes that natural convective heat transfer does not exist. This requirement leads, in some cases, to a limit on the possible range of densities or fineness to be used. Other apparatus designed for this range of temperatures are mentioned in Table 5.

TABLE 5
Apparatus Working at Low Temperatures

Authors	Year	Range of temp. (°C)	Plates position and dimensions (mm)	Ambience		Materials studied	Scope	Particularities of the apparatus	Ref.
				Gas	Pressure				
Cammerer, W.F.	1960	−150...50	 □ 500	Air	760	Mineral fibers; 40–330 kg m^{-3}; cellular plastics	Extension of the GHP method at low temperature		50
Achtziger, J.	1964	−180...50	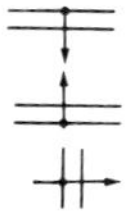 □ 500	Air	760	Low-density mineral fibers ($\rho = 8$ kg m^{-3})	Natural convection	Three positions of the thermal gradient	52
Karp, G.S., and Lankton, C.S.	1967	−100...25	 □ 350	Vacuum	10^{-5}	Multi-layer cryogenic insulation	Sources of errors; effects of pressure and temperature gradients	Measurement of very low thermal conductance	54
Brendeng, E., and Frivik, P.E.	1969 – 1974	−180 ...50	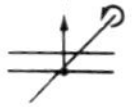 □ 400 □ 800	Different gases Air	Vacuum to 3 bars	Mineral fiber insulants	Mechanism of heat transfer; natural convection	Fully automatic control	55, 60, 74
Tye, R.P.	1971	−173...27	 ϕ 200	Different gases	Vacuum to 760	Cellular and granulated insulants	Thermal characteristics of of insulation material	Dynatech TCFG Model	56
De Ponte, F., and Di Filippo, P.	1971 – 1976	−173...27	 □ 300	Air	760	Cellular plastics and mineral fibers	Influence of thermal gradient; thermal characteristics of insulating material	The cold plates are cooled by an air flow coming from a cryo-generator; fully automatic control	58, 61
Marechal, J.-C., Doussin, and Marechal	1973	−150...80	 □ 500	Air	760	Cellular, fibrous, and granular insulation	Thermal and mechanical properties	Second lateral guard and outside guard for low temperature	37, 97, 99
Fournier, D., and Klarsfeld, S.	1974 –	−150...27	 ϕ 200	Dry N_2	760	Fibrous and cellular insulating materials	Use of insulants at low temperatures, taking into consideration the natural convection	Modified Dynatech TCFG GHP apparatus	59, 100

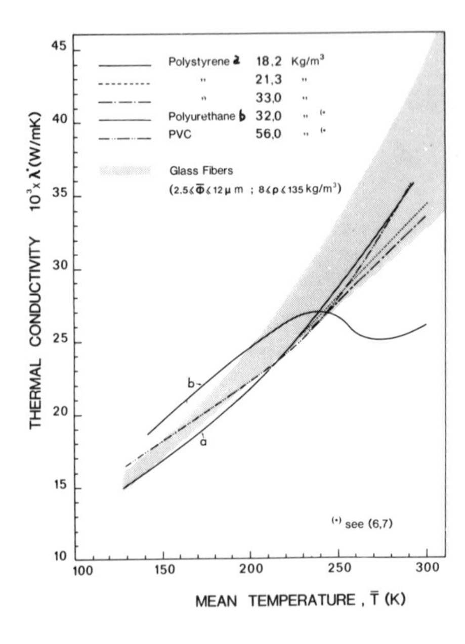

FIGURE 29. Thermal conductivity versus mean temperature for different cellular plastics.[56]

8.3. Apparatus Working at High and Very High Temperatures

The Dynatech TCFG apparatus already mentioned in Section 8.1 has been used at mean temperatures up to 900°C when equipped with special "high-temperature" plates. Figure 30 gives a sketch of the apparatus. The thermal conductivities, as a function of temperature, of four "refractory" products – alumina fibers, pure silica fibers, calcium silicate, and aerosil powders agglomerated with alumina fibers (product of the MIN-K type) – are presented in Fig. 31. These products cover the range of thermal conductivities of the more representative products used in the industry. Considering on the one hand, their relatively good mechanical resistance as a function of temperature, and on the

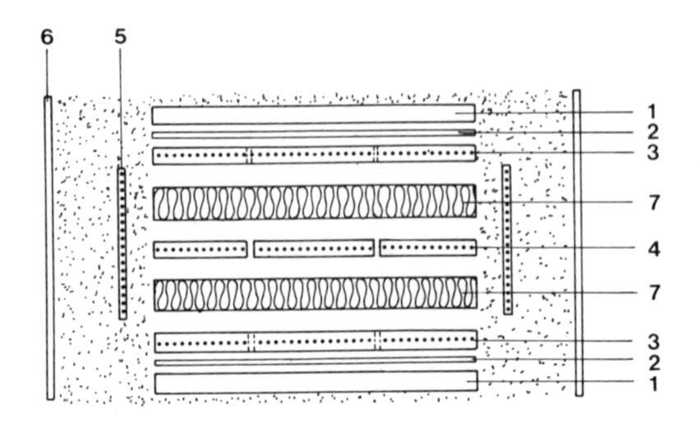

FIGURE 30. Guarded hot plate apparatus "model TCFG-N18." KEY: 1, refrigerated heat sink; 2, insulation; 3, auxiliary heater unit (cold plate); 4, guarded hot plate (central and guard ring); 5, outer-guard ring; 6, cooling shroud; 7, sample.

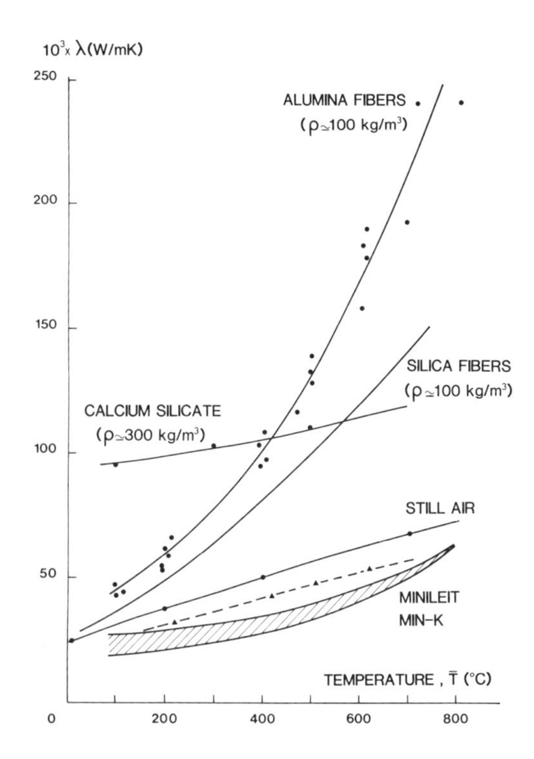

FIGURE 31. High-temperature thermal insulating materials.

other their low contact thermal resistances with the isothermal surfaces of the apparatus, these materials have been chosen as reference materials to delimit this zone. They are also used to check the working conditions of the apparatus and state their reproductibility.

8.3.2. Rapid Biguarded High-Temperature Thermal Conductivity Apparatus

The general principle of the apparatus[69] is similar to the one previously described for temperatures near ambient in Section 8.1. A biguarded hot plate, in which the central part is supplied with stabilized direct current, generates a heat flow through a single sample towards an auxiliary hot plate ensuring the required level of cold temperature (Fig. 32). Several requirements are necessary to allow the rapid measurement of the thermal conductivity of insulating products at high temperature and in stationary conditions. These requirements must permit increasing the number of measurements on an identical type of product in order to obtain statistical data. They suppose an apparatus with the metering area of the heating unit and the auxiliary heater unit ("hot" and "cold" plates) constantly maintained at a given temperature; with a system of change of the

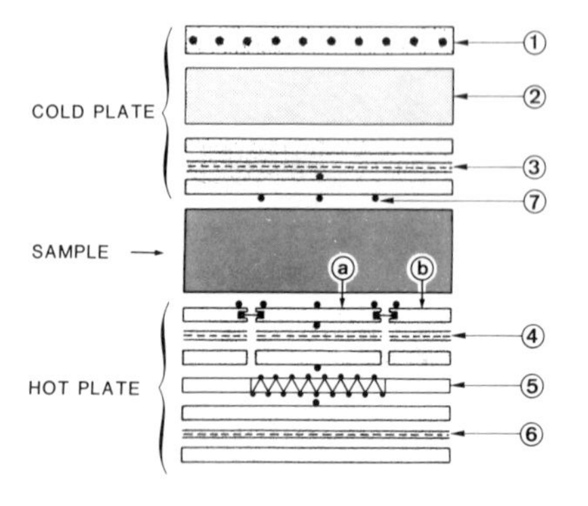

FIGURE 32. High-temperature biguarded hot plate apparatus (schematic diagram of the cold and hot plates). KEY: 1, refrigerated heat sink; 2, insulation plate; 3, auxiliary heater unit (top "cold plate"); 4, heating unit: central (a) and guard ring (b); 5, heat flowmeter (zero-balance controller; 6, lower guard plate; 7, thermocouple.

sample allowing continuous work and making a direct contact between the high-temperature plates and the ambient air impossible and permitting a preheating of the sample (Fig. 33); and with automatic controllers, reducing as much as possible the time necessary to reach a stationary regime, ensuring constant temperatures (T_C and T_H) by a simple display, and simultaneously allowing a precise measurement of the power dissipated in the central part of the hot plate (Fig. 34). To fit these requirements, the apparatus has the potential of accepting simultaneously three samples of identical thickness in a measurement tunnel: the central sample is under measurement; the left sample is to be measured while the last one has already been measured and is ready to be taken out of the apparatus once a new sample is introduced, the whole system working from left to right (Fig. 33). The "cold" unit in the top center of the tunnel can be slightly displaced vertically in order to ensure a contact, under controllable pressure, between the plates during the measurement ("low" position) and to allow the exit of the sample ("high" position) when it must be replaced.

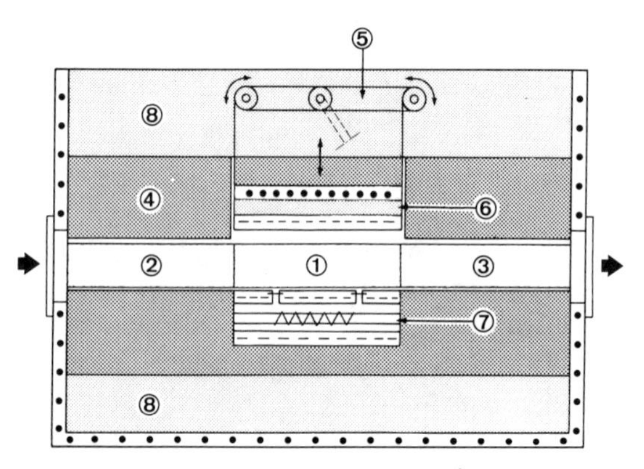

FIGURE 33. High-temperature biguarded hot plate apparatus; measurement tunnel. KEY: 1, sample under measurement; 2, sample to be measured; 3, measured sample; 4, tunnel in refractory bricks; 5, displacement system; 6, cold plate; 7, hot plate; 8, insulation.

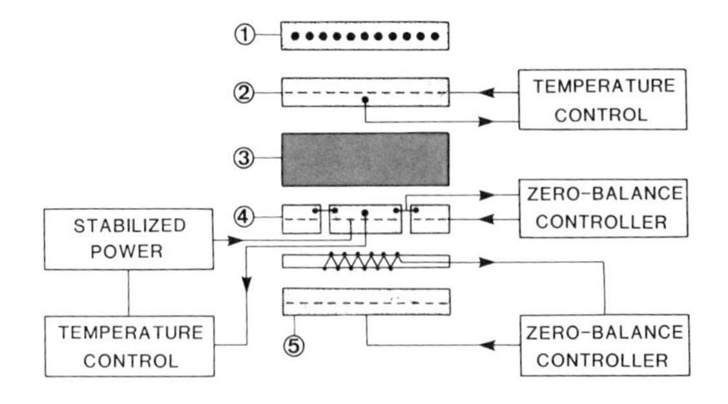

FIGURE 34. High-temperature biguarded apparatus. Temperature and zero-balance controllers. KEY: 1, refrigerated heat sink; 2, auxiliary heater unit; 3, sample; 4, heating unit: central and guard ring; 5, lower guard-plate.

The technical characteristics and the performances of the apparatus for a mean temperature of 400°C are given in Table 6.

The time needed to reach steady state varies between 3 and 10 hours depending on the type of mineral fiber insulant (Fig. 35). The reproducibility of the apparatus is of about ± 2.5%. The accuracy lies somewhere between ± 2% and 5%. This order of magnitude has been determined from comparative measurements with other types of apparatus working in a common or a complementary range of temperature (Figs. 36–38). Three types of fibrous insulants

TABLE 6
High-Temperature Biguarded Hot Plate Apparatus: Technical Characteristics and Performances

Specimen	
Surface	300 × 300 mm
Thickness	38–40 mm
Metering area	150 × 150 mm
Guarded HP + Cold Plate	
Total area: center + guard	300 × 300 mm
Metering area	150 × 150 mm
Range of possible temperature	100–700°C
Temperature heterogeneity	< ± 1°C
Temperature stability	± 0.25°C
Planarity	0.1 mm
Parallelism	0.1 mm
Thickness measurement	± 0.1 mm
Measurement devices	
Digital voltmeter	200,000 digits
Control thermocouples, power	Resolution: 1 μV
Time needed to reach steady-state (fibrous insulants)	3 to 10 h
Reproducibility	± 2%–3%
Accuracy	± 2%–5%

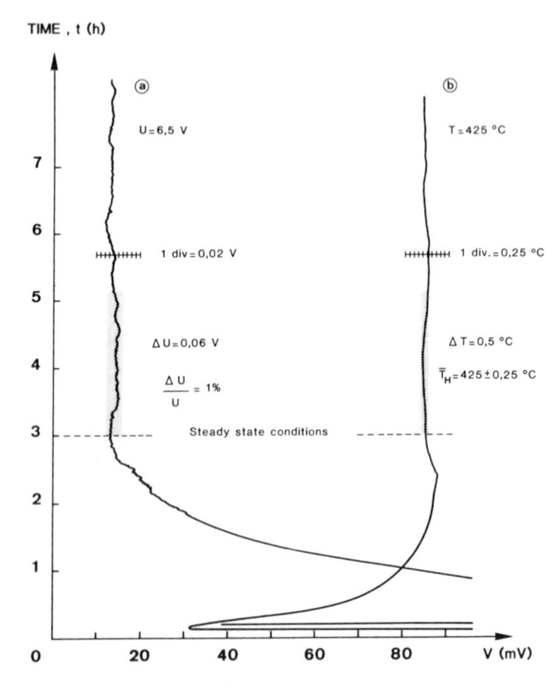

FIGURE 35. Running-up process (reaching the temperature of the hot plate). (a) Voltage, U, measured at the heating resistance connections during heating time. (b) Hot plate temperature (emf indicated by the control thermocouple).

have been taken as reference samples. Among those materials we find the RBGFB (resin-bonded glass-fiber board) reference material from the BCR (Community Bureau of Reference, Commission of the European Communities) used between 170 and 370 K.

The rapid biguarded high-temperature thermal conductivity apparatus made it possible to confirm the statistical nature of the results obtained on heavy-density products of given nominal characteristics and to confirm that the correlations thermal conductivity–density and specific surface ("fineness") existing at ambient temperature[(59)] remain valid at higher temperature, but that their importance increases considerably (Figs. 39, 40, and 41).[(69)]

Other installations used in this range of temperature are mentioned in Table 7.

8.4. *Computer-Aided Testing*

As in many metrological activities, the determination of the thermal conductivity in a stationary regime by the method of the guarded hot plate has

TABLE 7
Apparatus Working at High Temperatures

Authors	Year	Range of temp. (°C)	Plates position and dimensions (mm)	Ambience		Materials studied	Scope	Particularities of the apparatus	Ref.
				Gas	Pressure				
Rolinski, E.J., and Purcell, G.V.	1966	100...800	 □ 300	Vacuum Air Ar	760	Fibrous HT insulants	Mechanism of heat transfer; thermal characteristics	Dynatech T-3000 Model	62
Ferro, V., and Sacchi, A.	1968	50...1000	 □ 500	Air	760	Mineral fibers and granular insulants	Thermal characteristics	Guard zone of four independent parts; automatic controllers	63
Tye, R.P.	1969	27...600 	□ 300 ϕ 200	Vacuum Air Kr	10^{-4}–760 variable	MIN-K ~320 kg m^{-3}	Reference material	Outside guard annulus; Dynatech TCFG Model	64
Cammerer, W.F.	1972	200...900	 □ 500			Perlite concrete	Influence of density and crystallization water		65
Liermann, J.	1974	100...1000 	ϕ 200	He Air	10^{-3}–760	Graphite fibers. pure silica	Thermal characteristics		67
Boulant, J., *et al.*	1981	100...600	 □ 300	Air	760	Mineral fibers; insulants	Thermal characteristics; quality control and correlation between structure and thermal resistance	Rapid high-temperature thermal conductivity apparatus	69
Kamiuro, K., *et al.*	1982	27...800	 □ 400	Air	760	Alumino-silicate fibers	Radiative heat transfer	The temperature of the cold plate is maintained constant (~ 25°C)	70

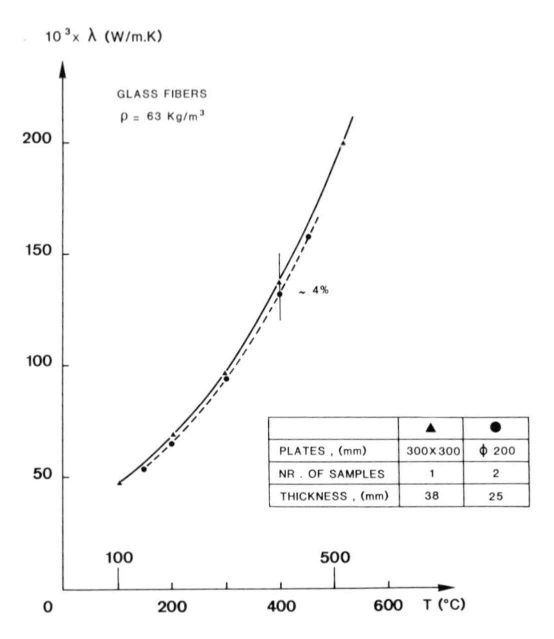

FIGURE 36. Thermal conductivity versus mean temperature, measured for a glass fiber reference specimen. ▲, High-temperature biguarded hot plate apparatus; ●, Dynatech TCFG guarded hot plate apparatus.

been more or less automated with the help of a computer in many laboratories in recent years.

Schematically, the guarded hot plate is then connected to a data acquisition and control system, and a desktop computer, which allow the automatic operation of the installation for the determination of the thermal conductivity as a function of temperature in a given range.

Such systems, set the temperature controllers, measure and evaluate the test data, check the thermal equilibrium, calculate the thermal conductivity, and store the results and advance the controller to the next set point. Computer-aided testing generally increase the efficiency of the installation and facilitates the work of the operating personnel.

9. REFERENCE MATERIALS AND INTERLABORATORY COMPARATIVE MEASUREMENTS

Despite the simplicity of the method, we cannot obtain high performances with guarded hot plate apparatus – far from it:

An examination of the documentation published by several laboratories relative to the experimental determination of the thermal conductivity, reveals

> that for the majority of building materials the values given differ greatly from each other; it has been unknown up to what extent such differences are the results of disparities inherent to the nature of the materials themselves or arise from the diversity of methods and testing conditions or from differences in the equipment employed in the determination (1963, RILEM,[26]).

The large dispersion of the results given by different laboratories, for the same material can be explained, especially in the past, by the lack of a precise method for computing the measurement error; by the very different performances of the sensors and the measurement instruments associated to the apparatus; and by the lack of a satisfactory standard dealing with the design of the apparatus and the measurements setup. To try to evaluate these differences between laboratories and especially to understand their causes, several round-robin tests, most of them being international, have been organized in the last 30 years.

Table 8 presents a summary of the main round-robin tests which have been organized by different associations between 1951 and 1981. Indeed, the analysis of the results shows that the dispersion may reach ± 50%. A closer look shows that most of the time a part of the dispersion can be attributed either to different measurement conditions (mean temperature, temperature difference, etc.) or to poorly defined reference materials.

The outlining of reference materials in the different ranges of thermal conductivities and temperatures of use is a major problem in order to validate

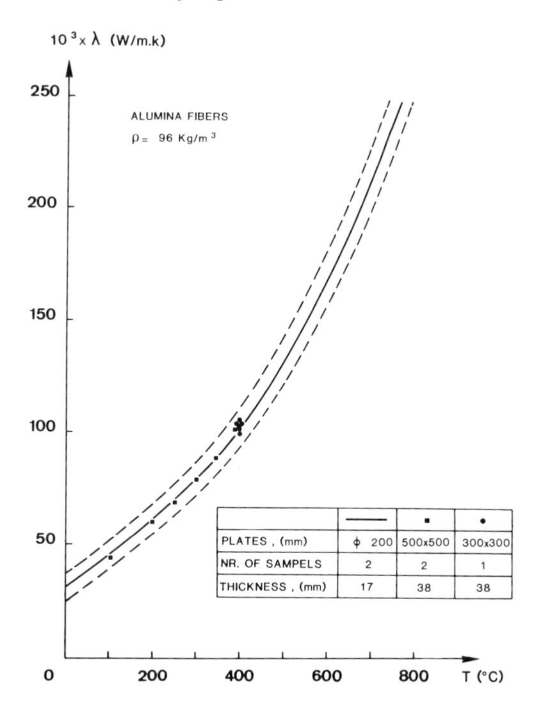

FIGURE 37. Thermal conductivity versus mean temperature, measured for an aluminum fiber reference specimen. —, Dynatech TCFG guarded hot plate apparatus (mean curve and total statistical dispersion zone); ■, CRIR, 500 × 500 mm, guarded hot plate apparatus; ●, CRIR, new high-temperature biguarded hot plate apparatus.

TABLE 8
Thermal Conductivity Round Robin for Guarded Hot Plate Apparatus

		Participants		Apparatus			Laboratory or (and) reference materials						
									Material				
Year	Leader	Number of apparatus	Number of countries	Dimensions (mm)	Range of temp. (°C)	$\bar{T}$ (°C)	Laboratory	Reference value	Nature and dimensions	Density (kg m^{-3})	Thickness (mm)	Results	Ref.
1951	ASHVE and ASTM	20	1	200–600	−6–55	Two or three	NBS (200 × 200 mm)	λ at 10°C	Cork	117	25	± 12% (total spread) ± 3% (75% of the results)	25
1963	RILEM	13	9	300–500	−10–800	?			Marble 300 × 300 500 × 500		38; 50	± 50%	26 40
									Cork 300 × 300 500 × 500		38; 50	± 10%	
1963	CERFIM	5	5						Mineral fibers	90	50	23.4%	27
1963	I.I.F.	20	16	200–1000	0–40	0 10 20	NPL	$\lambda = f(T)$	Resin-bonded glass fiber boards	88	38	0°C / 20°C 47% / 53% ± 2.5% 82.5% / 74% ± 5% 17.5% / 16% > ± 7.5%	
1981	CEC-BCR	4	4	−100...100				$\lambda = f(T)$	Resin-bonded glass fiber boards 1000 × 1000 600 × 600 500 × 500 300 × 300	88	35	2.4–1.2% (−100–100°C)	20 29
1981	ISO/TC 163 and NBS	105	20	300–1200	0–40	10 23			Resin-bonded glass fiber boards	160	25.4	In process	30
						0–40			Air gap		25.4		

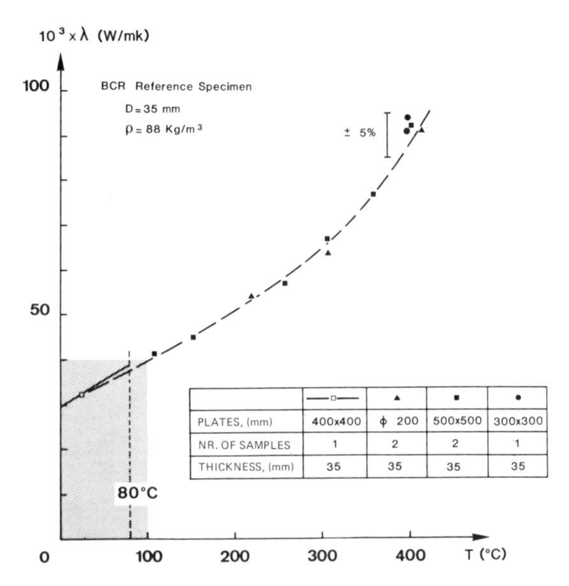

FIGURE 38. Thermal conductivity versus mean temperature measured for BCR reference specimen. □, CRIR standard biguarded hot plate apparatus (for measurements near ambient temperature); ▲, Dynatech TCFG guarded hot plate apparatus; ■, CRIR, 500 × 500 mm guarded hot plate apparatus; ●, CRIR, new high-temperature biguarded hot plate apparatus.

the measurement method and the present or future standards. An interesting example is the realization and the certification of such a material: "Reference Resin – Bonded Glass Fibers Board (RBGFB)" certified between −100 and 100°C by the Commission of European Communities, the Community Bureau of Reference (BCR).

Four European laboratories† took part in the measurements, using the guarded hot plate apparatus, in order to define the relationship between the thermal conductivity and the mean temperature, $\lambda(T)$. The standard reference material (SRM) was specially designed, manufactured and selected for this purpose (Fig. 39). It is presently available at Bruxelles, BCR.

The certification report on this standard reference material shows that by taking sufficient precautions[29] the dispersion observed in the results can be

† Forschungsinstitut für Wärmeschutz e.V. (FIW), Gräfelfing, Federal Republic of Germany; Instituto di Fisica Tecnica dell' Universita di Padova (IFT/P) in collaboration with the Instituto di Fisica Tecnica del Politecnico di Torino (IFT/T), Italy [the measurements reported by IFT/P were conducted on an apparatus of the Laboratorio per la Tecnica del Fredo di Padova, Consiglio Nazionale delle Ricerche (CNR), Italy]; Laboratoire National d'Essais (LNE), Paris, France; Propellants, Explosives and Rocket Motor Establishment (PERME), Waltham Abbey, in affiliation with the National Physical Laboratory (NPL), Teddington, United Kingdom. The statistical evaluation of the experimental results has been performed by the Joint Research Centre (JCR), Ispra, Italy.

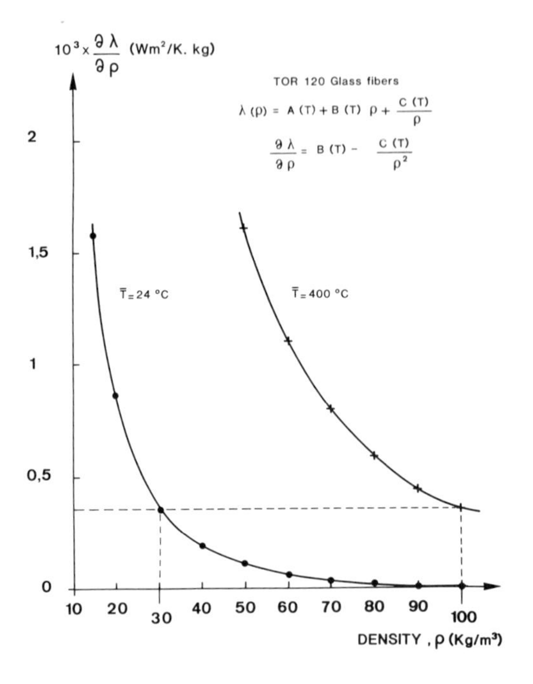

FIGURE 39. $\partial\lambda/\partial\rho$ versus density for TOR 120 glass fibers at two reference temperatures.

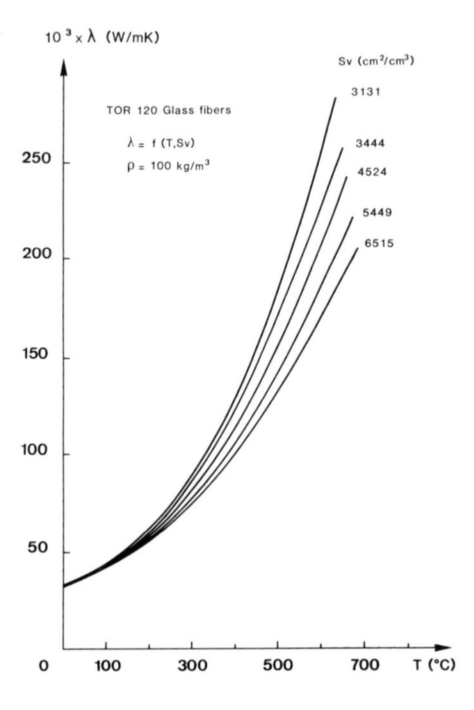

FIGURE 40. Thermal conductivity versus temperature for TOR 120 glass fibers of different specific surface.

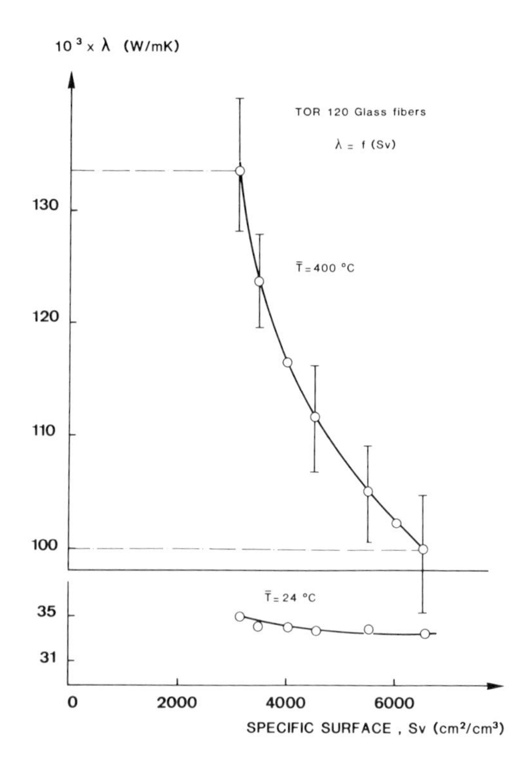

FIGURE 41. Thermal conductivity versus specific surface for TOR 120 glass fibers at two reference temperatures.

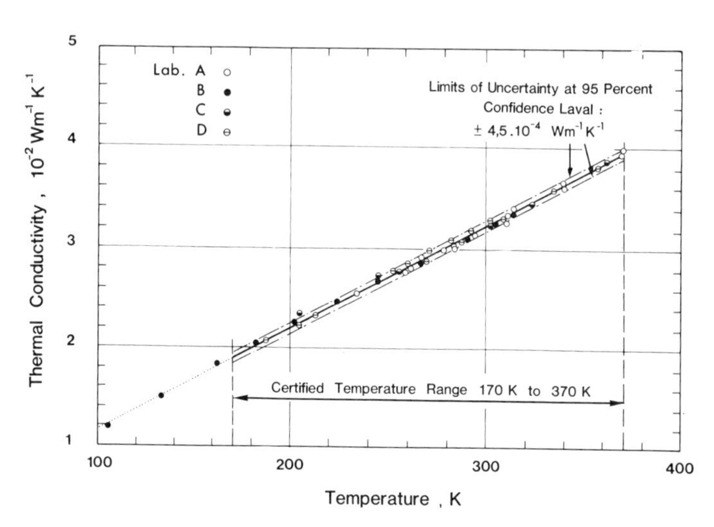

FIGURE 42. Reference material for the thermal conductivity of insulating materials RBGFB–BCR No. 64: Resin-bonded glass-fiber board.[29]

significantly reduced. The overall relative uncertainty gradually diminishes from 2.4% at the lowest temperatures to 1.2% for the higher temperatures, which corresponds to a constant limit of uncertainty of ± 0.00045 W/m K.

A similar approach is presently being undertaken at an international level, under the authority of the International Organization of Standardization (ISO), Technical Committee 163: "Thermal Insulation".[30]

The use of standard reference materials and of interlaboratory measurements is a very efficient way to improve the setup of this measurement method and the overall metrological performances of the apparatus. The recent ISO standard on the guarded hot plate measurement stipulates the use of reference materials and interlaboratory measurements, as follows:

> "After completion of new apparatus it is recommended that it be verified by undertaking tests on one or more of the various reference materials of different thermal resistance levels available."

It also includes recommended procedures and practices plus suggested specimen dimensions which together should enhance general measurements levels and assist in improving interlaboratory comparisons and collaborative measurement programes"

10. CONCLUSIONS

The principle of the guarded hot plate can be applied in different ways, leading to a great variety of apparatus. In most cases, these apparatus are classified according to their field of application, either by the range of measured thermal conductivities or by the temperatures of use.

The method of the guarded hot plate can be considered as a *reference method* not only to characterize the poorly conductive materials, in particular thermal insulants, but also to study and investigate thoroughly their behavior (research apparatus).

A great deal of empirical data, based on a broad practical experience, have been incorporated by the designers of these apparatus. More precise computations are, however, possible, in particular the computation of the errors due to lateral losses and to the nonequilibrium center guard, but these are difficult to apply to practical cases and require powerful computation means. These computations deal generally with "idealized" cases based on simplifying assumptions: homogeneous materials, purely conductive heat transfer, and "theoretical" boundary conditions. Much work still remains to be done in this field.

Considerable progress has been made on a metrological point of view concerning electrical measurements, temperature determination with the help of thermocouples and platinum resistances, and "zero balance" controllers and "set point" temperatures.

The present difficulties are linked to: the composite structure of the plates, the realization of "lateral" adiabatic boundary conditions to decrease the lateral losses in order to ensure a unidirectional density of heat flow, and also to the nature of the materials to be measured; in most cases these materials are industrial materials presenting a large heterogeneity for which the mechanism of heat transfer is very complex (coupling of conduction and radiation for highly porous materials), limiting the application of the notion of thermal conductivity itself.

The scattering of the measurement results determined on reference materials has considerably decreased in recent years, in particular around ambient temperature. However, an uncertainty equal to or lower than 2% will be considered negligible, which gives an idea of the limits of this method. Much remains to be done for high-temperature applications where the method is more difficult to implement.

The need to better characterize the thermal insulants used for building and industrial applications has contributed to increase the number of apparatus in service, in the industry for quality control, in "official" laboratories for the certification of these materials, and finally in research laboratories to investigate the mechanisms of heat transfer and think of new applications.

The international standard recently elaborated for the application of the method of the guarded hot plate (design, realization and use of apparatus) will contribute to an improvement of the metrological qualities of the apparatus.

ACKNOWLEDGMENTS. The author gratefully acknowledges the cooperation and assistance of his colleagues Jean Boulant, Yves Darche, Michel Degenne, Jean-Laurent Jaouen, Alex Seghers, and other members of the CRIR Thermal Laboratory who were involved in the research works dealing with the guarded hot plate and the heat fluxmeter methods.

He also takes this opportunity for expressing his gratitude to Francesco De Ponte, Daniel Fournier, K.H. Bode, and Walter Cammerer, with whom he participated to the Task Force Group of the ISO 163 Technical Committee; their works greatly helped him in writing the present synthesis.

Last of all, he wishes to express his thanks to Catherine Langlais, in charge of the "systems" section of the CRIR Thermal Laboratory, for her precious help in the redaction of this text.

REFERENCES

General Sources of Information

1. M. Jacob, *Heat Transfer,* Vol. 1 (7th printing), p. 151, John Wiley and Sons, New York (1959).

2. J.S. Cammerer, *Der Wärme- und Kälteschutz in der Industrie,* Springer-Verlag, Berlin (1962).
3. A.W. Pratt, *Heat Transmission in low-Conductivity Materials. Thermal Conductivity,* Vol. 1 (R.P. Tye, ed.), Academic Press, New York (1969).
4. ASTM Special Technical Publication, No. 217 (1957); No. 411 (1967); No. 544 (1974); No. 660 (1978); No. 718 (1980).
5. Conference on Thermal Conductivity, Vol. 1 a 17.
6. Institut International du Froid, conférences annuelles et congrès.
7. M. Combarnous and S. Bories, "Hydrothermal Convection in Saturated Porous Media," *Adv. Hydrosci.* **10**, 231–307 (1975).
8. S. Klarsfeld and M. Combarnous, "Analyse des transferts thermiques convectifs dans les isolants poreux perméables." *Rev. Gén. Therm.* **228**, 977–980 (1980).

Standardization

9. ANSI/ASTM C 177–1976, Standard Test Method for Steady-State Thermal Transmission Properties by Means of Guarded Hot Plate (First Edition ASTM C 177–45).
10. BSI 874–1973, Determining Thermal Insulating Properties with Definitions of Thermal Insulating Terms.
11. DIN 52-612–1979, Bestimmung der Wärmeleitfähigkeit mit dem Plattengerät.
12. NF X10-021–1979, Matériaux faiblement conducteurs. Détermination de la conductivité thermique par la méthode de la plaque chaude gardée avec échantillon symétrique.
13. UNI 7745, Determinazione della conduttivita termica dei materiali isolanti con il metodo della piastra calda con anello di guardia.
14. ISO/TC 163, Determination of Steady-State Specific Thermal Resistance and Related Properties by Means of Guarded Hot Plate Apparatus. (Draft Proposal).
15. What Property Do We Measure? ASTM, STP 544–1974, pp. 5–12.
16. R.P. Tye, "Thermal Insulation Evaluation: Present Status and Future Requirements," *J. Therm. Insulation* **2** (1), 109–130 (1979).

Reference Materials

17. R.W. Powell and R.P. Tye, "New Measurements on Thermal Conductivity, Reference Materials," *Int. J. Heat Mass Transfer* **10**, 581–596 (1967).
18. Reference Materials of Low Thermal Conductivity. ASTM, STP 544, pp. 307–309 (1974).
19. R. Doussin, Méthode et matériaux de référence pour la mesure de la conductivité thermique, Bulletin BNM (1974).
20. Reference Materials for Insulation Measurement Comparisons. ASTM, STP 660, pp. 7–29 (1978).
21. M. Bertasi *et al.,* Fibrous Insulating Materials as Standard Reference Materials at Low Temperatures. ASTM, STP 660, p. 30–49 (1978).
22. F. Cabannes, "Standard Reference Materials and Low Thermal Conductivity Experimental Methods", in Proc. 17th International Thermal Conductivity Conference, NBS, Gaithersburg, Maryland, June (1981).
23. B.G. Rennex *et al.,* "Development of Thick Low-Density Heat Transfer Standards," in Proc. 17th International Thermal Conductivity Conference, NBS, Gaithersburg, Maryland, June (1981).
24. D.R. Smith and J.G. Hust, "Effective Thermal Conductivity of Glass-Fiber Board and Blanket Standard Reference Materials," in Proc. 17th International Thermal Conductivity Conference, NBS, Gaithersburg, Maryland, June (1981).

Interlaboratory Comparative Measurements

25. H.E. Robinson and T.W. Watson, Interlaboratory Comparison of Thermal Conductivity Determinations with Guarded Hot Plate, ASTM, STP 119 (1951).
26. Détermination du coefficient de conductivité thermique par la méthode de la plaque chaude. Rapport sur l'essai en commun et spécifications provisoises, Bulletin RILEM No. 18, p. 41–71, March (1963).
27. R. Gasquet, "Méthodes de mesure de la conductivité thermique," *Rev. Gén. therm.* November (1963).
28. "Mesures comparatives internationales de conductivité thermique," *Bull. I.I.F.* **1**, 29–59 (1968).
29. H. Zibland, Certification Report on a Reference Materials for the Thermal Conductivity of Insulating Materials between 170 K and 370 K Resin bonded Glass Fiber Board, Commission of the European Communities, BCR Information, Report EUR 7677 EN (1981).
30. F. Powell, "Thermal Conductance Round Robin for Guarded Hot Plate and Heat Flow Meters and Status of ASTM C 16 Round Robin on Hot Boxes," in Proc. 17th International Thermal Conductivity Conference, NBS, Gaithersburg, Maryland, June (1981).

Apparatus Operating at and Near Ambient Temperature

31. R. Poensgen, "Ein Technisches Verfahren zur Ermittlung der Wärmeleitfähigkeit Stoffen," *Z. VDI* **56** (41), 1653–1658 (1912).
32. J.D. Verschoor and P. Greebler, "Heat Transfer by Gas Conduction and Radiation in Fibrous Insulations," *Trans. ASEM* **August**, 961–968 (1952).
33. A. Pascal, "La mesure de la conductivité thermique des matériaux du Bâtiment," *I.T.B.T.P. Bull.* **90**, 583–598 (Juin 1955).
34. Z. Zabawski, An Improved Guarded Hot Plate Thermal Conductivity Apparatus with Automatic Controls, ASTM, STP 217, pp. 3–16 (1957).
35. W. Fritz and K.H. Bode, "Zur Bestimmung der Wärmeleitfähigkeit Fester Stoffe," *Chem. Ing. Technik,* **11**, 118–1124 (1965).
36. H.M. Strong, F.P. Bundy and H.P. Bovenkerk, "Flat Panel Vacuum Thermal Insulation," *J. Appl. Phys.* **31**, (1), 39–60 (1960).
37. J.C. Marechal, Mesure de la conductivité thermique par la méthode du champ thermique unidirectionnelle. *Matér. Constr. (Paris)* **1**, (5), 443–456 (1968).
38. W. Fritz and W. Kuster, Beitrag zur Kenntnis der Wärmeleitfähigkeit poröser Stoffe. *Wärme Stoffübertrag.* **3**, 156–168 (1970).
39. D. Fournier and S. Klarsfeld, "Mesures de conductivité thermique des matériaux isolants par un appareil orientable à plaque chaude bi-gardée. I.I.F., Commission 2 & 6, Liège 1969," *Annexe 1969-7 au Bull. I.I.F.,* 321–331.
40. C. Bankwall, Guarded Hot Plate Apparatus for the Investigation of Thermal Insulations, National Swedish Institute for Building Research, Document D5 (1972).
41. M.H. Hahn, H.E. Robinson and D.R. Flynn, Robinson Line-Heat-Source Guarded Hot Plate Apparatus, ASTM, STP 544, pp. 167–192 (1974).
42. F.J. Powell and C.I. Siu, Development of the Robinson-Heat-Source Guarded Hot of Thermal Conductivity, I.I.E., Commission B1, 14ème Congrès I.I.F., Moscow.
43. R.P. Tye and S.C. Spinney, "Thermal Conductivity of Concrete: Measurement Problems and Effect of Moisture. I.I.F., Commission B1, Washington 1976," *Suppl Bull. I.I.F., Annexe 1976–2,* 110–118.
44. M. Degenne, S. Klarsfeld and M.P. Barthe, Measurement of the Thermal Resistance of Thick Low Density Mineral Fiber Insulation, ASTM, STP 660, pp. 130–144 (1979).

45. M.H. Hahn, "The 1000-mm Circular-Line-Heat-Source Guarded Hot Plate Apparatus," in Proc. 17th International Thermal Conductivity Conference, NBS, Gaithersburg, Maryland, June (1981).
46. M.H. Hahn and D. Ober, "Instrumentation and Auxiliaries for the 1000 mm Circular-Line-Source Guarded Hot Plate Apparatus," in Proc. 17th International Conductivity Conference, NBS, Gaithersburg Maryland, June (1981).

Theses

47. M.H. Hahn, "The Line-Heat-Source Guarded Hot Plate for Measuring the Thermal Conductivity of Building and INsulation Materials," doctoral dissertation, the Catholic University of America, Washington, D.C.
48. J.C. Castel and J. Huet, "Contribution à la détermination des caractéristiques thermophysiques des matériaux de Génie Civil," Thèse, I.N.S.A. de Rennes (1978).
49. J.M. Devisme, "Contribution à l'étude du comportement thermique des parois multicouches de bâtiment et à la mesure des caractéristiques thermophysiques des matériaux constructifs," thèse, Univ. Pierre et Marie Curie, Paris 6 (1980).

Apparatus Operating Between Ambient and Low and Very Low Temperatures

50. W.F. Cammerer, "Die Messung der Wärmeleitfähigkeit von Isolierstoffen bei tiefen Temperaturen," *Kältetechnik* Band **12** (4), 107–110 (1960).
51. J. Achtziger, "Wärmeleitfähigkeit messungen an Isolierstoffen mit dem Plattengerät bei tiefen Temperaturen," *Kältetechnik* **12** (12), 372–375 (1960).
52. J. Achtziger, "Einfluss der freien Konvection auf die Wärmeleitfähigkeit einer leichten Mineralfaser matte bei tiefen Temperaturen," *Kältetechnik* **16** (10), 308–311 (1964).
53. J.C. Rousselle, "A Guarded Hot Plate Apparatus for measuring Thermal Conductivity from – 80 to + 100°C," in Proc. 7th Conference on Thermal Conductivity, Gaitherburg, Maryland, pp. 513–520, November (1967).
54. G.S. Karp and C.S. Lankton, A Guarded Hot Plate Thermal Conductivity Apparatus for Multilayer Cryogenic Insulation, ASTM, STP 411, pp. 13–24 (1967).
55. E. Brendeng and P.E. Frivik, "On the Design of a Guarded Hot Plate Apparatus. I.I.F. Commission 2 & 6, Liège 1969," *Annexe 1969–7 Bull. I.I.F.*, 281–288.
56. R.P. Tye, "Measurements of Heat Transmission in Thermal Insulations at Cryogenic Temperatures Using the Guarded Hot Plate," in I.I.F., C.R. du 13ème Congrès, Washington, D.C., Vol. 1 (1971).
57. J.G. Bourne and R.P. Tye, Analysis and Measurement of the Heat Transmission of Multi-Component Insulation Penels of Thermal Protection of Cryogenic Liquid Storage Vessel, ASTM, STP 544, 297–305 (1974).
58. F. De Ponte and P. Di Filippo, "Thermal Conductivity Measurements of Insulating Materials down to 100 K," in *I.I.F. C.R. du* 13ème Congrès, Washington, DC, Vol. 1. (1971).
59. D. Fournier and S. Klarsfeld, Some Recent Expérimental Data on Glass Fiber Insulating Materials and their use for a Reliable Design of Insulations at Low Temperatures. ASTM, STP 544, pp. 223–242 (1974).
60. E. Brendeng and P.E. Frivik, New Development in Design of Equipment for Measuring Thermal Conductivity and Heat Flow. ASTM, STP 544, pp. 147–166 (1974).
61. G. Bigolaro, F. De Ponte and E. Fornasieri, Thermal Conductivity of Polyurethane at Low Temperatures, I.I.F., Commission B1, Washington, pp. 34–41 (1976).

Apparatus Operating at High and Very High Temperatures

62. E.J. Rolinski and G.V. Purcell, "Thermal Conductivity Measurements of Fibrous Insulations up to 2500°F (945°C)," in Proc. 3rd Conference on Heat Transfer, Chicago, pp. 133–140 (1966).
63. V. Ferro and A. Sacchi, "An Automatic Plate Apparatus for Measurements of Thermal Conductivity of Insulating Materials at Higher Temperature," in Proc. 8th Conference on Thermal Conductivity, pp. 737–760 (1968).
64. R.P. Tye, "The Thermal Conductivity of MIN K-2000. Thermal Insulation in Different Environments to High Temperature," in Proc. 9th Conference on Thermal Conductivity IOWA, pp. 341–351 (1969).
65. W.F. Cammerer, "Wärmeleitfähigkeit von Betonen und Massen im Temperaturbereich von 200 bis 900°C," *Ton Industrie-Z.* **96** (9), 274–276 (1972).
66. J.P. Brazel and R.P. Tye, "Thermal Characterization of Reusable External Insulation for the Space Shuttle," *High Temp. High Pressure* **4**, 639–674 (1972).
67. J. Liermann, Mesure absolue de la conductivité thermique des matériaux isolants. *Commiss. Energ. At.* **197**, 45–50 (1974).
68. A.H. Stripens, Heat Transfer in Refractory Fiber Insulations, ASTM, STP 660, pp. 293–309.
69. J. Boulant, C. Langlais and S. Klarsfeld, "Correlation between the Structural Parameters and the Thermal Resistance of Fibrous Insulants at High Temperature," in Proc. 17th International Thermal Conductivity Conference, NBS, Gaithersburg, Maryland, June (1981).
70. K. Kamiuto, I. Kinoshita, Y. Miyoshi and S. Hasegawa, "Experimental Study of Simultaneous Conductive and Radiative Heat Transfer in Ceramic Fiber Insulation," *J. Nucl. Sci. Technol.* **19**, June (1982).

Studies and Activities

71. C.J. Shirtliffe and H.W. Orr, "Comparison of Modes of Operation for Guarded Hot Plate Apparatus with Emphasis on Transient Characteristics," Proc. 7th Conf. on Thermal Conductivity, NBS Special Publ. 302 (1968).
72. F. De Ponte and P. Fillipo, Guarded Hot Plate Apparatus Design: The Hot Plate Imbalance Problems, University of Padova, Rapport no. 30 (1971).
73. F. De Ponte and P. De Filippo, Some Remarks on the Design of Isothermal Plates, University of Padova, Rapport No. 37 (1972).
74. F. De Ponte and P.E. Frivik, Automatic Control of Guarded Hot Plate Apparatus, I.I.F., Commission B1, E1, Freudenstadt (1972).
75. C.J. Shirtliffe, Establishing Steady-State Thermal Conditions in Flat Slab Specimens, ASTM, STP 544, pp. 13–33 (1974).
76. K.M. Letherman, "The Manufacture of Guarded Heater Plates by a Numerically Controlled Milling Machine," *J. Phys. E. Sci. Instrum.* **9**, 166 (1976).

Measurement Errors

77. E.V. Somers and J.A. Cyphers, "Analysis of Errors in Measuring Thermal Conductivities of Insulating Materials," *Rev. Sci. Instrum.* **22**, 583–585 (1951).
78. G.M. Dusinberre, "Further Analysis of Errors of the Guarded Hot Plate," *Rev. Sci. Instrum.* **23**, 649–650 (1952).
79. W. Woodside, Analysis of Errors Due to Edge Heat Loss in Guarded Hot Plates, ASTM, STP 217, pp. 49–62 (1957).
80. W. Woodside and A.G. Wilson, Imbalance Errors in Guarded Hot Plate Measurements, ASTM, STP 217, pp. 32–46 (1957).

81. W. Woodside, Deviations from One-Dimensional Heat Flow in Guarded Hot Plate Measurements. *Rev. Sci. Instrum.* **28,** 1033–1037 (1957).
82. I.G. Donaldson, "A Theory for the Square Guarded Hot Plate. A Solution of the Heat Conduction Equation for a Two-Layer Systems," *Q. Appl. Math.* **XIX,** 205–219 (1961).
83. A.W. Pratt, "Analysis of Errors Due to Edge Heat Loss in Measuring Thermal Conductivity by the Hot Plate Method," *J. Sci. Instrum.* **39,** 63–68 (1962).
84. I.G. Donaldson, "Computer Errors for a Square Guarded Hot Plate for the Measurements of Thermal Conductivity of Insulating Materials," *Br. J. Appl. Phys.* **13,** 598–602 (1962).
85. H.W. Orr, "A Study of the Effects of Edge Insulation and Ambient Temperature on Errors in Guarded Hot Plate Measurements," in Proc. 7th Conf. on Thermal Conductivity, Gaithersburg, Maryland, pp. 521–526 (1967).
86. R. Doussin, Influence du mode de construction des plaques chauffantes à anneau de garde et de l'isolation latérale sur la mesure de la conductivité thermique des matériaux isolants, I.I.F. Commission 2 & 6, Liège 1969, pp. 289–299.
87. R.P. Tye, "Effects of Edge Losses on Thermal Conductivity of Thermal Insulations at High Temperature," *Rev. Int. Hautes Temp. Réfract.* **7,** 308–312 (1970).
88. F. De Ponte and P. Di Fillipo, Design Criteria for Guarded Hot Plate Apparatus, ASTM, STP 544 (1974).
89. J.C. Marechal, "Métrologie et conductivité thermique," *Matér. Constr (Paris)* **37,** 61–65 (1974).
90. J.P. Bardon, "La mesure des températures de surface par contact. Erreurs liées aux transferts de chaleur parasites," *Rev. Gén. Therm.* **170,** 123–135 (1976).
91. K.H. Bode, "Wärmeleitfähigkeit messungen mit dem Plattengerät: Einfluss der Schutzringbreite auf die Mässungsicherkeit, *Int. J. Heat Transfer* 961–970 (1980).
92. M. Bomberg and K.R. Solvason, "Precision and Accuracy of the Guarded Hot Plate Method," in Proc. 17th International Thermal Conductivity Conference, NBS Gaithersburg, Maryland, June (1981).
93. K.H. Bode, (submitted for publication).
94. L.R. Troussart, "Three-Dimensional Finite Element Analysis of the Guarded Hot Plate apparatus and its Computer Implementation," *J. Therm. Insulation* **4,** 225 (April 1981).

Experimental Results

95. E. Cadiergues and J. Genevey, "La conductivité thermique des matériaux," *Suppl. Ann. I.T.B.T.P.* **52–53,** 472–490 (1952).
96. J.C. Marechal, "Conductivité thermique des matériaux du bâtiment," *Ann. I.T.B.T.P.* **185,** 436–442 (1963).
97. J.C. Marechal, "Propriétés mécaniques et thermiques des matériaux isolants," *Ann. I.T.B.T.P.* **301,** 22–42 (1973).
98. R.P. Tye, "Heat Transmission of Cellulose Fiber Insulation Materials," *J. Testing Evaluation* **2** (3), 176–179 (1974).
99. R. Doussin and J.C. Marechal, "Caractéristiques physiques des isolants légers, 1ère, 2ème parties," *Rev. Gén. Froid* **7,** 681–703 (1973).
100. D. Fournier and S. Klarsfeld, Données expérimentales récentes relatives à la conductivité thermique de matériaux isolants en fibres de verre et polystyrène expansé à basse température, I.I.F., Commission B1, Zurich pp. 183–187 (1973–4).
101. C. Langlais *et al.* Influence of moisture on heat transfer through fibrous insulating materials, ORNL/ASTM Symposium, December 1981 (to be published).

5

Hot-Wire Method for the Measurement of the Thermal Conductivity of Refractory Materials

W.R. DAVIS

1. INTRODUCTION AND BACKGROUND

1.1. General Introduction

The hot-wire method is a transient dynamic technique based on the measurement of the temperature rise of a linear heat source embedded in the test material. The heat source is assumed to have a constant output which is also uniform along the length of the test piece. From the change in the temperature over a known time interval the thermal conductivity of the test material can be derived. In its present form the application of the method is limited to materials whose thermal conductivity lies below 1.5–2.0 W/m K and which have a thermal diffusivity below 10^{-6} m^2/s (3×10^{-3} m^2/h). However, a recent modification may allow the limit to be raised to 25 W/m K. This paper is concerned specifically with refractory materials, although the line source technique has been used for a variety of other materials and in cryogenic research.

1.2. History

The method was described in 1888 by Schieirmacher,[1] but its first practical application was reported in 1949 by Van der Held and Van Drunen,[2] who used it to measure the thermal conductivity of liquids. In recent years its application to refractory materials has been developed in West Germany and submitted as a Provisional Standard DIN 51046.[3] In a modified form it has

W.R. DAVIS • British Ceramic Research Association Ltd., Queens Road, Penkhull, Stoke-on-Trent ST4 7LQ, England.

been accepted by PRE (Féderation Européene des Fabricants de Produits Refractaires), firstly as a Recommendation, PRE/R32,[4] and later as a Designated Standard in which form it has been submitted to ISO/TC33 for consideration as an ISO standard.

1.3. Mathematical Derivation

This has been discussed in detail by Van der Held and Van Drunen[2] and more recently by Carslaw and Jaeger.[5]

The formulas are derived most conveniently from the "source solution" of the heat conduction equation in rectangular coordinates. The temperature rise θ at time t at the origin of co-ordinates in an infinite solid due to a quantity of heat q being instantaneously generated at $t = 0$ at the point (x, y, z) is given (with slight changes in nomenclature) by Carslaw and Jaeger[5] as

$$\theta = \frac{q}{8\rho c(\pi\alpha t)^{3/2}} \exp\left(-\frac{x^2 + y^2 + z^2}{4\alpha t}\right) \tag{1}$$

where α is the thermal diffusivity $= \lambda/(\rho c)$, λ is the thermal conductivity, ρ is the density, and c is the specific heat.

Carslaw and Jaeger[5] go on to derive a formula for an instantaneous line source of heat. If a quantity of heat $Q'dz$ is instantaneously generated at $t = 0$ at all points on the infinite line parallel to the z axis and passing through the point $(x, y, 0)$, the temperature rise at the origin is obtained by replacing q in equation (1) with $Q'dz$ and integrating with respect to z, thus:

$$\theta = \frac{Q'}{4\pi\lambda t} \exp\left(-\frac{x^2 + y^2}{4\alpha t}\right) \tag{2}$$

Finally, the temperature rise at the origin resulting from a continuous line source of heat, at a rate for $t > 0$ of $Qdzdt$ at all points on the same line, is derived[5] by replacing Q' in equation (2) by Qdt and integrating with respect to t, giving

$$\theta = \frac{Q}{4\pi\lambda} \int_{r^2/4\alpha t}^{\infty} \frac{e^{-u}}{u} \cdot du, \qquad r^2 = x^2 + y^2 \tag{3}$$

$$= \frac{Q}{4\pi\lambda} \cdot \left[-\left(E_i \frac{-r^2}{4\alpha t}\right)\right] \tag{4}$$

which can also be written in the form

$$\theta = \frac{Q}{4\pi\lambda} \cdot E_1\left(\frac{r^2}{4\alpha t}\right) \tag{4a}$$

$E_1(r^2/4\alpha t)$ is an exponential integral of the form $E_1(x)$ given by

$$E_1(x) = -\gamma - \ln + x - \frac{x^2}{2 \cdot 2!} + \frac{x^3}{3 \cdot 3!} - \frac{x^4}{4 \cdot 4!}, \text{ etc.}$$

where γ is Euler's constant (0.577216 . . .). By manipulating equation (3), differentiating with respect to the lower limit $(r^2/4\alpha t)$, and putting $(r^2/4\alpha t) = b$ we have

$$\theta = (Q/4\pi\lambda) \cdot E_1(b) \tag{5}$$

Now since

$$\frac{dE_1(b)}{db} = -\frac{e^{-b}}{b}$$

then

$$\frac{d\theta}{db} = -\frac{Q}{4\pi\lambda} \cdot \frac{e^{-b}}{b} \tag{6}$$

and since

$$\frac{d\theta}{d \ln t} = \frac{d\theta}{db} \cdot \frac{db}{dt} \cdot \frac{dt}{d \ln t} \tag{7}$$

then

$$\frac{d\theta}{d \ln t} = \frac{Q}{4\pi\lambda} \cdot e^{-r^2}/4\alpha t \tag{8}$$

When the thermocouple is welded to the heater $r = 0$ and if we take the rise in temperature $\theta_2 - \theta_1$ $(= d\theta)$ over a time interval $t_2 - t_1$ $(= dt)$ we have

$$\frac{d\theta}{\ln t} = \frac{\theta_2 - \theta_1}{\ln t_2 - \ln t_1} = \frac{Q}{4\pi\lambda} \tag{9}$$

or

$$\lambda = \frac{Q}{4\pi} \cdot \frac{\ln(t_2/t_1)}{\theta_2 - \theta_1} \tag{10}$$

which can be written as

$$\lambda = Q/4\pi S \tag{11}$$

where S is the slope of the linear portion of the temperature–ln time curve. It should be noted that the thermal diffusivity term has disappeared.

This is the basic equation of the standard hot-wire test. If the temperature is plotted against ln t the resultant graph will be curved at the upper and lower limits with a linear portion in between. The equation is only valid for points on this linear portion. The reasons for these upper and lower limits are discussed in the following section.

1.4. Limitations

The available measurement time is set by the upper and lower curved portions of the temperature–ln time graph. The heat front is propagated radially in the form of a cylinder, from the heater wire to the outside of the test piece. The lower curved portion of the graph controls the earliest possible moment when the measurement can start. This can be denoted as t_{min} and is determined by the heat transfer coefficient between heater and test material. The transfer of heat between heater and test material can be adversely affected by poor embedding of the heater wire (see Sections 2.2 and 2.7). Because of the finite size of the test piece, the end of the measurement period, t_{max}, is fixed by the moment when the heat front reaches the exterior of the test piece. This is dependent on the test-piece geometry and the thermal diffusivity. For a standard shape 230 × 114 × 76 mm an approximation to t_{max} is given by the expression

$$t_{max} = 3.5 \times 10^{-4}/\alpha \text{ s}$$

For a typical magnesite brick having $\lambda = 12$ W/m K and $\alpha = 6 \times 10^{-6}$ m^2/s t_{max} will be of the order of 60 s. As the value of the thermal conductivity increases $t_{max} - t_{min}$ will decrease until the length of the linear portion of the graph is too short for accurate measurement. With typical test equipment this usually occurs at about 1.5–2.0 W/m K although with sensitive measuring equipment and careful technique it is possible to extend this to 6–8 W/m K. Jeschke[(6)] has discussed the above in more detail. The boundary effects and thermal resistance have also been discussed for heat probes by Blackwell[(21)] (see Section 4.2.1).

2. EXPERIMENTAL TECHNIQUES

2.1. Introduction

The method is a transient one and normally operates with a temperature rise in the specimen of up to a few tens of degrees. In this it differs from many other thermal conductivity methods in common use, which are equilibrium (steady state) methods with considerable temperature gradients (>100 K) across the test piece. In this respect the hot wire can be regarded as an isothermal technique and is therefore very suitable for materials having nonlinear conductivity–temperature graphs or for materials such as monolithics or chemically bonded refractories which undergo physical or chemical changes with temperature.

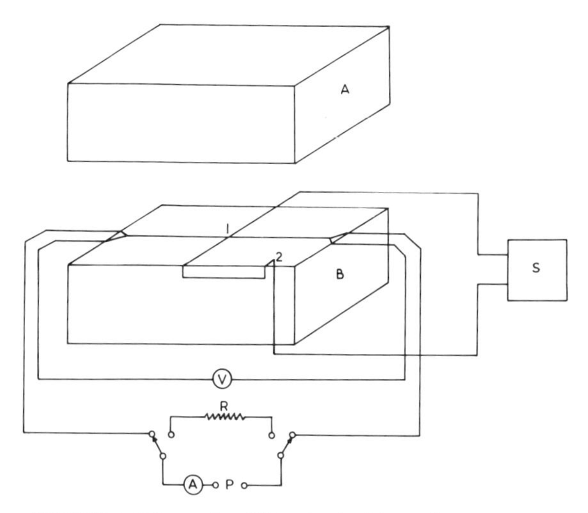

FIGURE 1. Schematic of the standard hot-wire apparatus.

The experimental setup is shown in Fig. 1 where the test pieces are two standard shapes with dimensions 230 x 114 x 76 mm. The latter dimension is sometimes 64 mm. With refractory insulating materials, having $\lambda < 0.5$ W/m K, the width of the specimen can be reduced to 76 mm without any great loss in accuracy although PRE/R32[4] and DIN 51046[3] specify 100 mm as the minimum. For lightweight insulating materials (<0.3 W/m K) the thickness can probably go below 50 mm, which is the lower limit specified.[3, 4]

2.2. *Specimen Preparation*

It is important that the contacting surfaces of the two units comprising the test assembly should be flat. PRE/R32[4] specifies grinding if necessary to give a deviation of less than 0.2 mm over 100 mm. Generally, testing with a straight edge will give good results. Grooves are cut in the bottom brick to take the measurement cross. These grooves should be only slightly wider than the diameter of the wire and the depth and width should not exceed 1.0 mm.

The measurement cross comprises a heater wire with a thermocouple welded to the midpoint. The heater wire is welded at each end to two "take-off" leads of similar but heavier gauge material. These supply the heater current (I) and measure the voltage drop (V) across the heater. The heater is usually platinum or platinum–20% rhodium being about 200 mm long. This length (L) should be measured to ±0.5 mm. The measurement thermocouple is Pt; Pt/Rh; being either type R (13% Rh) or Type S (10% Rh), spot-welded to the center of the heater wire with the limbs at right angles to the heater. The diameter of the couple should be smaller than that of the heater. DIN 51046[3] specifies

0.35 mm as the maximum size for heater and thermocouple. Base metal alloys can be used at room temperature and below 800°C having the advantage of low-temperature coefficient of resistivity and high thermoelectric output.

The leads of the measurement cross should be long enough to reach outside the furnace where the couple connections can be made either to compensating leads or to a thermally isolated junction box. A reference couple is placed near to the outside edge of the lower brick. DIN 51046[3] specifies 5 mm from the 230-mm edge and 10 mm or less from the 114-mm edge; PRE/R32[4] gives an alternative where three bricks are used in a vertical stack with the 230 x 114 mm faces in contact. The measurement cross is placed between the bottom and middle bricks and the reference couple between the middle and upper brick, vertically above the measurement couple.

The measurement cross can be cemented in with a paste made from finely ground test material and an organic binder. In many cases dry powdered material will give equivalent results. For soft insulating firebricks simply pressing the cross in position with a heavy weight on top of the upper brick will suffice. Poor embedding can give serious errors in the conductivity value, particularly when high-conductivity materials are being measured.

2.3. *Equipment*

2.3.1. *Power Supplies*

The heater can be energized by either direct or alternating current; in either instance a stabilized supply is essential. PRE/R32[4] gives a tolerance of 2% in the variation of current over the measurement period. DIN 51046[3] states ±0.5%. If possible the latter is to be preferred. If ac is used care should be taken that the stabilization circuits, commonly used in constant voltage transformers, do not produce a distorted waveform. If the ac ammeter or voltmeter is not of the "true rms" type, using the heating effect for measurement, considerable errors can arise with a nonsinusoidal waveform.

A simple means of measuring the heater current is to measure the voltage drop across a standard 1.00- or 0.100-Ω resistor placed in series with the heater. If a digital voltmeter is used for this purpose it can be switched to measure the voltage drop across the heater. There are a number of inexpensive digital voltmeters available with accuracies of the order of ±0.3% on ac and ±0.05% on dc.

Because of the high-temperature coefficient of resistivity of the platinum, or platinum–rhodium, heater its resistance will increase appreciably during the test run, even when the total temperature rise is only 20–25 K. If a constant current dc supply is used the power input will therefore tend to increase during the run. Conversely with a constant voltage supply the power will decrease. A third approach which is to be preferred is to use constant power. This is achieved by feeding the two voltages, corresponding to current (I) and voltage (V), into a multiplier module. The output from the module is $V \times I$, i.e., the power, and can be used as the output voltage in a standard regulator circuit to control the voltage supply. Commercial versions of the constant power supply are not as yet readily available.

The use of a dc supply can give rise to a spurious voltage effect in which a small positive or negative voltage is superimposed on the couple output. This arises from the finite size of the thermocouple junction. If the latter occupies, say, 0.2 mm of the 200-mm-long heater wire then for every volt drop across the heater there will be a corresponding millivolt drop across the couple. In practice the spurious voltage is lower than this but there is the possibility that it may affect the accuracy of the readings. It is therefore advisable to do a check run with the direction of current flow reversed. For this reason some experimenters prefer to use ac heating although the supply stability is inferior to that of a typical stabilized dc supply.

2.3.2. *Temperature Measurement*

As will be seen in Section 2.4 it is necessary to resolve temperature differences of the order of ±0.02 K. At 20°C with a type R (13% Rh) thermocouple this is roughly equivalent to a thermo-emf of ±0.1 μV. Using a flat-bed chart recorder of a standard width (200–250 mm) will require a full-scale deflection of 200–250 μV, i.e., 1 μV/mm minimum. A higher sensitivity than this is recommended, possibly 0.25–0.5 μV/mm. High sensitivity can be obtained with a standard recorder by the use of a high-stability low-noise instrumentation amplifier. Jeschke[6] advises the use of a nanovoltmeter with a printout. It is advisable to use screened leads for connections outside the furnace and to reduce interference effects as much as possible. Any dissimilar metal connections in the thermocouple leads between furnace and recorder should be thermally isolated or errors can arise from ambient temperature variations.

2.3.3. *Furnace*

This should be able to take one or more test assemblies with sufficient space between the heating elements and test assembly to result in a reasonably uniform temperature distribution. DIN 51046[3] gives a limit of 5 K between edge and middle of the test assembly. PRE/R32[4] requires a maximum difference of 10 K between any two points in the region of the test assembly. The test temperature should be known to better than ±5 K. The furnace temperature control should be good enough to give temperature variations below 0.1 K, over a 5-min period, measured on the outside of the test assembly. Because of the high sensitivity of the temperature-measuring equipment, thyristor control may produce interference problems. An alternative is the use of a motorized variable autotransformer. The furnace should be capable of reaching 1450°C using silicon carbide rod heaters, or 1600°C by the use of the more expensive molybdenum disilicide "hairpin" heaters.

2.4. Measurement and Calculation

2.4.1. Setting Up

To ensure uniform heating the test assembly is supported on two blocks either of the test material or of a refractory material having similar thermal conductivity characteristics. DIN 51046[3] specifies blocks 125 x 10 x 20 mm with the test assembly resting on the 125 x 10-mm faces. The bottom brick carrying the grooves for the measurement cross is placed in first and the cross then put in position. To ensure good embedding, either loose dry finely ground test material, or a paste from this powder with a minimum of organic binder are used. A jig, comprising a refractory material frame, can be used to hold the cross in position while the upper brick is being placed on the lower one. With a two-unit assembly the reference couple is set in before placing the upper brick. With a three-unit assembly the reference couple is placed between the upper and middle bricks vertically above the measurement couple. This brick can, in some instances, be replaced by a small block.

The voltage and current leads are connected up as in Fig. 1. If the test is to be done at high temperature the rate of heating should be low enough to ensure that there is no danger of spalling or thermal shock.

2.4.2. Measurement

The furnace and test assembly must be at thermal equilibrium before any measurements can be taken. This can be checked in the first instance using the measurement thermocouple to give a rough indication to about ± 1 K. The reference couple is then connected back-to-back with the measurement couple and the combination used to check the temperature difference between the inside and the outside of the assembly. This difference should not vary more than 0.05 K over a 5-min period.

A dummy load of equivalent resistance to the heater is connected to the power supply to set the heater current and to check on its stability. When thermal equilibrium has been reached the power supply is switched to the heater and the recorder started. The temperature rise of the hot wire is then recorded continuously for up to 15 min. The writing speed of the recorder should preferably be such that time measurements can be made to ± 1 s (± 2 s PRE/R32[4]). A typical test run will last for 10 min (600 s), but this can be varied to suit circumstances.

Determination of the power input to the heater is usually on the basis of experience, the expected thermal conductivity and the resistance of the heater. As a rough guide, for a total temperature rise of about 20 K, over a test period of 600 s, in a test material having a thermal conductivity of 0.3 W/m K the power input would be of the order of 6 W/m. With a 200-mm-long heater this gives an input of 1.2 W.

The power input can either be calculated from the product of the heating current (I) and the voltage drop (V) across the heater, divided by the length (L) of the heater, or from I^2R/L, where R is the resistance of the heater obtained from V/I. PRE/R32[4] suggests that the *total* temperature rise should be below 100 K. It is preferable to keep this rise as small as possible, say, of the order of 20 K. With a platinum or platinum–rhodium heater wire there will be a resistance change with temperature even for changes as low as 20 K. It is advisable therefore to take a number of readings of I and V during the test run, particularly at the beginning (t_1) and end (t_2) of the measurement period and to use these to calculate the mean power input (Q). Variation in the power input should be as small as possible; PRE/R32[4] suggests less than 2% but with available ac stabilized supplies a variation of under 1% should be possible. If the current variation does exceed 2% the test run should be repeated.

A minimum of three test runs should be made at each temperature. Where several test temperatures are required it is suggested that they should be made at intervals of 200 K.

2.4.3. Calculation

The temperature–ln t graph is curved at the upper and lower limits as discussed in Section 1.3. Assuming that the other criteria of sample preparation, size, and geometry have been satisfied and that the ambient temperature is stable to the specified limits, then if the input power has been correctly chosen there will be a linear portion between the two ends. The determination of the slope of this linear portion is the basis of the calculation. It can be done in one of three ways:

1. Plotting the temperature readings at fixed time intervals – say every 20 or 30 s on semilog graph paper. This averages out any measurement errors and is recommended where the material has not been previously tested or as a means of determining the correct input power.
2. Taking "spot readings," i.e., temperature θ_1 and θ_2 at known times t_1 and t_2 (usually 60 and 600 s or 120 and 600 s). Errors can arise if either of the spot reading pairs is not on the linear portion of the graph.
3. Using a computer to do a linear regression analysis on the temperature–ln t pairs. Linearity can be checked by using different time intervals,

e.g., 90–480 s, 120–540 s, etc. This technique will also give the correlation coefficient R. For reasonable accuracy (~5%) R should lie between 0.999 and 0.9999.

The conductivity is then calculated from equation (11), Section 1.3:

$$\lambda = Q/4\pi S \text{ W/m K}$$

S is the slope of the linear portion of the graph. For $t_1 = 120$ s and $t_2 = 600$ s $S = (\theta_2 - \theta_1)/\ln 5$. Q is the power input in watts per meter as defined in Section 2.4.2.

2.5. *Use of Nonrigid Materials*

2.5.1. *Powders and Granular Materials*

DIN 51046[3] and PRE/R32[4] both advise the use of refractory containers to give an assembly having dimensions similar to those of shaped materials. The test is carried out in the normal manner. Slight vibration or tamping may be necessary to ensure the absence of voids around the heater. When presenting the results the exact preparation procedure should be given as well as the bulk density of the material in the poured, untamped state.

2.5.2. *Ceramic Fibers*

Heat flow in the hot-wire test is radial so that with anisotropic materials the measured value will be the geometric mean of the conductivities in the two main directions [equation (15), Section 3]. With the panel test[10] the heat flow is linear so that a value can be obtained for the conductivity in a given direction. This is a major disadvantage of the hot-wire method and for this reason ceramic fibers are excluded from PRE/R32.[4] However, the parallel-wire modification may possibly overcome this difficulty (see Section 4).

2.6. *Monolithics and Conducting Materials*

The hot-wire method is ideally suited for testing monolithics and chemically bonded refractories. It operates with a low-temperature gradient across the test piece and the conductivity will therefore be measured at a specific temperature. In these materials combined water is released above 200°C with consequent changes in the conductivity. The panel test methods have considerable temperature gradients across the test piece so that with an insulating castable it is possible to have the material fully fired at the hot face while the cold-face temperature is only 150°C. The degree of firing will therefore vary through the test piece giving an average measured conductivity which is only accurate for that particular temperature gradient. One method of overcoming this is to prefire the material to an agreed temperature (800 or 1000°C) and do a normal conductivity test on the fired material. This, however, ignores any changes occurring at low temperatures, and as will be shown in Section 3 there can be considerable differences between the conductivity of a fired and an unfired material at temperatures below 500°C.

The hot-wire test suffers from none of these limitations and can be used with these materials at temperatures from room temperature upwards; the temperature intervals can be as low as 30–50 K if required. There is also no risk of overfiring the hot face to get a high mean temperature as the material is uniformly fired. If the thermal conductivity is known at a series of discrete temperatures there is no problem in calculating temperature distributions for a known thermal gradient although the conductivity may vary considerably from hot face to cold face.

With chemically bonded or acid-bonded refractories there may be corrosion problems if a base metal alloy measurement cross is used. Electrically conducting materials such as tar-bonded and carbon refractories cannot normally be tested but measurement is possible using a modification of the standard method to be described in Section 4.

2.7. *Errors*

Measurement errors such as those made in the determination of the heater length, power input, time, and temperature should normally be below ±2%. These are random errors and will give correspondingly random errors in the derived conductivity value. Systematic errors can also be present. These have been discussed by the author elsewhere[7] but can be summarized as follows:

1. Poor test-piece surface preparation and measurement cross embedding. These have been discussed in detail by Jeschke[6] and Eschner and coworkers.[8] It may be necessary to surface-grind the material. Flat surfaces and good embedding are particularly important when testing high-conductivity materials.
2. There should be minimum possible variation in the power input during a test run. For this reason it is good practice to keep the total temperature rise to as low a value as is consistent with the sensitivity and discrimination of the temperature-measuring equipment. This reduces the effect of changes in the resistance of the heater with temperature.
3. Ambient (room or furnace) temperature variations should also be kept as low as possible. A rising ambient temperature will give a low conductivity value. A falling ambient temperature will give a high value.

Any dissimilar metal junctions in the thermocouple leads outside the furnace should be isothermal.

4. Use of the fixed time–temperature method to obtain the slope of the temperature–ln t graph where one point is off the linear portion of the graph. Unless the material has been previously tested it is advisable to plot the first test run on semilog graph paper. This will also check whether the power input is high enough.

PRE/R32[4] quotes a repeatability of 8%; DIN 51046[3] gives 5%–10% for the same operator and same equipment. If care is taken with specimen preparation and embedding, using base-metal alloy measurement crosses at room temperature with insulating firebricks, it is possible to reduce this to about $\pm 2\frac{1}{2}\%$. With materials having a conductivity above 1 W/m K this will probably rise to ± 5%.

3. COMPARISON WITH OTHER METHODS

In a European cooperative test[8] in 1973 eight laboratories took part in measuring the thermal conductivity of a type 2300 Insulating firebrick. Five used the hot-wire method and three the panel test methods. Other methods such as radial-flow and split-column comparator were also used, with some laboratories using more than one method. The overall spread of the results was of the order of ± 25%, with the hot-wire tests having a spread of ± 12% and a mean level about 10% above the mean of the panel tests. Other cooperative tests described by Jeschke[6] on type 2300 and 2800 insulating firebricks have showed the hot-wire method giving results up to 30% higher than the values obtained by the panel test methods. In some cases the discrepancy can be explained by anisotropy, as it is not uncommon[7] for insulating bricks to have thermal conductivities differing by 30% depending on whether the panel test is done with the heat flow through the 76 (or 64) mm direction or through the 114 mm direction.

The anisotropy arises from the method of manufacture, which commonly uses combustible materials to produce the high porosity required in the fired material. This tends to give lenticular pores rather than spherical ones and the resultant structure will have a greater air path length in one direction than in the other. To determine how the hot-wire method operates under these conditions equation (1) in Section 1.2 has to be modified.

For an isotropic material with principal conductivities λ_1, λ_2, λ_3 in the directions of the x, y, z axes, respectively, the solution for an instantaneous point source of heat, corresponding to equation (1), is given[5] as

$$\theta = \frac{q}{8}\left(\frac{\rho c}{\pi^3 t^3 \lambda_1 \lambda_2 \lambda_3}\right)^{1/2} \exp\left[-\frac{\rho c}{4t}\left(\frac{x^2}{\lambda_1} + \frac{y^2}{\lambda_2} + \frac{z^2}{\lambda_3}\right)\right] \tag{12}$$

For an instantaneous line source of heat in a direction parallel to one of the principal axes of conductivity, we may, without loss of generality, assume it to be the z direction and to pass through the point $(x, y, 0)$. Substituting $Q'\,dz$ for q in equation (12) and integrating with respect to z gives

$$\theta = \frac{Q'}{4\pi t(\lambda_1 \lambda_2)^{1/2}} \exp\left[-\frac{\rho c}{4t}\left(\frac{x^2}{\lambda_1} + \frac{y^2}{\lambda_2}\right)\right] \tag{13}$$

For a continuous-line source of heat at a rate for $t > 0$ of $Q\,dz\,dt$ at all points on the line parallel to the z axis, the temperature rise at the origin of the co-ordinate system is obtained by substituting $Q\,dt$ for Q' in equation (13) and integrating with respect to t, giving

$$\theta = \frac{Q}{4\pi(\lambda_1 \lambda_2)^{1/2}} \cdot E_1\left[\frac{1}{4t}\left(\frac{x^2}{\alpha_1} + \frac{y^2}{\alpha_2}\right)\right] \tag{14}$$

Comparison of equation (14) and equation (4a) Section 1.3 indicates that for anisotropic materials, using an orthogonal-line source of heat, the measured value, λ_0, of conductivity is related to the principal values by

$$\lambda_0 = (\lambda_1 \lambda_2)^{1/2} \tag{15}$$

i.e., the measured value of the conductivity is therefore given by the geometric mean of the principal conductivities in the two directions perpendicular to the hot wire.

In the practical case, λ_1 and λ_2 would be conductivities in the 76- (or 64-mm) and 114-mm directions and λ_0 would be the measured value given by the hot-wire test. The heat flow in the panel test method is linear and for convenience is usually made in the 76- (64-) mm direction. If this direction happens to be the "low" conductivity direction then when this value is compared with the hot-wire "mean" value the latter will appear to be giving a high incorrect value.

Figure 2 shows measurements made on the same Type 2600 insulating firebrick by both the hot-wire and panel test methods. The conductivity was measured in both 76- and 114-mm directions by the latter method. As can be seen there is roughly 25% difference at 600°C between the values for the two directions. The line for the geometric mean value is also shown and this agrees to within 2% with the line for the hot-wire method. Six insulating bricks were tested in this manner(7) and five showed marked anisotropy. The sixth had only 2% difference between the two curves with the hot-wire curve parallel and 2.5% below the lowest curve. It is important to make these comparisons with the

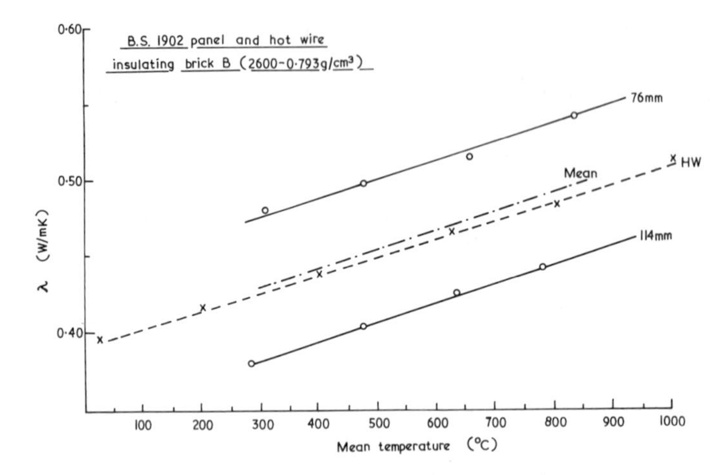

FIGURE 2. Comparison of hot-wire and panel test conductivity/temperature curves for an insulating firebrick.

same test piece as differences of ± 7.5% in conductivity values are not uncommon for a given batch of insulating bricks.[9] Values obtained from manufacturers data sheets should only be used as a general guide because of expected variations from batch to batch.

Further problems may arise in comparing test methods where the test materials are not temperature stable. Figure 3 gives the conductivity–temperature curves for a refractory concrete tested by the hot-wire and BS 1902 panel test[10] methods. With the latter the measurement calorimeter was removed overnight prior to each measurement at temperature. This raised the hot-face tem-

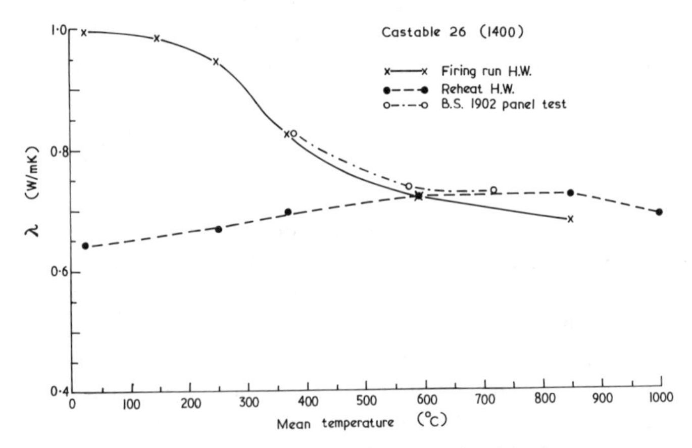

FIGURE 3. Comparison of hot-wire and panel test conductivity/temperature curves for a refractory concrete.

perature and reduced the temperature gradient across the test piece. It also allowed combined water to escape. The following morning the calorimeter was replaced, and when thermal equilibrium had been reestablished the conductivity was measured in the normal way. It can be seen that the hot-wire and panel test values agree quite well (to 5% at 600°C), although the panel test had gradients ranging from 280 to 550 K with corresponding variations in the degree of firing.

The third curve is for a "reheat" test, the test pieces having been held at 850°C for 12 h. Below 500°C there are marked differences between the "firing run" and "reheat run" curves. As stated in Section 2.6, one method of measuring the thermal conductivity of monolithics is to prefire the material and carry out a normal test. Unfortunately this fact or the prefiring temperature is not always stated in manufacturers' data sheets. Under these circumstances a valid comparison cannot be made. The panel and hot-wire tests of Fig. 3 could not be carried out on the same test pieces but were made on bricks from the same batch whose bulk densities agreed to 0.5%.

As reported by Davis and Downs,[7] with isotropic fired material there is usually good agreement between the panel and hot-wire tests provided that the tests are carried out on the same test pieces.

4. MODIFICATIONS

4.1. General

There have been a number of modifications suggested to what is now described as the "Standard Method," where the measurement thermocouple is welded to the midpoint of the heater. These are intended to simplify the method, as in the case of the "heat probe," or to extend its range and accuracy.

4.1.1. Heater and/or Thermocouple Modifications

The resistance change in a platinum heater wire has been suggested as a means of measuring the temperature rise. This can be done by using a Kelvin bridge technique[12] or by simply measuring the voltage drop across a known length of platinum wire when a measured current is passed through it. A digital voltmeter having a full-scale error of the order of 10^{-5} is used for this purpose. Julia and co-workers[13] describe such a technique for liquids, and a similar method has been used for measuring the thermal conductivity of molten slags.[14] The accuracy of this modification is dependent on reliable figures for the coefficient of resistivity at the test temperatures and freedom from heater wire corrosion by the test material. It offers the advantage of measuring the average temperature along the length of the heater as compared with the "spot" measurement provided by the welded couple.

A modification was suggested in 1960 by Haupin,[15] namely, that of using the thermocouple as the heater. Alternating current (60 Hz) was used to supply the heating power, a filter network blocking the ac from the temperature-measuring equipment. The low frequency required very bulky (10–20 H) iron-cored chokes and large (250 and 500 mfd capacitors). An improvement here would be to use a higher frequency (1–2 kHz), which would drastically reduce the physical size of the filter and eliminate hum pick up.

To ensure uniform heating the two legs of the thermocouple must be balanced for resistance and at the same time should have comparable wire diameters. This rules out the use of conventional thermocouple alloys such as Pt;Pt/Rh and Chromel–Alumel (type K). The two resistance alloys chosen should also have similar temperature coefficients of resistivity. These requirements combined with the need to calibrate each couple limits the method both as regards temperature and power input.

4.1.2. Thermal Conductivity or Heat Probes

In this variant the heater and thermocouple(s) are enclosed in a thin sheath or needle which is inserted into the test material. The method is intended basically for nonrigid materials which can be readily pierced by the "hypodermic needle" form of the probe. For this reason its application to refractories is mainly confined to powders, granular, or fibrous materials.

Amongst the first research workers to use a "heat probe" were Hooper and Lepper[16] in 1950. They enclosed an axial constantan alloy resistance wire and three or more series-connected base-metal thermocouples in an aluminium sheath, approximately 4.5 mm o.d. and 475 mm long. The thermocouples were not in direct contact with the heater but were electrically isolated and placed with the junctions near to its midpoint. Because of the finite size of the probe a correction factor, t_0, was calculated from the graph of the rate of change of temperature with time during the early part of the test run. This correction time, t_0, is subtracted from each measurement time, e.g., t_1 and t_2 in equation (10), Section 1.3. This t_0 correction has been suggested for the standard method.[2] However, for heater wire diameters below 0.4 mm it is only of the order of 1 s so it is normally ignored.

Since 1950 the heat probe has been applied to a wide range of materials including margarine and apples.[17] D'Eustachio and Schreiner[18] have described a probe 100 mm long and only 0.75 mm o.d. A comprehensive survey of probe theory and construction has been made by Wechsler,[19] and Pratt[20] also gives a review of design and theory. Blackwell[21] discusses sources of error which include thermal resistance at probe–sample boundaries, finite length of the probe, and boundary effects. The main conclusions are that the probe should be matched to the test material, otherwise there may be difficulties in obtaining a linear section on the temperature–ln time graph. (See Section 2.4.3.) Heat losses at the "free" end of the probe can also cause errors. Where the probe is used with materials of similar thermal conductivity and texture the above errors can be eliminated by the use of an "equipment factor," similar to the t_0 correction factor previously mentioned, or by the use of tables or nomograms. This in effect means the calibration of a specific probe for defined range of thermal conductivities and power inputs.

Because of their construction most heat probes are limited to a temperature range of about $-100°C$ to $+200°C$, although with specialized designs this range may be extended. More recently (1979) Bloomer[22] has used a heat-probe to measure the thermal conductivity of sedimentary rocks at ambient temperatures. This has been achieved by drilling a hole slightly larger than the probe diameter and ensuring good thermal contact with thermal grease (silicon grease with added MgO of Al_2O_3 powder). The probe dimensions were 60 mm long and 1.7 mm o.d., a bead thermistor being used as the temperature sensor. The probe was calibrated with a silica glass block and was semiautomatic in operation, readings being made and processed by a microprocessor. In an earlier work (1963) Wechsler[33] describes a direct graphical method for calculating thermal conductivity using a probe.

4.1.3. *The Parallel Wire*

It is generally accepted that the upper limit for the standard hot-wire test is of the order of 1.5 to 2.5 W/m K, which in effect rules out its application to most high-alumina firebricks and basic refractories. German and Dutch research workers have, over the last 4 or 5 years, investigated various modifications of the technique to raise this upper limit. In 1978 the "parallel wire" method was developed cojointly by research workers in Holland and West Germany in cooperation with the Clausthal Technical University.[11] This modification no longer has the measurement thermocouple welded to the midpoint of the heater but places the couple parallel to the heater wire at a distance of about 15 mm (Fig. 4). The method in itself is not new since many of the heat probes discussed in the previous section have the thermocouple separate from the heater. In these cases the distance is usually less than 2 mm.

4.2. *Mathematical Derivation*

The simplified equation (11) of Section 1.3 can no longer be used since $r > 0$. From equation (4a) we have the temperature θ at time t given by

$$\theta_{(t)} = \frac{Q}{4\pi\lambda} \cdot E_1\left(\frac{r^2}{4\alpha t}\right) \tag{4a}$$

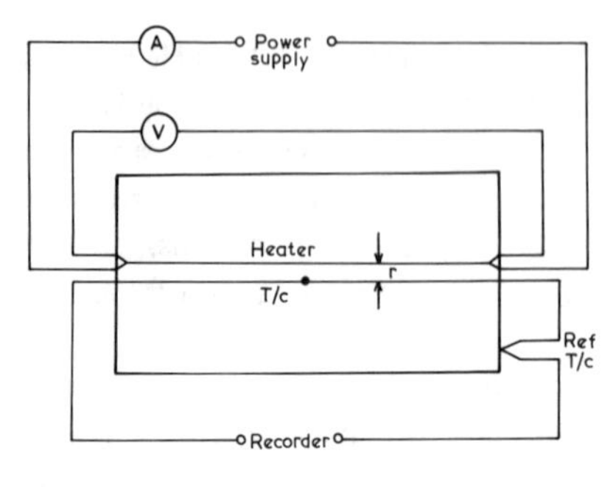

FIGURE 4. Schematic of the parallel-wire apparatus.

and at a time $2t$ by

$$\theta_{(2t)} = \frac{Q}{4\pi\lambda} \cdot E_1\left(\frac{r^2}{8\alpha t}\right) \tag{4b}$$

so that

$$\frac{\theta_{(2t)}}{\theta_{(t)}} = \frac{E_1(r^2/8\alpha t)}{E_1(r^2/4\alpha t)} \tag{16}$$

$E_1(r^2/4\alpha t)$ can be expressed in the form $E_1(x)$ or $-E_i(-x)$ values of which can be obtained from mathematical tables of exponential integrals.[12] By comparing values of $E_1(x)$ for x and $x/2$ with the ratio of the measured temperatures $\theta_{(2t)}$ and $\theta_{(t)}$ until the two ratios agree, the value of x, i.e., $(r^2/4\alpha t)$ and hence $(r^2/4\alpha)$ is obtained. λ_0 (the measured value of λ) is given by

$$\lambda_0 = \frac{Q}{4\pi} \cdot \frac{E_1(r^2/4\alpha t)}{\theta_{(t)}} \tag{17}$$

It should be noted that it is not necessary to know the value of r, the thermocouple–heater spacing, or the value of α the thermal diffusivity. However, if r is known then α can be determined.

With anisotropic materials we must again compare equation (14), Section 3 and equation (4a), Section 1.3. This indicates that, for these materials, using an orthogonal-line source of heat, the measured value, α_0, of the diffusivity, is related to the principal values α_1 and α_2 by

$$\alpha_0 = \frac{x^2 + y^2}{x^2/\alpha_1 + y^2/\alpha_2} \tag{18}$$

If the shortest path of heat flow from the heater to the thermocouple is also directed along one of the principal conductivity axes, e.g., the 76- or 114-mm direction then in equation (18) either x or y is zero and the measured diffusivity value α_0 is equal to the corresponding principal diffusivity value α_1 or α_2.

Knowledge of the thermal diffusivity α enables the specific heat, c, to be obtained from the equation

$$c = \lambda/\alpha\rho \tag{19}$$

where ρ is the bulk density. With anisotropic materials the value of λ is not the measured value λ_0 but λ_1 or λ_2 corresponding to the diffusivity values α_1 and α_2. From equation (15) we have

$$\lambda_0 = (\lambda_1 \cdot \lambda_2)^{1/2}$$

So it follows that

$$\lambda_1 = \lambda_0(\alpha_1/\alpha_2)^{1/2} \tag{20}$$

and

$$\lambda_2 = \lambda_0(\alpha_2/\alpha_1)^{1/2} \tag{21}$$

i.e., the conductivity values λ_1 and λ_2 in directions 1 and 2 (which may be the 76- and 114-mm directions) can be derived from the measured values of the conductivity, λ_0, and diffusivity, α_1, α_2.

This application of the parallel-wire method overcomes the main disadvantage of the hot wire, namely, that it cannot be used with anisotropic materials.

de Boer[8] suggests that the derivation of $(r^2/4\alpha t)$ from mathematical tables should be restricted to values of $\theta_{(2t)}/\theta_{(t)}$ which lie between 2.5 and 1.5 to obtain reasonable accuracy. However, this restriction can be overcome by the use of a computer, using an iterative process, to find the value of $(r^2/4\alpha)$ which leads to the best fitting line for the graph of the measured values of θ plotted against those calculated from equation (4a).

By placing the heater and thermocouple in the 230 × 114 mm and 230 × 76 mm faces values for α_1 and α_2 will be obtained. The two values obtained for λ_0 should agree. The use of equations (20) and (21) will then give the value of λ_1 and λ_2.

4.3. Applications

The experimental procedure is virtually the same as for the standard method. It will be found that the power output (Q) needs to be higher than that for the standard method. If a chart recorder is used there should be no problems in obtaining the $\theta_{(t)}$ and $\theta_{(2t)}$ values. If a printout is used the time intervals should be set according to the conductivity of the test material. With high-conductivity materials the usable time interval, before the heat front reaches the outside of the test array, may be as short as 60 s. It is advisable to take about 20 to 30 readings in this time.

The shape of the curve as drawn on the chart recorder differs from that obtained in the standard test. In the latter there is a rapid rise in temperature over the first few seconds followed by a levelling off. With the parallel wire the initial rate of rise is slow but increases with time.

Good embedding is again essential and considerable care must be taken with high-conductivity materials. This point is covered in some detail by de Boer and co-workers,[11] who also give details of the development of the method and of comparisons with the standard hot-wire method. It would appear that it is possible to raise the conductivity limit from 2.0 to 25 W/m K. Electrically conducting materials can be tested by the use of a sheathed heater and thermocouple. Because the heater and thermocouple are no longer in direct contact with the test material a correction factor must be applied to the time. This is done empirically by comparing the time–temperature curves obtained with bare and sheathed wires using a nonconducting test material. There is, however, an upper temperature limit of about 1050°C.

Repeatability would appear to be similar to that of the standard method. The accuracy of the determination of the thermal diffusivity is dependent on the error in measuring the heater–couple spacing (r). An expected error of ± 0.5 mm will give a corresponding error of about ± 7% in the diffusivity value.

5. SUMMARY

The hot-wire method for the measurement of thermal conductivity is a transient technique with radial heat flow. It has the advantage of operating with a small temperature gradient across the test assembly and so can be used at room temperature for quality control. However, the standard method cannot be used with anisotropic materials or with materials having a thermal conductivity above 2.0 W/m K. These limitations can be overcome by the use of the parallel-wire modification which raises the upper limit to 25 W/m K and provides a means of measuring the thermal diffusivity and specific heat at a defined temperature. Electrically conducting materials such as silicon carbide and tar-bonded refractories can also be tested. The standard method can be used at temperatures up to 1600°C with a probable accuracy of ± 5% if care is taken with test-piece preparation and heater–thermocouple embedding. Table 1 provides a summary of hot-wire transient heat flow experiments.

ACKNOWLEDGMENTS. The author wishes to thank Dr. D.W.F. James, Chief Executive of British Ceramic Research Association Ltd., for permission to publish this paper. He also wishes to acknowledge the assistance given by his colleague, Mr. F. Moore.

TABLE 1
Summary of Hot-Wire Transient Heat Flow Experiments Used for the Measurement of the Thermal Conductivity of Refractories and Other Materials

No.	Investigator	Ref.	Year	Material[a]	Method[b]	Temperature[c] measurement	Range (K)	λ Range W/m K	Accuracy[d] Max. error (%)	Scatter (%)	Comments	No.
1	Van der Held	2	1949	Liquids	Htr. sep. T/C	T/C	300	0.1–0.6	2	n.d.		1
2	Hooper	16	1950	SiO_2 (gran.) soils	Probe	T/C	300	0.1–2.0	n.d.	n.d.		2
3	Mann	17	1956	Insulation	Probe	T/C	300	0.03–0.6	n.d.	n.d.		3
4	Haupin	15	1960	I/Fb, Fb,	T/C as heater	T/C	300–800	0.8–1.8	n.d.	5	Uses htr as T/C	4
5	Mittenbühler	28	1962	Fb, B, powders	(T/C as heater	T/C	300–1300	0.1–6.0	5	11	(Uses htr as T/C	5
					Welded T/C/htr)	T/C					& welded T/C)	5
6	Mittenbühler	29	1964	(I/Fb, Fb, B, Chem-bond, Granules)	Welded T/C htr	T/C	300–1500	0.15–10.0	n.d.	±2		6
7	Wechsler	19	1966	Insulation, soils	Probe	T/C	200–350	0.03–4.0	n.d.	±5 (best)	Probe design	7
8	le Doussal	30	1971	Fb	Welded T/C/htr	T/C	300–1500	1.0–2.0	n.d.	n.d.	(Comparison with	8
9	Provost	31	1973	I/Fb	Welded T/C/htr	T/C	800–1100	0.1–0.2	n.d.	n.d.	other	9
10	Eschner	8	1974	I/Fb, Fb, B	Welded T/C/htr	T/C	300–1500	0.2–10.0	n.d.	5 to 9	techniques)	10
11	Hayashi	23	1973	C/clay, SiC/clay I/Fb, Fb, B	Welded T/C/htr	T/C	300	0.2–17.0	n.d.	n.d.	(Effects of sample size	11
12	Hayashi	24	1974	Graphite/clay plaster	Welded T/C/htr	T/C	100	0.2–3.0	n.d.	n.d.	& htr diam)	12
13	Hayashi	25	1974	I/Fb, Fb, B	Heater only	R (direct)	300–1500	0.1–10.0	2	±2		13
14	Morrow	12	1977	I/Fb, Fb, B	(Heater, welded T/C/htr heater as T/C)	T/C & R (bridge)	300–1400	0.1–5.0	n.d.	1.5 to 10		14
15	Julia	13	1977	Liquids	Heater only	R (direct)	270–350	0.1–0.7	±0.3	±0.1		15
16	Mills	14	1978	Molten slags	Heater only	R (direct)	>1400	n.d.	n.d.			16
17	Jeschke	6	1978	I/Fb, Fb, B	Welded T/C/htr	T/C	300–1500	0.2–3.0	n.d.	±4.5	Comparison	17
18	Jackson	26	1978	Insul. fibers	Welded T/C/htr	T/C	300–1900	0.03–0.8	n.d.	±5		18
19	Davis	27	1978	I/Fb, insul. fibers	Welded T/C/htr	T/C	300–1300	0.02–1.0	n.d.	n.d.	Comparison	19
20	Bloomer	22	1979	Sedim. rocks	Auto probe	Thermistor	300	1.0–2.1	n.d.	n.d.	Uses computer	20
21	de Boer	11	1980	I/Fb, Fb, B	Parallel wire	T/C	300–1500	0.3–21.0	n.d.	n.d.	Comparison	21
22	Davis	7	1980	I/Fb, Fb	Welded T/C/htr	T/C	300–1300	0.2–3.0	n.d.	±2	(Comparison with	22
23	Davis	32	1980	Monolithics, I/Fb, Fb	Welded T/C/htr	T/C	300–1300	0.2–10.0	n.d.	±5	panel test)	23
				I/Fb	parallel wire	T/C	300					

[a] I/Fb, Insulating firebricks; Fb, aluminosilicates; C/clay, carbon/clay mix; chem. bond, chemically bonded.
[b] T/C, Thermocouple.
[c] R, Resistance of heater used to measure temperature.
[d] Figures claimed by authors; n.d. no data.

NOTATION

θ	Temperature rise resulting from continuous line source of heat
t	Time
q	Heat quantity instantaneously generated at $t = 0$ at point x, y, z
$Q'dz$	Heat quantity instantaneously generated at $t = 0$ on line parallel to z-axis and passing through $(x, y, 0)$
Q	Heat generated per unit length of heater wire ($I^2 R$)
ρ	Bulk density
c	Specific heat
λ	Thermal conductivity
λ_0	Measured value of thermal conductivity
$\lambda_1 \lambda_2$	Thermal conductivity in directions 1 and 2
α or α_0	Mean or measured value of thermal diffusivity
$\alpha_1 \alpha_2$	Thermal diffusivity in directions 1 and 2
I	Current passing through hot wire
R	Resistance of hot wire per unit length
E_1 or E_i	Exponential integral
γ	Euler's constant

REFERENCES

1. A.L. Schieirmacher, *Wiedemann Ann. Phys.* **34** (1888).
2. E.F.M. Van der Held and F.G. Van Drunen, "A Method of Measuring the Thermal Conductivity of Liquids," *Physika* **15**(10), 865 (1949).
3. Deutsches Institut für Normung, "Testing of Ceramic Materials; Determination of Thermal Conductivity up to 1600°C by the Hot-Wire Method: Thermal Conductivity up to 2 W/m K" (in German). DIN 51046, Provisional Standard (Part 1) (August, 1976).
4. Fédération Européene des Fabricants de Produits Réfractaires (1978). "Determination of Thermal Conductivity up to 1500°C for Values of $\lambda \leqslant 1.5\,W \cdot m^{-1} \cdot K^{-1}$ by the Hot-Wire Method," 32nd PRE Recommendation (provisional) (1977).
5. H.S. Carslaw and J.C. Jaeger, *Conduction of Heat in Solids,* 2nd ed. Oxford U.P., Oxford (1959), pp. 256–261.
6. P. Jeschke, "Thermal Conductivity of Refractories: Working with the Hot-Wire Method," in *Thermal Transmission Measurements of Insulation,* ASTM STP 660 (R.P. Tye, ed.), American Society for Testing and Materials, Philadelphia (1978), pp. 172–185.
7. W.R. Davis and A.M. Downs, "The Hot-Wire Test – A Critical Review and Comparison with the BS 1902 Panel Test," *Trans. Br. Ceram. Soc.* **79**(2), 44 (1980).
8. A. Eschner, B. Grosskopf, and P. Jeschke, "Experiences with the Hot-Wire Method for the Measurement of the Thermal Conductivity of Refractories" (in German). *Tonind-Ztg.* 98(9), 212 (1974).
9. K.W. Cowling, A. Elliot, and W.T. Hale, "Note on the Relationship between Bulk Density and Thermal Conductivity in Refractory Insulating Materials," *Trans. Br. Ceram. Soc.* **53,** 461 (1954).
10. British Standards Institution, "Determination of Thermal Conductivity," BS 1902: Part 1A: Section 12 (1966).

11. J. de Boer, J. Butter, B. Grosskopf, and P. Jeschke, "The Hot-Wire Technique for the Determination of High Thermal Conductivity," *Refract. J* **(Sept–Oct)**, 22–28 (1980).
12. G.D. Morrow, "Improved Hot-Wire Thermal Conductivity Technique," Paper 9R-77F presented at the Bedford Springs Fall Meeting of the American Ceramic Society (1977).
13. Y.H. Julia, J.F. Renaud, D.J. Ferrand, and P.F. Malbrunot, "Device for Automatic Thermal Conductivity Measurements," *Rev. Sci. Instrum.* **48,** 1654 (1977).
14. K.C. Mills, B.J. Keene, J.S. Powell, and J.W. Bryant, "Proceedings of the Colloquium on Thermal Properties of Ceramics" Held at the National Physical Laboratory, Teddington, Middlesex, England, 19 October (1978).
15. W.E. Haupin, "Hot-Wire Method for Rapid Determination of Thermal Conductivity," *Bull. Am. Ceram. Soc.* **39,** 139 (1960).
16. F.C. Hooper and F.R. Lepper, "Transient Heat Flow Apparatus for the Measurement of Thermal Conductivities," *Trans. Am. Soc. Heat Vent. Engrs.* **56,** 309 (1950).
17. G. Mann and F.G.E. Forsythe, "Measurement of the Thermal Conductivity of Samples of Thermal Insulating Materials and of Insulation *in situ* by Heated Probe Method," *Mod. Refrig.* **59**(6), 188 (1956).
18. D. D'Eustachio and R.E. Schreiner, "A Study of a Transient Heat Method for Measuring Thermal Conductivity," *Trans. Am. Soc. Heat Vent. Engrs.* **18,** 331 (1952).
19. A.E. Wechsler, "Development of Thermal Conductivity Probes for Soils and Insulations," Cold Regions Research and Engineering Laboratories Technical Report 182. US Army Materiel Command (1966).
20. A.W. Pratt, "Heat Transmission in Low Conductivity Materials," in *Thermal Conductivity,* Vol. 1 (R.P. Tye, ed.), Academic Press, New York (1969), pp. 376–388.
21. J.H. Blackwell, *Can. J. Phys.* **31,** 472 (1953); and **34,** 412 (1956).
22. J.R. Bloomer and J. Ward, "A Semi-Automatic Field Apparatus for the Measurement of the Thermal Conductivities of Sedimentary Rocks," *J. Phys. E: Sci. Instrum.* **12**(11), 1033 (1979).
23. K. Hayashi, M. Fukui, and I. Uei, "On the Radius of the Cylindrical Specimen for the Measurement of Thermal Conductivity by the Hot-Wire Method" (in Japanese; English abstract), *Yogyo-Kyokai-Shi (J. Jpn. Ceram. Soc.)* **81**(12), 534 (1973).
24. K. Hayashi, M. Fukui, and I. Uei, "On the Length of the Specimen for the Measurement of Thermal Conductivity by the Hot-Wire Method" (in Japanese; English abstract), *Yogyo-Kyokai-Shi (J. Jpn. Ceram. Soc.)* **82**(4), 202 (1974).
25. K. Hayashi, M. Fukui, and I. Uei, "New Development of an Apparatus for Thermal Conductivity Measurement of Refractory Bricks by the Hot-Wire Method" (in English), *Yogyo-Kyokai-Shi (J. Jpn. Ceram. Soc.)* **82**(1), 13 (1974).
26. A.J. Jackson, J. Adams, and R.C. Miller, "Thermal Conductivity Measurements on High-Temperature Fibrous Insulation by the Hot-Wire Method," in *Thermal Transmission Measurements of Insulation,* ASTM STP 660 (R.P. Tye, ed.), American Society for Testing and Materials, Philadelphia (1978), pp. 154–171.
27. W.R. Davis, "Determination of the Thermal Conductivity of Refractory Insulating Materials by the Hot-Wire Method," in *Thermal Transmission Measurements of Insulation,* ASTM STP 660 (R.P. Tye, ed.), American Society for Testing and Materials, Philadelphia (1978), pp. 186–199.
28. A. Mittenbühler, "Determination of the Thermal Conductivity of Refractories by the Hot-Wire Method" (in German; English abstract), *Ber. Deut. Keram. Ges.* **39**(7) (1962).
29. A. Mittenbühler, "An Apparatus for the Measurement of the Thermal Conductivity of Refractory Bricks, Granular Materials and Powders" (in German; English abstract), *Ber. Deut. Keram. Ges.* **41**(1), 15 (1964).

30. H. le Doussal, "Interlaboratory Comparative Measurements of the Thermal Conductivity of Refractory Products" (in French), *Bull. Soc. Franc. Ceram.* **91**, 42 (1971).
31. G. Provost, "Thermal Conductivity; Theoretical Review; Summary of Comparative Tests, Practical Conclusions" (in French), *Bull. Soc. Franc. Ceram.* **99**, 17 (1973).
32. W.R. Davis, F. Moore, and A.M. Downs, "The Hot-Wire Method for the Determination of Thermal Conductivity; Castables and Modifications to the Standard Method," *Trans. J. Br. Ceram. Soc.* **79**(12), 158 (1980).
33. A.E. Wechsler and P.E. Glaser, "Thermal Conductivity of Non-Metallic Materials," Summary Report, Contract No NAS8-1567. Research Projects Laboratory, NASA, George C. Marshall Space Flight Center, Huntsville, Alabama.

6

B.S. 1902 Panel Test Method for the Measurement of the Thermal Conductivity of Refractory Materials

W. R. DAVIS

1. GENERAL INTRODUCTION

The B.S. 1902 panel test method[1] is basically similar to the ASTM C201-68 (1979) test,[2] the differences being mainly constructional. Both are equilibrium or "steady-state" methods involving the direct measurement of the amount of heat flowing linearly through a known area of the material under a known and steady temperature gradient. A water-flow calorimeter is used for the heat-flow measurement, the test piece being a plane parallel slab or panel of the material under test. Both methods have been used as the principal means of determining the thermal conductivity of refractory materials in the U.K. and U.S.A. over the past 30 years.

1.1. Range of Application

The method is commonly used over the temperature range of 400 to 1350°C (hot-face temperatures) for thermal conductivity values in the range 0.05–15 W/m K. Other limitations are mainly those of size. The test was originally designed to use the "standard square" – 230 x 114 x 76 (or 64) mm, two or three of such bricks being required. Where standard squares are not available the length and breadth of the samples should be sufficient to permit the cutting of two or more slabs 230 x 114 mm. The thickness should lie within the limits of 25 to 76 mm. Satisfactory measurements cannot be made with specimens of

W.R. DAVIS • British Ceramic Research Association Limited, Queens Road, Penkhull, Stoke-on-Trent, ST4 7LQ, England.

appreciable water content, the test pieces being oven dried at 110°C before testing. Unfired materials, such as monolithics, are usually prefired beforehand. The thermal conductivity of powders and nonrigid materials can be measured by the use of a refractory container approximating in size to that of the normal test panel (see Section 2.3.1).

The upper limit to the conductivity range is dependent on the maximum heat flux which can be handled by the calorimeter. This can be overcome to some extent by placing insulation between the calorimeter and the "cold face" of the test piece, but the thickness of such insulation is usually limited to about 10–15 mm. The lower limit to the conductivity range is set by the minimum heat flux measurable. This latter can be increased by the use of a thinner test piece (thickness $>$25 mm).

Because the heat flow is linear, anisotropic materials can be tested (see Section 3.3 paragraph 3). In such cases the direction of heat flow through the test material should be specified.

1.2. Thermometry

As already stated, the basis of the method is the measurement of the heat flux through a known area of the test piece material under a known temperature gradient. It is essential that the heat flow should be linear through the section covered by the calorimeter. Any divergence will result in an erroneous value for the area used in the calculation of the thermal conductivity. The following precautions are taken to ensure that the heat flow is truly perpendicular to the plane of the test slab:

1. The area of the test piece is considerably greater than that of the calorimeter.
2. The calorimeter is surrounded by a water-cooled guard ring which ensures that the "cold" (unheated) face of the rest of the test panel is at the same temperature as the area lying under the calorimeter. This transfers any lateral heat flow to the edges of the test panel, and, if the test panel is large enough, gives linear vertical heat flow in the central portion.
3. The "hot face" of the test piece is heated as uniformly as possible.

If the water calorimeter is placed directly onto the test panel then the temperature of the "cold face" of the latter cannot be much higher than that of the water in the calorimeter. With the higher conductivity materials (say above 1–2 W/m K) this would impose severe limitations on the hot-face temperature or require an impossibly high rate of water flow through the calorimeter. This restriction is overcome by interposing a thin slab of refractory insulating material between the cold face of the test piece and the calorimeter. This has the effect of raising the temperature of the "cold face" to a more reasonable value. Tests carried out on the same test piece, with and without this "backup insulation," have shown that the conductivity value obtained is the same in both cases.[10, 12]

The hot- and cold-face temperatures reported are the mean of the three thermocouple readings in each face. With proper adjustment of heater power and when thermal equilibrium has been reached the maximum temperature difference between the couple readings in a given face is usually of the order of 5°C (± 2.5°C). The couple thermo-emfs can be measured with either a potentiometer (± 1 μV) or high-accuracy digital meter (± 0.1°C).

1.3. History (Table 1)

In 1927 F.H. Norton described[3] a guarded water calorimeter method in which the testpiece was in the form of a cylinder, 108 mm diameter and 228 mm long. The test piece rested on top of the calorimeter being surrounded by insulation and heated on the top face by a gas furnace. Thermocouples were placed in holes drilled at four vertically spaced intervals from the top. A form of peripheral heating was used to reduce lateral heat flow. Hot-face temperatures ranged from 200 to 1450°C. A range of firebricks, fused alumina, and zirconia bricks were tested, having conductivities ranging from 0.3 to 6 W/m K. Claimed accuracy was ± 15% to ± 25%. In 1933 Wilkes[4] used a setup very similar to modern equipment. A three-brick panel was heated from the top with silicon carbide rod heaters and a guarded water calorimeter measured the heat flow. Mean temperatures (average of the hot- and cold-face temperatures) ranged from 200 to 1050°C. The tests were made on insulating and magnesite bricks with conductivities between 0.3 and 5 W/m K. The same author[5] described further experiments in 1934 with the same apparatus on a range of firebricks where the mean temperatures lay between 200 and 1350°C with conductivities 0.5 to 3 W/m K. The results were compared with those obtained by other methods.

Austin and Pierce[6] in 1935 used a similar apparatus to test firebricks and various types of silica brick. Here the mean temperature range was 200–1000°C with a conductivity range of 1–2 W/m K. Reproducibility is given as ± 5% and comparisons made with other methods. Reliability and reproducibility are discussed.

In a report for the ASTM Sub-Committee C-8 in 1936[7] Nicholls reviews various methods including the guarded water calorimeter, guarded hot plate, heat flowmeter, and radiation meter used in a series of cooperative tests by six laboratories on firebrick and silica brick samples over the mean temperature range 200–1350°C. Possible sources of error in each method are discussed in detail, although the wide variation of 30% in the test figures permitted no firm conclusions to be made regarding absolute accuracy.

A similar series of tests were carried out in the U.K. in 1937,[8] where four laboratories tested diatomite bricks. Results from three of the laboratories using

TABLE 1
Summary of Guarded Water–Calorimeter Thermal Conductivity Tests

No.	Investigator	Ref.	Year	Material[a]	Mean temp. range[b] (K)	Conductivity range (W/m K)	Reproducibility[c] (%)	No.
1	Norton, R.H.	3	1927	Fb, Al_2O_3 (f), ZrO_2	500–1700[d]	0.3–6	+15–± 25	1
2	Wilkes	4	1933	I/Fb, MgO	500–1300	0.3–5		2
3	Wilkes	5	1934	Fb	500–1600	0.5–3		3
4	Austin *et al.*	6	1935	Fb, SiO_2	500–1300	1–2	± 5	4
5	Nicholls	7	1936	Fb, SiO_2	500–1600	0.7–2		5
6	Oliver	8	1937	Diatomite Fb	400–800	0.09–0.14	5	6
7	Norton, C.L.	9	1942	Fb, I/Fb	400–1350	0.06–0.45	3–4	7
8	Patton *et al.*	10	1943	Fb, SiO_2, CC, I/Fb	500–1400	0.25–1.7	1–5	8
9	Clements *et al.*	11	1949	I/Fb, Sil.	700–1500	0.3–1.2	3	9
10	Watson *et al.*	12	1953	Al_2O_3 (f), I/Fb	400–1500	0.4–2	4.5	10

[a] Fb, aluminosilicate firebrick; I/Fb, insulating firebrick; MgO, magnesite brick; SiO_2, silica brick; Sil, siliceous brick; ZrO_2, Zirconia brick; Al_2O_3 (f), fused Al_2O_3 brick; CC, China clay brick.
[b] Arithmetic mean of hot and cold face temperature.
[c] Due to the high variability of test material accuracy figures cannot be quoted with any degree of confidence—these are therefore replaced by reproducibility values.
[d] Hot face temperatures (for this reference only).

guarded water calorimeter equipment had a scatter of 14%. Mean temperature range was 100–540°C. Reproducibility of better than 5% was claimed.

C.L. Norton[9] modified Wilkes's design in 1942 and the report gives considerable constructural detail and operational procedure. Test results are given for insulating bricks and firebricks over the mean temperature range of 100–1000°C with a conductivity range of 0.06 to 0.45 W/m K. Possible sources of error are discussed in detail and a reproducibility of 3%–4% is claimed.

Further tests in 1943[10] with this apparatus gave a reduction in the lateral heat flow and this arrangement formed the basis for the present ASTM C201-68 (1979) test. In these tests super-duty firebrick, silica brick, and China clay brick, having conductivities between 1.0 and 1.7 W/m K, were measured over the range 200–1100°C. Insulating bricks (~0.25 W/m K) were also tested between 200 and 700°C. Reproducibility of 1.5% is claimed.

Experiments carried out between 1949 and 1951 are reported by Clements and Vyse[11] and the equipment described was used in drafting the present B.S. 1902 test. Results are given for insulating bricks and siliceous bricks over the mean temperature range of 400 to 1250°C, conductivity range 0.3 to 1.2 W/m K. The total scatter of the results at a given temperature was better than 3%. Errors are discussed in some detail although a figure for absolute accuracy is not claimed due to the lack of reliable standards.

A cooperative test was made by Watson, Clements, and Vyse[12] in 1953 in which the ASTM C201-47 [now known as C201-68 (1979)] and the B.S. 1902 panel test equipments were compared. Tests were made on fused Al_2O_3 brick (2.0–1.5 W/m K) over the range 150 to 1200°C and high-temperature insulating brick (0.4 W/m K) over the mean temperature range 300–1100°C.

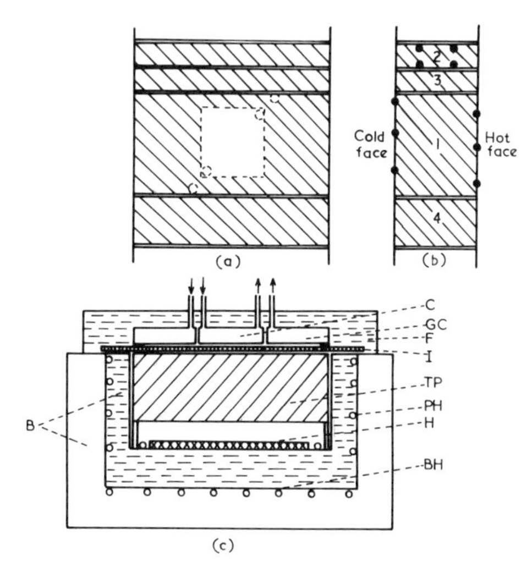

FIGURE 1. The B.S. 1902 test apparatus.

Tests were made with and without "backup insulation" with a maximum discrepancy of 2%. The results from the two sets of tests agreed to within 4.5%. Errors and equipment differences are discussed.

The present standard is designated as B.S. 1902: Part 1A: Section 12, 1966 and is currently (1983) under revision. This revision is a minor one and mainly concerned with arrangement and control of heaters and thermocouple layout.

The standard dealing with refractory insulating bricks, B.S. 2973: 1961, has Section 10, dealing with the measurement of thermal conductivity, identical to the B.S. 1902 standard and will be incorporated in it when the latter is issued in its revised form.

2. GENERAL OUTLINE OF THE METHOD

2.1. Description

A schematic outline of the equipment, roughly to scale, is given in Fig. 1. The test panel (TP) consists of a complete test brick [(1) in Fig. 1b) 230 × 114 × 76 (or 64) mm or approximating to these dimensions, with a half brick or slab (4) 230 × 55 × 76 (or 64) mm on one side and two slabs (2 and 3), 230 × 30 × 76 (or 64) mm on the other side, making up a panel roughly 230 × 230 mm. This panel rests on a refractory box in the furnace chamber.

The top and bottom faces of the panel are dressed flat and parallel and six thermocouples inserted into grooves as in Fig. 1b with the couples running parallel to the 230-mm dimension. The grooves are of such a depth that the couples lie half in and half out, being cemented into place with a commercial high-grade alumina cement. Four other couples are placed in slabs 2 and 3 as shown. These are used to detect lateral heat flow.

The main heaters (H) are Pt–20% Rh wire thinly coated with alumina cement. A set of nickel-chrome alloy heaters (BH) are used to boost the main heaters and reduce heat flow through the bottom of the furnace case at temperatures above 500°C. Peripheral heaters (PH) balance any lateral heat flow from the test panel, the thermocouples in slabs 2 and 3 being used for this purpose. The furnace casing has two-stage insulation, the inner lining being type 2800 insulating brick with a lightweight insulating brick layer between this and the casing.

The backup insulation is a 230 × 230 mm slab of insulating refractory (type 2600 or 2800) with a thickness of about 10 mm. This, however, can be altered to suit circumstances. This slab is bedded down onto the test panel with fine alumina powder to ensure good thermal contact. The calorimeter (C) and guard ring (GC) rest on thin strips (0.5 × 10 mm) of asbestos paper placed on the

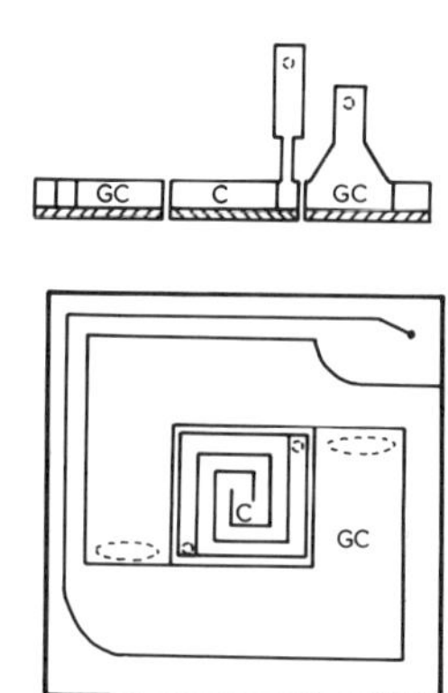

FIGURE 2. Design of calorimeter and guard ring.

insulating slab. This avoids the possibility of the calorimeter assembly touching at a few discrete points giving irregular temperature distribution.

To eliminate any ambient temperature effects the whole calorimeter assembly is covered with a 50 mm layer of ceramic insulating fiber. The design of the calorimeter and guard ring is shown in Fig. 2. Full constructional details of the apparatus are given in Ref. 1.

2.2. Experimental Procedure

Before assembly the thickness (L) of the test brick is measured over the thermocouple junctions, using vernier callipers, to better than ± 0.2 mm. The panel is assembled and gaps between the edges and the furnace wall are filled in with alumina cement. The peripheral heaters are switched on to dry out the panel and then dry alumina powder (250 μm) is poured over the warm panel and the backup slab is bedded down to give a uniform powder layer about 1 mm thick. Asbestos paper strips are placed on the insulating slab and the calorimeter–guard assembly positioned. The whole is covered with a 50-mm layer of ceramic insulating fiber.

The power input to the heaters is set to give the required hot-face temperature and the assembly is allowed to stabilize overnight (12–15 h). Readings are taken of the hot- and cold-face temperatures to ensure thermal equilibrium (less than 2°C change in 30 min). The water flow through the calorimeter/guard is set to give a difference of about 3°C between input and output water temperatures. Peripheral heat flow is checked with the lateral thermocouples and if necessary the peripheral heaters adjusted. If this is done then the measurements are delayed until thermal equilibrium is reestablished.

The heat flow rate is obtained by measuring the time taken for a known quantity of water to flow through the assembly, the input and output water temperatures being measured during this period. This gives the heat flow rate in calories per second.

The hot- and cold-face temperatures of the test pieces are measured before and after the heat flow measurement taking the mean of the three couples in each face to $\pm 0.5°C$.

The conductivity (λ) can then be calculated from

$$\lambda = \frac{QL}{A(t_1 - t_2)} \times 418.67 \quad \text{watts per meter Kelvin (W/m K)}$$

where Q is the heat flow in cal/s, L is the mean distance between junctions of hot and cold face couples (cm), t_1 is the average temperature of hot face (°C or K), t_2 is the average temperature of cold face (°C or K), and A is the area of calorimeter plus one-half the area of the gap between calorimeter and the guard (cm^2). The conductivity is quoted as having been determined as a "mean" temperature of $(t_1 + t_2)/2$ (°C or K). Three determinations are made at this temperature at intervals of 1 h. These usually agree to $\pm 2.5\%$. Measurements are normally made in ascending temperature steps usually agreed with the supplier although the "mean temperatures" will be dependent on the maximum permissible hot-face temperature.

2.3. *Variations in the Standard Method*

2.3.1. *Nonrigid Materials*

These comprise fiber materials, granules, and powders, and are tested by placing them in a rigid box of heat-resistant material, usually made from sillimanite slabs, having dimensions similar to the full test panel, 230 × 230 × 76 (or 64) mm. With powders and granular materials care must be taken to avoid air pockets. To remove these, light tamping or vibration is allowed but should be carried out in a specific manner, usually agreed with the supplier, to ensure homogeneity and with a known bulk density.

With fiber and granular materials the depth of the box is usually reduced to 25 mm to increase the heat flux. Because the thermal conductivity of these materials is dependent on bulk density, spacers are usually placed in the box to define the thickness and a known specified compressive load applied to produce the required bulk density.

2.3.2. *Electrically Conducting Materials*

In this case the thermocouples are insulated with close-fitting thin-wall pure alumina sheaths to avoid stray emfs or electrical short circuits between hot- and cold-face couples which would give erroneous readings.

2.3.3. Monolithics or Refractory Concretes

In service these materials are fired *in situ,* there being a loss of combined water during this process. With a lightweight insulating material ("castable") it is possible to have a temperature gradient such that the hot face is fully fired whereas the cold face is only in the "as-received–fully dried" condition. This can cause problems which are overcome by prefiring. This can be to a temperature slightly above the maximum hot-face temperature to be used in the conductivity test, the specimen being fired uniformly, or the complete panel can be gradient-fired in the conductivity apparatus to a temperature slightly above the hot-face temperature for any one set of measurements. This can be repeated at each temperature measurement. This technique gives conditions approximating to those encountered in service. The ASTM C201-68 test uses descending temperature steps so that the panel is usually uniformly prefired before testing.

2.3.4. Differences from the ASTM C201-68 (1979) Method

These are mainly constructional, the test panel being heated from above in the ASTM test. The B.S. 1902 test uses a smaller test panel with peripheral heaters to reduce lateral heat flow and there are also differences in calorimeter design. In the test procedure the ASTM test uses descending temperature steps as compared with ascending ones in the B.S. test. When compared, using the same temperature-stable test pieces, the two test methods usually agree to within ± 5%.

2.3.5. Modifications

The main modification consists of replacing the calorimeter assembly with a heat-flow meter. This usually comprises a small slab of material of known conductivity with multiple thermocouples set in the upper and lower faces connected in a differential mode. The output emf is converted to heat flow either by a simple calibration constant or graphically. Errors can arise because of lateral heat flow, a guard ring not being normally incorporated.

2.4. Summary of Application Range

The method is used for refractory materials in block form, usually the "standard square" – 230 x 230 x 76 (or 64) mm over the thermal conductivity limits of 0.05 to 15 W/m K. Temperature range is restricted by the temperature gradient across the test piece since the hot-face temperature must not exceed the maximum working temperature of the test material. This may cause problems with lightweight insulating materials where the temperature gradient may be as high as 900°C to give a "mean temperature" of 550°C. This problem can be overcome to some extent by the use of thinner test pieces. With the present apparatus maximum hot-face temperature is around 1350°C.

Nonrigid materials can be tested using a refractory container and electrically conducting refractories by the use of insulated thermocouples.

3. ERROR ANALYSIS

A more detailed review is given in Refs. 11 and 12. The work quoted there was carried out around 1955 and since then there have been considerable improvements in temperature measurement and control. However, the error analysis section is still valid.

3.1. Measurement Accuracies

The various measurements made during a test run can be summarized as follows:

a. Heat path length (e.g., thickness of test piece). This is measured with vernier callipers to ± 0.2 mm (roughly ± 0.25%) and is dependent on the specimen top and bottom faces being flat and parallel.
b. Calorimeter area is usually measured to 0.1 cm^2 – i.e., ± 0.2% – and can be regarded as an equipment constant.
c. Measurement of temperature gradient – taken as the mean of three cold-face and three hot-face temperatures to ± 0.5°C or better.
d. The evaluation of the heat-flow rate involves measurement of water volume, time, and temperature rise. The water is collected in a calibrated flask and with proper precautions the error should be negligible (<0.1%). Timing is usually to 0.2 s in 150 s, i.e., below 0.2% max. Temperature rise measurement is dependent on the type of detector used, e.g., mercury-in-glass restricted-range thermometers, multiple high-output thermocouples, platinum resistance thermometer. The overall temperature difference between the input and output water is normally set to about 3°C with an expected measurement accuracy of ± 0.01°C. The timing period is usually between 90 and 240 s, and five to seven water temperature readings are made at 30-s intervals.

3.2. Possible Error Sources

Some of these have already been discussed in the preceding section.

3.2.1. Departure from Linear Heat Flow

Linearity of heat flow is essential since any lateral heat flow in the measurement section will give an error in the derived conductivity value. It is difficult to achieve zero lateral heat flow in practice but it can be reduced to a low value by use of the peripheral heaters. The lateral thermocouples can be used to estimate the actual value, and this has been quoted as being typically below 0.5% of the total heat flow.[12] The error from this source usually gives a low value for the conductivity. The linearity of the heat flow through the panel is primarily dependent on attaining an even temperature distribution across both hot and cold faces. There is a gap of about 40 mm between the main heaters and the test pieces hot-face, and in practice this gives good heat distribution shown by the fact that the readings on the hot face usually agree to ± 2°C.

The alumina powder layer between the backup insulation and the test panel ensures good thermal contact provided that there are no voids or hollows, that the layer is of uniform thickness and that the asbestos strips ensure that there is no direct contact with the calorimeter to give a possible uneven temperature distribution.

With some insulating firebricks and silica bricks color changes occur on heating to high temperatures to give what could be termed "heat contour lines."[10] The presence of colored zones with perfectly straight boundaries indicates uniform temperature distribution and whether or not the heat flow is accurately perpendicular to the hot face.

3.2.2. Measurement of the Temperature Gradient

This has been covered to some extent in Section 3.1 (c) and involves the thickness and the surface temperatures. With steep temperature gradients a slight displacement of a thermocouple in or out of the grooves could give an error of 5°C or more. The finite size of the junction may also have an effect. The use of three couples in each face will reduce this error. Nicholls[7] suggests that if a steep temperature gradient exists at a surface then a thermocouple lying in that surface may not register the true temperature. This would apply if the calorimeter were in direct contact with the "cold" face. The presence of the backup insulation has the effect of a marked reduction in the temperature gradient. With modern temperature measuring equipment the actual temperature measurement error should be well below 0.5°C.

The effect of the backup insulation has been studied[10, 12] and it was concluded that any errors introduced by this technique lie within the ± 2% reproducibility region. It is difficult to estimate the maximum overall error that can arise in the temperature gradient measurement, but it has been given as about 1.5%.[11]

3.2.3. Measurement of Heat Flow [See Also 3.1 (d)]

The calorimeter system should not introduce any serious error. The design of the water channels in the guard ring and calorimeter results in "equivalent cooling areas" ensuring that the water flow rates in both are similar and can therefore be varied over a wide range without affecting the heat measurement. The thermal mass of the assembly gives a time lag of up to 60 s between a change in the temperature of the inlet water and the corresponding change in the outlet water temperature, but this is offset by taking readings every 30 s. The total water flow measurement time is usually between 90 and 240 s, 250 ml of water being measured in this time interval for a water temperature rise of about 3°C. With low flow rates, as with insulating refractories, this volume can be reduced to 100 ml. The total calorimeter error, based on an analysis of the individual errors in measuring rate of flow and temperature, and taking into account the possible exchange of heat between the calorimeter and the surrounding air, has been given as less than 1%.[11, 12]

A further possible source of error in the heat flow measurement is that due to direct radiation at high temperatures (say above 900°C hot-face temperature), the test material being regarded as semitransparent to infrared radiation. Calculation of the heat flow due to radiation is extremely complex, being affected by such parameters as effective refractive index, absorption coefficient, thermal gradient in the specimen, and the nature of the surfaces of the test piece and calorimeter (smooth or rough). At the present time the magnitude of the radiation effect, for refractory materials under practical experimental conditions, has not been reliably established.

3.2.4. Attainment of Thermal Equilibrium

The panel is normally held at the operating temperature for 12–15 h unless the peripheral heaters are adjusted, in which case up to a further 24 h may elapse. Equilibrium is defined as not more than 2°C change in the hot or cold face temperatures over a 30-min period.[1] These temperatures are also measured at the beginning and end of the heat flow measurements. With care any error arising from failure to reach thermal equilibrium should be very small.

3.2.5. Total Error

The individual errors listed above are of two kinds. The first are random errors which tend to cancel out when averaged, e.g., calorimeter and temperature measurement errors. The second type are constant or systematic errors arising from faulty setting up, e.g., positioning of backup insulation and/or calorimeter, lack of parallelism between hot and cold faces, faulty measuring equipment, contaminated thermocouples. In this respect the accuracy of a measurement is often confused with the precision with which it is made – equipment calibration should be checked at regular intervals – and reproducibility as absolute accuracy. The cooperative test in 1953[12] indicated that the ASTM and B.S. tests agreed to within 5%. However, no claims have been made for absolute accuracy although it is thought that the possible error in measuring thermal conductivity by either method is less than ± 5%.

3.3. *Comparison with Other Methods*

When comparing the B.S. 1902 method with another method such as the split-column comparative method it is essential that the comparison is a valid one. It is preferable that the same test piece(s) should be used in both tests and that the temperature gradient should be of the same order. This is particularly important where, as in the case of some basic refractories, the conductivity–temperature curve is nonlinear. It is also important where physical changes occur with rising temperature such as with monolithics (refractory concretes).

The thermal conductivity value obtained from the B.S. 1902 test is given as being at a "mean temperature" which is the arithmetic mean of the hot- and cold-face temperatures. This thermal conductivity value is actually equal to the integral of the true conductivity over the range defined by the hot and cold face temperatures divided by the temperature difference. The difference between this derived value and the true conductivity at the "mean temperature" will depend on the shape of the conductivity–temperature curve over the temperature interval. The error will be negligible where the curve is linear or slightly curved but considerable errors can arise where the curve exhibits a minimum as in the case of some magnesite refractories or an "elbow" as with ceramic insulating fibers.

Heat flow should be linear and in the same direction in both tests. Problems can arise where the heat flow is radial (as in the hot-wire and radial-flow methods) and the material is anisotropic. The actual conductivity value can also be important. The B.S. 1902 method can be used with confidence over the range 0.05 to 15 W/m K, and if comparison was made with the split-column comparative method on high-alumina firebricks with conductivity in the range 2–5 W/m K good agreement would be expected. However, if the comparison was made with carbon refractories (>20 W/m K) or insulating refractories (~0.3 W/m K), then in the first case the range would be outside that normally used with the B.S. 1902 test and in the second case well below that for the comparator test which gives best results when the conductivity value is of the same order as the steel standard (10–25 W/m K).

Where the comparison is valid, i.e., with isotropic materials in the range 0.1–1.5 W/m K good agreement (± 5%) is found with the hot-wire method.

ACKNOWLEDGMENTS. The author wishes to acknowledge the assistance given by his colleague A.M. Downs. Acknowledgment is also due to Dr. D.W.F. James,

Chief Executive, British Ceramic Research Association Ltd., for permission to publish this review.

NOTATION

λ	Thermal conductivity
Q	Heat flow (Heat flux)
L	Mean distance between hot and cold faces of the test piece
A	Area through which the measured heat flow passes
t_1	Average temperature of the test piece hot face
t_2	Average temperature of the test piece cold face

REFERENCES

1. British Standards Institution, "Methods of Testing Refractory Materials: Sampling and Physical Tests," B.S. 1902: Part 1A; Section 12, 1966 (to be reissued with minor revisions in 1984).
2. American Society for Testing and Materials, "Standard Test Method for Thermal Conductivity of Refractories," ANSI/ASTM C201: 68 (1979) supplemented by C182: 72 (1978), C202: 71 (1977), C417: 72 (1978), C767: 73 (1979).
3. F.H. Norton, "The Thermal Conductivity of Some Refractories," *J. Ceram. Soc.* **10**(1), 30 (1927).
4. G.B. Wilkes, "The Thermal Conductivity of Magnesite Brick," *J. Am. Ceram. Soc.* **16**(3), 125 (1933).
5. G.B. Wilkes, "The Thermal Conductivity of Refractories," *J. Am. Ceram. Soc.* **17**(6), 173 (1934).
6. J.B. Austin and R.H.H. Pierce, "The Reliability of Measurements of the Thermal Conductivity of Refractory Brick," *J. Am. Ceram. Soc.* **18**(2), 48 (1935).
7. P. Nicholls, "Determination of Thermal Conductivity of Refractories – Report for American Society for Testing and Materials, Sub-Committee C-8," *Bull Am. Ceram. Soc.* **15**(2), 37 (1936).
8. H. Oliver, "A Note of the Reliability of Thermal Conductivity Measurements for Insulating Materials," *Trans. Br. Ceram. Soc.* **37**, 49 (1937/8).
9. C.L. Norton, "Apparatus for Measuring Thermal Conductivity of Refractories," *J. Am. Ceram. Soc.* **25**(15), 451 (1942).
10. T.C. Patton and C.L. Norton, "Measurement of the Thermal Conductivity of Fire-Clay Refractories," *J. Am. Ceram. Soc.* **26**(10), 350 (1943).
11. J.F. Clements and J. Vyse, "A New Thermal Conductivity Apparatus for Refractory Materials," *Trans. Br. Ceram. Soc.* **53**, 134 (1954).
12. A.F. Watson, J.F. Clements, and J. Vyse, "A Co-operative Test on Thermal Conductivity," *Trans. Br. Ceram. Soc.* **53**, 156 (1954).

II

ELECTRICAL CONDUCTIVITY MEASUREMENT METHODS

7

High-Temperature Measurements of Electrical Conductivity

JEAN FRANÇOIS BAUMARD

1. INTRODUCTION

The electrical conductivity of solid materials at zero or near zero frequency encompasses probably the widest range covered by any common physical property. The difference between conductivity values of normal metallic conductors and good insulators is striking since the conductivity of a pure metal at low temperature may reach $10^{12}\ \Omega^{-1}\ m^{-1}$ while for insulators it may be much lower than $10^{-20}\ \Omega^{-1}\ m^{-1}$.[1] In order to meet the practical requirements for the measurements themselves over more than 30 decades of magnitude, a very important and specialized branch has developed. It would be neither possible nor desirable in this limited treatment to cover all of the many techniques that have been used through the years. Excellent, although often specialized, reviews of various methods may be found in the literature, for instance in Refs. 2–5. It must be realized first that one of the greatest hurdles facing the experimenter making meaningful conductivity measurements is the avoidance of many experimental errors that can arise. The determination of electrical conductivity is most often associated with, but must not be confused with, a resistance or impedance measurement. Too frequently the method is not selected with sufficient care to assure that spurious effects are eliminated for the material located in a cell assembly under given environmental conditions. It must be realized also that conductivity data will provide a very limited applicability if the solid material itself is insufficiently characterized with respect to chemical composition, microstructure, and thermal history, so that the distinction cannot be made

JEAN FRANÇOIS BAUMARD • Centre de Recherches sur la Physique des Hautes Temperatures, CNRS, 45045 Orleans Cedex, France.

between intrinsic and uncontrolled extrinsic properties. Along this development, it has been chosen primarily to focus on the high-temperature regime, let us say above 700–900 K, where the experimenter must be scrupulously careful in order to overcome difficulties as avoidance of multiple leakage paths, definition of a compositional equilibrium between the sample and the surrounding atmosphere, and determination of partial ionic and electronic conductivities. Mostly methods that are exemplary of the general principles that enable one to select a technique will be discussed to some extent.

The electrical conductivity is only one of the important electrical properties of solid materials, which also include dielectric constant (real part), loss tangent, electric strength, and so on. It is customary for the most general case of dielectrics to describe the current density and to define the electrical conductivity σ as

$$J = \sigma E + \frac{\partial D}{\partial t} \tag{1}$$

where E and D denote the applied electric field and the electric displacement, respectively. The latter are interrelated through

$$D = \epsilon\epsilon_0 E \tag{2}$$

where ϵ_0 and ϵ are the permittivity of free vacuum and the relative dielectric "constant" of the material, so that under an alternating field with ω frequency

$$J = (\sigma + i\omega\epsilon\epsilon_0)E \tag{3}$$

It may thus be expected that at least for specimens with a small or moderate conductivity, a simple measurement of the current density under a given applied field leads to combined effects of electrical conductivity and dielectric polarization. Although for many applications knowledge of a sample impedance is sufficient, it is often desirable to isolate individual contributions to equation (3). This is straightforward when σ and ϵ are real numbers and frequency independent, since the real part of the current density is then due to a true conductive process. But let us consider some ionic compound where mobile electrons contribute to a long-range transport through the crystallographic lattice, while lattice defects, for instance vacancies, induce dielectric relaxation phenomena around impurity sites, thus making the dielectric "constant" a dielectric function:

$$\epsilon(\omega) = \epsilon'(\omega) - i\epsilon''(\omega) \tag{4}$$

On assuming a constant conductivity $\sigma(\omega)$, equal to the dc conductivity $\sigma(0)$:

$$J = [\sigma(0) + \omega\epsilon''(\omega)]E + i\omega\epsilon'(\omega)E \tag{5}$$

The conduction component of the current may be used to define a *generalized* conductivity $\sigma(\omega) = \sigma(0) + \omega\epsilon''(\omega)$, that depends on frequency and may become indeed much greater than the dc conductivity itself. In this case, measurements down to low frequencies may be required to isolate both contributions of $\sigma(0)$ and ϵ''.

However, the term "conductive materials" generally refers to those solid compounds for which application of a dc field results in a long-range diffusive process of electronic and/or ionic charge carriers, implying partial dc conductivities. This chapter mainly deals with the determination of these quantities, and, unless otherwise specifically stated, dc conductivity and conductivity should be taken as synonymous. It should be kept in mind as a general trend that, owing to the thermally activated character of the various processes involved, all the materials become somewhat conductive when temperature increases, while localized relaxation processes will rapidly lie beyond the usual frequency range of interest for measurements, that extends up to 100 kHz to 1 MHz.

2. GENERAL CONSIDERATIONS ABOUT THE CONDUCTIVITY MEASUREMENTS

2.1. Sample Homogeneity

As mentioned earlier, the electrical conductivity is nothing but the limiting value of the ratio J/E at low frequencies, and it is measured generally in $\Omega^{-1}\ m^{-1}$ or $\Omega^{-1}\ cm^{-1}$ units, as (see Fig. 1):

$$\sigma = Il/VA \tag{6}$$

provided the current I is measured in amperes, the voltage V in volts, and the sample dimension in meters or centimeters, respectively. In this definition it is implicitly assumed that the sample chosen is *uniform* in its properties, a condition to get a homogeneous electric field. This in turn usually implies that the composition is itself uniform, and that no significant barrier exists within the solid. However, important exceptions may be found to this rule. First, a macroscopic nonuniformity may sometimes be taken into account with the aid of a specific model, so that integration over the entire specimen leads to an effective conductivity related to actual local conductivities with the aid of appropriate parameters. This happens, for instance, when electrical conductivity is used in a

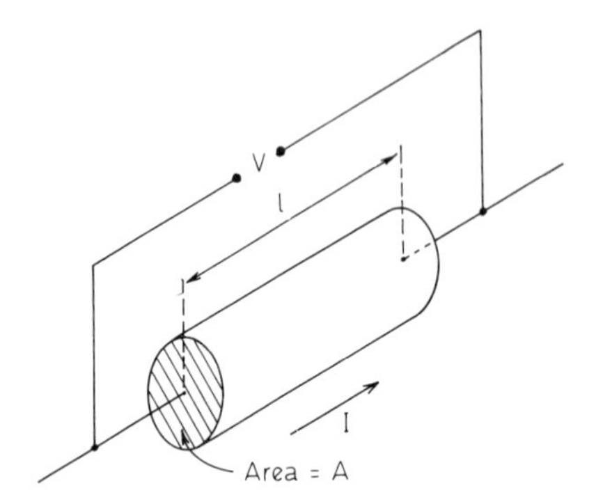

FIGURE 1. Schematic of a classical two-probe arrangement for electrical conductivity measurement.

transient technique to follow migration of point defects during a compositional change of oxide semiconductors.[7] The adjustable parameter is then a diffusion coefficient related to the mobilities of point defects.[7] An *effective* conductivity may also be defined in terms of Fig. 1 for a nonuniform specimen provided that the distribution of inhomogeneities is developed on a scale that remains much smaller than the sample dimensions. Such a situation is frequently encountered with composite materials.[8] Numerous equations have been derived over the years in an attempt to calculate the effective conductivity of heterogeneous solids, that involve the specific properties of each individual phase and the microstructure,[9, 10] as in the percolation models.[11]

Care should be taken that minor factors can deeply affect uniformity of electrical properties, mostly for polycrystalline semiconductor compounds. Segregation of impurities or subtle changes in the cation to anion ratio in the vicinity of grain boundaries may give rise to marked local changes in the resistivity.[12] When an electric field is applied, there is a buildup of charges, or interfacial polarization, that causes a frequency dependence, or dispersion, in both the measured conductance and capacitance of the specimen. Volger[13] has given a particularly attractive review of the layered model for electronic conductors, which will be referred to later on.

2.2. *The High-Temperature Measurements*

Probably the most salient feature of the high-temperature situation is the obvious possibility that ionic as well as electronic species can contribute to the charge transport phenomena, in a so-called mixed conductor:

$$\sigma = \sum_{\substack{i \text{ electronic} \\ \text{species}}} n_i q \mu_i + \sum_{\substack{j \text{ ionic} \\ \text{species}}} n_j z_j q \mu_j \tag{7}$$

where n denotes the concentration of a charge carrier with charge zq ($z = 1$ for electrons and electron holes) and mobility μ. It is common to define partial ionic and electronic conductivities σ_i and σ_e as

$$\sigma_i = \sum_{\substack{j\text{ ionic}\\ \text{species}}} n_j z_j q \mu_j \tag{7'}$$

$$\sigma_e = \sum_{\substack{i\text{ electronic}\\ \text{species}}} n_i q \mu_i \tag{7''}$$

the mobility and diffusion coefficient D of a carrier are related through the Nernst–Einstein relationship:[14]

$$zq\mu = kTD \tag{8}$$

Diffusion coefficients for ionic lattice species, although they may be strongly activated with temperature, are not generally large enough to provide a high mobility of ions.[15] Thus according to equations (7) and (8), a substantial part of the charge transport will take place by mobile ions only when their concentration greatly exceeds the electron and electron hole concentrations.† For instance, in recent years, there was an impressive effort of research in the field of solid electrolytes, that can be used in electrochemical devices as gauges or batteries, and where most of the charge transport is due to the large concentration and large mobility of a mobile ion.[16] In view of practical applications of these materials, even a small electronic conductivity may be deleterious.[17] This exemplifies that a complete characterization of the conductivity requires not only that total conductivity should be measured, but that, as far as possible, contributions of electronic and ionic conductivities should be explored. In general, these two quantities cannot be determined by a single experiment, and results from different approaches must be used in conjunction with a model to provide the necessary informations.

Another peculiar feature of the high-temperature measurements is the outstanding importance of environmental conditions. Generally speaking, reactivity of compounds is enhanced with temperature, so that the system constituted by the cell assembly, including specimen, electrodes, and leads, as well as the sample holder and the gas phase, will tend towards some complex compositional equilibrium when it is heated. Practically this means that the choice of materials that remain inert with respect to the sample becomes more

†This condition is necessary, but not sufficient, since vacant sites must be accessible to the ions in question. Quite often, the number of vacant sites is restricted to a few defects that become, in fact, the mobile species. For details, see Ref. 16.

and more restricted as temperature increases. Most often, beyond 1000–1200 K, direct contact of the sample with an insulating sample holder must be avoided to prevent solid-state reaction or contamination by diffusive processes, and the sample must be hung by its own leads.

Concerning the solid–gas equilibrium, the phase rule may be used to specify the number of independent variables that define the thermodynamic state of the compound. This follows from

$$V = C - P + 2 \tag{9}$$

where V denotes the number of independent variables, C the number of components, and P the number of phases. For a simple binary MO oxide in equilibrium with a given oxygen chemical potential, or oxygen partial pressure, the number of independent variables is 2. Thus the composition, or the oxygen-to-metal ratio, or alternatively the departure from stoichiometry x in $MO_{1 \pm x}$, are fixed once the temperature T and the oxygen chemical potential are themselves fixed.[18] For a ternary oxide compound, for instance a perovskite ABO_3, another independent variable, as AO activity, must be specified in order to determine the composition unequivocally. A small departure from stoichiometry is generally accommodated by lattice point defects, as interstitial extra ions or vacant sites.[18] In a semiconductor, internal electronic equilibrium then causes various charge transfers between (i) donors and acceptors, (ii) defects and electron energy bands, that control the number of mobile electrons and electron holes. In the above example of a MO oxide, some oxygen will probably be extracted from the lattice to reach equilibrium with a reducing gas phase. Oxygen deficiency may be accommodated by oxygen vacancies from which two free electrons associated with the formal -2 charge of O^{2-} ions will normally escape under the action of temperature and contribute to the electrical charge transport:

$$O^{2-} \rightleftharpoons \tfrac{1}{2}O_2\,(\text{gas}) + \square_O + 2e$$

The chemistry of point defects has been explored in detail in Ref. 18 to which the reader may refer. It should only be retained from these simple considerations that a precise measurement of the electrical conductivity under conditions where such a compositional change is possible is properly unsound unless the thermodynamic state of the compound can be defined with precision.

Obviously all this assumes that the specimen, at the temperature of interest, will attain equilibrium within a reasonable time scale. It must be noticed that the sample preparation, for instance single-crystal growth, or sintering of polycrystalline ceramics, is often operated at a temperature much higher than during the measurements. Complete reequilibration is sometimes hampered by diffusional kinetics, so that defects may be irregularly distributed or frozen in.

3. MEASUREMENT OF TOTAL CONDUCTIVITY: TWO- AND FOUR-PROBE ARRANGEMENTS

According to Fig. 1, the measurement in brief consists of applying electrodes to the sample, delivering some possibly time-dependent electric voltage to the cell assembly, and estimating the current density and its dependence with respect to the voltage. It should be noticed at this point that, although use of electrodes is widespread, other techniques that do not necessitate electrodes have sometimes been considered.[19, 20]

Use of electrodes signifies a charge transfer across boundaries between the electrodes and the material under investigation. In the measurement of total conductivity the electrical contacts must not impede the flow of carriers, i.e., no noticeable voltage drop must occur at these interfaces. Problems associated with the sample homogeneity have already been mentioned in Section 2.1. In turn, the electroding technique should not alter the physical property of interest. Again if some high-resistance layers are formed in the vicinity of electrodes during electrode preparation or during the time necessary to perform the measurement, a dc or low-frequency field will concentrate in the high-resistance thicknesses and will not be constant throughout the entire specimen.

In a classical two-probe arrangement as depicted in Fig. 1, it is evident that the total voltage drop V includes some potential differences due to the leads and contact resistances (or contact impedances under ac conditions). Consequently the two-probe techniques become inappropriate whenever these factors become significant in comparison with the voltage drop across the sample itself. This would be the case in investigations on bulky good conducting metals, for which the stray and sample resistances may be comparable in magnitude. The same remark holds for semiconductive compounds, since highly non-Ohmic behavior of contacts may be encountered for metal electrodes.[5] To eliminate these errors, additional probes are quite often necessary. In the well-known four-probe method, the potential drop is measured between two extra probes (Fig. 2), and in equation (6) the distance between them now replaces the sample length l:

$$\sigma = IL/VA \tag{10}$$

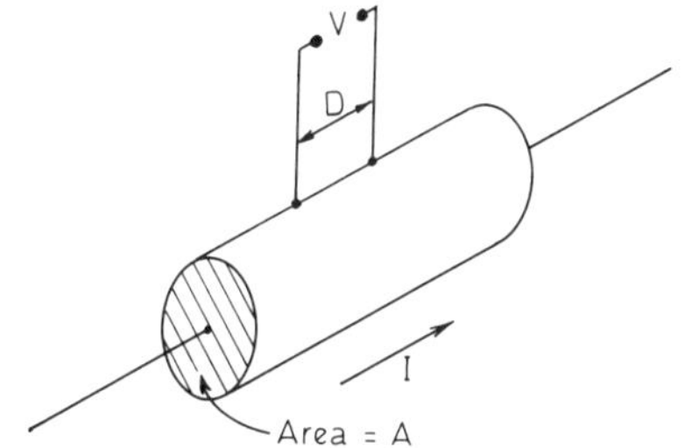

FIGURE 2. The four-probe arrangement.

To minimize the effect of contact resistance, some device measuring potential drop V with no or little current drawn from the probes in comparison with the current flowing through the sample is recommended. In theory this is the only way to reach an absolute measurement of conductivity, with point or bladelike voltage probes on the surface of the specimen. In practice, under usual conditions of high-temperature cell configurations, the contacts in question cannot be reduced down to a negligible surface. Lack of precision and reproducibility due to the determination of the geometrical factor is generally within 5%, and it can be mostly attributed to the determination of L. On the other hand, the precision on the geometrical factor is better in the two-probe arrangement, where both the surface of the contacts and the distance between electrodes can be defined accurately with regularly shaped specimens as cylinders.

A special four-probe technique was developed by Van der Pauw[21] and proves to be useful when working with flat samples, of arbitrary shapes, but with uniform thickness e (Fig. 3). Point contacts are placed on the circumference at successive arbitrary places denoted A, B, C, and D. Two voltage-to-intensity ratios must be measured, equivalent to two resistances R_1 and R_2. The first one R_1 is associated with the potential difference developed between C and D when current passes through terminals A and B. The second R_2 is defined as the ratio of the potential difference across DA to the current flowing through BC. The conductivity σ is then given by

$$\exp(-\pi e \sigma R_1) + \exp(-\pi e \sigma R_2) = 1 \tag{11}$$

It has also been shown that

$$\sigma = 2(\pi e)^{-1} \ln(2) [(R_1 + R_2) f(R_1/R_2)]^{-1} \tag{12}$$

where f is a function tabulated by Van der Pauw.[6, 21] The method, which was originally applicable to bidimensionally isotropic materials, was later extended to the case of anisotropic media.[22]

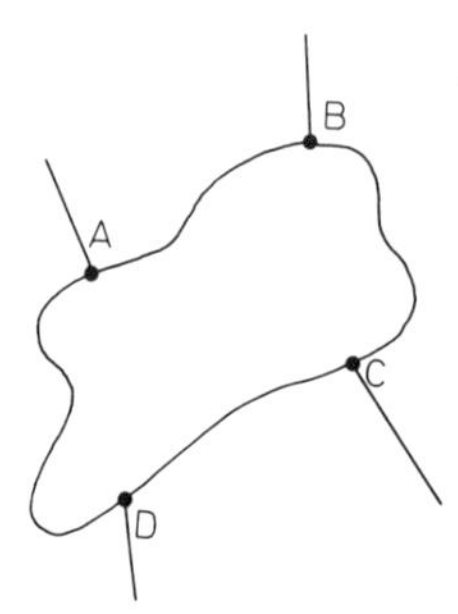

FIGURE 3. The arrangement of electrodes in the Van der Pauw method.

Two- and four-probe techniques are used with either dc or ac systems. In practice, possible occurrence of leakage paths with high-resistivity materials makes the four-probe technique more suited to compounds having moderate to large conductivities, e.g., higher than about $10^{-3}-10^{-6}\ \Omega^{-1}\ m^{-1}$. The two-probe method is often the only practical one when working on highly resistive materials. However, owing to its intrinsic simplicity, it is quite popular for many preliminary investigations.

4. MEASUREMENTS ON MIXED CONDUCTORS

For semiconductive compounds exhibiting mixed conduction, Hebb[(23)] made the observation that the fraction of current carried by ions depends on the type of electrodes used, and Wagner[(24)] elaborated a theory that incorporates effects arising from constraints as those imposed by the nature of electrodes in conductivity cells. In his book,[(18)] Kröger distinguishes four extreme situations:

a. reversible electrodes,
b. electrodes passing only electronic currents (semiblocking to ions),
c. electrodes passing only ionic currents (semi-blocking to electrons),
d. electrodes passing neither ionic nor electronic currents (blocking electrodes).

More recently, Wagner[(25)] has derived general equations for transport from irreversible thermodynamics and has discussed the case of semiblocking electrodes in detail. In practice cases (a) and (b) are found to be the most important. Reversible† electrodes are engaged in the measurement of total conductivity, and they may also be used in electrochemical cells to estimate the ionic transference number t_i, defined as the ratio of ionic to total conductivity:

$$t_i = \sigma_i/(\sigma_i + \sigma_e) \tag{13}$$

or conversely the electronic transference number t_e:

$$t_e = 1 - t_i = \sigma_e/(\sigma_i + \sigma_e) \tag{13'}$$

Polarization techniques employing electrodes semiblocking to ions (case b) provide a relatively unique way to determine small partial electronic conductivities for materials with prevailing ionic conduction. Analysis of cells with electron blocking electrodes is analogous to the ion blocking case, but practical

†The term "reversible" is used here in the sense that the exchange of carriers through the interface can be made without any limitation so rapidly that equilibrium is maintained between the electrode and the conductor for both directions of current.

implications are found to be limited to few examples in the literature.[25,26] Thus a separate discussion at this point seems unnecessary as well as for entirely blocking electrodes.

4.1. Conductivity Cells with Reversible Electrodes

Reversible electrodes have a double function since they act as current and/or voltage probes, and they locally fix the thermodynamic potential of one constituent in the mixed conductor, and thus the composition.[25] Under thermodynamic equilibrium, composition of a single-phase specimen will be uniform everywhere unless different reversible electrodes are disposed on the sample. By definition, the reactions taking place at the interface must not be hindered, and there should be no significant potential drop due to accumulation of either neutral or charged species. Separation of functions of reversible electrodes may often be achieved when the agent fixing the thermodynamic potential is nonconductor, for instance a gas as oxygen for an oxide semiconductor:

$$\text{Pt}(\text{O}_2) \parallel \text{oxide mixed conductor} \parallel \text{Pt}(\text{O}_2)$$

4.1.1. Cells with Identical Reversible Electrodes

Reversible electrodes are engaged in the basic two-probe arrangement to measure total electrical conductivity, but may also be useful to get partial conductivities. For instance, in the cell depicted just above, under the application of a dc bias, electronic species will flow and will be exchanged at the electrodes, mobile anions will migrate from the cathode to the anode, from which they escape as neutral species to the gas phase, while cations move towards the cathode, in the vicinity of which they combine with oxygen from the gas phase to form oxide. Duclot *et al.*[27] have developed a sensitive technique, named coulodilatometry, in which a special arrangement allows to measure a dimensional change of a sample due to cationic transport as a known quantity of faradays are discharged through the cell. The cationic transference number t_c is related to the variation in length Δl by the expression

$$t_c = aSz\Delta lM^{-1}Q^{-1} \tag{14}$$

where a is the specific gravity of the material, z the valency of cations, S the surface of the anode, M the molecular weight, and Q the amount of electrical charges expressed in faradays. This is in fact a modification of the method originally proposed by Tubandt,[28] in which transport fractions are found by measuring the amount of metal deposited at the cathode and non metal evolved at the anode once a known quantity of charges have been delivered to the sample. Although conceptually simple, the Tubandt technique has been pro-

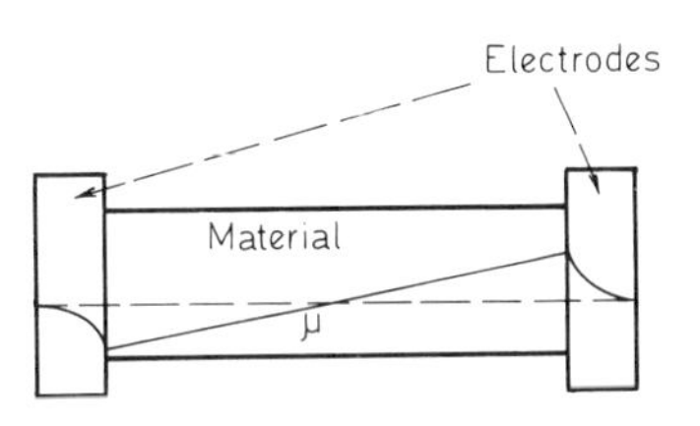

FIGURE 4. A schematic variation of the electrochemical potential of ions for a cell with actually reversible electrodes, in which a slow transport of discharge products leads to accumulation–depletion of these products in the vicinity of the electrode–material interfaces (full line). In the measurement of total conductivity with entirely reversible electrodes, this potential should be constant (dashed line). After Ref. 29.

gressively deleted owing to a number of reasons that can affect the validity of the calculated ionic transference numbers.[5]

In actual cells, however, in the measurement of total as well as partial conductivities, electrodes are rarely entirely reversible, and most tedious problems probably arise with ionic carriers. Slow charge transfer processes at the electrode–material interfaces tend to pile up ions with double layer formation.[29] The product of reaction must be eliminated as neutral species via some diffusional process; storage of reaction product within the electrode is expected to give rise to marked variations in the chemical potential normally imposed by an actually reversible behavior[30] (Fig. 4). Occurrence of such phenomena is likely to be observed under dc conditions where it can be detected from current voltage characteristics of the cell.[31,32] Marked departures from a simple Ohm law, even at low current densities, as well as instabilities in the current density with time under a constant applied field, generally indicate irreversible behavior. For gas–metal electrodes, use of thin coatings of porous metal is *a priori* preferable to foils, since diffusion of neutral products is expected to be much faster through open porosity than through metal itself.

One satisfactory solution to avoid influence of such dynamic irreversibilities is certainly to utilize a four-probe technique. Since negligible currents are drawn at the voltage probes, the analysis is made less difficult. However, the most interesting solution has been probably brought by impedance spectroscopy, adapted from aqueous electrochemistry by Bauerle[33] and largely developed by Kleitz and co-workers.[34] These authors have shown that impedance spectroscopy enables to isolate the bulk conductive properties with respect to interfacial phenomena, especially when electrodes are not too far from reversible behavior. For single crystalline specimens, when ac two-probe measurements are graphed on complex impedance plots (Z'' versus Z', where Z' is the real part and Z'' the negative imaginary part of the complex impedance $Z = Z' - iZ''$), data points are generally distributed over two arcs of circles, that are depressed below the real axis by a small angle, as in Fig. 5a. Appearance of a semicircular arc going through the origin in the complex impedance plot means that an equivalent circuit may be represented as a parallel RC combination (Fig. 5b). Thus succession of several arcs may be attributed to a same number of RC elements

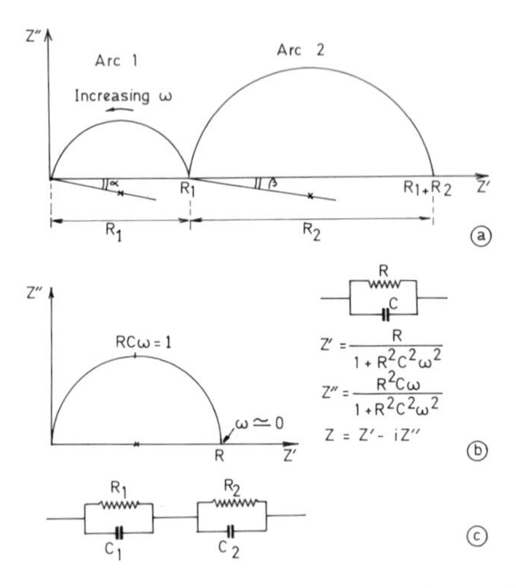

FIGURE 5. The analysis of electrical properties of mixed conductors by impedance spectroscopy: (a) A typical plot of the imaginary versus real parts of the impedance. (b) A simple electrical equivalent circuit that provides a semicircle in the complex impedance representation. (c) An electrical circuit often used to simulate the behavior reported on Fig. 5a.

with different time constants (Fig. 5c). The fact that the observed arcs are depressed below the real axis suggests that equivalent circuits are not an accurate representation, but a definitive explanation has not yet been established with certainty at the present time.[35,36] For simplicity, we will ignore it.

It was shown that low-frequency arc 2 (Fig. 5a) originates in electrode effects. The magnitude of this semicircle, that may be characterized by resistance R_2, will decrease the more as real electrodes approximate reversible behavior. It is to be noticed that data points will be distributed over an arc only if the electrode processes grossly resume to a resistance, including charge transfer resistance, shunted by some capacitor. This condition is not always fulfilled.[29]

Arc 1, at higher frequencies, is related to the bulk properties, and the conductivity calculated from R_1 and geometrical factor agrees with the results of four-probe dc measurements on the same material.[33] Easy separation of electrode and bulk effects explains the success of impedance spectroscopy for studies on solid-state electrolytes, but its range of applicability can probably be extended to many other materials. When operating on polycrystalline specimens, an extra arc is often observed and it is attributable to high-resistance or blocking layers distributed along grain boundaries.[37,38] For ionic conductors, this is more or less equivalent to the layered model developed by Volger for electronic conduction.[13] As many solid materials of practical interest are polycrystalline, influence of grain boundaries can thus be weighted and analyzed with the aid of several models published in the literature.[37,39]

4.1.2. Cells with Different Reversible Electrodes

In such cells, the thermodynamic potentials of the constituents at the two electrodes are fixed and different. This results in different compositions of the material in the vicinity of each electrode, that in turn gives rise to migration of mobile ionic and electronic species. As the latter quantities are charged, and under stationary conditions with no current flow in the external circuit, the emf of the cell, for a compound $M_\alpha X_\beta$ is given by

$$\begin{aligned} V &= q^{-1} \int_{\text{electrode 1}}^{\text{electrode 2}} z_X^{-1} t_i d\mu(X) \\ &= (2qz_X)^{-1} \int_{\text{electrode 1}}^{\text{electrode 2}} t_i d\mu(X_2) \end{aligned} \tag{15}$$

where $\mu(X)$ and $\mu(X_2)$ are the chemical potentials of X and X_2, and z_X is the valence of anions.[18] For purely ionic conductors, with $t_i = 1$, equation (15) simplifies to

$$V = (2qz_X)^{-1} [\mu(X_2)_2 - \mu(X_2)_1] = kT(2qz_X)^{-1} \ln [P_2(X_2)/P_1(X_2)] \tag{16}$$

where $P(X_2)$ denotes the partial pressure of X_2. In the case of mixed conductors, equation (15) is often approximated on assuming a mean ionic transference number $\bar{t}_i$:

$$V = \bar{t}_i kT(2qz_X)^{-1} \ln [P_2(X_2)/P_1(X_2)] \tag{17}$$

Although they do not allow to perform conductivity measurements directly, concentration cells were widely used in the past to calculate ionic and electronic contributions via equation (13) to the total conductivity, once the latter quantity was known.

Various configurations exist to investigate a metal compound, but the most popular arrangements are the following:

- An electronically conducting porous electrode in conjunction with a gas phase by which the activity of the nonmetallic component X_2 is imposed, for instance in the cell[40]:

$$Pt(O_2) \,||\, Al_2O_3 \,||\, Pt(O_2)$$

- An electronically conductive electrode containing another metal N in equilibrium with a compound $N_\gamma X_\delta$, where X is the same nonmetallic component as in $M_\alpha X_\beta$, for instance in the cell[41]:

$$Pt \,||\, Ni\text{–}NiO \,||\, BeO \,||\, Co\text{–}CoO \,||\, Pt$$

for which the emf was measured as a function of temperature. A major disadvantage is that the chemical potential at each electrode is then fixed at each temperature by the equilibrium between the metal and its compound, for instance in the above example by

$$Ni + \tfrac{1}{2}O_2\,(\text{gas}) \rightleftharpoons NiO$$

$$Co + \tfrac{1}{2}O_2\,(\text{gas}) \rightleftharpoons CoO$$

As a consequence, various kinds of metal–metal oxide couples should be used to cover a significant range of oxygen partial pressures and departures from stoichiometry at a given temperature. To circumvent this difficulty, the former arrangements with gas–metal electrodes are more frequently encountered, especially for oxides, since oxygen partial pressure may be varied over orders of magnitude on using diluted oxygen, then CO–CO_2 and H_2–H_2O gas buffers at lower oxygen activities.[42] Furthermore, any variation of t_i with oxygen partial pressure may be followed, since differentiation of equation (15) with respect to $P_2\,(X_2)$ yields, under constant $P_1\,(X_2)$.

$$dV/d\,[\ln P_2(X_2)] = kT(2qz_X)^{-1}t_i \tag{18}$$

A severe limitation to the accuracy of such cells lies in the mass transport through the electrolyte. Even if the net local current is zero under open circuit conditions, a charge flow of ionic species in the gradient of chemical potential occurs as long as it may be counterbalanced by an equivalent flow of electronic carriers. These flows, denoted, respectively, by J_i and J_e, may be calculated from[18]

$$J_i = -J_e = (\sigma_i + \sigma_e)(2qz_X)^{-1}t_i t_e \text{ grad } \mu(X_2) \tag{19}$$

The material behaves as pervious to X_2 molecules. Such a phenomenon will be important for compounds having comparatively high total conductivities and t_i, t_e of the same order of magnitude. The consequence is generally some polarization that affects the value of the emf delivered by the concentration cell.[43] In spite of this limitation, however, concentration cells prove to be useful when one wants to know whether the conductivity is entirely (or nearly entirely) ionic or electronic, or both in nature. For materials with low conductivities, polarization effects are probably not so important, and a better precision is expected provided that external leakage paths, such as surface conduction, may be eliminated or at least minimized, as in the conductivity measurements (see Section 5.4). Reference may be made for instance to the work of Brook *et al.*[31] on pure and doped α alumina where a special design of the cell was achieved in order to perform measurements on a very low conductivity material.

Estimation of accuracy in such cells cannot be reasonably worked out, according to the various factors mentioned. The method should be best suited to determine small electronic transference numbers in ionic conductors, or t_i and

t_e in poorly conducting materials. The developed emf should be large enough to avoid any confusion with parasitic thermoelectric voltages, and the lower detectable limit is probably of the order of 10^{-3} for t_i.

It must be emphasized that a measurement of the permeation rate, according to equation (19), under well-defined conditions of gradient in the chemical potential, may be interesting to reach the product $(\sigma_i + \sigma_e)t_i t_e$. Particularly, nearly pure ionic conductivity is often achieved by doping a compound to introduce a large number of lattice defects. The total conductivity is then primarily fixed by the dope content and does not depend on $\mu(X_2)$. Under these conditions, integration of equation (19) over the sample thickness l yields the permeation rate $J(X_2)$:

$$J(X_2) \cdot l = \int_0^l (J_i/2)\,dx = (4qz_X)^{-1} \int_0^l \sigma_e\,d\mu(X_2) \tag{20}$$

from which σ_e may be calculated once an appropriate model has been chosen for the variation of σ_e versus $\mu(X_2)$.[44] Derivation of electronic transference number t_e is straightforward from equation (20) after the total conductivity has been measured by any method. Permeation rate may be experimentally followed without any electrical measurement with the aid of the usual physico-chemical analysis apparatus, and this technique is undoubtedly more sensitive than emf measurements to get small electronic or ionic transference numbers.

4.2. Conductivity Cells with Semiblocking Electrodes

Reversible electrodes are quite extensively used in electrical conductivity work. However, the measurement of partial conductivities may also be carried out with the aid of semiblocking electrodes in asymmetric polarization cells. As permeation experiments, polarization methods were developed mainly to reach small electronic conductivities in compounds with a prevailing ionic conductivity.

Consider a binary metal compound $M_\alpha X_\beta$, with a predominant cationic conduction, sandwiched between a reversible electrode, generally metal M itself, and another electrode reversible to electrons and blocking to ions (Fig. 6). After application of a dc potential, or of a dc current, with the positive pole on the ion blocking electrode, a transient situation occurs during which cations, electrons, and/or electron holes move in the electric field. A depletion of cations

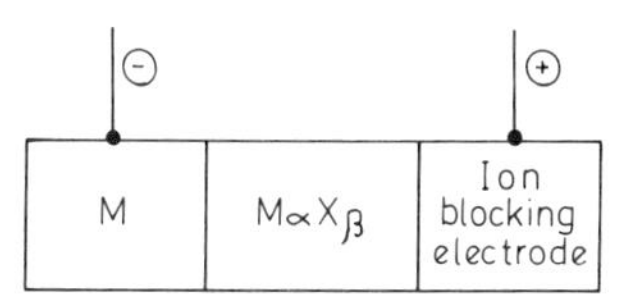

FIGURE 6. A representation of an asymmetric polarization cell.

is raised at the $M_\alpha X_\beta$-ion blocking electrode interface as no ions can be supplied by the electrode. A steady state will be reached when the gradient of cation concentration just balances the electric field. The electronic leakage current density is given by[45]

$$J_e = kT(ql)^{-1}\{\sigma_n[1-\exp(-qV/kT)] + \sigma_p[\exp(qV/kT)-1]\} \quad (21)$$

where V is the applied voltage, l the length of the sample between electrodes, and where σ_n and σ_p denote the electron and electron hole conductivities, respectively, at the equilibrium concentration fixed by the reversible electrode. In many cases equation (21) cannot be simplified by neglecting one term containing either σ_n or σ_p, and this equation must be evaluated in analyzing J–V plots. A typical curve is shown schematically in Fig. 7. As proposed by Patterson,[47] it is expedient to rearrange equation (21) as

$$J_e[1-\exp(-qV/kT)]^{-1} = kT(ql)^{-1}[\sigma_n + \sigma_p \exp(qV/kT)] \quad (22)$$

A graph of the left-hand term versus $\exp(qV/kT)$ should yield a straight line from which σ_n and σ_p are derived from the extrapolated intercept on the vertical axis and the slope. Let us notice that a similar analysis may be done for anionic conductors.[47]

Several conditions must be fulfilled in order to operate meaningful measurements.[45] The voltage applied to the sample must be kept below a critical value, and the volatile component X_2 should not be transferred between electrodes through the gas phase. Contact resistances must also be negligible. However, when these conditions are satisfied, an advantage of the method lies in the fact that a very low electronic conductivity, smaller than $10^{-15}\ \Omega^{-1}\ m^{-1}$ can be measured quite reproducibly, as demonstrated by the results of several authors on the same material, CuCl equilibrated with a copper electrode.[45]

Finally, let us mention that ac complex impedance measurements have also been performed with semi-blocking electrodes,[48] but mostly to investigate electrode processes. This will not be explored here.

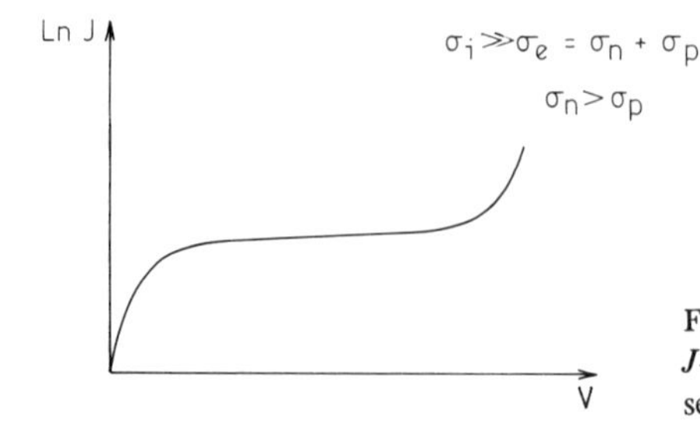

FIGURE 7. Schematic representation of a J–V plot for a dc polarization cell, as observed by Ilschner[46] for AgBr.

5. SPURIOUS EFFECTS IN HIGH-TEMPERATURE MEASUREMENTS OF ELECTRICAL CONDUCTIVITY

Various spurious effects may invalidate results obtained in the determination of total as well as partial conductivities. Some of them are even known to be prohibitive in the simple two- or four-probe configurations. As a matter of fact, a careful examination of possible spurious effects often restricts the choice among the experimental methods to the only one that avoids or minimizes the influence of undesirable factors. Accuracy of measurements is found to depend in a number of situations on the prior attention paid to extraneous voltages and leakage paths, which may be due to the electrical system, including the cell and its leads, the conductivity instrumentation, as well as to some interaction of this system with surroundings.

5.1. Noise

Noise may be generated internally or externally. Johnson noise, or Nyquist noise, is due to thermal agitation, and its rms value is given by

$$\langle V^2 \rangle = kT\Delta fR \tag{23}$$

where R is the equivalent resistance between voltage measuring terminals and Δf the frequency bandwidth. It should appear in measurements at very low level voltages as conductivity fluctuations. At lower frequencies "$1/f$" noise is known to become larger than the thermal noise and it also leads to conductivity fluctuations. It is frequently associated with poor contacts and recent investigations attribute it to fluctuations in the carrier mobility.[49] Generally those two sources of noise are not quite troublesome, because the effect of external noice is much more important. Anyway, their influence may be minimized with the aid of frequency-selective or time-averaging techniques.

Probably the origin of noise the most familiar to the experimenter is electrostatic pickup, when one is working on high-impedance circuitry. External noise also includes inductive pickup through the circuit under consideration, that results from various parts of the environment, such as transformers, power distribution lines, silicon-controlled rectifiers . . . at the power frequency and many harmonics. Ground loop currents are also troublesome when one deals with low-level signals. Ground loop currents are produced when (i) closed circuits formed by tying together system grounds to bring them to the same potential act as shorted paths when they are submitted to inductive pickup, or (ii) when ground potentials differ from one point to another in the circuitry.

There is no unique solution to minimize all these sources of external noise at the same time, because the problems generally depend on the technique adopted for measurements, the apparatus, the specimen characteristics, and the environmental conditions. As an example, an effective means to reduce electrostatic pickup is to shield all the circuitry, and to ground every shield to the same point. Alternatively, this induces additional stray capacitors between the sample and ground, that will undoubtedly affect high-frequency ac measurements. Ground loop effects may be avoided if signal processing is done with differential voltmeters. Mention may also be made of frequency selective techniques used in tuned detectors, selective or lock-in amplifiers, that eliminate signals with undesirable frequencies to a large extent.

5.2. Thermoelectric Voltages

Thermoelectric power induces extraneous voltages when the specimen is located in a temperature gradient, and this contribution to a small dc voltage is generally eliminated by performing measurements for both directions of dc current and averaging the results obtained. The problem is minimized by insuring that the cell assembly is heated in a zone of constant temperature.

5.3. Thermal Emission of Electrons

A frequent source of error in high-temperature measurements on poorly conducting materials is the thermal emission of electrons from various heated parts of the electrical circuit, and also from heated parts of the furnace. For semiconductors, the density of emitted current j is given by[50]

$$j = AT^2 n N_c^{-1} \exp(-\chi/kT) \tag{24}$$

where n denotes the concentration of free electrons, N_c the equivalent density of states in the conduction band, and χ the electronic affinity. Another similar expression holds for metal conductors that are engaged as electric leads.[51]

Thermal emission of electrons provides a source of carriers that will contribute to leakage paths between various parts of the circuit, making the conductivity apparently higher.[52] Three-probe arrangements may be found to eliminate this effect and they will be discussed in the next paragraph.

5.4. Surface Leakage Paths

It has been known for a long time that the behavior of low conductivity materials, having σ lower than 10^{-7}–$10^{-10}\ \Omega^{-1}\ \mathrm{m}^{-1}$ may be partially controlled in two- or four-probe configurations by surface effects, for instance adsorbed species as atmospheric moisture, making the current density larger in the vicinity of the surface than in the bulk. In order to circumvent this problem, a guard ring

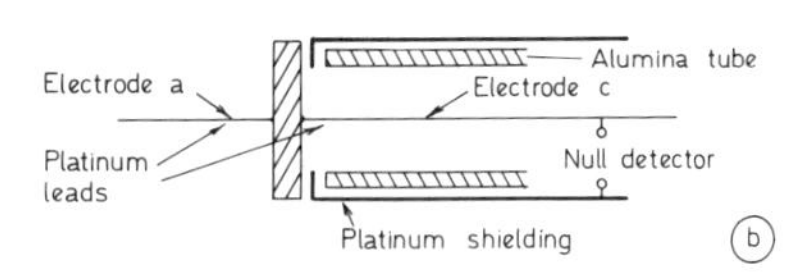

FIGURE 8. The three-probe arrangement used to minimize leakage paths on poorly conducting materials: (a) to eliminate the effects of surface conduction; (b) to eliminate effects of both surface and gas phase conduction.

collecting surface current may be added to the two-probe arrangement (Fig. 8a). The guard ring is most often adjusted to the same potential as central electrode c. The current is measured between electrodes c and a, and conductivity is now derived from the equation

$$\sigma = eI/VS \tag{25}$$

where e is the thickness of the sample, and $S = \pi(r_c + d/2)^2$ is the effective area of central electrode c. In practice, accuracy is limited, unless the geometry of central electrode can be well defined, for instance with evaporated metal layers. Extension of the guard ring to a cylindrical tube has been used[53] to minimize effects of both surface conduction and thermal emission of electrons. This arrangement is shown schematically in Fig. 8b. A platinum tube acts as a guard ring with respect to the sample for surface conduction and is kept at the same potential as inner electrode c. Thermal emission currents to the right-hand side of the cell are collected on the shield and their influence is reduced to zero if the current is measured on the central electrode. Such a configuration with a guard ring is commonly referred to as a three probe arrangement, and it may be used in dc or ac conditions.

For low-conductivity specimens, the four-probe technique necessitates a high-impedance voltmeter in order that no current flows through the potential probes. Since input impedance of most commercial ac voltmeters is limited to the range $1-100\,\mathrm{M\Omega}$, shunted by some capacitor, typically $10\,\mathrm{pF}$, this condition can be hardly satisfied unless dc techniques are used. Moreover with the four-probe technique the question of surface leakage currents remains unsolved. This is the reason why the latter is found to function best on materials having low to moderate resistivities.

Finally, for poorly conducting materials, some attention must be paid to stray capacitors between various parts of the system, especially when shielded cables are used in ac measurements. Generally this results in a capacitative coupling between the intensity and/or voltage leads, that should be taken into account as a correction when necessary.

6. CONTACTS AND ELECTRODING TECHNIQUES

Unless electrodeless techniques are used,[19, 20] an extremely important aspect of conductivity measurements is the making of suitable contacts between the specimen and the final electric leads to the voltage and intensity measuring system. These contacts must be appropriate to the material and the specific property being studied, e.g., total or partial conductivity, under a whole set of conditions such as maximum temperature and temperature range, nature of atmosphere and cell arrangement. Often they have to satisfy elementary mechanical criteria to withstand some handling during preparation and mounting, or simply to maintain a sample in position when it must be hung by its own leads.

Chemical aspects are first emphasized, since the electrode material must remain chemically inert with respect to the specimen and the atmosphere, and must not diffuse too appreciably in to the specimen. If the maximum temperature for measurements increases, such conditions are more and more difficult to fulfill. While at low or moderate temperatures a lot of electrode materials are available, choice is often restricted to noble metals as Ag, Au, Pt, Rh, and their alloys in oxidizing atmospheres, due to melting point and reactivity considerations.[6] Reactivity problems may sometimes be circumvented if the time necessary for measurements is small enough to make influence of chemical reactions negligible. Fast dynamic techniques[55] reduce exposure of the specimen and its immediate environment to high temperature to a very short period of time. However, the equilibrium value of the conductivity should be reached instantaneously and the thermal energy must be imparted uniformly through the entire specimen. These two conditions are satisfied by rapid resistance self-heating for highly conducting materials, mostly metals,[55] to which dynamic techniques were successfully applied.

Contacts may be classified in two categories, according as they provide some mechanical strength to the cell assembly or on the other hand necessitate some auxiliary setting to maintain electrodes at the proper location on the sample. The former category includes spot welded contacts, soldered contacts, painted, evaporated, or sputtered contacts, while the latter includes mainly pressure contacts and liquid phase contacts.

- Spot welding is convenient for metals or alloys studies on thin wires, since contacts can be small, and accurately located. Contamination by diffusion can be kept to a minimum since diffusion cannot take place over large distances in the short time required. It is even null if probes of the same nature as the specimen are used. Although they are probably not so extensively used, laser and ultrasonic welding may be classified, from a geometrical point of view, in the same group of techniques.

- Soldered contacts are useful particularly at low and moderate temperatures, below about 500 K, since some compound with a low melting point, such

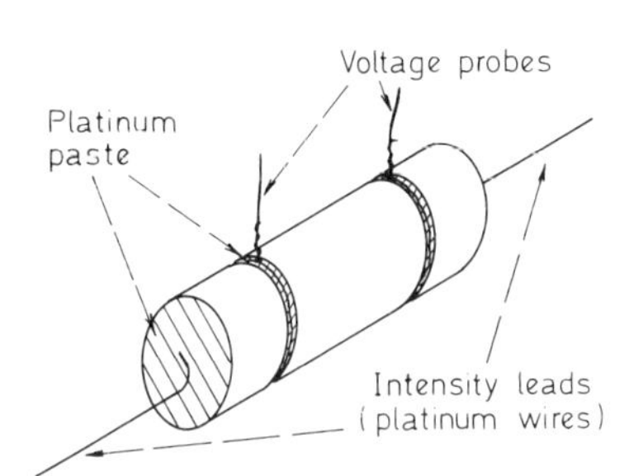

FIGURE 9. A typical arrangement of electrodes for four-terminal ac or dc measurements.

as tin, lead, indium, gallium, and their alloys, insures the junction between the sample and the leads after melting and solidification.

- Painted, evaporated, or sputtered contacts are used with both insulators and conductors, especially for medium- and high-temperature measurements. One of the most popular four-probe arrangements is depicted in Fig. 9. Small grooves are cut in a cylindrical sample to define the position of voltage probes. Voltage leads are tightly wrapped in the grooves around the sample. Intensity leads are held in position while the flat ends are coated with a paste of the same material as the leads, for instance with platinum paste. The small grooves are also coated with paste to improve the contact between the sample and the wires, and the whole assembly is fired. High-temperature annealing generally provides a good adhesion through some limited diffusive process and/or interlocking of the sample with the metal layers.

Comparisons between the respective qualities of contacts obtained with evaporated, sputtered, and painted coatings on mixed conductors have been reported in the literature with respect to their degree of reversibility.[31]

- Pressure is probably the simplest way, at least conceptually, to achieve contacts between a specimen and its leads. Pressure contacts may however develop high contact resistance on semiconductors, and they are used mostly with metals in four-probe arrangements, especially when a large number of samples are to be investigated in the same apparatus. Point or knife-edge probes allow quite an accurate determination of voltage probe spacing, making such contacts quite suitable for best precision. However, maintaining the probes in position is possible in fast dynamic techniques, but becomes unpractical under stationary high temperature conditions.

- Intermediary liquid phase contacts are used in specific cases, mostly for ionically conductive compounds, for which the electronic conductor plays the role of a reservoir for the ionic mobile carriers, for instance in the cells[56, 57]:

$$\text{liquid Na} \,\|\, \beta \text{ alumina} \,\|\, \text{liquid Na}$$

$$\text{liquid Ag}(O_2) \,\|\, \text{stabilized zirconia} \,\|\, \text{Pt}(O_2)$$

The main problems are associated with design and operation of a cell with impervious containers for liquid species. Recent developments of materials that exhibit both high electronic and ionic conductivities, such as intercalated compounds,[58] should be important for a proper selection of electrodes without handling such liquid phases. Very few electrode effects have been reported in ac measurements on the cell[59]:

$$C_{\text{graphite}} \,\|\, \beta \text{ alumina} \,\|\, C_{\text{graphite}}$$

and it has been suggested that some sodium may intercalate into graphite during electrode preparation, making them proper for a study of the bulk conductivity of this Na superionic conductor under ac conditions.

A concluding remark for this section is concerned with measurements on poorly conductive materials in two- or three-probe arrangements, for which influence of contact impedance must be lowered to a minimum level. Non-Ohmic behavior of a metal–electronic semiconductor can be attributed to rectifying effects due to the difference in work function of the two materials,[3] but also to other factors as inhomogeneous composition near the electrode or specimen surface, or poor wetting of the surface with formation of air gaps.[60] Fortunately, it is commonly observed that a high temperature annealing eliminates problems associated with these barriers, probably because they tend to reduce inhomogeneities and because actual chemical bonds can be realized through the formation of tiny intermediate phases or a short-range diffusion, that can deeply affect the band scheme of the junction.

7. A SHORT SURVEY OF INSTRUMENTATION

Impedance-measuring techniques, that allow the determination of conductivity may be categorized essentially, according to the instrumentation used, into bridge and voltage–current methods, that will be explored briefly.

In the bridge technique, circuit conditioning is required to achieve that a balance or null condition is detected and processed to indicate the measured value, as in the basic Wheatstone bridge. Many commercial instruments are available for dc and ac investigations, for which suppliers may be quite helpful in selecting the appropriate unit. However, the basic Wheatstone bridge has the important deficiency that it measures the total impedance of a cell, including leads and contacts impedance. Thus for samples with a high conductivity, configurations known as Kelvin double bridge or modified Kelvin double bridges[61] have been developed to improve the independence towards the contacts and leads effects. The procedure of balancing becomes unfortunately somewhat tedious, so that four probe measurements are generally preferred with voltage–current instrumentation. For samples with moderate to high resistances, ac transformer bridges seem to offer the desired versatility up to very high impedance levels, up to $10^{15}\ \Omega$, since they may be used in three terminal arrangements.[62] Conductive and reactive parts of impedance may be determined with a very good precision, making this apparatus judicious to operate complex impedance measurements.

Voltage–current techniques prove to be the most versatile in practice, since they may be used in any arrangement, including two, three, or four terminals, from poorly to highly conductive materials. Intensity is often measured through the voltage drop across a standard resistor placed in series in the circuit. In the case of ac measurements, analysis of the phase relationship between voltage and intensity may now be performed with a variety of electronic equipments, as phase sensitive detectors, lock-in amplifiers, network analyzers. For systematic investigations, automatic LCR meters even provide conductive and reactive parts of the impedance at a few selected frequencies.

NOTATION

A	Cross-sectional area of a sample
D	Electric displacement
D^*	Self-diffusion coefficient of a carrier
e	Thickness of a sample
E	Electric field
f	Frequency of the electric field (expressed in s^{-1})
i	Square root of -1
I	Intensity of the current flowing through the sample
j	Density of emitted current in thermal emission of electrons
J	Current density through a sample
J_e	Current density due to electronic charge carriers
J_i	Current density due to ionic charge carriers
$J(X_2)$	Permeation rate of X_2 molecules through a mixed conductor
k	Boltzman constant
l, L	Distance between voltage probes in a two- or four-probe measurement
n_j	Concentration of charge carriers j
$P(X_2)$	Partial pressure of X_2 gas
q	Electric charge on the electron
R	Sample resistance
t_e	Electronic transference number
t_i	Ionic transference number
T	Temperature
V	Voltage imposed or developed through two terminals
z_j	Valency of carriers with charge $z_j q$
Z	Impedance of a cell with real Z' and negative imaginary parts Z''
ϵ	Frequency-dependent dielectric function
ϵ', ϵ''	Real and imaginary parts of the dielectric function

μ_j	Mobility of charge carrier *j*
$\mu(X_2)$	Chemical potential of X_2 species
σ	Total conductivity of a material
σ_e	Electronic conductivity of a material
σ_i	Ionic conductivity of a material
σ_n	Partial electronic conductivity due to mobile electrons
σ_p	Partial electronic conductivity due to electron holes
ω	Pulsation of the electric field (rad s^{-1})

REFERENCES

1. C. Kittel, *Introduction to Solid State Physics,* 3rd Edition, p. 253, John Wiley and Sons, New York (1968).
2. W.C. Michels, *Electrical Measurements and Their Applications,* Van Nostrand, New York (1957).
3. L.P. Hunter, *Handbook of Semi-conductor Electronics,* McGraw Hill, New York (1956).
4. W. Crawford Dunlap, in *Methods of Experimental Physics* (K. Lark Horovitz and V.A. Johnson, eds.), Vol. 6, Part B, p. 32, Academic Press, New York (1959).
5. R.N. Blumenthal and M.A. Seitz, in *Electrical Conductivity in Ceramics and Glass* (N.M. Tallan, ed.), Part A, p. 35, Marcel Dekker, New York (1974).
6. J.F. Baumard, *Solid State Commun.* **20,** 859 (1976).
7. P.E. Childs, L.W. Laub, and J.B. Wagner, *Proc. British Ceram. Soc.* **19,** 29 (1971).
8. T.H. Geballe, in *Critical Materials Problems in Energy Production,* (Ch. Stein, ed.), p. 641, Academic Press, New York (1976).
9. L.K.H. Van Beek, in *Progress in Dielectrics,* Vol. 7, Heywood, London (1965).
10. J.M. Wimmer, H.C. Graham, and N.M. Tallan, Ref. 5, Part B, p. 620.
11. S. Kirkpatrick, *Rev. Mod. Phys.* **45,** 574 (1973).
12. A.J. Bosman and C. Crevecoeur, *Phys. Rev.* **144,** 763 (1966).
13. J. Volger, in *Progress in Semiconductors,* Vol. 4, John Wiley and Sons, New York (1960).
14. P. Kofstad, *Nonstoichiometry, Diffusion and Electrical Conductivity in Binary Metal Oxides,* John Wiley and Sons, New York (1972).
15. R.M. Dell and A. Hooper, in *Solid Electrolytes* (P. Hagenmuller and W. Van Gool, eds.), p. 291, Academic Press, New York (1978).
16. M. Pouchard and P. Hagenmuller, Ref. 15, p. 191.
17. B.C.H. Steele and R.W. Shaw, Ref. 15, p. 483.
18. F.A. Kröger, *The Chemistry of Imperfect Crystals,* 2nd Edition, North-Holland, Amsterdam (1974).
19. U. Strom and P.C. Taylor, *J. Appl. Phys.* **50,** 5761 (1979).
20. T. Sakurai, S. Mochizuki, and M. Ishigame, *High Temp. High Pressures,* **7,** 411 (1975).
21. L.J. Van Der Pauw, *Philips Res. Repts.* **13,** 1 (1958).
22. L.J. Van Der Pauw, *Philips Res. Repts.* **16,** 187 (1961).
23. M. Hebb, *J. Chem. Phys.* **20,** 185 (1951).
24. C. Wagner, *Z. Phys. Chem.* **B21,** 25 (1933).
25. C. Wagner, in *Progress in Solid State Chemistry,* Vol. 10, Part 1, Pergamon Press, London (1975).
26. J. Dellacherie, D. Balesdent, and J. Rilling, *J. Chim. Phys. (Paris)* **67,** 360 (1970).

27. M. Gauthier, M. Duclot, A. Hammou, and C. Déportes, *J. Solid State Chem.* **9**, 15 (1974).
28. C. Tubandt, in *Handbuch der Experimental Physik* (X. Wies and X. Harm, eds.), Vol. 12, Part II, p. 412, Akademische Verlagsgesellschaft, Leipzig (1933).
29. A.D. Franklin, *J. Amer. Ceram. Soc.* **58**, 465 (1975).
30. J. Vedel, in *Electrode Processes in Solid State Ionics* (M. Kleitz and J. Dupuy, eds.), p. 223, D. Reidel, Dordrecht, Holland (1976).
31. R.J. Brook, W.L. Pelzman, and F.A. Kröger, *J. Electrochem. Soc.* **118**, 185 (1971).
32. H. Yanagida, R.J. Brook, and F.A. Kröger, *J. Electrochem Soc.* **117**, 593 (1970).
33. J.E. Bauerle, *J. Phys. Chem. Solids* **30**, 2657 (1969).
34. E. Schouler, G. Giroud, and M. Kleitz, *J. Chim. Phys. (Paris)* **70**, 923 (1973).
35. D. Ravaine and J.L. Souquet, *J. Chim. Phys. (Paris)* **71**, 693 (1974).
36. A.K. Jonsher, *Phys. Stat. Sol. (b)* **83**, 585 (1977).
37. S.H. Chu and S.A. Seitz, *J. Solid State Chem.* **23**, 297 (1978).
38. W. Jakubovski and D.H. Whitmore, *J. Am. Ceram. Soc.* **62**, 381 (1979).
39. Da Yu Wang and A.S. Nowick, *J. Solid State Chem.* **35**, 325 (1980).
40. S.K. Mohapatra and F.A. Kröger, *J. Am. Ceram. Soc.* **60**, 141 (1977).
41. C.B. Alcock and G.P. Stravropoulos, *J. Am. Ceram. Soc.* **54**, 436 (1971).
42. K. Schwerdtfeger and E.T. Turkdogan, in *Physico-Chemical Measurements in Metals Research* (R.A. Rapp, ed.), Vol. 4, Part 1, p. 321, John Wiley and Sons, New York (1970).
43. J. Fouletier, P. Fabry, and M. Kleitz, *J. Electrochem. Soc.* **123**, 204 (1976).
44. A. Ounalli, B. Cales, and J.F. Baumard, *C.R. Acad. Sci. (Paris)* **C292**, 1185 (1981).
45. J.B. Wagner, Ref. 30, p. 185.
46. B. Ilschner, *J. Chem. Phys.* **26**, 1597 (1957).
47. J.W. Patterson, E.C. Bogren, and R.A. Rapp, *J. Electrochem. Soc.* **114**, 752 (1967).
48. R.D. Armstrong, Ref. 30, p. 261.
49. F.N. Hooge, *Physica* **83B**, 14 (1976).
50. J.C. Rifflet, P. Odier, A.M. Anthony, and J.P. Loup, *J. Am. Ceram. Soc.* **58**, 493 (1975).
51. J.P. Loup and A.M. Anthony, *Rev. Int. Htes Temp. Refract.* **1**, 193 (1964).
52. J.P. Loup and A.M. Anthony, *Rev. Int. Htes Temp. Refract.* **1**, 15 (1964).
53. R.J. Brook, J. Yee, and F.A. Kröger, *J. Amer. Ceram. Soc.* **54**, 444 (1971).
54. M.A. Seitz, R.T. MacSweeney, and W.M. Hirthe, *Rev. Sci. Instrum.* **40**, 826 (1969).
55. A. Cezairliyan, *High Temp. High Pressures* **11**, 9 (1979).
56. K.K. Kim, J.N. Mundy, and W.K. Chen, *J. Phys. Chem. Solids* **40**, 743 (1979).
57. L.M. Friedman, K.E. Oberg, W.M. Boorstein, and R.A. Rapp, *Met. Trans.* **4**, 70 (1973).
58. G.J. Dudley and B.C.H. Steele, *J. Solid State Chem.* **31**, 233 (1980).
59. D. Miliotis and D.N. Yoon, *J. Phys. Chem. Solids* **30**, 1241 (1969).
60. I. Warshawsky, *Rev. Sci. Instrum.* **26**, 711 (1955).
61. General Radio Instrumentation Catalog.

III

THERMAL DIFFUSIVITY MEASUREMENT METHODS

Thermal Diffusivity

An Introduction

Interest in the thermal diffusivity of materials is twofold. On the one hand, it represents a genuine transport property which characterizes thermal phenomena important in many engineering applications and fundamental materials studies. On the other hand, thermal diffusivity is a property directly related to thermal conductivity, another very important thermophysical property. As opposed to the measurement of thermal conductivity, which involves the measurement of heat fluxes that are difficult to control and measure accurately, the measurement of thermal diffusivity consists of the accurate recording of the time dependence of temperature due to a transient or periodic thermal perturbation at the specimen boundary. It is thus frequently easier to measure thermal diffusivity than thermal conductivity, as the two other thermophysical properties involved in the same relationship, density and specific heat, are either known or can be measured or calculated relatively simply and accurately.

Thermal diffusivity experiments are usually short and relatively simple. In most cases they require small specimens, which can be as small as disks a few milimeters in diameter and a fraction of a milimeter thick. Derivation of thermal diffusivity from recorded experimental data, however, frequently involves a sophisticated mathematical treatment, requiring, for some variants, large computers. This mathematical complexity, together with the need for fast and reliable temperature detectors in the high-temperature range, could account for the relatively late expansion of thermal diffusivity measurement methods. The present-day abundance of low-cost large-capacity mini- and microcomputers and the ease with which thermal diffusivity experiments can be interfaced and computer controlled make thermal diffusivity methods very attractive for application and further development in the nearest future.

Another important advantage of thermal diffusivity methods is that the temperature variation in the specimen can be kept quite small, maintaining thus the condition of constant temperature well satisfied. Beside satisfying the initial conditions of the mathematical model, which always incorporates property invariance with temperature, this feature of small temperature disturbance during the experiment enables studies of phase transitions throughout transition ranges with aid of thermal diffusivity measurements, something which is not feasible with any method involving appreciable temperature gradients during measurement.

According to the nature of the temperature perturbation at the specimen boundary, thermal diffusivity measurement methods may be divided into two basically separate large groups, transient heat flow methods and periodic heat flow methods. Transient heat flow methods include two distinctly different subgroups, the pulse or flash method, where duration of the perturbation compared with the experiment is short (1 ms or less), and the group of monotonic heating regime methods, in which the perturbation follows a certain, essentially linear temperature change over temperature intervals which can be very large. All periodic heat flow methods and variants may be accommodated within the same group, as all of them are based on the measurement of the attenuation or the phase shift of temperature waves along the path of their propagation through the material. They are, however, arbitrarily divided into two groups, according to the temperature range of application and the mode of energy supply. The first, called temperature wave techniques, constitute mainly variants applicable to lower and medium temperatures, and are suitable for implementation as multiproperty methods, which can generate a number of thermophysical properties in one experiment. The second group consistute the high-temperature variants, where the modulated energy input is effected by electron or photon bombardment of investigated materials.

Small and simple specimen shapes encountered in the pulse and the modulated energy beam input techniques represent an advantage in the study of materials where specimen manufacturing and handling is not simple, i.e., for rare substances, irradiated or radioactive materials, oxidized, two- or three-layer materials etc. The main requirements on the specimens then refer to their homogeneity, meaning that for a given shape they should be representative of the bulk of the material, and that boundary conditions of the method, such as the absorption of incoming energy in a very thin specimen surface layer or adiabatic conditions on the specimen boundary, are met in the measurement.

Less homogeneous and more complex materials need larger specimens, and for their study, a two-dimensional variant of the pulse technique, the radial temperature waves variant, or any of the monotonic heating variants should be employed.

Thermal diffusivity measurement methods have been used from very low to very high temperatures but their widest range of application has been from near room temperature to 2500 K. The greatest part of more recent data on thermal diffusivity in the literature has been produced by the laser pulse technique, and some thermophysical property data statistics state that even 75% of thermal diffusivity data published in the past decade were generated employing this technique. Modulated energy beam input techniques, including modulated electron beam or modulated light beam variants have been in expansion lately, covering a high proportion of the high-temperature thermal diffusivity measurements, both of metals and nonmetals and of metals in the molten state. The medium-temperature variants of the temperature wave technique have found widest application in the measurement of thermal diffusivity of liquid metals and fluids, the method being successfully applied to metals and nonmetals. The main attraction of this method lies in its suitability for multiproperty measurements, which include simultaneous measurement of thermal conductivity, thermal diffusivity, and specific heat on the same specimen. In spite of its somewhat lower precision, the method of monotonic heating offers unique advantages for coarse-matrix or large-grain complex materials, and where large temperature intervals in thermal diffusivity measurement should be traversed in relatively short periods of time with many data points, which is the case with materials which undergo certain chemical or structural changes with heating. The very wide temperature range of monotonic heating regime variants, particularly in the high-temperature region, render them attractive for wider use and further development and automation.

The four groups of thermal diffusivity measurement methods are briefly discussed in the following paragraphs and their mean features are summarized in Table 1.

The *pulse* method for the measurement of thermal diffusivity is based on the analysis of the temperature–time history of the rear face of a disk-shaped specimen whose front face has been exposed to a burst of radiant energy. The duration of the pulse should be approximately one hundredth of the time needed for the temperature of the rear face to reach 50% of its maximum value in order to avoid the need for the finite-pulse-effect correction. As a source of energy pulses the laser is most commonly used, although flash lamps may be sometimes adequate, particularly below room temperature. With correct specimen positioning and holding and fairly homogeneous laser flash beam that ensures unidirectional heat flow through the specimen, detection of the specimen rear face temperature is the main problem in this method of thermal diffusivity measurement. Thermocouples are usually used as detectors below 250 K. Above this temperature, photoconductive infrared detectors, photomultipliers and semiconductor detectors represent a better solution. The contact temperature detector may easily distort the temperature–time record and introduce a large systematic error in the measured thermal diffusivity. Radiant heat losses from the specimen become significant above 1000 K, and have to be adequately accounted for. The pulse method has been successfully applied to

TABLE 1
Thermal Diffusivity Measurement Methods

Measurement technique	Key features: Advantages	Key features: Disadvantages/limitations	Temperature range (K)	Principal specimen materials	Diffusivity range ($m^2 s^{-1}$)	Uncertainty (%)
Pulse method	Wide property range Wide temperature coverage Small and simple specimens Simple and rapid measurement Multilayer specimens Multiple specimen facility	Not well suited for coarse matrix materials Not convenient for translucent materials Sophisticated error analysis	100–3300 (300–2000)	All solids and encapsulated liquid metals Specimens: Disks 6–16 mm in diameter	10^{-7}–10^{-3}	1.5–5
Temperature wave method	Wide property range Very wide materials range Suitable for very high pressures Suitable for multiproperty measurement	Quasistationary temperature conditions necessary Complex mathematical treatment Complex error analysis	60–1300 (300–1300)	Solids, liquid metals, liquids, and gases Specimens: rods, cylinders, hollow cylinders, etc.	10^{-7}–10^{-4}	1–5 (solids) 1–3 (fluids) 5–9 (liquid metals)
Electron bombardment heat input method	High-temperature coverage Small specimens Applicable to electrical conductors and nonconductors AC techniques applicable Simultaneous specific heat movement possible	Complex experimental apparatus High-vacuum conditions required Arduous mathematical apparatus	330–3200 (1100–2200)	Metals, nonmetals, liquid metals Specimens: disks 6–9 mm, plates 0.25 cm^2 and more, hollow cylinders 6–8 mm i.d. 15–20 mm o.d.	10^{-7}–5×10^{-5}	2–10
Monotonic heating regime method	Wide materials range Continuous measurement in very wide temperature range Simple apparatus Simple measurement	Not very suitable for good thermal conductors Somewhat lower precision	4.2–3000	Ceramics, plastics, composites, thermal insulations Specimens: cylinders, plates, paralelepipeds	10^{-8}–10^{-5}	2–12

metals, ceramics, graphites and many other materials, and even to such heat sensitive materials as explosives, to homogeneous and certain heterogeneous materials. Apparatus with multispecimen facility as well as the apparatus with computerized error analysis and compensation have been developed and used successfully to high temperatures. With adequate low thermal inertia ambient furnace, thermal diffusivity measurement by the pulse method can be quite fast.

The *temperature wave* method has a large variety of variants and their modifications characterized by high informativeness. The sources of information on thermophysical properties are the mean temperature field, the amplitudes and phases of temperature waves. The amount of information carried enables the simultaneous measurement of more than one thermophysical property in one experiment. Variants are based on the following of plane, quasiplane, radial, or spherical temperature waves in the specimen material. Thermal diffusivity is determined from information on (a) the amplitude and phase of temperature oscillations at one frequency, (b) by measuring the amplitudes at two different frequencies, or (c) from data on temperature oscillation phases at two frequencies. Thus it is possible to cross-check data from one pair of measured quantities with data from another pair also obtained in the same experiment. The periodic nature of temperature oscillations permits use of mean values from a number of successive temperature cycles thus reducing the probable error of the measurement. Multiproperty variants, in which thermal conductivity, thermal diffusivity, and the specific heat of the investigated material are simultaneously measured are mostly applicable to electron bombardment and inductive radiofrequency energy supply, and the wire probe variant appropriate for liquids and gases. Temperature wave method has been used from very low temperatures to very high temperatures. In this presentation, because of many specific features inherent only to them, the high-temperature variants will be treated separately, in the next section. Temperature wave variants can be used for gases, liquids and solids, metals and nonmetals, over an extremely wide range of pressures. Except that they require quasistationary conditions and could be more time consuming, variants within this group are reliable and precise, and convenient for many applications.

The *electron bombardment modulated heat input* method is a modern high-temperature variant of the temperature wave method, which became feasible through developments in experimental and measurement techniques during the past few decades, electronics and opto-electronics in particular, as well as through the development and accessibility of computers and computing facilities. The method requires vacuum of 10^{-4} mbar and higher. According to the mode of modulated energy supply two main subgroups exist: electron gun heating variant for thin disk and plate specimens that can be electrically conducting or nonconducting, and direct anode heating variants for hollow cylinder specimens, appropriate for electrical conductors and liquid metals encapsulated within double-walled hollow cylindrical metallic containers. The combination of disk-shaped specimens mounted in the wall of a hollow cylinder with anode heating is also possible. Mathematical approaches for variants using plane or radial heat waves do not differ very much. In virtually all variants thermal diffusivity is calculated from the phase lag between either the modulation of the electron beam and the rear specimen face, or the signals from two detectors aimed at the front and the rear face. In its main domain of application, from 1200 to 3000 K, optical detectors are preferred. In the range between 350 and 1200 K, where the anode heating variant has successfully been applied thermocouples are used. All the precautions referring to the use of contact thermocouple detectors for dynamic temperature measurement hold here as well. The electron bombardment modulated heat input method has been applied with success in the measurement of the thermal diffusivity of refractory metals, insulators, semiconductors, dielectrics, and liquid metals.

The *monotonic heating regime* group of methods has two distinct subgroups, intended for measurements in narrow and in very wide temperature intervals. The first of them employs the so-called "regular regime," i.e., the quasistationary regime of temperature change which is established in the specimen some time after being brought in contact with a large thermal capacity thermostat at a given higher or lower temperature. The chapter contains a systematic presentation of the regular regime variants, with emphasis on theoretical aspects, the consistency of models and real experimental conditions for fundamental specimen geometries, the sources of systematic errors and possibilities of simultaneous determination of thermal conductivity and specific heat together with thermal diffusivity. Regular regime variants are applicable to a wide range of materials among medium and poor thermal conductors. They have also been applied to metals, but the accuracy was lower. Methods from this subgroup are adequate for lower temperatures. The second subgroup contains variants which give the temperature function of thermal diffusivity within wide temperature interval in one continuous experiment, with the boundary condition of ambient temperature varying according to a linear or almost linear law. In this group of variants measurements are performed in a quasistationary regime and thermal diffusivities are calculated from the heating rate and the temperature difference at two points of the specimen along the heat flow path. Variants and specimen geometries are reviewed. Monotonic heating regime variants are applicable from low to very high temperatures, and for the same categories of materials as the regular regime variants.

8

Pulse Method for Thermal Diffusivity Measurement

R.E. TAYLOR and K.D. MAGLIĆ

1. INTRODUCTION

Intensive development of transient-state techniques for the measurement of transport properties of materials has occurred within the last two or three decades, as a result of need to replace lengthy stationary state methods with faster, more productive methods, which require smaller specimens and can operate at higher temperatures. These developments were facilitated by the corresponding development of electronic recording and measuring equipment, and specifically during the last decade by the massive introduction of fast data acquisition systems and minicomputers into experimental research.

Unsteady-state methods, i.e., methods for the measurement of thermal diffusivity, are based on the analysis of the temperature response of the specimen subjected to transient thermal conditions. They could be grouped in accordance with the nature of this perturbance into (a) pulse methods, (b) periodic heat flow methods, and (c) monotonic heating regime methods. The first and the last group might be considered as two extremes of the same group, where the first duration of the heating period is very short compared with transient, while the last heating period in infinite compared with duration of the experiment. Still, because of many features inherent in experimental procedures which distinguish them, these two groups are treated separately.

The most recently developed of these groups, the pulse method, has gained very high popularity among thermophysical property investigators in the past two decades. Reasons for this lie in the ease with which initial and boundary

R.E. TAYLOR • Thermophysical Properties Research Laboratory and CINDAS, Purdue University, 2595 Yeager Road, West Lafayette, Indiana 47906. K.D. MAGLIĆ • Boris Kidrič Institute of Nuclear Sciences, Institute of Thermal Engineering and Energy Research, Belgrade, Yugoslavia.

conditions of the mathematical heat conduction model can be reproduced in a physical experiment, the simple shape of the specimen, and the wide range of materials, diffusivities, and temperatures to which the method is applicable. As a consequence, according to CINDAS statistics[1] about 75% of all the diffusivity results published in the primary literature in the early seventies were obtained with the pulse technique.

The pulse technique can be applied virtually to all solids,[2] and attempts have been made to use it for the measurement of thermal diffusivity of liquid metals.[3] Metals and alloys and opaque nonmetals[2,4,5] can be measured in a straightforward way, by exposing the specimen to an energy pulse,[2] while translucent materials and coatings,[6,7] temperature-sensitive materials,[8] and composites[9,10] have to be measured in a layered arrangement. Pulse technique has been successfully applied to anisotropic materials,[11,12] materials consisting of two and three layers,[13,14] and for the measurement of contact conductance between layers.[15]

Thermal diffusivity values ranging from 1×10^{-7} to 1×10^{-3} m^2/s are readily measurable by this method, and measurement can be made in the temperature range 100 to 3300 K. Specimens usually have the shape of a small disk 6 to 16×10^{-3} m in diameter, and less than 4×10^{-3} m thick. Some variants[16] employ specimens 3 to 5 times bigger and thicker.

The application of the method in its original and most frequently used form, employing a flash lamp or a laser energy pulse, is limited to materials which can be rendered thin enough to still be representative of the material and to satisfy the requirement that the half-time of the temperature transient (described in equation (8)] should stay within 0.040 and 0.250 s limits. Another limitation is connected with the requirement that the energy of the pulse has to be absorbed in a very thin layer of the front face of the specimen, without causing damage to the specimen due to large temperature excursion of the specimen material in this layer. These two limitations, however, can be overcome by application of the radial flash method[16] with bigger specimens, use of longer pulses, and corresponding extended pulse analysis[17] or multilayer techniques.[18,8] It is therefore difficult to ascribe limitations to the method other than those which arise from the temperature range of application and the range of diffusivities for which the method is applicable.

2. THE METHOD

The flash method was first described in 1960 by Parker *et al.*[19] In this method the front face of a small disk-shaped specimen is subjected to a very short burst of radiant energy coming from a laser or a xenon flash lamp, irradiation times being of the order of one millisecond or less. The resulting temperature rise of the rear surface of the specimen is recorded and measured and thermal diffusivity values are computed from temperature rise versus time data (Figs. 1a and 1b).

The method is based on mathematical solution of the heat conduction problem of the semi-infinite specimen initially at the constant temperature subjected to a flash of energy. According to Carslaw and Jaeger,[20] the temperature $T(x, t)$ at any time, t, after receiving a flash at $t = 0$, is given by

$$T(x,t) = \frac{1}{L}\int_0^L T(x,0)\,dx + \frac{2}{L}\sum_{n=1}^{\infty} \exp\left(\frac{-n^2\pi^2\alpha t}{L^2}\right) \times \cos\frac{n\pi x}{L}\int_0^L T(x,0)\cos\frac{n\pi x}{L}\,dx \tag{1}$$

Assuming the pulse of energy, Q, to be instantaneously and uniformly absorbed in a small depth g at the surface $x = 0$, then, at that instant, the temperature distribution is

$$T(x,0) = Q/dcg \qquad \text{for} \quad 0 < x < g$$

and

$$T(x,0) = 0 \qquad \text{for} \quad g < x < L \tag{2}$$

For these initial conditions equation (1) becomes

$$T(x,t) = \frac{Q}{dcL}\left|1 + 2\sum_{n=1}^{\infty} \cos\frac{nx}{L}\,\frac{\sin(n\pi g/L)}{(n\pi g/L)} \times \exp\left(\frac{-n^2\pi^2}{L^2}\alpha t\right)\right| \tag{3}$$

where d is density and c is specific heat.

For opaque materials, g is sufficiently small for $\sin(n\pi g/L) \simeq n\pi g/L$, and with this approximation the temperature at the rear face becomes

$$T(L,t) = \frac{Q}{dcL}\left|1 + 2\sum_{n=1}^{\infty} (-1)^n \exp\left(\frac{-n^2\pi^2}{L^2}\alpha t\right)\right| \tag{4}$$

Equation (4) indicates the maximum temperature of the rear face to be

$$T_{L\,\max} = \frac{Q}{dcL} \tag{5}$$

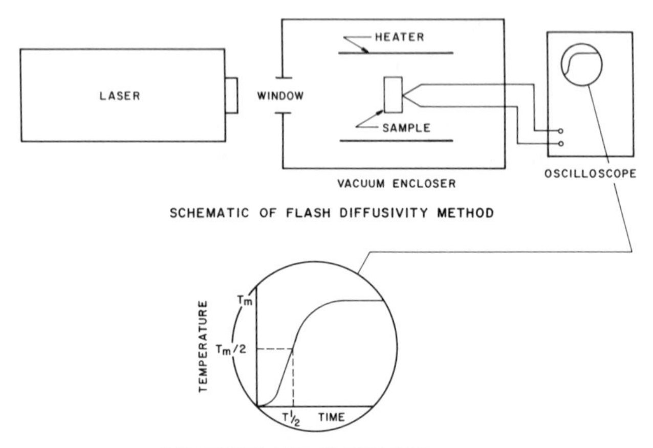

FIGURE 1. Schematic of flash diffusivity method.

and hence, at any time t, the rear face will rise to a fraction of its maximum rise which is given by

$$V = 1 + 2 \sum_{n=1}^{\infty} (-1)^n \exp\left(\frac{-n^2\pi^2}{L^2}\alpha t\right) \tag{6}$$

The value of V varies from 0 to 1.

From equation (6) thermal diffusion can be calculated as a function of percent temperature rise of the rear face:

$$\alpha = k_x L^2/t_x \tag{7}$$

where k is a constant corresponding to x percent rise and t_x is the elapsed time to x percent rise. Values of k_x may be found in Table 1.

The simplest and the most frequently used way to calculate thermal diffusivity is to use $t_{1/2}$ as the characteristic time, i.e., the time needed for the rear side temperature to reach 50% of its maximum value:

$$\alpha = 0.1388 L^2/t_{1/2}\ \mathrm{m^2/s} \tag{8}$$

Other percents are also used, and their use will be discussed in conjunction with measurement error analysis.

The mathematical analysis that forms the base of the flash method[20] employs series solutions that converge well after longer time, but not too well for times small compared to the temperature rise.[21, 22] It is convenient and satisfactory for experiments where heat losses are not significant, i.e., where the

TABLE 1
Values of k_x in Equation (7)

x (%)	k_x	x (%)	k_x
10	0.066108	60	0.162236
20	0.084251	66 (2/3)	0.181067
25 (1/4)	0.092725	70	0.191874
30	0.101213	75 (3/4)	0.210493
33 (1/3)	0.106976	80	0.233200
40	0.118960	90	0.303520
50 (1/2)	0.138785		

upper part of the response curve is not so much deformed due to heat losses to cause appreciable error in calculated thermal diffusivity. For the calculation and understanding of the temperature rise at relatively small times, James[17] recently proposed analysis based on series rapidly convergent at small t. It makes use of images to obtain an expression for the total temperature distribution as the sum of contributions from many sources. By this analysis thermal diffusivity is computed from the logarithmic relation between temperature rise and $1/t$, not depending on the data near the maximum or longer time.

The same analysis[17] is also applicable to accounting for heat losses from both surfaces of the specimen, step heating, and for two-layer specimen with contact resistance between layers. This approach[17] offers a new tool for the analysis of transient responses in the pulse method and establishes variants of the method convenient for materials sensitive to intensive heating of the front specimen surface.

3. EXPERIMENTAL

The flash diffusivity apparatus consists of a vacuum–inert gas environment chamber with a specimen holder and a furnace or chiller to ensure desired specimen reference temperature, vacuum pumping unit, system for the measurement of transient and reference temperature, and a laser as the energy flash source.

The vacuum chamber is usually equipped with two windows, for entrance of the laser pulse and for provision of a light path for contactless measurement of the temperature transient, and of the specimen reference temperature. Quartz glass is convenient for the first window, while the second is usually made of some ir-transparent material, like calcium fluoride or calcium aluminate, in order to enable use of a photodetector for the measurement of temperature transient in the ir range.

Vacuum environment is preferable for most materials from low temperatures to about 2500 K; at temperatures in excess of 2500 K and with graphite heaters or susceptors, inert gas atmosphere is necessary to prevent excessive evaporation of the heater material. It is advisable to protect both windows from direct radiation of hot furnace elements between measurements, particularly when the specimen temperature is measured with a pyrometer. It is also convenient to have a selective optical filter opaque at the laser light wavelength installed between the specimen and the optical pyrometer, to protect the pyrometer in the case of dislocation or breakdown of the specimen during operation of the apparatus. The vacuum chamber should be water cooled to ensure faster establishment and stability of the furnace and specimen temperature.

Thanks to the small dimensions of the specimen, the furnace can be small, with low thermal inertia, enabling rapid change of the specimen temperature, and rapid recording of thermal diffusivity data over large temperature intervals. Alternatively the thermal mass may be made large and a number of specimens positioned sequentially in the energy beam path at a particular temperature. In either case a large amount of data can be generated in a short period of time. The furnace may be horizontal or vertical. Both arrangements have advantages and disadvantages: horizontal is simpler for the positioning of the laser, specimen, and detector along the same optical axis, and for the alignment of all components using low-energy gas laser. A vertical arrangement (with a vertical furnace and the specimen in horizontal position) is preferable at very high temperatures (above 2300 K), as the relative motion of the furnace and the specimen holder elements will hardly effect alignment of the specimen, and the support of the specimen will be simpler, ensuring desired poor thermal contact between the specimen and its holder. Some deficiencies of the vertical system pertaining to a more complicated optical path can be largely overcome with aid of fiber optics.

Electrical resistance heating elements are used from room temperature to 3000 K, although at the highest temperatures grain growth and embrittlement of refractory metals reduce their life markedly. Above 2700 K, R. Taylor recommends high-frequency induction heating with a graphite susceptor.[(4)] Low ambient temperatures can be affected by fixing a copper jacket around the specimen and passing liquid-nitrogen-cooled gas through the jacket and specimen supports. The specimen temperature may be adjusted from liquid nitrogen to room temperature using this technique.

The flash diffusivity apparatus built by R.E. Taylor at Properties Research Laboratory, Purdue University[(1)] represents an advanced apparatus with a horizontal low thermal inertia furnace. It is shown in the photograph in Fig. 2. It consists of the base plate (1), specimen supports (2), tubular tantalum sheet furnace (3), and heat shields (4) (made of 25 and 50×10^{-6} m tantalum sheet, respectively); the specimen (5) is held by tantalum specimen support rods (6)

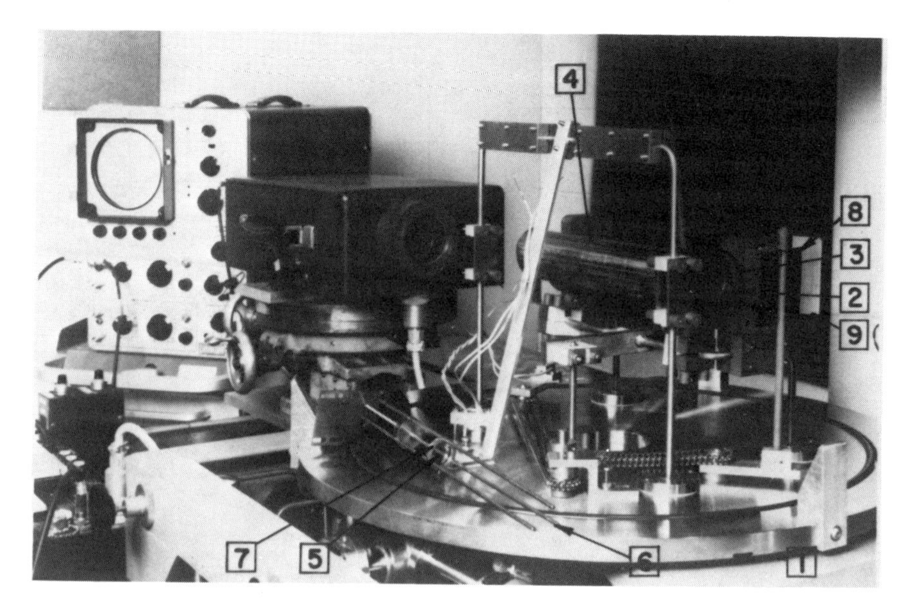

FIGURE 2. Close-up of base plate, furnace, and specimen holder of PRL apparatus.

and holder (7), which are shown lying on the base plate for purposes of clarity. Also visible in Fig. 2 are the heat beam reflector (8) and beam catcher (9) which protect the optical windows. Full description of the apparatus is given in Ref. 1.

The vertical furnace designed for operation from room to very high temperatures (3300 K) was built by R. Taylor[4] at the University of Manchester/ UMIST, U.K. Its chamber was designed both for work in vacuum and with inert gas (to 1 M Pa). A schematic diagram of the apparatus is given in Fig. 3. It employs high-frequency heating with a hf coil inside the chamber, and a graphite susceptor (insulated with graphite felt) as the heater. A ring-shaped specimen holder supports the specimen in the middle of susceptor as shown in Fig. 4. The light path from the laser and radiation from the specimen rear face are controllable via adjustment of the upper prism and bottom mirror. Thanks to the small mass of the susceptor, specimen holder, and thermal insulation, this furnace also has low thermal inertia, allowing fast coverage of a wide temperature range of investigation.

The thermal diffusivity apparatus realized by R.U. Acton of Sandia Laboratories,[23] has been built to meet the requirements of a large materials development program. Its furnace has large thermal mass and capability to measure diffusivity of as much as 20 specimens per setup for temperatures up to 2800 K. Tungsten wire mesh furnace of commercial vacuum tension testing

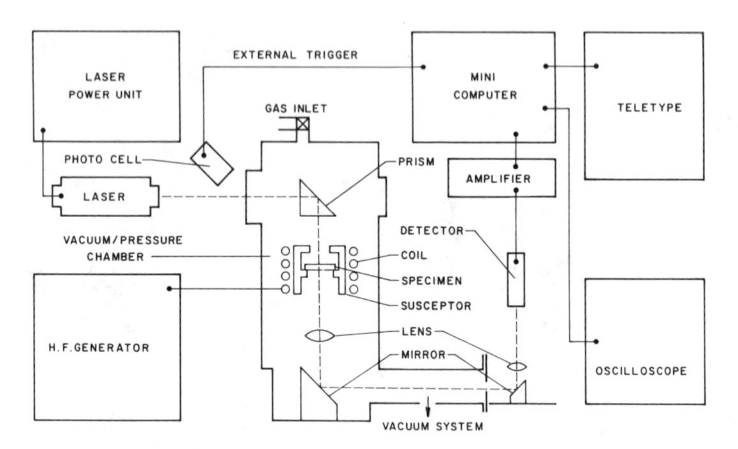

FIGURE 3. Schematic of a vertical furnace thermal diffusivity apparatus (U.M.I.S.T.).

apparatus has been adapted to receive a vertical graphite rod-shaped specimen holder, capable of receiving up to 20 specimens accommodated in horizontal holes disposed in two orthogonal vertical planes of the rod (Fig. 5). All specimens remain in the furnace hot zone, thus eliminating the time delay for equilibration of specimen at the desired temperature. With the aid of a water-cooled stainless steel extension rod manipulated from outside, specimens are brought in sequence in the horizontal optical path between laser and detector, and in this way 20 specimens may be tested during 30 min at selected temperature.

The three types of furnace and thermal diffusivity apparatuses briefly

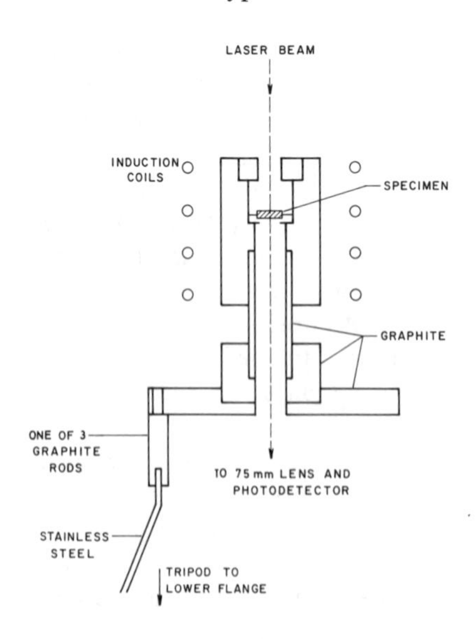

FIGURE 4. Specimen holder of U.M.I.S.T. thermal diffusivity apparatus.

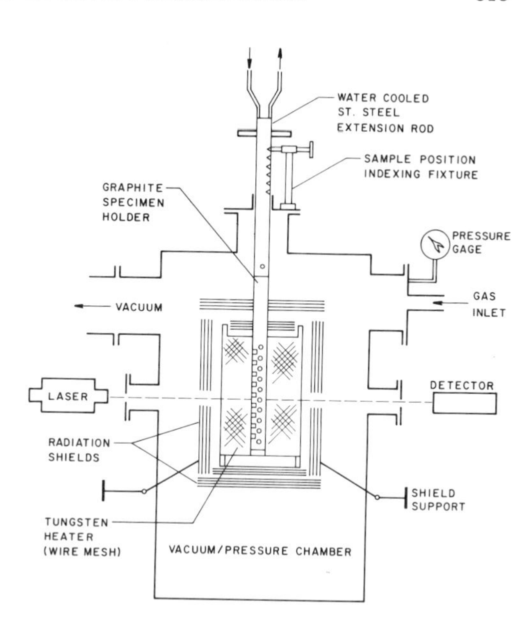

FIGURE 5. Schematic of multi-specimen thermal diffusivity apparatus (Sandia Laboratories).

outlined in the foregoing only illustrate a large variety of apparatuses designed and used by a multitude of researchers throughout the world. Their suitability for fast data generation and coverage of large temperature intervals singled them out for this illustration.

4. TRANSIENT TEMPERATURE MEASUREMENT

The detector of the transient temperature change may be a thermocouple, infrared detector, or automatic optical pyrometer. It must be capable of detecting 0.05 degree change above the ambient temperature. The time response of the detector and its associated amplifier is extremely important. While the response time of the detectors in optical instruments is often orders of magnitude faster than required for the flash method, the detector signal is fed into amplifiers and filters whose response times can be slow enough to affect transient readings. Therefore, the response time of the total circuit must be checked using choppers or other devices to prove that the response time is less than 0.1 of the half-time value. The use of electronic filters is discouraged as they tend to distort the shape of the time-temperature curve.

Thermocouples are used at low temperatures, below the sensitivity threshold of optical ir detectors. Thermocouples wires usually 25 to 50×10^{-6} m diameter are independently spot welded (in intrinsic arrangement), or the thermocouple is spring loaded against the specimen rear face.

In the case of intrinsic thermocouples, the response time (time to reach 95% of steady state value) can be defined as[24]

$$t_{95} = \frac{25}{\pi} \frac{D_T^2}{\alpha_s} \frac{\lambda_T}{\lambda_s} \tag{9}$$

where D_T is the diameter of the thermocouple wires, α_s is the thermal diffusivity of the specimen, λ is the thermal conductivity, and the subscripts T and s refer to the thermocouple and specimen, respectively. Thus the small-diameter thermocouple of low-conductivity material attached to a specimen of high-conductivity and high-diffusivity material yields the fastest response time. Equation (9) is misleading in that one might postulate the thermocouple response to be a smooth rise. Actually, the thermocouple response is a step change, followed by an asymptotic rise to the final value. This behavior is represented by equation (10) (Ref. 24):

$$\frac{T_T - T_0}{T_\infty - T_0} = 1-(1-a)\,e^{a^2 t^*} \operatorname{erfc}(a\sqrt{t^*}) \tag{10}$$

where t_0 and T are shown in Fig. 6, t^* is dimensionless time ($t^* = 4\alpha_s t/D_T^2$), and a is approximated by $1/[1 + 0.667\,(\sqrt{\lambda_T/\lambda_s})^{1/2}]$. In order to obtain the fastest response, one should use small-diameter thermocouple wire of an alloy with a low thermal conductivity attached to a substrate whose thermal diffusivity is large. For example, a 25×10^{-6} m constantan wire on a copper substrate requires 3 μs to reach 95% of steady state. However, if one considers the converse of this example, i.e., 25×10^{-6} m copper wire on a constantan substrate, it is found that 15 000 μs are required to reach 95% of steady state. This is 5000-fold slower than the first example. Thus, the proper selection of materials, based upon their thermal properties and geometries, is essential for accurate measurement of transient responses using intrinsic thermocouples. Intrinsic thermocouples, however, are sensitive to nonhomogeneous heating and temperature field at the specimen rear side, due to inhomogeneity of the laser beam and side heat losses, and should be used in thermal diffusivity measurement with caution. The use of intrinsic thermocouples in the flash technique is discussed in detail by Heckman.[25]

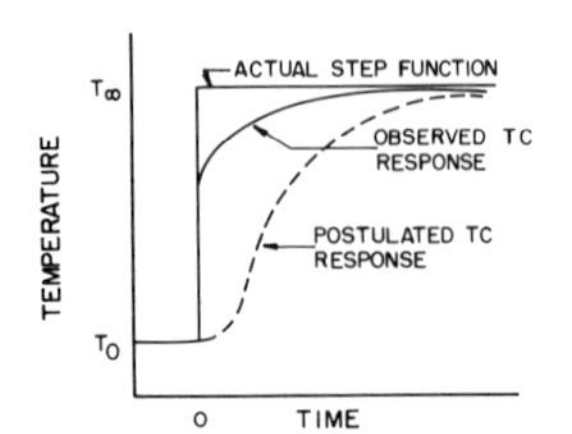

FIGURE 6. Thermocouple response characteristics.

Equations (9) and (10) relate to the minimum response time possible for a thermocouple. If the thermocouple is attached poorly to the specimen, the effective response time can be much larger. Thus, proper attachment of the thermocouple to the specimens is important. The preferred method is to spot-weld intrinsic thermocouples, i.e., nonbeaded couples where each leg is independently attached to the specimen about 1 mm apart. For specimen materials that cannot be spot-welded to, it may be possible to spring-load the thermocouple against the back surface. Realization of such spring-loaded thermocouple with low thermal inertia requires much skill, and before its use for testing materials its performance should be checked against thermal diffusivity standard material in the same thermal diffusivity range. For materials with low diffusivity values, it may be preferred to spot-weld thermocouples onto a thin metallic sheet and spring-load or paste this sheet onto the sample. Metal–epoxy and graphite pastes have been used successfully to bond layers together. This eliminates the problem of using thermocouples of relatively high diffusivity to measure samples of low diffusivity, which leads to very large response times [equation (9)].

Optical detectors are free from most deficiencies inherent to thermocouples. They can integrate temperature change from larger portions of the specimen rear face, in this way partially compensating for effects of nonuniformity of the laser beam. Their response times are in the microsecond range, and, due to their passive nature, they do not introduce additional thermal perturbance to the specimen. The most popular ir detectors are indium antimonide, sensitive to 5.5 μm, and more recently mercury cadmium telluride, which covers the same temperature range as indium antimonide but has to be cooled only to 230 K, and thermoelectric cooling is available. Photomultipliers and automatic pryometers are used above 1300 K.

When using remote temperature sensing, several precautions are required. The sensor must be focused on the center of the back surface. The sensor must be protected from the energy beam to prevent damage or saturation. When the sample is housed in a furnace, the energy beam may bounce or shine past the edges of the sample and enter the detector so proper shielding is required. For protection against lasers, dielectric spike filters, which are opaque at the selected wavelength, are very useful. The viewing window and any focusing lenses must not appreciably absorb in the wavelength region that the detector is sensitive to. This is particularly important for ir detectors. While measurements are being made at elevated temperatures, all windows must be monitored for deposit build-up and resulting absorption of energy. This can lead to loss of sensitivity of the optical detector and reduced or nonuniform sample heating from the pulsed energy source.

The signal conditioner includes the electronic circuitry to bias out the ambient temperature reading, spike protectors, amplifiers, and analog-to-digital converters. Precautions must be taken to prevent driving the components into saturation and to have adequate response time so that the transient signal is not distorted.

4.1. Experimental Data Reduction

Experimental data necessary for calculation of thermal diffusivity include base line, which represents equilibrium temperature of the specimen prior to laser discharge, time mark of the laser discharge and the transient temperature curve, extending at least ten lengths of the characteristic half-time, $t_{1/2}$.

The data collection system may be a digital data acquisition system, digital oscilloscope, or a regular oscilloscope. In the latter case, provisions must be made for photographing the trace. The timing measurements are provided by the data collection system and the accuracy of the time base must be verified.

The data processor may be a simple adjustable scale for measuring the elapsed time on a photograph plus hand-held calculator, a digital oscilloscope or a digital data acquisition system. It is important that the transient response curve by analyzed to verify the presence or absence of finite pulse time effects, radiation heat losses, and nonuniform heating. As a minimum this means computing the diffusivity using the 25%, 50%, and 75% rise times and ideally this means comparing the normalized rise curve with the theoretical model.

In the absence of a digital data acquisition system, a digital oscilloscope (preferably equipped with floppy disk memory) provides for fairly fast, simple, and accurate data acquisition. Calculations of basic corrections may be done with the aid of a hand calculator.

4.2. Test Specimen

The usual specimen is a circular disk with a front surface area less than that of the energy beam. Typically specimens are 6–16 mm in diameter. The optimum specimen thickness depends upon the magnitude of the estimated diffusivity and should be chosen so that the half-time falls within the 0.04- to 0.25-s range. This range is desirable because response times of detectors and electronics are usually satisfactory and heat losses are not excessive. Thinner specimens are desired at higher temperatures to minimize heat loss corrections. Typically, thicknesses are in the 1.5- to 4-mm range. Since the half-time is proportional to the square of the thickness, it may be desirable to use different thicknesses in different temperature ranges. In general, the same thickness will be too far from optimum for both cryogenic and high-temperature measurements.

4.3. Calibration

Both the micrometer used to measure specimen thickness and the time base of the data acquisition should be calibrated. However, it must be recognized that the most serious and insidious errors arise from inadequate time response of the detector and its associated electronics and from nonobservance of the boundary conditions assumed in the mathematical model, i.e., nonuniform heating, heat losses, or finite pulse-time effects. The response time can be checked by the measurement of a standard specimen whose thickness is estimated to cause about the same half-time as the unknown. The presence of the other effects is detected by the methods given in the section on error analysis.

While there are no standard reference materials (SRM) issued by the U.S. National Bureau of Standards (NBS) for thermal diffusivity, several materials can be used as reference materials. POCO graphite is available from NBS and the diffusivity values are being established through round robins.[26,27] Papers giving results are in various stages of completion.[28–30] ARMCO iron has also been used as a reference material. A data bank giving a summary of diffusivity values found in the literature and giving recommended values has been issued.[28] It must be emphasized that the use of reference materials to establish validity of the data on unknowns has often led to unwarranted statements on accuracy. The use of references is only valid when the properties of the reference (including half-times and diffusivity values) are nearly equal to that of the unknown and the temperature-rise curves are determined in an identical manner for the reference and unknown. One important check of the validity of data (in addition to the comparison of the rise curve with the theoretical model) is to vary the sample thickness. Since the half-times vary as L^2, decreasing the sample thickness by one-half should decrease the half-time to one-fourth of its original value. Thus, if one obtains the same diffusivity value with representative samples of significantly different thicknesses or, better yet, on the same sample with different thicknesses, then the results can be assumed valid.

4.4. *Errors in the Flash Method*

Experimental results are subjected to two general types of errors, namely, measurement errors and nonmeasurement errors. Measurement errors are associated with uncertainties that exist in measured qualities contained in the equation used to compute the diffusivity from experimental data. Nonmeasurement errors are associated with deviations of actual experimental conditions as they exist during the experiment from the boundary conditions assumed in the mathematical model used to derive the equation for computing the diffusivity.

5. MEASUREMENT ERRORS

Measurement errors include errors associated with determining the effective thickness of the sample and errors associated with measuring the time that the rear face temperature attains a certain percentage of the maximum rise. This latter measurement involves determining the temperature base line, the maximum temperature rise above the base line, and the time of initiation of the laser pulse as an integral part of the time determination. The response time of the detectors and amplifiers must be considered.

Determinations of sample thicknesses are often limited by the ability to machine flat and parallel surfaces, i.e., the samples may be tapered. Also an accurate thickness determination is difficult for softer materials or for materials with roughened surfaces, such as plasma-sprayed materials. However, in most cases the measurement uncertainty can be estimated with reasonable accuracy. The errors associated with two-dimensional heat flow caused by nonuniform thickness are smaller than the error in determining the effective thickness and can generally be neglected.

As an inherent part of determining the half time (or the time to reach any other percentage temperature rise), it is necessary to determine accurately the temperature base line and maximum temperature rise. The temperature base line can be obtained conveniently by measuring the detector response for a known period prior to charging the capacitor bank used to fire the flash tube (laser) and extrapolating the results to the time interval in which the rear face temperature is rising. This procedure is often complicated by the electronic disturbances associated with charging and discharging a high-voltage capacitor bank. These electronic disturbances may drive amplifiers into saturation or cause offsets. Also, in the case of optical detectors, part of the laser beam or flashlamp radiation may enter the detector and cause saturation or offsets. The presence of these disturbances depends upon the particular experimental setup and alignment and varies from day to day even within one laboratory, to say nothing of the differences between laboratories. The presence of offset or saturation can be detected and hopefully eliminated by trained researchers.

The problem of determining the maximum temperature rise also varies with each experiment. For the ideal case (Fig. 7, Experiment 1), the rear face temperature rise becomes greater than 99% of the maximum rise at four half times. From the shape of the ideal curve it should be noted that there is an initial temperature lag, then a rapid temperature rise, followed by an asymptotic increase to the maximum value. Thus the maximum temperature rise should occur after four half times and not at twice the half time. If the sample is heated nonuniformly, if there are heat leaks into the sample, or if the detector and/or amplifier response time is inadequate, the rear face temperature may continue to increase significantly after four half times (Fig. 7, Experiment 2). The presence of this situation, of course, shows that the experiment is nonideal. Therefore using mathematical procedures derived for the ideal case is bound to give results which are in error – the magnitude of this error depending upon the amount of derivation from ideality. Thus the problem of determining a usable

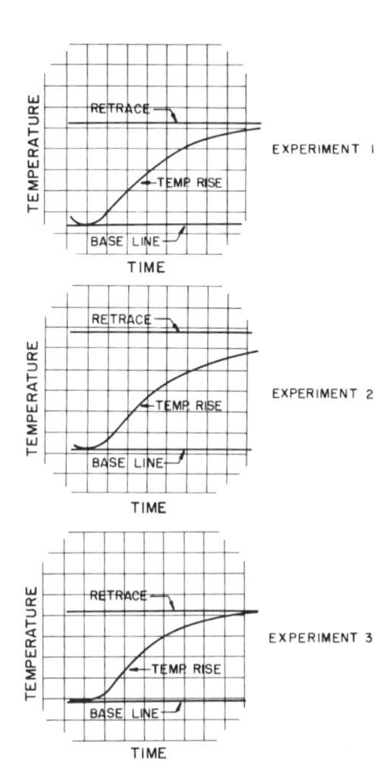

FIGURE 7. Reproductions of oscilloscope traces.

value for the maximum in the case of a continuing temperature rise is inseparable from the errors causing this rise.

On the other hand, the temperature maximum may be attained prior to four half times (Fig. 7, Experiment 3). Such a situation also shows the presence of nonideal behavior such as heat leaks from the sample to the holder, radiation losses, or nonuniform heating. While the maximum temperature is easy to determine in this case, the reasons for the departure from the ideal behavior must be known and either eliminated or corrected for.

The temperature rise curves are determined in a relative sense and not on an absolute basis. Therefore, in addition to measuring the response curves, it is necessary to measure the specimen reference temperatures. The initial (base line) temperatures are measured in an absolute sense and usually the measured diffusivity values are assigned to these temperatures since the temperature rises are small. However, a better measurement reference temperature is the initial temperature plus 1.6 times the temperature rise.[19] The error in the measurement of the specimen reference temperature may be avoided by adequate design of the furnace (favorable length-to-diameter ratio), or compensated for, by calibration of reference temperature thermocouple against thermocouple located in the center of the specimen.[30]

For the ideal case, the diffusivity is given by equation (8) and the measurement errors associated with the use of this equation can be expressed as

$$\frac{d\alpha}{\alpha} = \frac{2dl}{l} + \frac{dt_{1/2}}{t_{1/2}} \tag{11}$$

Often l is about 0.25 cm and can be measured to ± 0.0005 cm with a micrometer. The quantity $t_{1/2}$ is usually about 50 ms and can be measured to ± 1 ms with an oscilloscope using best practices. Thus the determinate errors are often about ± 2.4%, using an oscilloscope. Use of digital oscilloscope reduces this error to ± 1.5%, and with a digital data acquisition system (DDAS) the uncertainty in measuring t_x is less than 0.1 ms, reducing the determinate error to about ± 0.6% at $t_{1/2}$ and ± 0.5% at $t_{0.75}$.

In this estimate of measurement errors, it is assumed that detector and/or amplifier saturation and offsets are not present and that the curves are ideal. Thus the quoted measurement errors are the optimum that can be obtained in practice. Nonideal behaviors are treated separately in the next paragraphs, as nonmeasurement errors.

5.1. Nonmeasurement Errors

The major sources of nonmeasurement errors are (1) finite pulse time effect, (2) heat losses or gains, (3) nonuniform heating, and (4) in-depth absorption of the energy pulse. It should be noted that these effects are not properly classified as errors but are merely deviations from an ideal situation in which these effects are assumed to be negligible. It is entirely feasible to propose models which incorporate these effects and to generate mathematical expressions which properly account for them. This in fact has been done for the first three of these effects. Problems arise, however, in mathematically expressing the actual experimental conditions present, i.e., in knowing quantitatively the actual degree of nonuniformity of heating, the exact heat loss from each surface, etc.

Of these major effects, the finite pulse time effect (Fig. 8) is the easiest to handle and is the most advanced. Cape and Lehman[(21)] developed general mathematical expressions for including pulse time effects. Taylor and Clark[(31)] tested the Cape and Lehman[(21)] expressions experimentally. Larson and Koyama[(32)] presented experimental results for a particular experimental pulse characteristic of their flash tube. Heckman[(22)] generated tabular values for triangular-shaped pulses. Taylor and Clark[(31)] demonstrated that triangular-shaped pulses adequately approximated their commercial laser system, derived closed solution for practical cases, and demonstrated results experimentally. They used Armco iron and varied sample thickness and temperature. In this way, using the same sample the "finite pulse region" could be entered by decreasing sample thickness and the effect could be lessened by increased temperature (Fig. 9). The diffusivity

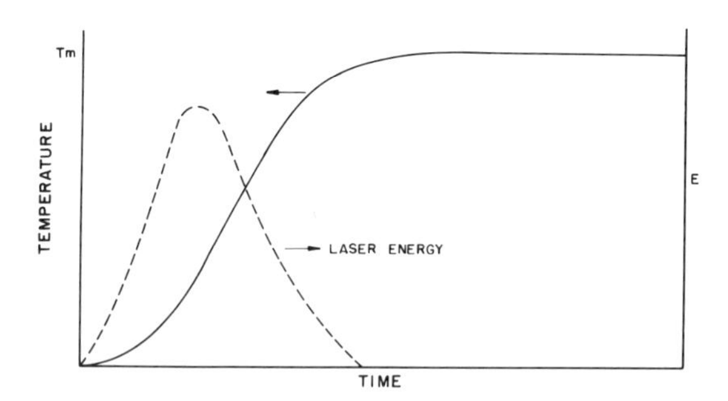

FIGURE 8. Finite pulse time effect.

values, obtained on the sample at thicknesses large enough for there to be finite pulse time effect (0.2535, 0.2530, and 0.1277 x 10^{-2} m, Fig. 9). are within 3% of the Thermophysical Property Research Center (TPRC) recommended curve,[33] which is believed accurate within 4%. Both original data and data corrected for the finite pulse time effect obtained on thinner samples (0.1181, 0.08788, 0.06375, and 0.03835 x 10^{-2} m thick) are shown in Fig. 9. All corrected values fall within 3% of the diffusivity without finite pulse effect and it was possible to make a correction as great as 21%. In addition, Taylor and Clark[31] showed how calculated diffusivity varied with percent rise. They also showed how to correct diffusivities at percent rises other than the half-rise value

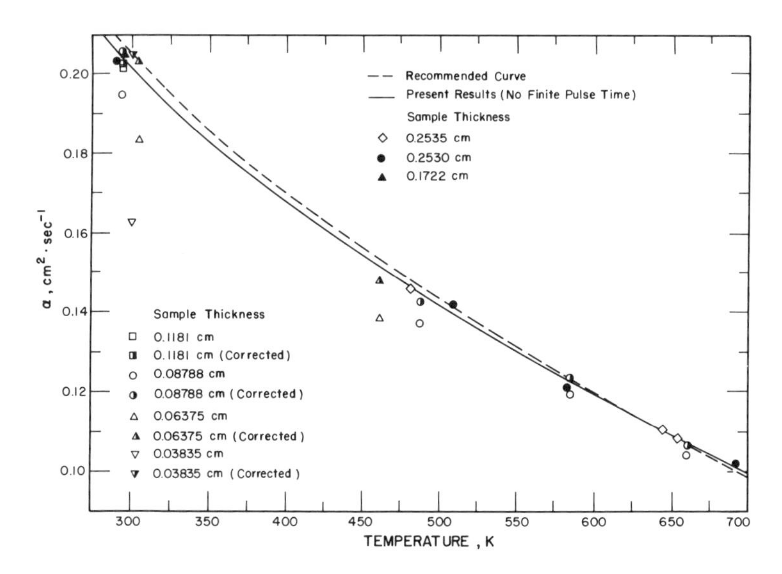

FIGURE 9. Thermal diffusivity of Armco iron (finite pulse time effect).

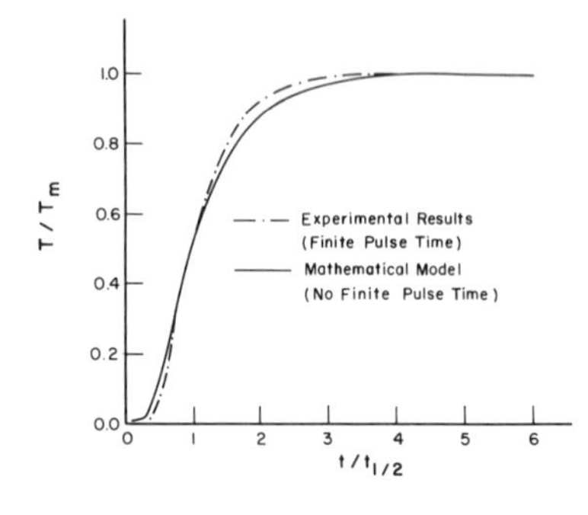

FIGURE 10. Rear-face temperature rise: comparison of mathematical model (no finite pulse time effect) to experimental values with finite pulse time.

so that the diffusivity values could be calculated over the entire experimental curve rather than at one point.

The presence of the finite pulse time effect (without other effects being present) and be readily determined from a comparison of the experimental curve to the theoretical model (Fig. 10). The distinguishing features are:

1. Experimental curve lags theoretical ideal curve from about 5% to 50% rise.
2. Experimental curve leads ideal theoretical curve from about 59% to about 98% rise.
3. A long, flat maximum is observed.

The calculated values of α increase with increasing percent rise and asymptotically approached the true value (Fig. 11).

Radiation heat losses have been extensively studied; Parker *et al.*,[34] Cape and Lehman,[21] Cowan,[35] Watt,[36] Heckman,[22] and Clark and Taylor.[37]

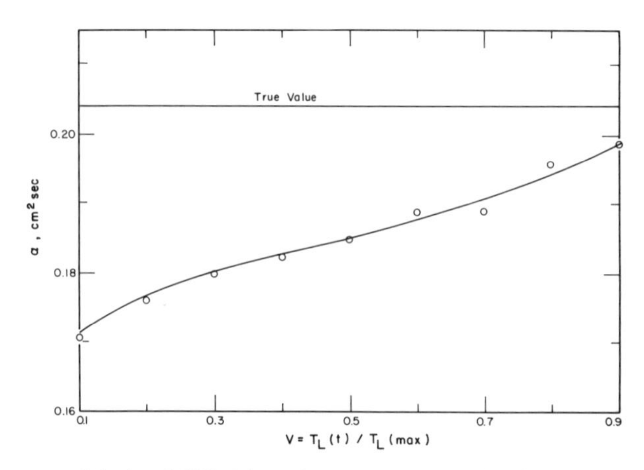

FIGURE 11. Calculated diffusivity values versus percent rise for a finite pulse time.

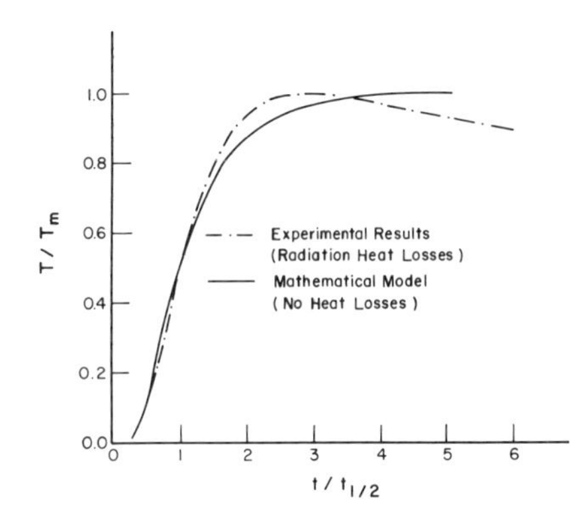

FIGURE 12. Rear-face temperature rise: Comparison of mathematical model (no heat loss) to experimental values with radiation hear losses.

Experimentally, the presence of heat losses is shown by the following features (Fig. 12):

1. Experimental curve slightly lags theoretical curve from about 5% to 50% rise.
2. Experimental curve leads theoretical curve from 50% to 100%.
3. A relatively short maximum is observed, followed by a pronounced smooth decline.

The calculated value of α increases with increasing percent rise at an increasing rate.

Radiation heat losses may be corrected for by the method of Cowan,[35] Heckman,[22] or Clark and Taylor.[37] Cowan's method involves determining the values of V_x, where V_x is the value of V [the nondimensionalized temperature rise, equation (6)] at $5t_{1/2}$ or $10t_{1/2}$ (Fig. 13). From V_x one can estimate the radiation loss parameter and correct $\alpha_{1/2}$.

The Clark and Taylor method is based upon the difference in the nondimensionalized curves representing various heat losses (Fig. 14). It can be seen that the ratio of time at a higher percent rise to a time at a lower percent rise decreases with increasing heat loss. A Cape and Lehman solution was generated for different loss parameters Y_r and Y_x on a t/t_c time scale. Division of t/t_c at a higher percent rise by t/t_c at a lower percent rise eliminates t_c, whereupon the results can be plotted. Plots of $\alpha t_{1/2}/l^2$ vs. the time ratios are shown in Fig. 15, generated by numerical evaluation of the Cape and Lehman solution. Three different ratios were plotted to enable the experimenter to select the most convenient. The procedure for finding thermal diffusivity is, then, to ratio the experimental times at two different fractional rises as per Fig. 14. For example, the time at $V = 0.8$ would be divided by the time at $V = 0.2$. Then from Fig. 15 a value of $\alpha t_{1/2}/l^2$ is found. Knowing $t_{1/2}$ and l^2, α is then calculated. It is desirable to have a number of ratios since, in practice, signal noise in either the

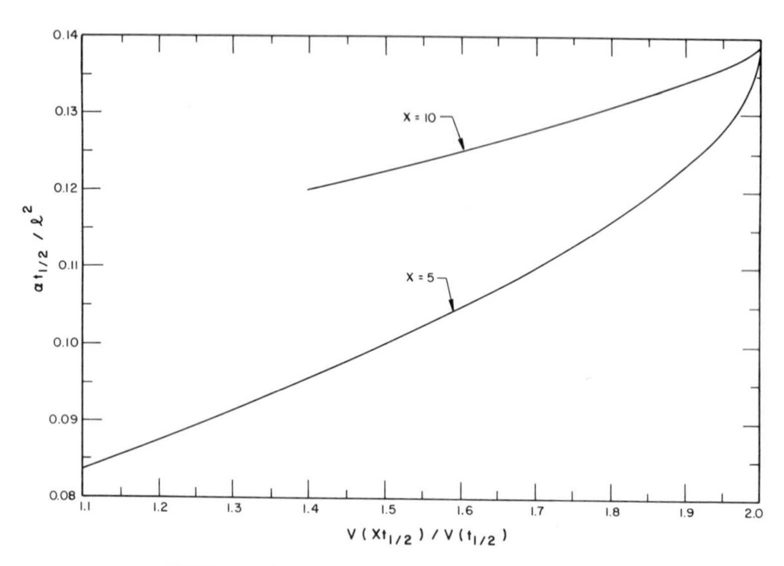

FIGURE 13. Data for Cowan's correction method.

low (0–25) percent range or higher (75–100) percent range or both can make effective ratioing difficult. Clark and Taylor tested their procedure under severe conditions (Fig. 16) and showed that corrections of 50% could be satisfactorily made. The Clark and Taylor method has the advantage of using the data collected during the initial rise rather than the cooling data, which are subject to greater uncertainties caused by nonuniform heating effects and conduction to the sample holder.

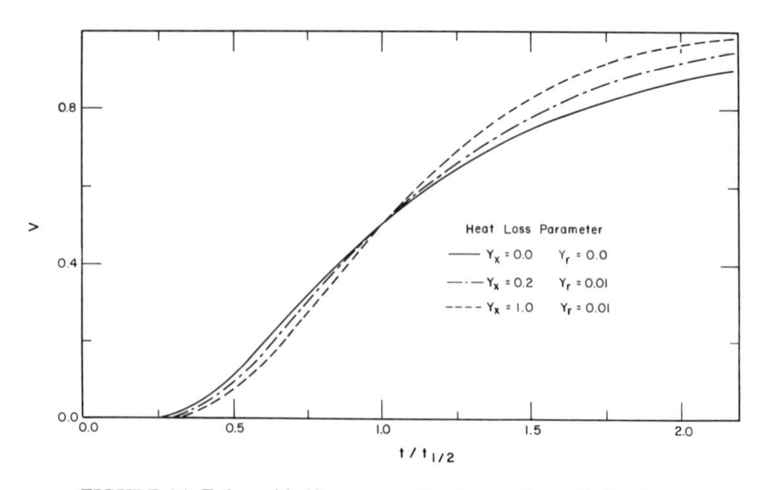

FIGURE 14. Enlarged halfmax normalized curve for radiation losses.

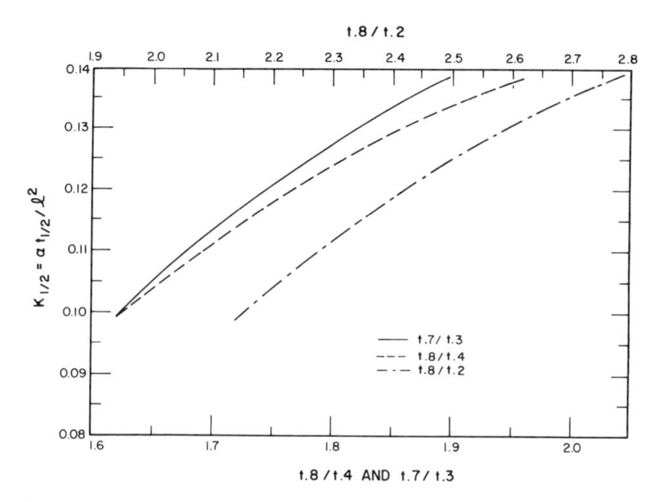

FIGURE 15. Data for Clark and Taylor ratio method for radiation heat losses.

However, further improvements in the correction procedure have been developed recently. One, under development by Taylor, compares the entire rise curve with curves calculated from a mathematical model, and then changes the heat loss parameter in the model until an optimum fit of experimental data to model is achieved. Using this fit, the value of α is obtained. Another technique, developed by James,(17) emphasizes the early time data and computes the diffusivity based on the logarithmic relation between $\theta t_{1/2}$ and $1/t$, where θ is the temperature rise. This method has the advantage of not depending on the

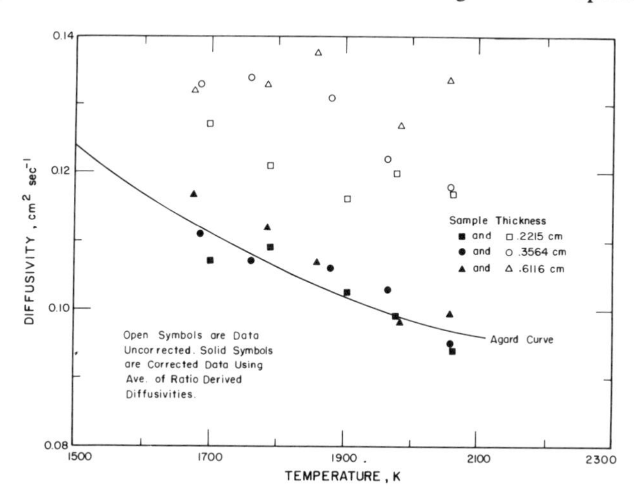

FIGURE 16. Thermal diffusivity of AXM-5Q graphite with corrections by ratio average method.

data near the maximum or longer times, and in fact the maximum is not even determined in this technique. The various methods have yielded corrected values of α within 2% even in the presence of unusually large heat losses where the uncorrected value of α was in error by more than 30%. It is interesting to note that one technique(16, 38) described later under radial heat flow is independent of heat loss corrections even where heat losses are very large.

Finite pulse time effects and radiative heat losses only occur in selective cases, i.e., finite pulse time effects occur with thin samples of high diffusivity materials and radiative heat losses occur at high temperatures with thicker samples. In contrast nonuniform heating can occur during any flash diffusivity experiment. However, very little has been published on the effects of nonuniform heating. Beedham and Dalrymple,(39) Schriempf,(40) and Taylor(41) have described the results for certain nonuniformities. These results show that when the heating is uniform over the central portion of the sample, reasonably good results can be obtained. However, a continuous nonuniformity over the central portion can lead to errors of at least ten times the determinate error.

5.2. *Application of Pulse Technique to Layered Structures*

In the case of layered structures (Fig. 17), the concept of effective diffusivity is not very meaningful. The heat-balance equation for transient conditions may be written as

$$\nabla \cdot \lambda \nabla T + \text{internal sources and sinks} = C_p \rho \frac{dT}{dt} \tag{12}$$

where λ is the thermal conductivity, C_p the specific heat at constant pressure, and ρ the density. If there are no internal sources and sinks,

$$\nabla \cdot \lambda \nabla T = C_p \rho \frac{dT}{dt} \tag{13}$$

For homogeneous materials whose thermal conductivity is nearly independent

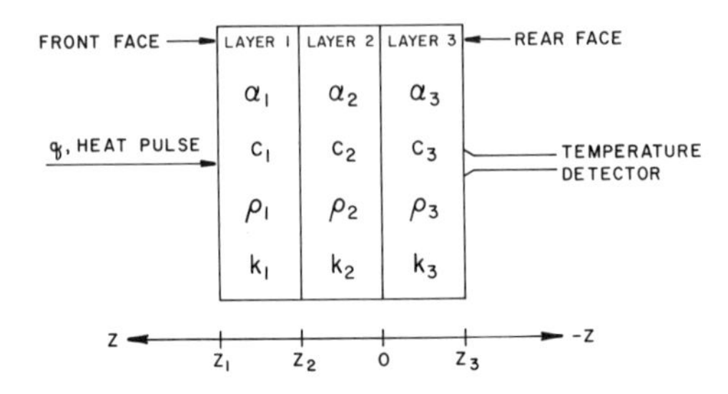

FIGURE 17. Layered composite samples.

of temperature, we may treat λ as a constant. Then $\nabla \cdot \lambda \nabla T$ becomes $\lambda \nabla^2 T$ and equation (13) can be written as

$$\lambda \nabla^2 T = C_p \rho \frac{dT}{dt} \tag{14}$$

or

$$\nabla^2 T = \frac{C_p \rho}{\lambda} \frac{dT}{dt} = \frac{1}{a} \frac{dT}{dt} \tag{15}$$

where $a = \lambda / C_p \rho$ is the thermal conductivity.

For one-dimensional heat flow

$$a \frac{d^2 T}{dx^2} = \frac{dT}{dt} \tag{16}$$

However, for layered structures, λ is discontinuous and λ may not be moved from behind the Dell operator. Nevertheless the individual layers still possess a unique thermal diffusivity. Layered structures have been studied by Larson and Koyama,[13] Gilchrist,[6] Bulmer and R. Taylor,[42] Murfin,[43] Ang,[44] Chistyakov,[45] H.J. Lee and R.E. Taylor,[14] T.Y.R. Lee, A.B. Donaldson, and R.E. Taylor,[18] and H. James.[17]

The nondimensionalized rear face temperature rise history following an instantaneous heat pulse to the front of a layered specimen composed of layers whose diffusivity ratios are less than 10:1, is the same as that for a homogeneous specimen providing there is no contact resistance between the layers. In fact deviations from the idealized model indicate interfacial resistance. It is also interesting to note that reversing the direction of heat flow does not affect the nondimensionalized response curve. The same half-time was noted, for example, when the stainless steel face of a two-layer specimen was irradiated and the temperature rise was measured on the rear of the copper layer as was observed when the copper was irradiated and the temperature rise was measured on the rear of the stainless steel face. This is true even though the diffusivities of the two layers are very different. The same situation also holds for three-layer specimens.

In references 14 and 18 computer programs were developed which calculate the thermal diffusivity of one layer of a two- or three-layer composite from the half-time measured in the conventional manner. Also programs were written and tested to compute the contact conductance between two layers whose thermal properties are known. T.Y.R. Lee and Taylor[9] also established the criteria for distinguishing between a resistive and capacitative layer.

Some of the layered specimens investigated by Lee, Donaldson, and Taylor[18] included stainless steel–epoxy layers with different thicknesses of each layer, an aluminium–epoxy and metal–metal layers. Diffusivity of either layer could be computed from the experimental data given the diffusivity of the other layer. Diffusivity of the resin was found to be in perfect agreement with data on a homogeneous sample while the diffusivity of the metal layer averaged within 2.5% of the value measured on homogeneous samples. It is also possible to compute the diffusivity of the poorly conducting layer assuming that the metal layer acts as a resistor or only as a capacitor (no temperature gradients), which leads to somewhat simpler mathematics.

The layered composite situations are important for a number of cases: a second layer may be placed on a homogeneous sample to limit temperature rises or prevent radiation from the energy source from penetrating the sample; one may wish to study coatings on a surface; one may study contact conductance between layers; one may form a cell to measure liquids or powders; or one may add a high-conductivity layer on the rear surface to increase temperature response characteristics of thermocouples. It is interesting that one measures shorter half-times using a thermocouple when a metallic layer is added to polymers than when the metallic layer is omitted, even though the heat pulse has a longer path length which enters as a square term under transient conditions.

The use of a relatively thick metal layer attached to the front of a sample is an effective way to reduce sizable temperature excursions. The back face temperature of this metal layer typically increases about 1°C as a result of the laser pulse and this layer serves as the front face of the sample layer. Thus the temperature excursion in the sample can be kept under 1°C. Using this technique, it is possible to measure heat-sensitive materials such as explosives and certain paints, to measure materials near phase transitions, and to measure liquids.

The one case of layered composites that the flash method has not been successfully applied to is that of measuring the properties of thin films of highly conducting materials applied to relatively thick layers of poorly conducting materials.

5.3. *Application of Flash Technique to Dispersed Composites*

A rigorous solution for the effective conductivity of a concentrated random array of particles of varying sizes has not been achieved. Several approximations have been developed, like those of Rayleigh,[46] Maxwell,[47] and Bruggeman,[48] which are useful in many cases. The Bruggeman equation, in particular, has proven rather effective in predicting the effective thermal conductivity of a dispersed composite containing a wide range of particle sizes.

Kerrisk[49,50] developed a criterion for the homogeneity of a heterogeneous material under transient conditions, which limits the particle size of a particulate phase to be much smaller than the sample thickness for the sample to

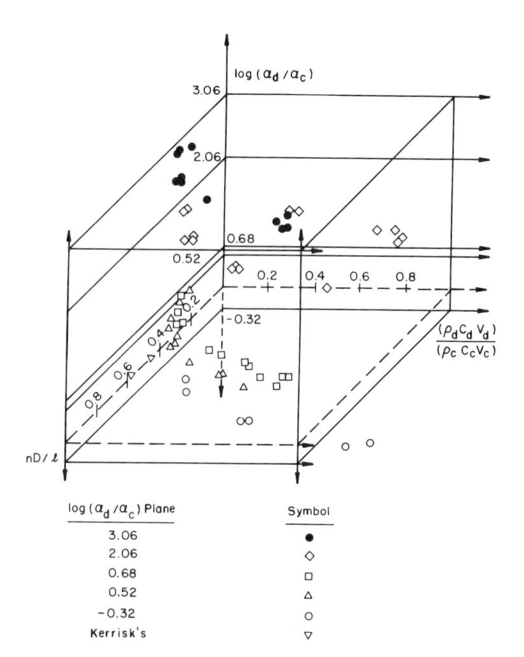

FIGURE 18. Three-dimensional representation of variables studied for homogeneity criteria.

be considered homogeneous, and for the usual relation between the thermal diffusivity and thermal conductivity to hold. H.J. Lee and Taylor(51) and T.Y.R. Lee and Taylor(9) demonstrated that the flash method was applicable to materials with particle sizes more than 50 times greater than Kerrisks' criteria. They also investigated the effects of ratios of various property values in addition to volume ratios. Particle-to-matrix diffusivity ratios from 0.48 to 1137, specific heat ratios of 0.04 to 1.16, and volume ratios of 0 to 0.34 were studied. The results were compared to results calculated using Rayleigh–Maxwell and Bruggeman equations. The largest deviation from the Bruggeman equation was 7%, in two cases, and averaged within 2% for the other 9 cases. A three-dimensional representation of the variables studied is given in Fig. 18. The x axis represents ratios of effective specific heats of the dispersed to continuous phase. As such it goes from 0 through 1 and could proceed to infinity but ratios only up to 1.16 have been investigated. The y axis represents the ratio of effective length through the particles to the sample length since D is the average particle diameter and n is the average number of particles per length l of the sample. As such, values along the y axis are limited to the range 0 to 1 with 0 being a homogeneous sample. Since Kerrisk only considered geometry, his criteria lie on the y axis only. The z axis is the logarithm of the ratio of the diffusivity of the dispersed to the continuous phase. Since a large range of ratios was considered, this axis was made logarithmic. More than 50 specimens were considered and

TABLE 2

Summary of Events (Developments in Experiments, Theory and Error Analysis) in the Pulse Thermal Diffusivity Method

No.	Investigator	Ref. No.	Year	Temperature sensor	Temperature range (K)	Material	Pulse source	Data reduction	Contribution
1	Parker *et al.*	19	1961	T/C	293–408		Flash lamp	Oscilloscope and camera	Original method
2	Cowan	35	1963		2500				Analysis of influence of convection and radiation heat losses
3	Cape and Lehman	21	1963						Theory of finite pulse time and heat loss
4	Taylor and Cape	56	1964						Study of finite pulse time
5	Watt	36	1966						Theoretical study of three-dimensional heat flow effects
6	Larson and Koyama	32	1967						Study of finite pulse time effect and correction for thin sample
7	Henning and Parker	24	1967						Study of intrinsic thermocouple dynamic response
8	Larson and Koyama	13	1968						Extension of pulse method to two-layer samples
9	Acton and Kahn	23	1970	IR	to 2800	Graphite	HeNe laser	Polaroid trace/ image digitizer/ computer	Multispecimen high-temperature facility
10	Walter *et al.*	55	1970	T/C, IR	300–1700		Electron beam	Oscilloscope/camera trace	Application of electron beam pulses (duration adjustable according to specimen)
11	Beedham and Dalrymple	39	1970						Study of influence of flash beam non-uniformity
12	Kerrisk	49, 50	1971, 1972						Study of criteria for application of pulse method to heterogeneous materials, part I and II
13	Donaldson	54	1972						Study of radial heat flow effects in pulse method
14	Schriempf	3	1972		286–573	Mercury	Nd–glass laser		Application of pulse method to liquid metals
15	Taylor, R.E.	1	1973	T/C, IR	80–2500	Metals and nonmetals	Nd–glass laser	DDAS (digital data acquisition system)	Application of DDAS to pulse method (reduction of measurement errors and application of corrections)
16	Ang *et al.*	44	1973		267–294	Mercury			Solution of three-layer problem of pulse diffusivity measurement on encapsulated liquids
17	Chistyakov	45	1973						Study of two-layer specimens with contact resistance between layers

18	Taylor, R.E., and Clark	31	1974						Study of laser pulse shapes and derivation of of analytical expressions for correction over entire temperature rise
19	Donaldson and Taylor, R.E.	38	1975	T/C	295	Armco iron	Nd–glass laser	DDAS	Introduction of radial heat flow variant
20	Clark and Taylor, R.E.	37	1975						Study of radiation heat losses and derivation of new correlation method based on rising portion of temperature–response curve
21	Murfin	43	1975						Analysis of two-layer conduction equation with contact resistance and application to pulse method
22	Lee, H.J., and Taylor, R.E.	51	1976						Study of application of pulse method to heterogeneous materials and effect of ratio of components
23	Heckman	25	1977	T/C					Theoretical study of behavior of intrinsic thermocouples in pulse thermal diffusivity measurement
24	Hanley *et al.*	5	1977	IR	300–1000	Eight representative rocks	Nd–glass laser	DDAS	Application of pulse method to rocks and minerals
25	Lee, T.Y.R., and Taylor, R.E.	9	1978						Study of application of pulse method to dispersed materials
26	Taylor, R.E., *et al.*	8	1979	T/C	224–457	Explosives	Nd–glass laser	DDAS	Application of pulse method to explosive materials (sandwiched three-layer specimens)
27	Chu *et al.*	16	1980	Pyrometer	1700–2000	Graphite	Nd–glass laser	DDAS	Diffusivity measurement at high temperatures with two-dimensional heat flow in the specimen
28	James	17	1980						Novel mathematical treatment for reduction of thermal diffusivity value from initial part of temperature response curve
29	Taylor, R.	4	1980	T/C	300–3000	Metals and nonmetals	Ruby laser	A/D converter and minicomputer	Detailed description of apparatus and procedure for high-temperature diffusivity measurement
30	Koski	53	1981						Review of use of minicomputers in pulse method and introduction of digital noise filters in transient signal filtering

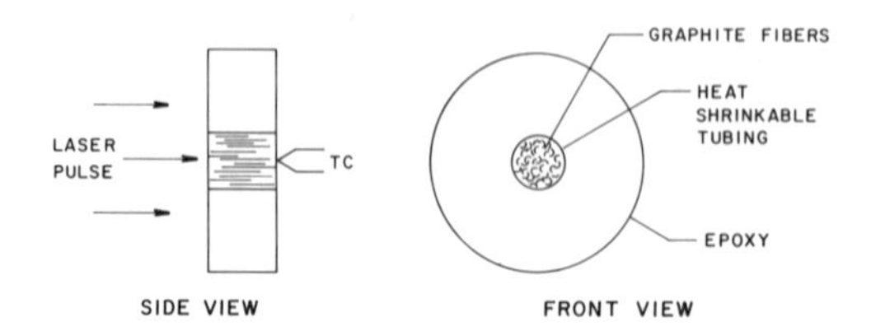

FIGURE 19. Sample arrangement for measuring diffusivity of graphite fibers.

they are indicated in the figure. In all cases when the volume percent of dispersed phase was less than 25%, the experimental value agreed within a maximum of 5% with the effective value computed using the Bruggeman equation. Above 25% dispersed phase the experimental values were generally lower than the value calculated from the Bruggeman equation but with one exception were all within 7% of the calculated value. The accuracy of the Bruggeman equation may be no better than 10% under these conditions since the Bruggeman equation may not accurately reflect dispersed particle–dispersed particle interactions. Therefore the deviations of the experimental diffusivity value from those calculated using the Bruggeman equation may not indicate that the experimental values are becoming inaccurate. This is particularly true since the deviation of the Bruggeman values is in the direction expected from this equation. In any event, it is obvious that the flash diffusivity method is applicable to a very wide range of heterogeneous materials – much wider than that indicated by Kerrisk.

H.J. Lee and Taylor[10] also found that it was possible to measure the diffusivity of fine fibers. They made bundles of fibres and embedded these in an epoxy matrix (Fig. 19). The heat pulse traveled through the fibers so rapidly compared to the epoxy that the temperature response due to the fibers could readily be separated from the response due to the epoxy.

6. SUMMARY

The current state-of-the-art concerning thermal diffusivity measurements using the flash technique is discussed in detail. In the original technique a flash of radiant energy is deposited uniformly over one surface of a homogeneous sample during a negligible time duration and the heat pulse diffuses unidirectionally to the opposite face. The diffusivity is calculated from the sample thickness and the time required for the rear face temperature rise to reach a known percentage of its maximum value. A method of nondimensionalizing the resulting time–temperature history of the rear face has been generated so that the experimental rise curve can be compared to mathematical models and deviations from these models detected.

It is possible to satisfactorily correct for radiation heat losses and for situations in which the time duration of the energy pulse is not negligible com-

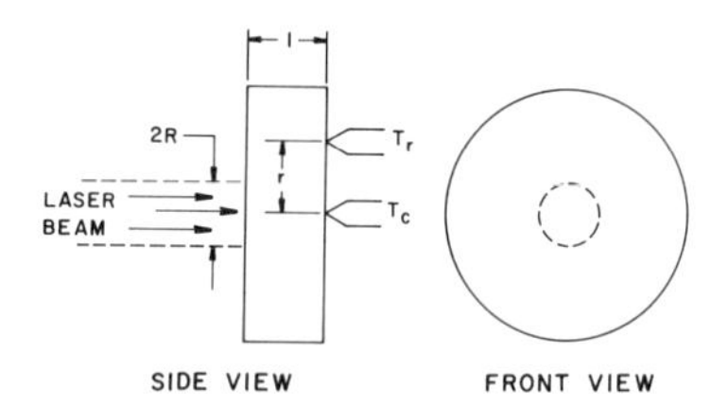

FIGURE 20. Radial heat flow experiment.

pared to the transient time. It is also possible to correct for nonuniform heating (at least in selected cases). It has been shown that very heterogeneous dispersed composites can be measured. Techniques for layered samples have also been developed, and this led to the ability to measure liquids, thick films, and contact conductance between layers. This technique also allows us to keep temperature excursions to less than one degree so that heat-sensitive materials can be measured or measurements can be made very near phase transitions.

The flash method has been extended to two-dimensional heat flow using a technique called the radial flash technique.[6, 38] This radial method involves measuring the rear face temperature rises at two locations (T_c and T_r) on the rear face. The sample is of larger diameter than the usual sample and it is subjected to a laser beam which irradiates only part of the front surface (Fig. 20). An interesting fact about the radial method is that diffusivity values in both the radial and axial directions may be obtained without regard to heat losses.[52] As such, the radial method has been used to verify the validity of the heat loss correction techniques. Of greater significance is the fact that the radial technique can measure the diffusivity of samples much larger than the energy source and thus can be used for nondestructive testing.

The flash method has been used to measure materials whose diffusivities range from 10^{-7} to 10^{-3} m^2 s^{-1} or a range of 10^4. Measurements have been made from at least liquid nitrogen temperatures to temperatures in excess of 2500 K. The method uses small, easy to fabricate samples and results can be obtained within seconds. No other technique has the versatility of this method.

REFERENCES

1. R.E. Taylor, "Critical Evaluation of Flash Method for Measuring Thermal Diffusivity," Report PRF-6764. Available from National Science Technical Information Service, Springfield, Virginia 22151 (1973).
2. R.E. Taylor, "Heat-Pulse Diffusivity Measurements," *High Temp. High Pressures* **11**, 43 (1979).
3. J.T. Schriempf, "A Laser-Flash Technique for Determining Thermal Diffusivity of Liquid Metals at Elevated Temperatures," Report of NRL Progress, 9–17 (1972).
4. R. Taylor, "Construction of Apparatus for Heat Pulse Thermal Diffusivity Measurements from 300–3000 K," *J. Phys. E: Sci. Instrum.* **13**, 1193–99 (1980).

5. E.J. Hanley, D.P. DeWitt, and R.E. Taylor, "The Thermal Transport Properties at Normal and Elevated Temperatures of Eight Representative Rocks," in *Proc. Seventh Symposium on Thermophys. Properties,* ASME, Washington, D.C., pp. 386–391 (1977).
6. K.E. Gilchrist, "Measurement of Thermal Conductivity of Ultrathin Single or Double Layer Samples," in *Proc. European Conf. Thermophys. Properties of Solids at High Temps.,* Baden-Baden, Nov. 1968, Zentralstelle für Atomkernenergie, Dokumentation (ZAED), Karlsruhe, pp. 368–392, (1968).
7. B. Marsicánin, K.D. Maglić, L. Jović, Z. Zivotić, and S. Hajduković, "Measurement of Thermal Diffusivity of Semitransparent Solid Materials Employing Laser Pulse Technique," in *Proc. of IV All-Union Heat and Mass Transfer Conference,* Minsk, Vol. **9**(11), pp. 479–488 (1972) (in Russian).
8. R.E. Taylor, H. Groot, and A.B. Donaldson, "Thermal Diffusivity of HNS High Explosives," in *Thermal Conductivity 16* (D.C. Larsen, ed.), Plenum Press, New York, pp. 251–260 (1983).
9. T.Y.R. Lee and R.E. Taylor, "Thermal Diffusivity of Dispersed Materials," *J. Heat Transfer* **100,** 720–724 (Nov. 1978).
10. H.J. Lee and R.E. Taylor, "Thermophysical Properties of Carbon/Graphite Fiber and MOD-3 Fiber-Reinforced Graphite," *Carbon* **13,** 521–527 (July 1975).
11. K.E. Gilchrist, "Thermal Conductivity of Pyrolitically Deposited Carbon Between 25 and 1100°C After Fast Neutron Irradiation," *High Temp. High Pressures* **4**(5), 497–501 (1972).
12. F.I. Chu, R.E. Taylor, and A.B. Donaldson, "Flash Diffusivity Measurements at High Temperatures by the Axial Flow Method," in *Proc. Seventh Symposium on Thermophysical Properties* (A. Cezairliyan, ed.), ASME, New York, p. 148 (1977).
13. K.B. Larson and K. Koyama, "Measurement by the Flash Method of Thermal Difusivity, Heat Capacity and Thermal Conductivity in Two-Layer Composite Samples," *J. Appl. Phys.* **39,** 4408–4416 (1968).
14. H.J. Lee and R.E. Taylor, "Determination of Thermophysical Properties of Layered Composites by Flash Method," in *Thermal Conductivity 14,* (P. Klemens and T.K. Chu, eds.), Plenum Press, New York, pp. 423–434 (1976).
15. R.E. Taylor and H.J. Lee, "Determination of Thermal Properties of Layer Composites by Flash Method," NTIS Report PB239-114 (1974).
16. F.E. Chu, R.E. Taylor and A.B. Donaldson, "Diffusivity Measurement at High Temperature by the Radial Flash Method," *J. Appl. Phys.* **51,** 336–348 (1980).
17. H. James, "Some Extensions of the Flash Method of Measuring Thermal Diffusivity," *J. Appl. Phys.* **51,** 4666–4672 (1980).
18. R.E. Taylor, T.Y.R. Lee, and A.B. Donaldson, "Thermal Diffusivity of Layered Composites," in *Thermal Conductivity 15* (V.V. Mirkovich, ed.), Plenum Press, New York, pp. 135–148 (1978).
19. W.J. Parker, R.J. Jenkins, C.P. Butter, and G.L. Abbott, "Flash Method of Determining Thermal Diffusivity, Heat Capacity and Thermal Conductivity," *J. Appl. Phys.* **32,** 1679 (1961).
20. H.S. Carslaw and J.C. Jaeger, *Conduction of Heat in Solids,* 2nd Ed., Oxford Univ. Press, Oxford, 258 pp. (1959).
21. J.A. Cape and G.W. Lehman, "Temperature and Finite Pulse-Time Effects in the Flash Method for Measuring Thermal Diffusivity," *J. Appl. Phys.* **34,** 1909 (1963).
22. R.C. Heckman, "Error Analysis of the Flash Thermal Diffusivity Technique," in *Proc. Fourteenth International Thermal Conductivity Conf.* (P.G. Klemens, and T.K. Chu, eds.), Plenum Press, New York, Vol. 14 (1976).

23. R.U. Acton and J.A. Kahn, "Thermal Diffusivity – A Multispecimen Automatic Data Reduction Facility," in *Proc. 10th Thermal Conductivity Conference,* Boston (1970).
24. C.D. Henning and R. Parker, "Transient Response of an Intrinsic Thermocouple," *J. Heat Transfer Trans. ASME* **39,** 146 (1967).
25. R.C. Heckman, "Intrinsic Thermocouple in Thermal Diffusivity Experiment," in *Proc. Seventh Symposium of Thermophysical Properties* (A. Cezairliyan, ed.), ASME, New York, p. 159 (1977).
26. E. Fitzer, "Thermophysical Properties of Solid Materials, Project Section II, Cooperative Measurements on Heat Transport Phenomena of Solid Materials at High Temperatures," AGARD Report No. 606 (1973). Available from NTIS.
27. M. Minges, "Analysis of Thermal and Electrical Energy Transport in POCO AXM-5Q1 Graphite," *Int. J. Heat Mass Transfer* **20,** 1161 (1977).
28. R.E. Taylor and H. Groot, "Thermophysical Properties of POCO Graphite," *High Temp. High Pressures* **12,** 147–160 (1980).
29. K.D. Maglić, M. Perović, and Z. Zivotić, "Thermal Diffusivity Measurements on Standard Reference Materials," *High Temp. High Pressures* **12** (1980).
30. K.D. Maglić, N. Perović, and Z. Zivotić, "Thermal Diffusivity and Electrical Resistivity of Cobalt," in *Proc. 16th Int. Thermal Conductivity Conference* (D. Larsen, ed.), Plenum Press, in press.
31. R.E. Taylor and L.M. Clark, III, "Finite Pulse Time Effects in Flash Diffusivity Method," *High Temp. High Pressures* **6,** 65 (1974).
32. K.B. Larson and K. Koyama, "Correction for Finite-Pulse-Time Effects in Very Thin Samples Using the Flash Method of Measurement Thermal Diffusivity," *J. Appl. Phys.* **38,** 465 (1967).
33. Y.S. Touloukian, R.W. Powell, C.Y. Ho, and M. Nicolaou, "Thermophysical Properties of Matter," *the TPRC Data Series, Vol. 10, Thermal Diffusivity,* IFI/Plenum Press, New York (1973).
34. W.J. Parker, "Flash Method of Measuring Thermal Conductivity," in *Proc. 2nd Conf. on Thermal Conductivity,* Ottawa, Ontario, pp. 33–45 (1962).
35. R.D. Cowan, "Pulse Method of Measuring Thermal Diffusivity at High Temperatures," *J. Appl. Phys.* **34,** 926 (1963).
36. D.A. Watt, "Theory of Thermal Diffusivity by Pulse Technique," *Br. J. Appl. Phys.* **17,** 231 (1966).
37. L.M. Clark, III and R.E. Taylor, "Radiation Loss in the Flash Method for Thermal Diffusivity," *J. Appl. Phys.* **46,** 714 (1975).
38. A.B. Donaldson and R.E. Taylor, "Thermal Diffusivity Measurement by a Radial Heat Flow Method," *J. Appl. Phys.* **46,** 4584–4589 (1975).
39. K. Beedham and I.P. Dalrymple, "The Measurement of Thermal Diffusivity by the Flash Method. An Investigation into Errors Arising from the Boundary Conditions," *Rev. Int. Hautes Temp. Refract.* 178–283 (1970).
40. J.A. Mackay and J.T. Schriempf, "Corrections for Nonuniform Surface Heating Errors in Flash-Method Thermal Diffusivity Measurements," *J. Appl. Phys.* **47,** 1668–1671 (1976).
41. R.E. Taylor, "Critical Evaluation of Flash Method for Measuring Thermal Diffusivity," *Rev. Int. Hautes Temp. Refract.* **12,** 141–145 (1975).
42. R.F. Bulmer and R. Taylor, "Measurement by the Flash Method of Thermal Diffusivity in Two-Layer Composite Samples," in *International Conference on Thermal Technique of Analysis,* The University of Manchester, Institute of Science and Technology, England (1974).
43. D. Murfin, "Development in the Flash Method for the Measurement of Thermal

Diffusivity," *Rev. Int. Hautes Temp. Refract.* 7, 284–289 (1975).
44. C.S. Ang, H.S. Tan, and S.L. Chen, "Three-Layer Thermal Diffusivity Problem Applied to Measurements on Mercury," *J. Appl. Phys.* **44**, 687–691 (1973).
45. V.L. Chisyakov, "Pulse Method of Determining the Thermal Conductivity of Coating," *Teplofiz. Vys. Temp.* **11**(4), 832 (1973) [English Transl. *High Temp.* **11**(4), 744–8 (1973)].
46. L. Rayleigh, "On the Influence of Obstacle Arranged in Rectangular Order upon Properties of a Medium," *Phil. Mag.* **34**, 481–507 (1892).
47. J.C. Maxwell, *A Treatise on Electricity and Magnetism,* 3rd Ed., (1) pp. 435–449, 1904.
48. D.A.G. Bruggeman, "Dielectric Constant and Conductivity of Mixtures of Isotropic Materials," *Ann. Phys.* (Leipzig) **24**, 636–679 (1935).
49. J.F. Kerrisk, "Thermal Diffusivity of Heterogeneous Materials," *J. Appl. Phys.* **42**, 267 (1971).
50. J.F. Kerrisk, "Thermal Diffusivity of Heterogeneous Materials, II. Limits of the Steady-State Approximation," *J. Appl. Phys.* **43**, 112 (1972).
51. H.J. Lee and R.E. Taylor, "Thermal Diffusivity of Dispersed Composites," *J. Appl. Phys.* **47**, 148 (1976).
52. A.B. Donaldson and R.C. Heckman, "The Measurement of Thermal Diffusivity by a Heat Loss Independent Technique," Paper presented at the 12th International Conference on Thermal Conductivity, Birmingham, Alabama, September 12–15 (1972), pp. 41–43 of Abstracts.
53. J.A. Koski, "Improved Data Reduction Methods for Laser Pulse Diffusivity Determination with the Use of Minicomputers," *Proceedings, 8th Symposium on Thermophysical Properties,* Washington, D.C., June 1981, Plenum Press (in press).
54. A.B. Donaldson, "Radial Conduction Effects in the Pulse Method of Measuring Thermal Conductivity," *J. Appl. Phys.* **43**, 4226–4228 (1972).
55. A.J. Walter, R.M. Dell, and P.C. Burgess, "Measurement of Thermal Diffusivities Using a Pulsed Electron Beam," *Rev. Int. Hautes. Temp. Refract.* (1970).
56. R.E. Taylor and J.A. Cape, "Finite Pulse-Time Effects in the Flash Diffusivity Technique," *Appl. Phys. Lett.* **5**(10), 210–221 (1964).

9

Temperature Wave Techniques

L.P. PHYLIPPOV

One of the main trends in developing new measuring techniques in thermophysics is the enhancement of information which can be obtained from a single experiment.[1] There are two main developments to this end: (1) Development and improvement of high-speed methods, and (2) evolution and implementation of multiproperty methods that generate a number of thermophysical properties data in one experiment. Both ways are connected with the use of nonstationary measurement methods. Nowadays two types of nonstationary methods are being successfully developed: (1) impulse and one-step (monotonic) heating methods, and (2) periodic heating methods. The present review deals with the second group. The author will review here methods of periodic heating that are used to measure thermal diffusivity and a set of other properties (temperature wave methods) including new indirect methods where manifestations of temperature waves in processes of various natures are being studied. The methods reviewed come close to periodic heating methods used for specific heat and thermal expansion measurements, which are the subject of a separate review. Some of the pertinent issues will be discussed in other reviews of the present compendium. This, for example, applies to electron bombardment heating in the plane wave method. This fact has influenced the review structure. On the whole, the range of methods and problems discussed is rather wide so that the review had to be condensed. For those interested in details a representative though not exhaustive bibliography is given at the end. While writing this review, the author has drawn on the wide experience in work in this field at Moscow University.

Generally, by temperature waves, periodic processes involving temporal and spatial changes of temperature are meant. These processes are analogous to familiar wave processes: it is possible to introduce concepts of harmonic (sinusoidal) waves, of amplitude and phase of such waves, and of phase velocity.

L.P. PHYLIPPOV • Moscow State University, Moscow, USSR.

At the same time the nature of temperature waves make them radically different from waves in the general meaning of the word. In contrast with electromagnetic and acoustic waves, temperature waves carry no energy, have no wave front separating the region that the wave has not yet penetrated,[(2)] and they are strongly damped (are absorbed) within one wavelength (a plane temperature wave is weakened more than 500 times within one wave length). However, the term "temperature waves" is commonly used and we shall employ it keeping in mind that it is arbitrary.

In thermophysical experiment one usually deals with stabilized (regular) periodic processes. When regular harmonic temperature waves exist, the temperature field may be given as

$$T = \bar{T}(x,y,z) + \theta(x,y,z)e^{i\omega t} \tag{1}$$

Here $\bar{T}(x,y,z)$ is the mean temperature field and θ is the complex amplitude of a temperature wave. The latter can be presented as follows:

$$\theta = \vartheta e^{i\varphi} \tag{2}$$

where $\vartheta = \vartheta(x,y,z)$ is the absolute value of amplitude and $\varphi = \varphi(x,y,z)$ is the wave phase.

In the experiment the temperature wave amplitude should be relatively small:

$$|\vartheta|/\bar{T} \ll 1 \tag{3}$$

As a rule, condition (3) is sufficient for the linearization of a thermal diffusivity problem. For linear approximations the superposition principle is true, and it is possible to get a solution of the problem for periodic processes of any form by summarizing solutions (1) for the harmonic components of the Fourier expansion of a given signal. This makes it possible to confine the general theory of temperature waves to the case of harmonic processes.

From the general concept of temperature waves it may be concluded that the distinctive characteristic of methods discussed consists in their high informativeness. The sources of information on thermophysical properties of a substance are mean temperature fields $\bar{T}(x,y,z)$, amplitudes $\vartheta(x,y,z)$, and phases $\varphi(x,y,z)$ of several harmonic components (or the frequency dependence of ϑ and φ for harmonic signals). The amount of information attainable makes periodic heating methods attractive for experiments of the multiproperty type which have already been mentioned as especially preferable. The same characteristics facilitate the use of methods which permit inner checking of the experiment when, for example, data received from the information on the amplitude of temperature oscillations are cross-checked against data from phase shift measurements in the same experiment.

Another main advantage of temperature wave methods is connected with their periodic character. During every period of such processes a whole cycle of temperature change is repeated. This makes it possible to considerably reduce chance errors and to increase the signal-to-noise ratio.

For relatively fast periodic processes it turns out to be convenient to use electronic instruments that not only permit one to design simple and precise and even automatized methods for temperature wave studies, but also to use well-established procedures for increasing the sensitivity of recording systems (selective amplifiers, synchronous detectors, etc.).[3]

There exist several characteristics that may serve as a basis for the classification of temperature wave experiments.[4] One of these is the type of symmetry of the functions $\theta(x, y, z)$. By this it is possible to distinguish between plane, quasiplane, radial, and spherical temperature waves. Another characteristic feature is the heating source: surface (outer), near-surface, or bulk. However, there exists as yet no generally accepted classification of these experiments.

Historically, the first were methods of measuring temperature waves in rods. The well-known Ångström method belongs to them, and we shall begin our analysis with these methods.

For a relatively small temperature dependence of the thermal conductivity, the theory reduces to the solution of the following equation:

$$\frac{\partial^2 \theta}{\partial z^2} + \frac{1}{r} \frac{\partial}{\partial r} r \frac{\partial \theta}{\partial r} - \frac{i\omega}{a} = 0 \tag{4}$$

where a is the thermal diffusivity. On the mantle surface of a cylindrical sample a boundary condition of the third kind is set

$$-\lambda \frac{\partial \theta}{\partial r} = \alpha \theta, \qquad r = R \tag{5}$$

(here λ is the thermal conductivity), the heat transfer coefficient α is supposed to be constant. In the Ångström method, the first boundary condition consists in setting the amplitude of temperature oscillations at the end surface:

$$\theta = \theta_0, \qquad z = 0 \tag{6}$$

For the solution of such problem see Ref. 5. For extremely small values of the Biot number (undimensional characteristics of heat transfer)

$$\text{Bi} \equiv \alpha R/\lambda \tag{7}$$

(5) is reduced to adiabatic condition. In this case a plane temperature wave is propagated along the rod.

$$\theta = \theta_0 e^{-(i\omega/a)^{1/2} z} \tag{8}$$

For finite values of Bi the surfaces of equal amplitudes and phases are curved, but for small Bi values they become similar to plane ones. It is in this approximation

$$\text{Bi} \ll 1 \tag{9}$$

that the theory of the Ångström method is considered. For these conditions (for metals and semiconductors) the problem is reduced to the following equation[3]:

$$\frac{d^2\theta}{dz^2} - \frac{2\text{Bi}}{R^2}\theta = \frac{i\omega}{a}\theta \tag{10}$$

with the boundary conditions $\theta(0) = \theta_0, z = 0$. The solution is[6]:

$$\theta = \theta_0 \exp\left[-\left(\frac{i\omega}{a} + \frac{2\text{Bi}}{R^2}\right)^{1/2} z\right] \tag{11}$$

From which follows

$$\vartheta = \theta_0 e^{-Kz} \tag{12}$$

$$K = \left(\frac{\omega}{2a}\right)^{1/2} \left\{\left[\left(\frac{2\text{Bi}}{\kappa^2}\right)^2 + 1\right]^{1/2} + \frac{2\text{Bi}}{\kappa^2}\right\}^{1/2} \tag{13}$$

$$\varphi = K'z \tag{14}$$

$$K' = \left(\frac{\omega}{2a}\right)^{1/2} \left\{\left[\left(\frac{2\text{Bi}}{\kappa^2}\right)^2 + 1\right]^{1/2} - \frac{2\text{Bi}}{\kappa^2}\right\}^{1/2} \tag{15}$$

with

$$\kappa^2 = \frac{\omega}{a} R^2 \tag{16}$$

The above solution is true in the case of a "semi-infinite" rod, i.e., in the case where the temperature wave practically does not reach the other end of the rod.

A sufficient condition for this is the inequality

$$L > 2\pi \left(\frac{2a}{\omega}\right)^{1/2} \tag{17}$$

(the sample length L is more than the length of the plane temperature wave).

In order to determine thermal diffusivity on the basis of (11) it is necessary to eliminate Bi. This can be done in three different ways: (1) by using information on amplitude and phase of temperature oscillations at one frequency; (2) by measuring amplitude at two different frequencies; (3) from data on temperature oscillation phases at two frequencies. To these one can add two more possibilities if use is made of the information on the mean temperature distribution along the sample $\bar{T}(z)$ that is determined by Bi; we accordingly get a combination of $\bar{T}$ and amplitude, $\bar{T}$ and phase.

In the "classical" Ångström method[7] use is made of the first of the above-mentioned methods. Thermal diffusivity is determined from (11) according to the formula

$$a = \frac{\omega}{KK'} \tag{18}$$

here

$$K = \frac{1}{l} \ln \frac{\vartheta_1}{\vartheta_2} \tag{19}$$

$$K' = \frac{\varphi_1 - \varphi_2}{l} \tag{20}$$

where ϑ_1, ϑ_2, φ_1, and φ_2 are the amplitudes and phases of temperature oscillation at two points at the distance l from one another. Examples for the use of this method are given in Refs. 8–23.

In the second method the following formula is used:

$$a = \frac{1}{2} \frac{[(\omega_1/K_1)^2 - (\omega_2/K_2)^2]^{1/2}}{(K_1^2 - K_2^2)^{1/2}} \tag{21}$$

It was employed in Ref. 24.

The third method, used in Ref. 25, is based on the formula that is analogous to (21):

$$a = \frac{1}{2} \frac{[(\omega_1/K_1')^2 - (\omega_2/K_2')^2]^{1/2}}{(K_1'^2 - K_2'^2)^{1/2}} \tag{22}$$

The Ångström method variants given above have equivalent accuracy, but the first one seems more convenient because it does not require measurements to be repeated at different frequencies. On the other hand a check on data reproducibility with a varied frequency may serve as a means of testing the quality of the experiment.

Apparatus for the Ångström method differ mainly in the means to produce periodic heating. Most common is the use of appliances that ensure the sinusoidal change of heating power and accordingly, of temperature (e.g., see Ref. 26). In this case the temperature sensor (thermocouple) signal does not need additional processing. An on-off-type of heating is easier to produce, but in this case the first harmonic of the temperature signal must be filtered out.

In the following, examples of recent applications of the Ångström method are given. In Ref. 20 a measuring system with automatized data acquisition and processing is described, the uncertainty of the experimental results being ~ 1.3%. In Ref. 21 is given a description of instruments that ensure measurements of amplitude ratios with an error of 0.5% and of phase differences with 0.2%; the net error (in thermal diffusivity) is ~ 2%.

The Ångström method can be employed for the study of both solid and liquid metals. The role of the metallic cell walls in experiments with liquid metals is discussed in Refs. 27–29.

In addition to the "classical" Ångström method described above modifications exist. In Ref. 30 a variant of the Ångström method is described which is characterized by the use of boundary conditions not of the third but of the fourth kind on the lateral surface of the sample in the case where the metallic specimen is surrounded by a medium with other thermophysical properties (a relatively poor heat conductor). In order to exclude the influence of the medium property, measurements have to be performed at different frequencies. This variant of the technique makes it applicable to investigations at very high (up to 3 Pa) pressures.[30–32]

Another modification of the Ångström method is based on the solution of equation (10) for rods of limited length with the condition of keeping the temperature at one end constant.[33, 34] However, a more radical means of making the sample short is that of employing relatively short oscillation periods and accordingly far smaller wavelengths. It is then possible in (13) and (15) to omit the term 2Bi in comparison to κ^2,

$$2\text{Bi} \ll \kappa^2 \tag{23}$$

The role of Bi in this case can be taken into consideration in the form of a small correction.[3] This means in fact a transition from quasiplane temperature waves in rods to plane ones. As the analysis shows,[3] inequality (23) can be realized for metals up to temperatures of 1000–2000 K on condition that heat exchange goes by means of radiation for periods 5–25 sec which corresponds to sample lengths in the range of 5–25 mm. The use of short samples is in principle more advantageous for it shortens the time necessary to reach regular conditions and lessens the demands on the quality of thermostating. In this case, however, it is necessary to take care that the errors due to the finite thickness of the thermocouples do not lead to a reduction in accuracy. The use of a system in which one electrode is formed by the metallic sample and the other by a 0.1-mm-diam wire welded to it perpendicularly makes it possible to use this method effectively at sample lengths of the order of centimeters.(3)

With short temperature waves, thermal diffusivity may be determined by means of one of two methods that are practically equivalent in accuracy: by measuring the ratio of the amplitudes and by determining the phases of temperature oscillation using the following formulas:

$$a = \frac{\omega l^2}{2 \ln (\vartheta_1/\vartheta_2)^2} \tag{24}$$

$$a = \frac{\omega l^2}{2(\varphi_1 - \varphi_2)^2} \tag{25}$$

Hence the redundant information received in the experiment provides a means for checking its consistency. The application of the short temperature wave technique to investigate the thermal diffusivity of solid and liquid metals is described in Refs. 3 and 29.

Moreover, under the condition (23), there is no necessity to use the solution for a semi-infinite sample since for the variable component of θ, adiabatic conditions may be assumed as a boundary condition on the other end of the sample. The possibility of choosing such a frequency range where the role of thermal leakage becomes insignificant is characteristic also for other types of periodic heating techniques. This is an important merit of the group of methods considered.

For the determination of the thermal diffusivity in a regime of plane temperature waves in samples of limited length two alternative formulas may be used:

$$\frac{\vartheta_1}{\vartheta_2} = \frac{[A_c^2(\kappa z_1/L) + B_c^2(\kappa z_1/L)]^{1/2}}{[A_c^2(\kappa z_2/L) + B_c^2(\kappa z_2/L)]^{1/2}} \tag{26}$$

and

$$\Delta\varphi = \arctan\frac{B_c(\kappa z_1/L)}{A_c(\kappa z_1/L)} - \arctan\frac{B_c(\kappa z_2/L)}{A_c(\kappa z_2/L)} \tag{27}$$

where distances z_1 and z_2 are taken from the unheated end of the sample;

functions A_c and B_c are determined by the following relation:

$$\cosh(\sqrt{i}\kappa) = A_c(\kappa) + iB_c(\kappa) \tag{28}$$

[they have been tabulated in Ref. 2, $\kappa = (\omega/a)^{1/2} L$].

Thermal diffusivity measurements by the phase variant of the method on the basis of (27) become especially convenient if a thermocouple is placed at $z_2 = 0$, i.e., at the free end of the sample, and the phase is measured from the power oscillation phase in small-inertia heating (see below). In this case the thermal diffusivity can be found from the formula

$$-\varphi = \frac{\pi}{4} + \arctan\frac{B_s(\kappa)}{A_s(\kappa)} \equiv \arctan\frac{A_s(\kappa)+B_s(\kappa)}{A_s(\kappa)-B_s(\kappa)} \tag{29}$$

Here $A_s(\kappa)$ and $B_s(\kappa)$ are determined by analogy to (28):

$$\sinh(\sqrt{i}\kappa) \equiv A_s(\kappa) + iB_s(\kappa) \tag{30}$$

Functions $A_s(\kappa)$ and $B_s(\kappa)$ have also been tabulated in Ref. 2.

The inverse of function (29) for $\kappa \geqslant 2$ can be approximated by a formula that is convenient for calculations[35]:

$$\kappa = 1.414\varphi - 1.11 \tag{31}$$

Another approximation of the same relation frequency used is[36]:

$$\kappa^2 = \frac{2\pi}{2.9}\varphi \tag{32}$$

For measurements by this method, readings from one sensor only are sufficient. Naturally, this sensor can consist of a thermocouple with wires welded to the sample near one another perpendicular to the second end. This minimizes the error due to inaccuracy of temperature sensor position. This kind of measurements (as part of a multiproperty study that will be discussed later) was described in Refs. 3 and 37. A similar technique was employed in Ref. 38 for samples in the form of tubes. The above-described experiment comes close to the method of plane temperature waves in plates which has evolved independently from the method of plane temperature waves in rods.[35,39]

In experiments with plates (disks) from fractions of millimeter to several millimeters thick one of the surfaces is subjected to periodic small-inertia heating by electron bombardment[34-36,40-46] or modulated heating by a powerful lamp.[47-49] Temperature oscillations of the opposite surface or, as in Refs. 43, 44, and 50, of both surfaces are registered. For thin plates periodic signal frequencies lie in the range of audiofrequencies, and to register the oscillations electronic devices are being used. The latter make the experiment elegant and suit it for automatization.[49]

A large group of experiments employing these techniques (more than 40 publications) was conducted by V.E. Zinov'ev and his colleagues (e.g., see Refs. 51–56). Paper 57 deals with the variant of the plane temperature wave technique developed for conditions when Bi proves to be relatively large (non-metallic materials at high temperatures); it is shown that the experiment conducted at two frequencies allows one to eliminate the role of Bi.

Due to the simplicity and the sensitivity of the plane temperature wave technique, it now competes quite successfully with the impulse heating method (flash technique), which is close to it in its concepts.[1] It complements well the Angstrom method in the range of temperatures higher than 1000 K where the use of noncontact means of determining temperature pulsations with the help of photosensors is especially convenient. Details on the plane temperature wave technique with modern devices where electron bombardment heating is used are the subject of the review by R. De Coninck and V.E. Peletsky in Chapter 10 of this volume.

All methods discussed above were meant for measuring thermal diffusivity. A possibility of employing similar experiments for thermal conductivity measurements is connected with the use of boundary conditions of the second kind with the possibility of determining the heat flow introduced in a sample during periodic heating (to be more precise, with the alternating component of a heat flow).† There exist three ways of measuring this flow. The first is the use of electron bombardment (for conductors) which allows the determination of the power dissipated on the anode-sample from the current and the potential difference, the second consists in employing low-inertia resistive heaters, and the third is based on the application of methods of calorimetry of radient flow in radiant heating. Each of these methods can be used when inequality (15) ensures that heat leakage is negligible. Otherwise the problem of accounting for heat exchange including the heated surface arises.

The problem of electron bombardment heating in experiments with plane temperature waves is treated separately in Chapter 10, so details are not given here. This type of heating was successfully used in the method of short cylindrical samples in Refs. 3, 37, and 58. Knowledge of the alternating component power introduced in a sample makes it possible to determine a set of thermophysical properties (thermal conductivity, thermal diffusivity, and as a result specific heat) with the aid of a single sensor (a thermocouple situated at the end opposite to the heated one). The phase difference between the first harmonic of a modulated power signal and a signal from the thermocouple allows determination of the thermal diffusivity according to (29); the amplitude of the tem-

†In Ref. 3 a possibility of using boundary conditions of the fourth kind in comparative experiments has also been discussed.

perature change signal gives information on specific heat. For this the following formula is used[3]:

$$\vartheta = \frac{\tilde{q}}{c_p \rho \omega L} F(\kappa) \equiv \vartheta_0 F(\kappa) \qquad (33)$$

Here $\tilde{q}$ is the alternating component of a signal of heating power, and

$$F(\kappa) \equiv \frac{\kappa}{[A_s^2(\kappa) + B_s^2(\kappa)]^{1/2}} \qquad (34)$$

ϑ_0 denotes the amplitude of the temperature oscillations at $\kappa \to 0$. The function $F(\kappa)$ for $\kappa^2 < 2$ differs only slightly from 1, hence the temperature oscillations amplitude ϑ determines the heat capacity of a unit volume $c_p \rho$. Measurements of thermal capacity in these conditions practically do not depend on thermal diffusivity measurements. For larger κ, where it is more convenient to determine thermal diffusivity because of the larger φ, calculation of $F(\kappa)$ does not present any difficulties.

In some experiments,[3] the electron source consisted of a plane spiral cathode placed parallel to a sample face. Between the cathode and the sample a direct voltage of several hundred volts was applied. It was periodically switched on and off so that a Π-type heating resulted. This technique was used to investigate the properties of metals and semiconductors in solid and liquid states at temperature range from 500 to 1400 K. In experiments with liquids a flat bottom cell-crucible was heated by electron bombardment. In order to prevent convective intermixing from the top downwards, a constant temperature gradient was created by a separate heater.[29]

The multiproperty method of measuring described above is in fact equivalent to the multiproperty technique in flat plates from the conceptual viewpoint. In the latter case the geometry is changed, the frequency range moves toward acoustic frequencies, and different instruments are used. The feasibility of measuring thermal conductivity together with thermal diffusivity in such experiments was shown in Refs. 3 and 4. One of the first experiments with the multiproperty technique is described in Ref. 59. Lately this method was further developed and implemented by modern apparatus in Refs. 60 and 61. Details can be found in the review by R. De Coninck and V.E. Peletsky, Chapter 10 of this volume.

In Ref. 62, heating by modulated radiation of a powerful lamp is used; in Ref. 63 a laser is applied. To stabilize the emissivity of the samples, their surface was blackened; for determining the heating power, a calibration experiment was conducted with a sample of known specific heat.

On the whole, a multiproperty technique for metallic plates appears to be one of the best modern methods. It inherently combines high sensitivity and precision, quick action and simplicity of measurement, and also the possibility of application to very high temperatures.

Now we review multiproperty measurement methods that use low-inertia resistive heaters. These methods are generally meant for nonmetallic materials. In the method developed in Refs. 64 and 65 a flat heater made of thin wire is tightly pressed between two identical samples of the material investigated. The temperature is registered by different thermocouples. A special feature of this method consists in the possibility of making measurements even before steady state has been reached and when the mean temperature has not had time to stabilize (massive samples are heated slowly): for this a differential compensating scheme is used. A multiproperty technique of a similar kind but for disks of finite thickness with temperature oscillations being registered on the outer surfaces of the samples is described in Ref. 66 and employed in Ref. 67. In Ref. 68 a comparative variant of the method is used: the heater is placed between discs of a standard material and of the one studied.

An important aspect of investigation of thermophysical properties of nonmetals at relatively high temperatures is the role of radiant (photon) heat transfer. For materials that are semitransparent in the infrared region, diffusion transfer of radiation by emission and absorption of photons within the material is important. Generally radiation transfer of heat presents difficulties for analytic description, since heat flow through a surface element at a given point is determined by the temperature field in the whole system and does not follow Fourier's law. The joint description of radiant–conductive thermal transfer constitutes one of the most complicated problems of modern thermophysics.[29, 69] The application of the temperature wave techniques is one of the most effective methods in the investigation of this problem. Variation of modulation frequency changes the relation between the length of the temperature waves and the length of the photon path and so changes the contribution of radiative transfer, allowing to distinguish the contribution of the latter. First experiments of this kind were based on crude solutions of the problem and had a rather qualitative character.[70, 71] A detailed theory of the problem as applied to particular experiments with temperature waves is given in Refs. 72 and 73. Reference 74 is characterized by an empirical approach to the problem, where the frequency dependence of the thermal diffusivity is regarded as a measure of radiative transfer. There are ample reasons to believe that in the study of radiative–convective transfer, the temperature wave techniques will play an important role.

There is one more type of experiment with plane temperature waves in nonconductive substances; it is the study of the pulsations of a foil strip that is heated by alternating current in the substance studied. This is an example of methods where a probe serves simultaneously to create a thermal disturbance and to register the response of the substance to this disturbance. The theory of the plane probe method[29, 75] concludes that amplitude and phase of temperature oscillations are determined by $b = \lambda/\sqrt{a}$. This property is called "thermal activity" in the Soviet literature. It really defines the activity of a material in transient processes and serves as a parameter that is of immediate importance for engineering and technology. The knowledge of b is also a convenient means of determining the thermal conductivity λ, provided the thermal diffusivity a is given, and as additional check of data on λ and a in multiproperty investigations.

The amplitude of temperature oscillations of a foil heated by alternating current with angular frequency ω is determined by the formula

$$\vartheta = \frac{\tilde{q}}{2s\sqrt{\omega}\,(2b^2 + 2bd + d^2)^{1/2}} \tag{35}$$

where

$$d = \sqrt{\omega} c_p \rho h \tag{36}$$

s and h are the area and thickness of the foil, respectively, c_p and ρ are its specific heat and density. The sensitivity of ϑ to b according to (35) is the greater the smaller the ratio d/b is, i.e., the weaker the influence of the thermal inertia of the probe is. (When $d \gg b$, ϑ is determined mainly by the specific heat of the probe. Such conditions may be used for measuring c_p[3, 29]).

To determine the temperature amplitude in a plane probe, Refs. 29 and 75 give a system which is based on concepts of the theory of nonlinear electric circuits. A cell with a probe constitutes one of the arms of a bridge circuit fed by alternating current from an audiofrequency generator. The bridge is balanced by an applied voltage of frequency ω. The probe temperature pulses at twice the frequency, 2ω, which results in its resistance changing twice during each cycle. The change of resistance with time makes the system nonlinear, which leads to the appearance of combined frequencies, particularly the frequency $2\omega + \omega$, three times the frequency of the applied voltage. The signal of the treble frequency can easily be separated from the bridge diagonal and then measured. The size of the signal of treble frequency is directly related to the amplitude of the temperature oscillations. Measurements of thermal activity of liquids by means of such a device are described in Refs. 3 and 29. In Refs. 76 and 77 another measuring system is used – one in which the probe is heated by the sum of direct and alternating current, which allows a convenient compensating scheme to be used.

A typical feature of experiments with liquids is the fact that the depth of penetration of temperature waves into the medium is very small (e.g., for water at a frequency of 50 Hz the wavelength is ~0.1 mm). Thus convective flow may be neglected, the probe may be placed arbitrarily in relation to the cell walls, and the temperature of the probe itself may be used as a reference temperature.

In Ref. 78 the role of radiative heat transfer under conditions corresponding to the experiments discussed is being considered. It is shown that values of λ pertain to purely conductive thermal transfer and are practically not distorted by radiative flow. This conclusion is of great importance, since the problem of radiative contribution to the thermal conductivity of nonmetallic liquids at temperatures higher than ambient is one of the most important both practically and in principle and is at the same time one of the least studied.[29, 79].

An application of plane probe method to the study of thermophysical properties of gases is hindered by the fact that the ratio d/b becomes too large, thus reducing the sensitivity of the scheme. This difficulty, however, can be avoided by the use of a differential two-probe circuit. Such a scheme as applied to the determination of the thermal activity of gases at high temperatures (when the probe is greatly overheated in respect to the walls) has been described in Ref. 80. The peculiarity of the scheme is its heating by the sum of direct and alternating currents which is convenient for a differential system, and the existence of an additional voltage of high (tens of kilocycles) frequency, that can balance the probe temperature without changing the alternating component at low frequency. One of the probes in this system is placed in a comparison cell with a gas of known properties (or with the same gas at a fixed pressure). The sensitivity of the differential scheme is approximately a hundred times higher than that of a common one. Measurements can be conducted at temperatures of thousands of degrees at pressures down to a few torrs.

An interesting and important application of thermal activity measurements of gases is the determination of pressure. When gases are near to the ideal state, values of b are directly proportional to the square root of pressure, so that the pressure change can easily be detected by the thermal activity change. The measurements can easily be made absolute; they cover without difficulties a wide range of pressures which usually requires the use of a number of different nanometers.

All the techniques described above are based on the outer heating of a studied material (or a medium). A method is now described where the inner heating of metallic samples is being used. It is a nonstationary variant of the Kohlrausch method. The stationary Kohlrausch method is one of the most widely used for measuring the thermal conductivity of metals at moderately high (up to $\sim 1000\,K$) temperatures.[3, 81] It is based on the measurement of the temperature at the middle of a sample in the form of a rod, which is heated by direct passage of current fed by massive clamps. Modulating the heating current and the measurement of the resultant alternating component of the temperature in the middle of the sample constitute the essence of the non-stationary variant of the said method. The theory of the method is given in Ref. 3, where it is shown that the phase of the sample temperature oscillation is determined by its thermal diffusivity:

$$\tan \varphi = \frac{B_c^2(\kappa) - A_c(\kappa)[1 - A_c(\kappa)]}{B_c(\kappa)} \tag{37}$$

and its amplitude depends on both specific heat and thermal diffusivity:

$$\vartheta = \frac{\tilde{q}}{c_p \rho \omega} \left\{ \frac{[1 - A_c(\kappa)]^2 + B_c^2(\kappa)}{A_c^2(\kappa) + B_c^2(\kappa)} \right\}^{1/2} \tag{38}$$

where $\tilde{q}$ is the power divided by the volume. At small κ it follows from (38) that ϑ is determined by the thermal conductivity of the sample alone. Thus the information received in the nonstationary experiment complements the data obtained in an ordinary experiment with the Kohlrausch method and turns the measurements into multiproperty ones. This kind of technique was used in Ref. 82 with thin rods. The error in thermal diffusivity was 5%; reproducibility of specific heat measurements was 1.3%.

Methods that are next viewed are based on the application of cylindrical (radial) temperature waves, i.e., of periodic processes which are characterized by temperatures changing along the radius of cylindrical samples. The thermal conductivity equation for the complex amplitude of cylindrical temperature waves in the absence of internal heating has the form

$$\frac{1}{r} \frac{d}{dr} r \frac{d\theta}{dr} = \frac{i\omega}{a} \theta \tag{39}$$

In experiments aimed at thermal diffusivity investigation only, a boundary condition of the first kind is used and the amplitude of temperature oscillations on the surface is to be determined. For uniform samples this condition will have the form

$$\theta = \theta_0 \qquad \text{for} \quad r = R \tag{40}$$

The solution of the corresponding problem gives the following formulas for the amplitude and the phase of temperature oscillations at the axis of a cylinder[3]:

$$\frac{\vartheta}{\theta_0} = [\text{bei}^2(\kappa) + \text{ber}^2(\kappa)]^{1/2} \tag{41}$$

$$\tan \varphi = \frac{\text{bei}(\kappa)}{\text{ber}(\kappa)} \tag{42}$$

where $\kappa = (\omega/a)^{1/2} R$, and $\text{bei}(\kappa)$ and $\text{ber}(\kappa)$ are Thomson (Kelvin) functions (tabulated in Ref. 90). The plots of functions (41) and (42) can be found in Refs. 3 and 29.

The analysis of these solutions makes it possible to ascertain that the thermal diffusivity can be determined both by measuring the ratio or temperature oscillation amplitudes $\vartheta(0)/\theta_0$ and by phase difference measurements. The sensitivity of both methods is approximately the same. For optimum values of $\kappa \approx 1.5-2$, the error of amplitudes of 1% and of phases of 1° results in an inaccuracy for a of 1%.

First experiments with cylindrical temperature waves were reported in Ref. 83 (the amplitude variant, molten glasses) and in Ref. 84 (the phase variant, polymers). Further development of this method was made in the study of solid and liquid metals (see also Ref. 37), where the simultaneous application of both amplitude and phase variants served as a means of inner checking of the experiment. The variant of the radial temperature wave method for hollow cylinders with the periodic heating of the outer surface was discussed in Refs. 3 and 29 and applied in Ref. 87.

What are the advantages of this technique as compared to the Angstrom method? In the cylindrical wave method there is no need of corrections for the heat exchange from the cylinder surface and end face corrections become small at radius/length ratios $<\frac{1}{3}$. The corrections for the cylindrical cell walls in investigations of melts become significantly simpler because the solid–liquid boundary is isothermal. Lastly, this arrangement has special advantages when it is used for multiproperty measurements which we are not going to consider.

For several variants of multiproperty type techniques that are based on electron bombardment heating (they are thoroughly described and analyzed in Refs. 3, 29, 88, 89) the most convenient one will be reviewed. This method employs a hollow cylindrical sample or a cell for liquid metals made of two coaxial thin-walled metal tubes which is placed in a vacuum chamber. Inside the sample, along its axis, a tungsten wire-cathode is stretched, the wire being heated by a current passing through it. Between the cathode and the anode-sample there exists a periodically changing voltage of several hundred volts. Temperature waves created by means of the electron bombardment of the inner surface are registered on the outer surface by means of thermocouples or in the case of a high-temperature device by means of a photomultiplier. The thermal diffusivity is determined from the phase difference between temperature oscillations by the formula

$$\varphi = \arctan \frac{\phi_2}{\phi_1}$$

$$\begin{aligned} \phi_2 \equiv\ & \mathrm{ber}'(\kappa_1)\,\mathrm{her}'(\kappa_2) + \mathrm{bei}'(\kappa_2)\,\mathrm{hei}'(\kappa_1) \\ & - \mathrm{ber}'(\kappa_2)\,\mathrm{her}'(\kappa_1) - \mathrm{bei}'(\kappa_1)\,\mathrm{hei}'(\kappa_2) \end{aligned} \qquad (43)$$

$$\phi_1 \equiv \mathrm{bei}'(\kappa_2)\,\mathrm{her}'(\kappa_1) + \mathrm{hei}'(\kappa_1)\,\mathrm{ber}'(\kappa_2) - \mathrm{bei}'(\kappa_1)\,\mathrm{her}'(\kappa_2) - \mathrm{hei}'(\kappa_2)\,\mathrm{ber}'(\kappa_1)$$

$$\kappa_1 \equiv (\omega/a)^{1/2} R_1, \qquad \kappa_2 \equiv (\omega/a)^{1/2} R_2$$

Here $\mathrm{bei}'(\kappa)$, $\mathrm{ber}'(\kappa)$, $\mathrm{hei}'(\kappa)$, and $\mathrm{her}'(\kappa)$ are the derivatives of the Thomson functions which are tabulated in Ref. 90. The specific heat may be determined in the same experiment from the formula

$$\vartheta = \vartheta_0 F, \qquad F \equiv \frac{R_2^2 - R_1^2}{\pi R_1 R_2} \cdot \frac{1}{(\phi_1^2 + \phi_2^2)^{1/2}} \tag{44}$$

where $\vartheta_0 \equiv \tilde{q}/Mc_p\omega$ and M is the sample mass.

Function F for $\kappa^2 < 2$ differs only little from 1 and specific heat measurements in such conditions are almost independent of thermal diffusivity.

The technique described and some of its modifications are widely used to investigate thermophysical properties of liquid metals in the temperature range from 1000 to 2000 K. The error in specific heat and thermal diffusivity amounts to about 5%.[3, 29, 91–99]

In Ref. 100 a similar variant of the multiproperty technique is described. It is based on heating by means of a low-inertia electric heater. The technique is used for the study of properties of nonmetallic liquid. In Ref. 101 the authors tested a technique that combines the determination of thermal diffusivity from the information on temperature oscillation at two points at different distances from a cylindrical heater with the determination of thermal conductivity from the values of steady components of power and temperature in the same experiment. The measurements were conducted with Teflon at pressures up to 3000 MPa. This is an interesting attempt of using information from the dc component of the measured signals.

Another type of multiproperty experiments with radial temperature waves is based on the use of modulated heating of samples in an inductive radiofrequency furnace. The theory of this method that takes into account the fact that heat is generated in the bulk near to the surface of the specimen is described at length in Ref. 3 and partly in Refs. 102 and 103. For the temperature oscillation phase shift on the sample surface relative to the power oscillation, the following formula holds:

$$\varphi = -\arctan \frac{F_2}{F_1} \tag{45}$$

$$F_1 \equiv F_1^0 + \eta + \eta^2(1 + \kappa^2 F_2^0) + \eta^3(\tfrac{5}{4} + \kappa^2 F_2^0) + \eta^4(\tfrac{3}{2} + \tfrac{5}{2}\kappa^2 F_2^0 - \kappa^4 F_1^0) + \cdots \tag{46}$$

$$F_2 \equiv F_2^0 - \eta^2\kappa^2 F_1^0 - \eta^3(1 + F_1^0)\kappa^2 - \eta^4\kappa^2 \times (\tfrac{3}{2}F_1^0 + \kappa^2 F_2^0 + 2) + \cdots \tag{47}$$

where

$$\eta \equiv \sigma/2R \tag{48}$$

and σ is the skin thickness [$\sigma = (\gamma/\pi\mu f)^{1/2}$, γ is the resistivity, μ the magnetic permeability, f the frequency of the induction furnace]. Functions $F_1^0(\kappa)$ and $F_2^0(\kappa)$ may be expressed through Thomson functions and their derivatives:

$$F_1^0(\kappa) \equiv \frac{\mathrm{ber}(\kappa)\,\mathrm{ber}'(\kappa) + \mathrm{bei}(\kappa)\,\mathrm{bei}'(\kappa)}{\mathrm{bei}'^2(\kappa) + \mathrm{ber}'^2(\kappa)} \cdot \frac{1}{\kappa} \tag{49}$$

$$F_2^0(\kappa) \equiv \frac{\mathrm{bei}(\kappa)\,\mathrm{ber}'(\kappa) - \mathrm{ber}(\kappa)\,\mathrm{bei}'(\kappa)}{\mathrm{bei}'^2(\kappa) + \mathrm{ber}'^2(\kappa)} \cdot \frac{1}{\kappa} \tag{50}$$

Formulas (46) and (47) are the result of expanding functions $F_1(\kappa, \eta)$ and $F_2(\kappa, \eta)$ into a series according to relative skin thickness. Evaluations show that for $\eta < 0.1$ the terms with η^4 do not exceed 1% of the total.

The relation $\varphi = \varphi(\kappa, \eta)$ is sensitive enough to the value of κ and can be used to determine the thermal diffusivity. The sensitivity of this technique is greater for smaller η, i.e., the nearer to the surface the heating occurs (hence it is advantageous to use generators with frequency of $\sim$1 MHz and higher). For the amplitude of temperature oscillations on the cylindrical sample surface the theory gives

$$\vartheta = \vartheta_0 2\kappa^2 (F_1^2 + F_2^2)^{1/2} \tag{51}$$

where ϑ_0 is the temperature oscillation amplitude at 0

$$\vartheta_0 = \frac{mQ}{Mc_p\omega} \tag{52}$$

Q is the power of inductive heating, m is the modulation coefficient. It follows from (51) that the thermal diffusivity can also be determined from the information on amplitudes by means of measuring temperature oscillation amplitudes at two modulation frequencies.

Both the phase and the amplitude variants were elaborated and used to measure the thermal diffusivity of metals in the temperature range from 1000 to 2400 K,[93, 104–106] For recording the temperature oscillations, a noncontact device was used (photomultipliers). The use of this as a multiproperty method came to pass by developing means of determining the inductive heating power Q in the same experiment. This was done by a device which consisted of several coils of wire encircling the sample near to its surface. From the emf $\mathscr{E}$ induced in this solenoid it is possible to determine the electric field near to the sample surface and thus to compute Q from the formula[103]

$$Q = \frac{2\mathscr{E}^2 \gamma \eta^3}{\pi} \cdot \frac{1 - \eta - \eta^2/4}{(R_s/R)^2 - 1 + 2\eta(1 + \eta^2/4) + \eta^2 4(1 - \eta - \eta^2/4)^2} \tag{53}$$

where R_s is the solenoid radius.

The multiproperty technique of measurements on the basis of the method of modulated heating was elaborated in detail and applied to a great number of investigations of properties of metals.[93, 107–115] The error in thermal diffusivity is 2%, and in thermal conductivity 5%. The practiced application of this method allows it to be considered as one of the most successful high-temperature multiproperty techniques for massive samples of metals such as Ti, Zr, Hf the study of which in the form of thin plates may be complicated by the occlusion of gases.

The last type of experiments with radial temperature waves reviewed here belongs to the group of probe methods, the method of wire probe. The principle of the method consists in the following: a thin wire probe placed into the substance to be investigated is heated by an audiofrequency current, the amplitude and the phase of the temperature oscillations giving information on the thermophysical properties of the substance. The theory of the method is given and discussed in Ref. 29. Expressions for phase and amplitude of temperature oscillations are as follows:

$$\tan \varphi = \frac{\kappa\eta\,[\mathrm{her}^2(\kappa) + \mathrm{hei}^2(\kappa)] - [\mathrm{her}(\kappa)\,\mathrm{hei}'(\kappa) + \mathrm{hei}(\kappa)\,\mathrm{her}'(\kappa)]}{\mathrm{her}(\kappa)\,\mathrm{her}'(\kappa) + \mathrm{hei}(\kappa)\,\mathrm{hei}'(\kappa)} \tag{54}$$

$$\vartheta = \vartheta_0 \left\{ \frac{\mathrm{her}^2(\kappa) + \mathrm{hei}^2(\kappa)}{[\mathrm{hei}(\kappa) + (1/\kappa\eta)\,\mathrm{her}'(\kappa)]^2 + [\mathrm{her}(\kappa) - (1/\kappa\eta)\,\mathrm{hei}'(\kappa)]^2} \right\}^{1/2} \tag{55}$$

where ϑ_0 is the amplitude of the probe temperature oscillations in vacuum, $\kappa \equiv (\omega/a)^{1/2} R$,

$$\eta \equiv \frac{(c_p\rho)_s}{2c_p\rho} \tag{56}$$

subscript s refers to the probe. In the given expression ω is the angular frequency of temperature change; when the heating is done by means of alternating current it is equal to twice the frequency of the current.

Formulas (54) and (55) form a system of two equations for two unknown parameters a and $c_p\rho$. Therefore the information on amplitudes and phases of temperature oscillations is sufficient for the determination of the two thermophysical properties. Alongside with such amplitude-phase variants of the technique there exist others where the same set is determined only on the basis of information on temperature oscillation phases measured at two frequencies (phase–frequency variant) or of amplitudes determined at two frequencies (amplitude–frequency variant).

The wire probe method was used by American authors[116–118] to measure the thermal conductivity of gases at high temperatures (the specific heat was considered as already known). The measuring scheme was the same as in the experiments with foil probes described above. Work along these lines was continued and further developed in Refs. 119 and 120. In Ref. 121 the same technique was applied to the measurement of the thermal conductivity of liquids.

The wire probe method as a multiproperty technique was elaborated in Refs. 122–124. The measuring system used consisted of an ac (60-Hz) bridge with three branches. The probe resides in the arm of the first branch, the second branch is used to balance the system at the frequency of the feeding current, in the third branch another similar probe is set in vacuum to get a standard signal of treble frequency that appears as a result of electric nonlinearity of the probes. Such a system ensures precise measurements of signal amplitude and phase. A comparison of the three variants mentioned above (phase-frequency, phase-amplitude, and amplitude-frequency) in Refs. 122–124 has led to the conclusion that they give results which are close to each other in sensitivity, but the first one is more convenient as it does not require the system to be readjusted when frequencies are changed. The method is best used as a relative technique, making an experiment with a liquid with known properties at one temperature only. The error in thermal conductivity measurements is about 1%–1.5%, that in specific heat being $\sim 2\%$.

The experimental technique described was successfully used to measure thermophysical properties of liquids and gases in the temperature range from 90 to 600 K at pressures of 30 MPa (e.g., see Refs. 123–127). Other variants of this method have been described and are being further developed. In Ref. 128 the heating is done by the sum of alternating and direct currents and the complex impedance of the bridge at the feeding voltage frequency is measured (this eventually provides the information on the amplitude and phase of temperature oscillations). On the whole there is every reason to believe that the cylindrical wave technique has a promising future. It is convenient for a wide range of states of matter, it is rapid and impervious to noise, allows automatized multiproperty measurements to be made, and yields thermal conductivity values which are free of the radiant heat transfer component.

Indirect methods are now reviewed:

One of the possible causes for the appearance of the temperature waves in a medium is the existence of periodic deformation processes. Such processes take place during bending oscillations of thin bars (reeds). The opposite sides of an oscillating reed are alternately cooled and heated in correspondence with the mechanical oscillations. Across the bar a periodically changing heat flow is propagating which leads to the dissipation of oscillation energy. This dissipation results in the so-called internal friction. The theory of internal friction of thermal origin was developed by Zener.[129] According to Zener, the amount of internal friction Q^{-1} is related to the frequency ν by the law:

$$Q^{-1} = Q_m^{-1} \frac{2\nu\nu_m}{\nu^2 + \nu_m^2} \tag{57}$$

and has a maximum at the frequency ν_m, which is connected with the thermal diffusivity in a simple way

$$\nu_m = \frac{\pi}{2h^2} a \tag{58}$$

where h is the thickness of the bar. The maximum value of internal friction Q_m^{-1} depends on the specific heat difference $c_p - c_v$:

$$Q_m^{-1} = \frac{1 - 2\sigma}{6} \cdot \frac{c_p - c_v}{c_v} \tag{59}$$

where σ is the Poisson coefficient. The same formula may have the form

$$Q_m^{-1} = \frac{1}{2} \frac{TE\alpha}{c_v} \tag{60}$$

where E is the Young modulus determined from the resonance frequency of the reed oscillations. The measurement of the ratio α^2/c_v offers particular advantage for multiproperty investigations.

Zener's theory was tested in Refs. 130–132, and in Refs. 133 and 134 it was used to measure a and $(c_p - c_v)/c_v$ of several metals at room and at lower temperatures. The possibility of using this technique at high temperatures up to 1400 K was studied in Ref. 135. In the apparatus described, the oscillations of a sample reed were generated electrostatically by the application of an alternating voltage to the sample and a small flat electrode near to it. The mechanical oscillations were recorded by a capacitor formed by the sample and another electrode. The internal friction can be determined in two ways: from the width of resonance peak and from the decrement of damping of free oscillations. The study of this technique showed that it ensures measurements of the thermal diffusivity and the ratio α^2/ω with an error of about 5%. With metals containing soluble inclusions like oxygen, nitrogen, carbon in the temperature dependence of Q^{-1} a relaxation due to solute diffusion appears in addition to thermal relaxation. In those temperature ranges where thermal relaxation is negligible it becomes possible to study the diffusion characteristics, such as activation energy of diffusion. There are reasons to believe that the acoustic technique may become a useful complement to direct thermophysical experiments.

In contrast with the acoustic techniques, nondirect optical methods of thermophysical property measurements deal with temperature waves that are inherent to the system rather than created externally. By this is meant a wave system that describes the "life" of fluctuations in liquids and dense gases. It is possible to distinguish two types of fluctuations: The first one consists of hyperacoustic waves similar to Debye waves in solids. The second type of movements are entropy (temperature) fluctuations that obey the thermal conductivity equations and thus are absorbed over distances of the order of a wavelength. Both fluctuations create density changes and produce inelastic light scattering, i.e., a modulation of scattered light waves. Hyperacoustic waves determine the fine structure of the Rayleigh diffraction line, Mandelstamm–Brillouin triplet. The study of this triplet provides information which is of great interest for thermophysics. For nonrelaxing liquids it is possible to derive data on adiabatic and isothermal compressibility and specific heat ratio c_p/c_v. Even more interesting is the information obtained in investigating the shape of the central component of the triplet that reflects temperature fluctuations as mentioned above. The width of the central component is proportional to the thermal diffusivity of fluids. The study of this effect became possible only recently, with the advent of so-called "optical mixing spectroscopy."[136,137] This is an electronic technique for detecting small frequency changes of optical signals. It ensures an optical resolution which is by several orders of magnitude higher than that of the best optical spectrometers. In such an apparatus a laser beam is diffracted by the medium to be investigated and recorded by a photomultiplier which also receives a nondiffracted laser radiation. The photomultiplier has the same function as the mixer in a superheterodine receiver. At its output appears a beating signal caused by the superposition of diffracted and nondiffracted light, i.e., a modulation which corresponds to the thermal fluctuations in the medium. The spectral analysis of these frequencies (of the time correlation function when a correlator is used at the output instead of a spectral instrument) allows one to determine the thermal diffusivity of a fluid. The error in these measurements amounts to several percent.[136] The application of this method is especially advantageous in investigations near the critical points. A direct study of thermal diffusivity and conductivity in this range is difficult,

TABLE 1

Summary of Temperature Wave Experiments (Examples of Use of the Main Modifications of the Method)

No.	Investigator	Ref.	Year	Material	Method	Temperature measurement[a, b]	Range (K)	a, λ, c_p $\left(\frac{m^2}{S}\right), \left(\frac{W}{m \cdot K}\right), \left(\frac{\tau}{g \cdot K}\right)$	Accuracy (%) Max. error	Accuracy (%) Scatter	Comments
1	Angstrom	7	1861	Cu, Fe, sand, clay	TW along rods (θ, φ variant)	Mercury thermometers	330	a 4×10^{-7}–10^{-4}	n.d.	~5	
2	Sidles and Danielson	9	1954	Cu, Ni, Th		TC	300–800	a 0.9×10^{-5}–1.2×10^{-4}	~5	n.d.	Use of X–Y recorder
3	Shanks *et al.*; Martin *et al.*	14 15	1967	Fe Pt		TC	300–1300	a $(0.4–2) \times 10^{-5}$	2–3	1–1.7	
4	Glanz	22	1968	MO_2–SiO_2		TC	400–1100	a $(0.7–1.3) \times 10^{-6}$	9	~5	
5	Leden *et al.*	20	1976	Cu		TM	300	a 10^{-4}	1	0.05	Automatized system
6	Savvides and Murray	21	1978	Si, Ge		TC	300–700	a $(1–3) \times 10^{-5}$	2	0.5	New apparatus
7	Rudnev *et al.*	28	1961	Liquid Na, Li		TC	620–1280	a $(2–7) \times 10^{-5}$	10	8	See also Ref. 27
8	King	25	1915	Cu, Sn	TW along rods (θ_{ω_1}, θ_{ω_2} variant)	TC	310–330	a 10^{-5}–10^{-4}	n.d.	~1	
9	Starr	24	1937	Ni	TW along rods (φ_{ω_1}, φ_{ω_2} variant)	TC	~300	a 1.5×10^{-5}	n.d.	~0.1	
10	Sundqvist and Bäckström	30, 31, 32	1977, 1978	Cu, Al, Ag, Au	TW along rods under high pressures (θ_{ω_1}, φ_{ω_1}, θ_{ω_2}, φ_{ω_2} variant)	TC	~310 P to 2.56 Pa	a $(1–2) \times 10^{-4}$	1–3	<1	
11	Tomokiyo and Okada	34	1968	Ge	TW in short rods with thermostatted second end	TC	300	a 3×10^{-5}	5	n.d.	
12	Phylippov	3, 29	1967	Solid Fe, liquid Sn	TW of small periods in short rods (θ variant, φ variant)	TP	400–900	a $(0.2–1.5) \times 10^{-5}$	4–9	~5	
13	El-Hifni and Chao	38	1956	Steels, Al	TW along tubes	TC	420–840	a $(0.4–2) \times 10^{-5}$	n.d.	n.d.	
14	Kraev and Stel'makh	40	1963, 1964	W, Ta, Mo, Nb	TW in plates (electron bombardment)	Photocell	1600–3000	a $(2–3) \times 10^{-5}$	5	n.d.	
15	Wheeler	50	1965	Pt, Ta, Mo, W, UO_2, UC, graphite, UO, boron, nitride, cellulose carbon		Photocell	1200–3000	$a < 4 \times 10^{-5}$	n.d.	5	
16	Brandt and Neuer	49	1978	Metals, insulators	TW in plates (xenon lamplight heating)	PbSe-detector, photomultiplier	700–2400	$a < 4 \times 10^{-5}$	1–5		No results adduced

17	Zinov'ev *et al.*	51–56	1968–1978	36 metals	TW in plates (electron bombardment)	Photoresistor	From 500–800 to melting points (up to 3300)	a 3×10^{-6} – 3×10^{-5}	3–5	0.1–0.3	
18	Petrunin and Yurchak	57	1971	Rocks, minerals	TW in plates (heating by light)	TC	350–1200	a $(0.4-1.7) \times 10^{-6}$	7	n.d.	
19	Phylippov and Nurumbetov	3	1967	Fe	TW along rods (multiproperty method)	TC	570–640	$a \sim 10^{-5}$ $\lambda - 50$	~5 4–7	~1 ~1	
20	Khusainova and Phylippov	58	1968	Mo		TC	500–1200	a $(4-6) \times 10^{-5}$ λ 130–160	4 7	~2 ~2	
21	Phylippov and Khusainova	29	1970	Liquid metals		TC	500–1100	$a \sim 2 \times 10^{-6}$ $\lambda \sim 30$	5–8 7–9	5 5	
22	Mebed *ct al.*	59	1973	Mo	TW in plates (multiproperty method)	Photomultiplier	1300–1500	a $(2-4) \times 10^{-5}$ c_p 0.3–0.4	3.5–4	2.5	
23	El-Sharkawy *et al.*	62, 60	1975, 1976	Graphite, corundum, fused quartz, Al_2O_3			1200–2200	a 1.5×10^{-6} – 1.5×10^{-5} $c_p \sim 1$	4 3.5–4	1.5 2.5	
24	Zaretskiy and Peletsky	61	1979	Fe		Photoresistor	600–1250	a $(3-10) \times 10^{-6}$ c_p 0.6–1.3	1–1.5 1.5–3	0.1–0.2 1	Electron bombardment heating
25	Ivlyev and Zinov'ev	63	1980	Fe		TP	700–1200	a $(3-10) \times 10^{-6}$ c_p 0.6–1.3	1.5 4.5	0.3 0.5	Laser heating
26	Basili *et al.*; Mebed *et al.*	67 68	1979 1979	$KNbO_3$ Nb_2O_5, La_2O_3, Ce_2O_3		TC	350–700	a $(0.1-1) \times 10^{-6}$ c_p 0.3–2.5	3.5 3.5	n.d. n.d.	Resistor heating
27	Phylippov	75, 29	1960	Organic liquids	Flat TW in fluids (probe method)	TP	~ 300	$b = \lambda/\sqrt{a}$	~3	~1	ac-heating, foil probe
28	El'darov and Orlova	77	1971	Organic liquids, water					~3	n.d.	ac + dc heating, wire probe
29	Phylippov *et al.*	80	1979	Gases (air, CO_2, Ar)			~ 1300		2–3	1.5	A possibility of gas pressure determination from $b(p)$ is shown
30	Phylippov *et al.*	82	1969	Ni	TW along rods mutli-property method (joule heating)	TC	~ 300	$a \sim 2 \times 10^{-5}$ $\lambda \sim 80$ $c_p \sim 0.4$	~5	~2 0.5 1	
31	Kirichenko	84	1961	Plastics	Radial TW		90–600	$a \sim 10^{-6}$	2–5	3–4	
32	Yurchak and Phylippov	85	1964	Fe, liquid Sn			300–900	a $(1-1.6) \times 10^{-5}$	7	3	

TABLE 1 (Continued)

No.	Investigator	Ref.	Year	Material	Method	Temperature measurement[a, b]	Range (K)	a, λ, c_p $\left(\frac{m^2}{S}\right)\left(\frac{W}{m \cdot K}\right), \left(\frac{\tau}{g \cdot K}\right)$	Accuracy (%) Max. error	Scatter	Comments
33	Phylippov and Yurchak	89	1966	Fe, liquid Pb	Radial TW, multi-property method		800–1140	a (0.5–1) × 10^{-5} c_p 0.1–0.8	2–4 3	1 2	
34	Banchila and Phylippov	94	1973	Liquid Sn, Pb		Photomultiplier	1100–2100	a (1.5–4) × 10^{-5} c_p (0.1–0.2)	5 3–4	5 3–4	Two variants of the method
35	Derman and Bogorodskiy	100	1970	Unorganic melts		TC	800–1500	$a \sim 10^{-6}$ $c_p\rho$ 2–12 (J/cm³ K)	6 4	n.d. n.d.	
36	Anderson and Bäckström	101	1972	Teflon		TC	~ 300; 0–30 kbar	a, Relative values	n.d.	n.d.	θ and φ variants of determining a
37	Phylippov and Pygal'skaya	102, 104	1964	W	Radial TW, inductive furnace heating	Photomultiplier	~ 1700	a 4 × 10^{-5}	4–8	1–2	θ and φ variants of the method
38	Phylippov and Makarenko	103	1968	Nb	Radial TW, inductive furnace heating, multiproperty method		~ 1900	a 2 × 10^{-5} $c_p \sim 0.3$	4–8 7–10	1–3 1–3	
39	Phylippov *et al.*	107–112, 93	1968–1971	W, Mo, Nb, Ta, V, Ti, Zi, Hf, Re			1100–2500	a (0.7–5) × 10^{-5} c_p 0.15–0.7	2 4	2 2	
40	Peterson and Bonilla	116	1965	He, N_2	Radial TW, wire probe method	TR	300–1300	λ 0.03–0.4	1.3–3	0.5–2	
41	Vargaftik *et al.*	119	1975	D_2O vapor			500–1000	λ 0.04–0.1	2.5	n.d.	
42	Siegel and Bonilla	121	1973	Liquid dowtherm			300–530	$\lambda \sim 0.1$	n.d.	n.d.	
43	Phylippov and Nefedov	122–127	1979, 1980	Fluid toluene, hexane, heptane, cyclohexane, CCl_4	Radial TW, wire probe multi-property method		300–600 0–30 MPa	λ 0.04–0.15 c_p 1.7–3	1–1.5 1.5–2	~ 1 ~ 1	Three variants of the method
44	Canneli and Canneli	133	1976	Nb	Acoustic		60–340	a (2.5–7) × 10^{-5} $\frac{c_p - c_v}{c_v}$ 10^{-3}–10^{-2}	n.d. n.d.	5 n.d.	
45	Kulish and Phylippov	135	1978	Mo, V, Ta, Nb			300–1400	a (1–6) × 10^{-5} $\frac{c_p - c_v}{c_v}$ (1–5) × 10^{-2}	3.5–10 n.d.	3.5–10 n.d.	
46	Ackerson and Hanley	138	1978	Fluid CH_4 at critical isochore	Optic		$T - T_c$ = 0.1–0.001	a 1 × 10^{-7} – 5 × 10^{-9}	~ 10	n.d.	

[a]TC, Thermocouple; TM, thermistor; TP, thermoprobe; TR, resistance thermometer (wire probe).
[b]Temperature oscillation registering.

since even very small temperature differences lead to the appearance of convective flow which distorts the results.[29] The optical method is free from this danger because temperature differences (with the exception of inherent fluctuations) do not exist. Results of thermal diffusivity investigations in the vicinity of critical points can be found, for example, in Ref. 138.

Let us draw some conclusions.

Periodic heating techniques form a wide range of means of investigation. They can be applied to the study of thermophysical properties of gases, liquids, and solids, both metals and nonmetals, over an extremely wide range of temperatures and pressures. They are very rapid, highly informative, and allow one to determine several thermophysical properties in the same experiment. Most of the periodic heating techniques are precise modern methods. Their wide use, improvement, and development can undoubtedly raise thermophysical investigations to a higher level.

Table 1 contains information on works that characterize various modifications of the method. These are either pioneering works, or examples of modern devices, or the latest publications pertaining to the theme.

REFERENCES

1. L.P. Phylippov, *Izv. Vyssh. Uchebn. Zavend. Energ.* **3**, 35–41 (1980).
2. A.V. Lykov, "Teoriya teploprovodnosty," "Vysshaya shkola," Moscow (1967).
3. L.P. Phylippov, "Izmerenie teplovikh svoistv tverdikh i shidkikh metallov pri vysokokh temperaturakh," Izdat. Mosc. Univ. (1967) ("Measurements of Thermal Properties of Solid and Liquid Metals at High Temperatures").
4. L.P. Phylippov, *Teplofiz. Vys. Temp.* (*High-temp. Thermophys.*) **2**(5), 817–828 (1964).
5. B. Abeles, G.D. Cody, and D.S. Beers, *J. Appl. Phys.* **31**(6), 1585–1592 (1960).
6. H.S. Carslaw and J.C. Jaeger, *Conduction of Heat in Solids.* 2nd Ed. Clarendon Press, Oxford (1959).
7. A.J. Angström, *Ann. Phys. Chemie* **144**, 513–530 (1861).
8. E.G. Shvidkovskiy, *Zh. Tekhn. Fiz.* **8**(10), 935–947 (1938).
9. P.H. Sidles and G.C. Danielson, *J. Appl. Phys.* **25**(1), 58–66 (1954).
10. L. Hugon and J. Jaffray, *Ann. Phys.* **10**, 377–385 (1955).
11. N.V. Zavaritskiy, *Zh. Ehksp. Teor. Fiz.* **5**(11), 1085–1097 (1957).
12. J. Gatacel and G. Weill, *J. Phys. Radium* **23**, 95A (1962).
13. K.G. Akhmetz'anov, N.Z. Pozdn'ak, and A.F. Dobrovolskiy, *Teplofiz. Vys. Temp.* **5**(1), 179–181 (1967).
14. H.R. Shanks, A.H. Klein, and G.C. Danielson, *J. Appl. Phys.* **38**(7), 2885–2892 (1967).
15. J.J. Martin, P.H. Sidles, and G.C. Danielson, *J. Appl. Phys.* **38**(8), 3075–3078 (1967).
16. H. Gonska, W. Kierspe, and R. Kohlhaas, *Z. Naturforsch.* **23**, 783–785 (1968).
17. H. Gonska and W. Kierspe, *Z. Angew. Phys.* **26**(5), 340–345 (1969).
18. J.C. Van Craeynest, J.C. Weilbacher, and J.C. Salbeaux, Proc. 8th Conf. on Thermal Conductivity (1969), Vol. 4, p. 587.

19. J.C. Weilbacher, J.C. Craeynest, J. Fauxinster, and J.J. Boy, *Rev. Int. Hautes Temp. Refract.* **10,** 77–78 (1973).
20. B. Leden, M.H. Hamza, and M.A. Sheirah, *Automatika* **12,** 445–456 (1976).
21. N. Savvides and W. Murray, *J. Phys., E: Sci. Instrum.* **11,** 941–947 (1976).
22. G. Glanz, *J. Nucl. Mater.* **27,** 331–334 (1968).
23. J.C. Van Craeynest and J.C. Weilbacher, *J. Nucl. Mater.* **26,** 132–136 (1968).
24. C. Starr, *Rev. Sci. Instrum.* **8,** 61–64 (1937).
25. R.W. King, *Phys. Rev. Ser. 2* **6,** 437–445 (1915).
26. D.A. Rigney, *Rev. Sci. Instrum.* **37,** 1376–1377 (1966).
27. I.I. Novikov, A.N. Solov'ev, E.M. Khabakhpasheva, V.A. Gruzdev, A.I. Pridantsev, and M.Ya. Vasenina, *At. Ehnerg.* **4,** 92 (1956).
28. N.I. Rudnev, V.S. L'ashenko, and M.D. Abramovich, *At. Ehnerg* **3,** 230 (1961).
29. L.P. Phylippov, "Issledovanie teploprovodnosti zhidkostei," Izdat. Moscow Univ., Moscow (1970) ("The Study of Liquid Thermal Conductivity").
30. B. Sundqwist and G. Bäckström, *High Temp. High Pressures* **9,** 41–48 (1977).
31. B. Sundqwist and G. Bäckström, *Solid State Commun.* **23,** 773–775 (1977).
32. B. Sundqwist and G. Bäckström, *J. Phys. Chem. Solids* **39,** 1133–1137 (1978).
33. F.K. Eder, *Monatsber. Dtsch. Akad. Wiss. Berlin* **2,** 86–91 (1960).
34. A. Tomokiyo and T. Okada, *Jpn J. Appl. Phys.* **7**(2), 128–134 (1968).
35. O.A. Kraev and A.A. Stel'makh, *Teplofiz. Vys. Temp.* **1**(1), 8–11 (1963).
36. J.B. Ainscough and M.J. Wheeler, *Brit. J. Appl. Phys. Ser. 2* **1,** 859–868 (1968).
37. L.P. Filippov, *Int. J. Heat Mass Transfer* **9,** 681–691 (1966).
38. M.A. El-Hifni and B.F. Chao, *Trans. ASME* **78**(1), 813–821 (1956).
39. R.D. Cowan, *J. Appl. Phys.* **32,** 1363–1370 (1961).
40. O.A. Kraev and A.A. Stel'makh, *Teplofiz. Vys. Temp.* **1**(1), 8–11 (1963); **2**(2), 302 (1964).
41. M. Serizawa, H. Kaneko, Y. Yokouchi, and M. Koizumi, Proc. 8th Conf. on Thermal Conductivity, Vol. 4, p. 549 (1969).
42. M. Serizawa, H. Keneko, Y. Yokouchi and M. Koizumi, *J. Nucl. Mater.* **34,** 224–226 (1970).
43. M.J. Wheeler, *High Temp. High Pressures* **1,** 13–20 (1969).
44. M.J. Wheeler, *J. Sci. Technol.* **33**(3), 102–107 (1971).
45. E.B. Zaretskiy and V.E. Peletskiy, *Teplofiz. Vys. Temp.* **17**(2), 310–313 (1979).
46. V.E. Peletskiy, E.S. Amasovich, E.B. Zaretskiy, J. Lierman, and P. Degas, *Teplofiz. Vys. Temp.* **17**(6), 1224–1231 (1979).
47. E. Chafik, R. Mayer, and R. Pruschek, *High Temp. High Pressures* **1,** 21–26 (1969).
48. M.R. Null and W.W. Lozier, Proc. 8th Conf. on Thermal Conductivity, Vol. 4, pp. 837–856 (1969).
49. R. Brandt and G. Neuer, *High Temp. High Pressures* **11** (1979).
50. J.M. Wheeler, *Brit. J. Appl. Phys.* **16,** 365–376 (1965).
51. V.E. Zinov'ev, R.P. Krentsis, and P.V. Gel'd, *Teplofiz. Vys. Temp.* **6**(5), 927–928 (1968).
52. P.V. Gel'd and V.E. Zinov'ev, *Izmer. Tekh.* **4,** 40–41 (1975).
53. V.E. Zinov'ev, P.V. Gel'd, G.E. Chuprykov, and N.I. Moreva, *Fiz. Tverd. Tela* **16**(2), 358–361 (1974).
54. L.P. Gel'd and V.E. Zinov'ev, *Izmer. Tekh.* **4,** 40–41 (1975).
55. P.V. Gel'd and V.E. Zinov'ev, *High Temp. High Pressures* **8,** 523–527 (1976).
56. E.M. Savitskii, P.V. Gel'd, V.E. Zinov'ev, N.B. Gernis, V.I. Sperelup, V.P. Polakova, and A.L. Sokolov, *Phys. Status Solidi A* **49,** 117–120 (1978).
57. G.I. Petrunin and R.P. Yurchak, *Vestnik Mosc. Univ. Ser. "Fizika"* **12**(5), 613–614 (1971).

58. B.N. Khusainova and L.P. Phylippov, *Teplofiz. Vys. Temp.* **5**, 929–930 (1968).
59. M.M. Mebed, R.P. Yurchak, and L.P. Filippov, *High Temp. High Pressures* **5**, 253–260 (1973).
60. A.A. El-Sharkawy, R.P. Yourchak, and S.R. Atalla, *Thermal Conduct.* **14**, 209–215 (1976).
61. E.B. Zaretskiy and V.E. Peletsky, *Teplofiz. Vys. Temp.* **17**(1), 124–132 (1979).
62. A.A. El-Sharkawy, S.R. Atalla, R.P. Yourchak, and L.P. Filippov, *Rev. Int. Hautes Temp. Réfract.* **12**, 168–170 (1975).
63. A.D. Ivlyev and V.E. Zinov'ev, *Teplofiz. Vys. Temp.* **18**(3), 532–539 (1980).
64. E.M. Kravchuk, *Inzh.-Fiz. Zh.* **5**(1), 59 (1962).
65. E.M. Kravchuk, *Inzh.-Fiz. Zh.* **6**(7), 3 (1963).
66. S.R. Atalla and El-Sharkawy, V. European Conference on Thermophys. Prop. of Solids at High Temp. Abstr. Moscow (1976), pp. 102–103.
67. R. Basili, A. El-Sharkawy, and S. Atalla, *Rev. Int. Hautes Temp. Réfract.* (1979), Vol. 16, pp. 340–345.
68. M.M. Mebed, M.A. Gaffar, and S. Sakindy, *Rev. Int. Hautes Temp. Réfract.* (1979), Vol. 16, pp. 340–345.
69. O.A. Sergeev and A.A. Men', *Teplofizicheskie Svoistva Poluprozrachnikh Materialov, Izdat. Standartov,* Moscow (1977).
70. J.F. Schatz and G. Simmons, *J. Appl. Phys.* **43**(6), 2586–2594 (1972).
71. J. Schatz and G. Simmons, *J. Geophys. Res.* **77**(35), 6966 (1972).
72. G.Ya. Belov, *Teplofiz. Vys. Temp.* **15**(5) (1977).
73. G.Ya. Belov, *Teplofiz. Vys. Temp.* **16**(4), 755–760 (1978).
74. R.P. Yurchak and A.A. Megakhed, *Vestnik Mosc. Univ. Ser. "Fizika"* **20**(6), 41–47 (1979).
75. L.P. Phylippov, *Inzh.-Fiz. Zh.* **3**(7), 121–123 (1960).
76. I.I. Novykov and F.G. El'darov, *Izmer. Tekh.* **1**, 50–53 (1970).
77. F.G. El'darov and T.I. Orlova, *Inzh.-Fiz. Zh.* **20**(6), 1087–1092 (1971).
78. S.N. Kravchoon and L.P. Phylippov, *Inzh.-Fiz. Zh.* **35**(6), 1027–1033 (1978).
79. N.B. Vargaftik, L.P. Phylippov, A.A. Tarzimanov, and R.P. Yurchak, "Teploprovodnost' gazov i zhidkostei," Izdat. Standartov., Moscow (1970) ("Thermal Conductivity of Fluids").
80. L.P. Phylippov, F.G. El'darov and S.N. Nefedov, *Inzh.-Fiz. Zh.* **36**(3), 466–471 (1979).
81. V.E. Mikrukov, "Teploprovodnost' i electroprovodnost' metallov i splavov," Izdat. "Metallurgiya" (1959) ("Thermal and Electric Conductivity of Metals and Alloys").
82. L.P. Phylippov, A.M. Yampolskiy, A.A. Kotlyar, N.V. Andreeva, and N.B. Shparo, *Electrovacuumnaya Tekhnika* **47**, 66–69 (1969).
83. A.F. Van Zee and C.L. Babcock, *J. Am. Ceram. Soc.* **34**, 244 (1951).
84. U.A. Kirichenko, *Inzh.- Fiz. Zh.* **4**(5), 12–15 (1961).
85. R.P. Yurchak and L.P. Phylippov, *Inzh.- Fiz. Zh.* **7**(4), 84–89 (1964).
86. Ya.I. Dutchak and P.V. Panasyuk, *Teplofiz. Vys. Temp.* **4**, 592 (1966).
87. R.P. Yurchak and L.P. Phylippov, *Teplofiz. Vys. Temp.* **2**(5), 696–704 (1964).
88. L.P. Phylippov and R.P. Yurchak, *Teplofiz. Vys. Temp.* **3**(6), 901–909 (1965).
89. L.P. Phylippov and R.P. Yurchak, *Vestnik Mosc. Univ. Ser. "Fizika"* **1**, 110–119 (1966).
90. E. Jahnke and F. Emde, *Funktionentafeln mit Formeln und Kurven,* 4 Aufl., B.G. Teubner, Leipzig und Berlin (1948).
91. S.R. Atalla, S.N. Banchila and L.P. Phylippov, *Teplofiz. Vys. Temp.* **10**(1), 72–76 (1972).

92. S.R. Atalla, S.N. Banchila, N.P. Dozorova, and L.P. Phylippov, *Vestnik Mosc. Univ. Ser. "Fizika"* **6**, 638–643 (1972).
93. L.P. Filippov, *Int. J. Heat Mass Transfer* **16**, 865–885 (1973).
94. S.N. Banchila and L.P. Phylippov, *Teplofiz. Vys. Temp.* **11** (3), 668–671 (1973).
95. I.I. Novykov and I.P. Mardykin, *Teplofiz. Vys. Temp.* **11** (3), 527–532 (1973).
96. S.N. Banchila and L.P. Phylippov, *Inzh.- Fiz. Zh.* **27** (1), 68–71 (1974).
97. I.I. Novykov, L.P. Phylippov, and V.I. Kostyukov, *Atom. Ehnerg.* **43,** 300–302 (1977).
98. I.I. Novykov, V.I. Kostyukov, and L.P. Phylippov, *Izv. Akad. Nauk SSSR Metals* **4,** 89–93 (1978).
99. S.N. Banchila, D.K. Palchaev, and L.P. Phylippov, *Teplofiz. Vys. Temp.* **17** (3), 507–510 (1979).
100. A.S. Derman and O.V. Bogorodskiy, *Izv. Akad. Nauk SSSR Ser. Fiz.* **34** (6), 1215–1216 (1970).
101. P. Andersson and G. Bäckström, *High Temp. High Pressures* **4,** 101–109 (1972).
102. L.P. Phylippov and L.A. Pygal'skaya, *Teplofiz. Vys. Temp.* **2** (3), 384–391 (1964).
103. L.P. Phylippov and I.N. Makarenko, *Teplofiz. Vys. Temp.* **6** (1), 149–156 (1968).
104. L.A. Pygal'skaya and L.P. Phylippov, *Teplofiz. Vys. Temp.* **2** (4), 558–561 (1964).
105. L.A. Pygal'skaya, R.P. Yurchak, I.N. Makarenko, and L.P. Phylippov, *Teplofiz. Vys. Temp.* **4** (1), 144–146 (1966).
106. L.A. Pygal'skaya, L.P. Phylippov, and V.D. Borysov, *Teplofiz. Vys. Temp.* **4** (2), 293–295 (1966).
107. I.N. Makarenko, L.N. Trukhanova, and L.P. Phylippov, *Teplofiz. Vys. Temp.* **8** (3), 667–669 (1970).
108. I.N. Makarenko, L.N. Trukhanova, and L.P. Phylippov, *Teplofiz. Vys. Temp.* **8** (2), 445–447 (1970).
109. A.V. Arutyunov, I.N. Makarenko, L.N. Trukhanova, and L.P. Phylippov, *Vestn. Mosk. Univ. Ser. III Fiz. Astronomiya* **3,** 340–343 (1970).
110. A.V. Arutyunov and L.P. Phylippov, *Teplofiz. Vys. Temp.* **8** (5), 1095–1097 (1970).
111. A.V. Arutyunov, S.N. Banchila, and L.P. Phylippov, *Teplofiz. Vys. Temp.* **9** (3), 535–538 (1971).
112. L.P. Phylippov and R.P. Yurchak, *Inzh.- Fiz. Zh.* **21** (3), 561–577 (1971).
113. A.V. Arutyunov, S.N. Banchila, and L.P. Phylippov, *Teplofiz. Vys. Temp.* **10** (2), 425–428 (1972).
114. A.V. Rumyantsev, I.B. Makarenko, and S.N. Banchila, *Inzh.- Fiz. Zh.* **36** (4), 581–587 (1979).
115. A.V. Arutyunov and S.N. Banchila, *Teplofiz. Vys. Temp.* **10** (1), 190–192 (1972).
116. B.L. Tarmy and Ch.F. Bonilla, in *Progress Int. Research on Thermodynamic and Transport Properties,* New York (1962), p. 404.
117. J.R. Peterson and Ch.F. Bonilla in *3rd Symp. on Thermophys. Properties,* New York (1965), pp. 265–276.
118. C. Lee and Ch.F. Bonilla in *Proc. of the 7th Conf. on Thermal Conductivity,* Washington D.C. (1968), pp. 561–578.
119. N.B. Vargaftik, Yu.K. Vynogradov, and N.A. Vanicheva, *Teplofiz. Vys. Temp.* **13** (6), 1291–1292 (1975).
120. N.B. Vargaftik and Yu.K. Vynogradov, *Teplofiz. Vys. Temp.* **11** (3), 523–526 (1973).
121. J.R. Siegel and C.F. Bonilla in *Proc. 6th Symp. on Thermophysical Properties,* Atlanta (1973), pp. 86–87.
122. L.P. Phylippov, S.N. Nefedov, and E.A. Kolykhalova, *Izv. Vyssh. Ucheb. Zaved. Pryborostr.* **22** (10), 78–82 (1979).

123. L.P. Phylippov, S.N. Nefedov, S.N. Kravchoon, and E.A. Kolykhalova, *Inzh- Fiz. Zh.* **38**(4), 644–650 (1980).
124. L.P. Phylippov., S.N. Nefedov, S.N. Kravchoon, and L.A. Bakhareva, *Izm. Tekh.* **6,** 32–35 (1980).
125. S.N. Nefedov and L.P. Phylippov, *Zh. Fiz. Khim.* **53**(8), 2112–2113 (1979).
126. S.N. Nefedov and L.P. Phylippov, *Inzh.- Fiz. Zh.* **37**(4), 674–676 (1979).
127. L.P. Phylippov, S.N. Nefedov, and S.N. Kravchoon, *Int. J. Thermophys.* **1**(2), 141–146 (1980).
128. A.A. Varchenco, in *Proc. of the 15th Conf. on Thermal Conductivity,* New York (1978), pp. 255–269.
129. C. Zener, *Phys. Rev.* **52(1),** 230–235 (1937).
130. I. Barducci, *Alluminio* **5,** 416–418 (1950).
131. B.S. Berry, *J. Appl. Phys.* **26,** 1221–1224 (1955).
132. G.B. Canneli, *Ric. Sci.* **38,** 1166–1170 (1968).
133. G. Canneli and G.B. Canneli, *J. Appl. Phys.* **47**(1), 17–22 (1976).
134. G. Canneli and G.B. Canneli, *Appl. Phys.* **17**(1) (1973).
135. A.A. Kulish and L.P. Phylippov, *Teplofiz. Vys. Temp.* **16**(3), 602–610 (1978).
136. B. Benedek, in *Polarisation,* Pr. Universitaires de France, Paris (1969).
137. *Photon Correlation and Light Beating,* (H.Z. Cummins and E.R. Pine, eds.), Plenum Press, New York (1974).
138. B.J. Ackerson and H.J.M. Hanley, *Chem. Phys. Lett.* **53,** 596–598 (1978).

10

Electron Bombardment Modulated Heat Input Method

R. DE CONINCK and V.E. PELETSKY

1. INTRODUCTION

The theory of thermal diffusivity and the techniques used for its measurement have been the subject of numerous investigations and review articles, and have constituted a large portion of the material in several books. However, a complete up-to-date review does not exist. These last few years in particular were marked by a basic and fast revolution of both the technical possibilities and the related procedures. Therefore, a comprehensive review of the topic seems called for, collecting the general body of actual knowledge in the field of thermal diffusivity with a particular emphasis on the most recent developments.

This chapter covers the electron bombardment modulated heat input method. The associated definitions and the principles of the procedure used are described together with the physical principles of heating by electrons, the measuring procedures and goals accomplished, and the necessary hardware and software. After an enumeration of the advantages and drawbacks, the main error sources are discussed.

Thermal diffusivity measurements with the aid of electrons have been put into practice only relatively recently. This is quite understandable considering the rather difficult and complex technical problems which had to be conquered, especially since most kinds of mechanical, electronic, and optoelectronic techniques are involved, while the mathematical procedures necessary are intricate and unmanageable without the help of a computer. The spectacular evolution of electronics and optoelectronics over the last few decades has especially made

R. DE CONINCK • Materials Science Department, SCK./CEN., B-2400 Mol, Belgium.
V.E. PELETSKY • Institute of High Temperatures, USSR Academy of Sciences, Moscow 127412, USSR.

these techniques more accessible and practical, making it possible to reach results and to develop procedures which were not attainable earlier.

Thermal diffusivity is an absolute property of matter, just as thermal conductivity is, while both are absolutely tied to each other. The knowledge of thermal diffusivity values is necessary in all phenomena involving temperature fluctuations enhancing possible thermal stresses.

The basic methods of measuring thermal conductivity are commonly part of the college physics syllabus. This might give rise to the expectation that obtaining accurate conductivity data is a relatively simple mission, whereas in reality this is definitely not the case, becoming obvious when referring to the chapters devoted to this matter in the present volume.

Thermal diffusivity is often not even mentioned in school books. Moreover, the measurement methods and associated theoretical treatments, as they generally are rather intricate, are often not fully recognized by nonspecialists.

Thermal conductivity is usually measured by a steady-state method. Thermal diffusivity measurements fall into the category of non-steady-state methods. A characteristic feature of these methods is that the temperature distribution in the specimen varies with time and that the rate of temperature change is measured instead of the rate of heat flow as is common for the steady-state methods. This implies avoidance of the very difficult measurement of heat flow, but unfortunately also implies the necessity of reckoning with many technical and mathematical problems.

Using electron heating has important and valuable advantages when measuring thermal diffusivity. It presents, unfortunately, some drawbacks too. However, when diffusivity values are required in as broad a temperature range as possible, with maximum precision and accuracy, and within a minimum of time, application of the electron bombardment modulated heat input method is presently, next to the flash method, the most obvious means.

2. PRINCIPLE OF THE PROCEDURE USED FOR THERMAL DIFFUSIVITY MEASUREMENT BY MODULATED ELECTRON BOMBARDMENT

2.1. General Introduction

Thermal diffusivity is measured by a dynamic or a non-steady-state method. The temperature distribution throughout the sample is varying with time, in contrast to what happens when applying a static or a steady-state method where the temperature distribution is fixed at a certain imposed value. Consequently, for a homogeneous, isotropic, heat conducting solid subjected to unidirectional heat flow, whose thermal properties possibly vary with temperature and position, the three-dimensional Fourier heat flow differential equation or time-dependent temperature distribution is involved:

$$\frac{\partial T_\theta}{\partial t} = a\left(\frac{\partial^2 T_\theta}{\partial x^2} + \frac{\partial^2 T_\theta}{\partial y^2} + \frac{\partial^2 T_\theta}{\partial z^2}\right) \tag{1}$$

T_θ is equal to

$$\frac{1}{\lambda_0}\int_0^T \lambda dT$$

in which λ_0 is the value of λ, the thermal conductivity when $T = 0$. T is the temperature, t the time, x, y, z are Cartesian coordinates, and a is the thermal diffusivity. The dimensions of a are $m^2\ s^{-1}$.

For any processes involving anisotropy, or inhomogeneity, or when very large temperature gradients are present or very short times are considered, or when work is done within the conductor, thermal diffusivity and thermal conductivity become ill defined and corresponding terms have to be added. Fortunately, experimental determinations of thermal diffusivity involve normally relatively small temperature excursions. This is certainly the case when using modulated electron heating and hence one can invariably assume constancy of thermal properties. Nevertheless, it is obvious that one must be watchful for conditions for which simplified equations could become quite inapplicable. This could well be the case for problems relating to some transient methods and possibly also for the flash method described elsewhere.

Thermal diffusivity is closely related to thermal conductivity, but also involves specific heat and thermal expansion. It was first expressed by Lord Kelvin as being a proportionality coefficient a, specific for each material, and related to λ by the equation

$$\lambda = a\rho C_p \tag{2}$$

ρ being the density and C_p specific heat.

Equation (1) can be integrated, taking into account equation (2), for one-dimensional heat flow in an homogeneous and isotropic body and where the temperature flow is governed by forces external to the substance only, to

$$\frac{dQ}{dt} = a\rho C_p \frac{dT}{dx}\, dy\, dz \tag{3}$$

where dQ/dt is the quantity of heat passing through the area $dy\, dz$ in the direction of x in a time dt. The rate of variation of temperature along x is given by dT/dx.

Equation (3) is easily transformed to equation (4) when considering the steady state only:

$$\frac{Q}{t} = \frac{\lambda(T_2 - T_1)yz}{x} \tag{4}$$

This equation represents the heat conducted in a time t, across a solid of section yz and thickness x. The constant λ is depending on the nature of the substance and is designated as the specific thermal conductivity, having the dimensions $\mathrm{W\,m^{-1}\,K^{-1}}$.

It has to be emphasized that, next to homogeneity and isotropy, ideal, not adversely interfering edge conditions are assumed while possible internal radiation, convection, or either internal sources or absorptions of heat (i.e., thermoelectric effects, radioactivity, chemical reactions, etc.), or "size effects" are supposed not to intervene.

A large number of dynamic techniques have been developed in recent years, and they show a great many advantages compared to steady-state techniques. A comprehensive survey of dynamic techniques related to thermal diffusivity was given by Danielson and Sidles[(1)] in 1969. Formerly, in 1964, Phyllipov[(2)] gave a literature survey of "stabilized periodic process methods," specifically intended to measure thermal diffusivity and specific heat. In 1973 a very extended survey of all known dynamic methods and associated results was presented by Touloukian *et al.*[(3)]

In principle, thermal diffusivity measurements are always carried out by applying a certain differential amount of energy to a specimen and by measuring in which way temperature evolves at another part of the specimen. This applied energy can vary as well continuously as repetitively or even can consistute of a single adjustable impact or withdrawal.

It is worthwhile to emphasize clearly that, although diffusivity and conductivity are simply related, these two properties require completely different experimental techniques. Thermal conductivity, measured directly by a steady-state method, requires the measurement of a thermal flux and a temperature gradient. Diffusivity requires the measurement of the time for a thermal disturbance to propagate a known distance. Only the temperature evolution has to be measured at two points or at two congruent surfaces of the specimen. This evolution can be translated mathematically into the value of the thermal diffusivity of the specimen under consideration. It is obvious that lengths and time intervals can be measured more easily and accurately than heat fluxes and temperature gradients.

Dynamic methods can broadly be divided into two categories, periodic or quasistationary, and transitory methods, depending upon whether the thermal energy is applied to the sample by modulating with a certain periodic function, or by a certain monotonous function, or a single, instantaneous stepwise change. For this last method a flash lamp or laser heating is generally used, as described elsewhere in this compendium. This pulse method, so far as the authors are aware, has only been produced a few times with electron heating – see Refs. 4 to 7 – while several authors use the modulation method without electron heating (laser, optical, etc) – see, e.g. Refs. 8–18.

On the other hand, it is worthwhile to refer to the work of Cutler and Cheney.[19] They demonstrate that the heat-wave method definitely is closely related to many other techniques for measuring thermal diffusivity, ranging from Angstrom's original method to several pulse heating techniques.

2.2. Periodic Methods

As should be apparent from the foregoing, a very large variety of different methods using electrons as a heating source might be imagined. Notwithstanding, only the periodic and the pulse or flash method have been brought to practice. This latter method is described elsewhere.

For the sake of simplicity, the specimens have to show two parallel surfaces. Therefore, either a semi-infinite plane-parallel plate or a semi-infinitely long hollow cylinder is most indicated.

The specimen is heated by transference to one of its surfaces of the energy of high-intensity electrons. The intensity, or the number of electrons, is adjusted, permitting one to tune to the desired temperature.

For the modulated electron heating diffusivity technique, the modulating signal has naturally to be periodic. This means a square wave or a triangular wave, but the most easily applicable, from the technical as well as from the mathematical point of view, is a sinusoidal wave. This periodic modulation signal, whose amplitude is preferentially small compared with the mean steady-state signal on which it is superimposed, drives the temperature of the bombarded surface to be modulated too. A phase shift with respect to the input wave is noted. The modulated heat quantity is transported through the sample toward the other surface, likewise creating a modulation of the temperature at that surface. This modulation is smaller in amplitude and is in turn lagged in phase versus modulation at the bombarded surface, at least for a finite velocity of temperature travel.

Both phase shifts, as well as the amplitude decrement, are unequivocally connected with the thermal diffusivity value of the specimen, as will be demonstrated in Section 4. In principle, it is thus sufficient to measure merely either a phase shift or an amplitude decrement to know thermal diffusivity. Moreover, when C_p, the specific heat, and ρ, the density of the specimen are known, thermal conductivity can be calculated using equation (2).

Periodic methods are often also called quasistationary or quasi-steady-state methods since it is clear that a steady-state condition prevails in the sense that the average thermal flux through the sample is constant for each adjusted value, and hence that the temperature at any point in the sample has an average constant value too.

The basic principle of the periodic method originates from the work of Angstrom, published some 120 years ago. However, his method has most probably never been used together with electron heating, as in this case electron heating does not show any advantage over other heating means. Moreover, the "Angstrom method" described elsewhere, is applicable to a long, thin cylindrical rod only, whereas utilization of samples in the form of a thin plate or disk is generally much more beneficent.

The possibility of deducing the thermal diffusivity of a thin solid plate heated by means of a high-energy electron beam impinging on one face of a flat plate from amplitude and/or phase measurements, was demonstrated theoretically by Cowan[20] in 1960.

So far as the authors are aware this represents the first proper, really relevant, and, in practice, usable mathematical expression which has ever been produced. The beam energy had to be modulated by either a square wave or a sine wave and the resulting temperature modulation of the faces observed photoelectrically. The most practical method seemed to be that which involved sine-wave modulation and measurement of the phase difference between the temperatures of the two faces of the plate. Cowan claimed: "With plate thicknesses of about 1 mm and frequencies of the order of 0.01 to 100 cps (depending on the value of a), it should be possible to measure thermal diffusivities to within 25% at any temperature from about 1000 K to the point where sublimation becomes troublesome." His work enhanced the breakthrough of practical thermal diffusivity determination in general and the use of electrons as a heat source in particular. Full disclosure and adequate elaboration occurred when in 1961 Cowan[21] presented in the open literature his revised mathematical model.

Nevertheless, the first thermal diffusivity apparatus by electron bombardment heating was developed probably independently of, although as a whole comparable with, the work of Cowan. It was reported in 1963 by Cerceo and Childers.[22]

Also Kraev and Stel'makh[23] describe in 1963 a periodic electron modulation thermal diffusivity apparatus. For their mathematical expression they use the Predvoditel and the Biot numbers – see Section 4 for more details – but in fact their expressions are also equivalent to those of Cowan.

Nearly simultaneously, Mustacchi and Giuliani[4] presented a study dealing with electron heating methods which was quite advanced for that time. For a reason which is unknown to the authors their study and apparatus has not been exploited any further. The same lot is shared by the experiments and the theory of Cerceo and Childers.

Also nearly simultaneously, Wheeler[24] gave a short description of a modulated electron beam thermal diffusivity apparatus and later on he[25, 26] presented a thorough and well-founded description of his modulated electron beam thermal diffusivity apparatus, altogether based on the theory of Cowan. He used original approximations and exposed a skillful iterative calculation method. His theory and technique has served as an example for many scientists accordingly. He assumed, as did Cowan, linear heat flow propagating axially through a disk-shaped specimen, neglecting heat losses from the edges of the sample, yet allowing for unequal heat losses from the front and back faces.

In the meantime, and also later on, interesting flash or pulse methods were presented, especially intended to be used with intensive light sources and lasers. As stated before, the pulse method has probably been used only rarely together with electron heating – see Refs. 4–7. These methods fall beyond the scope of this chapter, although similar techniques and theories are applicable.

Probably the first thermal diffusivity apparatus using hollow cylindrical specimens was reported in 1967 by Phyllipov.[27] His technique also made it convenient to measure liquid metals.

3. PHYSICAL PRINCIPLES INTERFERING WITH HEATING BY ELECTRON BOMBARDMENT

3.1. Introduction

The phenomenon that electrons are able to heat a substance, when applied under specific conditions, has been well known for a long time. Notwithstanding, since this phenomenon plays such an important role here, it will be described in more detail in the following, albeit restricted solely to those notions directly connected with thermal diffusivity measurements.

The most simple electron heating device consists of a diode, comprising primarily a source of electrons, the cathode, and an electron receiver or target, the anode.

In a more advanced electron heating setup, the target is remote and independent of the accelerating anode. Possibly a Wehnelt electrode is added as well as one or more beam-shaping electrostatic electrodes, or electromagnetic coils. The cathode–Wehnelt–anode electrode system and the shaping system serve the purpose of driving the electrons into a desired trajectory in order to obtain an electron beam with a well-defined configuration.

For the first type, the simple diode, the target is simultaneously the anode, which has to be electrically conducting hence. For the second type the electrons formed at the cathode hit a target which, in principle, is not necessarily a good electricity conducting material.

During the "historical period," before 1940, generally the first type was existing only as its construction was relatively simple and easy. Around 1955, the more sophisticated second type, called commonly the "electron gun," became the vogue as technology and especially as electron optics reached a more advanced level of development. To produce high temperatures the original "Pierce-type" guns have been rebuilt, taking into account the knowledge gathered from the techniques necessary for the construction of electron microscopes or cathode-ray tubes for oscilloscopes or for TV sets, etc.

For heating purposes, both types of electron generators make use either of a filamentary cathode, or of an indirectly heated block-cathode. Cold cathodes, or oxide-coated, thoriated, cesiated, or LaB_6-activated cathodes or the like, are generally not suitable when high currents are demanded or when working in arduous operating conditions. Also for some special applications a ring-shaped filament is used, while recently the "transverse gun" was developed. It can be considered to form an improved version of the diode-type generator or a combination of both types. The electron beam is deflected over an angle of 90 or even up to 270 degrees, forming an adequate solution for melting or the like purposes.

3.2. Energy Generated and Fraction Arriving at the Target

The voltage difference developed by an external power supply between the cathode and the anode determines largely the energy transferred by every electron to the target, most of which is converted to heat in the surface layers of this target.

The following facts with respect to the energy balance of each electron have to be taken into consideration.

An electron emitted by a thermionic cathode has an initial kinetic energy with a mean value of $E_0 = 2k_BT$ (k_B is Boltzmann's constant and T absolute temperature). After acceleration by the anode–cathode voltage difference V_a, it acquires an energy eV_a (e electron charge). On its way to the target, an electron may lose a part E_c of its energy as it collides with the molecules and atoms of the residual gases or products of evaporation of the heated substance. Finally as soon as it reaches the target, it releases an additional energy $e\varphi_a$ equal to the work function of the target material.

Thus, each electron may impart to the anode or target the following energy:

$$E^+ = 2k_BT + eV_a - E_c + e\varphi_a \qquad (5)$$

Whether or how this energy will be converted into heat in the target depends on the processes associated with the interaction between the electron and the material.

3.3. Energy Exchange between Electrons and Target

In any discussion concerned with electron energy exchange, it is absolutely necessary to state the energy range of the electrons interacting with the target.

In the very high-energy range (tens of MeV), some processes dominate (tunneling, Ramsauer, or Cerenkov effect, . . .) which have no or negligible influence in the range of say 1–10 keV, the range in which thermal diffusivity setups normally are operating. Here, the following interactions have to be considered.

The primary electron energy is partially converted into x rays. Most of the x-ray radiation power E_r is accounted for by bremsstrahlung, the energy of which can be defined by the formula

$$E_r = \eta e V_a \tag{6}$$

The value of η, which is known as the bremsstrahlung efficiency coefficient, is in turn calculated from the formula

$$\eta = k V_a Z \tag{7}$$

where Z is the atomic number of the target material, V_a the accelerating voltage in kV, and k an experimentally derived coefficient whose value ranges from 0.44×10^{-6} to 0.92×10^{-6} $(\mathrm{kV})^{-1}$ for voltages from 7 to 150 kV.

Calculations indicate that for values of V_a of the order of a few tens of kilovolts, the energy expended in x-ray radiation only constitutes some hundredths of a percent of the energy of the incident electron. This is certainly negligible compared to the total energy involved.

A more important phenomenon is the backscattering of electrons and, closely related, the secondary emission of electrons. These electrons may have many various energies while their number depends both on the target substance and on the primary beam energy. These two phenomena result simultaneously from elastic and inelastic reflection and from excitation of electrons of the target material by the primary electrons, together with multiple scattering in deeper lying layers of the material.

From the standpoint of thermophysical measurement instrumentation design some specific features of this phenomenon are important.

The total number of reemitted electrons may be either larger or smaller than the number of primary electrons, while the energy spectrum range can be comprised between a few eV and the initial electron energy.

As a consequence the energy of the bombarding electrons will be spread over the target and the surroundings. This certainly affects adversely power measurements while temperature measurements, or temperature fluctuation measurements may be affected also.

If the average energy lost per electron by secondary emission and/or back-scattering is denoted by E_s, the resulting energy balance E_i transformed into thermal excitation on the target or anode, can be written as follows:

$$E_i = E^+ - E_r - E_s \quad (8)$$

Multiplication of (8) by the number of electrons hitting the target per unit area and per unit time gives an expression for the power available for thermal excitation of the target

$$W_i = V_a J(1 + \zeta_0 - \zeta_c + \zeta_a - \zeta_r - \zeta_s) \quad (9)$$

with J the current density of the electron beam over the target and where the ζ parameters are correction factors for the effect of, respectively, the initial energy, losses in the interelectrode space, work function, x-ray radiation, and secondary electron emission.

The values of the correction factors ζ depend on the target substance, the beam parameters, and the vacuum level. Assuming accelerating voltages in the neighborhood of 5 to 10 kV and a residual gas pressure in the chamber not exceeding about 2×10^{-4} mBar, the only factor which has a certain importance is the secondary emission parameter ζ_s. Indeed its value may be higher than 40%[28] and thus will be determinative for the energy efficiency calculation or its measurement.

Nevertheless, in this respect it is certainly worthwhile to refer to a publication of Trouvé and Accary.[29] These authors state that for UC, Fe, and Zr the efficiency of heating by electron bombardment, for accelerating voltages between 4 and 10 kV, is 100%. They claim that, although approximately 40% of the electrons are reemitted by retrodiffusion, these electrons leave the surface with an energy which is nearly zero. Also tests carried out by one of the authors[30] proved that the loss in energy by secondary emission was in any case so low that it could not be determined by simple procedures.

3.4. Depth Distribution of the Power Deposited

The quantity W_i determines the energy imparted to the target during electron bombardment. For correct formulation of the boundary value problem and, accordingly, for correct determination of the target's properties during the thermal excitation, it is necessary to know how this power is distributed over the surface and throughout the volume of the target.

The thickness of the layer within which a primary electron will lose its energy completely can be equalized to the so-called total mean electron path p, using the formula recommended by Bronshtein and Fraiman.[31] They proclaim for primary electron energies ranging between 1 and 20 keV

$$p = 6 \times 10^{-6} A Z^{-1} \rho^{-1} V_a^{1.4} \tag{10}$$

with p in m, A the atomic weight, Z the atomic number, ρ the density, and V_a the primary electron energy (in keV). For example, for tungsten, and at $V_a =$ 10 keV, the primary beam loses its energy completely over a total length of 0.2 μm. The specimens usually employed in thermal diffusivity experiments are at least hundreds of times thicker. Consequently it can be accepted that the heat is released at the surface and that the simple mathematical description of heat transport [equations (1), (3), and (4)] can be applied, their assumed boundary conditions being respected.

As can be seen from formula (10), experiments should preferably be conducted utilizing small accelerating voltages. Should the voltage be increased to some hundreds of kilovolts, the total electron path would be extended to tens of microns.

3.5. *Control of Electron Density Deposition*

An important advantage of electron heating is coupled with the intrinsic possibilities allowing one to adequately and relatively easily control the space and time characteristics of the electrons. When using electron gun heating instead of simple diode heating, this is still more true.

An electron beam can be focused onto a rather small area, or even serve as a point heat source. Also it can rather easily be distributed or scanned over a particular specimen surface, thus allowing one to create a broad area at constant temperature. By applying electric and/or electromagnetic fields to the electron beam, it is possible to force the temperature of the bombarded surface to vary with time according to any prescribed function. These possibilities are especially appropriate for thermal diffusivity experiments.

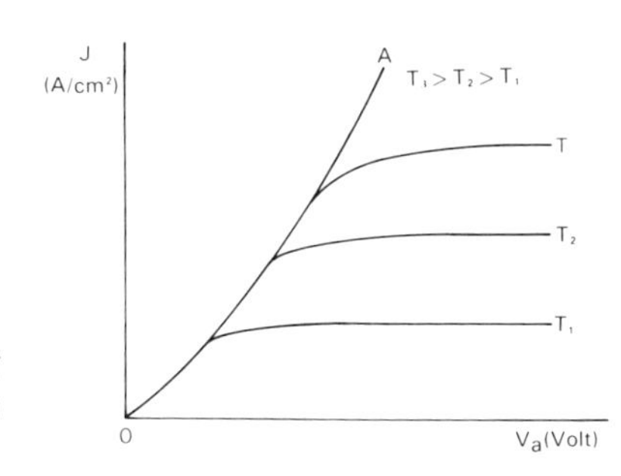

FIGURE 1. Current–voltage characteristic of a simple diode for three cathode temperatures $T_3 > T_2 > T_1$.

There are several ways to control the power of an electron heating device. The current–voltage characteristic for a simple diode with $T_3 > T_2 > T_1$, the cathode temperature as a parameter, has the typical form as represented in Fig. 1. Examination of this figure reveals that two situations can be considered.

A position on the curve OA corresponds to the space-charge-limited state. The magnitude of the anode current is in principle independent of the cathode temperature but is fully determined by the accelerating voltage between anode and cathode. As is well known, the electrons emitted by the cathode form a negative space charge around this cathode, creating a potential minimum and preventing an external field to reach the cathode surface. Generally, for lower accelerating voltages, only a part of the emitted electrons are entrained in the flow toward the anode. Current density J ($\mathrm{A\,m^{-2}}$) and voltage V_a (V) obey the general Langmuir–Compton law

$$J = P_e V_a^{3/2} \tag{11}$$

valid for each electrode configuration with P_e a constant called the perveance which depends on the geometry of the system solely. For a simple planar diode the Child law is obeyed:

$$J = 2.34 \times 10^{-6}\ V_a^{3/2}/x^2 \tag{12}$$

with x the interelectrode distance in m.

When designing circuits which will operate in this space charge current limited range, it is clear that a particular voltage variation will cause a reciprocal variation of the thermionic current. An accidental accelerating voltage modulation involves a reciprocal current modulation and thus an ac power modulation with a frequency twice as high.

The region on the right-hand side of curve OA is known as the saturation region. At sufficiently high accelerating voltages, the external field attracts all of the emitted electrons towards the anode. In this region voltage fluctuations change only very slightly the emission current intensity, at least for a fixed cathode temperature as a result of the Schottky effect. On the other hand, cathode temperature fluctuations immediately affect the beam power markedly.

The dependence of the thermionic emission current density J ($\mathrm{A\,m^{-2}}$) on the temperature T (K) of the cathode is expressed by the Richardson–Dushman equation

$$J = A_0 T^2 \exp\left(-e\varphi_a/k_B T\right) \tag{13}$$

where A_0 ($\mathrm{A\,m^{-2}\,K^{-2}}$) is the Richardson emission constant which is, based on the Fermi energy distribution theory, for very pure metals equal to $4\pi emk_B^2h^{-3} = 1.2 \times 10^6\ \mathrm{A\,m^{-2}\,K^{-2}}$ (e and m electron charge and mass respectively, and h

Planck's constant), $e\varphi_a$ (eV) is the work function, and k_B is Boltzmann's constant (8.62×10^{-5} eV K^{-1}). The above values are theoretical values which are not always confirmed experimentally in practice.

A variation of the cathode temperature changes the number of electrons leaving this cathode proportionally more than quadratically, varying the electron bombardment power consequently. This phenomenon can easily be used to modulate the total electron power. Unfortunately, requirements imposed on the stability of the emissive properties of the cathode and on the quality of the power supply used to heat the cathode are rather stringent. On the other hand, the power supply delivering the accelerating voltage should not be highly stabilized.

Notwithstanding, thermal inertia of the cathode can involve also nonsuitability for varying in time the power of an electron beam in accordance with a given pattern. This is the case when dealing with rapid variations. The notion "rapid" depends on the construction of the cathode of course. For a tungsten filament of, e.g., 0.5 mm ϕ already a frequency of 5 Hz can be called rapid.

For slow power variations, control of the cathode temperature is perfectly adequate and has indeed been exploited in practice. The smaller the thermal inertia of the cathode is, the greater the intensity of the modulation will be or also the higher-frequency response will be experienced.

It has to be remarked that this phenomenon either is searched for or is to be avoided, depending on one's aim, i.e., either modulate the electron power by modulation of the cathode temperature, or on the contrary keep the cathode temperature as constant as possible while modulation is carried out by another means.

The temperature modulation amplitude θ_0 of a filamentary cathode, energized by a current $i = i_0 + i_a \sin \omega t$ is expressed by

$$\theta_0 = \frac{2 i_0 i_a R \sin \varphi}{M C_p \omega} \tag{14}$$

R is the resistance of the filament with mass M, C_p the specific heat, and φ the phase shift between current and voltage across the cathode.

Formula (14) shows that θ_0 can be decreased by using a more massive cathode and/or by increasing the frequency. In the case of indirectly heated and sufficiently heavy cathodes, faulty modulation can be neglected. Should a directly heated and modulated filament be used, exactly the opposite features are requested.

Next to the possibility of varying the total heating power by changing either the temperature of the cathode or the accelerating voltage, depending on the working area chosen, an electron gun has the advantage of allowing for two more relatively simple and versatile modulation possibilities.

Introduction of a Wehnelt cylinder offers a most convenient means of modulation. Further, electromagnetic coils can be added which allow for a nearly unlimited range of scanning, or focusing–defocusing possibilities. Also electrostatic electrodes allow equal interactions but generally their construction and operation is much more delicate, and expensive.

Electromagnetic $X-Y$ scanning has been used by one of the authors[32] and by Schmidt *et al.*,[33] while Mayer and Neuer[34] present an interesting isothermal sample heating technique whereby the electron beam can be moved electromagnetically over any surface in such a way that controlled heat production is possible anywhere on the sample.

4. MATHEMATICAL PRINCIPLES FOR THE EXPERIMENTAL DETERMINATION OF THERMAL DIFFUSIVITY BY MODULATED ELECTRON HEATING

4.1. Introduction

As was stated before, electron bombardment allows one, in principle, to impose any kind of temperature distribution to be created on the surface of any sample. This implies that thermal diffusivity might possibly be measured using specimens of any configuration. However, for practical reasons, it is evident that preference is given to the simplest experimental setup and to the easiest mathematical description of the phenomenon, especially when being aware that the mathematics involved are already rather delicate and intricate even for that simplest case yet.

The simplest geometry for a specimen for this kind of measurements seems to be either a semi-infinite flat plate or a disk, or a semi-infinitely long cylinder. In the first case a modulated electron flux heats one surface of the disk, creating plane temperature waves which propagate through the specimen to the other colder face where the heat is dissipated to the surrounding space by thermal radiation. In the case of a cylindrical specimen generally radial temperature waves are created as well when using a solid cylinder as when using a hollow one.

The mathematics involved in thermal diffusivity measurements will consequently be treated in two ways. The first describes the plane heat wave method dealing with disk-type specimens; the second is the cylindrical specimen method. It is natural that each kind of specimen calls for equations which are adapted to each specific geometry. Nevertheless, it has to be remarked that these two ways do not differ basically. However, as the two theories were worked out and utilized by different scientists or groups of scientists, they diverged gradually from each other, especially regarding simplifying or approximating formulas.

4.2. *Plane Heat Waves – Mathematical Treatment*

The solution of the problem of the temperature distribution in an infinite plate, with one of its surfaces exposed to periodic modulated heating, is mainly ascribed to Cowan,[20, 21] as was mentioned in Section 2.2.

Cowan proposed three theoretical possibilities. He starts with the initial temperature distribution problem of a thin plate of which one surface is exposed to a steady electron beam, neglecting thermal conduction in directions parallel to the faces, thus assuming one-dimensional thermal flow.

A certain modulation, sufficiently small in amplitude, is superposed on this steady-state situation. After having found a solution for this relatively simple problem, which still is too general, however, he continues searching for a solution for three definite cases. The intensity of the electron beam is being altered by a step function, by square wave modulation, or by sine wave modulation.

Cowan argues that the step function method is limited to values of a_h (one of the heat loss parameters to be described further) greater than about 0.1 only. It is interesting to note that therefore this method may be of value in supplementing the sine wave method since for maximum accuracy the latter method requires approximate knowledge of a_h, if $a_h > 0.1$, which information is precisely obtainable from the step function method.

Later on Cowan[35] exposes a full mathematical treatment of the pulse method and he provides a solution which takes radiation into account, intended to be used with no matter which energy transfer mode, among others, electron heating. Nevertheless, as was already mentioned before, the flash or pulse method and moreover the step function change fall beyond the scope of this contribution.

For the square wave modulation method, initially Cowan does not present a plain solution, although the way in which a theoretical solution can be found is explained in summary.

It is interesting to note, when using square wave modulation instead of sine wave modulation, that the harmonics of the square wave will be highly attenuated and thus that the input will tend to be a pure sine wave. The phase relations will be almost those given for the sine wave modulation method, especially if the phase measurement is carried out over the sample, as described further. Harmonics can also be filtered out electronically from the signals, thus transforming the method integrally to the sine wave method. On the other hand, for, e.g., laser or thermal irradiation, square wave modulation can much more easily be applied than sine wave modulation as chopping of the irradiating energy is so much more easy to achieve then.

Combining Cowan's theoretical treatment and Wheeler's more practically oriented derivations, we can arrive at the following mathematical outline.

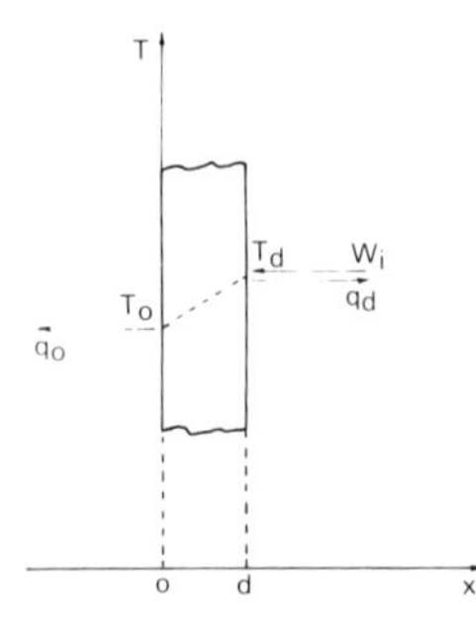

FIGURE 2. Schematic experimental specimen arrangement as used in Cowan's theoretical treatment.

Figure 2 illustrates the model used. Incident upon the plane $x = d$ of an infinite plate is an electron beam with energy density W_i, providing for uniform heating of that surface.

The heat applied to the plate is assumed to be partly transported through it, and simultaneously partly released by both of its surfaces in the form of radiation q_0 and q_d. Accepting that one-dimensional heat flow occurs, the differential equation of heat conduction – equation (1) in Section 2 – becomes

$$\frac{\partial T}{\partial t} = a \frac{\partial^2 T}{\partial x^2} \tag{15}$$

which is valid when putting that $T = T_d - T_0$ and that thermal conductivity λ, heat capacity C_p, and density ρ, and accordingly thermal diffusivity a, are constant with temperature T.

The rate of conduction $\lambda\ \partial T/\partial x$ must necessarily be equal to the difference between the effective power arriving at the surface and the power leaving it by thermal radiation. Moreover, it must be equal to the power leaving the non-bombarded surface by thermal radiation.

The boundary value conditions are thus obvious:

$$\left.\begin{aligned} \text{at } x = d\text{:} \quad \lambda(\partial T/\partial x)_d &= W_i - q_d \\ \text{at } x = 0\text{:} \quad \lambda(\partial T/\partial x)_0 &= q_0 \end{aligned}\right\} \tag{16}$$

In the case of sinusoidally modulated heating, the power–time relation can be expressed as

$$W_i(t) = W_i^0 + W_i^m \sin \omega t \tag{17}$$

where W_i^0 is the direct component, W_i^m the modulation amplitude and $\omega = 2\pi f$ is the angular frequency of modulation.

Consecutively the temperature distribution in the plate will have direct and alternating components also:

$$T(x,t) = T_x^0 + \Theta(x,t) \tag{18}$$

Substitution of (18) into the initial equation (15) leads to the formulation of the alternating temperature component:

$$\frac{\partial \Theta}{\partial t} = a \frac{\partial^2 \Theta}{\partial x^2} \tag{19}$$

with the boundary value conditions, taking into account equation (17)

$$\left.\begin{aligned} \text{at } x = d: \quad \lambda\,(\partial\Theta/\partial x)_d &= (\partial q_d/\partial T)\,\Theta(d,t) + W_i^m \sin \omega t \\ \text{at } x = 0: \quad \lambda\,(\partial\Theta/\partial x)_0 &= (\partial q_0/\partial T)\,\Theta(0,t) \end{aligned}\right\} \tag{20}$$

These boundary value conditions have been derived on the assumption that the temperature modulation is negligibly small or thus that $\Theta(x,t)/T_x^0 \ll 1$.

Equation (19) has been solved by Cowan, using Laplace transforms, the solution having the form

$$\Theta(x,t) = dW_i^m \lambda^{-1} P_x \sin(\omega t - \delta) \tag{21}$$

where the value of coefficient P_x depends on the x coordinate corresponding to the temperature Θ concerned.

In particular, for P_x the following relations were obtained: For $x = d$ (heated surface):

$$P_d^2 = \frac{2B^2 Q_0 + 2a_h B Q_1/(1 + r_h) + b_h Q_2/r_h}{2a_h^2 B^2 Q_0 + 2a_h b_h B Q_1 + (4B^4 + b_h^2) Q_2 + 4a_h B^3 Q_3} \tag{22}$$

and for $x = 0$ (rear surface):

$$P_0^2 = \frac{2B^2}{2a_h^2 B^2 Q_0 + 2a_h b_h B Q_1 + (4B^4 + b_h^2) Q_2 + 4a_h B^3 Q_3} \tag{23}$$

The following coefficients have been adopted in these equations:

$$B = d(\omega/2a)^{1/2} \tag{24}$$

[Remark: coefficient B can be written as $B = d(\mathrm{Tdl})^{-1}$ where (Tdl) is the thermal diffusion length which is equal to $(2a/\omega)^{1/2}$. B is called also thermal thickness, expressing the actual thickness in terms of the wavelength.]

$$\left.\begin{aligned} Q_0 &= \cosh^2 B \cos^2 B + \sinh^2 B \sin^2 B \\ Q_1 &= \cosh B \sinh B + \cos B \sin B \\ Q_2 &= \cosh^2 B \sin^2 B + \sinh^2 B \cos^2 B \\ Q_3 &= \cosh B \sinh B - \cos B \sin B \end{aligned}\right\} \tag{25}$$

as well as a_h, b_h, and r_h being the dimensionless heat loss parameters.

When accepting that the heat fluxes q_d and q_0 are determined for radiation into an environment which is at sufficiently low temperature, the following equation holds:

$$\frac{\partial q}{\partial T} = \frac{\partial(\epsilon\sigma T^4)}{\partial T} = 4\epsilon\sigma T^3 \tag{26}$$

with ϵ the total hemispherical emissivity and σ Stefan–Boltzmann constant.

Taking this into account the heat loss parameters can be transformed into practical useful expressions:

$$a_h = d(\partial q_0/\partial T + \partial q_d/\partial T)\lambda^{-1} = d4\epsilon\sigma T_m^3(1 + r_h)\lambda^{-1} \tag{27}$$

$$\text{with } T_m = \frac{T_0 + T_d}{2}$$

$$b_h = d^2 \left(\frac{\partial q_0}{\partial T}\right)\left(\frac{\partial q_d}{\partial T}\right)\lambda^{-2} = \frac{a_h^2}{4}, \qquad \text{assuming } r_h = 1 \tag{28}$$

and

$$r_h = \left(\frac{\partial q_d}{\partial T}\right)\Big/\left(\frac{\partial q_0}{\partial T}\right) = \left(\frac{T_d}{T_0}\right)^3 \simeq 1 \tag{29}$$

at least when the modulation amplitude is small.

The value of the phase shift δ in formula (21), which determines the phase difference between the specimen temperature modulation at a given x coordinate and the power modulation of the beam, is found from the formulas

$$\tan \delta_d = \frac{2B^3Q_1 + 2B^2a_hQ_2/(1 + r_h) + b_hBQ_3/r_h}{2a_hB^2Q_0 + b_h(1 + 2r_h)BQ_1/r_h + a_hb_hQ_2/(1 + r_h) + 2B^3Q_3} \tag{30}$$

for $x = d$, and from

$$\tan \delta_0 = \frac{b_h(\tan B - \tanh B) + 2a_h B \tan B \tanh B + 2B^2(\tan B + \tanh B)}{b_h(\tan B + \tanh B) + 2a_h B - 2B^2(\tan B - \tanh B)} \tag{31}$$

for $x = 0$.

These expressions are functions of frequency and of thermal diffusivity through the quantity B – see equation (24) – and also depend on the heat loss parameters a_h, b_h, and r_h.

The dependence of amplitude and phase on these quantities is obviously rather complicated and can be studied with satisfactoriness only by numerical calculus for various values of the parameters.

Measurements of amplitudes have to be avoided as the requirements regarding the sensitivity and the calibration of the temperature-sensing elements are so extremely high that they hardly can be attained in practice.

Phase comparisons are much easier to make. The measurement of δ_d does not seem to be interesting because of its high sensitivity to the heat loss parameter a_h at low frequencies, and its low sensitivity to B at high frequencies. Measurement of δ_0 seems to be more adequate.

In order to be able to calculate thermal diffusivity with maximum accuracy, it is necessary to know, at least approximately, either the values of two of the three heat loss parameters a_h, b_h, or r_h, or those of a_h and r_h, taking into account equation (28).

Formulas (30) and (31) indicate that the temperature modulation in the plate depends on both the intrinsic properties of the material itself (parameter B) and on the amount of heat loss at the edges (parameters a_h and r_h).

To determine thermal diffusivity it is thus sufficient to measure the phase shift δ_d or δ_0. When the heat losses are known one can calculate the value of parameter B from the respective formulas (30) or (31) whereafter one can find the value of the thermal diffusivity from (24).

Assuming that no heat losses occur through the edges of the specimen, then $a_h = 0$, $b_h = 0$, and $r_h = 1$, and then some of Cowan's formulas can be simplified. Equation (23) becomes

$$P_0^2 = (2B^2 Q_2)^{-1} \tag{32}$$

Equation (22) becomes

$$P_d^2 = Q_0(2B^2 Q_2)^{-1} = Q_0 P_0^2 \tag{33}$$

Equation (30) becomes

$$\tan \delta_d = Q_1/Q_3 \tag{34}$$

Equation (31) becomes

$$\tan \delta_0 = \frac{\tan B + \tanh B}{\tan B - \tanh B} \tag{35}$$

The preceding relations show that as parameter B increases, phase shift δ_d becomes less sensitive to it. Thus, the measurement of thermal diffusivity is in practice better done by the measurement of phase shift δ_0. At sufficiently large values of B (> 1.4), the following simple condition – see Ref. 23 – is met (at least when $a_h \approx 0$):

$$B = \delta_0 - \pi/4 = \delta_0 - 0.785 \tag{36}$$

where δ_0 is expressed in radians.

When $B < 1.4$, one must use the exact expressions and take into consideration heat loss parameters.

In cases where practical considerations (like temperature sensitivity of the transducers, specimen dimensions, or properties, etc.) do not allow the experiment to be carried out under circumstances where heat losses cannot be ignored, the theory still suggests that the heat loss parameters can be determined experimentally.

It should be noticed that the effect of parameter r_h on relation $\delta(B)$ – see equations (30) or (31) – is insignificant especially as r_h is generally almost equal to one. Therefore, in practice, one can introduce into the calculations the heat loss parameter $\bar{a} \approx (8\epsilon\sigma T^3)_{\text{mean}} d\lambda^{-1}$, without any perceptible loss in accuracy. This involves that $\bar{b} = \bar{a}^2/4$. Then, the phase shift $\bar{\delta}_0 = \delta_0(B, \bar{a})$ is a function of two variables only, and to determine them it is sufficient to carry out independent measurements at two frequencies at the same temperature. Such an approach was, e.g., successfully used by Chafik[11] (specimen heated by sine wave modulated light beam from a xenon arc lamp) and by Schmidt.[33] Schmidt determines a_h (using Cowan's equations) from the maximum of δ_d as a function of ω. He further measures $\Delta = (\delta_0 - \delta_d)$ also. His analysis is valid for $a_h < 1$ only.

Analysis of the effect of the heat losses on the possible phase shifts has prompted Cowan to conclude that the least sensitive to heat losses is Δ, and not either δ_0 or δ_d. Important in this case is the fact that parameter B can be lower and go down to approximately 0.1, thus extending the operational range of the experiment arrangement. Hence, when gun heating is used, with the irradiated and the nonirradiated surface of the specimen free for observation, measurement of temperature modulation shift between both surfaces should be done preferably.

This phase shift, $\Delta = (\delta_0 - \delta_d)$, may be calculated from formulas (30) and (31) or, if the temperature of the bombarded surface closely follows the harmonic law, from a formula proposed by Cerceo and Childers:[22]

$$\tan \Delta = \frac{\tanh B \tan B - (K/B)(\tanh B - \tan B)}{1 + (K/B)(\tanh B + \tan B)} \tag{37}$$

where $K = (\partial q_0/\partial T)(1/2\lambda)$ is a heat loss parameter standing for the heat released from the nonirradiated surface.

For very low values of K/B formula (37) is simplified to

$$\tan \Delta = \tanh B \tan B \tag{38}$$

As parameter B increases (higher frequency, greater thickness, low thermal diffusivity), phase shift Δ further approaches the value of B. Thus, the dimensionless parameter B can be defined in physical terms as the phase shift in radians between the temperature fluctuations on the two limiting surfaces of a thick plate, provided the heat source modulation ensures harmonic temperature fluctuations on the heated surface, while the other surface is adiabatically insulated.

As all mathematical procedures exposed hitherto are still rather cumbersome, and as a straight calculation is still not possible, some more practical procedures were worked out.

Wheeler proposes an interesting approach. He gives an approximating equation expressing a as a function of the directly measurable value of Δ:

$$a = \frac{2.9d^2\omega}{2\pi\Delta} = \frac{2.9d^2 f}{\Delta} \tag{39}$$

Parameter B can then be calculated from equation (24). With published or accepted data on the specific heat of the material under test, the value of λ can be calculated from equations (2). This value allows – equation (27) – the loss parameters to be calculated, allowing one to calculate in turn from (30) and (31) a new value for Δ. Iteration of this process is carried out until the values calculated for Δ, a, and λ become stable.

The theoretically exact values searched for are generally found after not more than two to three iteration steps. Unfortunately, ρ, C_p, and ϵ must be known and thus have to be measured separately or taken from literature.

Penninckx[36] reanalyzed Cowan's equations in the case generally accepted to be most interesting, that is, where the phase difference Δ is measured.

He depicts that the loss parameter a_h clearly hinders the straight use of Cowan's equations but that a can be expressed as a function of B, r_h, and d_0, where

$$d_0 = \frac{8\sigma\epsilon T_0^3}{d\omega C_p \rho} \tag{40}$$

Finally he proposes some simplified formulas for B which apply in confined but in practice fully normal areas. His chief merit is admittedly to present an equation relating the phase difference Δ to B as the sole unknown:

$$\tan\Delta = \frac{2\tan B\tanh B + B(\tan B - \tanh B)d_0}{2 + B(\tan B + \tanh B)d_0} \tag{41}$$

Unfortunately again one must know d_0 and therefore also ρ, C_p, and ϵ. This still can represent a source of rather large errors.

In Soviet literature, instead of parameter B, use is generally made of the so-called Predvodytelev number:

$$\mathrm{Pd} = \omega d^2/a = 2B^2 \tag{42}$$

or the quantity

$$Z = (\mathrm{Pd})^{1/2} = B\sqrt{2} \tag{43}$$

Parameters a_h and b_h are related to the Biot modulus $\mathrm{Bi} = (a/\lambda)d$, which is written as follows in the case of heat loss by radiation:

$$\mathrm{Bi} = \frac{4\epsilon\sigma T^3 d}{\lambda} \tag{44}$$

When the heat fluxes q_0 and q_d are determined for radiation into the environment which is at a distinctively lower temperature, then

$$\frac{\partial q}{\partial T} = \frac{\partial(\epsilon\sigma T^4)}{\partial T} = 4\epsilon\sigma T^3 \tag{45}$$

and thus, from (27), (28), and (29) it follows

$$a_h = \frac{d}{\lambda}4\sigma(\epsilon_0 T_0^3 + \epsilon_d T_d^3) = \mathrm{Bi}_{x=0} + \mathrm{Bi}_{x=d}$$

$$b_h = \frac{d^2}{\lambda^2}(4\sigma)^2\epsilon_0 T_0^3\epsilon_d T_d^3 = \mathrm{Bi}_{x=0}\mathrm{Bi}_{x=d} \tag{46}$$

$$r_h = \epsilon_d T_d^3/\epsilon_0 T_0^3 = \mathrm{Bi}_{x=d}/\mathrm{Bi}_{x=0}$$

As is known, in the case of thin specimens with high heat conductivity, the Biot modulus has low values up to extremely elevated temperatures. Analysis of formulas (30) and (31) indicates that with properly selected experimental conditions the space-time variations of the temperature across a thin plate closely follow a pattern meeting condition $\mathrm{Bi} = 0$.

4.3. Temperature Fluctuations in Hollow Cylindrical Bodies – Mathematical Treatment

As mentioned before, electrons accelerated and modulated in an electric field between a cathode and an anode, which is simultaneously the specimen, can be used to create temperature waves. This heating method can be applied using the former treatment dealing with plane heat waves, but is also successfully applicable to specimens in the form of a hollow cylinder with a directly heated filament-cathode in its axis. The solution of the corresponding heat transfer and boundary-value problem can primarily be attributed to Phyllipov.[27, 37]

For an adiabatic insulated outer cylinder surface, Phyllipov expressed the temperature as

$$\Theta(Z) = \Theta_0 \frac{xH_0^{(1)}(Z)I_1(Z_2) - H_1^{(1)}(Z_2)I_0(Z)}{(-i)^{1/2}H_1^{(1)}(Z_1)I_1(Z_2) - H_1^{(1)}(Z_2)I_1(Z_1)} \tag{47}$$

where

$$x = \frac{R_2^2 - R_1^2}{2R_1}\left(\frac{\omega}{a}\right)^{1/2}$$

R_2 being the outer radius of the cylinder and R_1 the inner radius and where $Z = r(-i\omega/a)^{1/2}$, with r the distance from the axis of the cylinder, $H_0^{(1)}(Z)$ the zeroth-order Hankel function of the first kind, $I_0(Z)$ the zeroth-order Bessel function of the first kind, and $I_1(Z)$ the first-order Bessel function of the first kind – see reference 37. Θ_0 is determined by the amplitude of heating power fluctuations through

$$\Theta_0 = \frac{Q}{MC_p\omega} = \frac{2R_1 W_i^m}{(R_2^2 - R_1^2)\rho C_p \omega} \tag{48}$$

where M and W_i^m are, respectively, the mass of the specimen and the modulation amplitude of the power.

Expression (47) permits one to calculate the phase difference between the input power and the sample temperature fluctuations at any point on the cylinder. Phyllipov[27] also showed that when the temperature is measured on the outer surface of the cylinder (at $r = R_2$) this phase difference can be determined from

$$\delta_0 = \arctan\frac{\varphi_2}{\varphi_1} \tag{49}$$

where

$$\varphi_1 = \mathrm{bei}'(x_2)\,\mathrm{her}'(x_1) + \mathrm{hei}'(x_1)\,\mathrm{ber}'(x_2) - \mathrm{bei}'(x_1)\,\mathrm{her}'(x_2) - \mathrm{hei}'(x_2)\,\mathrm{ber}'(x_1)$$

and

$$\varphi_2 = \mathrm{ber}'(x_1)\,\mathrm{her}'(x_2) + \mathrm{bei}'(x_2)\,\mathrm{hei}'(x_1) - \mathrm{bei}'(x_1)\,\mathrm{hei}'(x_2) - \mathrm{ber}'(x_2)\,\mathrm{her}'(x_1)$$

$$x_1 = R_1(\omega/a)^{1/2}, \qquad x_2 = R_2(\omega/a)^{1/2}$$

Here, ber$'$, her$'$ and bei$'$, hei$'$ are derivatives of the Thomson (Kelvin) function, see Ref. 38.

Thus, just as in the case of a semi-infinite plate, phase shift δ_0 turns out to be a function of the dimensions of the specimen and of the frequency-to-thermal diffusivity ratio – see equation (47). By measuring the phase shift one thus can equally determine thermal diffusivity.

Phylippov recommends calculating the phase shift δ_0 as a function of parameter x from formula (47) and for a ratio $\zeta = R_1/R_2$ of the cylinder radii. This corresponds physically with a flat-wall approximation of the same type as equation (24) when $\zeta \to 1$.

Indeed, when $\zeta \to 1, x$ tends to

$$x = \frac{R_2^2 - R_1^2}{2R_2}\left[\left(\frac{\omega}{a}\right)_{\zeta \to 1}\right]^{1/2} \approx (R_2 - R_1)\left(\frac{\omega}{a}\right)^{1/2} = B\sqrt{2} \qquad (50)$$

Under real conditions, the outer surface of the cylinder does not behave adiabatically. Heat losses will affect both the phase and the amplitude of the temperature fluctuations in the cylinder, as examined by Phyllipov and Yurchak.[37]

The formula they derived for the temperature of a cylinder losing heat from its outer surface, in accordance with the Newton–Richman law, is given by

$$\Theta(Z, \mathrm{Bi}) = \Theta_0 \frac{x}{(-i)^{1/2}} \times \qquad (51)$$

$$\frac{[H_0^{(1)}(Z)I_1(Z_2) - H_1^{(1)}(Z_2)I_0(Z)] - (\mathrm{Bi}/x\sqrt{i})[H_0^{(1)}(Z)I_0(Z_2) - H_0^{(1)}(Z_2)I_0(Z)]}{[H_1^{(1)}(Z_1)I_1(Z_2) - H_1^{(1)}(Z_2)I_1(Z_1)] - (\mathrm{Bi}/x\sqrt{i})[H_1^{(1)}(Z_1)I_0(Z_2) - H_0^{(1)}(Z_2)I_1(Z_1)]}$$

The variations in phase shift δ_0 due to heat exchange can be expressed by

$$\Delta\delta_0 = \mathrm{Bi}\,\psi(x) \qquad (52)$$

where the correction factor $\psi(x)$ is smaller than unity for normal values of x used in practical experiments. Moreover, $\psi(x)$ decreases with increasing x. Thus the effect of heat exchange on the phase shift does not, at any rate, exceed the Biot modulus (Bi) calculated from formula (44) with characteristic dimension

$$d = (R_2^2 - R_1^2)/2R_1 \tag{53}$$

It should be emphasized that taking heat exchange into consideration, even at small values of Bi, radically changes relation $\delta_0(x)$ for low x values. Instead of the phase shift tending to $\pi/2$ when $x \to 0$, which is typical for an isolated cylinder, heat losses lead to a substantial decrease of δ_0. As the frequency increases, the relative effect of heat exchange becomes less pronounced.

Just as for a semi-infinite plane specimen, a cylindrical specimen allows determination of both the heat loss parameter and the thermal diffusivity. The necessary information is provided for by measurements carried out at several frequencies. To determine the correction factor for heat loss in experiments with materials exhibiting low heat conductivity, Tkach and Yurchak[39] propose using relation $\varphi(2\omega) = f[\varphi(\omega)]$ between the phases of temperature fluctuations on the surface of a hollow cylinder at two frequencies, one being twice as high as the other. An important thing here is that, in view of the relatively weak effect of parameter Bi on the measurement results, one will have to be content with reduced accuracy. Also they propose relations $\Theta(\omega)/\Theta(2\omega) = f(\omega)$ or $\Theta(\omega)/\Theta(2\omega) = f[\varphi(\omega)]$ with Θ the amplitude and φ the phase difference. All these propositions are theoretically interesting but unfortunately rather delicate in practice.

In experiments with hollow cylindrical specimens thermal diffusivity can be determined from the phase shift between the temperature fluctuations of the inner and outer surfaces, or generalizing, from the phase shift between any two radii. This again can be called "conformal" with the semi-infinite plane specimen method; however, measurement of δ_0 seems to be the sole practical possibility when utilizing hollow cylindrical specimens.

5. EXPERIMENTAL TECHNIQUES OF THERMAL DIFFUSIVITY MEASUREMENT USING MODULATED ELECTRON HEATING

5.1. Survey of Literature References

Table 1 lists most of the major works dealing with modulated electron heating thermal diffusivity experiments. Immediately it can be remarked that all works enumerated refer to recent publications. Indeed, the first results date from 1963 only. This evidently confirms the relative novelty and modernness of the method and techniques put in practice.

Specimens in the form of a disk or a plate are mostly used and also specimens in the form of a hollow cylinder. Once a plain cylinder is used – see Ref. 41 – and a few times a disk-type specimen, which however is mounted in an opening of the wall of a hollow cylinder – see Refs. 81, 82, and 84.

TABLE 1
Survey of Major Publications Dealing with Thermal Diffusivity Modulated Electron Heating Methods

No.	Investigator(s)	Ref.	Year	Material(s)	Heating mode	Specimen geometry	Typical dimensions for ϕ and d (mm)	Temperature range (K)	Modulation frequency (Hz)	Measured magnitude	Uncertainty (%)	Remarks
1	Cerceo and Childers	22	1963	Al_2O_3, C	EG[a]–AH[b] combination	D[c]	19–25 ϕ 2.555 d	1190–1400	0.02–0.1	Δ[d]	± 10	
2	Kraev and Stel'makh	23	1963	W	AH	D	7–8 ϕ 0.2 d	1870–3220	280–1200	δ_0[e]	≃ 5	
3	Mustacchi and Giuliani	4	1963	Ni, Mo, Nb, UC	AH (triode-like)	D	12.5 ϕ 1–3 d	1200–1750 1200–2280	0.01–100	Either Δ or δ_0		Also amplitude and pulse method
4	Wheeler	24	1963	UC	EG	D	12.7 ϕ 1 d	1425–2375	0.48	Δ	≃ 3	
5	Kraev and Stel'makh	40	1964	Ta, Mo, Nb	AH	D	8–9 ϕ 0.2–0.3 d	1800–3150	130–530	δ_0	± 5	
6	Wheeler	25	1965	Pt, Ta, Mo, W, C, UO_2, U-49 at. % C, B_4C In oxide (probably In_2O_3)	EG	D or rectangular plate	0.3–1.3 cm^2 surf. 1–1.5 d	1200–2900	0.48	Δ	± 5	
7	Zankel	41	1967	Stainless steel, Ni, UC	EG + oven	Long cylinder	5.5–6 ϕ 60 length	330–975	0.05	Δ and amplitude	± 3.6	
8	Pascard	9	1967	UN, UC, UC-UN bin. syst. (U, Pu)C (U, Pu)N	EG	D	6 ϕ 1 d	probably 770–1970	1–5	Δ	10	
9	Ainscough and Wheeler	42	1968	2.7% enriched $UO_{2\cdot00}$	EG	D	1 cm^2 surf. 0.5 d	970–2020	0.5	Δ	7–14.5	
10	Zinov'ev and Krentsis	43	1968	Ti iodide	AH	Square plate	8 × 8 mm square 0.11–0.187 0.310 d	1000–1500	150–400	δ_0	7	
11	Zinov'ev *et al.*	44	1968	Ti iodide	AH	Square plate	8 × 8 mm square 0.11–0.187 0.0310 d	1000–1500	20–600	δ_0	7	
12	Wheeler	45	1969	Fe (99.95%)	EG	D	10 ϕ 0.1–0.2 d	550–1300	0.5	Δ	± 5	
13	Emel'yanov *et al.*	46	1969	Ta, Mo, Nb, V, Co	AH	D	10–20 ϕ 0.2–0.8 d	1100–2500	20–600	δ_0		

14	Schmidt *et al.*	33	1969	W	EG	D	6 ϕ $\simeq 2.7\ d$	1100–2850	0.5–1.2	Δ and $\delta_d{}^f$	± 7	
15	Wheeler	47	1970	Hf, Nb, Zircaloy 2	EG	D	6–10 ϕ 0.5–3 d	600–2050	0.5	Δ	± 5	
16	Schmidt and Caligara	48	1970	(U, Pu)-oxides	EG	D	6 ϕ 1–2 d	1250–2200	Probably 0.5–1.2	Probably Δ and δ_d	± 7	
17	Wheeler *et al.*	49	1971	Various UC sintered (U, Pu)C alloys	EG	D	10 ϕ 1 d	400–2200	0.5	Δ	± 5	
18	Wheeler	26	1971	Ta	EG	D	6 × 6 square 1 d	1050–2900	0.5	Δ	± 5	
				UO_2			4 × 4 square 0.5 d	620–1900				
19	Mardykin and Vertman	50	1972	Ce	AH	HC[g]	6–8 inner ϕ 15–20 outer ϕ 0.2 wall thickness	1150–1600	1 to, e.g., 0.005	δ_0	6–7	Also C_p
20	Zinov'ev *et al.*	51	1972	Gd	AH	Probably square plate		900–1500	168.8	δ_0	< 3.5	
21	Zinov'ev *et al.*	52	1972	Sc	AH	Probably square plate	? ϕ 0.225 d	900–1500	Probably 168.8	δ_0	4	
22	Tanaka and Suzuki	53	1972	Pyrolitic, natural, artificial and glassy C	EG	D	About 8ϕ and 1 d	1550–2200	0.5	Δ	Repeatability 1–5	
23	Tanaka and Suzuki	54	1972	Pyrolitic C	EG	D	8 ϕ 1.5 or 0.3 d	1660–2200	0.5	Δ	Repeatability 1–5	
24	Mebed and Yurchak	55	1972	Mo	AH (triode)	D		1500 mentioned only	5–7–10	δ_0	3.5–4.5	Also C_p
25	Gel'd and Zinov'ev	56	1972	Ir	AH	D		800–2600	280	Probably δ_0	4–5 scatter 1%–2%	
26	Atalla *et al.*	57	1972	In, Nd, Ce (liquid)	AH	HC	Ta, 6 inner ϕ 15 outer ϕ 0.1 wall thickness, length 70–80 mm	Probably 1150–2050	0.15–1	Phase ≠ between central T and power	6–7.5	Also C_p
27	Atalla	58	1972	La, Ce, Nd, In (liquid)	AH	HC	Ta, 6 inner ϕ 15 outer ϕ 0.1 wall thickness, length 70–80 mm	1100–2100	0.1–1	Phase ≠ between central T and power	3–6	Also C_p
28	Fridlender *et al.*	59	1973	TiC, CbC, TiN	AH	Plate	0.25–0.3 d	1780–2450–2670	112	δ_0	< 5	

TABLE 1 (Continued)

No.	Investigator(s)	Ref.	Year	Material(s)	Heating mode	Specimen geometry	Typical dimensions for ϕ and d (mm)	Temperature range (K)	Modulation frequency (Hz)	Measured magnitude	Uncertainty (%)	Remarks
29	Zinov'ev *et al.*	60	1973	Re (99.99%)	AH	D	5 × 5 mm square 0.3 d	750–3150	168.8	Probably δ_0	4	
30	Novikov and Mardykin	61	1973	Liquid Ce	AH	HC	Ta: 6 inner ϕ 15 outer ϕ 70–90 mm length	1210–1690			7	Also C_p
31	Banchila and Filippov	62	1973	Liquid Sn, Pb	AH	HC	Ta: 6 inner ϕ 13.8 outer ϕ 0.1 wall thickness ≃ 70 mm length	1130–1980 1150–2040	1.5–3	δ_0, phase ≠ between central or outer T and power	6	Also C_p
32	Mebed *et al.*	63	1973	Mo	AH (triode)	D		1300–2500	5–7–10	δ_0	3.5–4.5	Also C_p
33	Korshunov *et al.*	64	1973	TiC, ZrC	AH	D	8 ϕ 0.3 d	1325–2190 1460–2440	168.8	Probably δ_0	< 5	
34	De Coninck *et al.*	65	1973	U_2C_3	EG	D	6.5 × 6.5 square 1 d	1070–2200	0.5	Δ	5–10	
35	De Coninck *et al.*	66	1973	U_2C_3	EG	D	6.5 × 6.5 square 1 d	1070–2200	0.5	Δ	5–10	
36	Kulisch *et al.*	67	1973	Mo	AH (triode)	D		1420–1960	5–7	δ_0	4	
37	Zinov'ev *et al.*	68	1974	Fe and Fe–Si alloys	AH	D	8 × 8 square 0.2 d	950–1500	168.8	Probably δ_0	2.5–3 0.2 resolution	Also C_p
38	Tanaka	69	1974	LaB_6	EG	D		1280–2010	Probably 0.5	Probably Δ		
39	Zinov'ev *et al.*	70	1974	Pt	AH	D	36 mm^2 0.3 d	1000–1500	160	δ_0	± 5	
40	Neshpor *et al.*	71	1974	ZrC–NbC solid solutions	AH	D	8 ϕ 0.25–0.3 d	1780–2700	112–150	δ_0	4–6	
41	El-Sharkawy	72	1974	Graphite and Al_2O_3	AH + optical modulation	D	10–14 ϕ 1–3 d	1200–2100	0.38–0.77	Δ	5	Also C_p
42	De Coninck *et al.*	32	1975	UC	EG	D	7 ϕ 0.5–2 d	860–2500	0.5	Δ	Repeatability ≃ 2%	
43	Zinov'ev *et al.*	73	1975	Yb, Tm, Lu	AH	D		550–980 580–1400		Probably δ_0	4	
44	Novikov and Mardykin	74	1976	Y, Gd	AH	HC	15 outer ϕ 6 inner ϕ 70 mm length	1150–1700		Probably δ_0	5	

45	Schmidt	75	1976	(U, Pu)O, C, N	EG	D	6–8 ϕ 1–2 d	950–2000	Probably 0.5–1.2	Probably Δ and δ_d		
46	De Coninck *et al.*	76	1976	UC_2	EG	D	7–8 ϕ 1–2 d	840–2350	0.5	Δ	Repeatability $\simeq 5\%$	
47	Fridlender *et al.*	77	1977	ZrN, Z_r	AH	D	8 ϕ 0.2 d	1275–2070 1275–1700	112–190	Probably δ_0	< 5	
48	Benedict *et al.*	78	1977	(U, Pu)C	EG	D	6 ϕ $\simeq 2.7\ d$ probably	1110–1650	0.5–1.2 probably	Δ and δ_d	± 7 probably	
49	Kim *et al.*	79	1977	UO_2 (molten)	2 × EG (one of them modulated)	D (tungsten cell)	? ϕ 0.813–1.219 d	3187–3315	0.25–0.50–0.75	Δ		
50	Yurchak and Khromov	80	1978	Mo	AH–EG combination	D	10 ϕ 0.5–1.5 d	1220–2970	5–12	δ_0	3	Also C_p
51	Zaretsky and Peltsky	81	1979	Fe	AH	D (but HC aperture mounted)	15 ϕ $\simeq 0.41\ d$	670–1260	360	δ_0	± (1–1.5)	Also C_p
52	Zaretsky and Peltsky	82	1979	Ti iodide	AH	D (but HC aperture mounted)	15 ϕ $\simeq 0.5\ d$	650–1620	248–450	δ_0	± (1.5–2)	Also C_p
53	Banchila *et al.*	83	1979	Ga, In, Tl	AH	HC	Ta: 14 outer ϕ 6 inner ϕ 0.1 wall thickness 70 mm length	1200–1680 2050		δ_0 phase ≠ between central or outer T and power	4	Also C_p
54	Peletsky and Zaretsky	84	1982	Zr	AH	D (but HC aperture mounted)	Probably 15 ϕ $\simeq 0.4–0.5\ d$	1050–1230	Probably around about 400	δ_0	1.5	Also C_p
55	Atalla	85	1982	Austen. Stainless steel, C	AH	D	25–26 ϕ 2–3 d	400–1000	1–4	δ_0	0.5–0.8	

[a] EG, electron gun.
[b] AH, anode heating.
[c] D, disk-type.
[d] Δ, $\Delta = \delta_0 - \delta_d$.
[e] δ_0, phase shift between applied modulating wave function and temperature wave function of nonbombarded surface.
[f] δ_d, phase shift between applied modulating wave function and temperature wave function of bombarded surface.
[g] HC, hollow cylinder.

Electron gun heating as well as direct anode heating is utilized. The plain cylinder is simultaneously heated by a gun and by oven radiation. Also anode heating and optical modulation is used – see Ref. 72 – while Ref. 79 describes a heating method using two modulated electron guns. Also a combination of anode heating and electron gun heating has been used – see Refs. 22 and 80.

It should be remarked that mainly the measurement of Δ or of δ_0 is carried out, and in the Russian or related literature it is almost exclusively δ_0 that is measured. This is most probably due to the fact that in this latter literature solely anode heating is practiced. It is interesting to remark that amplitude decrements are nowhere measured, with Zankel[(41)] and Mustacchi and Giuliani[(4)] as the two sole exceptions. This is quite understandable as then one of the main advantages of the dynamic method is sacrificed.

Usually anode heating, in comparison with gun heating, is carried out at higher frequencies and on thinner specimens.

The disk-type specimens have mainly a diameter of approximately 6 to 9 mm and a thickness between 0.5 and 2 mm for the lower modulation frequencies (0.5 to 1 Hz), and between 0.1 and 0.3 for the higher frequencies (up to a few hundred Hz).

Experiments were primarily conducted on refractory metals, but also on insulators, semiconductors, dielectrics, etc. Also the liquid state is considered.

The high-temperature limit is either set by the specimen properties or is generally as high as 3000 to 3200 K.

The lower-temperature limit generally lies around 1100 K and only a few authors – see Refs. 41, 45, 49, 73, and 85 – succeeded in carrying out measurements down to some 350 or 550 K.

The uncertainty claimed probably seems to be somewhat optimistic in a few cases. Moreover, many times uncertainty, accuracy, precision, repeatability, and reproducibility are terms which are often either mixed up or not used neatly.

5.2. Experimental Designs

5.2.1. General Introduction

The specific features of experimental designs intended for the measuremeasurement of thermal diffusivity by modulated electron heating are described hereafter.

As mentioned before, one can recognize essentially two types of instruments depending on the type of specimen used, a disk or a hollow cylinder. However, these two types also can be called conformable as they only differ in the practical execution of the measuring chamber and in the final mathematical equations employed. Further a subdivision can be made when considering either direct anode-sample heating or electron gun heating.

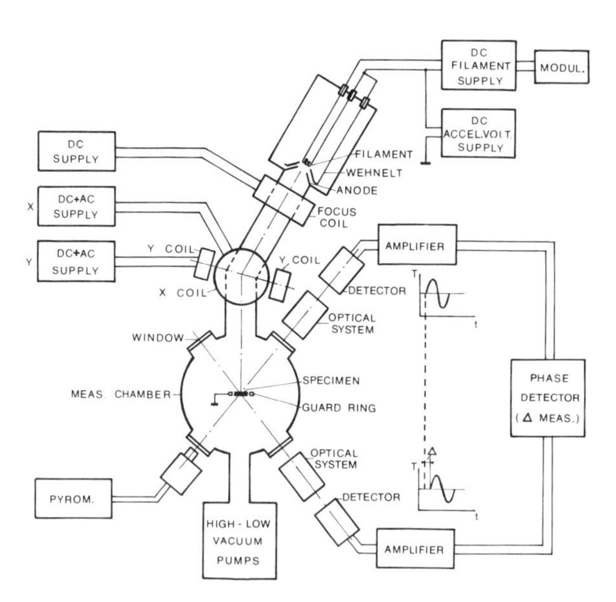

FIGURE 3. Schematic principle drawing of a sinusoidally modulated electron gun thermal diffusivity apparatus with disk-type specimen.

Also subdividing is possible considering the mode of measurement. Several phase and/or amplitude measurements are feasible in theory.

For the sake of simplicity, only an electron gun apparatus with a disk-type specimen will succinctly be described, illustrated by Fig. 3, but also in analogy and in contrast with a disk-type anode heating system.

5.2.2. Disk-Type Specimen Apparatus

An electron heating modulated thermal diffusivity apparatus comprises of course a measurement chamber which is brought to a vacuum of say better than 10^{-4} mBar by a vacuum pump group provided with its adequate vacuum-measuring instruments. The specimen and its specimen holder are contained within this chamber.

The electron power generator, either an electron gun or a direct anode heating system, is placed in the chamber or forms part of it. This generator is fed by the necessary power supplies and its power can be modulated as required. Viewing ports and/or electrical feed-throughs are provided. Ports are used for optical temperature measurement by a brightness pyrometer or by a two-color ratio pyrometer, or when measuring the temperature modulation of the specimen surface by radiation detectors. A phase detector enables the measurement of the phase difference either between both faces or between the applied power modulation and one of the faces, depending on the procedure adopted.

The construction of an anode heating setup as well as of an electron gun apparatus has mainly to be undertaken by preliminary trial and error tests. Therefore reference to former equipment is recommended. Both methods, anode heating or electron gun heating, have their advantages and their drawbacks. Selecting between both methods depends on the personal specific requirements and experimental capabilities. Each particular case has to be checked experimentally in view of an optimum solution.

From the technical point of view, in the case of disk-type specimens, anode heating is naturally much simpler than electron gun heating. The mechanical construction is much easier while lower voltages for the power supplies generally are sufficient.

Authors designing anode heating systems often implicitly accept that a planar cathode arranged in proximity of a flat specimen will ensure isothermal heating. Both the cathode and the specimen are considered to be components of a semi-infinite system of two planar electrodes. However, this has probably not yet been proved and serious doubts arise about it knowing that the cathode is always a flat wound or folded wire. It is self-evident that such a cathode does not show a homogeneous temperature distribution and that full symmetry can hardly be obtained. This means that the cathode will neither show equipotentiality, nor will it show a continuous surface. Further, the use of interelectrode distances of the same order of magnitude as the dimensions of the cathode and specimen invariably lead to a distortion of the equipotential planes of the accelerating electric field. This gives rise to focusing or defocusing effects, particularly pronounced at the specimen periphery. Moreover, anode heating systems usually operate with an accelerating voltage of the order of a few hundred volts to maximum 1500 to 2000 V. This entails that the electrons arrive at the target only weakly attracted and more or less intermingled as well in direction as in velocity.

Direct thermal and optical fluxes from the cathode to the specimen, and vice-versa, are practically unavoidable as the cathode is positioned in the immediate vicinity of the specimen. Under certain circumstances this thermal radiation is capable of raising the temperature of the cathode, besides inducing a noticeable delay before attaining a steady-state condition.

Molecular or ionic flows or depositions of vaporized–ionized materials will also easily affect the work function of the cathode and destabilize the heating and the emitting processes. Indeed, as a result of the interaction of the electrons with the specimen's evaporation or sublimation products or with the residual gases, active bombardment by positive ions of the cathode and cathode-adjacent elements can occur.

With this type of heating the heated surface of the specimen is practically masked for observation by optical means. Only thermocouples or pyrometry of the nonbombarded surface can be used.

In circuits with direct electron heating, exposure to the light emitted by the cathode can hardly be avoided. When, as is usually done, the temperature fluctuations of the back surface of the specimen are measured, a tubular light guide, e.g., a metal cylinder,[23, 40, 44, 67] is brought adjacent to that surface. This cylinder confines the luminous flux toward the photocell to the light emitted by the specimen only. Another way is to shield the cathode by solid screens and to direct the electrons to the specimen through an aperture having a diameter slightly smaller than that of the specimen.[80]

When using a disk generally an electron gun is used also. The electrons are accelerated to the working voltage and enter the equipotential zone of the vacuum chamber which accommodates the specimen. An important feature of this design is the remoteness of the cathode. Such an arrangement offers the advantage that intense thermal irradiation of the specimen surface by the cathode is precluded. Vice versa, exposure of the cathode by the heat of the specimen is rather low. This ensures higher stability. In addition, the specimen surface lends itself easily to temperature measurements by optical means being an important advantage. Moreover, even the small amount of direct light which still strikes the specimen can be reduced almost completely. Appropriate electromagnetic deflection of the beam, which is led through a bent guide tube – see, e.g., Refs. 25, 26, 32, and 49 – will allow only some stray light to pass.

The cathode of an anode heating system is normally a flat folded or wound tungsten filament. This applies also for a gun system but generally then the overall dimensions are much smaller. For a gun system sometimes also a block-cathode is used instead of a filament. This cathode is first anode-heated using an auxiliary tungsten filament. Such a block-cathode offers the advantage of delivering a high thermionic current and a high current density since its emitting surface is large and simultaneously its overall dimensions are small in comparison to a flat wound or folded wire.

The vacuum measuring chamber should preferably be made of a metal, e.g., stainless steel. This material has rather good vacuum degassing properties and is mechanically strong. Choosing a metal chamber has more advantages. Later adaptations are easily performed, a precise construction is possible, and electrostatic charges will not build up. A double wall can easily be composed. This allows for water cooling or even for hot water heating which improves considerably degassing efficiency during pumping down.

Electrical feed-throughs are commercially available and can be perfectly adapted to the chamber.

The viewing ports generally should be home-made as commercial ports seldom match the stipulations required. Shutters which protect the windows from vapor deposition in between the measurements are advisable – see, e.g., References 32 and 76. The windows themselves should be removable for cleaning purposes. Depending on the transmission characteristics of the pyrometer, of the detectors, or of other optical features in use, the windows have to be selected among materials like glass, quartz, sapphire, CaF_2, etc.

The purpose of the sample holder is of course to hold the sample in a well-defined and stable position with regard to the bombarding electrons. Since it is in physical, thermal, and electrical contact with the specimen, and as it often is as well bombarded by the electrons, either accidentally or intentionally, the material it is composed of has to meet several express requirements. Chemical compatibility, also at high temperatures, is necessary as well as a high melting point and a low evaporation rate, while both low thermal conductivity and high electrical conductivity are desirable.

Another purpose of the sample holder might be to provide simultaneously for thermal insulation of the specimen and/or thermal guarding of its edges. Moveover, the holder could be constructed in such a way as to counteract some specific parts to be bombarded by the electron flux.

It is obvious that the sample holder should be constructed so as not to disturb the results by its own diffusivity. This is in many cases a very delicate mission.

For solids several types of sample holders have been developed. Mostly the sample is simply resting on three or four tungsten wires 0.2 to 0.5 mm thick.

Wheeler's[25,26] specimens rest on three insulated, adjustable tungsten pins which are attached to a water-cooled tube. He provides for the possibility to slide the holder to another position where a second specimen is mounted. This allows one to leave the chamber closed in between two successive measurements.

Yurchak and Khromov[80] squeeze their sample between four quadrangularly positioned tungsten wires, two of them over, the other two beneath the sample.

Mebed and Yurchak[55,63] clamp their sample with its rim between four quadrangularly positioned molybdenum needles which can be moved with respect to the framework. This should allow one to center and hold samples having different diameters, but it is not clear whether this technique has any particular advantage, especially as it cannot be used for very thin or for easily deformable specimens.

Schmidt[33] uses an $X-Y$ movable table provided with three or four tungsten pins or wires on which the specimen rests.

Cerceo and Childers[22] suspend their sample by 0.5-mm tungsten wires and use radiation shields at the edges. Also De Coninck[86] suspends its samples by three 0.2-mm ϕ tungsten pins. However, these pins are protruding through the inner rim of a tungsten guard ring which has a slightly larger diameter and height than the specimen and which in turn is supported by three 0.5-mm tungsten wires. As the specimen and this ring are unitedly heated by the same electron beam which is scanned over the entire setup, radial heat losses are minimized and linear heat flow is promoted.

Zaretsky and Peletsky[81] use a completely different solution. Their sample, a thin disk of diameter 15 mm, is inserted into a niobium mandrel which is in turn inserted into a window on the lateral surface of a cylindrical, directly heated, niobium anode. In this way they combine the simplicity of a disk-shaped specimen with the advantages of a directly heated cylindrical specimen.

Next to the primary intention to measure solids, it is in principle also possible to carry out measurements on powders or granular materials and on liquids. Therefore a special encapsulating sample holder has to be designed while particularly a hollow-cylindrical-sample-holder and associated procedure is indicated for liquids.

Kim *et al.*[79] describe the measurement of molten UO_2 for which they constructed a disk-type tungsten cell which sandwiches a thin layer of UO_2.

For liquids, a deposition to control the internal pressure should possibly be provided. The preceding remark joins the idea of the possibility to carry out measurements under a certain pressure.

When powders or granulated materials or inhomogeneous or porous materials are to be measured, it is an absolute necessity that the inside diameter of the container be large compared to the average size of a single particle or of an individual zone, and that its surface is, e.g., some 100 times larger compared to the main individual cross-section area.

In the case of coarse-grained materials, the size of both the sample and the sample holder have to be large enough to sufficiently integrate the contribution of each grain into the overall thermal diffusivity of the specimen. No places may be found where accidentally either voids or particles are solely present.

For the modulated electron beam method, uniform temperature distribution or one-dimensional heat flow is obligatory. This is often sufficiently ascertained when the specimen is thermally guarded. A disk-shaped specimen lends itself easily to that purpose. A guard ring, whose internal diameter is a trifle larger than the overall diameter of the sample, and whose height is equal to or somewhat larger than that of the sample, offers a simple and reliable solution – see, e.g., Ref. 86. The value of the outer diameter of this ring is not critical. The material used for this ring is generally not critical either, provided its melting point is at least high enough to withstand the temperature envisaged for the measurement, while by preference its emissivity should be comparable to that of the specimen. This guard ring can be heated either with exactly the same electron beam density as used for the specimen, or by a slightly higher mean beam density in order to compensate for the heat losses.

Tungsten is a material which mostly meets sufficiently well all of the requirements imposed on the specimen holder and the guard ring. Only when an intricate mechanical construction is preconceived, tungsten might show the disadvantage of its uneasy machinability. Tantalum or a tantalum–tungsten alloy might be more adequate in that case.

The sample holder assembly can be mounted isolated from the chamber. The magnitude of the electron current passing through the specimen can then be measured if required. Consistently also an individual extra voltage can be applied to the specimen, if required.

The geometry and the dimensions of the specimen preferably should meet particular requirements.

In principle any configuration is admissible, but for the sake of simplicity, chiefly small disks or square plates are recommended.

A disk or a flat square are equally simple to make, but a disk shows the highest ratio of total surface to side surface for a given thickness. Radiation losses from the sides will therefore be minimum and proper axial linear heat transfer through the disk is conditioned by the choice of this ratio.

Thickness is governed not only by minimum radiation loss but also by minimum mechanical strength exigencies and by the modulation frequency. A smaller thickness calls for a higher frequency.

A good compromise which takes into account all these remarks seems for most materials to be a specimen with a diameter between 7 and 10 mm and a thickness between 0.4 and 1 mm and to use a modulation frequency between 0.5 and 5 Hz.

An important factor governing the design of an experimental system concerns the ability to make measurements of the main temperature and of the temperature modulation. This applies as well for anode heating as for gun heating.

The high-temperature limit, up to which measurements are possible, is in theory only determined by the restrictions imposed either by the characteristic properties of the material under investigation or possibly by those of the sample holder.

The low-temperature limit is not fixed by mean temperature measurement restrictions but only by the characteristics of the sensors used for the measurement of the temperature fluctuations.

Originally the low-temperature limit was approximately 1200 K. Meanwhile this limit could be decreased by a few authors. First Zankel(41) measured down to 330 K using thermocouples. Wheeler,(45, 49) using InSb detectors, developed a procedure making it possible in principle to carry out measurements down to approximately 400–450 K. Unfortunately his procedure does not seem to have been put into practice by others so far. Later, Zinov'ev(73) measured down to some 550 K and also Atalla(85) down to 400 K, both using thermocouples. Zaretsky and Peletsky(81,82) used PbS detectors and reached some 650 K.

When using thermocouples for the evaluation of the temperature fluctuations, the low-temperature limit will be the result of the low sensitivity of these couples. It is moreover especially difficult to screen off thermocouples from the primary, the secondary, and the stray electrons. This gives rise to spurious voltages which are very difficult to evaluate while possibly disparate electron bombardment of the specimen surface can occur.

One of the most critical operations when using thermocouples consists in joining the wires to the surface of the specimen, while exact positioning is also a delicate task. Spot-welding is mainly adequate for metallic specimens although often appreciable variations are noticed from one weld to another, despite numerous precautions. Frequently, what at room temperature appears to have been mechanically and electrically a good weld, betrays erratic response under operating conditions and particularly after exposure to higher temperatures.

With semiconductors often even more severe difficulties are experienced. The temperature response of the thermocouple does not always occur without a certain lag. Furthermore thermocouples often form eutectics with semiconductors well below the melting point of the semiconductor. On the other hand one can possibly utilize the thermal emf of the semiconductor itself to measure relative temperature differences. Regretfully, such a technique is usually not sufficiently reliable.

From all this it can be inferred that the main temperatures should be measured by radiation pyrometry and the temperature fluctuations by radiation detectors, as these offer so many advantages compared to thermocouples.

For the measurement of the main sample temperature several types of radiation pyrometers do exist. The total radiation pyrometer is normally not used since correction is often extremely worrisome due to the mostly defective knowledge of the total emissivity.

The brightness pyrometer, using the energy around a certain fixed wavelength, suffers somewhat from a lack of sensitivity because of the narrow energy band employed. On the other hand, the emissivity at a single wavelength is not susceptible to quite as much variation as is the total emissivity. The wavelength around 0.65 μm is commonly used for visible range brightness pyrometers, whereas 2.3 μm is a popular wavelength for an infrared brightness pyrometer.

The two-color or ratio pyrometer is still less dependent upon the emissivity. Here the energy around one wavelength is measured as well as the energy around a second wavelength and the ratio of these energies is taken as a measure for the temperature.

For the measurement of the temperature fluctuations radiation detectors which look through adequate vacuum-tight windows obviously constitute the ideal solution.

Photomultipliers or photocells are nowadays particularly well substituted by solid-state semiconductor radiation detectors such as Si detectors. PbS detectors are also in use on account of their ability to detect lower temperatures. For the lowest temperatures either InSb or GaAs or InGaAs or (HgCd)Te or PbSe or PnSnTe or the like detectors are necessary.

When using radiation detectors, the low-temperature limit will be bound to the sensitivity of these detectors. Indeed, the lower the temperature of the specimen, the more the radiated power shifts toward the infrared and, simultaneously, the lower the absolute value of the total black-body radiation becomes. This implies of course a smaller signal delivered by the detector. Unfortunately two more degenerating effects are acting. Indeed all detectors do show a rather sharp decrease of their detectivity above a certain wavelength and also their overall quantum efficiency decreases rather strongly with increasing wavelength.

For the evaluation of the experimental results it is neither necessary to know the heat flux nor the exact temperature difference. Therefore, phase shift measurements, although these might be difficult under certain circumstances, are sufficient. Consequently, the detectors do not need accurate calibration. All that is required is a high sensitivity and a linear response over a narrow interval.

After preamplification and amplification the signals picked up from the surfaces are either fed directly to a phase-measuring instrument, or else the leading signal is delayed or the lagged signal is advanced by a calibrated phase shifter before being fed to a zero detector.

For modulation frequencies above a few hertz adequate phase-measuring instruments or zero detectors are commercially available. For lower frequencies generally the phase-measuring instrument should be a home-made one. This cannot represent serious difficulties for a normally experienced electrician. As the signal is composed of a small ac modulation signal superimposed on a large dc signal, dc compensating circuits have to be provided for.

Phase shift measurements can be executed between the bombarded and the nonbombarded surface, or between the modulated beam and one of the two surfaces. For these latter phase measurements the use of a Wehnelt cylinder is particularly profitable as one of the signals is simply readily available at the modulator generator.

A Wehnelt cylinder offers the extra advantage of keeping the cathode always at a constant temperature throughout. This is beneficial for the stability of the power to be delivered as well as for the lifetime of the cathode.

With respect to both the pyrometer and the detector characteristics, additional attention has to be paid to the emissivity or to the reflectivity of the sample. Samples which exhibit a low emissivity are less easily measured for two reasons. Firstly, light reflection toward the detector is higher. This light can find its origin outside the apparatus, but obviously the most arduous light source is formed by the incandescent cathode. Secondly, the light emitted by a sample toward the detector is evidently lower.

In order to increase emissivity it might be interesting to roughen the surface by sandblasting or to apply a coating to it. Graphite is many times an adequate coating material. Anyhow, the coating integrity, e.g., physical adhesion, or chemical compatibility with the substrate specimen and negligible volatilization has to be ascertained.

A sample property which is indirectly related to emissivity is transparency. In connection with thermal diffusivity measurements sometimes very complex problems are experienced regarding acquisition of proper, representative results. Indeed thermal radiation can contribute to direct internal heat transfer between both specimen surfaces. This contribution generally increases with increasing temperature. This implies that the diffusivity values, obtained in the customary way, will not necessarily be the specific, absolute values searched for. Also mean temperature measurements are possibly inadmissibly erroneous.

A great many different situations are possible. A few particular cases need closer examination.

When the temperature modulation detectors operate at a wavelength for which the specimen is transparent, a signal will be detected which is small the more transparent the specimen is. Moreover this signal will be constituted of the mixture of both the signals generated by the bombarded and by the nonbombarded surface.

When the detector operates at a wavelength for which the specimen is opaque, but when simultaneously the specimen is transparent in a rather significant broad adjacent part of the radiation spectrum, thermal diffusivity measurements will deliver in principle regular, specific results, although also a radiation component will be present.

If necessary, differentiation between the two heat transport components can be controlled by measuring the full transmission spectrum or by carrying out a few consecutive measurements either using specimens with distinct thickness differences or using at least two different modulation frequencies. Indeed, diffusivity is by definition independent of thickness and frequency, but the measured phase difference is linearly dependent on d (except for $B \ll$) and on $\sqrt{\omega}$, whereas transmission obeys an inverse exponential law.

The formulas considered in Section 4 have been derived assuming that the temperature distribution is isothermal over the specimen surface and that no side or edge losses intervene. Absolute endeavor to respect this condition should be one of the main objects to be pursued when designing an electron bombardment modulated thermal diffusivity equipment. A deviation from one-dimensional heat flow can indeed give rise to rather large errors.

In electron heating systems the electron beam density shows theoretically a three-dimensional Gaussian distribution, at least when starting from a cathode with a circular surface. Parameters like the imperfect geometry of the cathode, its noneven temperature dissipation and/or thermionic emission, and its nonalignment, next to other misalignments, can affect this distribution. Consequently it should be realized that uniformity of the beam cannot be obtained unless taking a great deal of trouble only. The degree of nonuniformity and the corresponding effect on the measured diffusivity should be examined thoroughly experimentally. Next to providing for thermal guards around the specimen, attention has to be paid to a possible screening-off of the primary electrons by some parts of the setup (like thermocouples, earthing wires, etc.) or to extra heat losses at electrical contacts.

In order to minimize the error due to nonisothermality it is worth carrying out measurements whereby not the whole specimen surface is taken into account but only the central part of it. Therefore an optical system is indispensable. This system has to project an image of the surface of the specimen on an aperture which allows only that central part to pass through it before reaching the detector. It will consist either of normal lenses or, for the lower temperatures, of intrared lenses (e.g., Si, Ge, quartz, CaF_2, ZnSe, BaF_2, CdTe, etc.) or of a concave mirror disposition.

When temperature uniformity is sufficiently well satisfied, or when central radial symmetry is observed, it is not compulsory to restrict selection of the perceived zone to exactly the center of the specimen only. Indeed, no particular sensitivity to radial position is noticed as long as the area considered is comprised within the "homogeneous" zone. This is a great comfort since in that case the viewing area of the bombarded and of the nonbombarded face do not need exact reciprocal optical alignment.

With an electron gun almost perfect isothermal heating of the specimen is possible. In fact no matter which deliberately chosen temperature distribution can be imposed. Indeed, the beam produced by an electron gun can easily be refashioned and can moreover be swept cautiously over the specimen surface by electrostatic or electromagnetic means. Simple $X-Y$ scanning and focusing or defocusing is easily carried out. In this respect the sophisticated concept of Mayer and Neuer[(34)] is worth mentioning. They succeeded in compensating almost perfectly the heat loss from the edges of a disk-type specimen.

Electron heating devices might be designed in such a way that it is possible to determine the power transmitted to the sample. Besides thermal diffusivity, also heat capacity can then be measured and thus also direct calculation of thermal conductivity is possible.

To know exactly the power applied to the specimen is an extremely difficult challenge, and the results are often unreliable.

With electron beam heating this problem seems to be almost unsolvable, certainly when the beam is in addition scanned. For direct anode heating a solution can be proposed which is theoretically interesting but rather complex to realize, namely a planar cathode–anode configuration whose electrostatic field produces a focusing effect only toward the specimen. Unfortunately isothermal heating is then of course an additional problem. Moreover, for recapture of all secondary electrons onto the specimen, the distance between the wall of the chamber and the specimen should be as small as possible, whereas the potential of the chamber should be at least equal to that of the cathode, or possibly even negative with respect to it. It is evident that here very complex fulfillments have to be satisfied and that possible inadmissible electrical breakdown has to be faced.

5.2.3. Hollow-Cylinder-Type Specimen Apparatus

As mentioned before, the greater part equal conditions apply in practice as well for a directly heated disk-type apparatus as for a cylinder-type apparatus.

Gun heating is not recommended when a cylindrical geometry has been chosen. Heating of the outer surface requires more than one gun and simultaneous proper scanning of these beams. This obviously implies extremely serious problems. Gun heating of the inner surface of a hollow cylinder is technically not feasible.

Direct electron heating is the only possibility with cylindrical specimens. A filamentary cathode, stretched along the axis of the cylinder, will generate radial electron tracks.

The heat released by the inside surface will readily be uniform, at least provided the longitudinal measure of the cylinder and the distance over which it is bombarded are sufficiently large. Unfortunately faultless centering of the cathode over the full length of the cylinder is indispensable. Some authors[37, 62, 83] also heat the outer surface of their cylindrical specimen by anode heating.

An important advantage of direct electron heating together with a hollow cylinder specimen is the possibility to measure less erroneously the amount of energy induced into the specimen. This might facilitate the measurement of C_p, which is difficult to achieve with gun heating because of secondary electron emission.

In the case where the specimen is only a part of a cylindrical surface, as described in Ref. 81, one should take into consideration the possible lack of compensation for the secondary currents between the specimen and the cylinder. Also the unavoidable opening between the specimen and the cylinder may represent a source of systematic error in the determination of the heat capacity.

Adequate solutions to thermal guard cylindrical specimens are generally less easily experienced. Utilization of a relatively long cylinder is in many cases the only possibility.

Direct light emission from a central cathode to the inner wall of the cylinder is unavoidable. Besides, after all, optical measurement of the inner wall temperature is not possible in a simple and reliable way. Moreover, when a cylindrical specimen is anode heated from the outside, direct light emission toward this outside surface is inevitable.

A really positive aspect of the utilization of thin wall hollow-cylindrical-type specimens is found in a worthwhile extension of the application area of the electron heating diffusivity method since its perfect suitability to serve as a container for molten substances. Also the fact that specimens in the form of a tube can be measured directly in their original configuration is sometimes an important advantage. Indeed for certain construction tubes it is often not advisable to machine them. This is, among others, the case for, e.g., irradiated reactor fuel cladding tubes.

When measurements are carried out in the liquid state, internal convection streams might hamper acquisition of correct results or even lead to grave erroneous results. In order to attenuate as much as possible convection, and simultaneously in the hope not to affect too much the real diffusivity value, a few authors[50, 57, 58, 61, 62, 83] install several transverse tantalum foil partitions.

5.3. Enumeration of Advantages and Drawbacks and Some Limitations

The use of electron bombardment for the determination of thermal diffusivity proves to be a fast method. The time required for the stabilization of the mean temperature, after a new adjustment has been initiated, generally represents not more than one to two minutes. Subsequently, acquisition of each individual datum calls only for maximum a few minutes in total. Nevertheless, for temperatures below, e.g., 700°C, stabilization times are noticeably increasing and at, e.g., 500°C even some five or more minutes might be necessary.

Electron bombardment purveys the advantage to be highly flexible and versatile. Indeed, the power delivered by the electrons can easily be controlled and their path across the specimen surface can be commanded by a theoretically unlimited choice of signals. This leads to a wide variety of experimental possibilities.

Small specimens can be used. Therefore the method is particularly interesting when substances are to be measured which are expensive, exotic, or even whose manufacture or machineability is relatively difficult. Also because of the small sample dimensions very high temperatures can be reached in a short time and with a relatively low power.

Electron heating is the only practical technique which allows for high-temperature diffusivity measurements, at least next to the flash method described elsewhere, since classic steady-state methods entail severe difficulties when trying to calculate or to compensate the considerable radiation losses encountered.

Thermal diffusivity measurements with the aid of electron bombardment are, by the nature of things, applicable to electricity-conducting materials only. This limitation is evident knowing that it is imperative that the charges imparted by the bombarding electrons should flow away from the specimen easily. If this is not accomplished these charges will build up a more and more negative potential which can possibly even lead to the complete arrest of the bombardment. This problem exists likewise whether using beam heating or anode heating, although incidentally it is less acute with beam heating, particularly because of the higher voltages in use.

The electron current which passes through the sample can possibly distrub the temperature distribution of that sample by ohmic heating. A closer examination suggests itself.

For a disk-shaped specimen of diameter $2r$, thickness d, and resistivity ρ_e, and which is earthed by its whole nonbombarded surface, the resistance to earth is $R = \rho_e \, d/\pi r^2$.

The power P, discharged by the electrons in the specimen, must be at least equal to $\epsilon\sigma T^4 2\pi r(d + r)$, the power lost by thermal radiation.

The Ri^2 heating will be negligible provided it be less than, e.g., 1% of the beam power. Consequently: $\rho_e(d/\pi r^2)i^2 \leqslant 0.01P$, and as $i^2 = (P/V_a)^2$:

$$\rho_e \leqslant \frac{0.01 V_a^2 r}{2d(d + r)\epsilon\sigma T^4} \tag{54}$$

Therefore, for a specimen with $2r = 10$ mm, $d = 1$ mm, and $\epsilon = 0.5$, and using an accelerating voltage V_a of 10 kV, the resistivity has to be smaller than 1.4×10^4 Ω m at 1000 K, or 9.2×10^3 Ω m at 2000 K, or 1.8×10^2 Ω m at 3000 K. This however is an ideal, generally nonexisting situation. Commonly these requirements will be at least some ten times more severe since electrical contact will mostly be made through a few points along the edge of the sample only. However, even this will not present too often a problem.

A partial solution permitting to measure insulators lies in the installation of a supplementary heater to bring the material up to the temperature where conduction is low enough to be taken over by electron heating.

Another possibility is to deposit onto the surface of the sample a conducting layer. This layer has to be thin enough in order not to disturb the overall thermal diffusivity. A layer of a few Angstrom (10^{-10} m) deposited onto a specimen whose thickness is even as small as 0.1 mm (10^{-4} m) will reduce the overall electrical resistance by several decades but normally the thermal properties will only be disturbed by not more than a factor of the order of 10^{-4}–10^{-5}. Nevertheless, measurements on some ceramic materials will still not be possible below a certain temperature.

In connection with thermal diffusivity of layered composites, it is worthwhile to consult the work of Lee *et al.*,[87] while Brandt and Havránek[88] established an analytical solution allowing measurement of the thermal diffusivity of one of the layers of a two-layer sample.

Also, the calculations and measurements of Ang *et al.*[89] deserve attention although only the heat pulse technique is concerned. They expose a general solution for the three-layer conduction problem.

Dul'nev and Sigalov[90] treat the thermal diffusivity of inhomogeneous systems and carry out temperature-field calculations on quasihomogeneous-body models and Farooq *et al.*[91] present interesting considerations about three-layered cells.

Working under vacuum implies for one specimen an advantage and for another one a disadvantage. Oxidation will generally be prevented and moreover heat loss by convection will not exist. On the other hand, specimens with a high vapor pressure or a high sublimation rate may inflict corresponding difficulties, among others, possible vapor deposition on the sight-glasses. For these materials heating under normal pressures or under an inert atmosphere, by an arc-image, by direct thermal radiation, by light or by a laser or possibly even by a plasma, should then be advantageous.

High-temperature measurements often can give rise to difficulties or discrepancies by the fact that also other physical and moreover chemical instabilities can occur in the material under investigation. Oxidation or even complete disintegration can occur when the vacuum conditions are not met for that particular specimen. Contamination by vapors produced by some equally bombarded parts or incompatibility with the sample holder, next to phase transformations, diffusion of atoms or of defects which possibly leads to inhomogeneity or anisotropy, sintering, recrystallization, distortion, fatigue, cracking, vaporization, sublimation, or even ablation are phenomena which can occur.

These phenomena are in most cases drawbacks but they certainly are not solely confined to, or inherent in, the use of electrons as a heating means. In most cases they can readily be detected, or sometimes they are exactly one of the properties simultaneously searched for.

Utilization of thin specimens and relatively high modulation frequencies makes it possible to maintain the amplitudes of the temperature fluctuations to within a few degrees or even to a few tenths of a degree. This permits thermal diffusivity to be examined with high resolution as, e.g., is necessary in the neighborhood of a phase transition.

When one of these phenomena is deliberately pursued, it is sufficient to simply leave the specimen at the predisposed temperature for a time long enough to reach the desired effect. Examples of this possibility are given, e.g., in Refs. 32, 65, 76.

The measurement of thermal diffusivity is often, at least at high temperatures, an indirect means to determine thermal conductivity. Indeed at high temperatures the measurement of thermal diffusivity, specific heat and density of solid materials is less delicate than the measurement of thermal conductivity. However, it is moreover much more than an indirect means to measure conductivity. Next to its direct importance for technical applications it is an additional means to study transport phenomena.

Good thermal contacts are an absolute necessity for most thermal property measurement methods, particularly for the steady-state methods. This produces almost always severe difficulties which mainly cannot be solved adequately, neither at high, nor at low temperatures. In the case of electron beam modulated heating, good thermal contact between the sample and both the heat source and heat sink are conditions which simply do not come up for discussion. This is also the case for the contact between the sample and the temperature sensors, except when using thermocouples instead of radiation detectors.

The possibility to measure simultaneously heat capacity has been exploited since a few years, but nearly only together with direct electron heating so far – see last column in Table 1.

Only a few such experiments have been conducted with gun heating. The attempt of Wheeler[92] can serve as an example. He measured the heat capacity of platinum by a pulse-modified version of his modulated electron beam diffusivity apparatus. The results showed anomalies which he inputed to secondary electrons. Therefore he provided for compensation by secondary electrons from the sample holder. Unfortunately probably also an important fraction of the beam electrons passed alongside the specimen.

A drawback of the electron bombardment modulated method might lie in the fact that a rather arduous and elaborate mathematical procedure is compulsory. The use of a computer is an absolute necessity, except when simplified equations, and consequential lower accuracy, are judged to be satisfactory. Directly related to this is the necessity to know satisfactorily several parameters over the whole temperature range under consideration. This can cause extra embarassing or sometimes even unmanageable troubles. Many times, indeed, none of the parameters like λ, C_p, ρ, ϵ, ϵ_λ, Δd, are known for the material to be measured.

A significant advantage, when using an electron gun and a bent guide tube, is that next to diffusivity also simultaneously spectral emissivity measurements can be carried out over a broad temperature range. Examples thereof are given in Refs. 32, 65, and 76.

5.4. *Error Evaluation*

Experimental results are universally always subject to two not completely independent sources of error, measurement errors and nonmeasurement errors.

Measurement errors are associated either with direct, intrinsic, instrumental errors, or with uncertainties that exist in the values of those properties contained in the equation used to compute the desired result from experimental data.

Nonmeasurement errors are associated with experimental conditions which deviate from the conditions assumed in the theoretical model used to appropriately compute the desired result.

Numerous measurement errors are possible in the case of thermal diffusivity measurements. Next to pure instrumental errors, a large number of parameters can interfere adversely or can be the origin of a great deal of trouble. Indeed, the value of several of these parameters is either uncertain, or indefinite, or even unknown.

Particularly the following parameters are involved: the mean sample temperature (T), the phase difference Δ, the thickness of the sample (d) and its linear thermal expansion coefficient (Δd), the total emissivity (ϵ), the spectral emissivity for the wavelength at which temperature is measured with a brightness pyrometer (ϵ_λ), the density (ρ), the specific heat (C_p), and modulation frequency (f).

In order to assess the influence exerted by each, possibly erroneous individual parameter, some theoretical calculations have been carried out in compliance with the mathematical procedure adopted by one of the authors – see Refs. 25, 26, and 36. For two samples which had been measured in the past, one with a rather low diffusivity (a UC_2 sample – see Ref. 76), and another showing a high diffusivity (POCO AXM-5Q sample – see Ref. 93), a deviation of both plus and minus 10% was deliberately imposed to the accepted value of each parameter. The resulting influence on temperature (T), diffusivity (a), and on conductivity (λ) was analyzed.

Table 2 gives a survey of the average error, computed over the whole temperature range applicable.

Two classes of errors can be noticed. The small errors, ranging up to about 0.3% to 0.4%, concern those values all of which can be considered to be in reality zero, at least theoretically. Indeed, since all results were calculated by an iterative technique, each deviation imposed on one parameter carries along a certain, albeit small, departure of the value of all other results.

The large errors, ranging between say 0.5 and 20, are inherent in the thermal diffusivity determination method itself. It is obvious that this procedure will not provide veritable or substantial error predictions, although it clearly will infer a fair warning when only parameters are disposable whose reliability might look suspicious.

The next few paragraphs expose a more detailed analysis of possible errors associated with some of the various parameters involved.

When carrying out diffusivity measurements, each value must naturally be assigned to the mean measuring temperature. The degree of difficulty in determining this temperature grows, as a rule, with increasing temperature.

Thermophysical properties, and here in particular thermal diffusivity and conductivity, are meaningless when stated disengaged from temperature.

TABLE 2
Calculated Average Error for Temperature, Diffusivity, and Conductivity as a Result of a Plus or Minus 10% Deviation of the Interfering Parameters

Parameter with + or −10% deviation imposed	Calculated approximate mean error (%)					
	UC_2 sample (Ref. 76)			POCO AXM-5Q sample (Ref. 93)		
	T	a	λ	T	a	λ
T	11–13	0.2–0.6	5–15	10–11	0.1–0.7	≈ 2
Δ	0.02–0.07	11–15	11–15	0.01–0.1	≈ 15	≈ 15
d	0.01–0.06	≈ 18	≈ 18	0.01–0.08	≈ 20	≈ 20
Δd	0–0.005	0.2–0.5	0.1–0.3	0–0.005	0.03–0.4	0.02–0.2
ϵ	0.01–0.06	0.07–0.3	0.08–0.4	0–0.07	0.02–0.4	0.02–0.4
ϵ_λ	2–4	0.02–0.2	0.7–4	1–4	0.01–0.2	0.2–0.8
ρ	0.01–0.06	0.07–0.3	≈ 10	0.01–0.08	0.02–0.4	≈ 10
C_p	0.01–0.06	0.07–0.3	≈ 10	0.01–0.08	0.02–0.4	≈ 10
f	0–0.009	1–2	1–2	0–0.009	0.5–1	0.5–1

To reduce the temperature measurement uncertainty to tolerable limits constitutes, regretfully, many times a major embarrassment, and often the lack of agreement between the measurements of different scientists results primarily from this fact.

Unless thermal diffusivity does not change too much with temperature, imprecision of the temperature measurement is, proportionally, less important. When, on the contrary, thermal diffusivity changes very quickly with temperature, it has to be known with a high precision. This, e.g., is the case for measurements in the neighborhood of a phase transition. Otherwise both should be known with comparable precision.

The transmittance of the window through which temperature is measured should to be known as best it can and should be followed continuously during the whole measurement sequence. Transmission affects the measurements in exactly the same way as the emissivity of the specimens does.

The error equation for the brightness pyrometer, in accordance with Planck's radiation law, is

$$T_T = C_2 \lambda_B^{-1} \ln^{-1}\{\epsilon_\lambda \epsilon_T [\exp(C_2 \lambda_B^{-1} T_B^{-1}) - 1] + 1\} \tag{55}$$

where T_T is the true temperature (K), T_B is the brightness temperature or the reading of the brightness pyrometer (K), C_2 is the universal second radiation constant (1.438786×10^{-2} mK), λ_B is the wavelength selected about which to make the measurement (m), ϵ_λ is the spectral emissivity at λ_B (dimensionless), and ϵ_T is the transmission coefficient, at λ_B, of the window through which the pyrometer receives its radiation (dimensionless).

For $\epsilon_\lambda \epsilon_T = 0.5$ instead of unity (and using for $\lambda_B = 0.65\,\mu$m), the error will be about 8.5% and 18% at 2500 and 5000 K, respectively.

When temperature is not too high, this formula can be reduced to its simplified form when using Wein's radiation approximation:

$$T_T = [T_B^{-1} + \lambda_B C_2^{-1} \ln(\epsilon_\lambda \epsilon_T)]^{-1} \tag{56}$$

The supplementary error introduced by this simplification will be about 0.03% at 2500 K and about 1% at 5000 K.

The two-color or the ratio pyrometer is designed assuming ϵ_{λ_1} to be equal to ϵ_{λ_2}, the indexes 1 and 2 referring to the first and second wavelength, respectively. In that case the measurement of the pyrometer is, apart from its own instrumental inaccuracy, exactly conformal with the true temperature.

When ϵ_{λ_1} is not equal to ϵ_{λ_2} nonnegligible errors may be experienced. The relation between T_T, the true temperature, and T_R, the ratio or indicated temperature, is

$$\frac{\exp(-C_2\lambda_1^{-1}T_R^{-1})-1}{\exp(-C_2\lambda_2^{-1}T_R^{-1})-1} = \epsilon_{\lambda_1}\epsilon_{\lambda_2}^{-1}\epsilon_{T_1}\epsilon_{T_2}^{-1}\frac{\exp(-C_2\lambda_1^{-1}T_T^{-1})-1}{\exp(-C_2\lambda_2^{-1}T_T^{-1})-1} \tag{57}$$

For a commercial ratio pyrometer working at the wavelengths $\lambda_1 = 0.53$ and $\lambda_2 = 0.62\,\mu$m and for a difference of 10% between $\epsilon_{\lambda_1}\epsilon_{T_1}$ and $\epsilon_{\lambda_2}\epsilon_{T_2}$, the error of T_R versus T_T is about 6% and 13% at 2500 and at 5000 K, respectively.

In its simplified form this error function is straightforwardly calculable:

$$T_T = T_R[1 + \lambda_1\lambda_2(\lambda_2 - \lambda_1)^{-1}T_R C_2^{-1}\ln(\epsilon_{\lambda_1}\epsilon_{\lambda_2}^{-1}\epsilon_{T_1}\epsilon_{T_2}^{-1})] \tag{58}$$

The supplementary error introduced by this simplification will be about 0.003% and 0.25% at 2500 and at 5000 K, respectively.

A relatively simple method which enables one to correct for the possible difference in emissivity at the two wavelengths is described in Ref. 76. With a mirror adaptor mounted in a spectrophotometer it is possible to calculate the emissivity ratio. It has to be remarked that nowadays such an adaptor, or even a more sophisticated one which can serve the same purpose, is commercially available from most spectrophotometer constructors.

Measurement errors occurring when determining the phase difference can have several very complex origins, such as, among others, the response characteristics of the detectors, the amplifiers, or other networks. Simple means to verify these phase errors consist in bombarding a very thin (e.g., 0.02 mm) tungsten or tantalum foil. The phase difference will be close to zero, at least for modulation frequencies below, e.g., 1 Hz. One can also illuminate simultaneously both surfaces of a normal, high-reflecting sample by a high-intensity light source, modulated at the frequency under consideration.

Measurement errors can also be associated with determining the effective thickness and indirectly with the linear expansion coefficient.

The phase difference Δ is directly proportional to the specimen thickness. In order to avoid an extra source of error due to thickness, it is important to machine the specimen to flat and parallel surfaces.

Nonuniform thickness gives rise to two- or three-dimensional heat flow. Fortunately the error in consequence of this not one-dimensional flow is negligible in comparison to the error due to nonuniform thickness itself.

In connection with the influence of total emissivity and heat loss by radiation, a numerical analysis by Wheeler[94] showed that the accuracy of the diffusivity measurement is only faintly influenced by possible large errors in values assumed for radiation heat losses. When, e.g., total emissivity is in error by as much as ± 50%, the error in thermal diffusivity will be less than ± 0.5%. Our simple, practical calculation listed in Table 2 showed for this particular case a comparable conclusion.

In connection with errors due to energy transfer by radiation through transparent or semitransparent materials, see for instance Refs. 95 and 96.

About the influence of density and specific heat, no special comments have to be made. It is obvious that these parameters have in theory no influence on a, but that λ is directly proportional to them.

The influence of a deviation of the modulation frequency exerted on a and on λ is rather small and is negligible on T. This is normal as a and λ are proportional to less than the square root of frequency, while temperature is theoretically independent of it.

A higher frequency entails a more easy treatment of radiation losses, a shorter measuring time, a better signal-to-noise ratio, and a more convenient capacity to process the signals. On the other hand, a higher frequency implicates a decrease of the temperature amplitude while uncertainties might arise when the measured phase difference becomes larger than $\pi/4$ or some multiple thereof.

Measurements at lower frequencies are relatively more difficult from the technical point of view, and in order to avoid a small phase difference a larger specimen thickness is desirable.

Concerning the nonmeasurement errors the major source of error resides in the possible inhomogeneous temperature distribution over the sample.

Inhomogeneity of the temperature distribution can be provoked by a multitude of causes, several of which have been exposed formerly in Sections 5.2 and 5.3. A more exhaustive list should also quote the following causes: nonuniformity of some thermophysical properties of the sample, Thomson or Peltier effects, magnetic fields, radioactivity, external radiation, etc.

Each potential reason either contributes to inhomogeneity or constitutes a distinct source of error in itself. Nevertheless, most of these reasons generally involve theoretical rather than real measurable effects.

The open literature reports only a few investigations with respect to errors associated with inhomogeneous temperature distribution.

Pridmore-Brown[97] exposed an analysis of the influence exerted by a uniform beam and also by a beam concentrated into a spot at the center of a disk-shaped sample. For both cases the phase difference is calculated between the temperature at the center of the front and of the back face, and also between the average temperature of these faces.

The phase difference is found to be the same either at the center of a uniformly irradiated sample or when the average temperature is measured on a spot-irradiated sample. Nevertheless, in general thermal diffusivity will be overestimated for a nonisothermal specimen although the error will be very small.

Although concerned only with the laser flash method, the calculations of Beedham and Dalrymple[98] are to a large extent applicable to the modulated electron beam method also. They assess that a considerable degree of nonuniformity can be tolerated. For central hot spot heating, measurements over a small area at the center of the disk would not introduce any error. However, very large errors are calculated when measurements are carried out more toward the edges while averaging of the temperature fluctuations would lead to an increase of the diffusivity. Also, for a central cold spot, measurements of temperature responses over an extended area will lead to smaller errors.

Although also dealing with the flash method technique only, several features described by Taylor(99) apply unequivocally to the modulated electron heating technique as well.

Brandt and Neuer(16) examined by a simulation calculus possible errors due to nonisothermality. They showed that the diffusivity value obtained depends on the shape of the temperature distribution as well as on the coordinates of the area perceived by the detector. Higher values as well as lower values compared to the real value, can result. They also calculated that the error decreases with increasing frequency, at least between 0.5 and 5 Hz. When the phase shift is measured between the beam and one of the integral surfaces, the error introduced by inhomogeneity becomes negligibly small. For this reason integral phase shift measurements are proposed, although because of the possible adverse influence of edge losses, in practice measurements should be carried out on smaller areas.

Anomalous results which might be attributed to experimental uncertainties connected with inhomogeneous temperature distribution are also evaluated in Ref. 100.

To conclude this discussion on errors possibly due to inhomogeneous temperature distribution, the following statements can be made.

All thermal diffusivity measurements rely entirely on a mathematical treatment in which heat is transported through the sample in a one-dimensional way. When in practice, deviations from this one-dimensional assumption occur, rather large errors may result. In order to deal with one-dimensional heat transport some conditions have to be fulfilled. Either isothermal heating of the sample and conservation of linear heat flow perpendicular to the surface, or dictation of a desired energy profile over its surface is necessary. This is achieved either by $X-Y$ scanning of the electron beam, and guarding the edges, or by another scanning technique such as, e.g., described by Mayer and Neuer.(34)

One-dimensional heat transport is in practice never fully attainable. Nevertheless, unpublished tests from one of the authors have shown that in the case of a specimen in the form of a disk, a radially symmetrical nonuniformity of 1%–2% does not have any significant influence either on the repeatability or on the accuracy.

A few interesting initiatives dealing, among others, with error evaluations of thermal diffusivity measurements can be mentioned. Fitzer(93) and Minges(101, 102) came basically to equivalent final conclusions. They also propose possible calibration materials. Nowadays Minges is working out, in the framework of CODATA,* a complete new report, using newly analyzed and fresh data. In this connection, see also Hust.[103]

As a closing remark, a simple and generally adequate method of evaluating possible measurement errors consists in carrying out measurements under identical circumstances on a few specimens of the same materials but which have various thicknesses. Possibly also tests on specimens with different diameters can be carried out. Nonmeasurement errors can be ignored when sufficiently good mutual reproducibility of the results is observed, unless one or another intrinsic specimen property can be inculpated as, e.g., described when dealing with heat transfer by radiation. Finally, measurement errors can perfectly well be explored by making reproducibility–repeatability measurements in cooperation with a few colleagues or if possible, by measuring under ideal circumstances a calibrated material.

6. CONCLUSIONS

In assessing the role played by the electron bombardment modulated heat input method, the following conclusions can be drawn:

1. Electron heating allows one to shift the working temperature limitations, inherent in steady-state methods, to much higher values, while actually also the lower temperature range can be covered adequately.

2. The necessity to carry out experiments under vacuum represents for one substance a drawback and for another one an advantage.

3. It is a versatile, flexible, and fast method. As opposed to conventional heating systems, electron heating has made it possible to produce much easier, and in a more correct and adequate manner, the desired boundary-value conditions imposed by the theory. Materials of every sort and kind can be measured, ranging from excellent conductors to insulators.

4. The fact that an arduous mathematical procedure is necessary is disadvantageous although that does not represent a substantial problem any longer because of the current availability of computers.

5. As some ten different parameters are involved in the diffusivity determination there is every reason to accept introduction of as many errors. Fortunately most parameters do affect the results negligibly only.

6. Heating by electron bombardment allows the use of ac techniques instead of dc techniques. It promotes electronic measuring methods facilitating the acquisition and the processing of the data.

7. Measurements can be carried out utilizing thin specimens and/or small

*Committee on data for science and technology.

temperature differences. This makes it possible to study thermophysical properties of substances in the immediate vicinity of a phase transition and of other effects which change rapidly with temperature.

8. As disk-type specimens can be used with relatively small dimensions, electron modulated heating is an ideal solution when dealing with exotic, expensive, or not easily manufactured materials.

9. Electron anode heating forms the ideal solution for high-temperature thermal diffusivity determination of a hollow cylindrical sample. This makes it possible to measure certain samples in their original configuration, such as some particular tubes, as well as substances in their liquid state.

10. Electron heating, when properly worked out experimentally, permits the determination not only of thermal diffusivity but also of heat capacity and thus of thermal conductivity, utilizing a single experimental setup. Nevertheless, utilization of an electron gun presents in that case severe inconveniences in comparison to anode heating.

As a by-product, high-temperature spectral emissivity measurements can be carried out but now only using a remote and bent gun apparatus.

11. Although completely independent of any association with diffusivity measurements, it is worthwhile emphasizing an additional interesting extension possibility provided by an electron gun apparatus. It is not too difficult to increase the power delivered by a gun above the power needed for diffusivity measurements. Moreover, as adjustment and manipulation of this power can easily be managed, the possibility exists of devising an apparatus which can be used for thermal shock experiments on certain specimens. This has been exploited by one of the authors.[104]

12. Modulated electron heating equipments are not available commercially, so far as the authors are aware. Further developments and broader utilization of thermophysical experiments based on electron heating embraces an interesting and valuable domain covering science, technology, and applications. However, although several experimental setups are working satisfactorily well, some of them even approaching perfection, a lot of investigations are still necessary in order to acquire an apparatus working trouble-free and whose operation does not require any highly specialized dexterity.

NOTATION

Symbol	*Name*	*Unit of value*
a	Thermal diffusivity	$m^2\,s^{-1}$
$a_h, \bar{a}$	Heat loss parameters	Dimensionless
A	Atomic weight	Dimensionless

A_0	Thermionic emission constant	$4emk_B^2h^{-3} = 1.2 \times 10^6$ A m^{-2} K^{-2}
$b_h, \bar{b}$	Heat loss parameter	Dimensionless
B	"Thermal thickness" parameter	Dimensionless
B_i	Biot modulus	Dimensionless
C_2	Second radiation constant	1.438786×10^{-2} m K
C_p	Specific heat at constant pressure	J kg^{-1} K^{-1}
d	Thickness of disk-type specimen	m
d_0	Intermediary parameter – see equation (40)	Dimensionless
e	Electron charge	$1.6021917 \times 10^{-19}$ C
$e\varphi_a$	Work function of target material	eV (or J)
E	Energy (several indices are used. They are explained in the text)	J
f	Frequency	Hz
h	Planck's constant	6.626196×10^{-34} J s
$H_0^{(1)}(z)$	Zeroth-order Hankel function of first kind	–
i	Current through filamentary cathode	A
i_0	DC component of i	A
i_a	AC component of i	A
$I_0(z)$	Zeroth-order Bessel function of first kind	–
$I_1(z)$	First-order Bessel function of first kind	–
J	Current density	A m^{-2}
k	Experimental coefficient [see equation (7)]	(kV)$^{-1}$
k_B	Boltzmann constant	1.380622×10^{-23} J K^{-1}
m	Mass of electron	9.109558×10^{-31} kg
M	Mass	kg
p	Total mean electron path in target material	m
P_0, P_d	Intermediary parameters – see equations (22, 23)	Dimensionless
Pd	Predvodytelev number	Dimensionless
P_e	Perveance constant of electron gun	A V$^{-3/2}$ m^{-2}
P_x	Coefficient – see equations (21)	Dimensionless
q_0, q_d	Energy radiated at specimen face $x = 0$ and $x = d$, respectively	J
Q	Quantity of heat	J
Q_0, Q_1, Q_2, Q_3	Coefficients – see equation (25)	Dimensionless
r	Radius (disk-type or cylinder-type specimen)	m
R	Electrical resistance	Ω
R_1, R_2	Inner and outer radius of cylinder	m

r_h	Heat loss parameter	Dimensionless
t	Time	s
T	Absolute temperature (Several indices are used. They are explained in the text)	K
V_a	Anode–cathode accelerating voltage, or primary electron energy in equation (10)	V eV
W_i	Power available for thermal excitation of target [equation (16)]	J
$W_i(t)$	Power applied to specimen [see equation (17)] per unit area and per unit time	$W\,m^{-2}\,s^{-1}$
W_i^0	Direct component of power applied to specimen per unit area	$W\,m^{-2}$
W_i^m	Modulation amplitude component of power applied to specimen per unit area	$W\,m^{-2}$
x	Distance	m
x, y, z	Cartesian coordinates	m
Z	Atomic number of target materials [Also parameter used in equation (43)]	Dimensionless
δ_d	Phase difference between bombarded face and applied modulation	Radian
$\delta_0, \bar{\delta}_0$	Phase difference between non-bombarded face and applied modulation	Radian
Δ	Phase difference between both faces of sample $= \delta_0 - \delta_d$	Radian
Δd	Thermal expansion coefficient	Dimensionless
ϵ	Total hemispherical emissivity	Dimensionless
ϵ_s	Spectral emissivity	Dimensionless
ϵ_T	Transmission coefficient at a certain wavelength	Dimensionless
$\epsilon\lambda, \epsilon\lambda_1, \epsilon\lambda_2$	Spectral emissivity at wavelengths λ, λ_1, and λ_2 respectively	Dimensionless
ξ (index $0, a, c, r, s$)	Correction factors [see equation (9)]	Dimensionless
η	Bremstrahlung efficiency coefficient	Dimensionless
Θ	Alternating component of modulated temperature (several indices are used. They are explained in the text)	K
λ	Thermal conductivity	$W\,m^{-1}\,K^{-1}$

λ_B	Wavelength at which temperature is measured with a pyrometer	m
ρ	Density	kg m^{-3}
ρ_e	Electrical resistivity	Ωm
σ	Stefan–Boltzmann constant	5.66961 × 10^{-8} W m^{-2} K^{-4}
φ	Phase shift between current and voltage across filamentary cathode	Degree
ω	Angular frequency	s^{-1}

REFERENCES

1. G.C. Danielson and P.H. Sidles, "Thermal Diffusivity and Other Non-Steady State Methods," *Thermal Conductivity* (R.P. Tye, ed.), Vol. 2, pp. 149–201, Academic Press, New York (1969).
2. L.P. Phyllipov, "Use of Regular Conditions of the Third Kind for Measuring the Thermal Properties of Solid and Liquid Metals at High Temperatures," *Teplofiz. Vys. Temp.,* **2**(5) 817–828 (1964) (English transl. *High Temp.* pp. 733–741).
3. Y.S. Touloukian, R.W. Powell, C.Y. Ho, and M.C. Nicolaou, *Thermal Diffusivity in Thermophysical Properties of Matter,* Vol. 10, pp. 1a–50a. IFI Plenum, New York (1973).
4. C. Mustacchi and S. Giuliani, "Development of Methods for the Determination of the High Temperature Thermal Diffusivity of UC," European Atomic Energy Community–Euratom Report, EUR-337.e (August 1963), pp. 1–27.
5. A.J. Walter, R.M. Dell, and P.C. Burgess, "The Measurement of Thermal Diffusivities Using a Pulsed Electron Beam," *Rev. Int. Hautes Temp. Réfract.* **7**, 271–277 (1970).
6. A.J. Walter, R.M. Dell, K.E. Gilchrist, and R. Taylor, "A Comparative Study of the Thermal Diffusivities of Stainless Steel, Hafnium and Zircaloy," *High Temp. High Pressures* **4**, 439–446 (1972).
7. T. Kumada and K. Kobayasi, "Proposed Method of Measuring Thermal Diffusivity of a Small Specimen at Incandescent Temperatures," *J. Nucl. Sci. Technol.* **9**(3), 192–194 (1972).
8. B. Abeles, G.D. Cody, and D.S. Beers, "Apparatus for the Measurement of the Thermal Diffusivity of Solids at High Temperatures," *J. Appl. Phys.* **31**(9), 1585–1592 (1960).
9. R. Pascard, "Properties of Carbides and Carbonitrides," in *Nuclear Metallurgy,* Vol. 13 (K.E. Horton, R.E. Macherey, R.J. Allio, eds.), International Symposium on Plutonium Fuels Technology, Proc. Nuclear Metallurgy Symposium, Scottsdale, Arizona, October 4–6 (1967).
10. M.R. Null and W.W. Lozier, "Measurement of Thermal Diffusivity by the Phase Shift Method," *8th Conf. on Thermal Conductivity 1968, Purdue University (USA)* (C.Y. Ho and R.E. Taylor, eds.), Plenum Press, New York, pp. 837–856 (1969).
11. E. Chafik, R. Mayer, and R. Pruschek, "Messung der Temperaturleitzahl fester Stoffe bei hohen Temperaturen" ("Measurement of Thermal Diffusivity of Solids at High Temperatures" in German), *High Temp. High Pressures* **1**, 21–26 (1969); and Proc. 1st Europ. Conf. on Thermophysical Properties, Baden-Baden, 11–13 November (1968), pp. 81–94.
12. J. Kaspar and E.H. Zehms, "A Diffusivity Measurement Technique for Very High Temperatures," SAMSO-TR-70-416 or TR-0059 (9250-02)-1, p. 28, August (1970).

13. J.F. Schatz and G. Simmons, "Thermal Conductivity of Earth Materials at High Temperatures," *J. Geophys. Res.* 77(35), 6966–6983 (1972).
14. V.V. Mirkovich, Thermal diffusivity measurement of Armco iron by a novel method, *Rev. Sci. Instrum.* **48** (5), 560–565 (1977).
15. V.V. Mirkovich, "Thermal Diffusivity of Yttria-Stabilized Zirconia," *High Temp. High Pressures* **8**, 231–235 (1976).
16. R. Brandt, G. Neuer, "Thermal Diffusivity of Solids – Analysis of a Modulated Heating-Beam Technique," *High Temp. High Pressures* **11**, 59–68 (1979).
17. A.D. Ivliev and V.E. Zinov'ev, "Measurement of the Thermal Diffusivity and the Heat Capacity by the Method of Temperature Waves Using Laser Radiation and a Following Amplitude-Phase Receiver," *Teplofiz. Vys. Temp.* **18**(3), 532–539 (1980). (English transl. *High Temp.*, pp. 422–428).
18. J.B. Sundqvist, "Thermal Diffusivity Measurements Under Hydrostatic Pressure," *Rev. Sci. Instrum.* **52**, 1061–1063 (1981).
19. M. Cutler and G.T. Cheney, "Heat-Wave Methods for the Measurements of Thermal Diffusivity," *J. Appl. Phys.* **34**, 1902–1909 (1963).
20. R.D. Cowan, "Proposed Method of Measuring Thermal Diffusivity at High Temperatures," Los Alamos Scientific Laboratory Report LA-2460 (1960).
21. R.D. Cowan, "Proposed Method of Measuring Thermal Diffusivity at High Temperatures," *J. Appl. Phys.* **32**, 1363–1370 (1961).
22. M. Cerceo and H.M. Childers, "Thermal Diffusivity by Electron Bombardment Heating," *J. Appl. Phys.* **34**, 1445–1449 (1963).
23. O.A. Kraev and A.A. Stel'makh, "Thermal Diffusivity of Tungsten at Temperatures Between 1600 and 2960°C," *Teplofiz. Vys. Temp.* **1**(1), 8–11 (1963) (English transl. *High Temp.* pp. 5–8).
24. M.J. Wheeler, "Thermal Conductivity of Uranium Monocarbide," in: *Carbides in Nuclear Energy* (L.E. Russell, B.I. Bradbury, J.D.L. Harrison, H.J. Hedger, P.G. Mordon, eds.), Vol. 1, pp. 358–364, Macmillan & Co Ltd, London (1963).
25. M.J. Wheeler, "Thermal Diffusivity at Incandescent Temperatures by a Modulated Electron Beam Technique," *Brit. J. Appl. Phys.* **16**, 365–376 (1965).
26. M.J. Wheeler, "Rapid Measurement of Thermal Diffusivity Using a Modulated Electron Beam," *J. Sci. Technol.* **38**, 102–107 (1971).
27. L.P. Phyllipov, *Measurement of Thermal Properties of Solid and Liquid Metals at High Temperatures*, MGU Press, Moscow (1967) (in Russian).
28. H. Bruining, *Physics and Applications of Secondary Electron Emission*, Pergamon Press, London (1954).
29. J. Trouvé and A. Accary, "Bilan Electronique et Thermique au cours de la Fusion par Bombardement d'Electrons," *Rev. Int. Hautes Temp. Réfract.* **5**, 197–203 (1968) (in French: "Electron and Thermal Balance During Fusion by Electron Bombardment).
30. R. De Coninck, not published.
31. I.M. Bronshtein and B.S. Fraiman, "The Range of Kilovolt Electrons in Solid Bodies," *Fiz. Tverd. Tela* **3**(4), 1122–1124 (1961) (English transl. *Sov. Phys. Solid State* pp. 816–817).
32. R. De Coninck, W. Van Lierde, and A. Gijs, "Uranium Carbide: Thermal Diffusivity, Thermal Conductivity and Spectral Emissivity at High Temperatures," *J. Nucl. Mater.* **57**, 69–76 (1975).
33. H.E. Schmidt, M. van den Berg, and L. van der Hoek, "Zur Messung der Temperaturleitfähigkeit nach der Methode des modulierten Elektronenstrahls" ("Measurement of Thermal Diffusivity Using the Modulated Electron Beam Method" – in German), *High Temp. High Pressures* **1**, 309–325 (1969) and Proc. of the 1st European Conf. on

Thermophysical Properties of Solids at High Temperatures, Baden-Baden, 11–13 November (1968), pp. 95–124.
34. R. Mayer and G. Neuer, "Isothermal Sample Heating Using a Modulated Electron Beam," *Rev. Int. Hautes Temp Réfract.* **12**, 191–196 (1975).
35. R.D. Cowan, "Pulse Method of Measuring Thermal Diffusivity at High Temperatures," *J. Appl. Phys.* **34**, 926–927 (1963).
36. R. Penninckx, "Calculation of Thermal Diffusivity from Measurements with the Sine Wave Modulation Method: Modification of Cowan's Equations," *Appl. Phys. Lett.* **21** (2), 47–48 (1972).
37. L.P. Phyllipov and R.P. Yurchak, "A Method of Measuring the Specific Heat of Solid and Liquid Metals," *Teplofiz. Vys. Temp.* **3** (6), 901–909 (1965) (English transl. *High Temp.* pp. 837–844).
38. E. Jahnke, F. Emde, and F. Lösch, *Tafeln höherer Funktionen, Tables of Higher Functions* (B.G. Teubner, ed.), Verlags Gesellschaft, Stuttgart (1960).
39. G.F. Tkach and R.P. Yurchak, "Application of Periodic Temperature Oscillations to High-Temperature Studies of the Thermal Properties of Dielectrics," *Teplofiz. Vys. Temp.* **9** (1), 210–213 (1971) (English transl. *High Temp.* pp. 187–190).
40. O.A. Kraev and A.A. Stel'makh, "Thermal Diffusivity of Tantalum, Molybdenum and Niobium at Temperatures above 1800 K," *Teplofiz. Vys. Temp.* **2** (2), 302 (1964) (English transl. *High Temp.* pp. 270).
41. K. Zankel, "Contribution à l'Etude de la Diffusivité Thermique des Solides" (Contribution to the Study of Thermal Diffusivity of Solids" – in French), Report CEA-R-3150, 50 p. (July 1967).
42. J.B. Ainscough and M.J. Wheeler, "The High-Temperature Thermal Conductivity of Sintered Uranium Dioxide," *Brit. J. Appl. Phys. (J. Phys. D), Ser. 2* **1**, 859–868 (1968); and also: J.B. Ainscough and M.J. Wheeler, in *Proc. of the 7th Conf. on Thermal Conductivity* (D.R. Flyn and B.A. Pearcy, Jr., eds.), p. 467, National Bureau of Standards, Washington, D.C. (1968).
43. V.E. Zinov'ev and R.P. Krentsis, "Installation for Measuring the Thermal Diffusivity of Metals and Alloys at High Temperatures," in *Physical Properties of Alloys* (in Russian), Trudy UPI, No. 167, Sverdlovsk (1968), pp. 102–107.
44. V.E. Zinov'ev, R.P. Krentsis, and P.B. Gel'd, "Thermal Diffusivity and Conductivity of Titanium at High Temperatures," *Teplofiz. Vys. Temp.* **6** (5), 927–928 (1968) (English transl. *High Temp.* pp. 888–890).
45. M.J. Wheeler, "Thermal Diffusivity Measurement by the Modulated Electron Beam Method. Thermal Diffusivity of Iron Between 280°C and 1100°C," *High Temp. High Pressures* **1**, 13–20 (1969).
46. A.A. Emel'yanov, O.A. Kraev, A.A. Stel'makh, and R.A. Fomin, "Measurement of the Thermal Diffusivity of Metals in the Temperature Interval 1100–2500 K," *Zh. Prikl. Mekh. Tekh. Fiz.* **10** (3), 154–155 (1969) (English transl. *J. Appl. Mech. Tech. Phys.* pp. 482–483).
47. M.J. Wheeler, "Some Anomalous Thermal Diffusivity Results of Hafnium, Niobium and Zircaloy 2," *Rev. Int. Hautes Temp. Réfract.* 7, 335–340 (1970).
48. H.E. Schmidt and F. Caligara, "Wärmeleitfähigkeit und Wärmeleitungs-Integrale von Uran-Plutonium-Mischoxides" ("Thermal Diffusivity and Thermal Conductivity-Integral of Uranium–Plutonium Mixed Oxides," in German). *Deutsches Atomforum, Reaktortagung Berlin,* pp. 566–569 (1970).
49. M.J. Wheeler, E. King, C. Manford, and H.J. Hedger, "Thermal Diffusivity of Uranium and Uranium–Plutonium Monocarbides," *J. Br. Nucl. Energy Soc.* **10** (1), 55–64 (1971).

50. I.P. Mardykin and A.A. Vertman, "Thermal Properties of Liquid Cerium," *Izv. A N SSSR, Metally* **(1)**, 95–98 (1972) (English transl. *Res. Acad. Sci. USSR, Metallurgy*, pp. 68–72).
51. V.E. Zinov'ev, L.P. Gel'd, G.E. Chuprikov, and K.U. Epifanova, "High Temperature Transport Properties of Gadolinium," *Fiz. Tverd. Tela* **14**(9), pp. 2747–2749 (1972) (English transl. *Sov. Phys. Solid State* pp. 2372–2374).
52. V.E. Zinov'ev, L.I. Chipuna, and P.V. Gel'd, "Transport Properties of Scandium at High Temperatures," *Fiz. Tverd. Tela* **14**(9), 2787–2790 (1972) (English transl. *Sov. Phys. Solid State* pp. 2416–2418).
53. T. Tanaka and H. Suzuki, "Thermal Conductivity of Carbon Materials at High Temperatures," *Heat Transfer Jpn. Res.* **1**(4), 31–35 (1972).
54. T. Tanaka and H. Suzuki, "The Thermal Diffusivity of Pyrolytic Graphite at High Temperatures," *Carbon* **10**, 253–257 (1972).
55. M.M. Mebed and R.P. Yurchak, "Apparatus for Measuring Thermophysical Properties of Electrically Conducting Materials at Temperatures above 1000 K," *Zavod. Lab.* **38**(10), 1283–1285 (1972) (English transl. *Industrial Lab.* pp. 1620–1622).
56. L.P. Gel'd and V.E. Zinov'ev, "Temperature Conductivity of Iridium over a Wide Temperature Range," *Teplofiz. Vys. Temp.* **10**(3), 656–657 (1972) (English transl. *High Temp.* pp. 588–589).
57. S.R. Atalla, S.N. Banchila, and L.P. Filippov, "Investigation of the Complex of Thermal Properties of Liquid Metals at High Temperatures," *Teplofiz. Vys. Temp.* **10**(1), 72–76 (1972) (English transl. *High Temp.* pp. 60–63).
58. S.R. Atalla, "Experimental Investigation of Thermophysical Properties of Liquid Metals at Elevated Temperatures," *High Temp. High Pressures* **4**, 447–451 (1972).
59. B.A. Friedlander, V.S. Neshpor, B.G. Ermakov, and V.V. Sokolov, "Thermal Diffusivity and Conductivity of Pyrolytic Titanium Carbide, Columbium Carbide and Titanium Nitride at High Temperatures," *Inzh. Fiz. Zh.* **24**(2), 294–296 (1973) (English transl. *J. Eng. Phys.* pp. 210–212).
60. V.E. Zinov'ev, S.I. Masharov, and P.V. Gel'd, "Kinetic Properties of Rhenium at High Temperatures," *Fiz. Tverd. Tela* **15**, 1281–1284 (1973) (English transl. *Sov. Phys. Solid State* pp. 869–870).
61. I.I. Novikov and I.R. Mardykin, "Thermal Properties of Lanthanides at High Temperatures," *Teplofiz. Vys. Temp.* **11**(3), 527–532 (1973) (English transl. *High Temp.* pp. 472–476).
62. S.N. Banchila and L.P. Phyllipov, "New Measurements of the Complex of Thermal Properties of Liquid Tin and Lead," *Teplofiz. Vys. Temp.* **11**(3), 668–671 (1973). (English transl. *High Temp.* pp. 602–605).
63. M.M. Mebed, R.P. Yurchak, and L.P. Phyllipov, "Measurement of the Thermophysical Properties of Electrical Conductors at High Temperatures," *High Temp. High Pressures* **5**, 253–260 (1973).
64. I.G. Korshunov, V.E. Zinov'ev, P.V. Gel'd, V.S. Chernyaev, A.S. Borukhovich, and G.P. Shveikin, "Thermal Diffusivity and Thermal Conductivity of Titanium and Zirconium Carbides at High Temperatures," *Teplofiz. Vys. Temp.* **11**(4), 889–891 (1973) (English transl. *High Temp.* pp. 803–805).
65. R. De Coninck, W. Van Lierde, R. Penninckx, and A. Gijs, "Thermal Diffusivity Thermal Conductivity and Spectral Emissivity of U_2C_3 up to 2200 K," *Belgian Physical Society, General Scientific Meeting, University of Ghent (Belgium)*, 24–25 May (1973), Lecture III-15 (in Flemish).
66. R. De Coninck, W. Van Lierde, and A. Gijs, "Thermal Diffusivity and Conductivity of U_2C_3 up to 2200 K," *J. Nucl. Mater.* **46**, 213–216 (1973).

67. A.A. Kulish, R.P. Yurchak, and M.M. Mebed, "Apparatus for the Determination of Temperature Diffusivity of Electrically Conducting Material at High Temperatures," *Vestn. Mosk. Univ. Fiz. Astronomiya* **28**(2), 233–235 (1973) (English transl. *Bull. Moscow Univ., Astron. Phys.*, pp. 78–79).
68. V.E. Zinov'ev, Sh. Abel'ski, M.I. Sandakova, E.G. Dik, L.N. Petrova, and P.V. Gel'd, "Thermal Properties of Iron in Solid Solutions of Silicon in Iron near the Curie Point," *Zh. Eksp. Teor. Fiz.* **66**, 354–360 (1974) [English transl. *Sov. Phys. JETP* **39**(1), 169–171 (1974)].
69. T. Tanaka, "The Thermal and Electrical Conductivities of LaB_6 at High Temperatures," *J. Phys. C: Solid State Phys.* **7**, L177–L180 (1974).
70. V.E. Zinov'ev, L.P. Gel'd, T.G. Korshunov, and V.I. Chepkov, "Apparatus for Overall and Semiautomatic Measurements of Thermal Diffusivity and Heat Capacity of Metals at High Temperatures," *Pribory dlya issledovaniya fizicheskikh svoistv materialov,* Kiev, Naukova dumka, 1974, pp. 95–101 (in Russian).
71. V.S. Neshpor, B.A. Fridlender, S.S. Ordan'yan, and V.I. Grishchenko, "Thermal Diffusivity of Solid Solutions of Zirconium and Niobium Monocarbides at High Temperatures," *Teplofiz. Vys. Temp.* **12**(5), 1125–1128 (1974) (English transl. *High Temp.* pp. 992–995).
72. A.A. El-Sharkawy, S.R. Atalla, R.P. Yurchak, and L.P. Phyllipov, "An Apparatus for Simultaneous Determination of the Thermal Diffusivity, Capacity and Conductivity Coefficients of Solids in the Temperature Interval 1200–2200 K," *Proc. 4th European Thermophysical Properties Conf.*, Orléans (France), 4–6 September 1974, pp. 168–170 and *Rev. Int. Hautes Temp. Réfract.* **12**, 168–170 (1974).
73. V.E. Zinov'ev, P.V. Gel'd, and A.L. Sokolov, "High-Temperature Transport Properties of Thulium, Ytterbium and Lutetium," *Fiz. Tverd, Tela* **17**, 413–416 (1975) (English transl. *Sov. Phys. Solid State,* pp. 259–260).
74. I.I. Novikov and I.P. Mardykin, "The High-Temperature Thermal Diffusivity and Electrical Resistivity of Yttrium and Gadolinium, *At. Energ.* **40**(1) 63–64 (1976) (English transl. *Atomic Energy,* pp. 69–71).
75. H.E. Schmidt, "The Thermal Conductivity of Uranium–Plutonium Oxycarbonitrides," *Proc. 5th European Conf. on Thermophysical Properties of Solids at High Temperatures,* Moscow, USSR, May 1976 (Abstract only).
76. R. De Coninck, R. De Batist, and A. Gijs, "Thermal Diffusivity, Thermal Conductivity and Spectral Emissivity of Uranium Dicarbide at High Temperatures," *High Temp. High Pressures* **8**, 167–176 (1976).
77. B.A. Fridlender, V.S. Neshpor, M.A. Eron'yan, and A.V. Petrov, "Thermal Diffusivity of Zirconium Nitride in the Region of Homogeneity," *Teplofiz. Vys. Temp.* **15**(4), 779–784 (1977) (English transl. *High Temp.* pp. 779–784).
78. U. Benedict, G. Giacchetti, H. Matzke, K. Richter, C. Sari, and H.E. Schmidt, "Study of Uranium–Plutonium Carbide Based Fuel Simulating High Burnup," *Nucl. Technol.* **35**, 154–161 (1977).
79. C.S. Kim, R.A. Blomquist, J. Haley, R. Land, J. Fischer, M.G. Chasanov, and L. Leibowitz, "Measurement of Thermal Diffusivity of Molten UO_2," *Proc. 7th Symp. on Thermophysical Properties,* NBS, Gaithersburg, Maryland, USA, May 10–12 (1977) (A. Cezairliyan, ed.), NBS, The American Society of Mechanical Engineers.
80. R.P. Yurchak and A.V. Khromov, "Improved Apparatus for Measuring Thermal Properties of Electrically Conducting Materials at High Temperatures," *Zavod. Lab.* **44**(5), 557–558 (1978) (English transl. *Industrial Lab.* pp. 644–646).
81. E.B. Zaretsky and V.E. Peletsky, "A Device for Coordinated Study of Thermophysical Properties of Metals and Alloys," *Teplofiz. Vys. Temp.* **17**(1), 124–132 (1979) (English transl. *High Temp.* pp. 104–111).

82. E.B. Zaretsky and V.E. Peletsky, "Investigation of the Thermal Diffusivity of Titanium Iodide in a Large Neighbourhood of the fcc–bcc Transition," *Teplofiz. Vys. Temp.* **17**(2), 310–313 (1979) (English transl. *High Temp.* pp. 261–263).
83. S.N. Banchila, D.K. Palchaev, and L.P. Phyllipov, "Thermal Properties of Liquid Gallium, Indium, and Thallium at High Temperatures," *Teplofiz. Vys. Temp.* **17**(3), 507–510 (1979) (English transl. *High Temp.* pp. 507–510).
84. V.E. Peletsky and E.B. Zaretsky, "Thermophysical Properties of Group IV Transition Metals near Polymorphous Transformations," *Proceedings of the 8th Symposium on Thermophysical Properties,* Vol. 2, NBS, Gaithersburg, Maryland (USA) June 15–18, 1981, (J.V. Sengers, ed.), The American Society of Mechanical Engineers, New York (1982), pp. 83–86.
85. S.R. Atalla, "Thermal Properties of Stainless Steel and Graphite Specimens in the Temperature Range 400–1000 K," *Proceedings 8th European Thermophysical Properties Conference,* Baden-Baden, 27 September–1 October 1982 (To be published in *High Temp. High Pressures*).
86. R. De Coninck, private information, not published.
87. T.Y.R. Lee, A.B. Donaldson, and R.E. Taylor, "Thermal Diffusivity of Layered Composites," in: *Proc. of the 15th International Conference on Thermal Conductivity,* Ottawa, Ontario, Canada, August 24–26, Thermal Conductivity 15 (V.V. Mirkovich, ed.), Plenum Press, New York (1977), pp. 135–148.
88. R. Brandt and M. Havránek, "Determination of the Thermal Diffusivity of Two-Layer Composite Samples by the Modulated Heating Beam Method," *J. Non-Equilib. Thermodyn.* **3**, 213–230 (1978).
89. C.S. Ang, H.S. Tan, and S.L. Chan, "Three-Layer Thermal-Diffusivity Problem Applied to Measurements on Mercury," *J. Appl. Phys.* **44**, 687–691 (1973).
90. G.N. Dul'nev and A.V. Sigalov, "Thermal Diffusivity of Inhomogeneous Systems, 1. Temperature-Field Calculations," *Inzh. Fiz. Zh.* **39**(1), 126–133 (1980) (English transl. *J. Eng. Phys.* pp. 803–809).
91. M.M. Farooq, W.H. Giedt, and N. Araki, "Thermal Diffusivity of Liquids Determined by Flash Heating of a Three-Layered Cell," *Int. J. Thermodyn.* **2**(1), 39–54 (1981).
92. M.J. Wheeler, "Specific Heat Measurements by a Pulsed Electron Beam Method," *High Temp. High Pressures* **4**, 363–369 (1972).
93. E. Fitzer, ed., *AGARD Report No. 606,* Advisory Group for Aerospace Research and Development, NATO, "Thermophysical Properties of Solid Materials, Project 2 – Cooperative Measurements on Heat Transport Phenomena of Solid Materials at High Temperature," p. 114 (March 1973).
94. M.J. Wheeler, "The influence of Radiated Heat Losses on Thermal Diffusivity Measurements by the Modulated Electron Beam Technique," *Proc. 4th European Thermophysical Properties Conference,* Orléans (France), 4–6 September 1974, pp. 162–167; and *Rev. Int. Hautes Temp. Réfract.* **12**, 162–167 (1975).
95. D.W. Lee and W.D. Kingery, "Radiation Energy Transfer and Thermal Conductivity of Ceramic Oxides," *J. Am. Ceram. Soc.* **43**(11), 594–607 (1960).
96. A.A. Men' and O.A. Sergeev, "Investigation of Thermal Conductivity, Thermal Diffusivity and Emittance of Semitransparent Materials at High Temperatures," *High Temp. High Pressures* **5**, 19–28 (1973).
97. D.C. Pridmore-Brown, "Measurement of Thermal Diffusivity in Disks and Rods," *Hight Temp. Sci.* **2**(4), 305–310 (1970).
98. K. Beedham and I.P. Dalrymple, "The Measurement of Thermal Diffusivity by the Flash Method. An Investigation into Errors Arising from the Boundary Conditions," *Rev. Int. Hautes Temp. Réfract,* **7**, 278–283 (1970).

99. R.E. Taylor, "Critical Evaluation of Flash Method for Measuring Thermal Diffusivity," *Proc. of the 4th European Thermophysical Properties Conference,* Orléans (France), 4–6 September 1974, pp. 141–145; and *Rev. Int. Hautes Temp. Réfract.* **12**, 141–145 (1975).
100. A.J. Walter, R.M. Dell, K.E. Gilchrist, and R. Taylor, "A Comparative Study of the Thermal Diffusivity of Stainless Steel, Hafnium and Zircaloy," *High Temp High Pressures* **4**, 439–446 (1972).
101. M.L. Minges, "Evaluation of Selected Refractories as High Temperature Thermophysical Property Calibration Materials," *report AFML-TR-74-96,* pp. 66 (August 1975); and *Int. J. Heat Mass Transfer* **17**, 1365–1382 (1974).
102. M.L. Minges, "Analysis of Thermal and Electrical Energy Transport in POCO AXM-5Q1 Graphite," *Int. J. Heat Mass Transfer* **20**, 1161–1172 (1977).
103. J.G. Hust, "Graphite as a Standard Reference Material," in *Proc. of the 15th International Conference on Thermal Conductivity,* Ottawa, Ontario, Canada, August 24–26 (1977), Thermal Conductivity 15 (V.V. Mirkovich, ed.), Plenum Press, New York, pp. 161–167.
104. R. De. Coninck, A. Gijs, and M. Snykers, "Thermal Stock Tests on Some Proposed Limiter or First Wall Materials for Fusion Reactors," *Proc. 6th European Thermophysical Properties Conference,* Dubrovnik (Yugoslavia), 26–30 June, 1978, *Rev. Int. Hautes Temp. Réfract.* **16**(4), 294–308 (1979).

11

Monotonic Heating Regime Methods for the Measurement of Thermal Diffusivity

G.M. VOLOKHOV and A.S. KASPEROVICH

As is known, the estimation of thermal diffusivity involves the study of the space-time temperature dependence determined by the object–ambient-medium interaction.

The present review embraces a large group of the methods theoretically based on some particular solutions of the heat conduction equation under the first through third kinds of boundary conditions. These methods are all referred to as the methods of studying thermal diffusivity with monotonic heating. As the terminology adopted in the Soviet literature says, the unsteady-state developed (stabilized) thermal process characterized by an independence of a space-time temperature distribution on initial conditions is a regular regime. The first, second, and third kinds of regimes are distinguished. In the case of the first kind of regular regime, a sample is heated at constant ambient temperature. In the second kind of regular regime, heating proceeds at a constant rate (ambient temperature is a linear time function which implies a constant heat flux effect on a sample). The third kind of regular regime is a developed thermodynamic process. The term "monotonic regime" is used in studies implying heating rate close to linear and temperature-dependent thermophysical properties of a sample.

The third kind of regular regime being neglected, at all the above heat transfer regimes and no phase transitions, the time dependence of temperature used in thermal diffusivity calculations is a monotonic time function.

Despite a great variety of methods for calculating thermal diffusivity, they are all based on and primarily approximated by the well-developed solutions

G.M. VOLOKHOV and A.S. KASPEROVICH • Department of Physics, Belorussian State University, Minsk 220080, USSR.

of linear heat conduction equations for the simplest heat transfer cases (the feasibilities of modern computing technique are not meant here). One and the same theory underlies methods and devices of different engineering perfection for the study of different materials (solids, liquids, gases, plasma) in a wide range of ambient parameters. As it is impossible to describe in detail even the individual methods within the framework of a short review, the authors discuss the problems schematically with an emphasis on the theoretical premises of the method considered and their consistency with real experimental conditions. Experimental conditions usually differ from theoretical boundary ones. This involves essential difficulties in estimating systematic errors in many thermal diffusivity calculations. Most often these estimates are based on an approximate thermal analysis and are related to a certain experimental device, which does not allow any general conclusions on the limits of applicability of the method or on changes in the systematic error with alterations in the investigation procedure. In this connection, an attempt is made in some cases to estimate the components of the systematic error using more general solutions and to generalize the results applying the similarity theory.

1. MEASUREMENT OF THERMAL DIFFUSIVITY IN THE NARROW TEMPERATURE RANGE

To begin with, consider the methods of measuring thermal diffusivity in a narrow temperature range with a sample heated or cooled to some constant temperature slightly differing from the initial one. In this group of methods, equation (1),

$$\frac{\partial T}{\partial \tau} = a_{ij} \nabla^2 T \tag{1}$$

is solved using the following assumptions: (a) a body is isotropic or orthotropic, (b) the geometry is simple (e.g., an infinite plate, an infinite cylinder, a sphere, a parallelepiped), (c) heat transfer at a constant ambient temperature follows Newton's law, (d) the physical properties of the test samples are independent of temperature, (e) there are no internal sources or sinks. The available solutions for such cases are usually expressed in a series form[1-3] and may be generalized by one quite complicated expression.[1]

The analysis of heat transfer at a constant ambient temperature shows that the whole heating (cooling) process may be divided into three stages. For the first random, unsteady-state stage the initial temperature distribution is of great importance. At the second stage, known as a regular regime, the temperature changes follow a simple exponential law. The third stage corresponds to a steady state when a temperature at all points is equal to the ambient one.[1]

The simplest solution of (1) will be given for an infinite plate:

$$\Theta = \frac{T(x,\tau) - T_c}{T_0 - T_c} = \sum_{n=1}^{\infty} A_n \cos \mu_n \frac{x}{h} \exp\left(-\mu_n^2 \frac{a\tau}{h^2}\right) = f\left(\frac{x}{h}, \text{Bi}, F_0\right) \quad (2)$$

where h is the half-width of the plate, $\text{Bi} = \alpha h/\lambda$ is the Biot number, $F_0 = a\tau/h^2$ is the Fourier number, μ_n are the roots of the characteristic equation

$$\cot \mu = \mu/\text{Bi} \quad (3)$$

At small F_0 the solution of form (2) is usually represented in terms of the Gauss error function. As series (2) rapidly converges, then at $F_0 > 0.4$ the first term of series (2) may only be taken as

$$\Theta = A_1 \cos \mu_1 \frac{x}{h} \exp\left(-\mu_1^2 \frac{a\tau}{h^2}\right) = \varphi\left(\frac{x}{h}, \text{Bi}, F_0\right) \quad (4)$$

which corresponds to the second stage. At $F_0 > 3$ (theoretically at $F_0 \to \infty$)[1] the third steady-state stage sets in ($\Theta \to 0$). It follows from (4) that

$$\ln \Theta = f(\tau) \quad (5)$$

is the straight line with a slope defined by

$$m = -\frac{\partial\,[\ln(T_c - T)]}{\partial \tau} = \frac{\ln \Theta_1 - \ln \Theta_2}{\tau_2 - \tau_1} = \mu_1^2 \frac{a}{h^2} = \text{const} \quad (6)$$

where Θ_1 and Θ_2 are the excess relative temperatures at the instants τ_1 and τ_2, respectively. In (5) the dependence on τ is only shown since in estimating (6) $\ln(\Theta_1/\Theta_2)$ is independent of the coordinate. The quantity (6) is usually called a rate of the temperature change. For a regular stage, the theory of which is given in Refs. 4–8, the quantity (6) is constant. The rate of the temperature change may be related to the thermal diffusivity and Bi through the characteristic equations such as (3). The above comments are valid for all geometries enumerated at the beginning of the section. Therefore, thermal diffusivity and other properties can be studied using the regular regime methods.[4–8]

The rate of a temperature change in the case of the simplest geometries is expressed as follows:

For an infinite plate $2h$ thick:

$$m = \mu_1^2 \frac{a}{h^2} \tag{7}$$

For an infinite cylinder $2R$ in diameter and a sphere R in radius:

$$m = \mu_1^2 \frac{a}{R^2} \tag{8}$$

For a cylinder $2h$ in height and $2R$ in diameter:

$$m = \frac{\mu_{11}^2}{R^2} a_1 + \frac{\mu_{12}^2}{h^2} a_2 \tag{9}$$

For a parallelepipedon with dimensions $2h \times 2h_2 \times 2h_3$:

$$m = \frac{\mu_{11}^2}{h_1^2} a_1 + \frac{\mu_{12}^2}{h_2^2} a_2 + \frac{\mu_{13}^2}{h_3^2} a_3 \tag{10}$$

The numbers μ are found from the following characteristic equations:

For an infinite plate:

$$\cot \mu = \frac{\mu}{\mathrm{Bi}_h} \tag{11}$$

for an infinite cylinder:

$$\frac{\mathcal{T}_0(\mu)}{\mathcal{T}_1(\mu)} = \frac{\mu}{\mathrm{Bi}_R} \tag{12}$$

for a sphere:

$$\tan \mu = -\frac{\mu}{\mathrm{Bi}_R - 1} \tag{13}$$

where

$$\mathrm{Bi}_h = \frac{\alpha h}{\lambda_h}, \qquad \mathrm{Bi}_R = \frac{\alpha R}{\lambda_R}$$

Relation (11) is also valid for a parallelepipedon, and equations (11)–(12) are used for a finite cylinder. Tables of the roots of (11)–(12) can be found elsewhere.(1)

In accordance with the theory, determination of thermal diffusivity under a regular regime is a study of relation (4) for a temperature leveled over the volume of a sample heated or cooled from a certain initial state of equilibrium to a new one. The F_0 number determines the time for which the steady state is achieved.

Experiments can be performed at free convection ($\alpha \neq \infty$, $\mathrm{Bi} \neq \infty$) and also under the conditions when a constant temperature ($\alpha \to \infty$, $\mathrm{Bi} \to \infty$) is given and kept on the sample surface. In the former case comparative versions of the studies are realized; in the latter, the simple and reliable methods of direct determination of thermal diffusivity are used.[2,4,5] Indeed, at $\mathrm{Bi} = \infty$, equations (11)–(13) simplify to

$$\cos \mu = 0 \tag{14}$$

$$\mathcal{T}_0(\mu) = 0 \tag{15}$$

$$\sin \mu = 0 \tag{16}$$

Assuming that the ambient medium is isotropic ($a_1 = a_2 = a_3 = a$) and expressing μ_1 from (14)–(16) give the simple formulas for thermal diffusivities of the above geometries:

$$a = \frac{4h^2}{\pi^2} m_\infty \tag{17}$$

$$a = \frac{R^2}{5.783} m_\infty \tag{18}$$

$$a = \frac{R^2}{\pi^2} m_\infty \tag{19}$$

$$a = \frac{m_\infty}{\dfrac{5.783}{R^2} + \dfrac{\pi^2}{4h^2}} \tag{20}$$

$$a = \frac{m_\infty}{\dfrac{\pi^2}{4}\left(\dfrac{1}{h_1^2} + \dfrac{1}{h_2^2} + \dfrac{1}{h_3^2}\right)} \tag{21}$$

where m_∞ is the rate of a temperature change with $\alpha \to \infty$. So, in order to determine thermal diffusivity with $\alpha \to \infty$ over the whole sample surface, it is necessary to know and to maintain a certain constant temperature different from the initial one. This can be achieved by different means, the simplest of which may be described as follows. The test material is placed into a thermostat filled with vehemently stirred liquid. When a test sample is loose of fibrous material, it is placed into a sealed metal cylinder, sphere, or prism. Such a system is usually referred to as an a-calorimeter. Solid, waterproof materials can be conveniently tested without a protective shell, the thermoelectrodes being safely insulated. The general procedure of experiment and the processing of its results are as follows.[5] After the test sample of a certain geometry has been placed into the thermostat, the time changes of an excess temperature are recorded with a differential thermocouple

$$\Theta = f(\tau) \tag{22}$$

The data of (22) are used to plot (5), which gives the rate of temperature change (6). In accordance with the sample geometry, the thermal diffusivity is found from one of expressions (19)–(21). The experimental results are reliable if function (16) is linear.

Thermal diffusivity can be estimated using the alternative methods which imply the recording of the time of a temperature change for a prescribed value.[9]

If heat is transferred at an indefinite Bi value, then the rate of heating (or cooling) is a function not only of thermal diffusivity but also of the first root μ_1 of characteristic equations (11)–(13). Therefore, the coefficients in formulas (18)–(21) will also be indefinite. In this case, to find thermal diffusivity, temperature versus time has been measured at any two points of a sample. The processing of the results here becomes more tedious.[5, 6]

The initial section of the temperature curve inconsistent with relation (5) can be also employed to estimate thermal diffusivity using a semi-infinite body model:

$$\Theta = \operatorname{erf}\left(\frac{1}{2\sqrt{F_{0x}}}\right) \tag{23}$$

Solution (23) is obtained from (2) with appropriate ultimate transitions. The structure of general solution (23) does not allow an analytical formula for calculating thermal diffusivity. In these instants, the tabulated Gauss error functions[2] are used. More comprehensive information on the study of thermophysical properties at the initial stage of heat transfer can be found elsewhere.[10]

The functional relations of forms (7)–(10) enable one to implement a number of comprehensive methods of estimating a complex of thermophysical properties, namely, heat transfer coefficient, thermal diffusivity, and thermal conductivity.[11–14]

Thermal diffusivity at high temperatures has been studied when the boundary conditions of the first and third kinds are valid.[15–17]

The expression of general form (10) can be used to estimate thermal diffusivity of orthotropic materials.[18, 19] Three parallelepipeds of different sizes with the side ratios

$$h_1 : h_2 : h_3 = 1:2:3, \qquad 1:2:2, \qquad 1:1:2$$

are manufactured from the test material. For all three samples the rate of temperature change at $\alpha \to \infty$ is measured by the above method. Then, according to (10), with the ratio of the linear dimensions borne in mind, it is possible to write down the following equations:

$$m_1 = \left(\frac{\pi}{2}\right)^2 \left[\frac{a_1}{(h/2)^2} + \frac{a_2}{(2h/2)^2} + \frac{a_3}{(3h/2)^2}\right]$$

$$m_2 = \left(\frac{\pi}{2}\right)^2 \left[\frac{a_1}{(h/2)^2} + \frac{a_2}{(2h/2)^2} + \frac{a_3}{(2h/2)^2}\right]$$

$$m_3 = \left(\frac{\pi}{2}\right)^2 \left[\frac{a_1}{(h/2)^2} + \frac{a_2}{(h/2)^2} + \frac{a_3}{(2h/2)^2}\right]$$

From the solution to the above equations, the following formulas for thermal diffusivity can be obtained:

$$a_1 = 6.76h^2(27m_1 - 7m_2 - 5m_3) \times 10^{-3} \tag{24}$$

$$a_2 = 135.2h^2(m_3 - m_2) \times 10^{-3} \tag{25}$$

$$a_3 = 730.2h^2(m_2 - m_1) \times 10^{-3} \tag{26}$$

The main source of a systematic error in measuring thermal diffusivity by formulas (19)–(21) is the violation, in experiment, of the condition $\mathrm{Bi} = \infty$ required by theory as well as the dependence of the material properties on temperature. In order to determine thermal diffusivity as accurate as 2%, Bi must be approximately equal to 100.[4] In general, the systematic error due to the violation of the condition $\mathrm{Bi} = \infty$, corresponding to an ideal thermal contact, can be found from

$$\epsilon = \frac{\Delta a}{a} = \frac{\mu_{1\infty}^2 - \mu_1^2}{\mu_{1\infty}^2} \tag{27}$$

where μ_1, $\mu_{1\infty}$ are the roots of equations (11)–(13) and the roots of equations (14)–(16). In practice, μ_1 must be estimated to find (27). To do this, we shall use the method of Ref. 2. If over the surface of a sample with thermal conductivity λ_1 and thickness h there is a thin film (say, an air interlayer) of thickness δ and thermal conductivity λ_2 film heat capacity being neglected), then

$$\mathrm{Bi} = \frac{\alpha h}{\lambda_1} = \frac{h}{\lambda_1} : \frac{\delta}{\lambda_2} \tag{28}$$

Tentative estimation (28) and equations (14)–(16) yield μ_1 and then (27). These calculations as well as more complicated cases when film heat capacity is different from zero are given in Ref. 20.

Thus, systematic error (27) is a variable depending on (28). Therefore, in experiments with small-thermal resistance materials (e.g., metals) the systematic error of measurements may be very large.

Another source of the systematic error that is difficult to estimate lies in a temperature dependence of thermophysical properties. This portion of the error can be minimized if experiments run at low (5–10 K) temperature drops.[21, 22] Solid waterproof materials are difficult to test with the above procedure since the protective shells are a source of errors due to additional thermal resistances which are not easy to take into account.

It is often convenient to perform experiments when over some area of the surface (say, on the plate bases) a certain constant temperature is prescribed, and the other area of the surface (say, sides of a plate) is involved in heat transfer with the ambient medium at a different temperature. In those instances when this procedure does not work, thermal diffusivity has been calculated by formulas (17)–(18) obtained from the one-dimensional solutions to equation (1). The sample sizes are greatly increased to eliminate a systematic error due to the effect of sample end-to-ambient medium heat transfer.[11, 12, 23]

The strict analytical estimation of this error as well as the development of the methods based on a more general theory is possible using appropriate two- and three-dimensional solutions (1).[2, 20, 24, 25] A general form of the solutions is quite complicated and differs from solution (2) by a nonzero steady-state component.

For a better understanding of the practical application of the above solutions we shall consider an example of a finite cylinder $2h$ in height and $2R$ in diameter, with the bases kept at a certain constant temperature T_c and the side surface involved in heat transfer with the ambient medium at a certain initial temperature following Newton's law. The solution can be written as

$$\Theta = \frac{T(r,z,\tau)-T_0}{T_c-T_0} = \sum_{n=1}^{\infty} A_n \mathscr{T}_0\left(\mu_n \frac{r}{R}\right) \frac{\cosh \kappa\mu_n z/h}{\cosh \kappa\mu_n}$$

$$+ \sum_{n=1}^{\infty}\sum_{m=1}^{\infty} \frac{A_n \mathscr{T}_0(\mu_n r/R)\,\mu_m \cos \mu_n z/h}{\mu_n^2\kappa^2+\mu_m^2} \exp\left[-(\kappa^2 a_1\mu_n^2+\mu_m^2 a_2)\tau/h^2\right]$$

$$= f\left(\frac{r}{R}, \frac{z}{h}, \mathrm{Bi}, F_0\right) \qquad (29)$$

where μ_m, μ_n are the roots of characteristic equations (11) and (12) and the parameter $\kappa = h/R$ is the relative cylinder height. The first term in (29) is the steady-state component

$$\Theta_{st} = \varphi\left(\frac{r}{R}, \frac{z}{h}, \mathrm{Bi}_R\right) = \sum_{n=1}^{\infty} A_n \mathscr{T}_0\left(\mu_n \frac{r}{R}\right) \frac{\cosh \kappa \mu_n z/h}{\cosh \kappa \mu_n} \tag{30}$$

where

$$A_n = \frac{2\mathscr{T}_1(\mu_n)}{\mu_n[\mathscr{T}_0^2(\mu_n) + \mathscr{T}_1^2(\mu_n)]}$$

The analysis of the similar solutions gives correct and practically useful results.[20, 24, 25]

1. After a certain time interval, the regular regime sets in within the sample, the rate of temperature changes being expressed by

$$m = \frac{\ln(\Theta_{st} - \Theta_1) - \ln(\Theta_{st} - \Theta_2)}{\tau_2 - \tau_1} = \text{const} \tag{31}$$

and

$$\ln(\Theta_{st} - \Theta) = f(\tau) \tag{32}$$

is the straight line. The relative dimension $\kappa \to 0$ $(R \to \infty)$ corresponds to the case of an infinite plate [see solution (2)] and $\Theta_{st} \to 0$. In this case, the rate of a temperature change is given by (6). From solution (29) the constant of (31) may be written as

$$m = \left(\kappa^2 \mu_{m_1}^2 a_1 + \frac{\pi^2 a_2}{4}\right) \frac{1}{h^2} \tag{33}$$

2. At a certain linear dimension ratio of a sample, the core temperature varies in the same way as in the case of an infinite body, and the solutions of form (29) may be, under certain conditions, substituted by the simpler and more convenient one-dimensional solutions [e.g., solution (2)] with a high accuracy. Numerical calculation shows that at small Fourier numbers $(0 < a\tau/h^2 \leqslant 0.3)$ the temperatures at the finite cylinder center [see solution (28)] with the parameter $\kappa = \frac{1}{2}, \frac{1}{3}, \frac{1}{4}$ and at the infinite plate center $(\kappa = 0)$ coincide even at $\mathrm{Bi}_R = \alpha R/\lambda_2 \to \infty$. In the course of time the difference between these temperatures increases and achieves its maximum in a steady state. For example, if $\mathrm{Bi}_R = \infty$ and $\kappa = h/R = \frac{1}{4}$, then $\Theta_{st} = 0.013$, i.e., the error of measuring steady temperature at the finite plate center is about 1% because of the nonuniform heat flux.[20]

3. The solutions of form (29) can be used to determine thermal diffusivity

of isotropic and orthotropic materials as well as to study heat transfer parameters on the basis of the regular regime methods.[4] (See Appendix.)

A constant power source in a test material exchanging heat with the ambient medium makes it possible to develop a series of methods to determine not only thermal diffusivity but also thermal conductivity and heat capacity.[26–34] If, for example, over a certain plate surface a constant heat flux is prescribed and other surfaces are exchanging heat with the ambient medium according to Newton's law, then in this case also a regular thermal regime takes place, whose laws are similar to those considered above. The same is true for a hollow cylinder with a constant heat flux on its inner surface. The appropriate two- and three-dimensional solutions are discussed in Refs. 24–25. Consider first the simplest one-dimensional solutions.

The methods in Refs. 26–28 use a particular solution of equation (1):

$$\frac{\partial T(x,\tau)}{\partial \tau} = a\,\frac{\partial^2 T(x,\tau)}{\partial x^2}$$

for a plate $2h$ thick (the origin of the coordinates at the plate center) under the following boundary conditions:

$$T(x,0) = T_0 = \text{const}$$

$$\lambda\,\frac{\partial T(0,\tau)}{\partial x} = -q$$

$$T(h,\tau) = T_0$$

The solution of this problem convenient in practice at $F_0 = a\tau/h^2 > 1$ is of the form

$$\Theta(x,\tau) = T(x,\tau) - T_0 = \frac{q(h-x)}{\lambda} - \frac{2qh}{\lambda}\sum_{n=1}^{\infty}(-1)^{n+1} \times \frac{1}{\mu_n^2}\sin\mu_n\frac{h-x}{h}\exp\left(-\mu_n^2\frac{a\tau}{h^2}\right) \tag{34}$$

where

$$\mu_n = (2n-1)\frac{\pi}{2}$$

Solution (34) permits different versions of thermal diffusivity calculations to be used. In particular, it can be estimated applying *a priori* compiled tables[26] and making use of the described regular regime methods.[27] In this case the procedure is as follows. A steady component [the first term on the right-hand side of (34)] depends on the heat flux and thermal conductivity of the sample, i.e.,

$$\Theta_{st} = \frac{q(h-x)}{\lambda} \tag{35}$$

when measuring temperature difference between the surface and center of the plate

$$\Theta_{st} = \frac{qh}{\lambda} \tag{36}$$

In the regular regime which occurs at $F_0 > 0.3$, solution (36) may be in the form (37)

$$\Theta_{st} - \Theta = \frac{2qh}{\lambda\mu_1^2} \sin \mu_1 \frac{h-x}{h} \exp\left(-\mu_1^2 \frac{a\tau}{h^2}\right) \tag{37}$$

Hence, the rate of a temperature change

$$m_\infty = \frac{\ln(\Theta_{st} - \Theta_1) - \ln(\Theta_{st} - \Theta_2)}{\tau_2 - \tau_1} = \frac{\mu_1^2}{h^2} a \tag{38}$$

The formula for calculating thermal diffusivity

$$a = \frac{4h^2}{\pi^2} m_\infty \tag{39}$$

coincides with formula (17) but the rate of a temperature change must be taken in form (38). The experimental data are processed using the relationship

$$\ln(\Theta_{st} - \Theta) = f(\tau) \tag{40}$$

Thermal diffusivity is determined by the slope of straight line (40) (see Appendix).

The solution for a hollow cylinder with an inner radius R_1 and outer radius R_2 is of the same structure.[2] The steady component in this case is

$$\Theta_{st} = \frac{qR_1}{\lambda} \ln \frac{r_2}{r_1} \qquad (r_2 > r_1) \tag{41}$$

When measuring a temperature difference between the points $r_1 = R_1$ and $r_2 = R_2$

$$\Theta_{st} = \frac{qR_1}{\lambda} \ln \frac{R_2}{R_1} \tag{42}$$

Thermal diffusivity is estimated from

$$a = \frac{R_2^2}{\mu_1^2} m_\infty \tag{43}$$

But the roots μ_1 are found from the characteristic equation (see Table 2 in Appendix)

$$\mathcal{T}_1(\mu l) Y_0(\mu) - \mathcal{T}_0(\mu) Y_1(\mu l) = 0 \tag{44}$$

where

$$l = \frac{R_1}{R_2} \tag{45}$$

The procedure of thermal diffusivity calculation is the same as for a plate. Thermal diffusivity can be estimated by studying the curve of heating at the initial time moment.[28] The heating rate at an instant $\tau = 0$ and at a point $x = 0$ as it follows from solution (34) is determined by the expression

$$b = \frac{dT}{d\tau} = \frac{2qa}{\lambda h} \tag{46}$$

At the same time, this very heating rate can be expressed in terms of a steady state component (33)

$$\frac{\Theta_{st}}{\Delta\tau} = \frac{qh}{\lambda \Delta\tau} \tag{47}$$

From relationships (46) and (47)

$$a = \frac{h^2}{2\Delta\tau} \tag{48}$$

Values (35) and (42) being known, thermal conductivity can be calculated.

Heat capacity is found from the relationship

$$c\rho = \frac{\lambda}{a} \tag{49}$$

The approach[29–30] makes use of the solution to a one-dimensional problem for a semi-infinite body which, as a particular case, also follows from solution (34):

$$T(x, \tau) - T_0 = \frac{2q(a\tau)^{1/2}}{\lambda} \, i \operatorname{erfc} \frac{x}{2(a\tau)^{1/2}} \tag{50}$$

Using (50), studies can be performed in a pure unsteady-state regime. Specifically, when measuring an excess temperature between the heater plane and the base of the plate being in this case a model of a semi-infinite body, from (50) it follows that

$$\Delta T = \frac{2q(a\tau)^{1/2}}{\lambda\sqrt{\pi}} \tag{51}$$

From relationship (51), thermal activity is

$$\epsilon = \frac{\lambda}{\sqrt{a}} = \frac{2q\sqrt{\tau}}{\sqrt{\pi}\,\Delta T} \tag{52}$$

The experimental procedure and calculation of a set of thermophysical properties are discussed in Refs. 20, 29–34.

So, in the case of constant heat sources and sinks, estimation of thermal diffusivity by the regular regime method requires the knowledge of a steady component (36) or (42). The systematic error due to non-one-dimensional heat fluxes can be estimated from the solution and analysis of the appropriate two- and three-dimensional heat conduction problems.[2, 20, 24, 25] These solutions being tedious, only some consequences are considered here. From (36)

$$\lambda = \frac{qh}{\Theta_{st}} \tag{53}$$

Solution and analysis of the appropriate two-dimensional problem give the formula differing from (53) by its coefficient[20]

$$\lambda = \frac{qh}{\Theta_{st}} \frac{2}{\kappa} \sum_{\kappa=1}^{\infty} \frac{\tanh \kappa\mu_n}{\mu_n^2 \, \mathcal{T}_1(\mu_n)} \tag{54}$$

where μ_n are the roots of equation (15); $\kappa = 2h/2R$ is the relative height of the disk (plate). At $\kappa \to 0$ formula (54) coincides with (53). If $\kappa = \frac{1}{5}$, the relative error will be 0.8%; at $\kappa = \frac{1}{4}$ it is 1.4%. Thus, the systematic error increases with increasing height of the plate, which is observed in practice.

From relation (42)

TABLE 1

Summary of Monotonic Heating Regime Experiments in Measurements of Thermal Diffusivity of Solids

No.	Investigator	Ref.	Year	Material	Methods	Temperature measurement	Range	a (m^2 s × 10^7)	Accuracy (%) Max. error	Accuracy (%) Scatter	λ (W/m K)
1	Kondratiev, G.M.	4	1954	Disperse homo-geneous	Monotonic (regular heating)	t/c	300	0.25–2.5	No data	10–15	0.1–1
2	Kraev, O.A.	50	1956	Ceramics plastics	Regular heating	t/c	300–1300	5–10	No data	No data	5
3	Vasiliev, L.L.	46	1963	Disperse plastics	Adiabatic heating	t/c	10–400	0.8–5	No data	4	5
4	Platunov, E.S.	52	1964	Ceramics plastics	Regular regime	t/c	300–700	5–10	No data	3–8	5
5	Fraiman, Yu.E.	37	1964	Nonmetal	Linear heating	t/c	300–1500	0.8–5	No data	5–7	5
6	Volokhov, G.M.	27	1966	Disperse ceramics plastics	Monotonic heating with source	t/c	300	0.2–10	No data	2–5	0.1–5
7	Platunov, E.S.	22	1973	Metals	Regular heating	t/c	20–700	200–300	No data	5–8	100
8	Platunov, E.S.	22	1973	Refractory		Pyrometer	900–2000	5–10	No data	12	3
9	Shashkov, A.G. and Tyukaev, V.I.	57	1975	Heat insulators	Monotonic heating	t/c Pyrometer	3000	2–3	2	7–8	1

$$\lambda = \frac{qR_1}{\Theta_{st}} \ln \frac{R_2}{R_1} \tag{55}$$

The solution of the problem for a hollow finite cylinder yields[20]

$$\lambda = \frac{qR_1}{\Theta_{st}} \ln \frac{R_2}{R_1} \left[1 - \frac{2}{\kappa^2 \ln(1/\kappa)} \sum_{n=1}^{\infty} \frac{A_n}{\mu_n \cosh L\mu_n}\right] \tag{56}$$

where μ_n are the roots of the equation

$$\mathcal{J}_1(\mu_n\kappa)\, Y_0(\mu_n) = Y_1(\mu_n\kappa)\, \mathcal{J}_0(\mu_n)$$

$$\kappa = \frac{R_1}{R_2}, \qquad L = \frac{h}{R_2}$$

h is the cylinder half-height; L the relative height.

Without going into a detailed analysis of (56) note that in this case the systematic error can be estimated exactly.

2. MEASUREMENT OF THERMAL DIFFUSIVITY IN THE WIDE RANGE

Studies of thermal diffusivity as a function of temperature by successive thermostating of a sample at increasing temperatures cannot be considered convenient because of the great amount of time spent for the whole experiment. Therefore, the methods that seem promising are those which allow a temperature function of thermal diffusivity to be found within one continuous experiment. Solutions of the heat conduction equations [with heat flux at the surface prescribed or, in the case of heat transfer, with the ambient medium, whose temperature varies following a linear (or close to linear) law] are the theoretical basis for this group of methods.

Temperature fields in samples of very simple geometries heated symmetrically in a medium, whose temperature is a linear time function, with the assumption of the thermal properties independent of temperature are described by the expression[1]

$$\Theta = \frac{T(r,\tau) - T_0}{T_0} = \mathrm{Pd}\left\{F_0 - \Gamma\left[\left(1 + \frac{2}{\mathrm{Bi}}\right) - \frac{r^2}{R^2}\right]\right\}$$

$$+ \sum_{n=1}^{\infty} \frac{A_n}{\mu_n^2}\, \Phi\left(\mu_n \frac{r}{R}\right) e^{-\mu_n^2 F_0} \tag{57}$$

where $\mathrm{Pd} = bl^2/aT_0$, $F_0 = a\tau/l^2$ are the dimensionless complexes including the characteristic dimension of the sample (for a plate $l \equiv h$, for a cylinder and a sphere $l \equiv R$); r is the variable coordinate (for a plate $r = x$). $\Gamma = \frac{1}{2}, \frac{1}{4}, \frac{1}{6}$ for an infinite plate, a cylinder and a sphere, respectively, the numbers μ_n are found from characteristic equations (11)–(13). The analysis of expression (57) has revealed two temperature regimes, unsteady and quasisteady (regular regime of the second kind).[1] Under unsteady regime, for calculation of temperatures and heat fluxes expression (57) should be considered. At a certain value of $F_0 \geqslant F_{0_1}$ (for a plate $F_0 > 0.5$[1]), the series in (57) can be neglected. Then

$$\Theta = \mathrm{Pd}\left[F_0 - \Gamma\left(1 + \frac{2}{\mathrm{Bi}} - \frac{r^2}{R^2}\right)\right] \tag{58}$$

i.e., the temperature at any point of the sample changes following the linear law, and the temperature distribution along the coordinate obeys the quadratic law. Such a regime is called quasisteady.[1] The problems of the determination of the onset of a quasisteady regime are quite completely described in Refs. 1 and 22, and will be omitted from the present review. The simple and predicted relations are directly obtained from (58):

$$a = \Gamma\frac{b(r_2^2 - r_1^2)}{\Delta T} \tag{59}$$

i.e., to determine thermal diffusivity, it is necessary to know the heating rate b and the temperature difference between some two points of the sample, say, at the center and on the surface. The other formula also follow from (59):

$$a = \Gamma\frac{r_2^2 - r_1^2}{\Delta\tau} \tag{60}$$

where $\Delta\tau$ is the lag time, i.e., the time required for the temperature at a certain point of the sample to become equal to the temperature at another point of the same sample. Similar relationships hold for the case when a constant heat flux is prescribed on the surface.[1] From expression (59) it follows that the heat flux on the sample surface $q = 2\Gamma(\lambda/a)bR = \mathrm{const}$.

Formulas (59) and (60) are obtained from the solution of linear one-dimensional equation (1). Under real conditions the samples of finite dimensions are used, therefore, some corrections in (59), (60)[22] or the selection of the appropriate sample dimensions[20,25,35] are necessary. In Ref. 25, a general problem is considered on heating a finite cylinder in a medium with linearly changing temperature, whose side surface and bases are heated at different heating rates. The solutions (58) for an infinite plate and an infinite cylinder are obtained as a particular case of this solution. The formulas for thermal diffusivities at two-dimensional heat fluxes are more complex than relation (59). In a particular case of equal heating rates the formulas simplify. The factor Γ in (59) depends on the ratios of the linear dimensions of the cylinder. For several particular cases the values of these coefficients are the following:

Infinite plate	$\frac{1}{4}$	$\frac{1}{3}$	$\frac{1}{2}$	Infinite cylinder	3	2	1
0.5	0.4928	0.4752	0.4056	0.25	0.2496	0.2454	0.2006

For a plate, the temperature drop is measured between the center and base, and for a cylinder it is measured between the center and the side surface.

A method for determining thermal diffusivity under linear heating was originally suggested in Ref. 36 and its modifications were used by many authors.[20, 22, 37, 42] In this case the test material is placed in a medium whose temperature varies following the linear (or close to linear) law. A linear heating rate is preset by programmed temperature regulators.[37–40]

Thermal properties of the test material can continuously change during linear heating. A real thermogram is not linear but monotonic time function. Theoretical and experimental studies of many authors[20, 22, 37] have shown that thermal diffusivity as a function of temperature can be sought for from the formulas of form (60) if the experiment is performed at small temperature drops (5–10 K). The thermograms are processed after division into particular linear sections.

When in a sample there are heat sources of constant power and heating is linear, the capabilities of the quasisteady methods essentially increase since under such conditions they may be used for determination not only of the thermal diffusivity but also of the thermal conductivity and heat capacity.[37, 38, 41]

This variant is represented by the method of Refs. 37, 42, which is theoretically based on the solution of the equation

$$\frac{\partial T(r,\tau)}{\partial \tau} = \frac{1}{r}\frac{\partial}{\partial r}\left[r\frac{\partial T(r,\tau)}{\partial r}\right]$$

for a hollow cylinder with the inner radius R_1 and outer radius R_2 under the boundary conditions

$$T(r,0) = T_0$$

$$T_c = T_0 + b\tau$$

$$\lambda \frac{\partial T(R_2, \tau)}{\partial r} = \alpha[(T_0 + b\tau) - T(R_2, \tau)]$$

$$-\lambda \frac{\partial T(R_1, \tau)}{\partial r} = q$$

The formulas for calculating thermal diffusivity and thermal conductivity are the following:

$$a = \frac{b}{4\Delta T}\left(r_2^2 - r_1^2 - 2R_1^2 \frac{r_2}{r_1}\right) \tag{61}$$

$$\lambda = \frac{qR_1 \ln(r_2/r_1)}{\Delta T - \Delta T'} \tag{62}$$

The method provides for two experiments at $q = 0$ and $q = \text{const}$. In the former case (61) and in the latter (62) are calculated with $\Delta T'$ being a new temperature drop measured at $q = \text{const}$. The method has been applied for studying thermophysical properties of heat insulators for temperatures between 350 and 1000 K. The error of measurements is 5%–7%.

In some of the approaches the heat conduction equation is solved with a prescribed constant heat flux on the sample surface.

Different modifications of the above methods are known. In Refs. 43 and 44, a one-dimensional heat conduction equation for a plate $2h$ thick with the coordinate origin at its center is solved under the boundary conditions of the form

$$T(x, 0) = T_0$$

$$\lambda \frac{\partial T(h, \tau)}{\partial x} = q \tag{63}$$

$$\frac{\partial T(0, \tau)}{\partial x} = 0$$

Under these conditions a quasisteady regime temperature varies in time and space in the same way as with linear heating. Thermal diffusivity is estimated by the formula of form (59) and thermal conductivity is found from

$$\lambda = \frac{qh}{2\Delta T} \tag{64}$$

No special facilities have been used in the methods of Refs. 43 and 44 to prescribe a required regime. A constant heat flux is prescribed by using some similar samples with flat electric heaters between them.

Formula (64) does not take into account the specific heat of the source. Usually, for estimating the effect of the specific heat of the source the boundary condition of form (63) is supplied with a negative source:

$$q' = c\frac{\partial T}{\partial \tau} = cb$$

where c is the heat capacity of a heater. In Ref. 45, using the solution to the problem for an infinite cylinder, the formula for calculating thermal conductivity including heat capacity of the source is of the form

$$\lambda = \frac{(q - cb)R}{2\Delta T}$$

In Refs. 46–48, the solution of the heat conduction equation is obtained for an infinite cylinder having a constant-power heat source inside and ideal heat insulation ($\partial T(R, \tau)/\partial r = 0$) for its outer surface. This solution is used to develop a method for studying thermal conductivity and thermal diffusivity of poor heat conductors from 4.2 to 400 K. The thermophysical properties have been estimated not only under quasisteady conditions but also at the initial stage of heating with the aid of the compiled tables. The calculation error in this case is 2%–4%.

The quasisteady approach may also be implemented using a heat capacity standard (see, for instance, Ref. 49).

At high temperatures, the realization of the boundary conditions (constant heat flux or constant heating rate) involves considerable engineering difficulties, which has induced the development of the so-called monotonic regime (heating and cooling) methods, which are essentially a general case of the quasisteady heating method. The approximate solutions of the nonlinear heat conduction equation are a theoretical basis of the methods.[22, 50–56] Initially the monotonic heating methods were developed in Refs. 50 and 51. The temperature field was expressed in a power series form, and finally the formulas were obtained which differed from (59) and (60) by the corrections which included the changes of the heating rate and thermal diffusivity

$$a = \frac{R^2}{\Gamma \Delta \tau}(1 + \epsilon + \delta) \tag{65}$$

where

$$\epsilon = -\frac{1}{4\Delta T}\frac{d\Delta T}{d\tau}\Delta\tau, \qquad \delta = \frac{1}{4a}\frac{da}{dT}\Delta T$$

The experiment was carried out under a heating regime close to the linear one without using any special temperature regulators. At small temperature drops ($\Delta T < 10\,\mathrm{K}$) the corrections can be neglected. Then expression (65) coincides with (60).

The theory and experimental bases of the monotonic heating methods are most completely described in Ref. 22. The main theoretical assumptions made in the above work may be briefly summarized as follows. Within a small temperature drop ΔT, if it does not involve the phase transition points, thermal properties may be expressed by the series with any required accuracy

$$i = i_0(1 + \kappa_i\Theta + n_i\Theta^2 + \cdots) \qquad (i = \lambda, a, c, \rho) \tag{66}$$

Relative temperature factors κ_i and n_i in series (66) are a function of the reference temperature $T_0(\tau)$ but remain constant within the drop:

$$\kappa_i = \frac{1}{i_0}\frac{di_0}{dT}, \qquad n_i = \frac{1}{2i_0}\frac{d^2 i_0}{dT^2} \tag{67}$$

Some point convenient for direct temperature measurement – say the center of the sample or its surface – is chosen as a reference point with the temperature $T_0(\tau)$. Then

$$T(r, \tau) = T_0(\tau) + \Theta(r, \tau) \tag{68}$$

It is also assumed that the heating rate $b(r, \tau)$ is a monotonic time function and a series of form (66) is valid for it. Usually $|\kappa_i| < 3 \times 10^{-3}\,\mathrm{K}^{-1}$ and $|n_i| < 3 \times 10^{-6}\,\mathrm{K}^{-1}$, therefore, the conditions of the optimum convergence of (66) are determined by the temperature drops which occur during experiment. The linear one-dimensional heat conduction equation was solved by the successive approximation method with the above assumptions.[22] The formulas for calculation of thermal diffusivity were obtained from the solution. In the case of symmetric heating of a plate, a cylinder, and a sphere the formulas become

$$a = \frac{b_0 R^2}{\Gamma\Theta_R}(1 + \Delta\sigma_{i_\Theta}) \tag{69}$$

where b_0 is the heating rate at the reference point, R is the characteristic dimension of the sample, Θ_R is the temperature drop with respect to the reference point. Formula (69) differs from (60) by the correction

$$\Delta\sigma_{a_\Theta} = \frac{\Gamma}{\Gamma+4}\left(\kappa_{b_T} - 2\kappa_a - \frac{4}{\Gamma}\kappa_\lambda\right)\theta_R \tag{70}$$

where κ are the relative temperature factors of the heating rate, thermal diffusivity, and thermal conductivity of form (67). The formula for calculation of thermal diffusivity in terms of the lag time is of a similar form:

$$a = \frac{R^2}{\tau_R(\tau)}(1 + \Delta\sigma_{a_T}) \tag{71}$$

Corrections in formulas (69), (71) are complicated and their practical calculation is quite difficult. In Ref. 22 it is shown that at heating rates corresponding to small temperature drops (3–10 K), the corrections in (70) may be neglected and expressions (59) and (60) of quasisteady heating can be used for calculations.

On the basis of the monotonic heating method, the devices and instruments have been developed, which are designed for determining thermal diffusivity and thermal conductivity of heat insulators and metals over a temperature range between −150 and 2000°C.[22, 57] The accuracy of determining thermophysical properties was estimated through an approximate thermal analysis.[22] The value of the measurement error (5%–12%) depends on the test temperature range and the class of the test materials. For realizing the monotonic heating methods it is possible to use special facilities for prescribing the required thermal regimes. Different versions of thermal devices are considered in Refs. 22 and 58.

APPENDIX

1. A schematic drawing of a simple experimental device for determination of thermal diffusivity of solid, loose and fibrous materials at room temperature is shown in Fig. 1.[20] For solid material test two hollow plane parallel units 1, 3, are used, which can travel over a vertical plane along the guides (they are not shown in the figure). Loose and fibrous materials are tested in a copper or brass cylinder 2, fitted with a case 6, and removable covers 7, 8. Two differential chromel–alumel thermocouples 0.2 mm in diameter serve as a thermometer. Junctions of each of the thermocouples are soldered beforehand at the lower unit plane 1, and on the inner surface of the cylinder, 2, respectively. The other junction of DT_2 is mounted in a tube or needle fixed at the cylinder center by

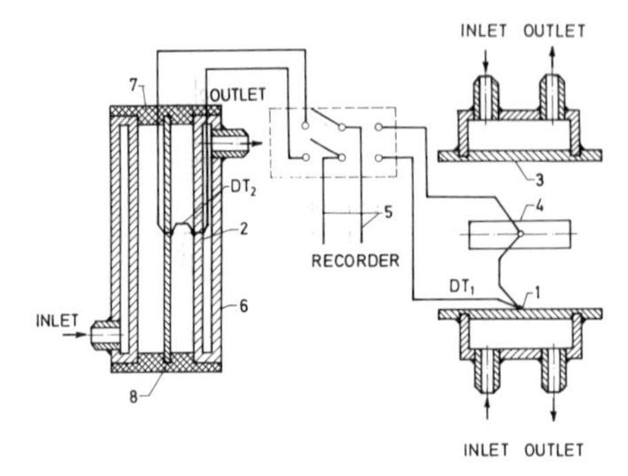

FIGURE 1. Device for determination of thermal diffusivity at room temperatures.

covers 7, 8. The junction of DT_1 is placed at the center of the test solid sample composed of two parallel plane square or cylinder plates 4. Cavities of the system are connected with a thermostat and filled with forced-through water of a constant temperature. So, on the plate bases and side surface of the cylinder, heat transfer occurs with the ambient medium of a constant temperature; the "free" surfaces exchange heat by convection with the surroundings at different temperature, say, room temperature – i.e., heat transfer with the surroundings of constant but different temperatures takes place. The analysis of the solutions of form (29) has revealed that for the present experimental procedure the sample dimensions should satisfy the following relations.

For a cylinder:

$$\kappa = \frac{2h}{2R} \geqslant 3$$

where h is the half-height of a cylinder and R is the inner radius of a cylinder.

For a plate:

$$\kappa = \frac{2h}{2R} \leqslant \frac{1}{4}$$

where h is the half-height of a plate and $2R$ is the diameter or the side of the square. In this case, heat transfer from the side surface of a plate and from the bases of a cylinder can be neglected, and formulas (17), (18) may be used in calculations.

In experiments with a solid sample, two parallel plane circular or square plates are made of the test material (the linear dimension ratio should satisfy the above formulas). Absolute dimensions of the sample depend on its structure and properties. Dense insulating materials (plastics, ceramics) can be of a small size, $h \approx 5$–8 mm. The sizes of the large-pore samples should be larger. A solid sample is initially outside the units filled with forced-through water of a constant temperature. Free ends of differential thermocouple DT_1 are connected with a self-recorder. After a stable temperature difference sets in, which can be inferred from the self-recorder readings, the sample is placed in the space between the units and pressed with their planes. Additional heat resistance can be considerably diminished by careful finish of the contact planes and strong compression of the samples. Thermal diffusivities are calculated by formula (17). When loose materials are tested, the cylinder is filled before the constant temperature water is supplied to its cavity. The results are processed in a way similar to the previous case. The calculation is performed according to formula (18). Thermal diffusivity of solid materials whose thermal conductivity lies between 0.1 and 5 W/m°C is accurate within 3%–5%.

2. A schematic drawing of a simple experimental device for determination of thermal diffusivities and thermal conductivities of solid, fibrous and loose materials is shown in Fig. 1. Its main elements are described in item 1 of the Appendix. This scheme is supplemented with cylindrical and plane constant-power heat sources and instruments for measuring the power (the instruments are not shown in Fig. 1). Both heaters should be of a low heat capacity. The cylindrical heater is wound on the frame of any thin material, say, thin copper or brass foil. The thermocouple junction DT_2 is fixed on the frame wall or at the center. The ready frame is covered with an insulating varnish or glue layer, and the heater of manganine or nichrome wire is wound thereafter. The tube heater is closed with plugs inserted into the recesses made in the covers of the cylindrical calorimeter. The heater is rigidly fixed to the lower cover, with the heater and thermocouple leads let through. Depending on the solid sample shape (a square, disk) the flat heaters are manufactured as a square or a circle from the thin wire of a high specific resistance or from the foil. The other junction of the differential thermocouple is glued into the heater.

The experimental procedure is as follows: after the hollow cylinder is filled with the test material and a flat test sample is placed together with the heater between the hollow units which press the sample, the thermostat should be switched on. After the temperature becomes uniform over the whole volume, which is shown by the readings of the temperature-difference records, the heaters are energized. The power to be supplied to the heater should be chosen according to the optimal temperature drop of 5 to 10 K. After the heaters are switched on, the excess temperature begins increasing and reaches its maximum in the steady state. During experiment relation (34) is obtained. Once the steady state is reached, steady-state temperature drops are determined, which can be used for plotting relation (40) and calculating the rate of temperature change (38). Thermal diffusivities are calculated from formulas (39) and (43). The values of μ are found from Table 2. The temperature behavior and the calculation methods are illustrated in Fig. 2.

TABLE 2
Values of the First Root of the Characteristic Equation
$\mathscr{J}_1(\mu l)\, Y_0(\mu) - \mathscr{J}_0(\mu)\, Y_1(\mu l) = 0$

l	μ_1	l	μ_1	l	μ_1
0.10	2.449	0.37	2.997	0.64	4.763
0.11	2.458	0.38	3.032	0.65	4.885
0.12	2.467	0.39	3.069	0.65	5.013
0.13	2.478	0.40	3.107	0.67	5.152
0.14	2.489	0.41	3.147	0.68	5.297
0.15	2.501	0.42	3.188	0.69	5.453
0.16	2.514	0.43	3.231	0.70	5.619
0.17	2.528	0.44	3.276	0.71	5.797
0.18	2.542	0.45	3.323	0.72	5.888
0.19	2.558	0.46	3.372	0.73	6.193
0.20	2.574	0.47	3.422	0.74	6.415
0.21	2.591	0.48	3.475	0.75	6.654
0.22	2.609	0.49	3.531	0.76	6.913
0.23	2.628	0.50	3.589	0.77	7.195
0.24	2.648	0.51	3.649	0.78	7.503
0.25	2.668	0.52	3.712	0.79	7.841
0.26	2.690	0.53	3.778	0.80	8.213
0.27	2.712	0.54	3.874	0.81	8.624
0.28	2.736	0.55	3.919	0.82	9.081
0.29	2.761	0.56	3.995	0.83	0.592
0.30	2.786	0.57	4.075	0.84	10.167
0.31	2.813	0.58	4.159	0.85	10.820
0.32	2.840	0.59	4.247	0.86	11.566
0.33	2.869	0.60	4.339	0.87	12.427
0.34	2.899	0.61	4.437	0.88	13.431
0.35	2.931	0.62	4.540	0.89	14.619
0.36	2.963	0.63	4.648	0.90	16.045

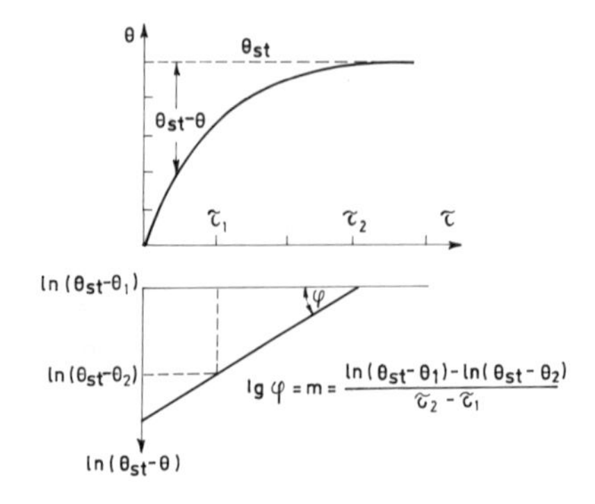

FIGURE 2. The method for calculating the rate of a temperature change with a heat source of constant power.

NOTATION

λ (W/m K)	Thermal conductivity
a (m^2/s)	Thermal diffusivity
c (J/kg K)	Specific heat
T (K)	Temperature
τ (s)	Time
Θ	Relative temperature
q (W/m^2)	Specific heat flux
α (W/m^2 K)	Heat transfer coefficient
$F_0 = a\tau/l^2$	Fourier number
$\mathrm{Bi} = \alpha l/\lambda$	Biot number
b (K/s)	Heating rate

REFERENCES

1. A.V. Luikov, *Analytical Heat Diffusion Theory,* Academic Press, New York (1968).
2. H.S. Carlslaw and D.C. Jaeger, *Conduction of Heat in Solids,* 2nd edition, Clarenden Press, Oxford (1959).
3. A.I. Tikhonov and A.A. Samarsky, *Mathematical Physics Equations,* Gostekhizdat Publ., Moscow (1951).
4. G.M. Kondratiev, *Regular Thermal Regime,* Gostekhizdat Publ., Moscow (1954).
5. G.M. Kondratiev, *Heat Measurements,* Mashgiz (1957).
6. G.N. Dulnev and G.M. Kondratiev, *Izv. Akad. Nauk SSSR, Otd. Tekhn. Nauk* 3 (1955).
7. G.N. Dulnev and G.M. Kondratiev, *Izv. Akad. Nauk SSSR, Otd. Tekhn. Nauk* 7 (1956).
8. N.A. Yaryshev, *Theoretical Fundamentals of Measuring Unsteady Temperatures,* Energya Publ., Leningrad (1967).
9. E.S. Fitzsimmons, *J. Am. Ceram. Soc.* **33**(11), 327 (1950).
10. A.F. Chudnovsky, *Heat Transfer in Dispersed Media,* Gostekhizdat (1954).
11. G.B. Simonov, *Stroit. Prom.* 8 (1952).
12. G.B. Simonov, *Zh. Teor. Fiz.,* vyp. 8 (1953).
13. A.F. Begunkova, *Zavod. Lab.* 10 (1952).
14. A.F. Begunkova, Sb. "Studies in the Region of Heat Measurements," Izd. LITMO, vyp. 20, Mashgiz, Leningrad (1956).
15. G.N. Tretiyachenko and L.V. Kravchuk, in *Heat and Mass Transfer,* Vol. 1, Izd. AN BESSR, Minsk (1962).
16. L.A. Plotnikov, *Zavod. Lab.* 9 (1950).
17. A.E. Paladino, E.L. Swarts, and W.B. Irndall, *J. Am. Ceram. Soc.* **40,** 340 (1957).
18. M.P. Emchenko, *Priborostroenie* 3 (1938).
19. M.P. Emchenko, *Tr. Leningr. Lesnoi Akad.* 83 (1959).
20. *Methods of Determining Thermal Conductivity and Thermal Diffusivity,* Energiya Publ., Moscow (1973).
21. A.V. Luikov, *Heat Conduction Theory,* GITTL, Moscow (1952).
22. E.S. Platunov, *Thermophysical Measurements at Monotonic Regime,* Energiya Publ., Leningrad (1973).
23. M.B. Das and M.A. Hossain, *Br. J. Appl. Phys.* **17,** 1 (1966).
24. G.M. Volokhov, *DAN BSSR, Ser. Fiz.-Tekhn. Nauk* 4 (1967).
25. G.M. Volokhov, V.P. Kozlov, and G.A. Surkov, *Heat and Mass Transfer,* Vol. 7, Minsk (1972).

26. V.O. Fogel and P.P. Alekseev, *Inzh. Fiz. Zh.* 2 (1962).
27. G.M. Volokhov, *Inzh. Fiz. Zh.* **9**, 5 (1966).
28. J. Hatta, *Rev. Sci. Instrum.* **28**, 50 (1979).
29. A.B. Verzhinskaya and L.N. Novichenok, *Inzh. Fiz. Zh.* 9 (1960).
30. A.B. Verzhinskaya, *Inzh. Fiz. Zh.* 4 (1964).
31. W.A. Plummer, D.E. Campbell, and A.A. Comstock, *J. Am. Ceram. Soc.* **45**(7), 310 (1962).
32. A.F. Chudnovsky, *Zh. Teor. Fiz.* **23**, 12 (1953).
33. J.C. Jager, *J. Geophys. Res.* **64**, 5 (1959).
34. W.E. Haupin, *Am. Ceram. Bull.* **39**, 3 (1960).
35. Yu.E. Fraiman and G.A. Surkov, *Inzh. Fiz. Zh.* **8**, 5.
36. A.V. Luikov, *Zh. Tekhn. Fiz.* **5**, 2–3 (1935).
37. Yu.E. Fraiman, *Inzh. Fiz. Zh.* 10 (1964).
38. A.V. Luikov *et al.*, in *Proc. 3rd All-Union Heat and Mass Transfer Conference*, Vol. 7, Nauka i Tekhn. Publ., Minsk (1968).
39. V.V. Vlasov, *Inzh. Fiz. Zh.* **7**, 3 (1964).
40. V.V. Vlasov and A.I. Fesenko, *Automation of Thermophysical Measurements,* Tambov (1972).
41. L.L. Vasiliev and S.A. Taneava, *Thermal Properties of Porous Materials,* Nauka i Tekhnika Publ., Minsk (1971).
42. L.L. Vasiliev and Yu.E. Fraiman, *Thermophysical Properties of Bad Heat Conductors,* Minsk (1967).
43. O. Krischer, *Chem. Ing. Techn.* **26**(1), 42 (1954).
44. L.A. Selinov, *Proc. Postov Engineering-Building Institute,* vyp. IV (1955).
45. T.S.E. Thomas, *Br. J. Appl. Phys.* **8**, 10 (1957).
46. L.L. Vasiliev, *Inzh. Fiz. Zh.* **6**, 9 (1963).
47. L.L. Vasiliev and G.A. Surkov, *Inzh. Fiz. Zh.* **7**, 6 (1964).
48. L.L. Vasiliev, S.A. Tanaeva, and A.D. Shnyrev, *Inzh. Fiz. Zh.* **17**, 6 (1969).
49. Th. Gast, Hellwege, and K. Knappe, *Kolloid Z.* 174 (1961).
50. O.A. Kraev, *Teploenergetika* 4 (1956).
51. O.A. Kraev, *Teploenergetika* 4 (1957).
52. E.S. Platunov, *Izv. Vuzov, Priborstr.* **1**, 1 (1961).
53. E.S. Platunov, *Teml. Vysok Temp.* **3**, 5 (1964).
54. E.S. Platunov, *Inzh. Fiz. Zh.* **9**, 4 (1965).
55. E.S. Platunov and S.E. Burovoi, *Izv. Vuzov, Priborstr.* **11**, 7 (1968).
56. E.S. Platunov, in *Proc. 3rd All-Union Heat and Mass Transfer Conference,* Nauka i Tekhnika Publ., Minsk (1968).
57. A.G. Shashkov and V.I. Tyukaev, *Thermal Properties of Decaying Materials at High Temperatures,* Nauka i Tekhnika Publ., Minsk (1975).
58. S.E. Burovoi, V.V. Kurenin, and E.S. Platunov, *Inzh. Fiz. Zh.* **30**, 4 (1976).

IV

SPECIFIC HEAT MEASUREMENT METHODS

Specific Heat

An Introduction

Six chapters are devoted to six different calorimetry techniques used for the measurement of specific heat mainly above room temperature and extending up to several thousand degrees. These techniques are: adiabatic, drop, levitation, modulation, pulse, and differential scanning calorimetry.

Although some of the calorimetry techniques, such as those that utilize the method of mixtures, date back over one hundred years, they have been used continuously after a succession of improvements and adaptations to emerging new requirements. A major surge in calorimetry came about in the mid-1950s and extended over approximately two decades. During this period, several conventional calorimetry techniques, such as adiabatic and drop, were perfected to their practical limits. Also during this period, several specialized calorimetry techniques such as levitation, modulation, and pulse, were developed for the measurements of specific heat at high temperatures and under other extreme conditions (high pressures, etc.).

In some of the calorimetry techniques (adiabatic, modulation, pulse, differential scanning), specific heat is obtained directly from the measured quantities; in others (drop, levitation), direct measurements yield enthalpy data which after differentiation yield specific heat.

The nature and the form of the specimen dictates, in some cases, the selections of the particular technique. For example, in adiabatic and drop techniques, the specimen is generally contained in a capsule which enables the use of a specimen practically in any form (powder, granules, chunks, etc.). Also, there is no restriction as to the specimen's electrical conduction characteristics. In some of the more specialized methods, such as in modulation and pulse, resistive self-heating is used which limits the technique to electrically conducting specimens in the form of wires, rods, or tubes.

Near room and at moderately high temperatures up to 1000 K, the measurement of specific heat does not present any serious problems. At higher temperatures and especially at temperatures above 2000 K, heat loss due to thermal radiation becomes a very important parameter. In addition, chemical reactions and specimen evaporation at high temperatures set an upper temperature limit for the applicability of the steady-state techniques for measurements of specific heat. Steady-state techniques have been used for measurements up to 3000 K and in a few cases to even somewhat higher temperatures; however, the limit for accurate measurements is about 2000 K. Pulse techniques with very short experiment duration (subsecond and submillisecond) can be used for measurements at very high temperatures (up to 10 000 K).

Almost all the calorimetry techniques have benefited immensely from the advances, during the last three decades, in general electronics and computer technologies. As a result, improvements in temperature and power controls as well as improvements in the measuring instruments and data reduction techniques contributed to the increased accuracy in specific heat determinations.

At present, it is possible to measure, with the best available techniques, specific heat with an uncertainty of not more than 1% in the temperature range 300–1000 K, not more than 2% in the range 1000–2000 K, and about 3% in the range 2000–3000 K. Above 3000 K, definitive measurements are not yet available; it may be said, however, that uncertainty in specific heat above 3000 K increases rapidly reaching an estimated value of 10% at 7000 K.

The six calorimetry techniques for the measurement of specific heat presented in this volume are discussed briefly in the following paragraphs. A summary table is also included (Table 1).

Adiabatic calorimetry is possibly the most versatile calorimetry technique for the direct measurement of thermal effects in substances, covering the range from cryogenic to moderately high temperatures. The technique has been used at temperatures as low as the helium point and as high as near 1900 K. However, for accurate measurements, the upper temperature limit is generally considered to be about 1300 K. This technique is used to measure specific heat of solids and liquids as well as heats of phase transitions, heats of solution and formation, and heat effects associated with structural changes in the specimen. This technique has high sensitivity, and as a result, can be used most effectively for measurements of specific heat near first- and second-order phase transitions.

Drop calorimetry, which is based on the classical method of mixtures, is generally used for measurements above room temperature. Although drop calorimeters have been built to operate at temperatures as high as near 3000 K, the accuracy of this technique deteriorates rather rapidly above 2000 K. The most accurate drop calorimeters available at present operate in the temperature range 300–1100 K. Since the specimen is generally placed in a capsule, the technique is applicable to both solids and liquids provided that no chemical reactions or alloying take place between the specimen and the container.

TABLE 1
Specific Heat Measurement Techniques

Technique	Advantages	Disadvantages	Temp. range[a] (K)	Specimen material	Uncertainty (%)
Adiabatic	Very versatile High sensitivity Solid and liquid specimen	Specimen in container Limitations in high temperatures	4–1300	All	1–3
Drop	Solid and liquid specimen	Specimen in container Slow measurement times Specific heat obtained from enthalpy data	300–2000	All	1–3
Levitation	No container for specimen Solid and liquid specimen	Specimen must be electrical conductor Specific heat obtained from enthalpy data	1000–2500	Electrical conductor (sphere)	2–5
Modulation	No container for specimen Multiproperty measurement capability	Solid specimen	80–3000	All	2–5
Pulse	No container for specimen Measurements at very high temperatures Solid and liquid specimen Multiproperty measurement capability	Specimen must be electrical conductor Sophisticated instru-mentation	1000–7000	Electrical conductor (wire, rod, tube)	2–10
Differential scanning	Quick and economical	Specimen in container Limitations in high temperatures	100–1000	All	1–3

[a] For each technique measurements at temperatures higher than that indicated have also been performed. However, the given value indicates the limit of the accurate measurements for that technique.

Accurate values of specific heat cannot be derived from drop calorimetry alone if nonequilibrium states are produced in the specimen during the experiment.

Levitation calorimetry is a variant of the method of mixtures. The main difference between levitation and drop calorimetry is the manner in which the specimen is held before the drop. While in drop calorimetry the specimen is usually placed in a capsule suspended in a furnace with a metal wire, in levitation calorimetry the specimen is levitated by a high-frequency electromagnetic field and is heated inductively. The key advantage of the levitation calorimetry is the fact that the specimen is not in contact with any other substance, other than the surrounding inert gas; thus the technique is immune to container problems. Levitation calorimetry is particularly attractive for measurements at temperatures above 2000 K, and can be used for electrically conducting specimens both in their solid and liquid phases. In some cases, excursions of the order of 1000 K above the melting point of the specimen have been achieved. Vapor pressure of the specimen usually sets the upper temperature limit for this technique.

Modulation calorimetry is applied to electrically conducting and non-conducting solids and is based on measurements performed by modulating the power used to heat the specimen. This method may be used from low temperatures up to near the melting point of the specimen. Since the amplitude of temperature oscillations can be as small as of the order of 0.001 K, the method provides high measurement sensitivity, and is suitable for measurements near solid–solid phase transformations. The technique is used for measurements at high temperatures as well as at high pressures.

Pulse calorimetry is developed to extend the limits of accurate measurements of specific heat of electrically conducting specimens to temperatures beyond the limits of other calorimetry techniques. Because of the extremely short experiment duration (subsecond to submillisecond), this technique is immune to most of the high-temperature problems that arise from heat losses, chemical reactions, evaporation, etc. Pulse calorimetry is generally used at temperatures above about 1000 K. In the case of subsecond-duration techniques, the upper temperature limit is the melting point of the specimen. In order to extend the measurements beyond the melting point and to several thousand degrees in the liquid phase, submillisecond-duration techniques are used.

Differential scanning calorimetry (DSC) and its predecessor, differential thermal analysis, have long been used for qualitative or semiquantitative studies of the behavior of materials. It has been only during the recent years that DSC has become a truly quantitative calorimetry technique. Commercial availability of differential scanning calorimeters and relatively rapid generation of reliable data over a reasonably wide temperature range have contributed to the widespread use of the DSC technique. The temperature range of the available DSC instruments is 100–1000 K with scanning rates of from a fraction of a degree to upwards of a hundred degrees per minute in both heating and cooling.

12

Adiabatic Calorimetry

D.N. KAGAN

1. INTRODUCTION

Adiabatic calorimetry is a most precise method of direct measurement of thermal effects accompanying thermodynamic transformations of substances. As distinct from many other experimental methods of investigation of thermodynamic characteristics, direct measurements by adiabatic calorimeters result in determination of both heat capacity and heats of phase transitions, chemical reactions, formation of alloys, processes of dissolution, and so on. High sensitivity and superb accuracy of the calorimetry, enabling the measurements to be performed even in the region of a phase transition of the second kind and, in particular, in the region of the critical state of the substance involved, are due to the adiabatic conditions of the calorimetry process. Heat loss, in this case, is excluded, and practically no need arises to introduce the corresponding corrections, which, as a rule, are far from being ideally precise. The adiabatic conditions of the calorimetry process are achieved at the expense of equality of the temperature of the calorimetric vessel surface and that of the adiabatic shield.

Heat exchange between the sample under test and the environment in the variable-temperature adiabatic calorimeters is removed by means of the shields whose temperature-control system perfectly ensures that the temperature is held equalized with that of the sample throughout the calorimetry process.

The achievements in the field of electronic control and measurement systems as well as in the sphere of computing technology obtained for the past 20 years have found their application in thermophysical research and, in particular, in adiabatic calorimetry, which has resulted in materially higher reliability of the method and in streamlined procedures for the experiments. The balance

D.N. KAGAN • Institute of High Temperatures, Korovinskoe Road, USSR Academy of Sciences, Moscow 127412, USSR.

of the temperature of the sample surface and that of the adiabatic shield is obtained best of all in the constant-temperature calorimeters in which the phase transition of the calorimetric substance is used for measuring the thermal effect. In this case the adiabatic conditions are provided by placing the calorimeter into a thermostat wherein the same phase transition takes place.

The variable-temperature adiabatic calorimeters can operate in continuous and stage heating modes. In the case of the continuous-heating operation the rate of heating usually lies within $0.1\ K\,s^{-1}$. When operating in the stage heating mode the rise of the temperature does not exceed a few degrees. These values are required for maintaining the equilibrium conditions in the sample. The lower restrictions are determined by gain of contribution of small heat loss with decrease of the heating rate. The said requirement concerning the equilibrium conditions in the sample confines the maximum dimensions of the sample (with consideration for its coefficient of thermal diffusivity) to 20–30 mm in diameter and 50–60 mm in length. The minimum dimensions of the sample are limited by the increase of the contributions of the heat capacity made by the calorimeter component parts, which leads to a lesser accuracy of the results.

The adiabatic calorimetry, in essence, is void of any limitations with respect to the materials to be tested and therefore is applicable to any substance in the condensed phase. It stands to reason that measurement of heat capacity of solids is found to be most suitable and correct. The use of the capsule in studies of a liquid phase causes deterioration of the correlation between the heat capacity of the substance under test and that of the calorimeter parts, which, in general, brings about some additional errors of the measurement results, but, in principle, imposes no restriction.

The temperatures in which adiabatic calorimetry is found applicable range from liquid-helium temperatures to a level of 1800–1900 K. The upper limitation is determined by a difficulty of meeting the adiabatic conditions at an extremely high intensity of heat exchange through radiation. The errors of the measurement results in good apparatus do not exceed 0.1%–0.3% at moderate and low temperatures; however, they tend to rise approximately by one order under high temperatures.

2. METHOD

Adiabatic calorimetry is considered one of the classical methods of measurements of true heat capacity and thermal effects of transformations of substances. The optimum condition for a calorimetry process is the absolute absence of heat flow across the boundary of the sample involved, i.e., whenever the following equation holds true at any point of the said boundary:

$$\frac{\partial T}{\partial \mathbf{n}} = 0 \qquad (11)$$

where **n** is the vector normal to the sample surface.

The uppermost accuracy of the calorimetry procedure can be obtained when no heat exchange with the environment takes place and, consequently, no need arises to introduce substantial corrections (such as the Reyno–Pfaundler correction in diathermal-shield calorimeters) to the main effect under measurement. The means for meeting condition (1) is enclosing the sample in an adiabatic shield with the temperature at each point of the shield being equal to that of the sample boundary throughout the period of the experiment. The temperature of the chamber enclosing the adiabatic shield either remains unchanged (isothermal chamber) or automatically follows the temperature of the adiabatic shield, the difference between the temperatures of the chamber and the shield remaining, as a rule, constant (follow-up chamber). The specific features of operation of the calorimeters with isothermal chambers are treated in Ref. 1. The peculiarities of the calorimeters with follow-up chambers are discussed in Refs. 2 and 3.

The adiabatic shield is designed not only for preventing loss of heat liberated in the process under consideration but also for compensating for heat flow produced by permanent side effects such as heat exchange over the fastenings or suspension parts of the calorimeter, heat evolution in the measuring system of the calorimeter, etc. The temperature compensation is confirmed by the absence of temperature change of the calorimeter in the initial and final periods of the experiment. This is precisely the condition for establishing the required negligible differences of the temperatures between the surface of the calorimeter and the adiabatic shield in some particular types of calorimeters. With a view to avoid heat loss in the process under study it is found imperative to hold this temperature difference constant throughout the time of the experiment (in thoroughly developed designs of the calorimeters this difference approximating zero).

Thus, in the practical case the condition for providing adiabatic conditions (adiabatization) of the calorimeter experiment can be expressed by the equation

$$\oint_s \frac{\partial T}{\partial \mathbf{n}} ds = 0 \qquad (2)$$

where ds is the element of the sample surface.

Nevertheless, in real cases some heat loss may occur due to feasible alterations of temperature gradients along the sample boundary in the main period and also because of inevitable imperfections of any control system meant for regulating the adiabatic shield temperature. However this heat loss is insignificant.

The corresponding correction is also small, and therefore the probable error, which may still remain after introducing the correction will also be negligible.

Depending on the desiderata of the investigations adiabatic calorimeters have found principal applications either in realization of the method of direct heating, i.e., for measuring true capacity and heat of phase transitions, or in reaction calorimetry, i.e., for measuring heat of reaction, heat of alloy mixing, heat of solution, etc.

The method of direct heating is based on the principle of imparting the amount of heat under measurement to the sample electrically. The possibility of reliable measurement of this value ensures a high sensitivity and precision in determination of heat capacity (including the phase transition regions). Heat may be supplied to the sample either periodically (by steps) or continuously. The first case is characterized by a temperature equilibrium in the sample in the initial and final periods of the experiment and some thermal imbalance in the main period of the experiment (in the period of heat application). In the second case some imbalance, i.e., availability of temperature gradients in the sample, exists during the entire period of the calorimetric experiment. For obtaining accurate data of the experiment it is necessary that these gradients remain practically unchanged during the measurements, i.e., all the sections of the calorimeter should get heated at a similar rate that will correspond to the so-called quasistationary mode of heating.[4,5] It is easy to show that this requirement can be best met for relatively small samples with a high coefficient of heat conduction and a small coefficient of heat emission (small Biot's criterion) and for fairly small rates of heating.

The analysis of temperature fields in direct-heating calorimeters of various types with an analytical determination of the time required for assuming the steady-state condition is given in Ref. 6. The detailed analysis[7] shows that deteriorations of the adiabatic conditions are identical for both methods, with the exception of the moments of switching-on and switching-off of the heater in the case of the stage (periodical) heating method. Thus, as far as accuracy is concerned, both the modifications are more or less equivalent, and selection of a certain method is determined by the problems to be solved in the experiment and also by the properties of the substance under investigation. For example, if the substance under test requires a considerable length of time for establishing equilibrium (multiphase systems, substances close to critical points, etc.) then preference should be given to the step heating method. For investigating substances possessing good heat conduction, the continuous-heating method has proven to be more convenient. The continuous application of heat requires less time for measurements and results in less heat loss, which is very essential at high temperatures in view of rise of radiation heat exchange.

Typically, the rate of heating in adiabatic calorimetry does not exceed

$0.1\,K\,s^{-1}$, and the rise of temperature in the stage-heating calorimeters lies just within a few degrees. All the relevant restrictions are considered below in this review.

In reaction calorimetry, when it is needed to measure the heat effect of the process the Nernst-type calorimeter is provided with an adiabatic shield. The advantage of this type of an adiabatic calorimeter resides in the fact that, first, the change of the temperature during the calorimetry procedure is minimal, which facilitates the determination of the reference temperature for the measured thermal effect, and, second, the contribution of the effective heat capacity of the reacting components, which is not always adequately known, is not substantial as compared with the known heat capacity of the calorimetric block.

In some cases for measurements of heat effects at any standard temperatures (say at 273 K, 298 K, 373 K, and so on) it is found extremely suitable to make use of constant temperature calorimeters where the phase transition of the calorimetric substance is employed at this temperature. The adiabatic conditions in this case are achieved in a particularly simple way – merely by submerging the calorimeter into a thermostat wherein the phase transition of the same substance takes place permanently. Both the variants of the mentioned reaction calorimeters will be discussed below.

The calorimeters in which the heat exchange between the sample and its environment comes to nought can operate as differential calorimeters, and specifically as microcalorimeters, as applied to biology, physics, chemistry. They may also be utilized for measuring enthalpy by the drop method [when in the Nernst-type calorimeter the requirements of the condition (1) are met by means of the adiabatic shield]. Besides, the adiabatization may be attained by way of rapid pulsed heating (when the share of heat loss is negligible as compared to the applied power due to the short duration of the process). These aspects of the calorimetry are discussed both in special works[(8–13)] and in some chapters of the present Compendium. As to the pulse heating method it should be noted that the use of extremely high rates of heating may cause, in some cases, a development of an heterogeneous structure or may distort the procedure of phase transformations in the substance concerned.

A typical diagram of a direct-heating adiabatic calorimeter with a follow-up chamber for measuring heat capacity and heats of phase transitions is presented in Fig. 1.

Sample 1, which is essentially a container filled with the substance under test, is heated by the use of electric heater 4. At a definite moment,[†] provided that the aforesaid requirements are met, the quasistationary heating condition is

[†] The time for attaining the steady-state condition for bodies of different geometry, which depends on the behavior of the highest-order terms in the series presenting the solution of the equation of heat conduction for the sample, is analyzed in Ref. 6.

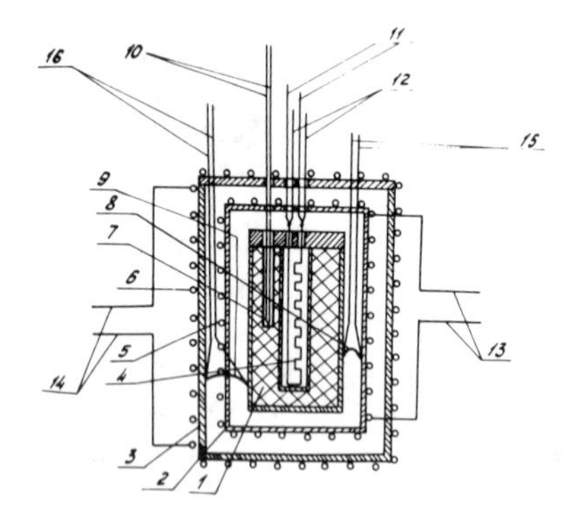

FIGURE 1. Typical diagram of direct-heating adiabatic calorimeter. 1, sample; 2, shield with controllable temperature; 3, furnace with controllable temperature; 4, calorimetric heater; 5, shield heater; 6, furnace heater; 7, temperature pickup for measuring reference temperature and rate of change of sample temperature; 8, differential thermocouple "sample-shield"; 9, differential thermocouple "sample-furnace"; 10, to system of measurement of temperature and its derivative; 11, to power supply source of calorimetric heater and to system of measurement of current; 12, to system of measurement of calorimetric heater voltage; 13, to power supply source of shield heater; 14, to power supply source of furnace heater; 15, to system of power control of shield heater power supply source; 16, same as under 15 for furnace heater.

obtained whose characteristic property is that the heating rate all over the sections of the sample is approximately equal and constant with time. The temperature of the sample and its derivative with respect to time are determined by readings of temperature pickup 7 (line 10) whose purpose may be served by a thermocouple or a resistance thermometer. This temperature sensor 7 inserted in the body of the sample makes possible the plotting of heating curves (thermograms) of the substance under test so as to determine the heat of the phase transformation. Quick-response shield 2 with heater 5 supplied with power from a controlled power source (line 13) ensures that the shield temperature "follows up" the temperature of the sample surface, i.e., provides the adiabatic conditions of heating. The feedback in the control system is ensured by differential thermocouple 8 transmitting to the control system (line 15) a signal of the difference between the temperatures of the shield and the sample surface. By means of controlling solely the temperature of the shield whose heaters might have substantial power (particularly at high temperatures) it is usually not possible to establish satisfactory adiabatic conditions. For this reason furnace (chamber) 3 may be equipped with a similar independent control system (6, 9, 14, 16) capable of following the temperature of the shield as well as the temperature of the sample surface (calorimeter) by the furnace temperature.

The indications of differential thermocouple 9 showing the difference between the temperature of the sample surface and that of the furnace serve as a feedback signal (line 15). Such a "two-layer" adiabatization system usually suffices for establishing the required accuracy of the calorimetric measurements; sometimes a need may arise to have more shields. The value of the power absorbed by the sample is determined by measuring the electric current (line 11) and voltage (line 12) of the calorimeter heater which is usually of a bifilar type so as to suppress the reactance when supplied with alternating current.

In practice, in the case of continuous application of heat, it is convenient to maintain the constancy of one of the values (input power or rate of temperature rise), while the other is measured as a function of time. Selection of this value determines to a large measure the design features of the apparatus. For instance, in Ref. 14 constant power to the calorimeter heater was maintained, and the temperature was measured as a function of time. This method suggests a lightweight adiabatic shield, provided that the thermometer is confined directly in the calorimeter. In Ref. 15, in contrast, the adiabatic shield represents a huge copper block whose temperature drift is held constant, while the power provided by the adiabatic regulator or controller in the calorimeter heater is measured as a function of time. In this case the calorimeter need not contain a thermometer. In some other research works both the variants have been used[(16)] in order to improve the reliability of the data sought.

The sample heat capacity value C_p ($\mathrm{J\,kg^{-1}\,K^{-1}}$) may be derived from the equation of the calorimeter energy balance.

(a) For continuous heating:

$$C_p = \frac{W - q}{M\,dT/d\tau} - \frac{A}{M} \tag{3}$$

where W (watt) is the power of the calorimeter heater; M (kg) is the mass of the sample; $dT/d\tau$ ($\mathrm{K\,s^{-1}}$) is the rate of change of the sample temperature; A ($\mathrm{J\,K^{-1}}$) is the calorimetric constant (heat value or thermal equivalent of the calorimeter) which is the total sum of the heat capacities of the calorimeter component elements (container for substance under test, heater, temperature sensor, suspension, etc.):

$$A = \sum_i C_{p_i}^c M_i^c \tag{4}$$

where $C_{p_i}^c$ and M_i^c are, respectively, the heat capacity and mass of each component part of the calorimeter; q (watt) is the intensity of heat exchange between the sample and environment.†

(b) For stage heating:

$$C_p = \frac{Q - q\Delta\tau}{M\Delta T} - \frac{A}{M} \tag{5}$$

where Q (J) is the amount of heat liberated by the heater; ΔT (K) is the variation

†From formula (3) it is seen that q is taken as a positive value when the heat exchange is characterized by loss of heat of the sample.

of the temperature of the sample and the calorimeter parts in the main period of the experiment; and $\Delta\tau$ (s) is the time during which the temperature was changing in the main period. The remaining designations in the expression (5) are interpreted as in formula (3).

As stated above, in spite of any measures taken for eliminating heat exchange between the sample and environment in the process of a calorimetric experiment, some insignificant heat exchange is probable by virtue of the following circumstances: change of temperature gradients in the main period over the surfaces of the sample, shield, and furnace when the number of the differential thermocouples is limited; errors of the thermocouples used; temperature lag of the adiabatic shield which may depend on the rate of heating, control condition, desired power, etc. Therefore in adiabatic calorimetry precision measurements it is necessary to introduce a correction for heat exchange, taking into account that the correction in its general aspect depends not only on the temperature but also upon its derivative with respect to time. Should it be assumed that the aforesaid factors insignificantly depend on the rate of heating, and are determined exclusively by the experiment temperature, then the heat exchange value q can be found by measuring the "run" of the temperature of the sample with the calorimeter heater switched off. After the heater has been turned off the temperature regulators still continue to maintain the adiabatic conditions, therefore the temperature of the sample will soon get equalized and tend to increase no further. The systematic change of the sample temperature in this period may occur only due to heat exchange with the environment. This period corresponds to the zero rate of heating. On condition that the temperature fields on the surfaces of the sample, shields, and furnace and the accuracy of the operation of the adiabatic system are independent of the rate of heating, the heat exchange will also be independent of this rate of heating. Therefore the actual heat exchange (value q) may be equated to the heat exchange determined at the zero rate of heating, i.e., in static conditions. The requirements for meeting these assumptions will be set forth in the sections devoted to the control systems and to the errors met in the adiabatic calorimetry.

The equation of the heat balance of the calorimeter in the temperature "run" period will take the following form:

$$q = -\frac{dT}{d\tau}^{*}(MC_p + A) \tag{6}$$

where the terms are to be interpreted similarly to those in formula (3).

The rate of change of the sample temperature with the calorimeter heater switched off [rate of "run" of the calorimeter temperature $(dT/d\tau)^*$]† and

†It is clear that at $q > 0$ (heat loss) the $(dT/d\tau)^*$ will, of course, be less than zero.

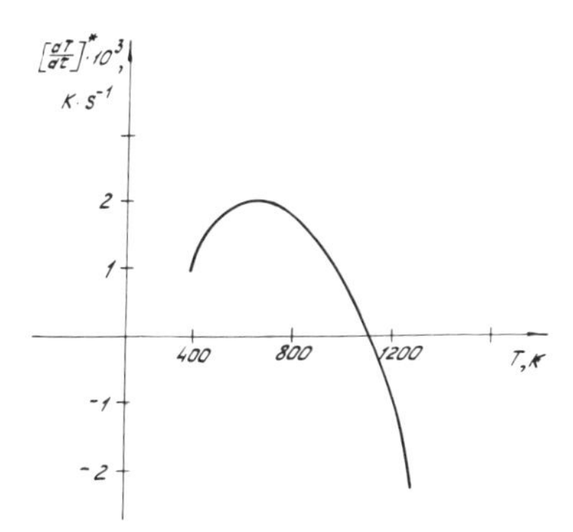

FIGURE 2. Characteristic of heat loss in adiabatic calorimeter.[17]

correspondingly the value of q shall be determined before and after the main period of the experiment, i.e., in the first and second auxilliary transitions. The obtained values are interpolated within the main period. The correctness of the stated assumption concerning the independence of the corrections for heat exchange of the rate of heating can be verified by reproducibility of the results under different power ratings of the calorimeter heater, i.e., at different rates of heating.

Experimentally found value of $(dT/d\tau)^*$ for one of the high-temperature adiabatic calorimeters[17] shown in Fig. 2, may be used as an example of the function $(dT/d\tau)^* = f(T)$.

The value of specific heat, C_p, can be derived by solving together equations (3) and (6). In adiabatic calorimetry practice, however, due to the negligible contribution the value of q makes, it is usually first determined from the equation (6) using an approximate value of C_p derived from equation (3) without the correction for heat exchange. Then the obtained value of q is introduced into equation (3) so as to find the more precise value of C_p. Usually two or three such iterations will suffice.

It should be emphasized that the overwhelming majority of the adiabatic calorimeters are operated in vacuum. This enables the heat loss (value q) to be decreased, and aids realization of conditions of equality of the rates of heating of all sections of the sample, i.e., minimum value of Biot's criterion.

The thermal effects of the phase transformations are determined from the analysis of the heating curves. The thermogram plotted in the "sample temperature–time" coordinates is at the same time the graph of the sample temperature versus the sample enthalpy, since at a constant power of the calorimeter heater the amount of heat applied to the sample (increase of the sample enthalpy) is proportional to the time. The process of absorption under liberation of heat is indicated on the thermogram by disturbed monotony of change of the temperature. In the study of phase transitions of the second kind accompanied by considerable rise of heat capacity as well as in the study of structural transformation

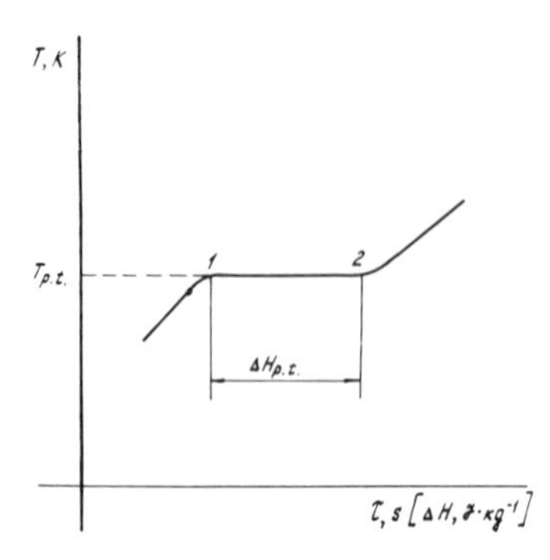

FIGURE 3. Thermogram of phase transition.

in solid alloys or melting of alloys where heat-absorbing processes cover certain range of temperatures, a considerable decrease of the value $dT/d\tau$ takes place on the thermogram within this temperature range. Phase transitions of the first kind in individual substances (isothermal processes) are shown on the thermogram by a section where $dT/d\tau \approx 0$ complying with the temperature and enthalpy of the phase transformation – T_{pt}, ΔH_{pt} (Fig. 3). Here the start and the finish of the transformation are characterized by irregularities of the thermogram. In the case of a properly organized process, that is, when the input power is minor and the phase transition is accompanied by a thermodynamic equilibrium in the sample, the illegible character of the thermogram bends and a slight inclination of isothermal section 1–2 may be caused by contamination of the sample, which in this event will represent an infinitely diluted solution. There are special methods to account for this fact, depending on the constitutional diagram of such a solution.[18]

The change of the enthalpy of the isothermal process under study – ΔH_{pt} ($J\,kg^{-1}$) – is determined by means of the following formula:

$$\Delta H_{pt} = \frac{(W - q) \cdot \Delta\tau_{pt}}{M} \tag{7}$$

where $\Delta\tau_{pt}$ (in s) is the time of the phase transformation, the remaining members of the equation being interpreted as in formula (3).

The equation of the thermal balance of the reaction calorimeter (which, as was stated above, is provided with a calorimetric block following the pattern of the Nernst-type calorimeter) expresses the following relationship for the reaction thermal effect under measurement Q_r ($J\,kg^{-1}$):

$$Q_r = \left[\frac{A}{M} + C_{eff}\right] \cdot \Delta T + \frac{q\Delta\tau}{M} \tag{8}$$

where C_{eff} ($J\,kg^{-1}\,K^{-1}$)† is the effective heat capacity of the substance under test consisting of reacting components and reaction products in the temperature range ΔT. The remaining members are interpreted as in formulas (3) and (5), the thermal value A incorporating the heat capacity and mass of the calorimetric block.

3. ADIABATIC CALORIMETERS

The sphere of applications of adiabatic calorimetry ranges from cryogenic helium temperatures up to very high temperatures even exceeding 1800 K. The upper limit is determined by the difficulty in meeting the conditions (1) under intensive radiation heat exchange.

The method has been developed rather intensively for the past 50 years. The chronological table illustrating the most important research works is presented below (as Table 1).

3.1. High Temperatures

One of the first adiabatic calorimeters was the Moser apparatus,[19] which was used to study the true heat capacities of nickel, brass, copper, and quartz within the temperature range of 300–1000 K.

Sample 1 (Fig. 4) is a block of the substance under test inserted into a thin-walled silver cylinder. The sample is provided with heater 2 and thermocouple 3 and is placed inside silver vessel 4 heated from outside with the aid of furnace 5. By manually selecting the current intensity in the external furnace 5 and in the calorimetric heater 2 a smooth rise of the temperature of the vessel (T_v) and that of the sample (T_s) (see Fig. 5) can be achieved, so that the temperatures of the vessel and the sample surface are practically equalized and maintained so during some lapse of time. The difference between the temperatures of the vessel and sample, i.e., $\Theta = T_s - T_v$ (Fig. 5), was measured by the use of differential thermocouple 6 (Fig. 4). The measurements were performed to determine the increment of the sample temperature (ΔT_s) and the amount of the heat liberated by the constant-power calorimetric heater (W) for the time $\Delta\tau$.

In calculating the heat capacity on the basis of the experimental results use was made of the formula (5), i.e., with heat supplied continuously for the particular time interval $\Delta\tau$, it is possible to resort to a processing procedure conforming to a stage application of heat. The heat exchange was taken into

†For the sake of generality the symbols Q_p^r and $C_{p\,eff}$ are omitted since in adiabatic reaction calorimetry for determination of heats of combustion reactions use is often made of bomb isochoric calorimeters.

TABLE 1
Adiabatic Calorimeters for Measuring Heat Capacity and Thermal Effects of Substance Transformation
(Chronology of Typical Variants)

No.	Investigator	Ref.	Year	Substance	Temperature range (K)	Property under determination	Error (%)	Kind of heating: C, continuous S stage
1	Eastman and Rodebush	40	1918	Na, K, Mg, Ca	60–290	C_p	>1	C
2	Simon and Swain	41	1935	Li	15–300	C_p	>1	C
3	Moser	19	1936	Ni, Cu, brass, quartz	300–1000	C_p	0.5–1	C
4	Sykes	20, 21	1935, 1938	Metals, alloys, steels	300–1000	C_p; heat effect of phase transitions; discovered phase transition of second kind in brass at temperature of 740 K	$\leqslant 2$ (δC_p) $\leqslant 1$ ($\delta \Delta H$)	C
5	Awbery, Snow, and Criffits	27	1940–1953	Steels, iron	300–1200	C_p; heat effects of phase transitions, $\Delta H_{\text{bcc—fcc}}$ at 1176–1183 K	$\leqslant 2$ (δC_p) $\leqslant 1$ ($\delta \Delta H$)	C
6	Sokolov	29	1948	Several substances	300–700	Study of second-kind phase transitions and phenomena near melting points	0.5	S
7	Popov and Galchenko	31	1951	Powdery and solid substances	300–1000	C_p	2	S
8	Eichen and Eugen	28	1951	Heavy water, H-propane	270–430	C_p	0.15; 0.2	C

9	Furukawa *et al.*	78, 79	1951	Benzoic acid, diphenyl	13–570	C_p	⩾(0.5–1)	C
10	Sinelnikov	80	1953	Quartz	300–1000	C_p	1–2	S
11	Strelkov *et al.*	1	1954	Several solids and liquids	12–300	C_p	0.2–2	S
12	Dauphinee *et al.*	43, 44,	1954, 1955	Na, K, Rb, Cs	30–330	C_p, heat and temperature of melting	0.2–2 (δC_p) 0.2 ($\delta \Delta H$)	C
13	Osborn *et al.*	81	1955	UF_5	5–300	C_p	0.2–1.5	C
14	Lazarev	25, 26	1955, 1956	Carbon and alloy steels; Al and Zn in liquid and solid phases	300–1300	C_p; thermal effects of phase transformations (melting, perlitic transformations)	2–3 (δC_p) 1–1.5 ($\delta \Delta H$)	C
15	West and Ginnings	2	1958	α-Al_2O_3	300–800	C_p	0.2	S
16	Backhurst	109	1958	Solid substances	900–1900	C_p	2–3	C
17	Stansbury	75	1959	Nickel	300–1300	C_p	⩽1	C
18	Henson, Simmons	84	1959	Graphite	300–800	C_p; accumulated energy in irradiated graphite	2–5	C
19	Sokolov, Shmidt	30	1960	Al_2O_3	300–1000	C_p	0.5	S
20	Gehring	86	1960	Liquid–metal systems with Sb, Ag	700–1200	C_p; ΔH (mixing)	1–2 (δC_p) 3 ($\delta \Delta H$)	S
21	Sklyankin	85	1960	Benzoic acid	4–410	C_p	0.2–2	S
22	Kiguradze	87	1960	Benzoic acid	80–273	C_p	0.3	S
23	Lyusternik	17, 32,	1959–1961	Carbon, lowalloy, stainless austenitic, chrome–nickel steels; alloys based on Zr, Nb and ternary alloys Pb–Sb–Sn	300–1400	C_p; heat of phase transformations	1 (δC_p) < 1 ($\delta \Delta H$)	C

TABLE 1 (Continued)

No.	Investigator	Ref.	Year	Substance	Temperature range (K)	Property under determination	Error (%)	Kind of heating: C, continuous S, stage
24	Martin	45, 46	1960	Ni, Na, Cu	20–300	C_p; ΔH (hexagonal closed packed–body centered cubic	0.2–2 (δC_p), < 1 ($\delta \Delta H$)	C
25	Frank *et al.*	47, 48	1961, 1963	Li, Na	3–30	C_p	1	C
26	Stingele	89	1961	Al	300–1000	C_p	0.5–1	C
27	Lukes	90	1961	Liquid–metal alloys	600–1100	$\Delta H_{mixing} = f(T)$	2–3	S
28	Shmidt	91	1961	Crashed and pulverized substances	300–1000	C_p	2–3	S
29	Kolesov *et al.*	92	1962	Several solids and liquids	12–340	C_p	0.2–2	C
30	Westrum *et al.*	83	1965	UC, UC_2	5–350	C_p	0.2–1.5	C
31	Filby and Martin	18	1965	K, Rb, Cs	(a) 3–26 (b) 20–320	(a) C_p (b) C_p; $\Delta H_{melting}$, $T_{melting}$	(a) 1 (δC_p) (b) 0.2–1 (δC_p)	(a) C (b) C and S
32	Martin and Snowdon	14,	1966	α-Al_2O_3, Na	300–475	C_p; $\Delta H_{melting}$, $T_{melting}$	0.1–0.2 (δC_p) < 0.1 ($\delta \Delta H$)	C
33	Gronvold	93	1967	Al_2O_3	300–1000	C_p	0.5	C
34	Braun *et al.*	16	1968	Ti, Zr, V, Cr, Mn, Fe Co, Ni, Cu, Ag, Au	300–1900	C_p; ΔH (melting, polymorphic transformations)	2–3 (δC_p), 0.5–1.5 ($\delta \Delta H$)	C
35	Martin	50	1970	Rb, Cs	Melting region	$\Delta H_{melting}$, $T_{melting}$	0.1 ($\delta \Delta H$)	C, S
36	Brooks *et al.*	110–113	1968–1973	Cu, Au, Pt	300–1200	C_p	0.5	C

37	Krusius	95	1971	Several metals	270–500	ΔH (superfine interactions in metals)	5	S
38	Mayer and Waluge	96	1971	Some solids and liquids	77–400	C_p	0.2–1	C
39	Robie and Hemingway	97	1972	Solutions and alloys	14–300	C_p; ΔH (mixing)	0.2–2 (δC_p); 1 ($\delta \Delta H$)	S
40	Buckingham and Edwards	15	1973	Liquid solutions near critical point of solution	200–800	C_p	0.4	C
41	Naito and Kamegashira	34	1973	Some solids and liquids	300–1000	C_p	<1	C
42	Orlova *et al.*	59	1974	Benzoic acid	4–300	C_p	0.15–1.8	S
43	Paukov *et al.*	42	1976	Sr, Ba	5–300	C_p	0.2–2	S
44	Kagan *et al.*	61, 63, 65, 68	1978–1982	Liquid alloys of alkali metals	400–1300	ΔH (mixing)	1–3	S
45	Gronvold	94	1978	In (solid and liquid)	300–1000	C_p	0.5	C
46	Orlova *et al.*	57, 60	1978–1980	Al_2O_3, superconductive materials	4–300	C_p	0.15–0.55	S
47	Cash *et al.*	114	1981	Solid metals	300–1200	C_p	0.5	C

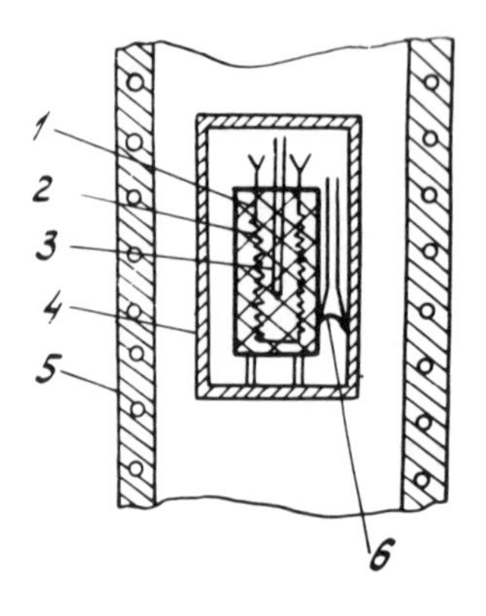

FIGURE 4. Diagram of Moser's calorimeter.[19] 1, sample; 2, heater; 3, thermocouple; 4, vessel; 5, furnace; 6, differential thermocouple "sample-vessel".

account on the basis of the records of the Θ value. The intensity of the heat exchange (q) was taken as

$$q = \alpha\Theta + \beta \tag{9}$$

The coefficients α, β and the calorimeter heat value A [refer to formula (5)] were determined by special calibrating experiments with an empty calorimeter at three different values of Θ (about 1, 3, and 5 K) using the following system of equations:

$$\begin{aligned} A\Delta T_{s1} &= (W - \alpha\Theta_1 - \beta)\Delta\tau_1 \\ A\Delta T_{s2} &= (W - \alpha\Theta_2 - \beta)\Delta\tau_2 \\ A\Delta T_{s3} &= (W - \alpha\Theta_3 - \beta)\Delta\tau_3 \end{aligned} \tag{10}$$

In order that the coefficients should remain unchanged both in the calibrating experiments and in the main ones, it was necessary to ensure the identity of the rates of heating, i.e., this may have an effect on distribution of the temperatures over the surface of the sample.

The meticulous work completed by Moser has a maximum error of 0.5%–1%, except in the region of phase transitions where it increased several times.

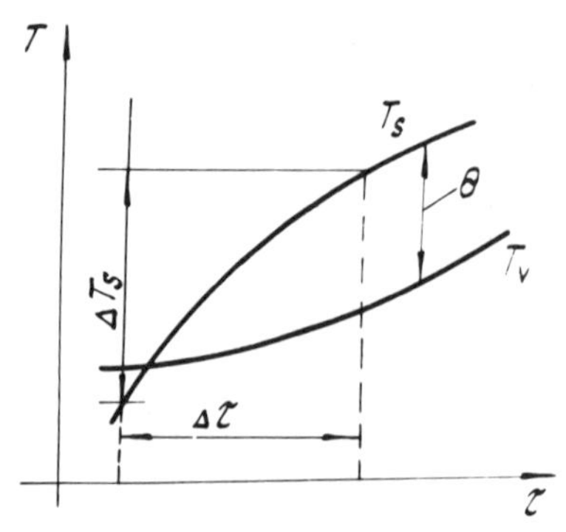

FIGURE 5. Thermogram of Moser's calorimeter.[19] T_s, temperature of sample; T_v, temperature of vessel; ΔT_s, increment of sample temperature; τ, time; $\theta = T_s - T_v$.

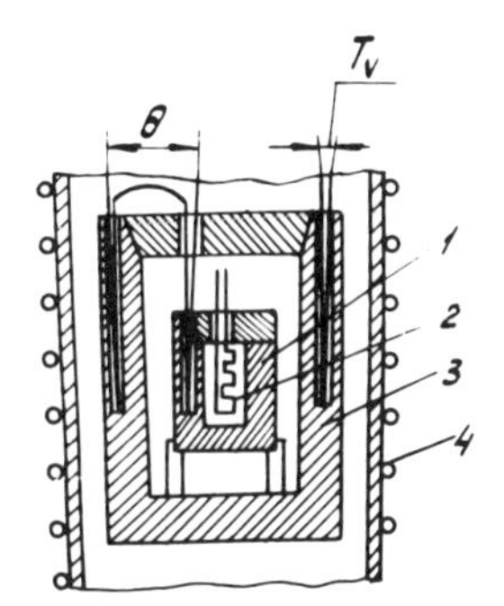

FIGURE 6. Diagram of Sykes's calorimeter.[20] 1, sample; 2, heater; 3, vessel; 4, muffle furnace.

A notable contribution to the development of adiabatic calorimetry was made by Sykes-type calorimeter.[20] Cylindrical sample 1 (Fig. 6) with internal heater 2 is housed in huge copper vessel 3 heated by muffle furnace 4. In the process of the experiments the vessel is heated so that its temperature T_v (Fig. 7) rises approximately at a constant rate.

Alternating the switching-on and switching-off of calorimeter heater 2 enables the sample to be heated up approximately at the same rate and with the least possible deviation of its temperature T_s from the vessel temperature T_v. The difference of the temperatures $\Theta = T_s - T_v$ and the temperature of the vessel (T_v) are measured with the help of thermocouples. Periods I, II, and III (Fig. 7) correspond to the switched-on condition of the heater. It is evident that at intersection points 1, 2, and 3 of the T_s and T_v temperature curves adiabatic conditions are established for heating the sample. The heat capacity was determined on the basis of the results obtained in the experiment by the use of the formula (3) corresponding to the continuous application of heat. The mean value of q was taken to be equal to zero for a small temperature range (5–10 K) being symmetric with respect to the intersection points of the T_s and T_v curves (points 1, 2, 3). The rate of heating of the sample ($dT_s/d\tau$) can be determined in the following way:

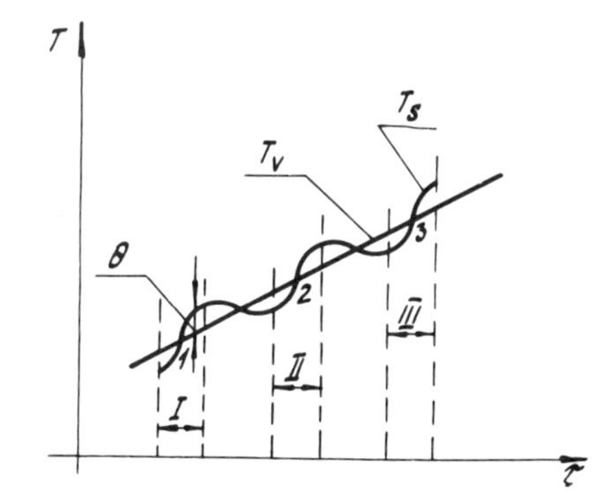

FIGURE 7. Thermogram of Sykes's calorimeter.[20] T_s, temperature of sample; T_v, temperature of vessel; $\theta = T_s - T_v$; τ, time.

$$\frac{dT_s}{d\tau} = \frac{dT_v}{d\tau} + \frac{d\Theta}{d\tau} \tag{11}$$

The calculation of the rate of heating of the vessel ($dT_v/d\tau$) presents no difficulties whenever its temperature rises smoothly. The value of $d\Theta/d\tau$ did not exceed 3% of the total rate of heating in the region of the monotonous variation of the heat capacity under measurement and lies within 10% in the case of abnormalities in the behavior of the heat capacity. This could be effected only with a highly experienced operator. Calculation of the value of $d\Theta/d\tau$ involves a number of principal difficulties. For the sake of differentiation the shortest range of the temperatures close to the intersection points of the curves should be considered preferable; however, this reduces the expected possible number of the measurements needed for calculating the derivative. It should also be borne in mind that the necessary alternative switching-on and -off operations of the heater give rise to temperature waves in the sample wall, therefore a certain length of time is required for establishing the desired quasistationary condition of heating after the heater has been actuated. At the same time, to diminish the value of Θ it is desirable to switch the heater on and off as frequently as possible. The process of the experiment is made complicated due to delay in measurements of the temperature. The apparatus of Sykes featured a materially high efficiency; however the above-mentioned causes made it impracticable to measure the values of C_p more frequently than in every 30–40 K interval. The error of the Sykes method amounts to some 2% in the maximum-temperature region, while in the phase-transformation regions the error increases to 5%–10%. This method was used to accomplish a number of calorimetric researches which dealt with measurements of true heat capacity and thermal effects of phase transitions of metals and steels up to 1000 K.[21–24]

The improvements of the Sykes method implemented by Lazarev[25, 26] made it possible to extend the range of application of this method to 1300 K. The distinctive characteristic of the Lazarev apparatus consists in the provision of a special additional calorimeter to account for heat loss whenever the temperature of the sample differs from that of the vessel. This allows measurements of heat capacities not only at the moments when $T_s = T_v$, but also in the case of some deviation of the temperatures from this equality. Two samples being absolutely similar in geometric dimensions and in blackness of the surface are placed in vessel 1 (Fig. 8). One of the samples (2) is furnished with internal heater 4. Vessel 1 is arranged in a furnace and can be heated therein so as to ensure a monotonous rise of its temperature T_v (Fig. 9). The rate of the temperature rise is sufficiently low that the requirements of the quasistationary condition are met. Both samples are subjected to heating inside the vessel. Sample 3 is heated only by the heat radiated from the vessel and serves as a calorimeter for measuring the intensity of the heat exchange (the so-called α-calorimeter).

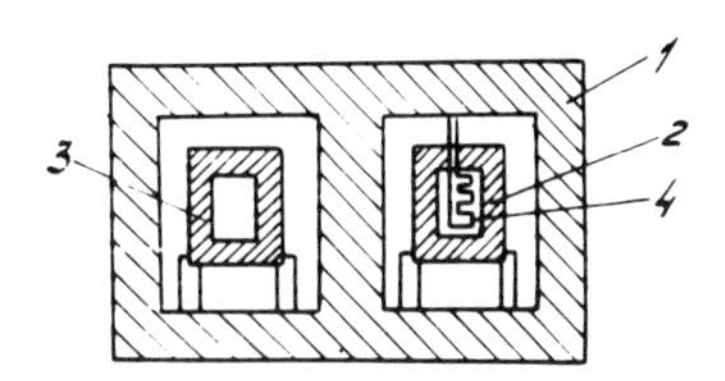

FIGURE 8. Diagram of Lazarev's calorimeter.[25,26] 1, vessel; 2, sample; 3, sample; 4, internal heater of sample.

Sample 2 is heated by calorimetric heater 4 in the conditions closely approximating the adiabatic conditions if the difference of the temperatures of the sample and the vessel ($\Theta_{s2} = T_{s2} - T_v$) is maintained close to zero. The temperature of the vessel was measured by the thermocouples, and the temperature differences Θ_{s2} and Θ_{s3} were determined by the use of the differential thermocouples which were alternately connected to a mirror galvanometer. In the process of the experiment the values T_v, Θ_{s2}, Θ_{s3}, W, τ are derived to obtain the required information for calculating the heat capacity. Sample 3 (α-calorimeter) may be fabricated either of the substance under test (absolute method) or of a substance with the known heat capacity (relative method), provided that the similar radiative characteristics of the surfaces of the samples are ensured. In both cases a simultaneous solution of the equations (3) and (6) makes it possible to find the heat capacity of sample 2 under test with due regard for a minor heat exchange between this sample and the vessel enclosing it. With the aid of sample 3 it is possible to determine the relationship between the heat-transfer coefficient and the temperature difference Θ for the respective absolute temperatures. Lazarev has investigated[25,26] the true heat capacity of several carbon and alloyed steels, aluminum, and zinc in solid and liquid phases with an error not in excess of 2%–3% at the maximum temperatures; however, in the regions of structural transformations the error tended to go up to 5%–10%. The same calorimeter allows direct measurements of thermal effects of phase transitions. Specifically, the heat of perlitic transformation in steel was measured. During heating in the temperature range of perlitic transformation a considerable amount of heat is absorbed, so that, with the heat supplied con-

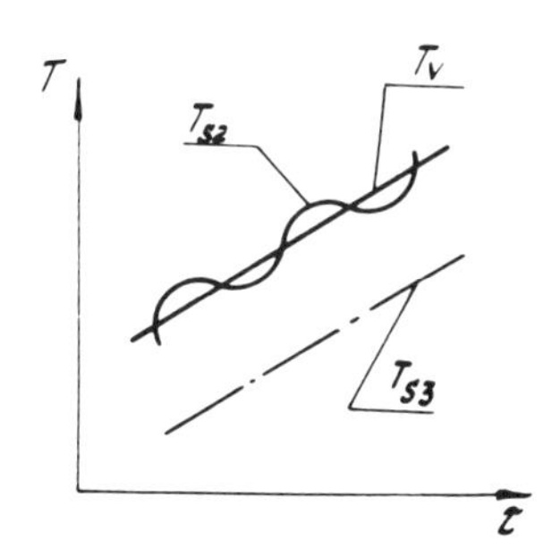

FIGURE 9. Thermogram of Lazarev's calorimeter.[25,26] T_v, temperature of vessel; T_{s2}, temperature of sample 2; T_{s3}, temperature of sample 3; τ, time.

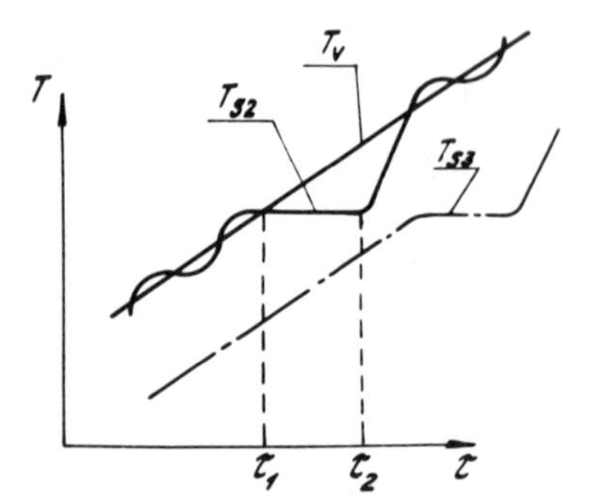

FIGURE 10. Thermogram of Lazarev's calorimeter at phase transition of substance under test.[25,26]

tinuously, the temperature of the sample practically remains constant. The thermograms for the absolute method take the form illustrated in Fig. 10. It should be noted that in the phase transformation zone ($\tau_1 < \tau < \tau_2$), as seen from Fig. 10, a sharp deterioration of adiabatic conditions is noted, and consideration for heat exchange with the environment (vessel) is growing in importance. From the standpoint of practical performance of the experiment this apparatus is more intricate than the Sykes-type calorimeter; however, it suffers from the same disadvantages associated with periodical switching-on and switching-off of the calorimetric heater.

An excellent adiabatic high-temperature calorimeter has been described by Awbery, Snow, and Griffits.[27] The authors have measured the heat capacity and heats of phase transitions of a wide variety of steels up to about 1200 K. Cylindrical sample 1 (Fig. 11) furnished with heater 2 and thermocouple 3 is continuously heated at a constant power W of the calorimetric heater within a considerable temperature range of some 50 K. In this case when the requirements of the quasistationary heating condition are met, an approximately constant gradient is attained in the sample (under smooth change of heat capacity and temperature), i.e., all the parts of the sample are characterized by a similar rate of temperature rise. The adiabatic conditions were established by means of thin-wall metal shield 4, whose temperature was continuously maintained in the process of heating within the level as close to the temperature of the sample surface as possible. To this end, quick-response furnace 5 enclosing and heating the shield was energized with a current which was controlled by the

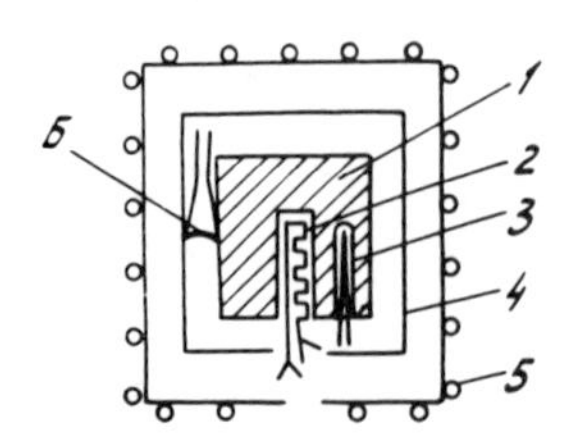

FIGURE 11. Diagram of calorimeter of Awbery, Snow, and Griffiths.[27] 1, sample; 2, heater; 3, thermocouple; 4, shield; 5, furnace; 6, differential thermocouple.

operator against the readings of a galvanometer using differential thermocouple 6 with one junction on the surface of sample 1 and the other on shield 4. The zero position of the galvanometer reading scale was adjusted prior to each experiment so that no heat exchange between the sample and the shield took place. This was noted by zero "run" of the sample temperature with the heater switched off.

In the Sykes's calorimeter[20, 21] considered above, as well as in other calorimeters closely approximating the former in the method,[22–26] the temperature of the massive vessel (T_v) surrounding the sample tended to rise monotonously (Figs. 7, 9), while the temperature of the sample proper (T_s) oscillated about the temperature T_v under the effect of switching-on and -off of the calorimeter heater, which, as was stated above, could be a source of systematic error. In the calorimeter devices by Awbery, Snow, and Griffits,[27] on the contrary, the monotonous rise was characteristic of the sample temperature T_s (Fig. 12), whereas the temperature of the quick-response shield (T_{sh}) tended to oscillate about the sample temperature T_s with a small amplitude (some 0.5 K). From the viewpoint of maintaining adiabatic conditions both methods, in principle, are interchangeable, but as to the requirement of holding the constancy of the value of $dT_s/d\tau$ within the entire volume of the sample, the latter calorimeter has the advantage. If at the end of the heating interval it was found that the temperature of the sample with the heater turned off exhibited certain "run" due to the heat exchange with the shield, then the respective correction was introduced for the measured heating rate. This procedure was as follows. Usually, subject of measurement was the time τ during which the sample was heated up by 50 K. If the rate of the "run" of the sample temperature measured with the heater switched off was equal to n ($K\,s^{-1}$), then the rise of the temperature for τ seconds in the absence of heat loss would be equal to $(50 + n\tau)$ (K). This value was the effective quantity of the temperature rise ΔT_{eff} in the experiment, and formula (5) for stage heating acquired the following form:

$$C_p = \frac{Q}{M\Delta T_{\text{eff}}} - \frac{A}{M} \tag{12}$$

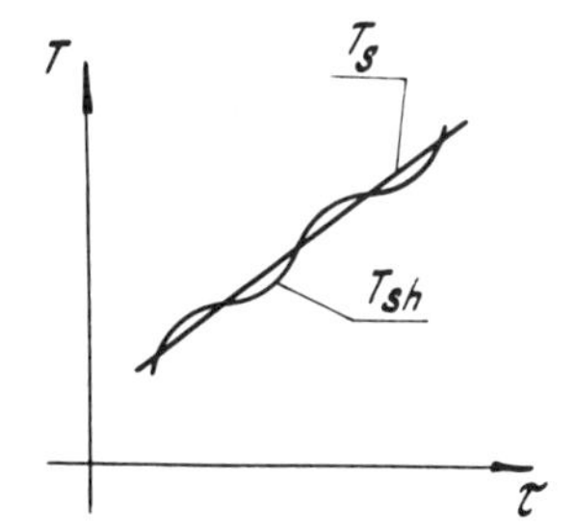

FIGURE 12. Thermogram of calorimeter of Awbery, Snow, Griffiths.[27] T_s, temperature of sample; T_{sh}, temperature of shield; τ, time.

where the remaining members of the formula are interpreted as in formula (5). In addition to the measurement of the heat capacity, the authors of Ref. 27 determined the thermal effect of the structural transformation of a body-centered cubic lattice of iron into a face-centered cubic lattice within the temperature range of 1176–1183 K. When iron is heated within this temperature range the heat is absorbed very intensively, which is accompanied by a sharp increase of the true heat capacity. In the process of heating the sample from 1168 to 1188 K, the amount of the heat applied to the sample was measured. Then the interpolated heat capacity values obtained by measurements below and above the phase transition range were integrated, thereby determining the amount of heat required only for heating the sample by 20 K. The transition heat was calculated as the difference between these two values accounting for the heat loss within the said range. The error of the obtained data on the heat capacity in the region of its monotonous variation did not exceed 2% at maximum temperatures. Averaging of the data in the 50 K range hampered the measurement of the true heat capacity in the regions of the phase transitions. Complications of manual control led to dispersion of individual values of heat of the high-temperature phase transition up to 10%. The recommended value was obtained by averaging many measurements. Measurement of the heat capacity close to the point of transformation involves additional errors due to rearrangement of temperature gradients in the sample.

One of the typical adiabatic calorimeters with continuous supply of heat was developed by Eicken and Eugen for the study of heat capacity of heavy water and H-propane within a range from room temperature to 430 K.[28]

The calorimeter vessel K (Fig. 13), 270 cm^3 in capacity, is made from Jena

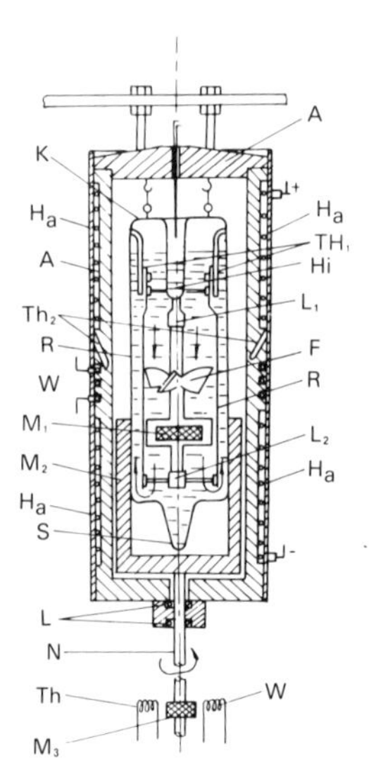

FIGURE 13. Adiabatic calorimeter of Eicken and Eugen.[28] K, calorimetric vessel; S, place of seal of vessel upon filling it with substance under test; $L_{1,2}$, step bearings; F, mixer; $M_{1,2,3}$, permanent magnets; N, shaft; R, glass tube; H_i, calorimetric heater; A, adiabatic shield; W, resistance thermometer; H_a, heater of adiabatic shield; T_h, junctions of differential thermocouple; L, bearings.

glass 20. The vessel is filled with a liquid under test and sealed at point S. The liquid under test is contained in a hermetically sealed vessel sturdy enough to operate, for example, with water under pressure of up to 6–8 kg/cm^2, i.e., approximately at temperatures up to 160°C. Inside the calorimetric vessel there is a glass mixer F which freely rotates in glass step bearings L_1 and L_2. Sealed in the mixer is a permanent magnet M_1. The mixer is driven by a permanent magnet M_2 located outside the calorimetric vessel. The magnet M_2 is fixed on shaft N made of German solver for minimizing heat transfer. The shaft N driven by a constant-speed electric motor rotates in ball bearings L. For better agitation of the liquid under test the calorimeter incorporates a glass guide tube R. Thus the specific features of the design of the calorimeter under consideration make it possible to preserve adequate purity of the liquid under test and to have it property mixed when conducting a calorimetric experiment. The calorimetric heater H_i fabricated of a constantan wire with a resistance of the order of 10 Ω is housed in a special sleeve in the upper part of the calorimetric vessel.

The adiabatic shield A is made of high-purity aluminum possessing a high heat conduction. In order to determine the reference temperature provision is made for five coils of a platinum resistance thermometer W wound of 100-Ω wire, 0.05 mm in diameter in the center of the adiabatic shield. Wound above and below the platinum thermometer is an electric heater H_a made of constantan wire with a resistance of approximately 350 Ω. This heater covers almost the entire side surface of the adiabatic shield. The necessary equality of the temperatures of the calorimetric vessel and the adiabatic shield is maintained in compliance with the indications of a three-junction differential copper-constantan thermocouple. Three junctions Th_1 of this thermocouple were accommodated in the sleeves of the calorimetric vessel, while three other junctions Th_2 were positioned in the apertures of the adiabatic shield nearby the resistance thermometer. The differential thermocouple was directly connected to the mirror galvanometer. The sensitivity of this thermocouple circuit was as high as 0.001°C. The electric measurements on the resistance thermometer were performed on the principle of a bridge circuit. The sensitivity of the temperature measurement corresponded to 0.001°C per mm of the scale length. In order to prevent interferences adversely affecting the operation of the galvanometers during rotation of the permanent magnets M_1 and M_2 (due to induced currents arising in the circuits of the thermocouples and resistance thermometer) provision was made for a minor permanent magnet M_3 fitted on the shaft running from the electric motor to the permanent magnet M_2. Located adjacent to the magnet M_3 are loop-shaped parts of the thermocouples and resistance thermometer, and the loops are so oriented with respect to the magnet M_3 that their induced currents are equal in magnitude but opposite in sign to the noise currents produced by the effect of rotation of the permanent magnetic M_1 and M_2. The power fed by the calorimetric heater was determined potentiometrically employing a compensating circuit. As shown by experience, the manual control of heating of the adiabatic shield sufficed to provide the equalization of the temperatures of the calorimeter and adiabatic shield with an error not exceeding 0.002°C. The heat loss corresponding to this possible difference of the temperature was always less than 0.01%.

As the result of the experiment the following relationship equation can be obtained: $Q = f(T)$, where T is the temperature and Q is the amount of heat applied in a definite time. Graphoanalytical treatment gives the derivative dQ/dT, which is essentially the heat capacity of the calorimetric system. The heat value of the calorimeter is determined either by calculations based on the known heat capacities of glass, magnet, and the remaining elements comprised in the calorimeter, or by resorting to a special calorimetric experiment with ordinary water whose heat capacity is well known. In the process of the experiment it could be seen that the indications of the differential thermocouples were different from zero with the calorimeter heater H_i switched off, which may be ascribed to the heat generation due to running of the mixer. The arbitrary "zero indication" of the differential thermocouples was determined experimentally at each temperature of the experiment with the calorimetric heater H_i switched off, with the mixer running, and under permanent heating of the adiabatic shield by the heater H_a. The heat flow determined in this way did not exceed 0.2%–0.3% of the total amount of heat supplied to the calorimeter.

Based upon the scatter of the data obtained in the experiments, the authors estimate the mean-square random error of the experimental data to be within 0.01%–0.02%. The total error in determination of values of heat capacity C_p of the liquid in the apparatus of Eicken and Eugen is about 0.15%.

One of the first calorimeters with automatic (two-position) system of control of adiabatic conditions is represented by the Sokolov apparatus[29] shown in Fig. 14. The temperature range was 700 K.† The calorimetric vessel and case A with double walled shield (B, C) were enclosed in furnace D. In addition to the radial, side heaters, provision was made for end heaters of the furnace: top heater F and bottom heater G. The cover of the furnace was secured to a porcelain tube E used to pass the lead wires of the heaters and thermocouples; and also three rods which support screens B and C. The calorimeter vessel of 20 ml in capacity, with shield A, was fabricated of platinum foil 0.08 mm thick. The vessel interior was provided with a special pocket for the heater and a sleeve to receive the thermocouples. Both the pocket and the sleeve were soldered to the bottom of the vessel using gold solder. The diameter of the shield was 2 mm larger than that of the calorimetric vessel. The calorimetric heater is made of platinum wire wound bifilarly around a mica plate. The adiabatic shields B and C are made of sheet silver, 0.6 mm thick. On the shield

†In a later work of this author[30] the temperature range was extended to 1000 K.

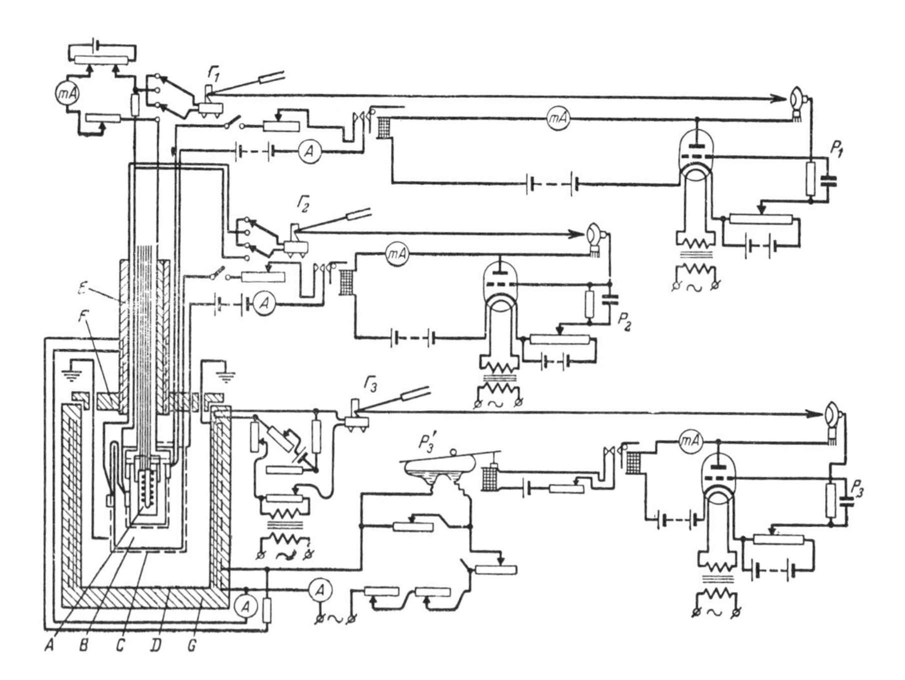

FIGURE 14. Circuit diagram of Sokolov's calorimeter with control system.[29,30] A, calorimetric vessel and case; B, C, two-ply shields; D, furnace; E, porcelain tube; F, top heater; G, bottom heater; $\Gamma_{1,2,3}$, mirror galvanometers; $P_{1,2,3}$, photorelays; P'_3, beam relays.

surfaces are arranged bifilarly wound nichrome heaters. The furnace heater is fabricated in the same manner. The three-junction thermocouples are used to measure the differences between the temperatures of the calorimeter and shield B, as well as between shields B and C. Figure 14 illustrates the schematic circuit diagram of the calorimeter adiabatic condition control system. The mirror galvanometers Γ_1 and Γ_2 coupled to the respective differential thermocouples are intended for controlling the temperatures of the shields B and C by means of photorelays P_1 and P_2. The electromagnetic relays (of the telephone type) in the anode circuits of the valves connect and disconnect the electric current of the storage batteries used to supply the power to the heaters of the shields. Galvanometer Γ_3 and relay P_3 control the temperature of the furnace through control beam relays P_3. Galvanometer Γ_3 is connected to the bridge arm of the furnace resistance thermometer. A part of the resistance on which a transformer turn is closed is connected in series with the galvanometer circuit, thereby preventing the galvanometer frame from sticking to the limit stops. The resistance thermometer of the furnace may be connected differentially with the similar thermometer wound around the external shield C. The differential thermocouple for measuring the difference between the temperature of the tube of furnace E and that of screen C is not shown in the diagram. This thermocouple operates in conjunction with the fourth galvanometer, and with the aid of its photorelay it controls the temperature of the lower part of the furnace tube, thus making it possible to prevent heat loss through the wires. The temperature of the sample was measured by means of a thermocouple composed of gold–palladium (60% Pd) and platinum–rhodium (10% Rh). The heat value of the calorimeter at room temperature was $A = 1.313$ cal/deg. The value of A tends to grow monotonously with rise of temperature; however, it offers a fairly reproducible anomaly with a peak at 600 K. The anomaly was apparently caused by phase transitions in mica or in sodium silicate contained in the calorimeter.

The Sokolov calorimeter was operated in a stage-heating mode and was employed for investigation of phase transitions of the second kind and miscellaneous phenomena close to melting points. According to the estimation by the author the error of the apparatus was within ± 0.5%.

The calorimeter with a similar system of automatic control of adiabatic conditions was developed by Popov and Galchenko.[31] The calorimeter was designed for measuring heat capacity of powdered and solid bodies up to 1000 K. The experiments were performed by two methods. One of the methods provided for compensation for heat loss due to side effects (nonuniformity of temperature over the surface, heat flow along the fastenings of the calorimeter, etc.) by way of maintaining some permanent difference between the shield temperature and that of the sample. The said difference was set against the zero run of the calorimeter in the auxiliary periods. In the other case a difference was maintained on the differential thermocouple which differed from the fictitious "zero" difference (i.e., this difference might actually be equal to zero). Then a correction for heat loss detected in this case was introduced into the heat capacity concerned. This correction was determined in conformity with different values of the measured heat capacity at different indications of the thermocouple, and was not in excess of 0.5% of the heat supplied. The maximum error of the experiment at 1000 K was about 2%.

An automatic adiabatic calorimeter for measuring true heat capacity and heats of phase transformations of steels and alloys in the temperature range of 300–1400 K was developed by Lyusternik.[17, 32, 33] The calorimeter was based on the principle of continuous heating. The circuit diagram of the calorimeter is shown in Fig. 15. Cylindrical sample 1 of the material under test is subjected to continuous heating within a considerable temperature range by means of calorimetric heater 2 whose power is maintained constant. The rise of the temperature of the sample is measured by platinum–rhodium: platinum thermocouple 3 (0.2 mm diam) mounted in the pyrometric line of the sample. The current (I), voltage (V), and the sample temperature (T_s) are measured potentiometrically. The sample is jacketed with a double adiabatic shield in the form of quick-

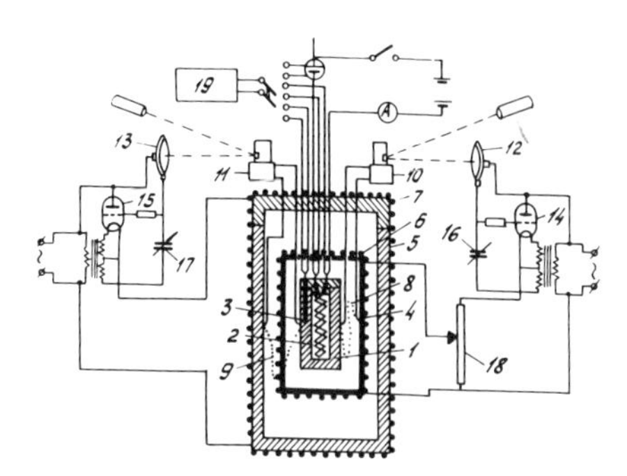

FIGURE 15. Lyusternik's adiabatic calorimeter.[17,32,33] 1, cylindrical sample; 2, calorimetric heater; 3, platinum–rhodium: platinum thermocouple; 4, quick-response screen; 5, furnace; 6, 7, heaters of shields; 8, 9, differential thermocouples; 10, 11, mirror zero galvanometers; 12, 13, photocells; 14, 15, thyratrons; 16, 17, variable capacitors; 18, 19, potentiometers.

response screen 4 and furnace 5. The heaters of both the shields (6 and 7) are supplied from the proportional-type photothyratron control systems, whereas the pickups (sensors) serve differential thermocouples 8 and 9, respectively (platinum–rhodium and platinum, 0.2 mm in diameter). The thermocouples sense and read the difference between the temperatures of the said shields and that of the sample. The difference between the temperature of the furnace and that of the sample was not in excess of 1 K, while that between the shield and the sample was no more than 0.02 K. The value of the heat loss was determined by the use of formula (6) and did not exceed 3% of the power supplied within any section of the temperature interval. The correction for possible heat exchange using the measurements of the calorimeter temperature run with the heater switched off (as was stated in Section 2) is found tolerable only if the heat exchange is independent of rate of heating, but is governed only by the reference temperature. This means that the distribution of the temperatures over the surfaces of the sample and shield, and the adiabatization quality, shall not vary in the transition from the main period (heating) to the auxiliary period (period of measurement of the temperature run with the heater turned off). Only in this case the differential thermocouple will correctly react to heat exchange between the sample and environment. Therefore in Refs. 17, 32 and 33 the rates of heating in the experiments were changed more than by ten times in order to check the reliability of the correction introduced for the heat exchange (the usual working rate of heating was $0.05\,\mathrm{K\,s^{-1}}$). The appropriate measures were taken to ensure that the heat emission of the calorimetric heater was constant along its length, so that the temperature field was constant over the surface of the sample. The same measures were taken for the heaters of the shield and the furnace. The temperature gradients in the process of heating did not exceed $4 \times 10^{-3}\,\mathrm{K\,mm^{-1}}$ for the sample and shield, and $4 \times 10^{-2}\,\mathrm{K\,mm^{-1}}$ for the furnace.

The regulators of temperature of the shield and furnace have no principal differences. These are photothyratron systems with phase control of anode current. The systems operate in the proportional control mode. The differential thermocouples 8 (9) are connected to a mirror zero galvanometer 10 (11). The light beam reflected from the galvanometer mirror is incident on the photocathode of photocell 12 (13). Deflection of the mirror of the galvanometer showing temperature differences detected by the differential thermocouples results in different illumination of the photocathode. The photocell is connected in the grid circuit of thyraton 14 (15) together with variable capacitor 16 (17) using a phase bridge circuit. The resistance of the photocell changes with variation of the illumination intensity, and so does the phase of the cutoff voltage at the thyratron grid. As the illumination intensifies, the duration of the operation of the thyratron grows within each positive half-cycle, thereby causing the anode current to increase. The furnace heater is connected directly to the anode circuit of the thyratron, while the shield heater is coupled across potentiometer 18. The range of the working power of the regulators is set in advance by altering the anode voltage and capacitance in the grid circuit. The power consumed by the furnace tends to go high with rise of the experiment temperature. In the case of proportional control mode this can be ensured only at the expense of an increase of lagging of the furnace temperature behind the temperature of the sample, i.e., due to gradual deviation of the temperature difference under control from zero. Change of the power from 50 W (at 400 K) to 300 W (at 1200 K) results in a shift of the light spot on the galvanometer scale for 12 mm, which corresponds to the 1.5 K lag of the furnace temperature behind the temperature of the sample at 1200 K. That is why it is necessary to select a new zero position of galvanometer 11 in every 200–300 K. This is achieved by adjusting the regulator so that at the moment of the onset of the subsequent stage of heating the furnace power required for the given temperature is realized at a zero current in the circuit of galvanometer 11, which will correspond to a zero (minimum) heat exchange. The absence of heat exchange in this case is indicated by the run of the sample temperature which is measured by thermocouple 3. Having the regulator so adjusted for each range of the temperatures it is possible to ensure that even at maximum temperatures the temperature difference under control is not above ± 1 K. Quick-response shield 4 installed between the sample and the furance for finer adjustment of adiabatic conditions is, consequently, under the conditions where fluctuation of the temperature difference is not in excess of 2 K. Therefore the power consumed by the shield heater does not exceed 5 W and remains virtually unchanged with change of the experiment temperature. This makes it unnecessary to adjust the regulator of the shield temperature in the process of the experiments and at the same time enables one to continuously maintain the equality of the temperatures of the surfaces of the sample and shield within the limit of 0.02 K up to 1400 K.

The rate of heating of the sample was obtained by measuring the time with a stopwatch, for an increase of temperature of 20 μV as read by the Jager–

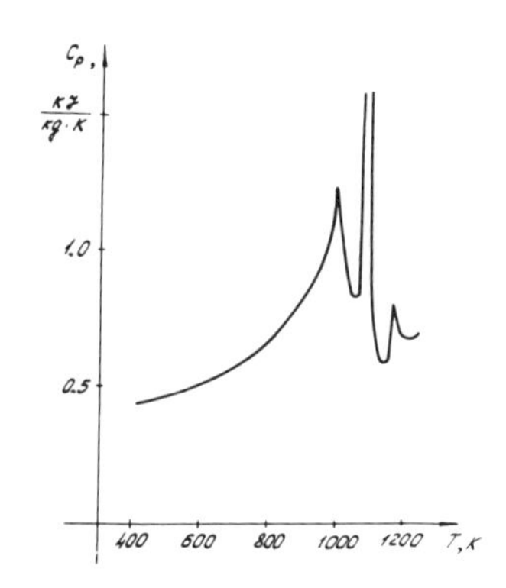

FIGURE 16. Heat capacity of stainless chromium steel 4X13.[33]

Düsselhorst potentiometer. The derivative $dT/d\tau$ was determined by means of graphical differentiation.

The calorimeter under consideration was used to investigate the heat capacity and heats of phase transformations of carbon steels, low-alloy chromium, stainless, austenitic chrome–nickel stainless steels, as well as alloys based on zirconium and niobium, ternary alloys of the lead–antimony–tin kind, within the temperature range of 300 to 1400 K with an error of 0.8%–1%.

In Fig. 16 is presented the heat capacity of stainless chromium steel 4X13 (C–0.4%; Cr–13%; Ni $\leqslant$ 0.5%; Cu $\leqslant$ 0.3%; Mn $\leqslant$ 0.6%; Si $\leqslant$ 0.6%; S, P $\leqslant$ 0.03%) as measured in Ref. 33. The heat capacity curve has its peak with a maximum value $C_p = 1.26\,\text{kJ}\,\text{kg}^{-1}\,\text{K}^{-1}$ at 1018 K due to the fact that steel loses ferromagnetic properties (Curie point). In the region of 1100 K the heat capacity tends to increase sharply because of pearlitic transformation in steel 4X13. The measured heat effect of the pearlitic transformation amounts to $9.21\,\text{J}\,\text{kg}^{-1}$. In the range of 1175–1185 K there is one more minor peak of the heat capacity that is apparently due to absorption of heat in the process of solution of carbide (Cr Fe_7C_3 in γ-iron. The thermal effect of this process is not over $0.63\,\text{J}\,\text{kg}^{-1}$.

One of the highest-temperature adiabatic calorimeters is manifestly the apparatus devised and developed by Braun, Kohlhaas, and Vollmer[16] shown in Fig. 17. This calorimeter with continuous heat supply operates from 300 to 1900 K. The calorimeter is designed for measuring the heat capacity of solids (2% error) and liquids (3% error), heats of phase transformations in solids (0.5% error), and melting heat (1.5% error). The operating conditions are (1) power applied is constant, heat capacity of sample is inversely proportional to rate of heating; (2) rate of heating is maintained constant, heat capacity is proportional to power applied; (3) difference of temperatures between calorimeter and shield is registered with heater switched off. The metals studied are Ti, Zr, V, Cr, Mn, Fe, Co, Ni, Cu, Ag, Au.

At the 7th All-Union Conference on calorimetry held in Moscow in 1977 professor Edgar Westrum informed the author of this Chapter of this Compendium that he was engaged in development of a high-temperature adiabatic calorimeter with a temperature range of 1800–1900 K. Unfortunately at the present no publications concerned have been found as yet.

The work of Naito and Kamegashira[34] was devoted to a study of two important methodological problems: (1) methods of disposition of the calorimeter heater; (2) presence of heat exchange gas in the apparatus. To solve the first problem two calorimeters were tested. In one of them the heater was placed in a special seat located along the axis of a cylindrical sample. The heater of the other calorimeter was made in the form of a coil, and was wound externally. The experiments showed that the heat loss of the second calorimeter is less. Besides, variation of heat loss after filling the calorimeter with the substance under test in the second case was approximately ten times less compared with the first case. The second problem is of great importance for high-temperature calorimeters (where utilization of cements and varnishes is not possible). The researches disclosed that the availability of heat exchange gas somewhat increases the heat exchange between the calorimeter and the adiabatic shield, but materially improves the thermal contact of the heaters with the objects under control. Use of argon as a heat exchange gas at atmospheric pressure resulted in considerable (some 5 K) temperature gradient on the shield. The drop of the pressure down to 5–8 torr caused a decrease of the gradients to the values characteristic of vacuum calorimeter, without appreciably deteriorating the thermal contact of the heaters with the shield and the calorimeter. The investigations were performed up to the temperature of 1000 K.

3.2. Low Temperatures

The specific characteristics of low temperatures, primarily a decrease of intensity of heat exchange and the applicability of resistance thermometers, enable the measurements by the method of adiabatic calorimetry to be done more accurately than in the region of high temperatures (except ultralow temperatures). This aspect of adiabatic calorimetry has been developed best of all, and a considerable number of research works are devoted to this problem.[35,36] First of all a work that is especially worthy of consideration is a fundamental review by Westrum, Furukava, and McCullough published in the first volume of *Experimental Thermodynamics*.[37] Much attention has been given to analysis of errors and correction methods in the works.[1,38,39] Now we shall discuss some works of common interest which have not been reflected in the relevant summarizing literature.

Among the first research works dealing with adiabatic low-temperature calorimetry were classical investigations by Eastman and Rodebush[40] covering the range from liquid nitrogen to room temperatures, and by Simon and

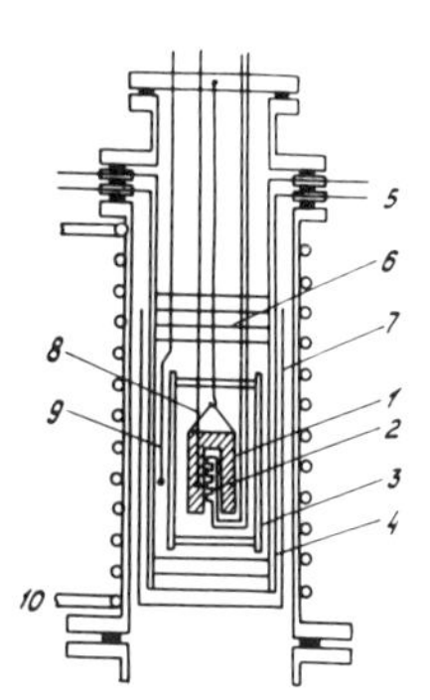

FIGURE 17. Continuous-heating adiabatic calorimeter (300–1900 K) of Braun, Kohlhaas, Vollmer.[16] 1, sample; 2, heater; 3, 7, thermal shields; 4, tube furnace; 5, leads of furnace electrodes; 6, shields; 8, 9, thermocouples; 10, cooling water.

Swain[41] (from the boiling point of hydrogen to room temperatures). A rather typical calorimeter is one developed by Strelkov[1] and repeatedly modified in the aspect of the methodology in a number of works.[38, 42] The range of temperatures for measurement of heat capacity of solids and liquids is 12–300 K. The diagram of the calorimeter and the experimental thermogram are shown in Fig. 18. Container 3 fabricated from 0.3-mm-thick copper has a capacity of 85 cm^3. Thin-walled sleeve 1 houses a platinum thermometer whose resistance varied from 110 Ω (300 K) to 0.3 ohms (12 K). Heater 2 is made of constantan wire arranged bifilarly in a copper capillary tube. The latter is mounted on vertical cruciform fins. Cover 4 is provided with capillary tube 6 for filling the calorimeter with heat exchange gas (helium) and substance to be tested (if it is in liquid state at room temperature). The same cover has sleeve 5 to admit the lead wires of calorimetric heater 2. The calorimeter is enclosed in an adiabatic shield with sectionalized heaters. The circuit diagram of the adiabatic control system employing the proportional-integral principle is presented in Fig. 19. The photothyratron system makes use of photoresistors as sensors in

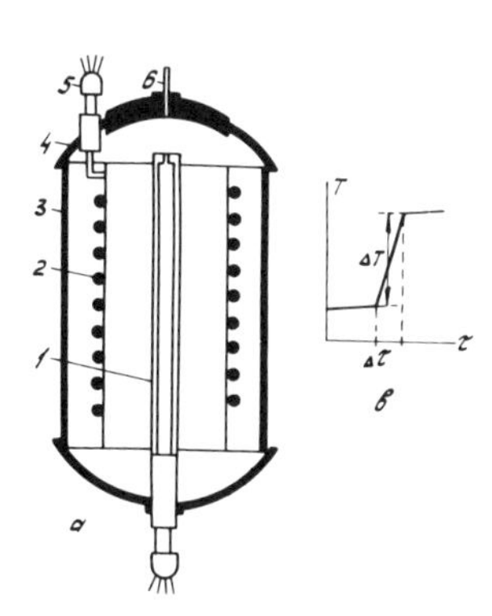

FIGURE 18. Low-temperature vacuum adiabatic calorimeter of Strelkov.[1] (a) Cut-away view of calorimeter: 1, platinum resistance thermometer; 2, heater; 3, container; 4, cover; 5, wire leads of heater; 6, capillary tube for filling container with helium gas (for better heat exchange in calorimeter). (b) Determination of heat capacity.

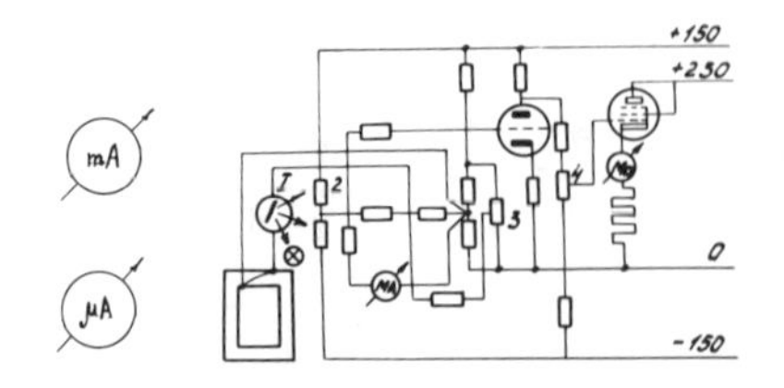

FIGURE 19. Schematic diagram of automatic control of one of the elements.[1] 1, galvanometer; 2, photoresistors; 3, zero corrector; 4, setting of rated power of heater.

the differential thermocouple unbalance amplification circuit. The adiabatic shield is surrounded with an isothermal chamber in the form of a double-wall Dewar flask. The thermostatic media in use are: hydrogen or nitrogen at pressure $p \leqslant 760$ torr (with temperature ranges of 12–60 K and 60–195 K, respectively), dry ice with alcohol (195–253 K), ice with NaCl (253–273 K), and ice (273–300 K).

To minimize introduction of heat into the calorimeter the wire leads are pressed to fit tight to the adiabatic shield thereby making its temperature equal to the temperature of the calorimeter. With the same objective in mind the vacuum line atop the upper part of the calorimeter contains a black body simulator which prevents radiation from heated parts of the apparatus towards the calorimeter. The heat value of the calorimeter was determined when filled with helium. The portion of the heat consumed for heating up the helium, as compared with the heat value of the calorimeter, was about 1% at low temperatures and tended to be less at higher temperatures.

The calorimeter is designed for operation in stage heating mode (Fig. 18b). The run of the calorimeter before and after the main period was within 2×10^{-5} K s^{-1}. The temperature rise in the main period amounted to 0.05 to 0.5 K. The random error did not exceed 0.2%, but errors of temperature measurement increased the total error to 1%–2%.

An excellent complex of adiabatic low-temperature calorimeters has been built at the National Research Council (NRC, Canada) by Dauphinee, Martin, Preston-Thomas, and colleagues.[14, 18, 43–50] These are precision calorimeters used to measure heat capacity (and in some cases heat of structural transformations and melting) of many metals including alkali metals with an error of 0.1–0.2%, covering the temperature ranges of 3–26 K,[18, 47, 48] 20–330 K[18, 43–46, 50] and 300–475 K.[14, 49] The calorimeters mentioned have undergone a detailed analysis in the summarizing work devoted to thermophysical properties of alkali metals.[51] A number of methodological peculiarities of the works in this series[52–56] typical of adiabatic low-temperature calorimetry has been considered in the work.[37] One of the latest calorimeters of the National Research Council[14, 49] has been mentioned in Ref. 51 in a brief-outline form, therefore we shall dwell at length on it. The work[14] describes the apparatus developed by Martin and Snowdon and designed for measuring heat capacity by the continuous-heating method within the temperature range of 300 to

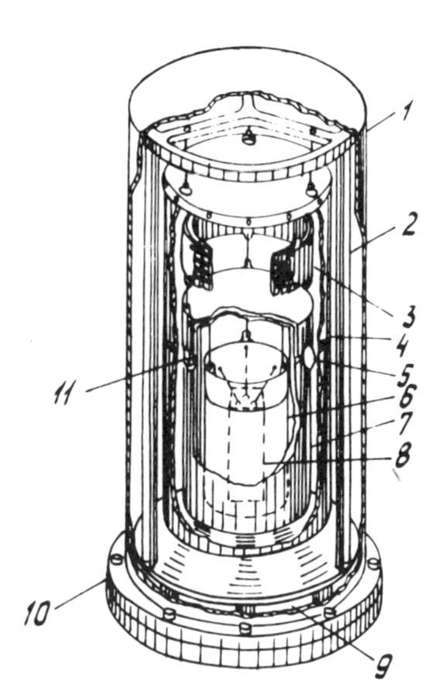

FIGURE 20. The calorimeter apparatus of Martin and Snowdon.[14,49] 1, outer case; 2, support; 3, thermal anchoring ring for wires; 4, outer shield (uncontrolled); 5, radiation shield over vent; 6, adiabatic shield; 7, inner shield; 8, calorimeter; 9, to diffusion pump underneath; 10, O-ring seal; 11, heater terminal.

475 K. This apparatus should hardly be considered a low-temperature calorimeter, as is customarily assumed; however it has generalized the best properties of the existing low-temperature calorimeters (see Fig. 20). For maintaining adiabatic conditions provision is made for three shields of which two are controllable and one (outer) is passive. The calorimeter is modified in the usual fashion for minimizing heat exchange through gas and wire leads (evacuation and proper thermal contact of wires with intermediate shields). The construction of the calorimeter is also of interest (shown in the figure with a dotted line). The calorimeter consists of a container with the sample under test and an outer case. The container is made from stainless steel. In the center of the container there is a recess to receive a platinum resistance thermometer. Wound on the side of the container is a heater. The outer case is fabricated from silver foil. Change of temperature gradients on the outer case of the calorimeter of such construction, during the main period, generally speaking, depends upon thermal diffusivity of the foil material and rate of heating; however, the gradients remain unchanged at different degrees of filling of the calorimeter. The rates of heating were varied from 10^{-3} to 10^{-2} K s^{-1}. The measurement of the heat capacity of a standard (reference) substance α-Al_2O_3 has proven the design precision of the apparatus (± 0.1–300 K, ± 0.2–475 K). The calorimeter was used for precision measurements of heat capacity of sodium[49] in the region of the melting point, which demonstrated identity of its values of heat capacity C_p in solid and liquid phases.

The work[15] presents a description of a unique apparatus intended for measuring heat capacity of liquids near critical point by the continuous-heating method. A substantial length of time required for establishing the equilibrium and the intensive increase of heat capacity necessitate operation at very low rates of heating. In[15] the heating rate was 10^{-6} K s^{-1} (i.e., 0.4×10^{-2} K h^{-1}), which involved extraordinary requirements concerning the adiabatization. The fact that the calorimeter was fabricated with particular care, and that, for the most part, high-sensitivity 20-kΩ thermistors were used as temperature pickups made it practical to minimize the heat capacity error to 0.4% with a random error of some 0.2% (at about 300 K). The thermistors were selected in view of such characteristics and then were properly calibrated against a platinum resistance thermometer.

One of the recent devices which dealt with adiabatic low-temperature calorimetry is the apparatus developed by Orlova.[57] This calorimeter is meant for measuring heat capacity of solids in the temperature range of from 4 to 300 K.

The author employs the method of vacuum adiabatic calorimeter with stage heating. The calorimeter is schematically shown in Fig. 21. The calorimeter is fabricated of stainless steel (0.1 mm in thickness and about 20 cm^3 in capacity). In sleeves 6 inside the calorimeter are placed a thermometer (in the center) and three sections of a constantan heater with a total resistance of 600 Ω. To improve heat transfer from the heater to the sample use is made of platinum cylinder 5. The adiabatic condition of the calorimeter is obtained by controlling the temperature of the triple-walled shield enveloping the calorimeter (shields 1, 2, 3) by means of the heaters attached to the respective surfaces. Shield 2 was used as an active one only at calorimeter temperatures exceeding 100 K. When the departures of the calorimeter temperature from the cryogenic bath temperature exceed 200 K, it becomes necessary to use shield 1 also as an active one. The perfectness of adiabatization essentially depends upon the equality of the temperature fields of the calorimeter (empty and filled with substance) and main shield 3. To minimize the temperature gradients, the external surface of the

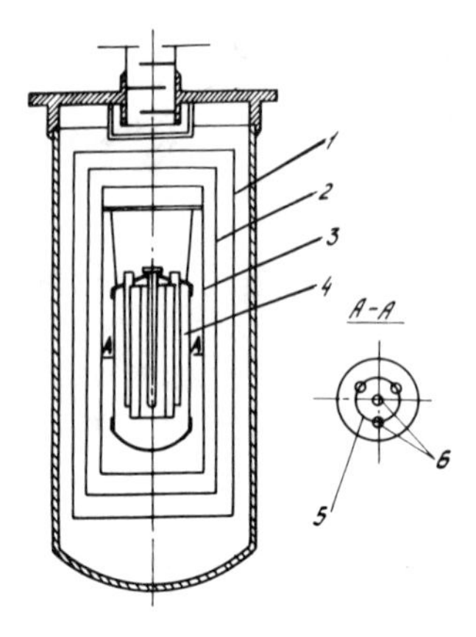

FIGURE 21. Vacuum adiabatic calorimeter with discrete application of heat.[59] 1, 2, 3, shields; 4, calorimeter; 5, platinum cylinder; 6, wells with thermometer (in center) and heater sections.

calorimeter was plated with copper (deposited electrolytically in a layer of 0.3 mm) and, besides, the difference between the temperatures of shield 2 and shield 3 was altered in such a way that the power supplied to the main shield should be minimum (5×10^{-3} W), and approximately constant within the entire temperature range. The required differences between the temperatures of the surfaces of the calorimeter and shield 3 and also between those of shields 2 and 3 were maintained automatically with an error not exceeding $(2\text{–}3) \times 10^{-4}$ K. In this case the value of the equilibrium temperature "run" of the calorimeter was not more than 3×10^{-7} K s^{-1}.

The temperature was measured by means of a germanium resistance thermometer (in the region of 4.2–20 K) and a platinum resistance thermometer ($R_0 \approx 100\,\Omega$) calibrated against the IPTS-68 scale. The germanium resistance thermometer ($R_0 \approx 125\,\Omega$) was calibrated with reference to the scale of a gas thermometer, where the scale of the latter gradually turned into the IPTS-68 scale at 13.81 K. Parasitic emf was eliminated by reversing the supply current. The error of the temperature measurements lay within ± 0.003 K, the temperature differences ΔT being $\pm 5 \times 10^{-5}$ K. The power produced by the calorimeter heater was measured and the heating duration was determined with an error not exceeding 0.001%. The calorimeter was evacuated, filled with thermal exchange gas (helium) and sealed in a special appliance making it possible to completely exclude the difference between the masses of the solder in the experiments performed with an empty calorimeter and with the calorimeter filled.

The pressure of the thermal exchange gas (10 and 30 torr, respectively) was selected using the condition of equality of masses of the gas in an empty calorimeter and in a filled one.

The root-mean-square deviation of the experimental points from the smoothed curve is 0.2%–0.3% in the range of 4–14 K, and tends to drop down to 0.1% with increase of the temperature up to 40 K. The level of a probable unaccounted systematic error is evaluated to be 0.25% in the regions of the temperatures of 4–14 K, and 0.03% in the region of 14–300 K.

It is emphasized that this apparatus is fully automated (Fig. 22) and is operated in conjunction with data acquisition system Hewlett-Packard HP 3050B and computer HP 9825.[57] The adiabatic condition of the calorimeter is maintained by a proportional-integral-differential regulator (PID-controller) with differential thermocouples Au + 0.07 atom % Fe–chromel used as temperature sensor. The following processes are automatically controlled (main channels are shown in Fig. 22): achievement of the temperature of the beginning of the experiment, selection of electric current across thermometers depending on the temperature of the experimental point, selection of the resistance coil depending on the thermometer resistance, changing over the sense of current in the circuits of the thermometers and heater, plotting of curve of the temperature "run" [i.e., a curve $T(\tau)$ before and after supplying the heat to the calorimeter],

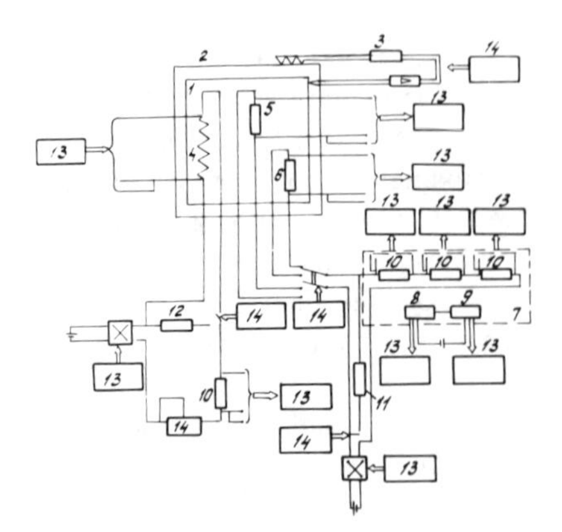

FIGURE 22. Functional diagram of calorimetric apparatus.[57] 1, calorimeter; 2, shield; 3, proportional-integral-differential controller; 4, heater; 5, resistor of germanium thermometer; 6, resistor of platinum thermometer; 7, thermostat; 8, thermostat thermometer; 9, resistor in circuit of thermostat thermometer; 10, standard resistor; 11, resistor of auxiliary thermometer; 12, resistor of auxiliary heater; 13, HP 3495: channels 0–23; 14, HP 6940.

selection of temperature step ΔT, change of current across the heater depending on the required step of the temperature, measurement of the heat applied. The temperature was measured every 20 s during the temperature "run". That made it possible to select the desired duration of the temperature "run," depending upon the dynamics of control of adiabatic conditions as well as the time required for equalizing the temperature after completing the heating process. The results are recorded on a magnetic tape. Upon completing the experiment the accumulated data are formed into a data file for the substance under test.

In calculation of heat capacity of a sample use is made of smoothed values of the heat value of an empty calorimeter. The heat value of the empty calorimeter is smoothed by the use of the spline technique as described in Ref. 58. The tables of the splines for different calorimeters are stored on a magnetic tape and can be retrieved depending on the calorimeter selected for an experiment. The experimental values of C_p of the sample are smoothed in a way similar to that used in the case of the heat value of an empty calorimeter, using the spline technique on the same computer.

The necessary measurements were accomplished to determine the heat capacity of synthetic corundum and benzoic acid to recommend them as a standard sample,[57, 59, 60] as well as a number of superconducting materials within a temperature range of 4–300 K.

3.3. *Applications in Reaction Calorimetry*

As pointed out above in the section devoted to discussion of the method, the adiabatic calorimeter is a reliable instrument of reaction calorimetry. This is of particular importance for determining minor heat effects. This aspect has been in our opinion inadequately generalized in the relevant literature, particularly as applied to high temperatures, so it seems to be useful to discuss this problem separately.

The author of this chapter of the Compendium (at the High-Temperature Research Institute of the Academy of Sciences of the USSR) has developed an adiabatic high-temperature calorimeter for measuring integral heats of mixing of liquid binary and multicomponent systems of alkali metals.[61] This calorimeter is believed to be typical enough and is considered suitable for analyzing basic peculiarities of high-temperature adiabatic reaction calorimetry. At the same time the apparatus has been designed with due regard for the latest achievements in this field of science.

Reaction calorimetry places a number of general requirements which must be met to accurately measure heat effect of mixing processes. These requirements are: the absence or provision of minimum value of vapor volume (to avoid adverse effect of side thermal effect of condensation or evaporation), constancy of pressure and temperature in the process of mixing, and necessity of reliable homogeneity of mixture in calorimeter.[62] Determination of heats of mixing of interacting systems of alkali metals involves some additional difficulties due to small values of excess thermodynamic functions of these systems,[63] and substantial corrosive and chemical activity of alkali metals, especially at high temperatures. From the experimental standpoint a positive value of heats of mixing,[64] i.e., endothermal effect of alloying reaction, is also inconvenient. The experiments involving alkali metals demand complete tightness of the calorimetric container, and the level of temperatures excludes a presence of fittings and other accessories in the hot zone. The calorimeter meets the aforesaid requirements and ensures measurements of heats of mixing of alkali–metal systems with an error of 1%–2% up to temperatures of some 1300 K. Provision is also made for measurement (within one experiment) of the heats of mixing of binary systems of several compositions or multicomponent systems comprising at most four components.

The diagram of the calorimeter is diagrammatically illustrated in Fig. 23. Calorimetric vessel 12 contains inner calorimetric vessels 13 each filled with an exactly known mass of component of alloy 18. The remaining space of larger calorimetric vessel 12 is also filled with one of the components of alloy 18. Each vessel may be filled either with pure metal or with an alloy whose initial concentration is well known. Bellows-type pistons 14 and compensator 22 ensure absence of vapor in the vessels. Initially the vessels are so filled that bellows-type pistons 14 are held compressed, while bellows-type compensator 22 is expanded. In its lower part each internal vessel 13 is provided with diaphragms 17 rated at the difference of pressures of saturated vapours of the respective components. Located beneath the daiphragms are dividers 21. All the bellows spaces communicate through ducts 23 with the controllable pressure system. One of these lines is shown in the figure where 1 represents a flow-control valve, 2 a two-position valve. The sensitive element of tensometer 4 is fixed on a cold part of capillary 3. Calorimetric vessel 12 is placed inside cylinder 15, which serves as

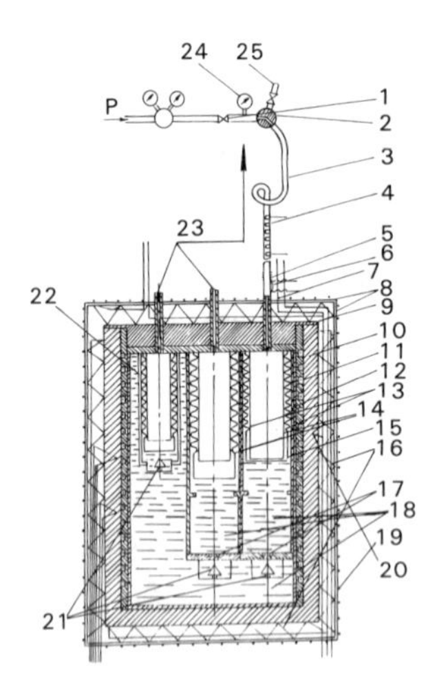

FIGURE 23. Diagram of high-temperature reaction adiabatic calorimeter for measuring heats of mixing of alkali–metal alloys.[61] 1, Flow-control valve; 2, two-way valve; 3, capillary tube; 4, tensometer sensor; 5, protective heater of capillary tube; 6, differential thermocouples; 7, heater of end shields; 8, adiabatic shields (radial and end); 9, block cover; 10, calorimetric block; 11, calibrating calorimetric heater; 12, calorimetric vessel; 13, inner calorimetric vessels; 14, bellows-type pistons; 15, body of calibrating calorimetric heater; 16, differential thermopiles (radial and end sections); 17, diaphragm; 18, components of alloy under test; 19, heater of radial shield; 20, measuring thermocouples; 21, dividers; 22, bellows compensator; 23, to system of controllable pressure; 24, manometer; 25, line to receiver.

a case for calibrating heater 11. As noted above, adiabatic reaction calorimeters are designed in compliance with the principle realized in the Nernst-type calorimeter, i.e., provided with a heavy and robust block (Fig. 23, Refs. 9, 10). An increase of the heat value of the calorimeter at the expense of the substance with the known heat capacity is necessary, first, for bringing the mixing process to an isothermal stage, and, second, for reducing the contribution of reagents and reaction products (whose heat capacities are sometimes inadequately known) to the thermal value of the calorimeter. This contribution does not exceed 3%–5%, which leads to a change of the experiment temperature of the order of 1 K. The calorimeter vessel is enclosed in thin-walled adiabatic shield 8 whose temperature is maintained so as to be equal to the temperature of the block surface. The purpose of the control system pickup is served by a multijunction battery of differential thermocouples 16, their junctions being uniformly distributed over the surfaces of the block and shield. The multijunction thermocouple battery is useful not only for enhancing the control sensitivity but also for ensuring a compensation for the mean difference of the surface temperature between the calorimeter block and adiabatic shield. The independent control systems feed heater 19 of the radial screen and heater 7 of the end face screens corresponding to two sections of the thermopiles. Change of temperature in the process of mixing as well as its uniformity are sensed by four thermocouples 20 built into the body of block 10 and distributed over its surface. Heat loss through the capillary tubes and fasteners is compensated for by protective heaters 5 and differential thermocouples 6.

The initial pressure of inert gas in each bellows chamber is somewhat higher than the pressure of saturation of the respective component of the alloy. The mixing takes place as a result of rupture of diaphragm 17 in one of inner

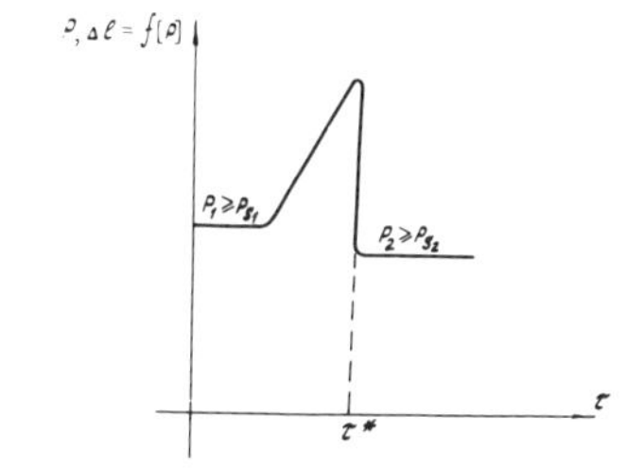

FIGURE 24. Barogram (tensogram) of calorimetric experiment.[61]

calorimetric vessels 13 under the effect of increased pressure exerted by virtue of an additional supply of gas to the bellows chamber of this vessel via flow-control valve 1. The moment of the rupturing of the diaphragm and, consequently, the beginning of the experiment is indicated by the tensogram of tensometer sensor 4 or by the barogram of manometer 24. Figure 24 illustrates change of pressure in the first inner vessel, where P is the manometer reading, $\Delta l = f(P)$ is the reading of the tensometer, τ is the time, τ^* is the start of the experiment, P_1 is the initial pressure, P_2 is the experiment pressure, and P_{S_1} and P_{S_2} are, respectively, the pressure of the saturated vapors of the component in the vessel and that of the obtained binary mixture. As soon as the diaphragm is ruptured, bellows piston 14 moves down, the liquid metal from vessel 13 flows into vessel 12. At the same time the piston of bellows compensator 22 goes upward. By operating flow-control valves 1 and two-way valves 2 which communicate with the receivers through line 25, it is possible to impart reciprocating motion to the bellows–piston system. The liquid metal at every cycle passes through the holes in the zone of dividers 21 thereby creating jet flow which makes for reliable mixing of the components. Upon completing the required procedure for measuring the temperature change in the main period of the experiment, as well as the "run" of the calorimeter, it is advisable to repeat the operation discussed above with the second internal vessel, and then the third one, for the sake of accounting for heat loss. It has been found that each time either a new-composition binary solution or a three four-component solution is obtained, and a series of the experiments, as can be seen, is conducted in a single assembling of the calorimeter. In view of probable nonadditivity of heat capacity of alloys, generally speaking, the heat value of the calorimeter should be measured before and after each experiment. However, it is well to point out that in the case of minor contribution of heat capacity of the substances under investigation, only a great deviation from additivity may be noted, which is not characteristic of alkali–metal systems.

The heat effect is measured in the process of formation of alloy at pressure P_2 (see Fig. 24) which remains almost unchanged. Taking into consideration the foregoing about the change of the temperature it may be asserted that the measured heat of mixing corresponds to the isobaric–isothermal process, i.e., presents enthalpy of alloy formation.

All the component elements of the calorimetric vessels, including the bellows chambers and capillary tubes (3, 12, 13, 14, 17), are fabricated from stainless steel. The body of calibrating heater 15, calorimetric block 9, 10 and adiabatic shield 8 are made of copper. The thermocouples are all platinum–rhodium (10% Rh): platinum wire, 0.2 mm diam, and the heaters 7, 11, 19 are wound of tungsten–rhenium (20% Re) wire, 0.35 mm in diam. It should be noted that substituting elements made of Armco iron or molybdenum for elements made from copper enables the temperature range of the apparatus to be enlarged.

The construction features and basic dimensions of the calorimeter are seen from Fig. 25. Three inner calorimetric vessels 1 are arranged axial-symmetrically in outer calorimetric vessel 2. In the same fashion are positioned bellows compensators 3. Some sections of the cylindrical part of each inner vessel are cut away so as to allow the liquid involved to freely flow into the cavities of the outer vessel. Bellows 4 provides reciprocating motion in the annular gap formed by inner and outer reinforcing cylinders 5. The internal sides of the outer cylinders have vertical grooves 6 along which the metal can flow down under

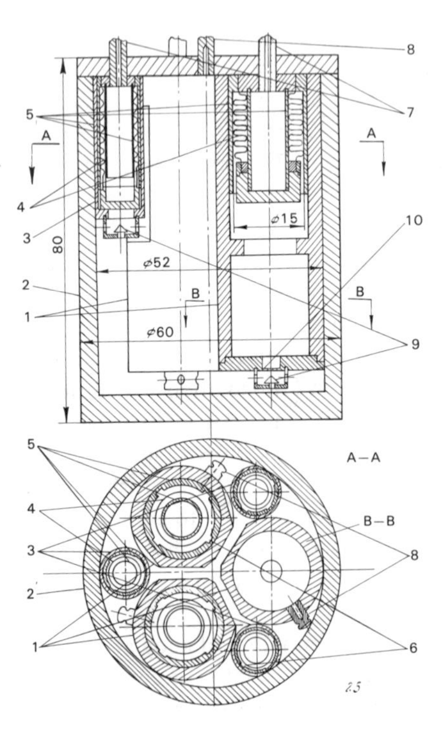

FIGURE 25. Construction of a high-temperature reaction adiabatic calorimeter for measuring heats of mixing of alkali–metal alloys.[61] 1, inner calorimetric vessels; 2, outer calorimetric vessel; 3, compensators; 4, bellows; 5, reinforcing cylinders; 6, grooves for flowing down of liquid metal; 7, capillary tubes for evacuation and for delivery of gas; 8, capillary tubes for filling vessels with alloy components; 9, dividers; 10, diaphragm.

compression of the bellows. Shown in Fig. 25 are the capillary tubes 7 used for evacuating and for delivering gas into the bellows spaces, and capillary tubes 8 (first squeezed and then sealed by argon-arc welding) used for filling the vessels with liquid alkali metals. The filling procedure ensures that neither extraneous gas nor vapor phase of the metal under test are present in each vessel. The chambers with dividers 9 are furnished with several ports to provide multistream flow under reciprocating motion of the pistol system after diaphragm 10 has been ruptured. In assembling the calorimeter, firstly all the parts of inner vessels 1 are welded in consecutive order, and then, upon filling with alloy components and sealing capillary tubes 8, these vessels are mounted and fixed by welding on the cover of the outer vessel 2, which is then welded to the vessel body. Thereupon the remaining space of outer vessel 2 is filled with liquid metal and the respective capillary tube 9 is sealed. Filling each vessel is accompanied with the relevant operations of weighing so as to exactly determine the final concentration of the obtained alloy. Then the frame of calibrating heater 15 (Fig. 23) is placed in position by the method of close-sliding fit together with calorimetric block 10 (Fig. 23). Then cover 9 of the copper block (Fig. 23) is installed. The difference in the coefficient of linear thermal expansion of stainless steel and copper provides reliable thermal contact in heating. Initially a calibrating heater is mounted on frame 15 (Fig. 23); the respective junctions of differential thermocouples 16 (Fig. 23) and also measuring thermocouples 20 (Fig. 23) are arranged on the block (see Refs. 9, 10, Fig. 23). Then radial and end adiabatic shields 8 (Fig. 23) are fitted into position and fastened. After that the external junctions of differential thermocouples 16 (Fig. 23) (passed through the holes in the shields) are arranged and fixed. Heaters 7, 19 of the end and radial shields (Fig. 23) are attached from the outside of the adiabatic shield.

Due to the fact that the heat of mixing of systems of alkali metals is positive[64, 65]† the process of production of an alloy is endothermic and is followed by decrease of the calorimeter temperature. Therefore the initial temperature of calorimeter 1 (Fig. 26) shall be somewhat higher than the temperature of furnace 2. Both the temperatures are maintained by means of three independent standard thyristor control systems: one for heater 3 of furnace 2, and the remaining two for heaters 5 of the calorimeter (radial and end shields). The latter two systems are also used for preserving the adiabatic conditions in the calorimetric experiment. A signal from a temperature sensor (6 or 7) supplied to controller-comparator 8, where the unbalance between preset and actual signals is determined and preamplified. Then this signal is measured in control unit 9 and in the form of a positive feedback is fed to the base of a transistor preamplifier. In the collector circuit of this amplifier is shaped and also measured a unified (0–5 mA) control signal depending on the selected

†Exception being the Cs–Rb alloy with a negligible negative heat of mixing.[64]

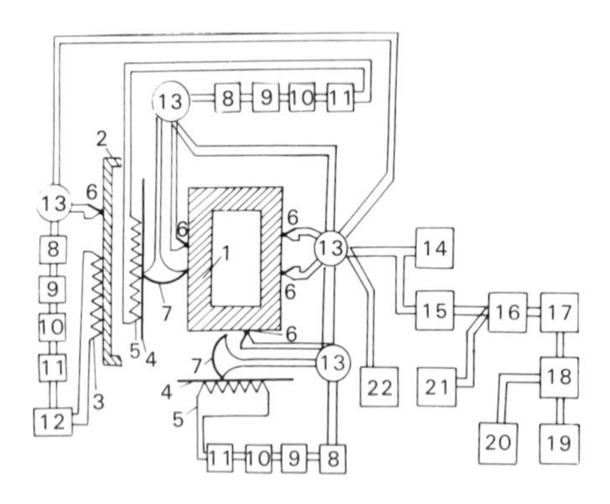

FIGURE 26. Diagram of control of adiabatic conditions and measurement of temperature.[61] 1, Calorimetric block; 2, thermostat unit of furnace; 3, heater of furnace; 4, adiabatic shields (radial and end); 5, heaters of adiabatic shields (radial and end sections); 6, measuring thermocouples; 7, differential thermopiles (radial and end sections); 8, controller-comparator and dc preamplifier; 9, intermediate amplifier and controller (setting of control law and shaping of control signal); 10, thyristor control unit; 11, thyristor power amplifier; 12, step-down transformer; 13, switches; 14, potentiometer; 15, photocompensation amplifier; 16, resistance box (voltage divider); 17, digital voltmeter; 18, transcriber; 19, printer; 20, frequency meter; 21, multichannel recording potentiometer; 22, potentiometer.

control mode (as a rule, proportional-integral-differential control). This signal taken from the winding of the pulse transformer is supplied via thyristor control unit 10 to the control electrode of thyristor power amplifier 1 whose anode circuit is coupled with the corresponding heater. In this way the phase principle of continuous control in accordance with the selected law is realized. For a comparatively low-resistance load (furnace heater) the thyristor amplifier of the control system is connected to the heater through step-down transformer 12. This circuit, in general, complies with the general principles set forth in Refs. 37 and 38.

After achieving the required temperature level of the furnace both the regulators of the adiabatic shield are adjusted to a similar temperature exceeding by 5–10 K the temperature of the furnace. These temperature levels are obtained in the remote control mode of all the three regulators, each of them being coupled to one of five measuring thermocouples 6. Three measuring thermocouples of the calorimeter are arranged symmetrically in the lower half of the radial part of calorimetric block 1 beneath the compensators (see also Fig. 25). The fourth thermocouple (Ref. 6) is located in the center of the lower end face block 1. Thermocouple 6 of the furnace is placed in a special thermostat unit 2, which serves as the frame of the furnace heater. Bringing the unbalance on the comparator to zero in the process of obtaining the needed temperature level is a means of measuring the current temperature by the comparator. Temperature data are simultaneously recorded using standby measuring thermocouples of the calorimeter. Upon stabilization of the temperature the adiabatic regulators are changed over (switch 13) to automatic control with input from differential thermobatteries 7. In this period the "run" of the calorimeter is being measured.

The application of multijunction differential thermopiles as well as differential thermocouples and protective heaters on the connecting elements of the calorimeter results in the unbalance on the comparator, selected for minimal (close to zero) "run" of the calorimeter, being practically absent. After the beginning of the main period (time τ^* in Fig. 24) the calorimeter temperature will start to decrease. In these circumstances the temperature of the shields in the adiabatization conditions will also tend to fall, until the mixing process has been completed. Due to the smallness of the temperature change in the experiment, the "run" will be similar in both the auxiliary periods. It is emphasized that the power to the heaters is not high because of comparatively small difference of the temperatures between calorimeter 1 and heater 2; therefore sufficiently reliable control of adiabatic conditions can be obtained in the case of a single-wall shield. This can be easily verified by varying the difference of the temperatures of the calorimeter and the furnace.

For automatic recording of the calorimeter temperature provision is made for a compensating circuit having discrete and analog outputs.[66, 67] The temperature in the main period of the experiment shows fairly slow variation, therefor using switches 13 it is possible to introduce into the circuit any measuring thermocouple 6, checking in this way uniformity of the temperature. Prior to commencing the calorimetric procedure the electromotive force of thermocouple 6 is fully compensated by potentiometer 14. Connected in series with the latter is compensating amplifier 15, which is responsible for compensation and measurement of slight changes of the electromotive force in the auxiliary and main periods. The current output signal of amplifier 15 is measured with the aid of digital voltmeter 17 for drop of voltage across resistance box 16 (voltage dividers). Automatic recording of the voltmeter readings is effected by a digital printer (transcriber 18 and printer 19). To ensure a higher accuracy of the time intervals† in measurements, transcriber 18, controlling the operation of voltmeter 17, is connected to an external crystal oscillator from frequency meter 20. For monitoring the experimental conditions, a signal from the output of photoamplifier 15 is also supplied to one of the channels of multichannel recording potentiometer 21. Amplifier 16 serves for matching the impedances of amplifier 15 and potentiometer 21. In the beginning and at the end of the main period of the experiment the signal of thermocouples 6 is changed over by switch 13 to potentiometer 22. The readings of the potentiometer are considered reference for determining the experiment temperature, and for processing the printed and recorded information.

†It is found very essential in determination of heat loss [see the formula (6)], and what is more, when the calorimeter is utilized for measuring of heat capacity [see formula (3)].

We have also developed a constant-temperature, adiabatic calorimeter for directly obtaining precise data (at 373.15 K) of the enthalpy of formation of alkali–metal alloys.[68] The constant-temperature adiabatic calorimeters represent a fairly specific group and find applications in measurements of thermal effects at some standard (reference) temperature. These calorimeters employ the phase transition of calorimetric substances taking up or liberating heat due to the process under study. Phase transitions of calorimetric substances provide constancy of the calorimeter temperature, and the same phase transition in a thermostat enclosing the calorimeter ensures adiabatic effect of the calorimetric process.

The following example will show specific features and sources of error of the calorimeters of this type. The calorimeter construction is schematically shown in Fig. 27. Calorimeter 3 filled with boiling distilled water is placed in thermostat 2 also filled with boiling water. The amount of heat liberated (absorbed) in the reaction under study is determined by change of intensity of vaporization of calorimetric liquid as compared with the constant level (natural "run" of the calorimeter) established by calorimetric heater 1. Maintaining intensive boiling in the calorimeter is necessitated by the requirement of saturation temperature throughout the experiment (i.e., both in the main period and in the "running" periods), as well as of its constancy in the calorimeter volume. Reliable conditions of thermodynamic equilibrium in the "liquid–vapor" system of the calorimetric substance and also a substantial difference between the volumes of the two phases will provide a very high sensitivity and accuracy of the experiment.

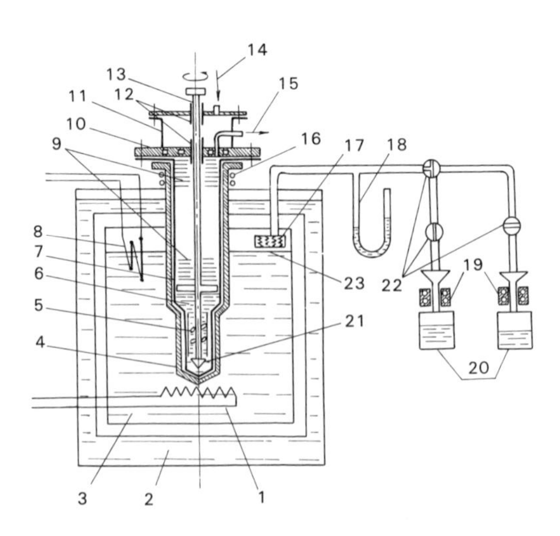

FIGURE 27. Reaction constant-temperature adiabatic calorimeter.[68] 1, Calorimetric heater; 2, thermostat; 3, calorimeter; 4, receptacle case; 5, mixer; 6, alloy under test; 7, reaction vessel; 8, differential thermopile; 9, shields; 10, channels for thermostatic liquid; 11, inert gas chamber; 12, packing glands; 13, rod; 14, delivery of inert gas; 15, evacuation; 16, protective heater; 17, separator; 18, differential oil manometer; 19, condenser coolers; 20, condensers; 21, striker; 22, distributing valves; 23, level of calorimetric liquid.

Intensity of boiling and variations thereof in the main period are determined by measuring the amount of the vapor escaping the calorimeter. The vaporization heat of the calorimetric liquid is well known, and determination of the calorimeter constant which should be practically equal to this value is essentially a check of the apparatus for accuracy of the method used and for reliability of the construction. The vapor is discharged from the calorimetric vessel through a heated tube and enters alternately in detachable condensers 20 which are all weighed on an analytical balance in equal lapses of time. Placed in the upper part of the calorimetric vessel is differential thermocouple 8 used to determine an overheat of the calorimetric liquid as compared to the saturation temperature, and a separation device 17 of the vapor discharge line.

Tightly inserted in receptacle case 4 of the calorimeter is reaction vessel 7. Prior to a calorimetric experiment two (or more) hermetically sealed glass vials with the components of the expected alloy are placed on the bottom of the said reaction vessel. The cover of the reaction vessel is provided with seals 13 to pass a rod terminating in striker 21 to break the glass vials (when the rod moves down) and mixer 5 for uninterrupted agitation of the alloy components after the vials are broken. The radiation heat exchange in the vertical direction is suppressed by end shields 9 spaced apart through the length of the rod. The lower set of the shields is attached directly to the rod, while the upper one is located under the cover of the reaction vessel. Soldered to the vessel cover is chamber 11 for an inert gas to prevent ingress of air through the lower seal. Inside the cover there are lines 10 for continuous flow of the thermostat (boiling) water. The part of the receptacle case projecting beyond the surface of the calorimeter is heated by protective heater 16. This heater, like the thermostatic conditions for the cover, serves to decrease heat exchange due to radiation and heat conduction along the walls of the reaction vessel and the receptable case. All the parts of the calorimeter are made of stainless steel. The pressure of the vapor of the substance under test at the experiment temperature does not exceed 10^{-2} torr, therefore the thermal effect associated with evaporation or change of composition of the vapor phase is negligible.

The detailed description of the calorimetric procedure is given in Ref. 68. As far as quality of adiabatization is concerned, the adiabatization in this calorimeter, as in any adiabatic calorimeter, cannot be absolute. The sources of nonadiabatic condition also either may be permanent (this is taken into account when measuring the "run") or may arise under the action of heat effect in the main period (and, consequently, cannot be taken into account by measurements of the "run" in static state). The permanent heat flows are caused by relationships between overheat of the liquid and height of the boiling layer, the density of vaporization centers, boiling conditions, etc. In principle, there may be several causes of deterioration of adiabatic conditions in the main period. During the main period at some change of hydrodynamics of the boiling liquid there may take place either disappearance or generation of vaporization centers in the calorimeter. This brings about a change of overheat, and, correspondingly, causes heat exchange between the calorimeter and the thermostat. It should be noted that this is an irreversible process, therefore if it arises (that is, sensed by differential thermocouple 8) it is accounted for in measurements of the "run" in the second auxiliary period of the experiment. A short-time loss of the adiabatic condition may occur in the study of exothermic reactions when a positive thermal impulse in the main period leads to some rise of pressure (and, correspondingly, of the saturation temperature) at the expense of intensification of evaporation. However, as is seen from evaluations in the actual processes, this effect is sufficiently negligible ($\ll 1$ cal) in corrections, and may be regarded as systematic error of calorimetry. In endothermic reactions, such as processes of formation of alkali–metal alloys, the mentioned effect is absent. In case of a change of atmospheric pressure in the main period, the saturation temperature tends to change in the calorimeter and in the thermostat simultaneously, therefore, no deterioration of adiabatic conditions takes place.

The sensitivity of the calorimeter amounts to several calories, which in an unfavorable case of measurement of small thermal effects of formation of alkali–metal alloys with account made for volume of liquid phase in the reaction vessel (60–70 cm^3), corresponds to an error not exceeding 3%–5%.

4. AUTOMATIC CONTROL OF ADIABATIC CONDITIONS

As shown in the section devoted to development of adiabatic calorimetry, all modern calorimetric apparatuses are equipped with automatic control systems for maintaining adiabatic conditions. This is an indispensable condition for improving the accuracy of the obtained data and for facilitating the procedure of calorimetric experiments. The adiabatic condition automatic control systems[37, 38] comprise the following basic constituent elements:

1. Preamplifier of weak DC signals of differential temperature pickups determining difference in temperatures between calorimeter surface and adiabatic shield.
2. Controller used for presetting temperature difference necessary for compensating for all heat loss, i.e., corresponding to zero "run" of calorimeter.†
3. Comparator intended for determining deviation of actual temperature difference from that preset.
4. Control element forming response to the above deviation in the form of control effect in compliance with the respective control law.

†In properly developed calorimeters this difference is, of course, approximating zero.

5. Power amplifier used to amplify output signal of the control element to a value sufficient for controlling temperature of adiabatic shield.

The transfer function of the automatic control of adiabatic conditions, determining the effect exerted by the temperature difference sensed by the differential temperature pickup on the power of the heaters of the shields, depends on the principle used in design of the control system. At present the following laws of control of adiabatic conditions are adhered to[36, 38, 69]:

a. proportional (P-control);
b. proportional-integral (PI- or isodrome control);
c. proportional-integral-differential (PID-control).

It should be noted that two-position (relay) regulators widely used before due to their design simplicity[29, 31, 36, 70] find presently decreasing applications. Their principal shortcoming is an oscillation tendency dependence between oscillation amplitude and range of control. From the section devoted to errors in adiabatic calorimetry it will be evident that the shortcomings mentioned are also very much inherent in proportional controllers.

Figure 28 illustrates a typical proportional-integral (floating) control circuit for controlling the temperature difference between the calorimeter surface and adiabatic shield.[38]

The diagram presented in Fig. 29 shows a proportional-integral-differential circuit designed for providing adiabatic condition with a higher accuracy. As distinct from the circuit given in Fig. 28 the proportional-integral-differential circuit is additionally provided with differentiator10 for realizing the proportional-integral-differential control law.

In Fig. 30 is demonstrated the diagram of a circuit connected up for most precise calorimetric experiments.[38] Unlike the circuit shown in Fig. 29, here provision is made for two additional circuits. The first additional circuit (2, 16, 15, 13) is intended for establishing invariance of temperature difference under control with respect to heat liberation in the calorimeter. The necessity of this circuit consists of the following. Any real control system is incapable of instan-

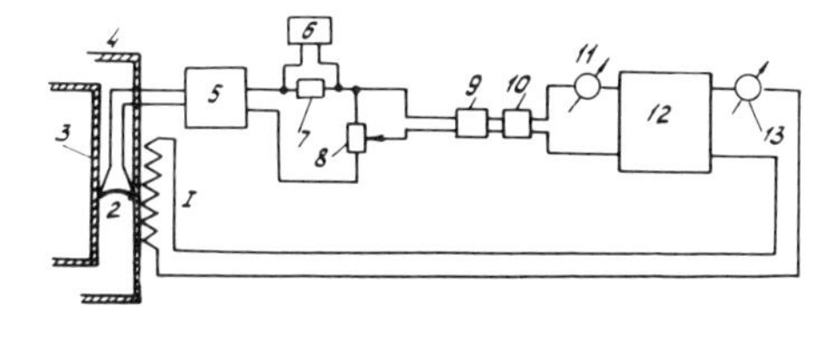

FIGURE 28. Block diagram of adiabatic-process system with proportional-integral control.[38] 1, heater of adiabatic shield; 2, differential thermocouple (thermopile); 3, surface of calorimeter; 4, adiabatic shields; 5, compensating dc preamplifier with current output; 6, electronic recording potentiometer; 7, standard resistor; 8, potentiometer; 9, intermediate amplifier; 10, regulator; 11, milliammeter (0–5 mA); 12, power amplifier (magnetic, thyratron or thyristor); 13, ammeter.

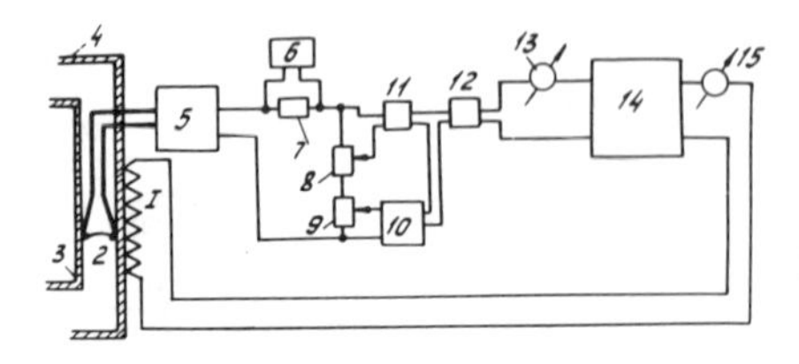

FIGURE 29. Block diagram of adiabatic process with proportional-integral-differential control.[38] 1, Heater of adiabatic shield; 2, differential thermocouple (thermopile); 3, calorimeter surface; 4, adiabatic shields; 5, compensating dc preamplifier; 6, recording potentiometer; 7, standard resistor; 8, 9, potentiometers; 10, differentiator; 11, intermediate amplifier; 12, regulator; 13, milliammeter (0–5 mA); 14, power amplifier (magnetic, thyratron, thyristor); 15, ammeter.

taneously removing the difference between the temperature of the calorimeter and the shield from the preset value (zero "run" of the calorimeter), since this deviation appears primarily due to the effect of heat evolution in the calorimeter and the character of its variation with time. This is particularly essential in stage-heating calorimeters. Besides, this also may govern the distribution of temperatures over the surface of the calorimeter. That is why the analysis of heat loss in the main period of the experiment, undertaken in adiabatic calorimetry through two "running" periods, cannot be absolutely exact, and needs, at least, obligatory checking with varying intensity of heat evolution and its dependence on time. To minimize the effect of the said factor in precision control systems, the power of the adiabatic heater is controlled depending on the value and the character of change not only of the deviation of the controlled temperature difference from the preset value (deviation control) but also of the intensity of heat evolution in the calorimeter. In this event the operation accuracy becomes invariant relative to thermal agitation, as the causes inviting variation of heat exchange in the main period are being compensated for. When power is supplied to the calorimetric

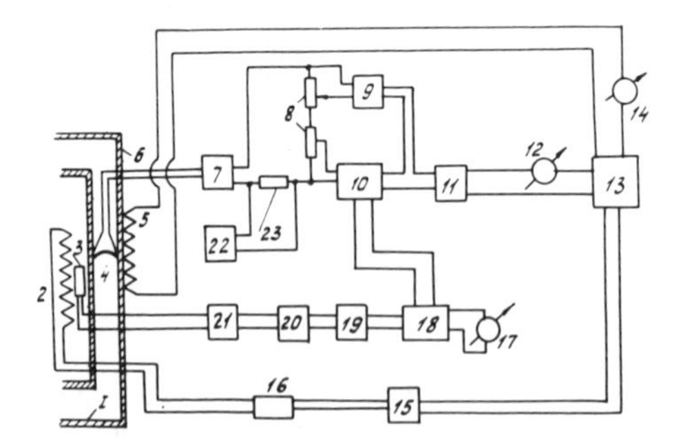

FIGURE 30. Block diagram of precision adiabatic system.[38] 1, calorimeter surface; 2, calorimeter heater; 3, calorimeter temperature pickup; 4, differential thermocouple (thermopile); 5, heater of adiabatic shield; 6, adiabatic shields; 7, compensating dc preamplifier; 8, potentiometers; 9, differentiator; 10, intermediate amplifier; 11, regulator; 12, milliammeter (0–5 mA); 13, power amplifier (magnetic, thyratron or thyristor); 14, ammeter; 15, excitement compensator – integral–differential circuit; 16, system of supply of power to calorimetric heater; 17, milliammeter; 18, remote control unit; 19, intermediate amplifier and regulator; 20, dc preamplifier; 21, measuring potentiometer or bridge; 22, recording potentiometer; 23, standard resistor.

system, an additional power is fed at the same time to the adiabatic heater. In this case the transfer function of the heat agitation compensator can be so selected that the controlled difference of temperatures becomes independent of intensity of heat evolution in the calorimeter. The transfer function in question is determined through the transfer functions of the calorimetric system and adiabatic shield. Proceeding from this it is possible to show[38] that the purpose of excitement compensator 15 shall be served by an integral-differential circuit[71] for the calorimeter with a "follow-up" chamber (or this circuit connected in series with an integrator[72] for a calorimeter with an isothermal chamber). For the circuit illustrated in Fig. 30 if, for example, power amplifier 13 is a magnetic amplifier, then it should incorporate several control windings. One of the windings receives an output signal from the deviation proportional-integral-differential controller (9, 10, 11). A signal from system 16 is so fed to another winding via excitation compensator 15 as to avoid adverse effect of heat evolution on the temperature difference under control. System 16 allows transmission of this signal synchronously with the start and finish of heat liberation in the calorimeter. The method of design of the thermal excitation compensator in the form of an integral-differential circuit is described in Refs. 38 and 72, and in simpler form (without consideration for dynamic properties of adiabatic shield)† in Refs. 69 and 73.

The other circuit (3, 21, 20, 19, 18. 17. 9) is designed for bringing the calorimeter to the exactly preset temperature level in order to eliminate a temperature "run" in the initial period. The required preliminary working condition is obtained by means of calorimetric heater 2 and the proportional-integral-differential controller of adiabatic conditions (9, 10, 11), when both the additional circuits are cut in. Measuring potentiometer (or bridge) 21 is tuned so as to achieve a balance at the preset temperature. The unbalance signal from the potentiometer is applied via DC preamplifier 20 to the intermediate amplifier and to proportional-integral controller 19, and further to remote control unit 18 and amplifier 10, the latter being an algebraic adder of signals from 7 and 18. The proportional-integral-differential controller (9, 10, 11) keeps the total signal constant, corresponding to the zero "run" at the preset temperature of the calorimeter.

Prior to the main period of the experiment the output signal of controller 19 is replaced, with the aid of unit 18 and measuring instrument 17, by a permanent remote-control signal of the same magnitude.

†Referred to an inertial link of first order if resorting to the terminology peculiar to the theory of automatic control.

5. ESTIMATION OF ERRORS OF ADIABATIC CALORIMETRY

In a real calorimetric experiment performed in adiabatic conditions there exist two chief sources of errors associated with heat exchange between the calorimeter and adiabatic shield. On the one hand, the temperature difference under control is not a difference of mean integral temperatures of surfaces involved in heat exchange, since the differential temperature sensors can average temperatures only at some sections of these surfaces. Therefore in measuring temperature gradients on the surfaces during the experiment (i.e., in case of differences between these gradients in the main and "running" periods of the experiment) the total heat exchange may take place even when a constant difference of the temperatures is maintained at a level corresponding to the zero (minimum) "run," This source of the errors has not been sufficiently studied as yet, and practically all the experimental data on heat capacity contain this error to a certain extent. This error may be minimized by means of a differential multijunction thermopile that should be placed between the surfaces of the calorimeter and the shield. However, there is also a restriction factor consisting of heat exchange through the wires and an increase of uncertainty in the heat value of the calorimeter. In Refs. 39 and 74 the authors endeavored to analyze the effect of change of the temperature gradients over the surface of a particular calorimeter depending on the rate of heating and the degree of filling of the calorimeter.

On the other hand, any real control system is incapable of instantaneously eliminating deviations of the temperature difference under control from the preset value for the zero "run" of the calorimeter. This difference between the temperature of the calorimeter surface and that of the shield arises under the control effect [intensity of heat evolution in the calorimeter and characteristics of its change with time $W = f(\tau)$] and external disturbing effects (fluctuation of supply voltage of the adiabatization system, change of room temperature, and so on).

These factors may lead to unexpected heat losses in the main period, where, the higher the intensity of heat evolution, the greater will be the loss of heat. The unaccounted heat flow from the calorimeter to the shield, of course, leads to increase of C_p values. The effect of input power on the experimental results is well exemplified by the work of Standsbury *et al.*,[75] which is one of the first works where automatic control of adiabatic conditions in a calorimeter with continuous heating was utilized. Figure 31 illustrates the heat capacity of nickel obtained in this work. The three upper curves correspond to the measurements made at heating rates of 0.07, 0.05, and 0.03 $\mathrm{K\,s^{-1}}$ (from top to bottom). The lowest curve stands for corrected values.

For expressing the probable error arising from control and external effects use is made[38, 76] of the so-called integral criterion for estimation of quality of

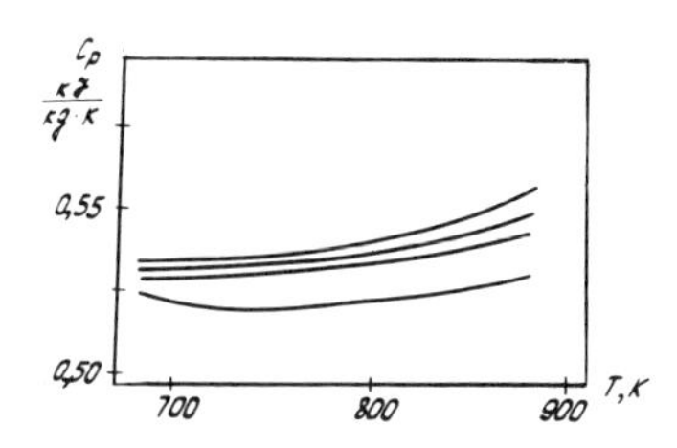

FIGURE 31. Heat capacity of nickel at different rates of heating.[75]

control (δ):

$$\delta = \frac{\Delta Q}{Q} \tag{13}$$

where

$$Q = \int_{\tau_1}^{\tau_2} W(\tau)\,d\tau$$

is the total heat effect of the process under study;

$$\Delta Q = \int_{\tau_1}^{\tau_2} q\,d\tau = \int_{\tau_1}^{\tau_2} S\alpha\Theta(\tau)\,d\tau$$

are heat losses in the process under study; S is the area of the calorimeter surface; α is the effective coefficient of heat exchange between the calorimeter and shield which is determined by radiation, heat conduction via residual gas (and also by convection in some cases), and by heat conduction through the lead wires of the differential thermopiles; $\Theta = f(\tau)$ is the difference between the temperatures of the calorimeter and shield counted from the initial level, established in the zero "run"; τ_1 and τ_2 are the beginning and the end of the main period; and q is the intensity of heat exchange (see formula 3).

The theory of automatic control enables one to obtain for a particular model of the calorimeter the quantitative estimates of the criterion δ depending on the control law, dynamic characteristics of the elements incorporated in the closed control circuit (shields, calorimeter, temperature sensors, etc.),† and control and external disturbances.[38, 76]

It is possible to show that

$$\delta = \delta_1 - \delta_2 \tag{14}$$

where δ_1 is a relative error resulting from introduction of energy into the calo-

†Using the terminology characteristic of the automatic control theory—depending on the transfer function of the regulator, calorimeter, adiabatic shield, etc.

rimeter (control effect), and δ_2 is a relative error caused by fluctuations of the voltage in the supply network of the automatic regulator (i.e., external disturbing effect).

Such an analysis reveals advantages and disadvantages of control systems of various types. For example, absolutely adiabatic conditions with proportional control is impossible in principle, since for the case of an isothermic chamber this may result in $\delta_1 = 1$, i.e., 100%, while for a follow-up chamber, it will lead to $\delta_2 = \infty$, i.e., in the case of any external disturbing effect the error may be infinite. Introduction of differential component into the control law will considerably extend the range of stability of the control system, therefore the proportional-integral-differential control possesses an advantage over the proportional-integral control in that it enables the integral criterion δ to be decreased by several orders of magnitude.†

The apparent error δ_2 in precision calorimetry can be minimized by decreasing random disturbing effects. For instance, utilization of highly stabilized power supply sources for the regulators of the adiabatic shields[2, 3] together with realization of the PID-control law requirements may bring δ_2 down to a negligible value.

The probable error δ_1 can be minimized by introducing into the control system a compensator of disturbance as an invariant component, i.e., enabling the accuracy of maintaining adiabatic conditions to become invariant in relation to the intensity of heat evolution in the calorimeter.[15, 73] It can be readily seen (as was mentioned in the section which dealt with the adiabatization schemes) that the functions of such a compensator can be performed by an integral-differential circuit for an adiabatic calorimeter with a follow-up chamber and an integrator connected in series with this circuit for an adiabatic calorimeter with an isothermal chamber.

Thus, meeting the listed requirements (measurement of controlled difference of mean-integral temperature of calorimeter and adiabatic shield by the use of a multiply differential temperature pickup, application of the proportional-integral-differential control law with introduction on an adiabatic invariant and use of high-stabilization power supply sources of adiabatization systems) is an imperative for realization of the possibility to estimate a correction for heat exchange in the main period in compliance with the calorimeter "run" measured in the initial and final periods, i.e., to derive the correction and its error in a fairly simple form. However, in this case it is also necessary to perform a direct experimental check of reproducibility of results under variation of power of heat effect in the calorimeter. Generally, if it is impractical to reduce criteria δ_1 and

†Formally due to the fact that the proportional-integral-differential control allows an essential increase of the transfer coefficient of the regulator and a decrease of time of isodrome as compared to the proportional-integral control.

δ_2 to zero (or, at least, to have them minimized) and also to reduce to zero the factor associated with a change of distribution of temperatures over the relevant surfaces, then for accurate estimation of the heat exchange correction and an error that will still remain, as could be seen, it is necessary to know the transfer functions of all the elements of the closed adiabatic-condition control circuit of a particular calorimetric system.

In precision calorimetric measurements an important role is played by provision of suitable reproducibility of conditions of calibrating and practical experiments. The temperature gradients over the sample surface may change. If for filling the calorimeter with the substance to be tested it is necessary to extract the container out of the adiabatic shield, the following may also change: positioning of the container with respect to the adiabatic shield, location of the junctions of the differential thermopile and response time of the junctions, quantity of sealing and thermocontact substances (solder, welding additive agents, cement), mass of heat-exchange gas in the container and mass of the capillary tubes for filling the container with gas. The listed changes lead, from experiment to experiment, to an increase of probable errors (primarily random errors), and attempts to eliminate them will materially complicate the procedure of preparation for the experiment. An example of removal of sources of such errors is given in Ref. 38, where, with this in view, in a calorimeter similar, in principle, to that described in Ref. 1, the procedure of preparation of the experiment and design features of some components have been materially changed.

In case the mentioned requirements are met, i.e., when the value q depends only on the temperature of the experiment and therefore can be determined in static conditions (in "running" conditions), it is possible to derive the expression (15) for heat capacity by using formulas (3) and (6). Subject to consideration, first of all, is the most complicated case from the viewpoint of calculation of error, i.e., measurement of heat capacity under continuous delivery of heat,

$$C_p = \frac{W}{M(dT/d\tau)_{\text{eff}}} - \frac{A}{M} \tag{15}$$

where

$$\left(\frac{dT}{d\tau}\right)_{\text{eff}} = \left[\frac{dT}{d\tau} - \left(\frac{dT}{d\tau}\right)^*\right] = \left[\frac{dE_T}{d\tau} - \left(\frac{dE_T}{d\tau}\right)^*\right]\frac{dT}{dE_T}$$

$$= \frac{[dE_T/d\tau - (dE_T/d\tau)^*]}{dE_T/dT} \tag{16}$$

is the effective rate of heating of the sample, i.e., the rate of heating with the correction for heat exchange accounted for.

E_T is the electromotive force of the calorimeter measuring thermocouple† as a function of the temperature.

The remaining components of the equation are as in formula (3).

The standard procedure of the theory of error gives the sources of relative error of heat capacity (δC_p):

$$\psi \delta W, \qquad \psi \delta \left(\frac{dT}{d\tau}\right)_{\text{eff}}, \qquad \psi \varphi \delta A, \qquad \delta M \tag{17}$$

where

$$\varphi = A \left(\frac{dT}{d\tau}\right) \Big/ W \tag{18}$$

is a portion of power W introduced by the calorimetric heater and consumed for heating up the parts of the calorimeter. This value may be considered a known constant value with a sufficient degree of accuracy (for calculation of error), namely,

$$\psi = \frac{1}{1-\varphi} \tag{19}$$

The remaining symbols in (17) stand for relative errors of the value included in formula (15).

As seen from formulas (17) and (18), it is expedient to possibly minimize the value φ when developing new designs of calorimetric appatatuses. In measuring heat capacity of solids the value of φ is usually equal, or less than 3%. In the experiments with liquid samples, the presence of a container results in a substantial increase of this coefficient, which may reach (at high temperatures and high vapour pressure of the substance under test) values materially exceeding 10%, which of course, impairs the accuracy of the data obtained.

The error of measurement of power (δW) determined basically from the systematic errors of the respective electrical measurements is, as a rule, $\delta W \leqslant 0.10\%$.

Determination of the effective rate of heating of the sample $(dT/d\tau)_{\text{eff}}$ is the most complex part of the calorimetric experiment and introduces the main contribution into the total error. Using again the common operation of the theory of error, from formula (15) the sources of relative error of the effective rate of heating $[\delta (dT/d\tau)_{\text{eff}}]$ can be derived:

$$\xi \delta \left(\frac{dE_T}{d\tau}\right), \qquad \xi \kappa \delta \left(\frac{dT_E}{d\tau}\right)^*, \qquad \delta \left(\frac{dE_T}{dT}\right) \tag{20}$$

†The case of usage of a resistance thermometer as a temperature sensor may be considered in a similar way.

where

$$\kappa = \left(\frac{dE_T}{d\tau}\right)^* \bigg/ \frac{dE_T}{d\tau} \tag{21}$$

is the ratio of the rate of "run" of the calorimeter to the rate of heating. For the heating rates adopted in adiabatic calorimetry lying within 0.1 K s^{-1} the value of κ, as seen from Fig. 2, even for high-temperature calorimeters does not exceed 2%,† and in calculation of the error within a relatively narrow temperature range this value may be considered a known constant quantity.

$$\xi = \frac{1}{1-\kappa} \tag{22}$$

The remaining members in formula (20) stand for relative errors of the values contained in formula (16).

For measuring the function $E_T = f(\tau)$ and its derivative $dE_T/d\tau$ the measuring circuit (to line 10 in Fig. 1) is usually supplemented with a system with an analog output (recording potentiometer or bridge) and with a discrete output (digital reading voltmeter, transcriber, digital printer). The analog output provides mainly for a qualitative picture of the process, while the discrete output gives the table of measured values E_{Ti} and the associated times τ_i. In the modern measuring circuits the discrete output is directly coupled to computers,[57] wherein the measured values E_{Ti} and τ_i are converted into the values of derivative $dE_T/d\tau$. Besides, the subsequent algorithm of calculation of C_p can also be completely realized. Measurement of the values of E_{Ti} or their increments $\Delta E_{Ti} = E_{Ti+1} - E_{Ti}$, determined to obtain the derivative $dE_T/d\tau$ (measurements of electrical quantities) are very accurate. It should be particularly emphasized that the action of the main sources of systematic error in measurements of small differences of temperatures (emf) at one point by use of one temperature sensor is practically eliminated. This also holds true for measurements of time τ_i and its intervals $\Delta\tau_i = \tau_{i+1} - \tau_i$, for the stabilized crystal oscillators producing command for measuring E_{Ti} provide a high accuracy of the values of τ_i and $\Delta\tau_i$. The frequency of readings may be substantial enough to be confined to linear interpolation for calculation of the derivative.

In the regions wherever heat capacity involves peculiarities, one of the interpolation formulas (Stirling, Bessel, Newton-Gregory) can be introduced into the algorithm of calculation of the derivative by the discrete output tables.

†Expressions (20) and (21) are used for evaluation of the base limitation of the heating rate $dE_T/d\tau$. In case of an excessive drop of the heating rate a rise of the coefficient κ will indicate an increase of contribution made by heat loss to the error; this is also clear from the formulas (3) and (5). The lower limit of the heating rate, as well as the upper one, shall be checked in the experiment of the result of this rate of heating.

From the foregoing it transpires that the main source of error of values $dE_T/d\tau$ and $(dE_T/d\tau)^*$ is the random error caused by instability of operation of the entire calorimetric apparatus, comprising the working cell proper with a calorimetric heater and temperature sensor a control system for regulating the heater power for maintaining the adiabatic conditions, a system of measurement of temperatures, etc. It is random deviations that are deciding in restrictions of frequency of readings of the values E_{Ti} and τ_i (when a change of measured value between two readings is approximately equal to scattering).

The well-known methods of statistical processing of the measurement results (for example, see Ref. 77) make it possible, in each particular case, to obtain the characteristics of distributions of random deviations of the values $dE_T/d\tau$ and $(dE_T/d\tau)^*$, and primarily their variance, for introduction into the calculation of the total error. The analysis of many calorimetric researches† involving Student's statistics shows that the general standard of distribution of individual deviations of the heating rates usually lies within 2%–3%, and the respective value for the smoothed quantities, within 0.3%. For the rate of the calorimeter "run" due to nonideal adiabatic conditions [i.e., value $(dE_T/d\tau)^*$], the said characteristics of distribution of random deviations (in relative units) are materially higher because of minor values of $(dE_T/d\tau)^*$. However, the respective contribution to the total error of the value $(dT/d\tau)_{\text{eff}}$ proves not to be higher, since the value $\delta(dE_T/d\tau)^*$ is introduced with a small coefficient as one of the sources of total error [see expressions (20) and (21)].

The relative error $\delta(dE_T/dT)$ associated with calibration of the temperature sensors is systematic for the calorimetric experiment. For a more complicated case (in the region of high temperatures where the purpose of the temperature sensor is served by platinum–rhodium: platinum thermocouples) this value, according to special investigations,[101-104] is $\delta(dE_T/dT) \approx 0.2\%$.

The expression $dE_T/dT = f(T)$ either is presented in the form of a comprehensive table or is centered into the computer memory in the form of a data file. Consequently, the error of this quantity is also affected by the error of the reference temperature. The corresponding additional error of the value dE_T/dT is about 0.05% within the value d^2E_T/dT^2[101-104] and the error of the reference temperature ΔT (not exceeding 1.2–1.5 K in the region of 1300 K) taken into account.

In calculating the error it is essential to take into consideration the differences of the temperature under measurement and its derivative from the mean value for the sample. This difference may result from the inertia of the temperature pickup, as well as from temperature gradients in the sample volume, and nonuniformity of the rates of heating at different points of the sample. The said factors will have an effect both on determination of the reference temperature

†In connection with this, in addition to the works mentioned above, Refs. 78–100 are also of interest.

and directly on determination of the value of $(dT/d\tau)_{\text{eff}}$. For estimating the effect of the temperature gradients in the volume of the sample it is possible to write down directly on the basis of the heat conduction equation (dimensional estimation) that

$$\Delta T \approx \frac{c\rho}{\lambda} l^2 \frac{dT}{d\tau}$$

where c, ρ, λ are, respectively, heat capacity, density, and thermal conductivity of the sample, l is the characteristic dimension, and ΔT is the temperature drop in the sample. For measurements with a relative error δc it is necessary to meet the inequality

$$\frac{1}{c} \cdot \frac{\partial c}{\partial T} \Delta T < \delta c$$

or

$$\frac{\partial c}{\partial T} \cdot \frac{\rho l^2}{\lambda} \cdot \frac{\partial T}{\partial c} < \delta c \tag{23}$$

The inequality (23) involves additional upper limitations for the tolerable rate of heating. Calculations of the temperature fields[4–6, 105–107] and special experiments disclose that for rates of heating commonly characteristic of adiabatic calorimetry (not in excess of $0.1\,\text{K}\,\text{s}^{-1}$) and the associated dimensions of the sample (not over 25 mm in diameter) the temperature drops (particularly in studies of metals) do not exceed fractions of a degree throughout the sample, and the difference between the average temperature and the temperature within the zone of the temperature sensor is not above 0.1–0.2 K. Utilization of thin temperature sensors (0.1–0.2 mm) ensures their low inertia response, with a lag in temperature for the aforesaid rates of heating lying within 0.2–0.3 K. The effect of thermal lagging of calorimetric temperature pickups has been discussed in detail in Ref. 6. These factors have been taken into account in the foregoing value of the error of the reference temperature.

When the method-prescribed conditions of heating are met (see Section 2) a quasistationary condition takes place (regular condition of the second kind, see Refs. 4–6), when all the sections of the sample are heated practically at the same rate of heating, special evaluations show that probable differences may yield an additional systematic error of the value $(dT/d\tau)_{\text{eff}}$ reaching 0.1%.

All the listed factors of the relative error $\delta\,(dT/d\tau)_{\text{eff}}$, both systematic and random, will serve [with the respective coefficients from expression (20)] as individual independent sources of heat capacity relative error δC_p [each additionally with coefficient ψ taken from expression (17)].

The calorimeter constant A is determined either by calculation as a total effective heat capacity of all the parts of the calorimeter, or through a special experiment. The latter is much more typical for studies of liquids when one of the main constituent parts of the calorimeter is represented by a container for the substance under test, the total heat capacity of this part being included in the value of constant A. In calculation of the value of constant A, the error (value δA), which in the present case is, primarily, systematic, is determined by errors in the literature data on the materials used for fabrication of the calorimeter component parts (with the reference error taken into account), and also by improper account for the effective part of the mass of these component parts. Normally, it is found rather difficult to determine exactly the portions of the wires (suspension, temperature sensors, supply circuits, control arrangement), wire insulation, fastening rods, etc. which take part in the process of heating at the expense of the power of the calorimetric heater. The data on heat capacity of construction materials have, as a rule, an error within 1%. The use of thin wires (0.1–0.2 mm) and fastenings, as well as possibly accurate determination of their natural mass, will reduce the error of the effective mass to a value not exceeding 0.1%. The experimental determination of the value of A is, in essence, an independent calorimetric experiment, the results of which are used for deriving this value from expression (3) when $C_p = 0$. In this case, the error δA, generally speaking, contains both random and systematic components considered above [see expressions (17) and (20)]. However, in the practice of calorimetric research, the heat value of the calorimeter is, as a rule, determined by performing many experiments; therefore the contribution of random errors to the value of δA is considerably minimized.

Measurement of the mass of the sample (M) adds, as a rule, just a negligible value to δC_p, an exception being the cases of study of the substances intensively reacting with oxygen that necessitate special conditions and facilities for filling the containers. This is redoubled by the fact that at high temperatures, when evaporation is great, the containers shall be tight. Alkali metals, for example, can be placed into the category of such substances. Nevertheless in this case the value of δM lies within 0.05%.

Upon analyzing all the error sources associated with determination of heat capacity by the results of the experiment from formula (8) it is necessary to take into account the independent reference error δC_p^T:

$$\delta C_p^T = \frac{1}{C_p} \left(\frac{dC_p}{dT} \right) \Delta T \tag{24}$$

From this it will be obvious that the error of measurement of the sample temperature is taken into account independently several times [in determination of errors $\delta (dT/d\tau)_{\text{eff}}$, δA, δC_p^T]. The error δC_p^T depends very much on the temperature ranges of the measurements. At high temperatures the main factor is the ΔT temperature error. At the same time in the Debye region the value of error δC_p^T is mainly determined by substantial coefficients $1/C_p$ and dC_p/dT. The reference error for different classes of substances and temperature ranges usually varies within the limits of $0.05 \leqslant \delta C_p^T \leqslant 0.1\%$.

At present there are methods of construction of graphs of total error distribution with account for randomization of systematic components of the error.[108] Such a procedure is found reasonable in the case of a great number of independent sources of systematic errors. This distribution is characterized with all the necessary components of distribution of random values such as variance, confidence probability, range of deviations corresponding to this probability, Student criterion, etc. Taking into account the confidence probability of 95% adopted in experimental thermodynamics, the heat capacity measured in an adiabatic calorimeter with continuous supply of heat, as proven by lots of relevant investigations, is characterized with the following ranges of relative total error.

In the high-temperature calorimeter under consideration $\delta C_p \leqslant 0.9\%–1\%$ for solid samples ($\varphi \leqslant 0.02$) and $\delta C_p \leqslant 1.6\%–2\%$ for samples in the form of containers with liquid substance under test ($\varphi \leqslant 0.3–0.4$). In the region of moderate and low temperatures a sharp decrease of heat exchange and use of resistance thermometers as temperature sensors instead of thermocouples result in appreciable improvement of accuracy. This is due to the fact that, on the one hand, resistance thermometers are more precisely graduated, i.e., error $\delta(dE_T/dT)$ goes down. This is of importance both for the error of determination of the value of C_p and for the reference error. On the other hand, the readings of the resistance thermometer capable of responding to the averaged temperature of the sample is less liable to random variations of the calorimetric system, as a result of which the errors $\delta(dE_T/d\tau)$ and $\delta(dE_T/d\tau)^*$ reduce significantly. As stated above, it is a random component of these errors that makes the uppermost contribution to the total error. Besides, due to a lesser intensity of heat exchange at low temperatures, the operation of the system of controlling and maintaining adiabatic conditions is markedly facilitated. All this enables the error, in some cases, to be decreased to 0.1%–0.2%.† In this case a leading role is played by the fact that in the region of low temperatures the container is often found unnecessary or its walls are extremely thin ($\varphi \leqslant 0.02–0.03$).

Analysis of formula (7) shows that the error of measurement of heat effects of phase transformations, as a rule, is less than the error of measurement of heat capacity, since the main component of the heat capacity error, i.e., error of determination of heating rate – $\delta(dE_T/d\tau)$, is absent. With due regard for nonirothermality of the thermogram in the process of melting and uncertainty

†Except for lowest temperatures when such factors as small values of heat capacity under measurement, complexity of measurements of the temperatures, and decreasing of sensitivity of the differential thermocouples lead to an increase of the error.

of the moments of beginning and end of the process, the error of determination of heats of phase transitions in adiabatic calorimetry usually is not in excess of 0.2%–0.5%.

In stage-heating adiabatic calorimeters determination of derivative $(dT/d\tau)_{eff}$ is not necessary, which also makes for better accuracy. But in this case the analysis of the temperature field in the sample is more complex, for the heating condition due to transient processes is often not quasistationary. The asymptotic nature of the establishment of equilibrium after application of heat determines the duration of the experiment. This enhances the effect of heat losses, and also causes ambiguity in the value of ΔT [see formula (5)]. It is also essential to take into consideration the fact that the arithmetical mean of the temperature between two equilibrium positions may be considered the reference temperature only when the temperature dependence of heat capacity in the given range of temperatures is not more complex than the linear dependence. As seen from the analysis of the unique research works, the error of results in adiabatic calorimeters with stage heating and continuous heating are close in value.

Most of the factors considered for errors of stage-heating calorimeters are also characteristic of reaction adiabatic calorimetry.

The thermal value of the calorimeter (A) in the present case is determined primarily by the heat capacity and mass of the calorimetric block, which is fabricated, as a rule, from a standard material (copper, for example); therefore the value of δA is very small. This applies both to the case of determination of A according to the literature information, and to measurement of A in a special experiment, because of reduction of relative contribution by an unidentified effective mass of wire leads, suspension, supports, etc. It can be seen from formula (8) that the error of heat capacity of reacting components ($\delta C_{p_{eff}}$) will be included as a source of error in the total error with the coefficient

$$\gamma = C_{eff} \bigg/ \frac{A}{M} \tag{25}$$

which is close in sense to the coefficient φ from formula (18). From this it is inferred that it would be good practice to use the Nernst-type calorimeters in the reaction adiabatic calorimetry, where, as was mentioned above, this ratio is small. Usually $\gamma \leqslant 0.1$. It is of great importance, because the effective heat capacity C_{eff} of the substance under test consisting of reacting components and reaction products may involve a substantial error with regard to variable composition. An analysis similar to that presented above shows that even in the region of high temperature ($T \leqslant 1300$ K), the heat effect of the process (chemical reaction, production of alloy, etc.) in reaction adiabatic calorimetry can be measured with an error $\delta Q^r \leqslant 2\%$. In the region of room temperatures the error can be diminished by one order of magnitude due to the factors considered for a continuous-heating calorimeter.

The analysis of all the error sources of a constant-temperature reaction adiabatic calorimeter, including the errors still remaining after having introduced the relevant corrections,[68] with due regard for randomization of systematic errors, shows that the error does not exceed 3%–5% even for such small thermal effects as enthalpies of formation of alkali–metal alloys. In determinations of heats of chemical reactions the error will decrease materially.

REFERENCES

1. P.G. Strelkov, E.S. Itskevich, V.N. Kostryukov, G.G. Mirskaya, and B.N. Samoilov, "Thermodynamic Investigations on Low Temperatures. Measurements of Heat Capacity of Solids and Liquids between 12 and 300°K," *Zh. Fiz. Khim.* **28**(3), 459–468 (1954).
2. E.D. West and D.C. Ginnings, "An Adiabatic Calorimeter for the Range 30 to 500°C," *J. Res. NBS* **60**(4), 309–315 (1958).
3. F.E. Karasz and I.M. O'Reily, "Wide Temperature Range Adiabatic Calorimeter," *Rev. Sci. Instrum.* **37**(3), 255–260 (1966).
4. G.N. Kondratiev, *Regular Heat Condition,* Gostekhteorizdat, Moscow (1954).
5. E.S. Platunov, *Thermophysical Measurements in Monotonous Conditions,* Energia, Moscow (1973).
6. B.N. Oleynik, *Precision Calorimetry,* Izdatelstvo Standartov, Moscow (1973).
7. E.D. West, "Heat Exchange in Adiabatic Calorimeters," *J. Res. NBS* **67A,** 331–341 (1963).
8. L.I. Anatichuk and O.Ya. Luste, *Microcalorimetry,* Visha Shkola, Lvov (1981).
9. R.C. Wilhouit and D. Shiao, "Thermochemistry of Biologically Important Compounds. Heats of Combustion of Solid Organic Acids," *J. Chem. Eng. Data* **9**, 595–601 (1964).
10. V.A. Sokolov, E.I. Banashek, and G.M. Rubinchik, "Enthalpy of Korund for the Range 678–1300°K," *Zh. Neorg. Khim.* **8**(9), 2017–2021 (1963).
11. N.A. Landiya, A.A. Chuprin, and G.D. Chachanidze, "Automatic Control of Shield Temperature of the High Temperature Adiabatic Calorimeter in Drop Method, *Teplofiz. Vys. Temp.* **3**(6), 910–916 (1965).
12. A.C. Macleod, "High Temperature Adiabatic Drop Calorimeter and the Enthalpy of α-Alumina," *Trans. Faraday Soc.* **63**(2), 300–307 (1967).
13. E.V. Matizen and R.I. Efremova, "Adiabatic Calorimeter for Measurement of Enthalpy and Heat Capacity by Drop Method at High Temperatures," *Metrologia* **8,** 56–61 (1972).
14. D.L. Martin and R.L. Snowdon, "Continuous-Heating Adiabatic Calorimeter for the Range 300 to 475°K – The Specific Heat of α-Al_2O_3," *Can. J. Phys.* **44,** 1449–1465 (1966).
15. M.J. Buckingham, M. Edwards, and J.A. Lipa, "A High Precision Scanning Ratio Calorimeter for Use Near Phase Transitions," *Rev. Sci. Instrum.* **44**(9), 1167–1172 (1973).
16. M. Braun, R. Kohlhaas, and O. Vollmer, "Zur Hochtemperatur-Kalorimetrie von Metallen," *Z. Angew. Phys.* **25**(6), 365–372 (1968).
17. V.A. Kogan and V.E. Lyusternik, "Study of Heat Capacity and Melting Heat of Alloys Based on Pb," *Inzh. Fiz. Zh.* **4,** 105–108 (1961).

18. J.D. Filby and D.L. Martin, "The Specific Heats below 320°K of Potassium, Rubidium and Caesium, *Proc. R. Soc. London* **A284** (1936), 83–107 (1965).
19. H. Moser, "Messung der wahren spezifischen Wärme von Silber, Nickel, β-Messing, Quarzkristall und Quarzglas zwischen +50 und 700°C nach einer verfeinerten Methode," *Phys. Z.* **37**(21), 737–753 (1936).
20. C. Sykes, "Methods for Investigating Thermal Changes Occurring During Transformation in a Solid Solution," *Proc. R. Soc. London* **A148**(800), 422–446 (1935).
21. C. Sykes and H. Evans, "Specific-Heat/Temperature Curves of Commercially Pure Iron and Certain Plain Carbon Steels," *J. Iron Steel Inst.* **138**(2), 125–162 (1938).
22. P.L. Gruzin, G.V. Kurdyumov, and R.I. Entin, "About the Nature of the 'Third Transition' at Heating of Tempered Steel," *Metallurg* **8**(128), 15–23 (1940).
23. P.L. Gruzin, "Sykes's Method of Definition of the Metals and Alloys Heat Capacity at High Temperatures," *Zavod. Lab.* **7**(11), 1266–1270 (1938).
24. Ya.M. Feenberg and U.A. Berman, "Definition of True Heat Capacity of Solids in the Wide Temperature Range," *Zavod. Lab.* **9**, 1057–1062 (1953).
25. A.I. Lazarev, in *Trudi Instituta Tochnoi Mehaniki i Optiki* Vol. 20, pp. 12–19, Leningrad (1955).
26. A.I. Lazarev, thesis, "New Methods and Apparatus for Study of Transition Heats and True Heat Capacity of Metals at High Temperatures," LITMO, Leningrad (1955).
27. J.H. Awbery, W. Snow, and H. Griffits, *Physical Constants of Some Commercial Steels at Elevated Temperatures,* Int. Publ., London (1953).
28. A. Eucken and M. Eigen, "Untersuchung der Assoziationsstruktur in Schwerem Wasser und *n*-Propanol mit Hilfe Thermisch-kalorischer Eigenschaften, insbesondere Messungen der spezifischen Wärmen," *Z. Electrochem. Angew. Phys. Chem.* **55**(5), 343–354 (1951).
29. V.A. Sokolov, "Calorimeter for Definition of True Heat Capacity and Hidden Heats at High Temperature," *Zh. Tekh. Fiz.* **18,** 813–822 (1948).
30. N.E. Schmidt and V.A. Sokolov, "Adiabatic Calorimeter for Measurement of True Heat Capacity of Low Heat Conductivity Substances for the Range 30–750°C. Heat Capacity of Korund," *Zh. Neorg. Khim.* **5**, 797–801 (1960).
31. N.N. Popov and G.L. Galchenko, "Definition of True Heat Capacity of Powder Substances at High Temperatures," *Zh. Obshch. Khim.* **21**(12), 2220–2235 (1951).
32. V.E. Lyusternik, "Automatic Calorimeter for Quantitive Thermal Analysis of Fireproof Steels," *Prib. Tekh. Eksp.* **4,** 127–129 (1959).
33. B.E. Neimark and V.E. Lyusternik, "Influence of Tempering on Temperature Conductivity of Carbon Steels," *Teploenergetika* **5**, 16–18 (1960).
34. K. Naito, N. Kamegashira, N. Yamada, and J. Kitagawa, "Studies on a Dynamic Adiabatic Calorimeter 1: Heat Leakage at High Temperatures," *J. Phys. E: Sci. Instrum.* **6**(9), 836–840 (1973).
35. *Experimental Thermochemistry* (F.D. Rossini and H.A. Skinner, eds.), Int. Publ., New York–London (1956, 1962).
36. S.N. Skuratov, V.P. Kolesov, and A.F. Vorobjev, *Thermochemistry,* Moskovski Universitet, Moscow (1964, 1966).
37. E.F. Westrum, G.T. Furukawa, and J.P. McCullough, in *Experimental Thermodynamics,* Vol. 1, Butterworths, London (1968).
38. O.A. Sergeev, *Metrological Bases of Thermophysical Measurements,* Izdatelstvo Standartov, Moscow (1972).
39. A.V. Voronel, S.R. Garber, V.M. Mamnitsky, and V.V. Shchekochikhina, in *Tr. Metrol. Inst. SSSR* **92**(152), 71–88, Izdatelstvo Standartov, Moscow (1967).
40. E.D. Eastman and W.H. Rodebush, "The Specific Heats at Low Temperatures of

Sodium, Potassium, Magnesium and Calcium Metals, and of Lead Sulfide," *J. Am. Chem. Soc.* **40**, 489–497 (1918).
41. S. Simon and R.C. Swain, "Untersuchungen über die spezifischen Wärmen bei tiefen Temperaturen," *Z. Phys. Chem.* **B28**(3), 189–198 (1935).
42. L.M. Khriplovich and I.E. Paukov, "Heat Capacity, Entropy and Enthalpy of Strocium at 4.9–302°K," *Zh. Fiz. Khim.* **50**(2), 567–568 (1976).
43. T.M. Dauphinee, D.K.C. MacDonald, and H. Preston-Thomas, "A New Semiautomatic Apparatus for Measurement of Specific Heats and the Specific Heat of Sodium Between 55 and 315°K," *Proc. R. Soc. London* **A221**, 267–276 (1954).
44. T.M. Dauphinee, D.L. Martin, and H. Preston-Thomas, "The Specific Heats of Potassium, Rubidium and Caesium at Temperatures Below 330°K," *Proc. R. Soc. London* **A233**(1193), 214–222 (1955).
45. D.L. Martin, "The Specific Heat of Sodium from 20 to 300°K: The Martensitic Transformation," *Proc. R. Soc. London* **A254**(1279), 433–443 (1960).
46. D.L. Martin, "The Specific Heat of Lithium from 20 to 300°K: The Martensitic Transformation," *Proc. R. Soc. London* **A254**(1279), 444–454 (1960).
47. J.P. Franck and D.L. Martin, "The Superconducting Transition Temperature of Lead," *Can. J. Phys.* **39**, 1320–1329 (1961).
48. J.D. Filby and D.L. Martin, "The Specific Heats Below 30°K of Lithium Metal of Various Isotopic Compositions and of Sodium Metal," *Proc. R. Soc. London* **A276**(1365), 187–203 (1963).
49. D.L. Martin, "Specific Heat of Sodium from 300 to 475°K," *Phys. Rev.* **154**(3), 571–575 (1967).
50. D.L. Martin, "Specific Heat Below 3°K, Melting Point, and Melting Heat of Rubidium and Cesium," *Can. J. Phys.* **48**(11), 1327–1333 (1970).
51. *Thermophysical Properties of Alkali Metals* (V.A. Kirillin, ed.), Izdatelstvo Standartov, Moscow (1970).
52. T.M. Dauphinee, "An Isolating Potential Comparator," *Can. J. Phys.* **31**(4), 577–591 (1953).
53. T.M. Dauphinee and H. Preston-Thomas, "Copper Resistance Temperature Scale," *Rev. Sci. Instrum.* **25**(9), 884–886 (1954).
54. T.M. Dauphinee and H. Preston-Thomas, "A d.c. and Square Wave a.c. Resistance and Voltage Comparator," *J. Sci. Instrum.* **35**, 21–24 (1958).
55. D.L. Martin, "The Specific Heats of Lithium Fluoride, Sodium Chloride and Zinc Sulphide at Low Temperatures," *Phil. Mag.* **46**(378), 751–758 (1955).
56. D.L. Martin, "The Specific Heat of Copper from 20 to 300°K. The Specific Heat of a Lithium–Magnesium Alloy. The Martensitic Transformation," *Can. J. Phys.* **38**(1), 17–31 (1960).
57. M.P. Orlova, Ya.A. Korolev, and L.M. Kheifetz, "Automatization of Heat Capacity Measurements for the Range 4.2–237°K," *Zh. Fiz. Khim.* **54**(1), 246–249 (1980).
58. J. Alberg, E. Nilson, and J. Yolsh, *Theory of Splines and Applications,* Mir, Moscow (1972).
59. N.P. Rybkin, M.P. Orlova, A.K. Baranjuk, N.G. Nurullaev, and L.N. Rojnovskaya, "Exact Calorimetry at Low Temperatures," *Izmer. Tekh.* **7**, 29–32 (1974).
60. M.P. Orlova and Ya.A. Korolev, "Korund – A Model Substance for Low Temperature Calorimetry," *Zh. Fiz. Khim.* **52**(11), 2756–2759 (1978).
61. D.N. Kagan, "The High Temperature Reaction Adiabatic Calorimeter for Measurement of Heats of Mixing of Liquid Alkali Metals Alloys," *Zh. Fiz. Khim.* **56**(5), 1318–1322 (1982).
62. V.P. Belousov and A.G. Morachevsky, *Heats of Mixing of Liquids,* Himia, Leningrad (1970).

63. D.N. Kagan, E.E. Shpilrain, and G.A. Krechetova, "Analysis of Types of Binary Alkali Metals Systems Based on Excessive Thermodynamic Functions Data, *Teplofiz. Vys. Temp.* **18**(3), 639–643 (1980).
64. T. Yokokawa and O.J. Kleppa, "Heat of Mixing in Binary Liquid–Alkali–Metal Mixtures," *J. Chem. Phys.* **40**(1), 46–54 (1964).
65. E.E. Shpilrain, D.N. Kagan, and G.A. Krechetova, "Study of Thermodynamic Properties of Liquid Binary Alkali Metals Systems at High Temperatures," *Teplofiz. Vys. Temp.* **16**(5), 951–959 (1978).
66. E.E. Shpilrain, D.N. Kagan, and S.N. Ulyanov, "Measurement of Heat Capacity and Heats of Phase Transition with the Pulse-Differential Method," *Teplofiz. Vys. Temp.* **18**(6), 1184–1191 (1980).
67. E.E. Shpilrain, D.N. Kagan, and S.N. Ulyanov, "Installation for Calorimetric Researches," *Teplofiz. Vys. Temp.* **19**(5), 1040–1044 (1981).
68. E.E. Shpilrain, D.N. Kagan, and G.A. Krechetova, *Caloric Properties of Liquid Binary Systems of Alkali Metals at High Temperatures. Review of Thermophysical Properties of Substances,* No. 4, Teplofizicheski Zentr, Moscow (1978).
69. E.D. West and D.C. Ginnings, "Automatic Temperature Regulation and Recording in Precision Adiabatic Calorimetry," *Rev. Sci. Instrum.* **28**(12), 1070–1074 (1957).
70. M.M. Popov, *Thermometry and Calorimetry,* Moskovski Universitet, Moscow (1954).
71. V.Ya. Rotach, *Calculation of Adjustment of Industrial Control Systems,* Gosenergoizdat, Moscow (1961).
72. B.Ya. Kogan, *Electronic Simulators and Their Application for Study of Automatic Control Systems,* Fizmatgiz, Moscow (1963).
73. O.N. Tikhonov, *Solution of Problems on Automatisation,* Nedra, Leningrad (1969).
74. D.C. Ginnings and E.D. West, in *Experimental Thermodynamics,* Vol. 1, Butterworths, London (1968).
75. E.E. Stansbury, D.L. McElroy, M.L. Picklesimer, G.E. Edler and R.E. Pawel, "Adiabatic Calorimeter for Metals in the Range 50 to 1000°C," *Rev. Sci. Instrum.* **30**(2), 121–127 (1959).
76. V.A. Besekersky and E.P. Popov, *Theory of Automatic Control,* Fitmatgiz, Moscow (1966).
77. D.J. Hudson, *Statistics. Lectures on Elementary Statistics and Probability,* Geneva (1964).
78. G.T. Furukawa, D.C. Ginnings, R.E. McCoskey, and R.A. Nelson, "Calorimetric Properties of Diphenyl Ether from 0 to 570°K," *J. Res. NBS* **46,** 195–199 (1951).
79. G.T. Furukawa, R.E. McCoskey, and G.J. King, "Calorimetric Properties of Benzoic Acid from 0 to 410°K," *J. Res. NBS* **47,** 256–262 (1951).
80. N.N. Sinelnikov, "Vacuum Adiabatic Calorimeter and Some New Data about $\beta = \alpha$ Transition of Quartz," *Dokl. Akad. Nauk SSSR* **92,** 369–372 (1953).
81. D.W. Osborne, E.F. Westrum, and H.R. Lohr, "The Heat Capacity of Uranium Tetrafluoride from 5 to 300°K," *J. Am. Chem. Soc.* **77,** 2737–2742 (1955).
82. E.F. Westrum, in *Advances in Cryogenic Engineering* (K.D. Timmerhaus, ed.), Vol. 7, Plenum Press, New York (1962).
83. E.F. Westrum, E. Suits, and H.K. Lonsdale, in *Advances in Thermophysical Properties at Extreme Temperatures and Pressures,* ASME, New York (1965).
84. R.W. Henson and J.H.W. Simmons, "An Adiabatic Rise Calorimeter Stored Energy in Irradiated Graphite," Atomic Energy Research Establishment Report AERE M/R 2564 (1959).
85. A.A. Sklyankin and P.G. Strelkov, "About Reproducing and Accuracy of Contemporary Numerical Values of Entropy and Enthalpy of Condensed Phase at Standard Temperature," *Zh. Prikl. Mekh. Tekh. Fiz.* **2,** 100–111 (1960).

86. E. Gehring, "Kalorimetrische Untersuchungen an Alüssigen Systemen mit Antimon und Silber, Diss., München (1960).
87. O.A. Kiguradze, "Working out and Study of Standard Means of Measuring of Heat Capacity of the Heat-Isolating Solid Materials at the Temperature Range 80–273°K, VNIIM, Leningrad (1971), thesis.
88. A.G. Boganov, "High Temperature Adiabatic Calorimeter for Measurement of Heat Capacity of Solids," *Peredovoi Nauchnotekh. Proizvod. Opit* **4,** 3–22 (1960).
89. A. Stingele, "Ein neues Hochtemperaturkalorimeter zur Messungen wahrer spezifischer Wärmen bis 700°C. Die Spezifische Wärme des Aluminium bis zum Schmelzpunkt, Diss., München (1961).
90. H.L. Lukas, "Bestimmung der Mischungswärme und ihrer Temperaturabhändigkeit binären metallischen Schmelzen mit einem Adiabatischen Hochtemperaturkalorimeter," Diss., Stuttgart (1961).
91. E. Smidt, "Über die Entwicklungen eines adiabatischen Kalorimeters zur genauen Messung von Spezifischen warmen körniqes und pulverförmiger Stoffe," n. 994, Köln-Opladen, West Deutsch. Verl. (1961).
92. V.P. Kolesov, E.A. Seregin, and S.M. Skuratov, "Small Adiabatic Calorimeter for Measurement of True Heat Capacity for the Range 12–340°K," *Zh. Fiz. Khim.* **36,** 312–315 (1962).
93. F. Gronvold, "Heat Capacity of Indium from 300 to 1000°K. Enthalpy of Fusion," *J. Thermal Anal.* **13**(3), 419–428 (1978).
94. F. Gronvold, "Adiabatic Calorimeter for the Investigation of Relative Substances in the Range from 25 to 775°C. Heat Capacity of α-Aluminium Oxide," *Acta Chem. Scand.* **21**(7), 1695–1713 (1967).
95. M. Krusius, "Calorimetric Measurements of Hyperfine Interactions in Some Metals," Thesis, Helsinki (1971).
96. J. Mayer and T. Waluga, "An Adiabatic Low Temperature Calorimeter for Specific Heat Measurement of Solids and Liquids in Temperature Range 77–400°K," Inst. Fiz. Jadrowej. IFJ Report 750/PL, p. 18, Krakow (1971).
97. R.A. Robie and R.S. Hemingway, *Calorimeters for Heat of Solution and Low-Temperature Heat Capacity Measurements,* U.S. Government Printing Office, Washington, D.C. (1972).
98. H. Weber, "Isothermal Calorimetry for Thermodynamic and Kinetic Measurements," Dissertation 5098, Zürich (1974).
99. W. Gautschi, "Parameter-Schatzung auf Warmeflussmessungen," Abhandl. Diss. ETH 5652, Zürich (1976).
100. W. Hemminger and G. Höhne, *Grundlagen der Kalorimetrie,* Akad. Verl., Berlin (1980).
101. W.F. Roeser and H.T. Wensel, "Reference Tables for Platinum to Platinum–Rhodium Thermocouples," *J. Res. NBS* **10**(2), 275–287 (1933).
102. M. Tanaka and K. Okada, in *Res. Electrotechn. Lab.,* No. 404, Tokyo (1937).
103. A.M. Sirota and B.K. Maltsev, "About Thermocouple Au–Pt," *Izmer. Tekh.* **8,** 27–28 (1959).
104. V.E. Lyusternik, Reproduction of PtRh-Pt Thermocouple Calibration in the Wide Range of Temperatures," *Teplofiz. Vys. Temp.* **1,** 11–16 (1963).
105. G. Greber and S. Erk, *Fundamental of Heat Exchange,* ONTI, Moscow–Leningrad (1936).
106. H.S. Carslaw and J.C. Jaeger, *Condition of Heat in Solids,* 2nd Ed., Clarendon Press, Oxford (1959).
107. A.V. Lyikov, *Theory of Heat Conduction,* Vishaya Shkola, Moscow (1967).
108. S.G. Rabinovich, "Method of Calculation of a Measurement Result Error," *Metrologia* **1,** 3–12 (1970).

109. J. Backhurst, "The Adiabatic Vacuum Calorimeter From 600 to 1600°C," *J. Iron Steel Eng.* **189,** 124–134 (1958).
110. C.R. Brooks, W.E. Norem, D.E. Hendrix, J.W. Wright, and W.G. Northcutt, "The Specific Heat of Copper From 40 to 920°C," *J. Phys. Chem. Solids* **29,** 565–574 (1968).
111. E.E. Stansbury and C.R. Brooks, "Dynamic Adiabatic Calorimetry," *High Temp. High Pressures* **1,** 289–307 (1969).
112. G. Cordoba and C.R. Brooks, "The Heat Capacity of Gold From 300 to 1200 K; Experimental Data and Analysis of Contributions," *Phys. Status Solidi A* **6,** 581–595 (1971).
113. Yeh Ching-Chuan and C.R. Brooks, "The Heat Capacity of Platinum from 350 to 1200 K; Experimental Data and an Analysis of Contributions," *High Temp. Sci.* **5,** 403–413 (1973).
114. W.M. Cash, E.E. Stansbury, C.F. Moore, and C.R. Brooks, "Application of a Digital Computer to Data Acquisition and Shield Temperature Control of a High-Temperature Adiabatic Calorimeter," *Rev. Sci. Instr.* **52,** 895–901 (1981).

13

Heat-Capacity Calorimetry by the Method of Mixtures

DAVID A. DITMARS

1. INTRODUCTION

Thermophysical-property data are indispensable components in the design and technical evaluation of energy-producing and energy-converting systems. They assume yet further enhanced importance in light of both the present world energy crisis and the proliferation of new composite construction materials in modern technology. Because of this, it is necessary that technical and managerial personnel faced with decisions related to the measurement of thermophysical properties have at hand the most current yet all-inclusive information available on the state of measurement technology in this area. This chapter aims to fulfill this need for one well-defined measurement technique: the measurement of the heat capacity of non-reacting solid or liquid systems by the method of mixtures (drop calorimetry).

2. FUNDAMENTAL CONSIDERATIONS

2.1. Need for Heat-Capacity Data

The need for heat-capacity data can arise from several stimuli which range from the very practical to the primarily theoretical:

(A) Heat-capacity data over an extended temperature range are frequently required in calculating the transient and steady-state response of candidate systems in engineering design. Sometimes these data have already been published and are available in raw or evaluated form by consulting one of the prominent

DAVID A. DITMARS • National Bureau of Standards, Washington, D.C. 20234.

sources for thermophysical or thermodynamic property data (e.g., Touloukian and Ho[1]; Stull and Prophet[2]; Gurvich *et al.*[3]; Rabinovich[4]; Hultgren *et al.*[5]). Often, the required data will be nonexistent or of dubious accuracy. For composite materials it is sometimes sufficient to estimate these data additively from already existing data for the various components. If data are to be measured directly, it is important that the composition of the specimen(s) being measured be indeed representative of the composition of the actual material of interest as well as its state in actual use.

(B) If a change of phase (solid–solid or solid–liquid) occurs in the application of a specific material then relative-enthalpy data (which are related to heat-capacity data in a way described later in this chapter) obtained through the calorimetric method of mixtures may be useful in quantitatively characterizing the energy involved in this change of phase. Again, for the enthalpy data on a specific material to be useful, they must represent the same reproducible differences between high-temperature and low-temperature thermodynamic states as obtained in the engineering application of the material. In some cases due to certain sudden temperature changes during these calorimetric measurements, the terminal thermodynamic state after an enthalpy measurement may not be reliably reproducible. Then, either this nonreproducibility must somehow be quantified or another calorimetric procedure must be employed such as solution calorimetry.

(C) In order to carry out thermochemical calculations such as estimating the Gibbs energy change of a proposed chemical reaction, predicting high-temperature equilibria from low-temperature measurements or correcting vapor-pressure measurements, it is necessary to have the thermodynamic functions over a wide temperature range for the substance(s) of interest. If low-temperature (from room temperature to near 0 K) heat-capacity data exist, then relative enthalpy measurements from room temperature to as high a temperature as is technically feasible will suffice for derivation of the thermodynamic functions.

(D) Modern transient techniques of thermophysical property measurement commonly have multiproperty capability. For instance, instead of measuring only thermal conductivity, λ, it frequently proves more convenient to measure the quantity thermal diffusivity, a

$$a = \frac{\lambda}{\rho C_p} \tag{1}$$

where ρ is the density and C_p is the heat capacity of the material of interest. If the thermal diffusivity and the heat capacity are known independently (the latter through calorimetry), then it is possible to calculate from these data and the density, the thermal conductivity.

(E) Direct calculation of the heat capacity of arbitrary solids, a task as yet beyond present solid-state theory, will ultimately depend on more sophisticated knowledge of solid-state interactions than those embodied in the classical Debye theory (harmonic approximation) or in calculations based on an imperfect knowledge of band structures. Contributions to the heat capacity of solids which cause it to differ from the Debye approximation may arise in varying degree from electronic contributions, anharmonic contributions, vacancy or defect contributions as well as from gross atomic displacements within an otherwise stable lattice. Most of these effects become more pronounced at high temperatures ($\gg \Theta_D$). Highly precise and accurate heat-capacity data obtained through conventional calorimetry as well as through transient-measurement techniques provide a fertile area in which new approaches to solid-state heat-capacity theory can be generated or tested.

2.2. *Basic Physical Principles*

It is possible to derive from calorimetric enthalpy data thermodynamic functions such as heat capacity, entropy, and Gibbs energy. This section provides a summary description of the calculation of these functions. In all calorimetric experiments employing the method of mixtures the directly measured quantity is the enthalpy difference (relative enthalpy) between an initial equilibrium thermodynamic state of a system at an elevated temperature and a terminal state at the calorimeter temperature (usually near room temperature). The emphasis, however, is on ***relative*** enthalpy and if circumstances so dictate, it may be desirable that the initial state be at the lower temperature or that the calorimeter be at some temperature much different from room temperature.

Consider the thermodynamic definition of enthalpy:

$$H = U + PV \tag{2}$$

where U is the internal energy, P is pressure, and V is volume. When the system proceeds from an initial to a final equilibrium state,

$$dH = dU + PdV + VdP \tag{3}$$

However, the heat dQ transferred to this process,

$$dQ = dU + PdV \tag{4}$$

Therefore

$$\frac{dH}{dT} = \frac{dQ}{dT} + V\frac{dp}{dT} \tag{5}$$

where T is the thermodynamic temperature. Under isobaric conditions,

$$\left(\frac{dH}{dT}\right) = \left(\frac{dQ}{dT}\right)_p \equiv C_p \tag{6}$$

The change in enthalpy, dH, under isobaric conditions is equal to the heat transferred to or from the system,

$$H_2 - H_1 = Q = \int_1^2 C_p \, dT \tag{7}$$

State 2, the initial state, is assumed to be characterized by some arbitrary temperature T; state 1, by a convenient calorimeter reference temperature T_r (commonly 298.15 or 273.15 K). With this notation,

$$S_T = S_{T_r} + \int_{T_r}^{T} \frac{C_p}{T} \, dT \tag{8}$$

and

$$-\frac{(G_T - H_{T_r})}{T} = S_T - \frac{(H_T - H_{T_r})}{T} \tag{9}$$

where S is the entropy and G, the Gibbs energy function. One can only calculate from drop calorimetric data the relative entropy increments to temperatures above or below T_r. The entropy at T_r is customarily obtained from low-temperature adiabatic calorimetric studies, assuming the third law.

The relationship between the measured relative enthalpy and the heat capacity are also shown in Fig. 1. Usually, differences in experimental technique lead to more or less of a mismatch in all derivatives of the low- and high-temperature heat-capacity functions at T_r. This situation must be resolved by

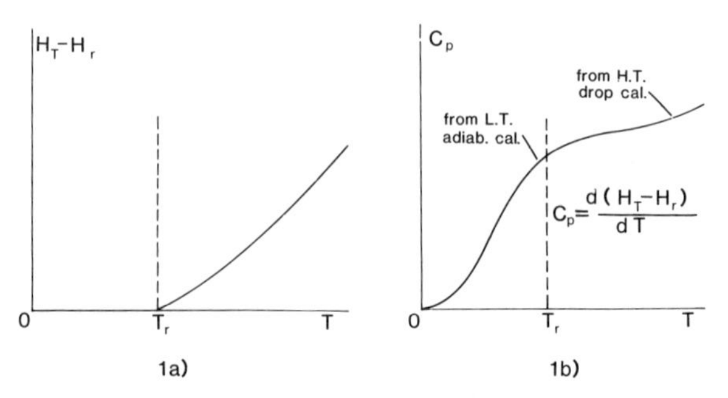

FIGURE 1. Derivation of high-temperature heat capacity from enthalpy measurements; correlation with low-temperature data.

detailed consideration of the sample analysis and the sources of inaccuracy in the individual measuring techniques, leading to the assignment of weight to each datum prior to final smoothing.

3. THE METHOD OF MIXTURES

3.1. General Considerations

The term "method of mixtures" was originally derived from early calorimetric experiments in which specimens were dropped into a water bath and the relative enthalpy calculated from the difference between the final and the initial bath temperatures. In fact, if one is only interested in relative enthalpies rather near to room temperature and is willing to accept one to two percent inaccuracy, then some version of this method may still be an appropriate approach.

More often, one is interested in obtaining relative enthalpies over a temperature range of several hundred degrees with inaccuracies from one to several *tenths* of a percent. This entails a substantial increase in construction time, in sophistication of instrumentation and certainly in cost.

Considerably more scientific and technical awareness on the part of the operator will also be required if the apparatus is to fulfill its potential. These more complex calorimeters may still be thought of in an extended sense as employing a "method of mixtures" in that they all depend in their basic principle on the fairly rapid translation of the specimen from some exterior, temperature-controlled zone into a calorimeter held at a well-characterized equilibrium state. The change in state of the calorimeter then becomes a measure of the heat transferred from the specimen to the calorimeter. To obtain a relative enthalpy datum from such an experiment, means must be at hand for evaluating the heat loss from the specimen during its translation, the loss or gain of heat from the environment by the calorimeter ("heat leak") as well as the dependence of the calorimeter state parameters upon the net heat transferred to the calorimeter. This latter dependence of the state parameters is commonly referred to as the "calorimeter calibration" and accurately measured electric energy is most often used as the calibrating heat source. For brevity, in the remainder of this chapter, we shall refer to calorimeters which operate on the above principle as "*drop calorimeters*".†

†The very special discipline of "levitation calorimetry," while conceptually a variant of the method of mixtures, is characterized by a highly specialized technique of sample heating (electronic levitation) and is treated at length in Chapter 14 of the present volume.

3.2. Points to Consider Before Choosing a Drop Calorimeter

3.2.1. Availability

Drop calorimetry has been historically and continues to be largely a "do-it-yourself" task as far as apparatus is concerned. From time to time there have been attempts to produce drop calorimeters commercially; however, these have not shown continuous economic viability due to a restricted market. The situation is vastly different from the one obtaining in the DSC (differential scanning calorimetry) field described in Chapter 17; there are at present time a number of commercially available DSC calorimeters. These instruments rely heavily in their basic measuring principle on heat-capacity standard reference materials which can only be realized through drop calorimetry, or adiabatic calorimetry. In contrast, the acquisition of a drop calorimeter may involve a good deal of time spent in thermal design and shop construction even after all the raw materials and components have been assembled. The overall time from conception of a design to testing the finished calorimeter may vary from two to three years dependent on the shop facilities available as well as on the number and expertise of personnel active in the project. This has had one beneficial effect: experimenters who go to these lengths are usually anxious to document their efforts and thus much valuable technical information and experience find their way into the literature. The ultimate purpose of this chapter and of the complementary presentation in Volume II is not only to offer guidance in the selection, but through the presentation of designs and operating guidelines to shorten considerably this overall time.

3.2.2. Temperature Range and Overlap with Other Techniques

Drop calorimeters are in general limited in their temperature range (i.e., the maximum specimen temperature at which they can produce acceptable data) only by *specimen* limitations, e.g., specimen decomposition or a specimen's reaction with its high-temperature environment. Modern refractory materials together with present-day vacuum, heating, and temperature-control technology (including electron-beam, laser, and induction heating as well as conventional resistive heating) permit the design of furnaces to heat specimens to temperatures exceeding 2000 K.

The lower end of the usual drop-calorimeter temperature range (near room temperature) can overlap for a few tens of degrees the uppermost part of the low-temperature adiabatic calorimeter range. Since the drop calorimeter shows its greatest inaccuracy at the lower end of its temperature range, greater weight is most often given to heat-capacity data from low-temperature adiabatic calorimetry when an overlap occurs. The drop-calorimeter temperature range includes completely those temperatures attainable by high-temperature adiabatic calorimetry of both the equilibrium and the continuous-heating variety. High-temperature adiabatic calorimeters, if designed with very careful attention to controlling heat-flow paths, and particularly, to heat exchange by radiation, have inaccuracies comparable to or slightly better than that of the most accurate drop calorimeters, below about 600–700 K. At higher temperatures inaccuracy due to radiation heat loss becomes a limiting factor in adiabatic calorimetry. From about 1000 K to about 1600–1700 K, at which temperature transient, electric pulse-heating techniques become possible, drop calorimetry is the only widely available technique for measuring relative enthalpy (and hence, heat capacity). It should be noted that these electric pulse techniques have only been applied successfully to solids having high electrical conductivity. This situation may change in the near future, at least as regards measurements on solids, due to recent developments in the area of laser-flash calorimetry (Takahashi *et al.*[6]). At the time of this writing, this transient method appears to have an inaccuracy from two to four times as great as that of the best drop calorimetry.

3.2.3. Characteristics and Some Advantages and Disadvantages of Drop Calorimeters

Drop calorimeters have certain features in common, only some of which they share with other calorimeter types described in this volume. These can be conveniently grouped as they relate to construction and maintenance of the calorimeter. In addition there are factors peculiar to each type of drop calorimeter which will be considered below in the description of the individual types of calorimeters.

3.2.3a. Construction and Maintenance Factors. It is imperative that the laboratory planning the installation of a drop calorimeter have close access to a high-quality machine shop with full capabilities for metal turning, drilling, milling, and joining (as by welding, brazing, or soldering). During the construction process, one should not underestimate the value of close rapport between the machinist and the designer. While in theory it is possible to specify everything through formal technical drawings, it is an exceptional designer who cannot benefit from the practical experience of a skilled machinist. This frequently leads to improvements in design or, at least, uncovers cost-cutting modifications.

3.2.3b. Sample Factors. Drop calorimetry is a suitable method for relative enthalpy measurement on a very broad spectrum of materials: metallic or nonmetallic, either massive or in any state of subdivision, and solid or liquid (or a mixture thereof), with virtually no restriction on either the range of material transport properties or on radioactive behavior. In the customary applications of drop calorimetry, the specimens are encapsulated, sometimes hermetically while under reduced gas pressure. Such encapsulation prevents loss of specimen material as well as its reaction with the environment. The only absolute requirement is that the sample be nonreactive with its container, if one is used. It is highly desirable that the sample have a low vapor pressure, both to avoid mass loss and to circumvent the necessity of correcting for a heat of vaporization. If the sample container is properly designed (essentially, this implies sufficiently thick walls and low surface emissivity) then all the heat "lost" (i.e., not transferred to and measured in the calorimeter) during the translation from the furnace to the calorimeter will come from the container and the true sample heat can be calculated from the difference between the gross measured heat of the sample–container combination and that of the empty container alone, measured in a second experiment. If one is dealing with a massive, unencapsulated sample of a simple geometry and for which good emissivity data are available, it may also be possible to calculate directly the heat lost during translation (Kirillin *et al.*[7]). If a sample container is used, it should have a total heat capacity less than that of the sample by a factor of several. Otherwise, any uncertainty in the heat content of the container may make a sizable contribution to systematic error. For most metals, oxides, and refractory materials, this implies a minimum sample volume of from two to ten cubic centimeters.

The initial cooling rate of the sample in the calorimeter may approach 1000 to 2000 K/min. Experience has shown this rate to be sufficiently rapid that some solid–solid transitions cannot be completed in the time available and the sample becomes "frozen" in varying degrees in metastable states. Thus for some samples exhibiting these transitions, the cooled sample in the calorimeter may be in an uncertain thermodynamic state. This effect is likely to show up as imprecision in the enthalpy data. There are two remedies available to the experimenter: alter the calorimeter design to reduce the heat-transfer coefficient between sample and calorimeter (thus, reducing the cooling rate), or perform a second experiment (solution calorimetry) on *each* measured specimen to refer all specimens to a common thermodynamic state. The latter remedy leads to a protracted and laborious procedure for each enthalpy datum which of course cannot be repeated since the specimen in question will have been consumed.

3.2.3c. Operational Factors. Drop calorimetry is a method well-suited to the accurate enthalpy measurement of those solid–solid and solid–liquid phase changes which show evidence of proceeding to a repeatable thermodynamic state each time the sample in question is cooled in the calorimeter. This is so because the initial state of the sample can be quite precisely controlled by allowing sufficient equilibration time in the furnace. In this regard, one must use caution to allow sufficient equilibration time, since it may be required (as near a melting transition) that substantial amounts of heat be transferred via a very small temperature gradient.

Drop calorimetry can be a quite tedious method due to the number of manual operations which must be performed for each measurement, and due to the relatively long fore- and after-periods connected with each measurement, during which the calorimeter heat leak and the furnace temperature are controlled or monitored. Automation or computer control are applicable at certain phases of each measurement, for instance in controlling and recording furnace temperature or calorimeter temperature or in recording calibrating electric power. The actual translation of the sample to the furnace together with the opening and closing of such radiation gates as may exist can be automated to a certain extent, but is frequently left as an operator function. A drop calorimeter operating at peak efficiency under the most favorable circumstances ordinarily is capable of completing a few measurements per working day, say four or five, with full attendance of one operator, but the long-term average is likely to be less. It is not a technique well-suited to acquiring a mass of data in a short time.

Although in principle is may be possible to calibrate a drop calorimeter using a heat-capacity standard reference material such as α-Al_2O_3 (also called in the literature synthetic sapphire, alumina, or corundum) or Mo, a more accurate calibration can be made electrically. There are specific requirements, however, which must be met for an accurate calibration. The design of the calibrating heater should be such that during operation, it closely simulates the effect of inserting a hot sample into the calorimeter. The spatial configuration of the heater may be a very important consideration, as will be the design and placement of heater-current and potential leads. Additionally, there are certain basic instrumentation requirements for any laboratory engaged in electrical calibrations: a stable source of dc current, voltage and resistance standards whose calibration is traceable to a major standards laboratory, calibrated dc-voltage measuring instruments, and a precise counter for time-interval measurement. All this instrumentation should be carefully chosen to be compatible in producing the desired level of accuracy in the measured power. As regards the calorimeter operating environment, drop calorimeters can be highly sensitive pieces of research equipment and their correct functioning will almost always be degraded by vibration or drastically fluctuating temperature or humidity. The nature of these apparatuses also has implications for the quality of operating personnel. In order to realize the greatest chance of consistently accurate performance, it is desirable that the operator or at least someone who supervises him closely be conversant in the fundamental physical and chemical principles underlying the particular apparatus as well as the thermodynamic description of the material under investigation.

4. CLASSIFICATION AND DESCRIPTION OF DROP CALORIMETERS

4.1. Prior Reviews

The very general nature of the present review precludes setting before the reader the mass of technical detail which he must consider before arriving at a specific design of drop calorimeter; this detailed information will form the substance of Volume II. Some of this material can be found in the reviews of calorimetric techniques which already exist in the literature (Sturtevant,[8] McCullough and Scott,[9] Ginnings,[10] Kubaschewski and Alcock[11]). Consultation of these reviews is advised; for even though one may sometimes find only a specific facet of calorimetry treated, such material will frequently suggest to the calorimetrist alternative solutions to problems of apparatus design.

In this section, various types of drop calorimeters are described and illustrated by functional, schematic diagrams. Distinctive features of each type are emphasized. Since each feature and variant cannot be covered in the limited framework of the present review, we present in the section following this one a comprehensive list of bibliographic citations (with comments) wherein specific apparatuses are discussed in detail. These are taken, in the main, from the literature covering the period 1967 to May 1979. Citations from the literature before 1967 were included as well if they seemed to be of special historic interest or illustrated important technical points.

4.2. Classification of Drop Calorimeters: Operating Principles

A calorimeter can be thought to consist of an active measuring part and its immediate, enclosing surroundings. In this context, one useful method of categorizing most drop calorimeters is based on the temperature distribution within the calorimeter. Thus, one has *isoperibol* calorimeters in which the enclosure is controlled at a constant temperature. In this case, the calculation of the heat exchange between the measuring part and the enclosure is an important part of each sample heat measurement. If, on the other hand, the average temperature of the enclosure is controlled to equal at all times that of the measuring part, heat exchange between the two is minimized and we have an *adiabatic drop* calorimeter or *adiabatic receiving* calorimeter. If the measuring part is a two-phase system and heat transfer between the sample and the measuring part results in a change in the phase distribution at constant temperature, one has an *isothermal* calorimeter. The phase change can be either solid–liquid or liquid–vapor. Note that even in these "isothermal" calorimeters, however, small temperature differences on the order of millidegrees can exist and correct evaluation of each measurement may depend on an understanding of the effect of these differences.

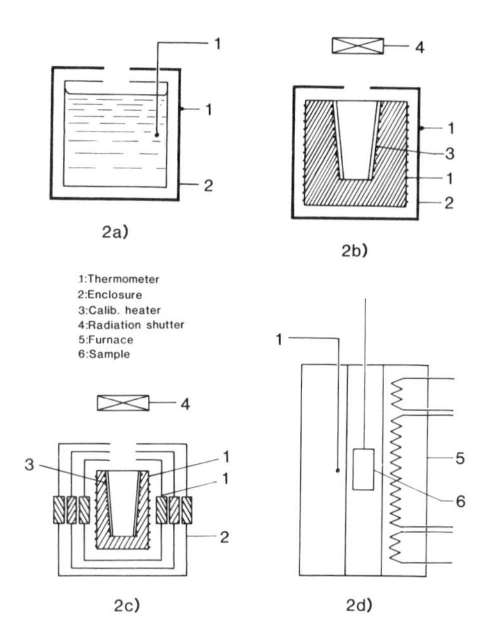

FIGURE 2. Schematic configuration for nonisothermal calorimeters in the method of mixtures: (a) liquid-bath calorimeter, (b) Isoperibol metal-block calorimeter, (c) adiabatic receiving calorimeter with radiation shielding, (d) furnace.

4.2.1. Liquid-Bath Calorimeter

In this class of drop calorimeter (Fig. 2a) the main measuring part consists of a liquid bath, so chosen in size to give a conveniently measurable temperature rise for the least amount of heat anticipated from a single measurement. The enclosure usually consists of another liquid bath – this one, thermostatted. Both baths can be stirred to promote temperature equilibration but in this case, the work of stirring must be accounted for in the inner bath. During the operation of this type of calorimeter, it is important to reduce or eliminate loss of mass or heat through either splashing or vaporization.

The construction and instrumentation of this type of calorimeter can be quite simple. Ordinarily, water is used as the calorimetric substance, though other liquids are also possible, provided their heat capacities are known independently. An electrical calibration is possible in principle but probably unnecessary at the modest level of accuracy attainable with this type of calorimeter. Alternatively, one could use a solid of known heat capacity in this arrangement to measure the liquid heat capacity. The accuracy attained with this apparatus will depend mainly on the accuracy with which the true sample temperature can be measured and the extent of one's ability to accurately account for heat lost during the sample drop as well as for heat exchange between the inner bath and the enclosure.

With reasonable care, an uncomplicated apparatus of this type should show an inaccuracy level in heat measurements no greater than one to two percent. For many applications, this level of inaccuracy is tolerable. This type of calorimeter which employs direct immersion is, however, limited in the maximum feasible sample temperature. This is likely to be about 300°C. Instead of using the simple direct immersion method, one can drop the sample into a cavity immersed in the calorimetric fluid. A liquid calorimeter working in this fashion with the added feature of an adiabatic shield has been used in enthalpy measurements to 2700 K. (See Section 4.3.)

A single furnace has been shown schematically in Fig. 2d. This could be used in conjunction with either of the two calorimeters shown as well. Furnaces with electric resistive heating has been used as well as those operating on inductive (eddy-current) or electron-bombardment principles. Furnace temperatures are measured with platinum resistance thermometers or thermocouples (usually Pt/Pt–10 Rh pairs) to about 800 K. Above this temperature, thermocouples are available for service to above 2000 K. Optical pyrometry can be used for temperature measurement at 1200 K and above.

4.2.2. Isoperibol Metal-Block Calorimeters

This widely applied drop-calorimeter type is schematically shown in Fig. 2b. Here, the use of a solid calorimetric substance eliminates any errors due to liquid mass loss. Copper is the metal most frequently chosen for the block material due to its ready availability, easy machinability, and high thermal diffusivity. Nickel and aluminium have also been used for this purpose. The advantage conferred by the high thermal diffusivity of silver (50% greater than that of copper) is usually vitiated by the initial cost consideration.

The higher metal thermal diffusivity reduces temperature gradients (compared to those in a liquid medium) and allows rapid temperature equilibration after addition of a hot sample. These calorimeters are useful for enthalpy measurements at any sample temperature above about 370 K. The designs in this class of calorimeter vary from exceedingly simple to highly complex, depending on the measurement accuracy desired. In many designs, the metal block has been gold- or nickel-plated in order to reduce radiant heat exchange between block and enclosure. The constant-temperature enclosure usually depends on a stirred, thermostatted liquid bath. At temperatures such that radiation from the furnace could be a significant component of the calorimeter heat-leak, one or more radiation gates (retractable, to allow entry of a sample) are interposed between furnace and calorimeter. Resistance thermometers – either of the capsule variety inserted in a hole in the block or of the distributed variety, wound over the exterior surface of the block – are the usual means of measuring temperature. Thus, precision resistance-measuring apparatus will be required for operation of the calorimeter. These calorimeters are almost universally calibrated electrically by a heater situated so as to produce nearly the same temperature effect on the block as a warm sample. The overall inaccuracy of an enthalpy measurement depends again on the inaccuracy in the temperature assigned to the sample as well as on the experimenter's ability to accurately account for heat loss during the time of translation into the calorimeter and for heat exchange between the block and its enclosure. As a rough guide, if a block calorimeter is carefully designed and instrumented, enthalpy inaccuracy of not more than 0.2% to 0.4% should be attainable up to 1200 K. From 1200 K to about 3000 K, overall inaccuracy of the enthalpy measurements will increase from about 0.4% to about 2% to 3%. This increase in inaccuracy comes largely from uncertainty in the heat loss during translation of the sample at the higher temperature to the calorimeter.

4.2.3. *Adiabatic Receiving Calorimeters*

In adiabatic receiving calorimeters (Fig. 2c) the massive isothermal enclosure of the isoperibol, metal-block calorimeter is replaced with a usually less-massive heated one whose average temperature is at all times matched closely to that of the measuring part contained within. The measuring part is similar in many respects to that of the usual block isoperibol calorimeter described above, though a liquid bath has been used on occasion (see Section 4.3). The measurement of a good "average temperature" in the above context depends strongly on the thermal design of the calorimeter and on the surface distribution of the temperature-sensing elements. In light of the currently available control and recording technology, it must be considered inefficient and impractical to attempt the manual operation of any adiabatic calorimeter. Therefore, more automatic control instrumentation (than, say, is needed for an isoperibol calorimeter) will be required to maintain "adiabatic" conditions at all times within the calorimeter.

Temperatures of the calorimeter measuring portion are obtained through conventional resistance thermometry or through quartz-crystal thermometry. The latter method uses the frequency dependence on temperature of a quartz-crystal oscillator imbedded in the calorimeter. These methods require appropriate instrumentation for precision resistance or frequency measurement. In the temperature range from 1200 to 2000 K a well-designed adiabatic-receiving calorimeter should be capable of enthalpy measurements with an inaccuracy not exceeding 0.6%.

4.2.4. *Phase-Change (Isothermal) Calorimeters*

4.2.4a. Solid–Liquid Systems. In this class of calorimeter, heat from a sample is transferred to the calorimeter working fluid, bringing about at constant temperature a change either in phase or in phase distribution within this fluid.

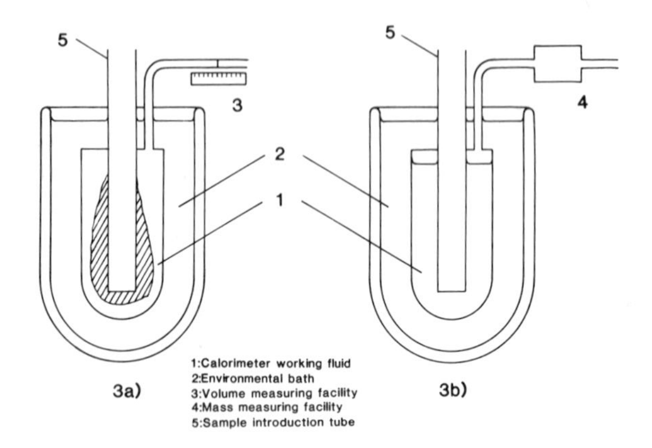

FIGURE 3. Schematic configuration for isothermal, phase-change calorimeters in the method of mixtures: (a) solid–liquid system, (b) liquid–gas system.

The temperature of the phase change having been well-established beforehand, no further temperature measurement within the calorimeter is necessary. The temperature of the phase change becomes the reference temperature for enthalpy measurement. One is then only concerned with the sample temperature in the furnace. Illustrated in Fig. 3 are the measuring principles of solid–liquid and of liquid–gas phase-change calorimeters.

Figure 3a is essentially a sensitive dilatometer. Heat transferred between a sample in the entry tube and the calorimeter working fluid will cause either melting or freezing of the working fluid with the attendant change in volume (measured at 3). This change in volume becomes a measure of the heat transferred after an electrical calibration of the calorimeter. In Fig. 3b, heat transferred to the working fluid causes vaporization, and the mass vaporized (measured either as a gas flow at 4 or after subsequent condensation) is the measure of the heat transferred, given the heat of vaporization of the working fluid.

Since both types of calorimeter illustrated in Fig. 3 rely ultimately on the repeatable thermodynamic properties of the working fluid for their accuracy, it is of considerable importance that all phases of these fluids have the same composition. One way to achieve this is to require that the calorimetric working fluid of Fig. 3a be both gas- and solute-free. The fluid employed in Fig. 3b should likewise be free of other species in solution. The two most common working fluids for solid–liquid phase change calorimetry have been water and diphenyl ether (DPE) due to their ready availability and the ease with which they can be purified. The equilibrium temperatures of these two types are roughly 273 and 300 K, respectively. Other fluids have been proposed and used (see Section 4.3) but none so widely. The environmental baths are usually water ice in the case of the ice calorimeter and a stirred, thermostatted liquid bath in in the case of the DPE calorimeter. The temperature fluctuations permissible in these baths is small (0.001 K or less) since the steady-state heat exchange ("heat leak") between the working fluid and the environmental bath usually takes place over a temperature difference of at most a few millidegrees. The solid phase is most often formed prior to an experiment by introducing dry ice into the sample tube. Usually, one is concerned as to the spatial distribution of the solid phase (the "mantle"): whether it covers the sample tube (and any fins) completely; when are direct paths melted through it to the bulk liquid phase; has supercooling occurred in freezing? For this reason, partial- or all-glass construction is frequently used in these calorimeters, and it is useful to have access to glass-working facilities.

Mercury is used as the means to determine the total volume change of the working fluid. It is introduced into the space with the working fluid, confining and maintaining the total amount of the working fluid constant. The mercury communicates with an external mass-accounting system via a small capillary and facilitates measurement of volume changes of the working fluid. The mercury does not significantly contaminate either water or diphenyl ether and permits a very accurate measure of this volume change to be made in the form of a measured mass change of mercury at the external system. With this feature, the "calibration factor" (which expresses the equivalence between the heat absorbed by the calorimeter and the change of mercury mass) is given by

$$\Gamma = H_f/[(V_s - V_l)/\rho_{\mathrm{Hg}}]$$

where Γ is the calibration factor expressed in heat units per unit mass of mercury, H_f is the heat of fusion of the calorimeter working fluid, V_s and V_l are the specific volumes of the solid and liquid phases, respectively, and ρ_{Hg} is the density of mercury at the calorimeter equilibrium temperature. Γ is most conveniently measured by a direct electrical calibration. For ice calorimeters, it has a value of about 270.5 J g^{-1} and for diphenyl ether calorimeters, 79.1 J g^{-1}. The diphenyl ether calorimeter thus has about four times the sensitivity of the ice calorimeter. One ice calorimeter in use for many years at NBS has consistently shown an imprecision of about ± 0.5 J in drop calorimetric heat measurements and this imprecision has been independent of the amount of heat measured. The heat leak of this calorimeter varied from day to day but has generally been less than 300 μW. Under ideal conditions, the estimated overall inaccuracy of heat measurement at temperatures up to 1200 K has been 0.2%.

Operation of either of these types of solid–liquid phase-change calorimeter usually involves continual handling in the open of mercury. This demands good laboratory technique and constant caution on the part of the operator to avoid the creation of a toxic mercury hazard. Besides the usual furance-control and temperature-measuring instrumentation, one must have a good analytical balance (preferably one that reads directly to at least 0.1 mg) and a cathetometer or comparable scale-reading device for observation at the mercury-accounting system of the effect of the calorimeter heat leak. The overall time for acquisition of data is likely to be somewhat longer than that for the block calorimeters because of the additional time (about one day) needed to form solid mantles from the liquid working fluid and the continual care needed to keep all parts of the mercury accounting system clean.

4.2.4b. Liquid–Vapor Systems. Liquid–vapor phase-change calorimeters (Fig. 3b) have been used rather infrequently in recent years for drop calorimetry and then have almost invariably employed low-boiling-point liquefied gases (such as nitrogen, helium, or argon). There appears to exist only one example of boiling-water calorimeter (see Section 4.3). The claims and evidence advanced for its inaccuracy, not exceeding 1.5% on enthalpy measurements at 3100 K, would certainly make it competitive with other high-temperature drop-calorimetric techniques.

Liquefied gas calorimetry has two notable advantages: (A) The apparatus construction is quite simple, consisting principally of a closed Dewar vessel into which a sample tube is inserted; (B) use of an inert gas such as argon provides a desirable environment for high-temperature systems which might otherwise react with the atmosphere or with other calorimeter parts. On the other hand, one does need the availability of liquefied gases, and the instrumentation necessary to measure the gas flow from the calorimeter can be moderately involved. One reported investigation used, besides automatic data acquisition components, a laminar flow unit, a pressure transducer, and a capacitance-type electronic manometer. The limited amount of data reported from this type of calorimeter show it to have apparently an inaccuracy about twice that of other types of drop calorimeters used over the same temperature range.

4.2.5. Other Drop Calorimeters

The calorimeter types described in Sections 4.2.1.–4.2.4. have historically comprised the bulk of drop calorimeters. Types employing different principles are usually advanced as solutions to some peculiar technical requirement of the system under study. Seven bibliographic citations to these miscelleneous types are presented below but are not discussed in detail here. It is significant that five of these most recent citations feature the application of a Calvet-type twin heat-flow calorimeter in drop calorimetry for the express purpose of reducing sample size and operating time for an individual experiment. This has import for future developments in drop calorimetry which are discussed below.

4.3. Bibliographic Citations for Apparatus Descriptions

The coverage of the calorimetric literature from 1966 to May 1979 is quite complete. A computerized search for this period was conducted using seven different commercially available data bases with all conceivable key words which might lead to literature references. (See Table 1.) Citations prior to 1966 were included only if of special value. All citations are given in complete form in Section 5.

5. CURRENT ACTIVITY AND FUTURE PROSPECTS

The drop-calorimeter types described in Sections 4.2.1–4.2.5 above have provided a large fraction of our current data base covering the high temperature heat capacity of the elements and their technologically important compounds. When applied in a sophisticated manner, they are still unsurpassed in their precision and accuracy of measurement. However, they have the disadvantage of being slow producers of data and of being poorly suited to accommodate extreme conditions of temperature and pressure. It is in precisely this context that conventional drop calorimetry meets its greatest challenge. Sometimes one is faced with the necessity of choosing one material from among many. If heat capacity is an important factor in the choice, it is often more cost-effective to have in a relatively short time a large set of less-accurate data than a lesser set of highly accurate data. This accounts in part for the current efforts to develop more reliable techniques in the areas of continuous-heating and differential-scanning calorimetry, multi- (thermophysical) property measurements and transient heat measurements.

It is not likely that conventional drop calorimetry will be supplanted. It *is* likely that in the future it will function at a much lower level of activity than heretofore. Even within this contracted sphere of activity, emphasis can be anticipated on design and instrumentation features which increase the efficiency of the drop method. Recent efforts to speed up this sluggish technique have relied heavily on comparative measurements and have concentrated on using smaller measuring samples, discarding thermally ponderous calorimeters in favor of smaller, more sensitive instruments such as those of the Calvet twin type.

Although the use of a container in conventional drop calorimetry does confer certain advantages, it is a significant hinderance at very high temperatures when sample and container may react with one another. This has led to the introduction of the containerless liquid-sample experiments of levitation calorimetry. These are in turn limited to good electrical conductors. One approach to the problems of high-temperature heat-capacity measurements on non-electrical conductors which might react with a containment is to perform the

TABLE 1
Bibliographic Citations for Apparatus Descriptions

Method	Citations	Remarks
A. All-liquid calorimeters	Barton, 1938[12]	Elementary presentation of method of mixtures
	Eitel, 1952[13]	Calorimeter and furnace for moderate accuracy
	Andrews, 1926[14]	Ingenious use of twin calorimeter principle Calibrated with silver as standard reference material
	Conway and Hein, 1965[15]	Liquid bath calorimeter with adiabatic shield Sample temperature to 2700 K
B. Metal block isoperibol calorimeters	Southard, 1941[16]	One of the best early descriptions of a copper block calorimeter
	Dworkin and Bredig, 1960[17]	Claims 1%–2% accuracy for heats of fusion at $T > 700$ K
	Kirillin *et al.*, 1961[18]	Uses 2300 K furnace; detailed apparatus cross section
	Fomichev *et al.*, 1962[19]	
	Chekhovskoi and Sheindlin, 1963[20]	Graphite tube furnaces only; for use to approximately 3400 K
	Sheindlin *et al.*, 1963[21]	
	Guseva *et al.*, 1966[22]	
	Fredrickson *et al.*, 1969[23]	Electron-beam heated furnace.
	Marchidan and Ciopec, 1969[24]	A thorough exposition of an elementary design.
	Marchidan and Ciopec, 1971[25]	Calibrates calorimeter with α-Al_2O_3
	Sheindlin *et al.*, 1970[26]	Calorimeter with furnace for use at 3700 K.
	Oetting, 1970[27]	Presents an exhaustive analysis of method of accounting for heat transfer in isoperibol calorimetry. A "must" for the most accurate work.
	Chekhovskoi and Berezin, 1970[82]	
	Lindroth and Krawza, 1971[28]	

	Stout *et al.*, 1973[29]	
	Chekhovskoi *et al.*, 1974[30]	
	Blachnik *et al.*, 1974[31]	
	Buchnev *et al.*, 1974[32]	
	Proks *et al.*, 1977[33]	Exemplary description of calorimeter
	Chemykhim *et al.*, 1978[34]	Metal block calorimeter for liquid metals; liquid specimen poured directly from aluminium oxide crucible into calorimeter
C. Adiabatic receiving calorimeters	Olette, 1958[35]	Uses massive stirred liquid enclosure as the adiabatic shield
	Levinson, 1962[36]	Furnace temperatures to 2800 K with graphite tube furnace; describes dropping mechanism
	West and Ishihara, 1965[37]	Induction heating of sample; relies on multiple radiation shields around measuring part; sample automatically lifted into calorimeter.
	Dennison *et al.*, 1966[38]	
	Macleod, 1967[39]	
	Tydlitat *et al.*, 1971[40]	
	Spedding and Henderson, 1971[41]	Uses spherical furnace design.
	Grønvold, 1972[42]	Follows closely design of West and Ishihara[37]
	Efremova and Matizen, 1972[43]	
	Arkhipov, 1974[44]	
	Ditmars *et al.*, 1977[45]	More information on operation of calorimeter of West and Ishihara[37]
D. Phase-change calorimeters, solid–liquid (water)	Swietoslawski, 1946[46]	Detailed consideration of processes internal to the ice calorimeter
	Ginnings and Corruccini, 1947[47]	First "modern" ice calorimeter; great detail on electrical calibration
	Ginnings *et al.*, 1950[48]	Major revision of 1947 calorimeter above; much construction detail
	Foley and Giguère, 1951[49]	
	Leake and Turkdogan, 1954[50]	Detailed construction and operation details

TABLE 1 (Continued)

Method	Citations	Remarks
D. Phase-change calorimeters, solid–liquid (water) (cont.)	Oriani and Murphy, 1954[51]	Detailed construction and operation details
	Deem and Lucks, 1958[52]	All-metal construction of ice calorimeter
	Lucks *et al.*, 1960[53]	Presents two simple furnace designs
	Spedding *et al.*, 1960[54]	
	Smith *et al.*, 1961[55]	An excellent primer for ice calorimeter construction, operation
	Colwell and Halsey, 1962[56]	Ice calorimeter for enthalpies below 273.15 K
	Kokes *et al.*, 1962[57]	The simplest possible student calorimeter
	Neel *et al.*, 1962[58]	
	Welty and Wicks, 1962[59]	
	Opdycke *et al.*, 1966[60]	First attempt to continuously monitor mass of mercury in calorimeter
	Maglić and Perrot, 1968[61]	
	Ditmars, 1976[62]	Information on electrical calibration and heat leak
	Zakurenko and Kuz'michev, 1978[63]	Special reading unit for calorimeter
D. Phase-change calorimeters, solid–liquid (diphenyl ether)	Giguère *et al.*, 1955[64]	
	Jessup, 1955[65]	Describes construction, filling, operation, calibration
	Hultgren *et al.*, 1958[66]	
	Peters and Tappe, 1968[67]	Exhaustive description and analysis of a large DPE calorimeter
	Davis and Pritchard, 1972[68]	Another exhaustive study; painstaking attention to detail
D. Phase-change calorimeters, solid–liquid (other)	Lakhanpal and Parashar, 1970[69]	Uses cetyl alcohol as dilatometric fluid
D. Phase-change calorimeters, liquid–vapor	Gilbreath and Wilson, 1970[70]	Not a drop calorimeter but presents useful instrumentation for liquid nitrogen vaporization as a measure of heat

	Shpil'rain *et al.*, 1972[71]	Uses water as working fluid; calorimeter calibrated with tungsten as a standard reference material
	Stephens, 1974[72]	Uses argon as working fluid; thorough apparatus description
	Stephens, 1977[73] Stephens, 1978[74]	Not a drop calorimeter but uses CCl_3F (freon-11) as a working fluid to measure heat of radioactive decay
E. Miscellaneous types	Orehotsky and Schröder, 1970[75]	Calorimeter combining features of radiation and drop calorimetry
	Konicek *et al.*, 1971[76] Suurkuusk and Wadsö, 1974[77]	Fully automated drop calorimeter for small ($\leqslant 1$ g) solid or liquid samples near room temperature; based on twin heat-flow calorimeter; 0.01% precision, claims 0.1% accuracy
	Fontana and Winard, 1972[78]	Small block calorimeter contained within a furnace so that enclosure can be thermostatted at temperatures to 1300 K
	Arkipov *et al.*, 1975[79]	Receiving part designed as a Calvet microcalorimeter; for small samples (~ 500 mg) at 350 to 1100 K; features "rapid" measurement (~ 15 min per datum)
	Tamura *et al.*, 1975[80]	Features "transposed" method in which room temperature sample is dropped into a high-temperature Calvet twin, heat-flow calorimeter
	Morizur *et al.*, 1976[81]	Calvet conduction calorimeter used for measurements to 3000 K; claims 6% accuracy

measurements in the microgravity environment of outer space. At this writing several proposals to accomplish this are being actively pursued.

In the area of temperature measurement – one of the major sources of inaccuracy in drop calorimetry – intensive development is going on in platinum resistance thermometry for use at temperatures up to the gold point (~ 1337 K). It is likely that the next revision of the International Practical Temperature Scale will adopt the platinum resistance thermometer as the defining instrument of this temperature scale up to the gold point.

NOTATION

a	Thermal diffusivity
°C	Celsius temperature
C_p	Heat capacity at constant pressure
G	Gibbs energy function
H	Enthalpy
H_f	Enthalpy of fusion
J	Joule
K	Kelvin
k	Thermal conductivity
l	Subscript indicating liquid phase
mg	Milligram
P, p	Pressure
Q	Amount of heat
r	Subscript indicating a reference enthalpy, temperature, etc.
S	Entropy
s	Subscript indicating solid phase
T	Thermodynamic temperature
U	Internal energy
V	Volume
W	Watt
Γ	Ice calorimeter calibration constant
λ	Thermal conductivity
ρ	Density
Θ_D	Debye characteristic temperature

REFERENCES

1. Y.S. Toloukian and C.Y. Ho, eds., *Thermophysical Properties of Matter, The TPRC Data Series,* in thirteen volumes, IFI, Plenum Press, New York (1970).
2. D.R. Stull and H. Prophet, eds., *JANAF Thermochemical Tables,* Second Edition, NSRDS – NBS 37 (1971), for sale by the Superintendent of Documents, U.S. Govt. Printing Office, Washington, D.C. 20402. (Semiannual supplements of this compilation are issued by and obtainable from Dow Chemical U.S.A., Midland, Michigan 48640).
3. L.V. Gurvich, ed., *Thermodynamic Properties of Chemical Substances,* a Handbook in

Two Volumes (in Russian), Academy of Science, Moscow, USSR (1962).
4. V.A. Rabinovich, ed., *Thermophysical Properties of Matter and Substances,* 2nd ed., State Service for Standard and Reference Data (GSSSD), Moscow (1970). (English translation available from NTIS.)
5. R. Hultgren, P.D. Desai, D.T. Hawkins, M. Gleiser, K.K. Kelley, and D.D. Wagman, *Selected Values for the Thermodynamic Properties of the Elements,* American Society for Metals, Metals Park, Ohio (1973).
6. Y. Takahashi, H. Yokokawa, H. Kadokura, Y. Sekine, and T. Mukaibo, "Laser-Flash Calorimetry I, Calibration and Test on Alumina Heat Capacity," *J. Chem. Thermodyn.* **11**, 379–394 (1979).
7. V.A. Kirillin, A.E. Sheindlin, and V.Ya. Chekhovskoi, "Enthalpy and Heat Capacity of Some Solid Materials at Extremely High Temperatures," in *Proc. Int. Symp. High Temp. Tech.* held at Pacific Grove, California, pp. 471–484, IUPAC (1963).
8. J.M. Sturtevant, Calorimetry, in *Physical Methods of Organic Chemistry, Part I,* A. Weissburger, ed., pp. 523–582, Interscience, New York (1959).
9. J.P. McCullough and D.W. Scott, eds., *Experimental Thermodynamics,* Vol. I, IUPAC, Plenum Press, New York (1968).
10. D.C. Ginnings, ed., *Precision Measurement and Calibration – Heat,* NBS Special Publication 300, Vol. 6, for sale by the Superintendent of Documents, U.S. Govt. Printing Office, Washington, D.C. 20402 (1970).
11. O. Kubaschewski and C.B. Alcock, *Metallurgical Thermochemistry* 5th ed., Pergamon Press, New York (1979).
12. A. Barton, in *A Textbook on Heat,* pp. 29 ff., Longmans, Green & Co., New York (1938).
13. W. Eitel, in *Thermochemical Methods in Silicate Investigation,* pp. 133 ff., Rutgers University Press, New Jersey (1952).
14. D.H. Andrews, G. Lynn, and J. Johnston, "The Heat Capacities and Heat of Crystallization of some Isomeric Aromatic Compounds," *J. Am. Chem. Soc.* **48,** 1274–1279 (1926).
15. J.B. Conway and R.A. Hein, "Enthalpy Measurements of Solid Materials to 2400 C by Means of a Drop Technique," in *Advances in Thermophysical Properties at Extreme Temperatures and Pressures,* Third Symposium on Thermophysical Properties, Purdue Univ., Lafayette, Indiana, ASME (March, 1965).
16. J.C. Southard, "A Modified Calorimeter for High Temperatures," *J. Am. Chem. Soc.* **63,** 3142–3146 (1941).
17. A.S. Dworkin and M.A. Bredig, "The Heat of Fusion of the Alkali Metal Halides," *J. Phys. Chem.* **64,** 269–272 (1960).
18. V.A. Kirillin, A.E. Sheindlin, and V.Ya. Chekhovskoi, "Experimental Determination of the Enthalpy of Corundum (Al_2O_3) in the Temperature Range 500 to 2000°C" (in Russian), *Inzh.-Fiz. Zh.* **4** (2), 3–17 (1961).
19. E.N. Fomichev, V.V. Kandyba, and P.B. Kantor, "Calorimetric Apparatus for the Determination of Enthalpy and Specific Heat of Substances" (in Russian), *Izmer. Tekh.* **(5),** 15–18 (1962).
20. V.Ya. Chekhovskoi and A.E. Sheindlin, "Laboratory Furnace with a Graphite Heating Element for Investigations to Temperatures of 3100 C" (in Russian) *Prib. Tekh. Eksp.* **(1),** 197–199 (1963).
21. A.E. Sheindlin, V.Ya. Chekhovskoi, and L.A. Reshetov, "High Temperature Laboratory Furnace With a Graphite Heating Element for Investigations to Temperatures of 3000 C" (in Russian), *Prib. Tekh. Eksp.* **(2),** 153–156 (1963).
22. E.A. Guseva, A.S. Bolgar, S.P. Gordienko, V.A. Gorbatyuk, and V.V. Fesenko, "Deter-

mination of the Enthalpy of Self-Bound Silicon Carbide in the 1300–2000 K Temperature Range," *High Temp.* (Engl. transl.) **4** (5), 609–612 (1966).

23. D.R. Fredrickson, R. Kleb, R.L. Nuttall, and W.N. Hubbard, "A Drop Calorimeter With an Electron-Beam Heated Furnace," *Rev. Sci. Instrum.* **40** (8), 1022–1025 (1969).
24. D.I. Marchidan and M. Ciopec, "Drop Calorimeter for Determining the Heat Capacity of Solids in the Temperature Interval 25–1000°C" (in Romanian), *Acad. Repub. Pop. Rom. Stud. Cercet. Chim.* **17** (9), 737–762 (1969).
25. D.I. Marchidan and M. Ciopec, "High-Temperature Enthalpy and Related Thermodynamic Functions of $UO_{(2+X)}$," *Rev. Roum. Chim.* **16** (8), 1145–1154 (1971).
26. A.E. Sheindlin, I.S. Belevich, and I.G. Kozhevnikov, "An Apparatus for Determining the Enthalpy and Specific Heat of Solids at Temperatures up to 2700 K" (Engl. transl.), *High Temp.* **8** (3), 563–565 (1970).
27. F.L. Oetting, "On the True Temperature Rise in Isoperibol Calorimetry," *J. Chem. Thermodyn.* **2**, 727–739 (1970).
28. D.P. Lindroth and W.G. Krawza, "Heat Content and Specific Heat of Six Rock Types at Temperatures to 1000°C," U.S. Bur. Mines Report Invest. 7503, 25 pp. (1971).
29. N.D. Stout, R.W. Mar, and W.O.J. Boo, "The High-Temperature Enthalpy and the Enthalpy of Fusion of Boron by Drop Calorimetry," *High Temp. Sci.* **5**, 241–251 (1973).
30. V.Ya. Chekhovskoi, V.D. Tarasov, and I.A. Zhukova, "Apparatus for Measuring the Enthalpy and Specific Heat of High-Melting-Point Materials at High Temperatures" *High Temp.* (Engl. transl.), **12** (6), 1088–1091 (1974).
31. R. Blachnik, R. Igel, and P. Wallbrecht, "Thermodynamische Eigenschaften von Zinnchalcogeniden," *Z. Naturforsch.* **29a**, 1198–1201 (1974).
32. L.M. Buchnev, V.I. Volga, B.K. Dymov, and N.V. Markelov, "Enthalpy of Carbon Materials at 500–3250 K" *High Temp.* (Engl. transl.), **11** (6), 1072–1076 (1974).
33. I. Proks, M. Eliasova, I. Zlatovsky, and J. Zauska, "High-Temperature Drop Calorimeter for the Determination of Increases in Enthalpy," *Silikaty (Prague)* **21** (3), 253–264 (1977).
34. V.I. Chemykhin, I.N. Zedina, and S.E. Vaisburd, "A Calorimeter for Measuring Liquid-Metal Enthalpies," *Inzh.-Fiz. Zh.* **34** (5), 870–874 (1978).
35. M. Olette, "An Adiabatic Dropping Calorimeter for Enthalpy Measurements at High Temperatures. The Heat Content of Silicon from 1200 to 1550 C," in *The Physical Chemistry of Steelmaking* (Proc. Conf. at Endicott House; Dedham, Massachusetts, 28 May–3 June 1956; J.F. Elliot, ed.), pp. 18–26, John Wiley, New York (1958).
36. L.S. Levinson, "High Temperature Drop Calorimeter," *Rev. Sci. Instrum.* **33**, 639–642 (1962).
37. E.D. West and S. Ishihara, "A Calorimetric Determination of the Enthalpy of Graphite from 1200 to 2600 K," in *Advances in Thermophysical Properties at Extreme Temperatures and Pressures*, pp. 146–151, ASME, New York (1965).
38. D.H. Dennison, K.A. Gschneider, Jr., and A.H. Daane, "High-Temperature Heat Contents and Related Thermodynamic Functions of Eight Rare-Earth Metals," *J. Chem. Phys.* **44** (11), 4273–4282 (1966).
39. A.C. Macleod, "High-Temperature Adiabatic Drop Calorimeter and the Enthalpy of α-Alumina," *Trans. Faraday Soc.*, No. 530, **63** (Pt 2), 300–310 (1967).
40. V. Tydlitat, A. Blazek, J. Halousek, K. Pietsch, and K. Prihoda, "A Copper Drop-Calorimeter with Adiabatic Shield for Enthalpy Measurement up to 1700 K," *Czech. J. Phys. (B)* **21** (8), 817–822 (1971).
41. F.H. Spedding and D.C. Henderson, "High-Temperature Heat Contents and Related Thermodynamic Functions of Seven Trifluorides of the Rare Earths," *J. Chem. Phys.* **54** (6), 2476–2483 (1971).

42. F. Grønvold, "Inverse Drop Calorimeter for Enthalpy Determinations in the Range 300 to 1500 K." *Acta Chem. Scand. Ser. A* **26**, 2216–2222 (1972).
43. R.I. Efremova and E.V. Matizen, "Adiabatic Calorimeter for the Determination of the Enthalpy of Corundum at High Temperatures by the Method of Mixtures" (in Russian), *Metrologiya* **(8)**, 56–63 (1972).
44. V.A. Arkhipov, E.A. Cutina, V.N. Dobretsov, and V.A. Ustinov, "Enthalpy and Specific Heat of Neptunium Oxide in the Temperature Interval 350–1100 K" (in Russian), *Radiokhimiya* **16** (1), 123–126 (1974).
45. D.A. Ditmars, A. Cezairliyan, S. Ishihara, and T.B. Douglas, *Enthalpy and Heat Capacity Standard Reference Material; Molybdenum SRM 781, from 273 to 2800 K*, NBS Special Publication 260-55, 80 pp. (1977), for sale by the Superintendent of Documents, U.S. Govt. Printing Office, Washington, D.C. 20402.
46. W. Swietoslawski, *Microcalorimetry*, Reinhold, New York (1946).
47. D.C. Ginnings and R.J. Corruccini, "An Improved Ice Calorimeter – The Determination of its Calibration Factor and the Density of Ice at 0°C," *J. Res. Natl. Bur. Stand. (U.S.)* **38**, RP 1796, 583–591 (1947).
48. D.C. Ginnings, T.B. Douglas, and A.F. Ball, "Heat Capacity of Sodium Between 0°C and 900°C, the Triple Point and Heat of Fusion," *J. Res. Natl. Bur. Stand. (U.S.)* **45**, RP 2110, 23–33 (1950).
49. W.T. Foley and P.A. Giguère, "Hydrogen Peroxide and its Analogues, IV. Some Thermal Properties of Hydrogen Peroxide," *Can. J. Chem.* **29**, 895–903 (1951).
50. L.E. Leake and E.T. Turkdogan, "The Construction and Calibration of a Bunsen Ice Calorimeter, *J. Sci. Instrum.* **31**, 447–449 (1954).
51. R.A. Oriani and W.K. Murphy, "The Heat Capacity of Chromium Carbide (Cr_3C_2)," *J. Am. Chem. Soc.* **76**, 343–345 (1954).
52. H.W. Deem and C.F. Lucks, "An Improved All-Metal, Bunsen-Type Ice Calorimeter," *Instrument Society of America, Proc. 13th Annual Instrument-Automation Conference*, Vol. 13, Pt. II, Paper No. PPT-4-58 (1958).
53. C.F. Lucks, H.W. Deem, and W.D. Wood, "Thermal Properties of Six Glasses and Two Graphites," *Am. Ceram. Bull.* **39** (6) 313–319 (1960).
54. F.H. Spedding, J.J. McKeown, and A.H. Daane, "The High-Temperature Thermodynamic Functions of Cerium, Neodymium, and Samarium," *J. Phys. Chem.* **64** (3), 289–290 (1960).
55. D.F. Smith, C.E. Kaylor, G.E. Walden, A.R. Taylor, Jr., and J.B. Gayle, "Construction, Calibration, and Operation of Ice Calorimeters," U.S. Bur. Mines Report Invest. 5832 (1961).
56. J.H. Colwell and C.D. Halsey, Jr., "The Properties of α-Sulfur Trioxide, *J. Phys. Chem.* **66**, 2182–2184 (1962).
57. R.J. Kokes, M.K. Dorfman, and T. Mathia, "Calorimetry," *J. Chem. Educ.* **39**, 90–91 (1962).
58. D.S. Neel, C.D. Pears, and S. Oglesby, Jr., "The Thermal Properties of Thirteen Solid Materials to 5000°F or Their Destruction Temperatures," Technical Documentary Report No. WADD 60-924, Southern Research Institute, Birmingham, Alabama, AD No. 275 536 (available from NTIS) (1962).
59. J.R. Welty and C.E. Wicks, "Ice Calorimeter for the Precise Measurement of Heat Content From 0 to 1500°C" U.S. Bur. Mines Report Invest. 6028 (1962).
60. J. Opdycke, C. Gay, and H.H. Schmidt, "Improved Precision Ice Calorimeter," *Rev. Sci. Instrum.* **37** (8), 1010–1013 (1966).
61. K. Maglić and J.E. Parrott, "Thermal and Thermoelectric Properties of Sintered Iron Disilicide for High-Temperature Thermoelectric Applications," Bundesministerium für Bildung und Wissenschaft, Forschungsbericht K 70-01, pp. 263–282 (1970).

62. D.A. Ditmars, "Measurement of the Average Total Decay Power of Two Plutonium Heat Sources in a Bunsen Ice Calorimeter," *Int. J. Appl. Radiat. Isot.* **27**, 469–490 (1976).
63. O.E. Zakurenko and V.M. Kuz'michev, "Reading Unit for Ice Calorimeter," *Prib. Tekh. Eksp.* **6** 167–168 (1978).
64. P.A. Giguère, B.G. Morissette, and A.W. Olmas, "A 27°C Isothermal Calorimeter," *Can. J. Chem.* **33,** 657–664 (1955).
65. R.S. Jessup, "A New Bunsen-Type Calorimeter," *J. Res. Natl. Bur. Stand. (U.S.)* **55**(6), RP 2635, 317–322 (1955).
66. R. Hultgren, P. Newcombe, R.L. Orr, and L. Warner, "A Diphenyl Ether Calorimeter for Measuring High-Temperature Heat Contents of Metals and Alloys," *The Physical Chemistry of Metallic Solutions and Intermetallic Compounds,* N.P.L. Symposium No. 9, col. I., paper 1 H, H.M.S.O., London (1958).
67. H. Peters and E. Tappe, "Theoretische und experimentelle Analyse eines grossen Bunsen-Kalorimeters mit Diphenyloxidfüllung, I. Mitteilung: Theoretischer Teil, II. Mitteilung: Experimenteller Teil," *Montasber. Dtsch. Akad. Wiss. Berlin* **10**(2), 88–111 (1968).
68. J.V. Davies and H.O. Pritchard, "The Properties of Diphenyl Ether Calorimeters," *J. Chem. Thermodyn.* **4,** 9–22 (1972).
69. M.L. Lakhanpal and R.N. Parashar, "Cetyl Alcohol as a Dilatometric Fluid for Isothermal Phase-Change Calorimeter," *Indian J. Chem. Sec. A* **8**(4), 368–369 (1970).
70. W.P. Gilbreath and D.E. Wilson, "A Precision Liquid Nitrogen Boil-Off Calorimeter and its Application to Surface Energy Measurements," *Rev. Sci. Instrum.* **41**(7), 969–973 (1970).
71. E.E. Shpil'rain, D.N. Kagan, and L.S. Barkhatov, "Thermodynamic Properties of the Condensed Phase of Alumina near the Melting Point," *High Temp.-High Pressures* **4,** 605–609 (1972).
72. H.P. Stephens, "Determination of the Enthalpy of Liquid Copper and Uranium with a Liquid Argon Calorimeter," *High Temp. Sci.* **6,** 156–166 (1974).
73. H.P. Stephens, "Design of a Vaporization Calorimeter for Assay of Radioactive Samples, Sandia Laboratories Report SAND-76-0635, 21 pp. (1977).
74. H.P. Stephens, "Construction and Testing of a Vaporization Calorimeter for Assay of Radioisotopic Samples," Sandia Laboratories Report SAND-77-0105, 38 pp. (1978).
75. J.L. Orehotsky and K. Schröder, "A High-Temperature Heat-Exchange Calorimeter," *J. Phys. E* **3,** 889–891 (1970).
76. J. Konicek, J. Suurkuusk, and I. Wadsö, "A Precise Drop Heat-Capacity Calorimeter for Small Samples," *Chem. Scr.* **1,** 217–220 (1971).
77. J. Suurkuusk and I. Wadsö, "Design and Testing of an Improved Precise Drop Calorimeter for the Measurement of the Heat Capacity of Small Samples," *J. Chem. Thermodyn.* **6,** 667–679 (1974).
78. A. Fontana and R. Winand, "Mise au Point Experimentale d'un Calorimetre a Haute Temperature Mesure de la Chaleur Specifique de ZrF_4 Solide en Fonction de la Temperature et Estimation de la Temperature et Estimation de la Chaleur Sensible de ZrF_4 Liquide," *J. Nucl. Mater.* **44,** 295–304 (1972).
79. V.A. Arkhipov, V.N. Dobretsov, L.S. Perkatora, and V.A. Ustinov, "Drop Calorimeter for Measurement of Enthalpy in the Temperature Range 350–1100 K" (in Russian), *Zh. Fiz. Khim.* **49**(5), 1331–1333 (1975).
80. S. Tamura, R. Yokokawa, and K. Niwa, "The Enthalpy of Beryllium Fluoride from 456 to 1083 K by Transposed-Temperature Drop Calorimetry," *J. Chem. Thermodyn.* **7,** 633–643 (1975).

81. G. Morizur, A. Radenac, and J.-C. Cretenet, "Calorimètre à Chute 3000 K. Application à la Détermination de la Chaleur Spécifique du Tungstène," *High Temp.-High Pressures* 8, 113–120 (1976).
82. V.Ya. Chekhovskoi and B.Ya. Berezin, "Experimental Apparatus for Measuring the Enthalpy and Specific Heat of Refractory Metals" *High Temp.* (Engl. transl.), 8(6), 1244–1246 (1970).

14

Levitation Calorimetry

VITALIY YA. CHEKHOVSKOI

1. INTRODUCTION

Development of new branches of technology dealing with high-temperature processes called for new information on properties of refractory metals and their alloys in solid and liquid states at temperatures exceeding 2000 K. However, a high chemical activity of metals and, particularly, their melts at temperatures above 1500–2000 K presented a most serious handicap in experimental study of their thermodynamic properties. This circumstance gave rise to new methods of research making it possible to eliminate adverse effects of chemical activity of metals and their melts on the measurement results over tha range of high temperatures. One of these original methods is the so-called "levitation" calorimetry, which first found practical application in 1968–1970. The levitation calorimetry method employs the principle of mixtures and levitation heating of specimens held suspended in a high-frequency electromagnetic field of an inductor. The application of such an electromagnetic crucible makes it practicable to prevent chemical interaction of metals, alloys, and their melts with the surrounding elements of the experimental apparatus in investigation of enthalpy and heat capacity. Besides, the levitation calorimetry method enables the limits of high-temperature research of enthalpies and heat capacities of metal melts to be extended to 3400 K. There is every ground to expect that the method of levitation heating will allow still greater temperature ranges. Taking into account a comparatively short history of application of the levitation calorimetry method it is believed that the potentialities of the levitation calorimetry have not yet been exhausted.

VITALIY YA. CHEKHOVSKOI • Institute of High Temperatures, Korovinskoe Road, USSR Academy of Sciences, Moscow 127412, USSR.

The present chapter covers the following aspects of investigation of thermodynamic properties of metals and their melts by the levitation calorimetry method: peculiarities of levitation heating of specimens; methods of measurement of temperatures of solid and liquid-state metals; methods of measurement of spectral emissivity of metal melts, including that at the melting point; description of the methods and techniques used for measurement of enthalpy by the use of the existing experimental equipment; analysis and estimation of probable sources of errors yielded in measurements of enthalpy of metals in solid and liquid states.

2. PECULIARITIES OF INVESTIGATIONS OF THERMODYNAMIC PROPERTIES OF REFRACTORY METALS AND THEIR MELTS AT TEMPERATURES ABOVE 1500–2000 K

The classical method of mixtures commonly involving a massive calorimeter and an electric resistance furnace has found wide use and is essentially one of the principal methods for research concerning enthalpies and heat capacities of substances in solid and liquid states up to the temperatures of 1500–2000 K. This was mainly because of two important features of the method: (1) a high accuracy of research results, and (2) versatility. The latter is illustrated by applicability of the method of mixtures over an extremely wide range of temperatures for researching a great variety of substances: metals and nonmetals in solid and liquid states as well as porous, powderlike, or massive specimens.

However, the further increase of the experimental temperatures in the classical method encountered an insurmountable difficulty. This is associated with the fact that the chemical activity of refractory substances and metals in the solid state, and particularly in the liquid state, at the temperatures exceeding 1500–2000 K tends to increase substantially. There are no unreactive containers. This factor together with some other difficulties arising in high-temperature experiments seriously limit the temperature range appropriate for application of the method of mixtures. Exceeding these limits may bring about erroneous experimental results due to changes of composition of the object under test because of chemical interaction with the crucible, the suspension wire of the specimen under test in the furnace, etc.

A search for a way out of this situation gave rise to a crucibleless method for investigation of enthalpy of a number of refractory metals almost up to their melting temperatures, e.g., molybdenum up to 2834 K,[1] niobium up to 2587 K,[2] chromium up to 2096 K,[3] tungsten up to 3594 K,[4] silicon carbide up to 2843 K,[5] etc. In this case the specimen under test was heated without a crucible in a furnace with a tungsten heater. The heat loss of the specimen for the time of its drop from the furnace to the calorimeter was calculated by the use of the Stefan–Boltzmann formula, the integral hemispherical emissivity of the substance under investigation being known. The absence of the ampoule ensures that the experiment range is nearly half as much as compared with that when the ampoule is used, for this obviates the need for measuring the enthalpy of an empty ampoule. The heat lost by a dropping specimen is usually not in excess of few per cent of the heat brought by the specimen into the calorimeter. Therefore the error in calculations of this loss often proves to be less than that associated with measurement of the enthalpy of an empty ampoule.

Nevertheless, the method of mixtures cannot be employed for research on enthalpies and heat capacities of molten metals and other substances at temperatures exceeding 2000 K, since crucible materials which do not react with chemically active melts have not been found as yet. It is known that some attempts[6] were made to use the method of mixtures and the so-called autocrucibles in the form of cooled copper casting molds in which the metal was fused by arc heating or by other methods of heating. Such a solution of the problem will not do for precision investigations of enthalpies by the method of mixtures, since there is an uncertainty in the fusion temperature because of the great temperature difference between the melt and the cold walls. The same disadvantage associated with an uncertainty of "melt temperature" is characteristic of application of a zone crucibleless melting by electron-beam or induction heating of a metal rod.

The problem of measuring enthalpies and heat capacities of refractory metals and their melts by the method of mixtures at temperatures exceeding 1500–2000 K has been solved most satisfactorily by application of a crucibleless induction heating of the specimens under test in suspended state in a high-frequency electromagnetic field of an inductor. With this kind of heating the metal specimen is protected from contamination, for a crucible is absent and the specimen has no contact with other parts of the installation. This allows the temperature range for application of the method of mixtures to be substantially extended up to 3600 K[7] so as to obtain new data on thermodynamic properties of metals and their alloys in solid and in liquid states, which is so essential for development of new directions in high-temperature engineering and technology. Levitation heating has a certain advantage also in that only a short period of time is required to achieve the desired equilibrium high temperature of the specimen under test, both in solid state and in liquid state. This kind of heating is applicable not only to metals and their alloys but also to other substances which possess a high electrical conductivity, for example, carbides, borides, nitrides, etc. Levitation heating in a vacuum aids in cleaning the melts at the expense of evaporation of gases and some volatile impurities. Besides, it is found that levitation heating is a unique means for investigating refractory metals in supercooled states as shown in Refs. 8 and 9.

The combined application of the method of mixtures and the levitation heating of a specimen is referred to as levitation calorimetry. Levitation calorimetry originated and found application almost at the same time in the Soviet Union[10] and in the USA.[11] As a result of the investigations reported so far, data were obtained on heats and enthalpy of fusion, as well as heat capacities of many liquid refractory metals, i.e., Mo, Nb, V, Pt, Ru, Rh, Cu, Ti, Zr, Ir, Ni, Co, Pd, Fe, and U. The results obtained from the experimental studies of heat capacities of refractory metals, entropies, and heat of melting materially differ from the commonly cited values of these properties in the widely known handbooks. It is hoped that this fact will give impetus to further development of theories for molten metals and for melting, to determination and correlation of the data obtained with other parameters, as well as to semiempirical generalizations of the experimental data.

Levitation heating represents a versatile means in physicochemical and physical metallurgy investigations of metal melts, metal alloys, and other substances at high temperatures. In particular, it finds application for investigating[12]

- phase diagrams;
- surface tensions and densities of melts;
- interactions of melts and their admixtures with gases;
- rates of evaporation and gassing in vacuum;
- brightness temperatures and spectral emissivities of molten metals at melting points and at higher temperatures;[13]
- distribution of elements between metal and slag.

The levitation heating is also widely used for technological purposes, namely, for producing metal alloys including the alloys of the metals whose melting temperatures differ materially, for instance, niobium and gallium[14] (the melting temperature of niobium is 2750 K, while that of gallium is 303 K), for synthesis of chemical compounds, as well as for production of metallic films.

It is extremely difficult or essentially impossible to obtain most of the above-stated parameters for molten refractory metals by any method other than levitation heating. The investigations of the spectral emissivities of molten metals provide valuable information which is indispensable for determination of the true temperatures of melts, for levitation calorimetry, and also for other relevant investigations.

Concurrent with levitation calorimetry another method for investigating enthalpies and heat capacities of metals and their melts – the method of electrical explosion of a wire[15–17] – is being developed. The merits of this latter method reside in the fact that due to the quick heating of the wire from room temperature to a definite temperature of melt with heating rates from 10^4 to $10^8\ K\,s^{-1}$ the metal will not chemically react with the environment. However, this method is characterized by other sources of experimental error, which could hardly be predetermined. The question has been left open concerning the uncertainty of the physical state of a specimen subjected to very rapid heating which may differ from the equilibrium state.[18] This will require a special verification of the data obtained in any specific instance.

3. LEVITATION HEATING

Metals may be held in a suspended state in a high-frequency electromagnetic field possessing a potential well, i.e., the area wherein the field intensity reduces from the edges towards the center.[19] Interaction of a high-frequency electromagnetic field with a metal induces in the latter the currents which force the field out of the space occupied by the metal, and at the same time cause the metal to heat up. By virtue of physical interaction of the induced currents and the electromagnetic field, the metal is expelled from the area of the higher field intensity to the area of the lesser strength of the field. The metal may take a stable position in the potential well, if the resultant of the electromagnetic forces acting on the metal from beneath is equal to the force of gravity. For the sake of a stable position of the metal in a suspended state, provision should be made for an inductor consisting of several turns arranged under the specimen to hold it in the suspended state, and other turns placed above the specimen to stabilize the position of the specimen.[20] The current in both groups of inductor turns shall be opposite in direction. In one study[21] it is suggested that the upper stabilizing turns be replaced by a water-cooled copper ring which is not directly connected with the source of voltage. Induced in this ring are the electric currents whose fields stabilize the position of the specimen in the inductor.

It should be noted that for a stable position of the molten metal in the inductor some additional conditions are required, aside from those listed above. Now we shall briefly consider these conditions.[19] The intensity of the field and the density of the current induced by this field tend to diminish exponentially from the surface of the specimens inward, penetrating into the metal to a depth which is referred to as a penetration depth, which can be approximately determined by use of the formula

$$\Delta = (2\rho/\mu\omega)^{1/2} \tag{1}$$

where μ is the magnetic permeability of the metal, ρ is the specific resistance, and $\omega = 2\pi f$ is the circular frequency of the electromagnetic field.

As a consequence of interaction of the induced current with the associated electromagnetic field, some mechanical forces arise in the surface layer of the metal. These forces exert an electromagnetic pressure in a metal, similar to hydrostatic pressure, and equal to $\mu H_0^2/4$. The electromagnetic pressure in a levitated drop of metal equals the hydrostatic pressure of the metal with a height h, and a section equal to the unit area, i.e.,

$$h\gamma g = \mu H_0^2/4 \tag{2}$$

where H_0 is the amplitude of the intensity of the magnetic component of the field on the surface of the metal under the specimen, γ is the metal density, and g is the acceleration due to gravity.

Due to the fact that there are specific areas on the surface of the molten metal where induced currents are absent due to attenuation of the magnetic field in these places, then only the metal surface tension forces are capable of compensating for attenuation of the electromagnetic shell in the lower part of the molten metal. Using the Laplace formula, it can be written down that the difference of the surface tension forces in the lower surface and upper surface of the melt drop is balanced with a melt column of height h and a section equal to the unit area, i.e.,

$$h\gamma g = \sigma\{[1/r_{1l} + 1/r_{2l}] - [1/r_{1u} + 1/r_{2u}]\} \tag{3}$$

where r_{1l}, r_{2l}, r_{1u}, r_{2u} are the main radii of curvature of the lower and upper surfaces of the melt drop, and σ is the surface tension.

If the radius of curvature of the upper part of the melt drop is much greater than that of the lower part of the drop, when the drop takes the form of a body of revolution, then $r_{1l} = r_{2l} = r_l$, and from (3) we shall derive the relationship between the height of the drop and the curvature radius r_l:

$$h = 2\sigma/(\gamma g r_l) \tag{4}$$

In Ref. 19 it is noted that as the mass of the levitating metal grows, some folds are formed in its lower part. The direction of these folds coincides with the directions of the field lines, while the depth of the folds may be commensurable with the electromagnetic field penetration depth Δ. Such folds will be almost transparent for the field. A similar situation will take place if the field shaping a drop proves to be able to compress the lower part of the melt to a diameter equal to 2Δ. In both cases, with further decrease of the curvature radius, this part of the drop will remain transparent for the field of the given frequency, and the melt will flow out. After substituting the expression of the field penetration depth (1) for r and expression (2) in formula (4), we shall obtain the critical height of the molten metal drop, i.e.,

$$h_c = 2\sigma/(\gamma g \Delta) = a^2/\Delta = \mu H_c^2/(4\gamma g) = \sigma(2\mu\omega/\rho)^{0.5}/(\gamma g) \tag{5}$$

a new one. The F_0 number determines the time for which the steady state is achieved.

Experiments can be performed at free convection ($\alpha \neq \infty$, $\mathrm{Bi} \neq \infty$) and also under the conditions when a constant temperature ($\alpha \to \infty$, $\mathrm{Bi} \to \infty$) is given and kept on the sample surfacc. In the former case comparative versions of the studies are realized; in the latter, the simple and reliable methods of direct determination of thermal diffusivity are used.[2,4,5] Indeed, at $\mathrm{Bi} = \infty$, equations (11)–(13) simplify to

$$\cos \mu = 0 \tag{14}$$

$$\mathscr{T}_0(\mu) = 0 \tag{15}$$

$$\sin \mu = 0 \tag{16}$$

Assuming that the ambient medium is isotropic ($a_1 = a_2 = a_3 = a$) and expressing μ_1 from (14)–(16) give the simple formulas for thermal diffusivities of the above geometries:

$$a = \frac{4h^2}{\pi^2} m_\infty \tag{17}$$

$$a = \frac{R^2}{5.783} m_\infty \tag{18}$$

$$a = \frac{R^2}{\pi^2} m_\infty \tag{19}$$

$$a = \frac{m_\infty}{\dfrac{5.783}{R^2} + \dfrac{\pi^2}{4h^2}} \tag{20}$$

$$a = \frac{m_\infty}{\dfrac{\pi^2}{4}\left(\dfrac{1}{h_1^2} + \dfrac{1}{h_2^2} + \dfrac{1}{h_3^2}\right)} \tag{21}$$

where m_∞ is the rate of a temperature change with $\alpha \to \infty$. So, in order to determine thermal diffusivity with $\alpha \to \infty$ over the whole sample surface, it is necessary to know and to maintain a certain constant temperature different from the initial one. This can be achieved by different means, the simplest of which may be described as follows. The test material is placed into a thermostat filled with vehemently stirred liquid. When a test sample is loose of fibrous material, it is placed into a sealed metal cylinder, sphere, or prism. Such a system is usually referred to as an a-calorimeter. Solid, waterproof materials can be conveniently tested without a protective shell, the thermoelectrodes being safely insulated. The general procedure of experiment and the processing of its results are as follows.[5] After the test sample of a certain geometry has been placed into the thermostat, the time changes of an excess temperature are recorded with a differential thermocouple

$$\Theta = f(\tau) \tag{22}$$

The data of (22) are used to plot (5), which gives the rate of temperature change (6). In accordance with the sample geometry, the thermal diffusivity is found from one of expressions (19)–(21). The experimental results are reliable if function (16) is linear.

Thermal diffusivity can be estimated using the alternative methods which imply the recording of the time of a temperature change for a prescribed value.[9]

If heat is transferred at an indefinite Bi value, then the rate of heating (or cooling) is a function not only of thermal diffusivity but also of the first root μ_1 of characteristic equations (11)–(13). Therefore, the coefficients in formulas (18)–(21) will also be indefinite. In this case, to find thermal diffusivity, temperature versus time has been measured at any two points of a sample. The processing of the results here becomes more tedious.[5,6]

The initial section of the temperature curve inconsistent with relation (5) can be also employed to estimate thermal diffusivity using a semi-infinite body model:

$$\Theta = \operatorname{erf}\left(\frac{1}{2\sqrt{F_{0x}}}\right) \tag{23}$$

Solution (23) is obtained from (2) with appropriate ultimate transitions. The structure of general solution (23) does not allow an analytical formula for calculating thermal diffusivity. In these instants, the tabulated Gauss error functions[2] are used. More comprehensive information on the study of thermophysical properties at the initial stage of heat transfer can be found elsewhere.[10]

The functional relations of forms (7)–(10) enable one to implement a number of comprehensive methods of estimating a complex of thermophysical properties, namely, heat transfer coefficient, thermal diffusivity, and thermal conductivity.[11–14]

Thermal diffusivity at high temperatures has been studied when the boundary conditions of the first and third kinds are valid.[15–17]

The expression of general form (10) can be used to estimate thermal diffusivity of orthotropic materials.[18,19] Three parallelepipeds of different sizes with the side ratios

$$h_1 : h_2 : h_3 = 1:2:3, \qquad 1:2:2, \qquad 1:1:2$$

are manufactured from the test material. For all three samples the rate of temperature change at $\alpha \to \infty$ is measured by the above method. Then, according to (10), with the ratio of the linear dimensions borne in mind, it is possible to write down the following equations:

$$m_1 = \left(\frac{\pi}{2}\right)^2 \left[\frac{a_1}{(h/2)^2} + \frac{a_2}{(2h/2)^2} + \frac{a_3}{(3h/2)^2}\right]$$

$$m_2 = \left(\frac{\pi}{2}\right)^2 \left[\frac{a_1}{(h/2)^2} + \frac{a_2}{(2h/2)^2} + \frac{a_3}{(2h/2)^2}\right]$$

$$m_3 = \left(\frac{\pi}{2}\right)^2 \left[\frac{a_1}{(h/2)^2} + \frac{a_2}{(h/2)^2} + \frac{a_3}{(2h/2)^2}\right]$$

From the solution to the above equations, the following formulas for thermal diffusivity can be obtained:

$$a_1 = 6.76h^2(27m_1 - 7m_2 - 5m_3) \times 10^{-3} \tag{24}$$

$$a_2 = 135.2h^2(m_3 - m_2) \times 10^{-3} \tag{25}$$

$$a_3 = 730.2h^2(m_2 - m_1) \times 10^{-3} \tag{26}$$

The main source of a systematic error in measuring thermal diffusivity by formulas (19)–(21) is the violation, in experiment, of the condition $\text{Bi} = \infty$ required by theory as well as the dependence of the material properties on temperature. In order to determine thermal diffusivity as accurate as 2%, Bi must be approximately equal to 100.[4] In general, the systematic error due to the violation of the condition $\text{Bi} = \infty$, corresponding to an ideal thermal contact, can be found from

$$\epsilon = \frac{\Delta a}{a} = \frac{\mu_{1\infty}^2 - \mu_1^2}{\mu_{1\infty}^2} \tag{27}$$

where μ_1, $\mu_{1\infty}$ are the roots of equations (11)–(13) and the roots of equations (14)–(16). In practice, μ_1 must be estimated to find (27). To do this, we shall use the method of Ref. 2. If over the surface of a sample with thermal conductivity λ_1 and thickness h there is a thin film (say, an air interlayer) of thickness δ and thermal conductivity λ_2 film heat capacity being neglected), then

$$\text{Bi} = \frac{\alpha h}{\lambda_1} = \frac{h}{\lambda_1} : \frac{\delta}{\lambda_2} \tag{28}$$

Tentative estimation (28) and equations (14)–(16) yield μ_1 and then (27). These calculations as well as more complicated cases when film heat capacity is different from zero are given in Ref. 20.

Thus, systematic error (27) is a variable depending on (28). Therefore, in experiments with small-thermal resistance materials (e.g., metals) the systematic error of measurements may be very large.

Another source of the systematic error that is difficult to estimate lies in a temperature dependence of thermophysical properties. This portion of the error can be minimized if experiments run at low (5–10 K) temperature drops.[21, 22] Solid waterproof materials are difficult to test with the above procedure since the protective shells are a source of errors due to additional thermal resistances which are not easy to take into account.

It is often convenient to perform experiments when over some area of the surface (say, on the plate bases) a certain constant temperature is prescribed, and the other area of the surface (say, sides of a plate) is involved in heat transfer with the ambient medium at a different temperature. In those instances when this procedure does not work, thermal diffusivity has been calculated by formulas (17)–(18) obtained from the one-dimensional solutions to equation (1). The sample sizes are greatly increased to eliminate a systematic error due to the effect of sample end-to-ambient medium heat transfer.[11, 12, 23]

The strict analytical estimation of this error as well as the development of the methods based on a more general theory is possible using appropriate two- and three-dimensional solutions (1).[2, 20, 24, 25] A general form of the solutions is quite complicated and differs from solution (2) by a nonzero steady-state component.

For a better understanding of the practical application of the above solutions we shall consider an example of a finite cylinder $2h$ in height and $2R$ in diameter, with the bases kept at a certain constant temperature T_c and the side surface involved in heat transfer with the ambient medium at a certain initial temperature following Newton's law. The solution can be written as

$$\Theta = \frac{T(r,z,\tau)-T_0}{T_c-T_0} = \sum_{n=1}^{\infty} A_n \mathscr{T}_0\left(\mu_n \frac{r}{R}\right)\frac{\cosh \kappa\mu_n z/h}{\cosh \kappa\mu_n}$$

$$+\sum_{n=1}^{\infty}\sum_{m=1}^{\infty}\frac{A_n \mathscr{T}_0(\mu_n r/R)\,\mu_m \cos\mu_n z/h}{\mu_n^2\kappa^2+\mu_m^2}\exp\left[-(\kappa^2 a_1\mu_n^2+\mu_m^2 a_2)\tau/h^2\right]$$

$$= f\left(\frac{r}{R},\frac{z}{h},\mathrm{Bi},F_0\right) \qquad (29)$$

where μ_m, μ_n are the roots of characteristic equations (11) and (12) and the parameter $\kappa = h/R$ is the relative cylinder height. The first term in (29) is the steady-state component

$$\Theta_{st} = \varphi\left(\frac{r}{R}, \frac{z}{h}, \mathrm{Bi}_R\right) = \sum_{n=1}^{\infty} A_n \mathscr{T}_0\left(\mu_n \frac{r}{R}\right) \frac{\cosh \kappa \mu_n z/h}{\cosh \kappa \mu_n} \tag{30}$$

where

$$A_n = \frac{2\mathscr{T}_1(\mu_n)}{\mu_n[\mathscr{T}_0^2(\mu_n) + \mathscr{T}_1^2(\mu_n)]}$$

The analysis of the similar solutions gives correct and practically useful results.(20, 24, 25)

1. After a certain time interval, the regular regime sets in within the sample, the rate of temperature changes being expressed by

$$m = \frac{\ln(\Theta_{st} - \Theta_1) - \ln(\Theta_{st} - \Theta_2)}{\tau_2 - \tau_1} = \text{const} \tag{31}$$

and

$$\ln(\Theta_{st} - \Theta) = f(\tau) \tag{32}$$

is the straight line. The relative dimension $\kappa \to 0$ $(R \to \infty)$ corresponds to the case of an infinite plate [see solution (2)] and $\Theta_{st} \to 0$. In this case, the rate of a temperature change is given by (6). From solution (29) the constant of (31) may be written as

$$m = \left(\kappa^2 \mu_{m_1}^2 a_1 + \frac{\pi^2 a_2}{4}\right) \frac{1}{h^2} \tag{33}$$

2. At a certain linear dimension ratio of a sample, the core temperature varies in the same way as in the case of an infinite body, and the solutions of form (29) may be, under certain conditions, substituted by the simpler and more convenient one-dimensional solutions [e.g., solution (2)] with a high accuracy. Numerical calculation shows that at small Fourier numbers $(0 < a\tau/h^2 \leqslant 0.3)$ the temperatures at the finite cylinder center [see solution (28)] with the parameter $\kappa = \frac{1}{2}, \frac{1}{3}, \frac{1}{4}$ and at the infinite plate center $(\kappa = 0)$ coincide even at $\mathrm{Bi}_R = \alpha R/\lambda_2 \to \infty$. In the course of time the difference between these temperatures increases and achieves its maximum in a steady state. For example, if $\mathrm{Bi}_R = \infty$ and $\kappa = h/R = \frac{1}{4}$, then $\Theta_{st} = 0.013$, i.e., the error of measuring steady temperature at the finite plate center is about 1% because of the nonuniform heat flux.(20)

3. The solutions of form (29) can be used to determine thermal diffusivity

of isotropic and orthotropic materials as well as to study heat transfer parameters on the basis of the regular regime methods.[4] (See Appendix.)

A constant power source in a test material exchanging heat with the ambient medium makes it possible to develop a series of methods to determine not only thermal diffusivity but also thermal conductivity and heat capacity.[26–34] If, for example, over a certain plate surface a constant heat flux is prescribed and other surfaces are exchanging heat with the ambient medium according to Newton's law, then in this case also a regular thermal regime takes place, whose laws are similar to those considered above. The same is true for a hollow cylinder with a constant heat flux on its inner surface. The appropriate two- and three-dimensional solutions are discussed in Refs. 24–25. Consider first the simplest one-dimensional solutions.

The methods in Refs. 26–28 use a particular solution of equation (1):

$$\frac{\partial T(x,\tau)}{\partial \tau} = a\,\frac{\partial^2 T(x,\tau)}{\partial x^2}$$

for a plate $2h$ thick (the origin of the coordinates at the plate center) under the following boundary conditions:

$$T(x,0) = T_0 = \text{const}$$

$$\lambda\,\frac{\partial T(0,\tau)}{\partial x} = -q$$

$$T(h,\tau) = T_0$$

The solution of this problem convenient in practice at $F_0 = a\tau/h^2 > 1$ is of the form

$$\Theta(x,\tau) = T(x,\tau) - T_0 = \frac{q(h-x)}{\lambda} - \frac{2qh}{\lambda}\sum_{n=1}^{\infty}(-1)^{n+1} \times \frac{1}{\mu_n^2}\sin\mu_n\frac{h-x}{h}\exp\left(-\mu_n^2\frac{a\tau}{h^2}\right) \tag{34}$$

where

$$\mu_n = (2n-1)\frac{\pi}{2}$$

Solution (34) permits different versions of thermal diffusivity calculations to be used. In particular, it can be estimated applying *a priori* compiled tables[26] and making use of the described regular regime methods.[27] In this case the procedure is as follows. A steady component [the first term on the right-hand side of (34)] depends on the heat flux and thermal conductivity of the sample, i.e.,

$$\Theta_{st} = \frac{q(h-x)}{\lambda} \tag{35}$$

when measuring temperature difference between the surface and center of the plate

$$\Theta_{st} = \frac{qh}{\lambda} \tag{36}$$

In the regular regime which occurs at $F_0 > 0.3$, solution (36) may be in the form (37)

$$\Theta_{st} - \Theta = \frac{2qh}{\lambda\mu_1^2} \sin \mu_1 \frac{h-x}{h} \exp\left(-\mu_1^2 \frac{a\tau}{h^2}\right) \tag{37}$$

Hence, the rate of a temperature change

$$m_\infty = \frac{\ln(\Theta_{st} - \Theta_1) - \ln(\Theta_{st} - \Theta_2)}{\tau_2 - \tau_1} = \frac{\mu_1^2}{h^2} a \tag{38}$$

The formula for calculating thermal diffusivity

$$a = \frac{4h^2}{\pi^2} m_\infty \tag{39}$$

coincides with formula (17) but the rate of a temperature change must be taken in form (38). The experimental data are processed using the relationship

$$\ln(\Theta_{st} - \Theta) = f(\tau) \tag{40}$$

Thermal diffusivity is determined by the slope of straight line (40) (see Appendix).

The solution for a hollow cylinder with an inner radius R_1 and outer radius R_2 is of the same structure.[2] The steady component in this case is

$$\Theta_{st} = \frac{qR_1}{\lambda} \ln \frac{r_2}{r_1} \qquad (r_2 > r_1) \tag{41}$$

When measuring a temperature difference between the points $r_1 = R_1$ and $r_2 = R_2$

$$\Theta_{st} = \frac{qR_1}{\lambda} \ln \frac{R_2}{R_1} \tag{42}$$

Thermal diffusivity is estimated from

$$a = \frac{R_2^2}{\mu_1^2} m_\infty \tag{43}$$

But the roots μ_1 are found from the characteristic equation (see Table 2 in Appendix)

$$\mathcal{T}_1(\mu l) Y_0(\mu) - \mathcal{T}_0(\mu) Y_1(\mu l) = 0 \tag{44}$$

where

$$l = \frac{R_1}{R_2} \tag{45}$$

The procedure of thermal diffusivity calculation is the same as for a plate. Thermal diffusivity can be estimated by studying the curve of heating at the initial time moment.[28] The heating rate at an instant $\tau = 0$ and at a point $x = 0$ as it follows from solution (34) is determined by the expression

$$b = \frac{dT}{d\tau} = \frac{2qa}{\lambda h} \tag{46}$$

At the same time, this very heating rate can be expressed in terms of a steady state component (33)

$$\frac{\Theta_{st}}{\Delta\tau} = \frac{qh}{\lambda \Delta\tau} \tag{47}$$

From relationships (46) and (47)

$$a = \frac{h^2}{2\Delta\tau} \tag{48}$$

Values (35) and (42) being known, thermal conductivity can be calculated.

Heat capacity is found from the relationship

$$c\rho = \frac{\lambda}{a} \tag{49}$$

The approach[29–30] makes use of the solution to a one-dimensional problem for a semi-infinite body which, as a particular case, also follows from solution (34):

$$T(x, \tau) - T_0 = \frac{2q(a\tau)^{1/2}}{\lambda}\, i\, \mathrm{erfc}\, \frac{x}{2(a\tau)^{1/2}} \tag{50}$$

Using (50), studies can be performed in a pure unsteady-state regime. Specifically, when measuring an excess temperature between the heater plane and the base of the plate being in this case a model of a semi-infinite body, from (50) it follows that

$$\Delta T = \frac{2q(a\tau)^{1/2}}{\lambda\sqrt{\pi}} \tag{51}$$

From relationship (51), thermal activity is

$$\epsilon = \frac{\lambda}{\sqrt{a}} = \frac{2q\sqrt{\tau}}{\sqrt{\pi}\,\Delta T} \tag{52}$$

The experimental procedure and calculation of a set of thermophysical properties are discussed in Refs. 20, 29–34.

So, in the case of constant heat sources and sinks, estimation of thermal diffusivity by the regular regime method requires the knowledge of a steady component (36) or (42). The systematic error due to non-one-dimensional heat fluxes can be estimated from the solution and analysis of the appropriate two- and three-dimensional heat conduction problems.[2, 20, 24, 25] These solutions being tedious, only some consequences are considered here. From (36)

$$\lambda = \frac{qh}{\Theta_{st}} \tag{53}$$

Solution and analysis of the appropriate two-dimensional problem give the formula differing from (53) by its coefficient[20]

$$\lambda = \frac{qh}{\Theta_{st}} \frac{2}{\kappa} \sum_{\kappa=1}^{\infty} \frac{\tanh \kappa\mu_n}{\mu_n^2\, \mathscr{T}_1(\mu_n)} \tag{54}$$

where μ_n are the roots of equation (15); $\kappa = 2h/2R$ is the relative height of the disk (plate). At $\kappa \to 0$ formula (54) coincides with (53). If $\kappa = \frac{1}{5}$, the relative error will be 0.8%; at $\kappa = \frac{1}{4}$ it is 1.4%. Thus, the systematic error increases with increasing height of the plate, which is observed in practice.

From relation (42)

TABLE 1
Summary of Monotonic Heating Regime Experiments in Measurements of Thermal Diffusivity of Solids

No.	Investigator	Ref.	Year	Material	Methods	Temperature measurement	Range	a (m^2 s × 10^7)	Accuracy (%) Max. error	Accuracy (%) Scatter	λ (W/m K)
1	Kondratiev, G.M.	4	1954	Disperse homo-geneous	Monotonic (regular heating)	t/c	300	0.25–2.5	No data	10–15	0.1–1
2	Kraev, O.A.	50	1956	Ceramics plastics	Regular heating	t/c	300–1300	5–10	No data	No data	5
3	Vasiliev, L.L.	46	1963	Disperse plastics	Adiabatic heating	t/c	10–400	0.8–5	No data	4	5
4	Platunov, E.S.	52	1964	Ceramics plastics	Regular regime	t/c	300–700	5–10	No data	3–8	5
5	Fraiman, Yu.E.	37	1964	Nonmetal	Linear heating	t/c	300–1500	0.8–5	No data	5–7	5
6	Volokhov, G.M.	27	1966	Disperse ceramics plastics	Monotonic heating with source	t/c	300	0.2–10	No data	2–5	0.1–5
7	Platunov, E.S.	22	1973	Metals	Regular heating	t/c	20–700	200–300	No data	5–8	100
8	Platunov, E.S.	22	1973	Refractory		Pyrometer	900–2000	5–10	No data	12	3
9	Shashkov, A.G. and Tyukaev, V.I.	57	1975	Heat insulators	Monotonic heating	t/c Pyrometer	3000	2–3	2	7–8	1

$$\lambda = \frac{qR_1}{\Theta_{st}} \ln \frac{R_2}{R_1} \tag{55}$$

The solution of the problem for a hollow finite cylinder yields[20]

$$\lambda = \frac{qR_1}{\Theta_{st}} \ln \frac{R_2}{R_1} \left[1 - \frac{2}{\kappa^2 \ln(1/\kappa)} \sum_{n=1}^{\infty} \frac{A_n}{\mu_n \cosh L\mu_n} \right] \tag{56}$$

where μ_n are the roots of the equation

$$\mathcal{T}_1(\mu_n \kappa) Y_0(\mu_n) = Y_1(\mu_n \kappa) \mathcal{T}_0(\mu_n)$$

$$\kappa = \frac{R_1}{R_2}, \qquad L = \frac{h}{R_2}$$

h is the cylinder half-height; L the relative height.

Without going into a detailed analysis of (56) note that in this case the systematic error can be estimated exactly.

2. MEASUREMENT OF THERMAL DIFFUSIVITY IN THE WIDE RANGE

Studies of thermal diffusivity as a function of temperature by successive thermostating of a sample at increasing temperatures cannot be considered convenient because of the great amount of time spent for the whole experiment. Therefore, the methods that seem promising are those which allow a temperature function of thermal diffusivity to be found within one continuous experiment. Solutions of the heat conduction equations [with heat flux at the surface prescribed or, in the case of heat transfer, with the ambient medium, whose temperature varies following a linear (or close to linear) law] are the theoretical basis for this group of methods.

Temperature fields in samples of very simple geometries heated symmetrically in a medium, whose temperature is a linear time function, with the assumption of the thermal properties independent of temperature are described by the expression[1]

$$\Theta = \frac{T(r,\tau) - T_0}{T_0} = \mathrm{Pd} \left\{ F_0 - \Gamma \left[\left(1 + \frac{2}{\mathrm{Bi}} \right) - \frac{r^2}{R^2} \right] \right\}$$

$$+ \sum_{n=1}^{\infty} \frac{A_n}{\mu_n^2} \Phi \left(\mu_n \frac{r}{R} \right) e^{-\mu_n^2 F_0} \tag{57}$$

where $\mathrm{Pd} = bl^2/aT_0$, $F_0 = a\tau/l^2$ are the dimensionless complexes including the characteristic dimension of the sample (for a plate $l \equiv h$, for a cylinder and a sphere $l \equiv R$); r is the variable coordinate (for a plate $r = x$). $\Gamma = \frac{1}{2}, \frac{1}{4}, \frac{1}{6}$ for an infinite plate, a cylinder and a sphere, respectively, the numbers μ_n are found from characteristic equations (11)–(13). The analysis of expression (57) has revealed two temperature regimes, unsteady and quasisteady (regular regime of the second kind).[1] Under unsteady regime, for calculation of temperatures and heat fluxes expression (57) should be considered. At a certain value of $F_0 \geqslant F_{0_1}$ (for a plate $F_0 > 0.5$[1]), the series in (57) can be neglected. Then

$$\Theta = \mathrm{Pd}\left[F_0 - \Gamma\left(1 + \frac{2}{\mathrm{Bi}} - \frac{r^2}{R^2}\right)\right] \tag{58}$$

i.e., the temperature at any point of the sample changes following the linear law, and the temperature distribution along the coordinate obeys the quadratic law. Such a regime is called quasisteady.[1] The problems of the determination of the onset of a quasisteady regime are quite completely described in Refs. 1 and 22, and will be omitted from the present review. The simple and predicted relations are directly obtained from (58):

$$a = \Gamma \frac{b(r_2^2 - r_1^2)}{\Delta T} \tag{59}$$

i.e., to determine thermal diffusivity, it is necessary to know the heating rate b and the temperature difference between some two points of the sample, say, at the center and on the surface. The other formula also follow from (59):

$$a = \Gamma \frac{r_2^2 - r_1^2}{\Delta\tau} \tag{60}$$

where $\Delta\tau$ is the lag time, i.e., the time required for the temperature at a certain point of the sample to become equal to the temperature at another point of the same sample. Similar relationships hold for the case when a constant heat flux is prescribed on the surface.[1] From expression (59) it follows that the heat flux on the sample surface $q = 2\Gamma(\lambda/a)bR = \text{const}$.

Formulas (59) and (60) are obtained from the solution of linear one-dimensional equation (1). Under real conditions the samples of finite dimensions are used, therefore, some corrections in (59), (60)[22] or the selection of the appropriate sample dimensions[20, 25, 35] are necessary. In Ref. 25, a general problem is considered on heating a finite cylinder in a medium with linearly changing temperature, whose side surface and bases are heated at different heating rates. The solutions (58) for an infinite plate and an infinite cylinder are obtained as a particular case of this solution. The formulas for thermal diffusivities at two-dimensional heat fluxes are more complex than relation (59). In a particular case of equal heating rates the formulas simplify. The factor Γ in (59) depends on the ratios of the linear dimensions of the cylinder. For several particular cases the values of these coefficients are the following:

Infinite plate	$\frac{1}{4}$	$\frac{1}{3}$	$\frac{1}{2}$	Infinite cylinder	3	2	1
0.5	0.4928	0.4752	0.4056	0.25	0.2496	0.2454	0.2006

For a plate, the temperature drop is measured between the center and base, and for a cylinder it is measured between the center and the side surface.

A method for determining thermal diffusivity under linear heating was originally suggested in Ref. 36 and its modifications were used by many authors.[20, 22, 37, 42] In this case the test material is placed in a medium whose temperature varies following the linear (or close to linear) law. A linear heating rate is preset by programmed temperature regulators.[37–40]

Thermal properties of the test material can continuously change during linear heating. A real thermogram is not linear but monotonic time function. Theoretical and experimental studies of many authors[20, 22, 37] have shown that thermal diffusivity as a function of temperature can be sought for from the formulas of form (60) if the experiment is performed at small temperature drops (5–10 K). The thermograms are processed after division into particular linear sections.

When in a sample there are heat sources of constant power and heating is linear, the capabilities of the quasisteady methods essentially increase since under such conditions they may be used for determination not only of the thermal diffusivity but also of the thermal conductivity and heat capacity.[37, 38, 41]

This variant is represented by the method of Refs. 37, 42, which is theoretically based on the solution of the equation

$$\frac{\partial T(r,\tau)}{\partial \tau} = \frac{1}{r}\frac{\partial}{\partial r}\left[r\frac{\partial T(r,\tau)}{\partial r}\right]$$

for a hollow cylinder with the inner radius R_1 and outer radius R_2 under the boundary conditions

$$T(r,0) = T_0$$

$$T_c = T_0 + b\tau$$

$$\lambda \frac{\partial T(R_2, \tau)}{\partial r} = \alpha[(T_0 + b\tau) - T(R_2, \tau)]$$

$$-\lambda \frac{\partial T(R_1, \tau)}{\partial r} = q$$

The formulas for calculating thermal diffusivity and thermal conductivity are the following:

$$a = \frac{b}{4\Delta T}\left(r_2^2 - r_1^2 - 2R_1^2 \frac{r_2}{r_1}\right) \tag{61}$$

$$\lambda = \frac{qR_1 \ln(r_2/r_1)}{\Delta T - \Delta T'} \tag{62}$$

The method provides for two experiments at $q = 0$ and $q = \text{const}$. In the former case (61) and in the latter (62) are calculated with $\Delta T'$ being a new temperature drop measured at $q = \text{const}$. The method has been applied for studying thermophysical properties of heat insulators for temperatures between 350 and 1000 K. The error of measurements is 5%–7%.

In some of the approaches the heat conduction equation is solved with a prescribed constant heat flux on the sample surface.

Different modifications of the above methods are known. In Refs. 43 and 44, a one-dimensional heat conduction equation for a plate $2h$ thick with the coordinate origin at its center is solved under the boundary conditions of the form

$$T(x, 0) = T_0$$

$$\lambda \frac{\partial T(h, \tau)}{\partial x} = q \tag{63}$$

$$\frac{\partial T(0, \tau)}{\partial x} = 0$$

Under these conditions a quasisteady regime temperature varies in time and space in the same way as with linear heating. Thermal diffusivity is estimated by the formula of form (59) and thermal conductivity is found from

$$\lambda = \frac{qh}{2\Delta T} \tag{64}$$

No special facilities have been used in the methods of Refs. 43 and 44 to prescribe a required regime. A constant heat flux is prescribed by using some similar samples with flat electric heaters between them.

Formula (64) does not take into account the specific heat of the source. Usually, for estimating the effect of the specific heat of the source the boundary condition of form (63) is supplied with a negative source:

$$q' = c\frac{\partial T}{\partial \tau} = cb$$

where c is the heat capacity of a heater. In Ref. 45, using the solution to the problem for an infinite cylinder, the formula for calculating thermal conductivity including heat capacity of the source is of the form

$$\lambda = \frac{(q - cb)R}{2\Delta T}$$

In Refs. 46–48, the solution of the heat conduction equation is obtained for an infinite cylinder having a constant-power heat source inside and ideal heat insulation ($\partial T(R, \tau)/\partial r = 0$) for its outer surface. This solution is used to develop a method for studying thermal conductivity and thermal diffusivity of poor heat conductors from 4.2 to 400 K. The thermophysical properties have been estimated not only under quasisteady conditions but also at the initial stage of heating with the aid of the compiled tables. The calculation error in this case is 2%–4%.

The quasisteady approach may also be implemented using a heat capacity standard (see, for instance, Ref. 49).

At high temperatures, the realization of the boundary conditions (constant heat flux or constant heating rate) involves considerable engineering difficulties, which has induced the development of the so-called monotonic regime (heating and cooling) methods, which are essentially a general case of the quasisteady heating method. The approximate solutions of the nonlinear heat conduction equation are a theoretical basis of the methods.[22, 50–56] Initially the monotonic heating methods were developed in Refs. 50 and 51. Th temperature field was expressed in a power series form, and finally the formulas were obtained which differed from (59) and (60) by the corrections which included the changes of the heating rate and thermal diffusivity

$$a = \frac{R^2}{\Gamma \Delta \tau}(1 + \epsilon + \delta) \tag{65}$$

where

$$\epsilon = -\frac{1}{4\Delta T}\frac{d\Delta T}{d\tau}\Delta\tau, \qquad \delta = \frac{1}{4a}\frac{da}{dT}\Delta T$$

The experiment was carried out under a heating regime close to the linear one without using any special temperature regulators. At small temperature drops ($\Delta T < 10\,\mathrm{K}$) the corrections can be neglected. Then expression (65) coincides with (60).

The theory and experimental bases of the monotonic heating methods are most completely described in Ref. 22. The main theoretical assumptions made in the above work may be briefly summarized as follows. Within a small temperature drop ΔT, if it does not involve the phase transition points, thermal properties may be expressed by the series with any required accuracy

$$i = i_0(1 + \kappa_i\Theta + n_i\Theta^2 + \cdots) \qquad (i = \lambda, a, c, \rho) \tag{66}$$

Relative temperature factors κ_i and n_i in series (66) are a function of the reference temperature $T_0(\tau)$ but remain constant within the drop:

$$\kappa_i = \frac{1}{i_0}\frac{di_0}{dT}, \qquad n_i = \frac{1}{2i_0}\frac{d^2 i_0}{dT^2} \tag{67}$$

Some point convenient for direct temperature measurement – say the center of the sample or its surface – is chosen as a reference point with the temperature $T_0(\tau)$. Then

$$T(r,\tau) = T_0(\tau) + \Theta(r,\tau) \tag{68}$$

It is also assumed that the heating rate $b(r,\tau)$ is a monotonic time function and a series of form (66) is valid for it. Usually $|\kappa_i| < 3 \times 10^{-3}\,\mathrm{K}^{-1}$ and $|n_i| < 3 \times 10^{-6}\,\mathrm{K}^{-1}$, therefore, the conditions of the optimum convergence of (66) are determined by the temperature drops which occur during experiment. The linear one-dimensional heat conduction equation was solved by the successive approximation method with the above assumptions.[22] The formulas for calculation of thermal diffusivity were obtained from the solution. In the case of symmetric heating of a plate, a cylinder, and a sphere the formulas become

$$a = \frac{b_0 R^2}{\Gamma\Theta_R}(1 + \Delta\sigma_{i_\Theta}) \tag{69}$$

where b_0 is the heating rate at the reference point, R is the characteristic dimension of the sample, Θ_R is the temperature drop with respect to the reference point. Formula (69) differs from (60) by the correction

$$\Delta\sigma_{a_\Theta} = \frac{\Gamma}{\Gamma+4}\left(\kappa_{b_T} - 2\kappa_a - \frac{4}{\Gamma}\kappa_\lambda\right)\theta_R \tag{70}$$

where κ are the relative temperature factors of the heating rate, thermal diffusivity, and thermal conductivity of form (67). The formula for calculation of thermal diffusivity in terms of the lag time is of a similar form:

$$a = \frac{R^2}{\tau_R(\tau)}(1 + \Delta\sigma_{a_T}) \tag{71}$$

Corrections in formulas (69), (71) are complicated and their practical calculation is quite difficult. In Ref. 22 it is shown that at heating rates corresponding to small temperature drops (3–10 K), the corrections in (70) may be neglected and expressions (59) and (60) of quasisteady heating can be used for calculations.

On the basis of the monotonic heating method, the devices and instruments have been developed, which are designed for determining thermal diffusivity and thermal conductivity of heat insulators and metals over a temperature range between −150 and 2000°C.[22, 57] The accuracy of determining thermophysical properties was estimated through an approximate thermal analysis.[22] The value of the measurement error (5%–12%) depends on the test temperature range and the class of the test materials. For realizing the monotonic heating methods it is possible to use special facilities for prescribing the required thermal regimes. Different versions of thermal devices are considered in Refs. 22 and 58.

APPENDIX

1. A schematic drawing of a simple experimental device for determination of thermal diffusivity of solid, loose and fibrous materials at room temperature is shown in Fig. 1.[20] For solid material test two hollow plane parallel units 1, 3, are used, which can travel over a vertical plane along the guides (they are not shown in the figure). Loose and fibrous materials are tested in a copper or brass cylinder 2, fitted with a case 6, and removable covers 7, 8. Two differential chromel–alumel thermocouples 0.2 mm in diameter serve as a thermometer. Junctions of each of the thermocouples are soldered beforehand at the lower unit plane 1, and on the inner surface of the cylinder, 2, respectively. The other junction of DT_2 is mounted in a tube or needle fixed at the cylinder center by

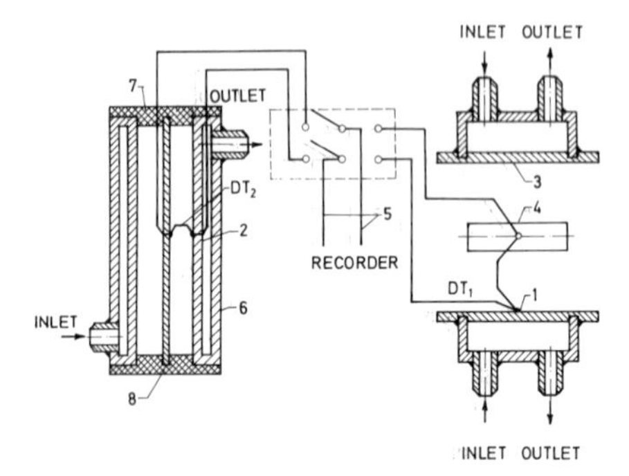

FIGURE 1. Device for determination of thermal diffusivity at room temperatures.

covers 7, 8. The junction of DT_1 is placed at the center of the test solid sample composed of two parallel plane square or cylinder plates 4. Cavities of the system are connected with a thermostat and filled with forced-through water of a constant temperature. So, on the plate bases and side surface of the cylinder, heat transfer occurs with the ambient medium of a constant temperature; the "free" surfaces exchange heat by convection with the surroundings at different temperature, say, room temperature – i.e., heat transfer with the surroundings of constant but different temperatures takes place. The analysis of the solutions of form (29) has revealed that for the present experimental procedure the sample dimensions should satisfy the following relations.

For a cylinder:

$$\kappa = \frac{2h}{2R} \geqslant 3$$

where h is the half-height of a cylinder and R is the inner radius of a cylinder.

For a plate:

$$\kappa = \frac{2h}{2R} \leqslant \frac{1}{4}$$

where h is the half-height of a plate and $2R$ is the diameter or the side of the square. In this case, heat transfer from the side surface of a plate and from the bases of a cylinder can be neglected, and formulas (17), (18) may be used in calculations.

In experiments with a solid sample, two parallel plane circular or square plates are made of the test material (the linear dimension ratio should satisfy the above formulas). Absolute dimensions of the sample depend on its structure and properties. Dense insulating materials (plastics, ceramics) can be of a small size, $h \approx 5$–8 mm. The sizes of the large-pore samples should be larger. A solid sample is initially outside the units filled with forced-through water of a constant temperature. Free ends of differential thermocouple DT_1 are connected with a self-recorder. After a stable temperature difference sets in, which can be inferred from the self-recorder readings, the sample is placed in the space between the units and pressed with their planes. Additional heat resistance can be considerably diminished by careful finish of the contact planes and strong compression of the samples. Thermal diffusivities are calculated by formula (17). When loose materials are tested, the cylinder is filled before the constant temperature water is supplied to its cavity. The results are processed in a way similar to the previous case. The calculation is performed according to formula (18). Thermal diffusivity of solid materials whose thermal conductivity lies between 0.1 and 5 W/m°C is accurate within 3%–5%.

2. A schematic drawing of a simple experimental device for determination of thermal diffusivities and thermal conductivities of solid, fibrous and loose materials is shown in Fig. 1. Its main elements are described in item 1 of the Appendix. This scheme is supplemented with cylindrical and plane constant-power heat sources and instruments for measuring the power (the instruments are not shown in Fig. 1). Both heaters should be of a low heat capacity. The cylindrical heater is wound on the frame of any thin material, say, thin copper or brass foil. The thermocouple junction DT_2 is fixed on the frame wall or at the center. The ready frame is covered with an insulating varnish or glue layer, and the heater of manganine or nichrome wire is wound thereafter. The tube heater is closed with plugs inserted into the recesses made in the covers of the cylindrical calorimeter. The heater is rigidly fixed to the lower cover, with the heater and thermocouple leads let through. Depending on the solid sample shape (a square, disk) the flat heaters are manufactured as a square or a circle from the thin wire of a high specific resistance or from the foil. The other junction of the differential thermocouple is glued into the heater.

The experimental procedure is as follows: after the hollow cylinder is filled with the test material and a flat test sample is placed together with the heater between the hollow units which press the sample, the thermostat should be switched on. After the temperature becomes uniform over the whole volume, which is shown by the readings of the temperature-difference records, the heaters are energized. The power to be supplied to the heater should be chosen according to the optimal temperature drop of 5 to 10 K. After the heaters are switched on, the excess temperature begins increasing and reaches its maximum in the steady state. During experiment relation (34) is obtained. Once the steady state is reached, steady-state temperature drops are determined, which can be used for plotting relation (40) and calculating the rate of temperature change (38). Thermal diffusivities are calculated from formulas (39) and (43). The values of μ are found from Table 2. The temperature behavior and the calculation methods are illustrated in Fig. 2.

TABLE 2
Values of the First Root of the Characteristic Equation
$\mathcal{T}_1(\mu l)\ Y_0(\mu) - \mathcal{T}_0(\mu)\ Y_1(\mu l) = 0$

l	μ_1	l	μ_1	l	μ_1
0.10	2.449	0.37	2.997	0.64	4.763
0.11	2.458	0.38	3.032	0.65	4.885
0.12	2.467	0.39	3.069	0.65	5.013
0.13	2.478	0.40	3.107	0.67	5.152
0.14	2.489	0.41	3.147	0.68	5.297
0.15	2.501	0.42	3.188	0.69	5.453
0.16	2.514	0.43	3.231	0.70	5.619
0.17	2.528	0.44	3.276	0.71	5.797
0.18	2.542	0.45	3.323	0.72	5.888
0.19	2.558	0.46	3.372	0.73	6.193
0.20	2.574	0.47	3.422	0.74	6.415
0.21	2.591	0.48	3.475	0.75	6.654
0.22	2.609	0.49	3.531	0.76	6.913
0.23	2.628	0.50	3.589	0.77	7.195
0.24	2.648	0.51	3.649	0.78	7.503
0.25	2.668	0.52	3.712	0.79	7.841
0.26	2.690	0.53	3.778	0.80	8.213
0.27	2.712	0.54	3.874	0.81	8.624
0.28	2.736	0.55	3.919	0.82	9.081
0.29	2.761	0.56	3.995	0.83	0.592
0.30	2.786	0.57	4.075	0.84	10.167
0.31	2.813	0.58	4.159	0.85	10.820
0.32	2.840	0.59	4.247	0.86	11.566
0.33	2.869	0.60	4.339	0.87	12.427
0.34	2.899	0.61	4.437	0.88	13.431
0.35	2.931	0.62	4.540	0.89	14.619
0.36	2.963	0.63	4.648	0.90	16.045

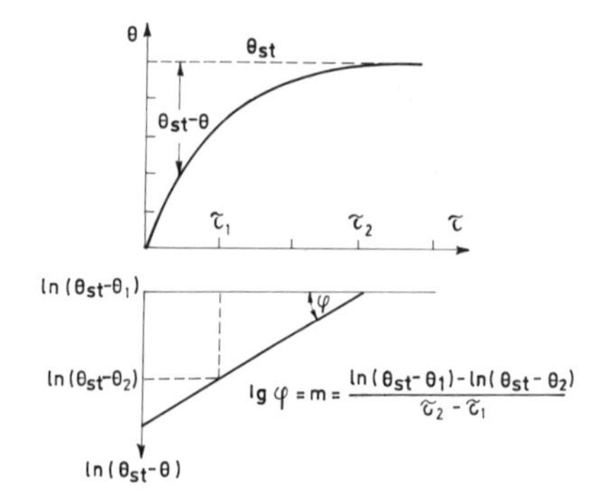

FIGURE 2. The method for calculating the rate of a temperature change with a heat source of constant power.

NOTATION

λ (W/m K)	Thermal conductivity
a (m^2/s)	Thermal diffusivity
c (J/kg K)	Specific heat
T (K)	Temperature
τ (s)	Time
Θ	Relative temperature
q (W/m^2)	Specific heat flux
α (W/m^2 K)	Heat transfer coefficient
$F_0 = a\tau/l^2$	Fourier number
$\mathrm{Bi} = \alpha l/\lambda$	Biot number
b (K/s)	Heating rate

REFERENCES

1. A.V. Luikov, *Analytical Heat Diffusion Theory,* Academic Press, New York (1968).
2. H.S. Carlslaw and D.C. Jaeger, *Conduction of Heat in Solids,* 2nd edition, Clarenden Press, Oxford (1959).
3. A.I. Tikhonov and A.A. Samarsky, *Mathematical Physics Equations,* Gostekhizdat Publ., Moscow (1951).
4. G.M. Kondratiev, *Regular Thermal Regime,* Gostekhizdat Publ., Moscow (1954).
5. G.M. Kondratiev, *Heat Measurements,* Mashgiz (1957).
6. G.N. Dulnev and G.M. Kondratiev, *Izv. Akad. Nauk SSSR, Otd. Tekhn. Nauk* 3 (1955).
7. G.N. Dulnev and G.M. Kondratiev, *Izv. Akad. Nauk SSSR, Otd. Tekhn. Nauk* 7 (1956).
8. N.A. Yaryshev, *Theoretical Fundamentals of Measuring Unsteady Temperatures,* Energya Publ., Leningrad (1967).
9. E.S. Fitzsimmons, *J. Am. Ceram. Soc.* **33**(11), 327 (1950).
10. A.F. Chudnovsky, *Heat Transfer in Dispersed Media,* Gostekhizdat (1954).
11. G.B. Simonov, *Stroit. Prom.* 8 (1952).
12. G.B. Simonov, *Zh. Teor. Fiz.,* vyp. 8 (1953).
13. A.F. Begunkova, *Zavod. Lab.* 10 (1952).
14. A.F. Begunkova, Sb. "Studies in the Region of Heat Measurements," Izd. LITMO, vyp. 20, Mashgiz, Leningrad (1956).
15. G.N. Tretiyachenko and L.V. Kravchuk, in *Heat and Mass Transfer,* Vol. 1, Izd. AN BESSR, Minsk (1962).
16. L.A. Plotnikov, *Zavod. Lab.* 9 (1950).
17. A.E. Paladino, E.L. Swarts, and W.B. Irndall, *J. Am. Ceram. Soc.* **40**, 340 (1957).
18. M.P. Emchenko, *Priborostroenie* 3 (1938).
19. M.P. Emchenko, *Tr. Leningr. Lesnoi Akad.* 83 (1959).
20. *Methods of Determining Thermal Conductivity and Thermal Diffusivity,* Energiya Publ., Moscow (1973).
21. A.V. Luikov, *Heat Conduction Theory,* GITTL, Moscow (1952).
22. E.S. Platunov, *Thermophysical Measurements at Monotonic Regime,* Energiya Publ., Leningrad (1973).
23. M.B. Das and M.A. Hossain, *Br. J. Appl. Phys.* **17**, 1 (1966).
24. G.M. Volokhov, *DAN BSSR, Ser. Fiz.-Tekhn. Nauk* 4 (1967).
25. G.M. Volokhov, V.P. Kozlov, and G.A. Surkov, *Heat and Mass Transfer,* Vol. 7, Minsk (1972).

26. V.O. Fogel and P.P. Alekseev, *Inzh. Fiz. Zh.* 2 (1962).
27. G.M. Volokhov, *Inzh. Fiz. Zh.* **9,** 5 (1966).
28. J. Hatta, *Rev. Sci. Instrum.* **28,** 50 (1979).
29. A.B. Verzhinskaya and L.N. Novichenok, *Inzh. Fiz. Zh.* 9 (1960).
30. A.B. Verzhinskaya, *Inzh. Fiz. Zh.* 4 (1964).
31. W.A. Plummer, D.E. Campbell, and A.A. Comstock, *J. Am. Ceram. Soc.* **45**(7), 310 (1962).
32. A.F. Chudnovsky, *Zh. Teor. Fiz.* **23,** 12 (1953).
33. J.C. Jager, *J. Geophys. Res.* **64,** 5 (1959).
34. W.E. Haupin, *Am. Ceram. Bull.* **39,** 3 (1960).
35. Yu.E. Fraiman and G.A. Surkov, *Inzh. Fiz. Zh.* **8,** 5.
36. A.V. Luikov, *Zh. Tekhn. Fiz.* **5,** 2–3 (1935).
37. Yu.E. Fraiman, *Inzh. Fiz. Zh.* 10 (1964).
38. A.V. Luikov *et al.,* in *Proc. 3rd All-Union Heat and Mass Transfer Conference,* Vol. 7, Nauka i Tekhn. Publ., Minsk (1968).
39. V.V. Vlasov, *Inzh. Fiz. Zh.* **7,** 3 (1964).
40. V.V. Vlasov and A.I. Fesenko, *Automation of Thermophysical Measurements,* Tambov (1972).
41. L.L. Vasiliev and S.A. Taneava, *Thermal Properties of Porous Materials,* Nauka i Tekhnika Publ., Minsk (1971).
42. L.L. Vasiliev and Yu.E. Fraiman, *Thermophysical Properties of Bad Heat Conductors,* Minsk (1967).
43. O. Krischer, *Chem. Ing. Techn.* **26**(1), 42 (1954).
44. L.A. Selinov, *Proc. Postov Engineering-Building Institute,* vyp. IV (1955).
45. T.S.E. Thomas, *Br. J. Appl. Phys.* **8,** 10 (1957).
46. L.L. Vasiliev, *Inzh. Fiz. Zh.* **6,** 9 (1963).
47. L.L. Vasiliev and G.A. Surkov, *Inzh. Fiz. Zh.* **7,** 6 (1964).
48. L.L. Vasiliev, S.A. Tanaeva, and A.D. Shnyrev, *Inzh. Fiz. Zh.* **17,** 6 (1969).
49. Th. Gast, Hellwege, and K. Knappe, *Kolloid Z.* 174 (1961).
50. O.A. Kraev, *Teploenergetika* 4 (1956).
51. O.A. Kraev, *Teploenergetika* 4 (1957).
52. E.S. Platunov, *Izv. Vuzov, Priborstr.* **1,** 1 (1961).
53. E.S. Platunov, *Teml. Vysok Temp.* **3,** 5 (1964).
54. E.S. Platunov, *Inzh. Fiz. Zh.* **9,** 4 (1965).
55. E.S. Platunov and S.E. Burovoi, *Izv. Vuzov, Priborstr.* **11,** 7 (1968).
56. E.S. Platunov, in *Proc. 3rd All-Union Heat and Mass Transfer Conference,* Nauka i Tekhnika Publ., Minsk (1968).
57. A.G. Shashkov and V.I. Tyukaev, *Thermal Properties of Decaying Materials at High Temperatures,* Nauka i Tekhnika Publ., Minsk (1975).
58. S.E. Burovoi, V.V. Kurenin, and E.S. Platunov, *Inzh. Fiz. Zh.* **30,** 4 (1976).

IV

SPECIFIC HEAT MEASUREMENT METHODS

Specific Heat

An Introduction

Six chapters are devoted to six different calorimetry techniques used for the measurement of specific heat mainly above room temperature and extending up to several thousand degrees. These techniques are: adiabatic, drop, levitation, modulation, pulse, and differential scanning calorimetry.

Although some of the calorimetry techniques, such as those that utilize the method of mixtures, date back over one hundred years, they have been used continuously after a succession of improvements and adaptations to emerging new requirements. A major surge in calorimetry came about in the mid-1950s and extended over approximately two decades. During this period, several conventional calorimetry techniques, such as adiabatic and drop, were perfected to their practical limits. Also during this period, several specialized calorimetry techniques such as levitation, modulation, and pulse, were developed for the measurements of specific heat at high temperatures and under other extreme conditions (high pressures, etc.).

In some of the calorimetry techniques (adiabatic, modulation, pulse, differential scanning), specific heat is obtained directly from the measured quantities; in others (drop, levitation), direct measurements yield enthalpy data which after differentiation yield specific heat.

The nature and the form of the specimen dictates, in some cases, the selections of the particular technique. For example, in adiabatic and drop techniques, the specimen is generally contained in a capsule which enables the use of a specimen practically in any form (powder, granules, chunks, etc.). Also, there is no restriction as to the specimen's electrical conduction characteristics. In some of the more specialized methods, such as in modulation and pulse, resistive self-heating is used which limits the technique to electrically conducting specimens in the form of wires, rods, or tubes.

Near room and at moderately high temperatures up to 1000 K, the measurement of specific heat does not present any serious problems. At higher temperatures and especially at temperatures above 2000 K, heat loss due to thermal radiation becomes a very important parameter. In addition, chemical reactions and specimen evaporation at high temperatures set an upper temperature limit for the applicability of the steady-state techniques for measurements of specific heat. Steady-state techniques have been used for measurements up to 3000 K and in a few cases to even somewhat higher temperatures; however, the limit for accurate measurements is about 2000 K. Pulse techniques with very short experiment duration (subsecond and submillisecond) can be used for measurements at very high temperatures (up to 10 000 K).

Almost all the calorimetry techniques have benefited immensely from the advances, during the last three decades, in general electronics and computer technologies. As a result, improvements in temperature and power controls as well as improvements in the measuring instruments and data reduction techniques contributed to the increased accuracy in specific heat determinations.

At present, it is possible to measure, with the best available techniques, specific heat with an uncertainty of not more than 1% in the temperature range 300–1000 K, not more than 2% in the range 1000–2000 K, and about 3% in the range 2000–3000 K. Above 3000 K, definitive measurements are not yet available; it may be said, however, that uncertainty in specific heat above 3000 K increases rapidly reaching an estimated value of 10% at 7000 K.

The six calorimetry techniques for the measurement of specific heat presented in this volume are discussed briefly in the following paragraphs. A summary table is also included (Table 1).

Adiabatic calorimetry is possibly the most versatile calorimetry technique for the direct measurement of thermal effects in substances, covering the range from cryogenic to moderately high temperatures. The technique has been used at temperatures as low as the helium point and as high as near 1900 K. However, for accurate measurements, the upper temperature limit is generally considered to be about 1300 K. This technique is used to measure specific heat of solids and liquids as well as heats of phase transitions, heats of solution and formation, and heat effects associated with structural changes in the specimen. This technique has high sensitivity, and as a result, can be used most effectively for measurements of specific heat near first- and second-order phase transitions.

Drop calorimetry, which is based on the classical method of mixtures, is generally used for measurements above room temperature. Although drop calorimeters have been built to operate at temperatures as high as near 3000 K, the accuracy of this technique deteriorates rather rapidly above 2000 K. The most accurate drop calorimeters available at present operate in the temperature range 300–1100 K. Since the specimen is generally placed in a capsule, the technique is applicable to both solids and liquids provided that no chemical reactions or alloying take place between the specimen and the container.

TABLE 1
Specific Heat Measurement Techniques

Technique	Advantages	Disadvantages	Temp. range[a] (K)	Specimen material	Uncertainty (%)
Adiabatic	Very versatile High sensitivity Solid and liquid specimen	Specimen in container Limitations in high temperatures	4–1300	All	1–3
Drop	Solid and liquid specimen	Specimen in container Slow measurement times Specific heat obtained from enthalpy data	300–2000	All	1–3
Levitation	No container for specimen Solid and liquid specimen	Specimen must be electrical conductor Specific heat obtained from enthalpy data	1000–2500	Electrical conductor (sphere)	2–5
Modulation	No container for specimen Multiproperty measurement capability	Solid specimen	80–3000	All	2–5
Pulse	No container for specimen Measurements at very high temperatures Solid and liquid specimen Multiproperty measurement capability	Specimen must be electrical conductor Sophisticated instrumentation	1000–7000	Electrical conductor (wire, rod, tube)	2–10
Differential scanning	Quick and economical	Specimen in container Limitations in high temperatures	100–1000	All	1–3

[a] For each technique measurements at temperatures higher than that indicated have also been performed. However, the given value indicates the limit of the accurate measurements for that technique.

Accurate values of specific heat cannot be derived from drop calorimetry alone if nonequilibrium states are produced in the specimen during the experiment.

Levitation calorimetry is a variant of the method of mixtures. The main difference between levitation and drop calorimetry is the manner in which the specimen is held before the drop. While in drop calorimetry the specimen is usually placed in a capsule suspended in a furnace with a metal wire, in levitation calorimetry the specimen is levitated by a high-frequency electromagnetic field and is heated inductively. The key advantage of the levitation calorimetry is the fact that the specimen is not in contact with any other substance, other than the surrounding inert gas; thus the technique is immune to container problems. Levitation calorimetry is particularly attractive for measurements at temperatures above 2000 K, and can be used for electrically conducting specimens both in their solid and liquid phases. In some cases, excursions of the order of 1000 K above the melting point of the specimen have been achieved. Vapor pressure of the specimen usually sets the upper temperature limit for this technique.

Modulation calorimetry is applied to electrically conducting and non-conducting solids and is based on measurements performed by modulating the power used to heat the specimen. This method may be used from low temperatures up to near the melting point of the specimen. Since the amplitude of temperature oscillations can be as small as of the order of 0.001 K, the method provides high measurement sensitivity, and is suitable for measurements near solid–solid phase transformations. The technique is used for measurements at high temperatures as well as at high pressures.

Pulse calorimetry is developed to extend the limits of accurate measurements of specific heat of electrically conducting specimens to temperatures beyond the limits of other calorimetry techniques. Because of the extremely short experiment duration (subsecond to submillisecond), this technique is immune to most of the high-temperature problems that arise from heat losses, chemical reactions, evaporation, etc. Pulse calorimetry is generally used at temperatures above about 1000 K. In the case of subsecond-duration techniques, the upper temperature limit is the melting point of the specimen. In order to extend the measurements beyond the melting point and to several thousand degrees in the liquid phase, submillisecond-duration techniques are used.

Differential scanning calorimetry (DSC) and its predecessor, differential thermal analysis, have long been used for qualitative or semiquantitative studies of the behavior of materials. It has been only during the recent years that DSC has become a truly quantitative calorimetry technique. Commercial availability of differential scanning calorimeters and relatively rapid generation of reliable data over a reasonably wide temperature range have contributed to the widespread use of the DSC technique. The temperature range of the available DSC instruments is 100–1000 K with scanning rates of from a fraction of a degree to upwards of a hundred degrees per minute in both heating and cooling.

12

Adiabatic Calorimetry

D.N. KAGAN

1. INTRODUCTION

Adiabatic calorimetry is a most precise method of direct measurement of thermal effects accompanying thermodynamic transformations of substances. As distinct from many other experimental methods of investigation of thermodynamic characteristics, direct measurements by adiabatic calorimeters result in determination of both heat capacity and heats of phase transitions, chemical reactions, formation of alloys, processes of dissolution, and so on. High sensitivity and superb accuracy of the calorimetry, enabling the measurements to be performed even in the region of a phase transition of the second kind and, in particular, in the region of the critical state of the substance involved, are due to the adiabatic conditions of the calorimetry process. Heat loss, in this case, is excluded, and practically no need arises to introduce the corresponding corrections, which, as a rule, are far from being ideally precise. The adiabatic conditions of the calorimetry process are achieved at the expense of equality of the temperature of the calorimetric vessel surface and that of the adiabatic shield.

Heat exchange between the sample under test and the environment in the variable-temperature adiabatic calorimeters is removed by means of the shields whose temperature-control system perfectly ensures that the temperature is held equalized with that of the sample throughout the calorimetry process.

The achievements in the field of electronic control and measurement systems as well as in the sphere of computing technology obtained for the past 20 years have found their application in thermophysical research and, in particular, in adiabatic calorimetry, which has resulted in materially higher reliability of the method and in streamlined procedures for the experiments. The balance

D.N. KAGAN • Institute of High Temperatures, Korovinskoe Road, USSR Academy of Sciences, Moscow 127412, USSR.

of the temperature of the sample surface and that of the adiabatic shield is obtained best of all in the constant-temperature calorimeters in which the phase transition of the calorimetric substance is used for measuring the thermal effect. In this case the adiabatic conditions are provided by placing the calorimeter into a thermostat wherein the same phase transition takes place.

The variable-temperature adiabatic calorimeters can operate in continuous and stage heating modes. In the case of the continuous-heating operation the rate of heating usually lies within $0.1\,K\,s^{-1}$. When operating in the stage heating mode the rise of the temperature does not exceed a few degrees. These values are required for maintaining the equilibrium conditions in the sample. The lower restrictions are determined by gain of contribution of small heat loss with decrease of the heating rate. The said requirement concerning the equilibrium conditions in the sample confines the maximum dimensions of the sample (with consideration for its coefficient of thermal diffusivity) to 20–30 mm in diameter and 50–60 mm in length. The minimum dimensions of the sample are limited by the increase of the contributions of the heat capacity made by the calorimeter component parts, which leads to a lesser accuracy of the results.

The adiabatic calorimetry, in essence, is void of any limitations with respect to the materials to be tested and therefore is applicable to any substance in the condensed phase. It stands to reason that measurement of heat capacity of solids is found to be most suitable and correct. The use of the capsule in studies of a liquid phase causes deterioration of the correlation between the heat capacity of the substance under test and that of the calorimeter parts, which, in general, brings about some additional errors of the measurement results, but, in principle, imposes no restriction.

The temperatures in which adiabatic calorimetry is found applicable range from liquid-helium temperatures to a level of 1800–1900 K. The upper limitation is determined by a difficulty of meeting the adiabatic conditions at an extremely high intensity of heat exchange through radiation. The errors of the measurement results in good apparatus do not exceed 0.1%–0.3% at moderate and low temperatures; however, they tend to rise approximately by one order under high temperatures.

2. METHOD

Adiabatic calorimetry is considered one of the classical methods of measurements of true heat capacity and thermal effects of transformations of substances. The optimum condition for a calorimetry process is the absolute absence of heat flow across the boundary of the sample involved, i.e., whenever the following equation holds true at any point of the said boundary:

$$\frac{\partial T}{\partial \mathbf{n}} = 0 \tag{11}$$

where **n** is the vector normal to the sample surface.

The uppermost accuracy of the calorimetry procedure can be obtained when no heat exchange with the environment takes place and, consequently, no need arises to introduce substantial corrections (such as the Reyno–Pfaundler correction in diathermal-shield calorimeters) to the main effect under measurement. The means for meeting condition (1) is enclosing the sample in an adiabatic shield with the temperature at each point of the shield being equal to that of the sample boundary throughout the period of the experiment. The temperature of the chamber enclosing the adiabatic shield either remains unchanged (isothermal chamber) or automatically follows the temperature of the adiabatic shield, the difference between the temperatures of the chamber and the shield remaining, as a rule, constant (follow-up chamber). The specific features of operation of the calorimeters with isothermal chambers are treated in Ref. 1. The peculiarities of the calorimeters with follow-up chambers are discussed in Refs. 2 and 3.

The adiabatic shield is designed not only for preventing loss of heat liberated in the process under consideration but also for compensating for heat flow produced by permanent side effects such as heat exchange over the fastenings or suspension parts of the calorimeter, heat evolution in the measuring system of the calorimeter, etc. The temperature compensation is confirmed by the absence of temperature change of the calorimeter in the initial and final periods of the experiment. This is precisely the condition for establishing the required negligible differences of the temperatures between the surface of the calorimeter and the adiabatic shield in some particular types of calorimeters. With a view to avoid heat loss in the process under study it is found imperative to hold this temperature difference constant throughout the time of the experiment (in thoroughly developed designs of the calorimeters this difference approximating zero).

Thus, in the practical case the condition for providing adiabatic conditions (adiabatization) of the calorimeter experiment can be expressed by the equation

$$\oint_s \frac{\partial T}{\partial \mathbf{n}} ds = 0 \tag{2}$$

where ds is the element of the sample surface.

Nevertheless, in real cases some heat loss may occur due to feasible alterations of temperature gradients along the sample boundary in the main period and also because of inevitable imperfections of any control system meant for regulating the adiabatic shield temperature. However this heat loss is insignificant.

The corresponding correction is also small, and therefore the probable error, which may still remain after introducing the correction will also be negligible.

Depending on the desiderata of the investigations adiabatic calorimeters have found principal applications either in realization of the method of direct heating, i.e., for measuring true capacity and heat of phase transitions, or in reaction calorimetry, i.e., for measuring heat of reaction, heat of alloy mixing, heat of solution, etc.

The method of direct heating is based on the principle of imparting the amount of heat under measurement to the sample electrically. The possibility of reliable measurement of this value ensures a high sensitivity and precision in determination of heat capacity (including the phase transition regions). Heat may be supplied to the sample either periodically (by steps) or continuously. The first case is characterized by a temperature equilibrium in the sample in the initial and final periods of the experiment and some thermal imbalance in the main period of the experiment (in the period of heat application). In the second case some imbalance, i.e., availability of temperature gradients in the sample, exists during the entire period of the calorimetric experiment. For obtaining accurate data of the experiment it is necessary that these gradients remain practically unchanged during the measurements, i.e., all the sections of the calorimeter should get heated at a similar rate that will correspond to the so-called quasistationary mode of heating.[4,5] It is easy to show that this requirement can be best met for relatively small samples with a high coefficient of heat conduction and a small coefficient of heat emission (small Biot's criterion) and for fairly small rates of heating.

The analysis of temperature fields in direct-heating calorimeters of various types with an analytical determination of the time required for assuming the steady-state condition is given in Ref. 6. The detailed analysis[7] shows that deteriorations of the adiabatic conditions are identical for both methods, with the exception of the moments of switching-on and switching-off of the heater in the case of the stage (periodical) heating method. Thus, as far as accuracy is concerned, both the modifications are more or less equivalent, and selection of a certain method is determined by the problems to be solved in the experiment and also by the properties of the substance under investigation. For example, if the substance under test requires a considerable length of time for establishing equilibrium (multiphase systems, substances close to critical points, etc.) then preference should be given to the step heating method. For investigating substances possessing good heat conduction, the continuous-heating method has proven to be more convenient. The continuous application of heat requires less time for measurements and results in less heat loss, which is very essential at high temperatures in view of rise of radiation heat exchange.

Typically, the rate of heating in adiabatic calorimetry does not exceed

TABLE 1
Works on Development of Modulation Method

Subject covered	Author(s), year	Ref.
General theory, supplementary-current method	Corbino, 1910	1
Third-harmonic method	Corbino, 1911	2
Use of thermionic-current oscillations	Smith and Bigler, 1922	3
Development of the third-harmonic method	Filippov, 1960	4
	Rosenthal, 1961	5
Bridge circuit for wire samples	Kraftmakher, 1962	6
Photodetectors	Lowenthal, 1963	7
Electron-bombardment heating	Filippov and Yurchak, 1965	8
Liquid metals	Akhmatova, 1965	9
Account of temperature gradients	Holland and Smith, 1966	10
Account of thermal resistances, measurements at low temperatures	Sullivan and Seidel, 1966	11, 12
Modulated-light heating	Handler, Mapother and Rayl, 1967	13
Induction heating	Filippov and Makarenko, 1968	14
Nonconducting materials	Glass, 1968	15
Measurement of the temperature coefficient of specific heat	Kraftmakher and Tonaevskii, 1972	16
Nonadiabatic regime	Varchenko and Kraftmakher, 1973	17
High pressures	Bonilla and Garland, 1974	18
Microcalorimeters for organic materials	Schantz and Johnson, 1978	19
	Smaardyk and Mochel, 1978	20
	Tanasijczuk and Oja, 1978	21
Improvement of the modulated-light method	Ikeda and Ishikawa, 1979	22
Measurement of specific heat using temperature fluctuations	Kraftmakher and Krylov, 1980	23, 24
Frequencies up to 10^5-10^6 Hz	Kraftmakher, 1981	25

Table 1 compiles the works on development of the modulation method for measuring specific heat.[1–25] Together with the term "modulation method,"[6] the term "ac calorimetry"[11,12] is widely used now. In this review, as well as in other review papers,[26–28] we will use the term "modulation method".

2. THEORY OF MODULATION METHOD

The basic theory of the modulation method is quite simple. If the power heating the sample is modulated by a sine wave ($p_0 + p \sin \omega t$) then the sample temperature starts to oscillate around a mean value T_0. The heat balance equation takes the form

$$mcT' + P(T) = p_0 + p \sin \omega t \tag{1}$$

where m and c are the mass and specific heat of the sample, $P(T)$ is heat loss from the sample, $T' = dT/dt$, and ω is the modulation frequency ($\omega = 2\pi f$). Assuming $T = T_0 + \Theta$, $\Theta \ll T_0$, and taking $P(T) = P(T_0) + P'\Theta$ (P' is the heat transfer coefficient), one obtains the steady-state solution of the equation (1):

$$\begin{aligned} T &= T_0 + \Theta_0 \sin(\omega t - \varphi) \\ P(T_0) &= p_0 \\ \Theta_0 &= (p/mc\omega) \sin \varphi \\ \tan \varphi &= mc\omega/P' \end{aligned} \tag{2}$$

The condition $\tan \varphi \gg 1$ ($\sin \varphi \approx 1$) is the criterion of the adiabatic regime of measurements. If the angle φ is close to 90° the corrections for heat losses can be neglected. Adiabatic conditions mean that the amplitude of oscillations of the power heating the sample (p) is much larger than the amplitude of oscillations of the heat loss from the sample due to temperature oscillations ($P'\Theta_0$). Although the heat trasnsfer coefficient grows rapidly with increasing temperature, the regime of the measurements can be kept adiabatic by increasing the modulation frequency. Under adiabatic conditions, for calculating the heat capacity of the sample it is enough to determine the quantities p, ω, and Θ_0.

With separate heater the thermometer, one has to take into account thermal resistances in the system, as well as a finite thermal conductivity of the sample. Sullivan and Seidel[11,12] have considered a heater, sample, and thermometer interconnected by thermal conductances (Fig. 1a). An ac power applied to the heater provides an oscillating heat input to the sample. The heat flows through the sample and out to the heat sink via the thermal link. At first

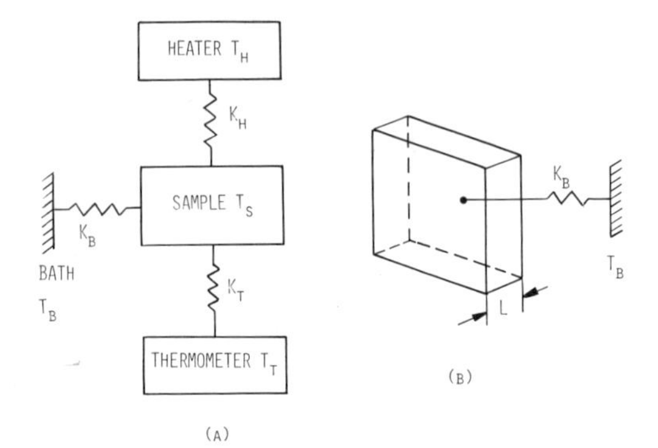

FIGURE 1. Calculation of the temperature oscillations in the system heater–sample–thermometer.[12]

the thermal conductivity of the sample was assumed to be infinite. The heat balance equations for the system (for the designations see Fig. 1) are

$$
\begin{aligned}
C_h T'_h &= p_0 + p \sin \omega t - K_h(T_h - T_s) \\
C_s T'_s &= K_h(T_h - T_s) - K_b(T_s - T_b) - K_t(T_s - T_t) \\
C_t T'_t &= K_t(T_s - T_t)
\end{aligned}
\tag{3}
$$

The temperature variations are sufficiently small so that the parameters C and K may be considered constant. The steady-state solution of these simultaneous equations consists of two terms, one a constant that depends upon K_b and the other an oscillatory term inversely proportional to the heat capacity. The expression for T_t is

$$T_t = T_b + p_0/K_b + (pB/\omega C) \sin (\omega t - \varphi) \tag{4}$$

where $C = C_s + C_h + C_t$, and B is a complicated expression involving quantities from equations (3); φ is a phase angle approximately equal to 90° under the conditions discussed below.

If (1) the heat capacities of the heater and thermometer are much less than that of the sample, (2) the sample, heater, and thermometer come to equilibrium in a time much shorter than the modulation period, and (3) the period of modulation is much shorter than the sample-to-bath relaxation time, then to first order in $1/\omega^2\tau_s^2$ and $\omega^2(\tau_h^2 + \tau_t^2)$,

$$
\begin{aligned}
B &= [1 + 1/\omega^2\tau_s^2 + \omega^2(\tau_h^2 + \tau_t^2)]^{-1/2} \\
\tan \varphi &= [1/\omega\tau_s - \omega(\tau_h + \tau_t)]^{-1}
\end{aligned}
\tag{5}
$$

where the relaxation times are defined as $\tau_s = C/K_b$, $\tau_h = C_h/K_h$, and $\tau_t = C_t/K_t$.

For determining the effect of the finite thermal conductivity of the sample, Sullivan and Seidel have considered a sample in the form of a slab of thickness L (Fig. 1b) which is heated uniformly on one side ($x = 0$) by a sinusoidal heat flux. The other side of the slab ($x = L$) is coupled uniformly to the bath through the thermal conductance K_b. The temperature oscillations of the sample can be expressed in the form

$$\Theta_0 = (p/\omega C) \, [1 + 1/\omega^2\tau_s^2 + \omega^2\tau^2 + 2K_b/3K_s]^{-1/2} \tag{6}$$

where various time constants have been lumped into $\tau, \tau^2 = \tau_h^2 + \tau_t^2 + \tau_{\text{int}}^2, \tau_{\text{int}}$ is the internal relaxation time of the sample depending on its thermal diffusivity and thickness, and K_s is the thermal conductance of the sample in the direction of heat flow.

Under operating conditions, τ_s is by two to three orders larger than τ so that condition $\omega^2\tau_s^2 \gg 1 \gg \omega^2\tau^2$ is satisfied by the proper choice of the modulation frequency. The term $2K_b/3K_s$ is very small due to the small thickness of the sample and can be neglected. The most modulation measurements were performed under conditions when the above assumptions are well satisfied so that the expression for calculating the heat capacity of the sample takes the form

$$mc = p/\omega\Theta_0 \tag{7}$$

3. MODULATION OF HEATING POWER

3.1. Direct Heating

Samples with sufficiently high electric conductivity, in the form of a wire, a foil, or a rod, can be conveniently heated by passing an electric current through them. Three methods of modulation of the heating power have been used:

(1) Heating by alternating current. The modulation frequency is twice the frequency of the current. The amplitude of the power oscillations is equal to the effective power and can be easily determined. If the current is the only means of heating the sample and the measurements are made in a wide temperature range, then the amplitude of the temperature oscillations strongly changes with the temperature.

(2) Heating by direct current with a small ac component. The modulation frequency is equal to the frequency of the ac component. In this case the mean temperature and the amplitude of temperature oscillations can be controlled independently. An important advantage of this regime is that the measurements can be made by the equivalent-impedance method when specific heat of the sample is calculated from the parameters of a balanced bridge or potentiometer circuit which do not depend on the amplitude of the ac component.

(3) Heating by high-frequency current modulated with a necessary frequency. Such a manner of heating is useful when sample temperature and its oscillations are measured with a thermocouple connected electrically with the sample. The low-frequency signal corresponding to the temperature oscillations can be easily separated from a high-frequency voltage which, in this case, can branch off to the thermocouple circuit.

3.2. Electron-Bombardment Heating

In the electron-bombardment heating the heating power can be modulated in several ways:

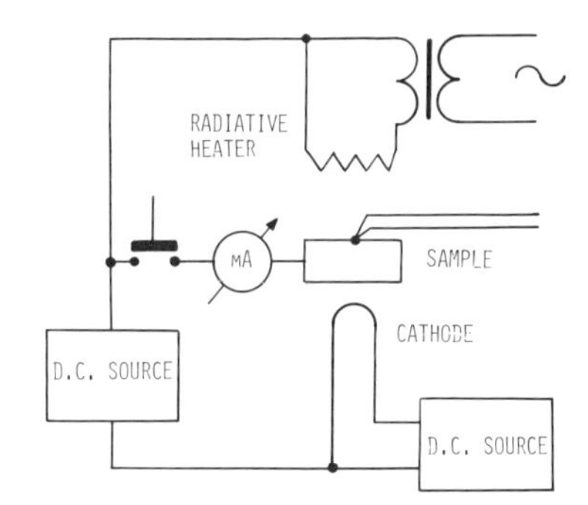

FIGURE 2. Electron-bombardment heating. Use of a radiative heater for changing the mean temperature of the sample.[29]

(1) In the saturation regime the accelerating voltage is modulated, the electron beam current remaining constant.

(2) The temperature of the cathode is varied or a control electrode is used to produce the modulation of the electron beam current, the accelerating voltage remaining constant.

(3) The accelerating voltage is periodically switched on and off so that the form of the heating-power pulses is rectangular. A disadvantage of this method is that when the mean heating power is increased the amplitude of power oscillations increases too. A separate source of heating can be used for eliminating this effect[29] (Fig. 2).

3.3. *Use of Electrical Heaters*

Separate heaters for the periodical heating have been used mainly at low temperatures and in the microcalorimeters for organic substances (Fig. 3). The main requirements to a heater are its small heat capacity and good thermal contact with the sample. Therefore for this purpose microresistors or specially deposited thin films are often used. An alternating current is used for heating and high accuracy in the determination of the heating-power oscillations can be achieved when modern measuring equipment is employed.

3.4. *Modulated-Light Heating*

This elegant method was first used in the measurements of specific heat of nickel near the Curie point[13] and is widely employed now. A sample in the form of a foil is placed in a furnace which controls the mean temperature (Fig. 4). The sample is illuminated by the light from an incandescent lamp passed through a chopper. The temperature oscillations, of the order of 10^{-2} K, are measured with a thermocouple connected with a lock-in amplifier. A photocell produces the reference voltage for the amplifier. The output signal from the lock-in amplifier is a dc voltage proportional to the input ac voltage, i.e., inversely

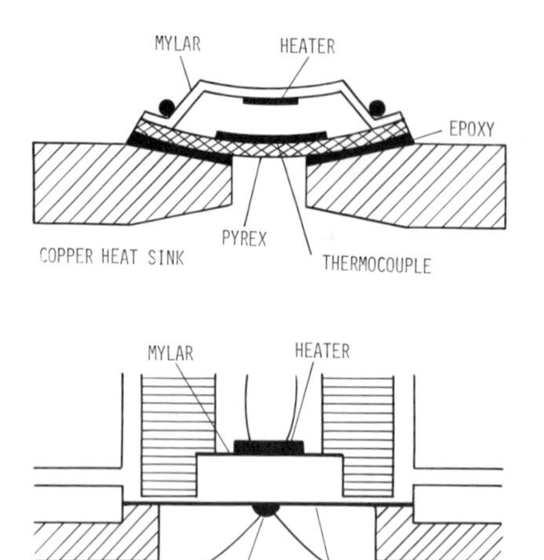

FIGURE 3. Use of separate heaters in microcalorimeters for organic liquids.[20,21]

proportional to the specific heat of the sample. It is fed to the Y-input of a recorder. The X-input if fed from a thermocouple which measures the mean temperature in the furnace. The measurements are made with the furnace temperature gradually changing. The modulation frequency is selected based on the thickness and thermal diffusivity of the sample.

In the majority of works using this method the absolute magnitude of the heating-power oscillations was not determined and only the measures were taken to make it independent of the sample temperature. For this purpose the samples were usually covered with a thin layer of graphite.

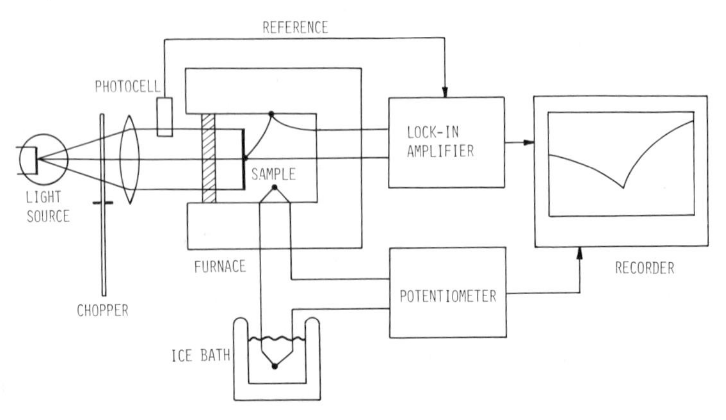

FIGURE 4. Modulated-light heating. The amplitude of temperature oscillations in a nickel foil is recorded automatically as a function of the mean temperature.[13]

3.5. Induction Heating

Induction heating of the samples is achieved by placing them in an induction furnace whose high-frequency current is modulated with a low frequency. To measure the supplied power, a separate coil is used; voltage induced in the coil is measured. Electrical conductivity of the sample, which must be known in order to calculate the heating-power oscillations, is measured by means of current and potential probes.[14]

4. REGISTRATION OF TEMPERATURE OSCILLATIONS

4.1. Use of Temperature Dependence of Sample Resistance

Determination of the temperature oscillations of the sample from the oscillations of its electrical resistance has been used by many workers. The advantage of this method is that measurements can be carried out also at relatively low temperatures when the brightness of the samples is small; also, high modulation frequencies can be employed at high temperatures since the problem of the inertia of a temperature sensor is here eliminated. The only, but very serious, drawback of the method is the necessity of knowing the temperature dependence of both the sample resistance and its temperature derivative. Therefore the method should not be applied when these quantities exhibit anomalies. Sample resistance oscillations can be used to detect temperature oscillations in different ways: by the supplementary-current method, by the third-harmonic method, and by the equivalent-impedance method.

4.1.1. Supplementary-Current Method

The method was first used by Corbino.[1] He observed the appearance of a dc voltage across a sample when a supplementary ac current was passed through it with a frequency twice that of the heating current, i.e., with the frequency of sample termperature oscillations. We have used a modified version of this method: a supplementary current is passed through the sample with a frequency close to the frequency of the temperature oscillations. This results in a low-frequency, rather than dc, voltage appearing across the sample, with the amplitude proportional to the amplitude of the sample resistance oscillations and to the magnitude of the supplementary current. This method has the advantage that the low-frequency signal is related only to the oscillations of the sample resistance, which accounts for the high noise-immunity of the method. The supplementary-current method was used for measuring the temperature coefficient of specific heat when it was necessary to measure the amplitude of the

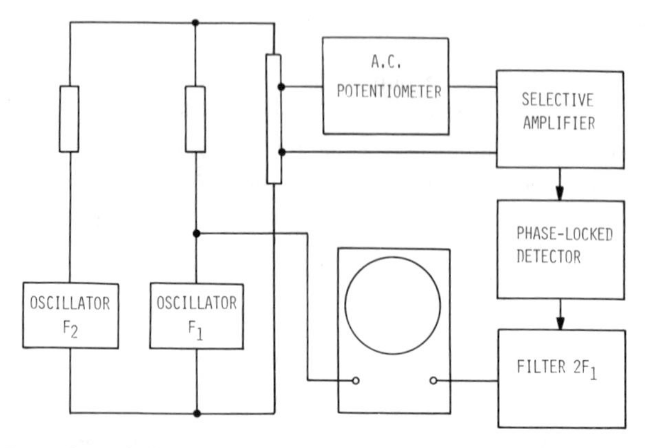

FIGURE 5. Detection of the resistance oscillations by a supplementary current of higher frequency.[30]

second-harmonic temperature oscillations in the presence of a much stronger signal of the fundamental frequency.[16]

Gerlich *et al.*[30] used a different version of the supplementary-current method. A current I_1 with the frequency $f_1 = 1\,\text{Hz}$ was passed through the sample, setting up temperature oscillations at 2 Hz. The resistance oscillations were detected by means of a supplementary current with the frequency $f_2 = 320\,\text{Hz}$ (Fig. 5). The main component of the voltage at the frequency f_2 across the sample was compensated by a potentiometer, while the component associated with the temperature oscillations was amplified with the aid of a selective amplifier and then fed to a phase-locked detector (the voltage at frequency f_2 was used as a reference). A signal at the frequency $2f_1$ obtained at the detector was fed through a selective filter into an oscilloscope. The second input of the oscilloscope was connected to a voltage with the frequency f_1. The Lissajous figure on the oscilloscope screen was photographed and its shape was used to calculate the specific heat of the sample from the known temperature coefficient of resistance.

4.1.2. *Third-Harmonic Method*

Corbino[2] has shown that when an alternating voltage is applied to a sample the current through it contains a third-harmonic component connected with sample temperature oscillations. Corbino measured the third-harmonic current by compensating the fundamental frequency current and observing the Lissajous figures on the oscilloscope tube.

If the sample is fed by ac current through a sufficiently high resistance, the amplitude of the third-harmonic voltage appearing across the sample under adiabatic conditions is

$$V_3 = I^3 R_0 R'/8mc\omega = V^3 R'/8R_0^2 mc\omega \tag{8}$$

where I and V are the amplitudes of the fundamental frequency current and voltage across the sample.

Phyllipov[4] used a bridge circuit with the sample as one of the arms (Fig. 6). The bridge was balanced at the fundamental frequency and the third-harmonic signal at the bridge output was measured with a selective amplifier. An oscilloscope indicated the balance of the bridge at the fundamental frequency. A variable capacitor shunting one of the arms was used to compensate the additional fundamental frequency voltage associated with sample temperature oscillations and its phase lag of 90° with respect to the heating current. Rosenthal[5] studied the frequency dependence of the temperature oscillations in wire samples over a wide frequency range. Other workers measured the third-harmonic voltage with the aid of phase-locked detectors.

4.1.3. Equivalent-Impedance Method

Radio engineers have known for a long time that a temperature sensitive resistor through which a direct current is flowing can be represented by an equivalent impedance which is, in particular, a function of the specific heat.[31-33] However, some time elapsed before thermophysicists realized the advantages of this method. The method of equivalent impedance has proved to be very convenient for measuring the specific heat of wire samples at high temperatures.[6]

Consider a wire sample with a current $I_0 + i \sin \omega t$ $(i \ll I_0)$ passing through it. Sample temperature oscillations about the mean value are described by the following expressions:

$$\begin{aligned} \Theta &= \Theta_0 \sin(\omega t - \varphi) \\ \Theta_0 &= (2I_0 i R_0/mc\omega) \sin\varphi \\ \tan\varphi &= mc\omega/(P' - I_0^2 R') \end{aligned} \tag{9}$$

Here R_0 and R' are the sample resistance at the mean temperature and its temperature derivative, and φ is the phase shift between the ac component and the temperature oscillations.

In order to derive an expression for equivalent impedance, we shall write the resistance of the sample in the form

$$R + R'\Theta = R_0 + R'\Theta_0 \cos\varphi \sin\omega t - R'\Theta_0 \sin\varphi \cos\omega t \tag{10}$$

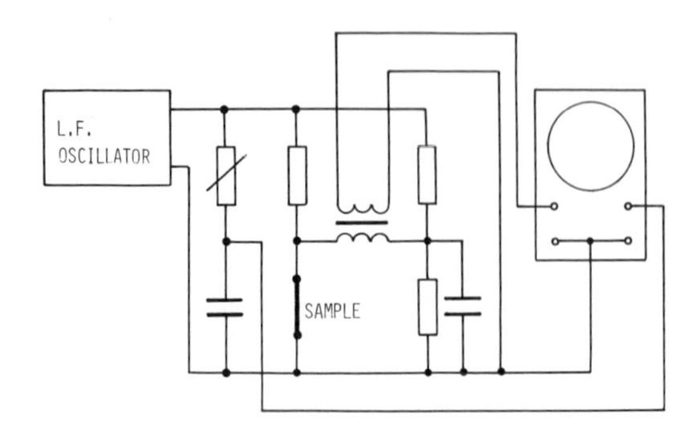

FIGURE 6. Circuit for measuring the third-harmonic signal.[4]

The oscillating component of the resistance consists of two parts, one in phase with the ac component of the current, and the other quadrature lagging. An ac voltage appears across the sample (small terms are neglected):

$$V = iR_0 \sin \omega t + I_0 R' \Theta_0 \cos \varphi \sin \omega t - I_0 R' \Theta_0 \sin \varphi \cos \omega t \tag{11}$$

The impedance of the sample, which describes the amplitude and phase relation between ac component of the current and the ac voltage across the sample can therefore be written as $Z = R_0 + A - jB$. The quantities A and B can be readily determined using equations (9) and dividing the ac voltage across the sample by the ac component of the current:

$$Z = R_0 + (2I_0^2 R_0 R'/mc\omega) \sin \varphi \cos \varphi - j(2I_0^2 R_0 R'/mc\omega) \sin^2 \varphi \tag{12}$$

An analogous impedance is displayed by a circuit of resistance R and capacitance C connected in parallel:

$$Z_1 = R/(1 + \omega^2 R^2 C^2) - j\omega R^2 C/(1 + \omega^2 R^2 C^2) \tag{13}$$

At sufficiently high modulation frequencies $\omega^2 R^2 C^2 \ll 1$, $A \ll R$, and $\sin^2 \varphi \approx 1$. Then $R = R_0$, $B = \omega R^2 C$, and

$$mc = 2I_0^2 R'/\omega^2 RC \tag{14}$$

It is easy either to estimate the errors due to the assumptions made or to replace the approximate expressions by the exact ones.

The specific heat of the sample is thus directly related to the parameters of the equivalent impedance R and C. Consequently the specific heat can be measured by use of a bridge circuit in which one arm is shunted by a variable capacitor (Fig. 7). A selective amplifier tuned to the modulation frequency permits high sensitivity to be obtained even when the temperature oscillations of

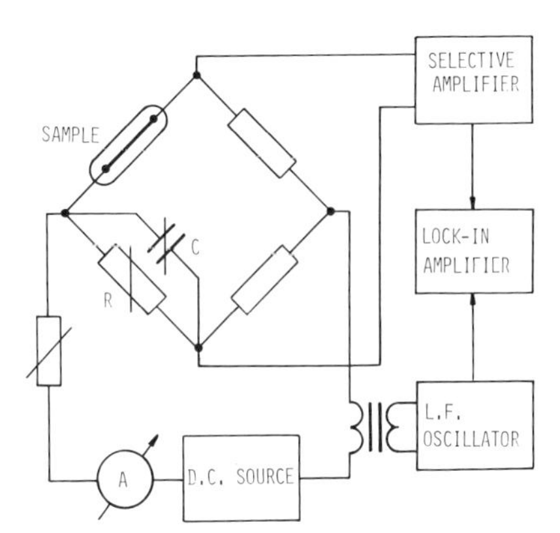

FIGURE 7. Bridge circuit for measuring specific heat of wire samples.[6]

the sample are small. At very small temperature oscillations it is convenient to use phase-locked detection.

The bridge circuit does not allow measurements to be made at relatively low temperatures since the cold-end effects become significant. In this case it is necessary to use very long samples or to heat the current leads clamped to the ends of the samples. More convenient is a potentiometric scheme which makes it possible to measure the specific heat of the central part of a sample.[26,34] The sample is in the form of a wire or a strip and is heated by a current with dc and ac components (Fig. 8). The central part of the sample is defined by fine potential probes. The voltages across the resistors R_1 and R_3 are compared with the aid of a selective amplifier tuned to the ac component frequency. The resistor R_2 and capacitor C_2 are adjusted until the ac current in the potentiometer circuit has an appropriate amplitude and the phase coincides with that of the

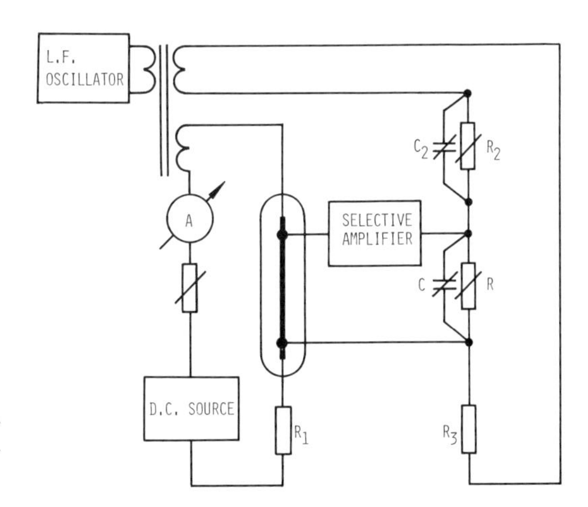

FIGURE 8. Potentiometer circuit for the elimination of cold-end effects.[34]

ac current heating the sample. An exact phase equality in this case is very important. The selective amplifier is then switched on to measure the equivalent impedance of the central part of the sample by the compensation method. As in the case of the bridge circuit, the conditions of the balance (R and C values) do not depend on the magnitude of the ac current flowing through the sample. The balance is indicated by an oscilloscope, and a phase-locked detector can be also used.

4.2. Photoelectric Detectors

Determination of temperature oscillations by measuring the oscillations of sample brightness is now being widely used. Lowenthal[7] used this method to measure the specific heat of tungsten, tantalum, molybdenum, and niobium in the 1200–2400 K range. The samples were heated by an ac current and a photomultiplier was used as the temperature sensor (Fig. 9). The dependence of the photomultiplier current on sample temperature was assumed to have the form $I = AT^n$, with n weakly dependent on temperature. The amplitude of the temperature oscillations can then be determined from the expression

$$\Theta_0 = TV/nV_0 \tag{15}$$

where V_0 and V are the dc and ac components of the photomultiplier output signal.

Phyllipov and Yurchak[8] determined the amplitude of temperature oscillations considering the current-temperature dependence in the form $I = B\exp(-A/T)$. In this case

$$\Theta_0 = T^2 V/AV_0 \tag{16}$$

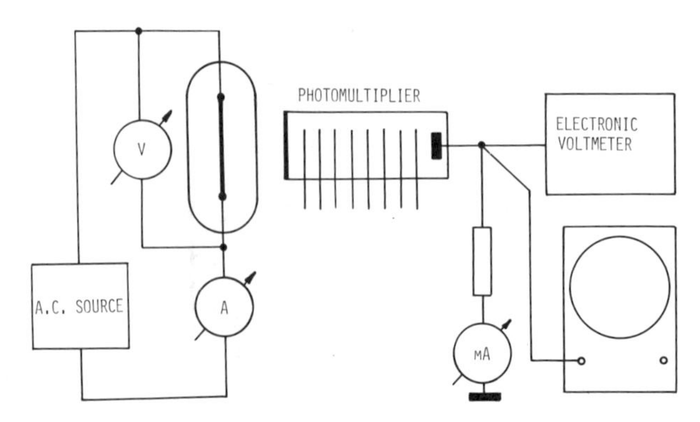

FIGURE 9. Measurement of temperature oscillations with a photomultiplier.[7]

Since the emittance of the sample is temperature dependent this expression is not accurate and can be used only when the dependence of the quantity A on the temperature can in some way be established.

Akhmatova[9,35] measured the specific heat of molten tin, gallium, and copper. The method used in this case was based on the comparison of the amplitudes of temperature oscillations of two niobium capillaries, one filled with the metal under study and the other empty.

4.3. *Thermocouples and Resistance Thermometers*

Thermocouples and resistance thermometers provide the most reliable method of measuring the sample temperature and the amplitude of its oscillations, and should be used whenever possible.

Thermocouples are widely used now together with the modulated-light heating or when a separate heater is used. They are made from thin wires or strips, $10^{-3}-10^{-2}$ mm thick. In some cases the thermocouples were formed by depositing thin films, $10^{-5}-10^{-4}$ mm thick. Thus a good thermal contact with the sample and low inertia can be achieved which allows to perform measurements at modulation frequencies of the order of 10 Hz.

Measurement of specific heat at modulation period of a few seconds was carried out as follows (Fig. 10). The sample was heated by an amplitude-modulated mains current. The modulation was obtained by passing through the

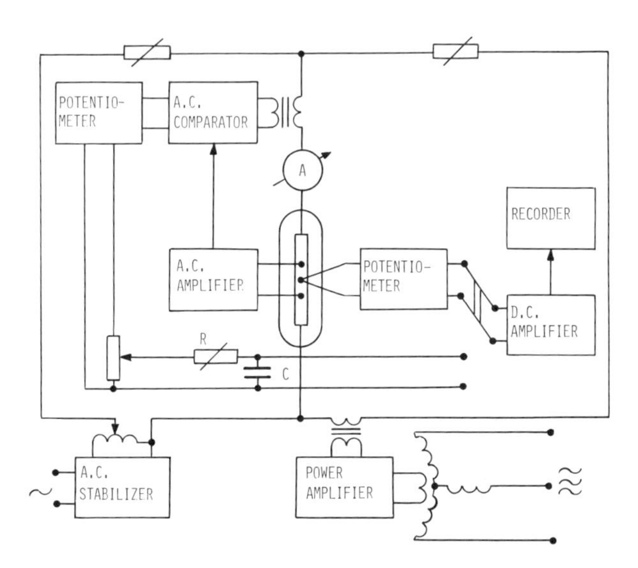

FIGURE 10. Measurement of the specific heat of bulk samples using a thermocouple.[26]

sample, together with the mains current, a supplementary current of the same frequency but with a time-varying phase. The supplementary current was generated by a rotor winding turning at a constant speed in a rotating magnetic field of a stator fed from a three-phase supply. The interaction of the mains and supplementary currents produced beats which resulted in modulating the heating power at a frequency depending on the speed of rotation of the rotor. The dc component of the thermocouple signal was measured with a potentiometer and the amplified ac component was recorded. The oscillations of the heating power were measured with a photocompensation ac comparator. The comparator works on the principle of automatically equalizing torques, one of which is created by ac currents in the windings of an electrodynamometer system, while the other is due to a dc current in a permanent-magnet system. The moving parts of both systems are mounted on the same shaft. The torque of the electrodynamometer system is proportional to the power dissipated in the central part of the sample. The output signal of the comparator, determined by the current in the permanent-magnet system, is also proportional to this power. The dc signal at the output of the comparator is compensated by a potentiometer while the ac component is fed to the input of an integrating *RC* circuit. The output signal from this circuit was measured with the system used for determining sample temperature oscillations. The parameters of the *RC* circuit were chosen in such a way that its output signal was close to the ac signal from the thermocouple. This method permits the elimination of the effect of the inertia of the measuring system at low modulation periods (the comparator is assumed to have a sufficiently short response time). The relation between the oscillations of power and those of sample termperature, is the same as that between the ac signals at the input and output of the integrating circuit. This allows measurements to be made under conditions where the modulation of the heating power is nonsinusoidal.

If the adiabatic conditions hold, the amplitude of the sample temperature oscillations is $\Theta_0 = p/mc\omega$ and the ac signal from the thermocouple with a sensitivity E' is $V_1 = pE'/mc\omega$. The amplitude of the ac component at the output of the comparator is Kp, K being a proportionality coefficient. If the time constant of the integrating circuit is much larger than the modulation period, then the output signal is $V_2 = Kp/\omega RC$ (the input resistance of the amplifier is much larger than R). Therefore the specific heat can be calculated from the expression

$$mc = V_2 RCE'/V_1 K \tag{17}$$

The relationship between V_1 and V_2 is determined by analyzing data from the recorder. It is also possible to employ direct mutual compensation of V_1 and V_2. The coefficient K is easily determined under static conditions.

4.4. *Use of Phase-Locked Detectors*

Phase-locked detectors are widely used now for registration of weak periodic signals in the presence of noise. The method is based on the fact that the frequency of the signal is known. A periodic signal is detected with a parametric detector controlled by a reference voltage. The reference voltage is taken from the same oscillator which controls the process under study. Therefore the frequency of the expected signal always coincides with the reference frequency while the phase difference between them remains constant. In the modulation method the reference voltage is supplied either by the source of the modulated power or by a special sensor (for example, a photodetector when the chopped-light heating is used). The output signal from the phase-locked detector is averaged over a time sufficiently long for required suppression of the noise. The efficient bandwidth of the phase-locked detector is inversely proportional to the time of averaging. Thus, a phase-locked detector is always tuned to the signal frequency and has a variable bandwidth.

Amplifiers employing phase-locked detection are called lock-in amplifiers. Due to the narrow bandwidth they have high sensitivity and noise-immunity. With the thermocouples it is possible to measure temperature oscillations of the order of $10^{-3}-10^{-2}$ K, the record value being 10^{-4} K.[18] Some lock-in amplifiers employ two phase-locked detectors which are controlled by reference voltages whose phases are shifted by 90°. This makes it possible to determine the amplitude of the measured signal and the phase shift of the signal with respect to the reference voltage. Such double phase-locked detector is necessary, for example, in bridge measurements when the signal may contain both in-phase and quadrature components.

5. NONADIABATIC REGIME OF MEASUREMENTS

Usually the modulation method is employed under conditions when the changes in the heating power are much larger than the changes of the heat losses of the sample due to its temperature oscillations. Such regime, which we shall call adiabatic, can be achieved by simply increasing the modulation frequency. However, in some cases the increased modulation frequency is undesirable (for example, when the heat is generated not in the whole volume of the sample or when a temperature sensor with a long time constant is used). As it is shown below, measurements in the nonadiabatic regime when the correction for heat losses is large make it possible to determine the specific heat with the same accuracy as in adiabatic regime. This is due to the fact that although the heat losses under nonadiabatic conditions are large they can be accurately measured and taken into account. Moreover, in a nonadiabatic regime it is possible to determine the specific heat and the temperature coefficient of the resistance of the sample using its heat transfer coefficient. In the case of radiative heat transfer the above coefficient is determined by the temperature dependence of the total emittance of the sample.

Consider a wire sample through which the current $I_0 + i \sin \omega t$ $(i \ll I_0)$ passes. The oscillations of the sample temperature about its mean value are determined by the expression

$$\begin{aligned} \Theta &= (2I_0 i R_0/mc\omega) \sin\varphi \sin(\omega t - \varphi) \\ \cot\varphi &= (P' - I_0^2 R')/mc\omega \end{aligned} \tag{18}$$

where φ is the phase shift between the ac component of the current through the sample and the temperature oscillations.

One can also obtain the expression for the temperature oscillations in terms of the voltage across the sample:

$$\begin{aligned} \Theta &= (2V_0 v/R_0 mc\omega) \sin\psi \sin(\omega t - \psi) \\ \cot\varphi &= (P' + I_0^2 R')/mc\omega \end{aligned} \tag{19}$$

where φ is the phase shift between the ac component of the current through the sample and the temperature oscillations.

Thus, it is possible to determine any of the three quantities c, R', and P' in terms of the other two:

$$\begin{aligned} c &= 2I_0^2 R'/(\cot\psi - \cot\varphi)m\omega \\ c &= 2P'/(\cot\psi + \cot\varphi)\, m\omega \\ R' &= P'(\cot\psi - \cot\varphi)/I_0^2(\cot\psi + \cot\varphi) \end{aligned} \tag{20}$$

Since I_0, m, and ω can be determined with a high degree of accuracy, the measurements reduce to finding $\cot\psi$ and $\cot\varphi$. One of the possible methods of doing this is to directly measure the phase shift between the signal from a photocell viewing the sample and either the current through or the voltage across the sample. The second possibility consists in measuring the electrical impedance of the sample. The sample responds to an ac current as a complex resistance since there is a phase shift between the ac components of the voltage and current:

$$\mathbf{Z} = |Z| \exp(-j\phi), \qquad \phi = \varphi - \psi \tag{21}$$

$$\cot\phi = (1 + \cot\psi \cot\varphi)/(\cot\psi - \cot\varphi) \tag{22}$$

By equating the amplitudes of the sample temperature oscillations in (18) and (19) one gets

$$|Z| = v/i = R_0 \sin\varphi / \sin\psi \tag{23}$$

In the case of the measurements of the impedance of the sample by a bridge, one of the bridge arms consists of the resistance R shunted by a capacitance C. The modulus of the impedance of this circuit is $R(1 + \omega^2 R^2 C^2)^{-1/2}$, and the phase angle is $\arctan(-\omega RC)$. From the requirements for balance of the bridge one obtains:

$$\begin{aligned} \cot\psi &= (R/R_0 - 1)/\omega RC \\ \cot\varphi &= (\cot\psi - \omega RC) R_0/R \end{aligned} \tag{24}$$

Thus, $\cot\psi$ and $\cot\varphi$ can be found from the measurable parameters R, R_0, and C.

The equation for calculating the specific heat by the temperature coefficient of resistance can be written in a form similar to that for the adiabatic case:

$$c = (2I_0^2 R'/m\omega^2 R_0 C)/[1 + (R/R_0 - 1)^2/\omega^2 R^2 C^2] \tag{25}$$

where the correction coefficient approaches unity with increasing frequency. If the frequency response of the zero indicator is flat and if the amplitude of the temperature oscillations is kept constant within the frequency band used, the error of calculations by equation (25) is independent of frequency (except at lowest frequencies where the error increases). In contrast to this case, when calculating the specific heat and the temperature coefficient of resistance from the heat transfer coefficient, the error is minimal when the angles ψ and φ are close to 45°. This result is quite understandable since the use of the heat transfer coefficient as the basic quantity is justified only under substantially nonadiabatic conditions (in an adiabatic regime the amplitude and the phase of temperature oscillations are independent of the heat transfer coefficient).

The choice of either R' or P' as a basic parameter is determined by the accuracy with which each of them is known. It is useful to employ the heat transfer coefficient P' at extremely high temperatures when point defects begin to give a substantial contribution to the temperature derivative of the resistance.

In the case of radiative heat exchange $P = S\sigma\epsilon T^4$ and

$$P' = P[4 + (T/\epsilon) d\epsilon/dT] = nP/T \tag{26}$$

Usually the second term in the brackets is several times smaller than the first one and decreases with increasing temperature. Especially favorable is the

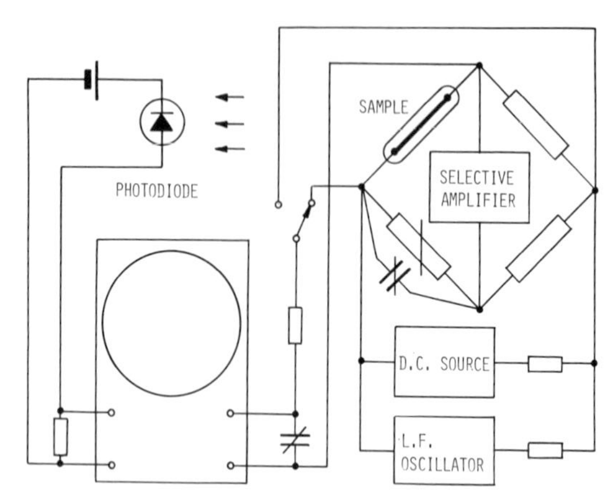

FIGURE 11. Experimental arrangement for measurements in the nonadiabatic regime.[17]

case when the total emittance is large and weakly temperature dependent; here it is possible to take $n = 4$ as in the case of a perfectly black body. One may expect that it would be useful to artificially increase the blackness of the samples used in measurements.

An experimental test of the measurements in the nonadiabatic regime was made on vacuum incandescent lamps with tungsten filaments (110 V, 8 W). The equivalent electrical impedance was measured at 6 Hz by an ac bridge (Fig. 11). A selective amplifier was used as a zero indicator. The same bridge was also used for measuring the dc sample resistance. The quantities $\cot\psi$ and $\cot\varphi$ were measured also by the photoelectric method. The signal from the photodiode came to the Y-input of an oscilloscope. The voltage across either the sample or the bridge arm in series with the sample came to the X-input through a phase-shifting circuit RC. The value of C was adjusted so that the Lissajous figure on the oscilloscope was transformed into a straight line. In contrast to the impedance measurements, in this case different modulation frequencies from 3 to 8 Hz were used for different heating currents in order to maintain the phase shifts ψ and φ which are temperature dependent close to 45°.

There is also a way to maintain the adiabatic regime of measurements at low modulation frequency. It consists in surrounding the sample with a thermal shield which temperature oscillates at the same frequency.[36] With a proper choice of the amplitude and phase of these oscillations, the heat exchange corrections can be made insignificant.

The non adiabatic regime of measurements and use of a thermal shield are applicable only in the case of radiative heat exchange when the heat transfer coefficient does not depend on the modulation frequency. Hence, they are inapplicable in measurements of specific heat under high pressures. In such measurements, one has to introduce large corrections in order to obtain a true value of specific heat.[37,38]

6. DIRECT MEASUREMENT OF TEMPERATURE COEFFICIENT OF SPECIFIC HEAT

The method of measuring the temperature coefficient of specific heat (TCSH) is based on the modulation method with strictly sinusoidal oscillations of the heating power. The temperature dependence of the specific heat brings about a deviation of the form of temperature oscillations from the sinusoidal one. The second-harmonic component appears in the temperature oscillations with the magnitude depending on the TCSH. This component is measured by harmonic analysis.

Consider first the simplest case of a purely sinusoidal heating power. The heat balance equation, with the temperature dependence of the specific heat being taken into account, is

$$mc(1+\alpha\Theta)\Theta' + P + P'\Theta + P''\Theta^2/2 = p_0 + p\cos\omega t \tag{27}$$

where $\alpha = (1/c)dc/dT$ is the TCSH, P, P', and P'' are the heat loss from the sample and its temperature derivatives.

Since $\alpha \ll 1$ this nonlinear equation can be readily solved by the successive approximations method. With high-order terms neglected, the solution has the form

$$\begin{aligned} \Theta &= \Theta_1\cos(\omega t - \varphi_1) + \Theta_2\cos(2\omega t - \varphi_2) \\ \Theta_1 &= p\sin\varphi_1/mc\omega, \qquad \Theta_2 = \alpha\Theta_1^2/4\cos\varphi_2 \\ \tan\varphi_1 &= mc\omega/P', \qquad \tan\varphi_2 = P''/2mc\omega\alpha \end{aligned} \tag{28}$$

At sufficiently high modulation frequencies $\varphi_1 \approx 90°$, $\varphi_2 \approx 0$, and the expressions become simplified:

$$\Theta_1 = p/mc\omega, \qquad \Theta_2 = \alpha\Theta_1^2/4 \tag{29}$$

Thus, to determine the TCSH one has to measure the fundamental component and the second harmonic of the sample temperature oscillations. The α value is usually of about $10^{-4}\ K^{-1}$ at temperatures above the Debye temperature. It increases several times due to increasing specific heat caused by point-defect formation. At the second-order phase transitions α is 10^{-2} to $10^{-1}\ K^{-1}$ and one may be successful in measuring the second harmonic even with temperature oscillations less than 1 K. In all cases a high selectivity of the registering apparatus is needed to guarantee the accurate measurement of the second-harmonic signal in the presence of a much stronger fundamental-frequency signal.

On deriving equations (28) and (29) it was supposed that the strictly sinusoidal modulation of the heating power is achieved by some means. However, it is difficult to realize such a mode in practice. To measure the TCSH of wire samples it is convenient to heat them by alternating current. Let us suppose that the internal resistance of the power source for all harmonics of the current is high enough so that no changes of the heating current are produced by the oscillations of the sample temperature and electrical resistance. But an additional second-harmonic signal arises due to the main current flowing through the oscillating resistance of the sample. To take this into account, the right side of equation (27) should be written in the form $I^2 R(1 + \beta\Theta) \cos^2(\omega t/2)$, where I is the amplitude of the heating current, R is the sample resistance, and $\omega/2$ denotes the frequency of the heating current; $\beta = (1/R)\, dR/dT$ is the temperature coefficient of resistance. Since $\beta \ll 1$ we obtain instead of equation (29)

$$\Theta_2 = \Theta_1^2(\alpha - \beta)/4 \tag{30}$$

Of great importance is the method of the registration of the sample temperature oscillations. If they are determined using some parameter with a nonlinear temperature dependence, the results can be severely distorted. For example, it is not suitable to measure the temperature changes by the changes of the brightness. It would be better to use the temperature dependence of the resistance of the sample making the necessary corrections for the nonlinearity which, far from the phase transition points, is not strong. Taking this correction into account we obtain instead of equation (30)

$$\Theta_2 = \Theta_1^2[\alpha - \beta - (1/\beta_0)d\beta_0/dT]/4 \tag{31}$$

where $\beta_0 = (1/R_{273})dR/dT$.

Thus, on heating the sample by the alternating current and on registering its temperature oscillations by the oscillations of the resistance, equation (31) for calculating the TCSH is no longer as simple as equation (29). However, in many cases such method of measurements can be used.

The method has been tested experimentally with a bridge circuit fed by 50 Hz mains current (Fig. 12). Two gas-filled incandescent lamps with tungsten filaments were employed as samples. Since the expected value of the ratio Θ_2/Θ_1 is 10^{-3} to 10^{-2} the harmonic content in the heating current must be sufficiently small. The third and fifth harmonics of the feeding current are especially dangerous since their interaction with the main current brings about the doubled-frequency modulation of the power generated in the sample, and

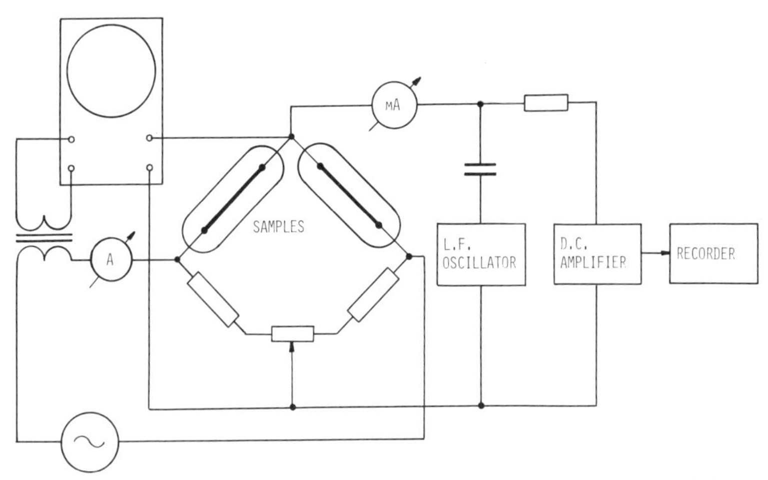

FIGURE 12. Measurement of the temperature coefficient of specific heat.[16]

produces an additional second-harmonic signal. To suppress the harmonics which are present in the feeding current, as well as those arising from the sample temperature oscillations we used LC-filters (not shown in Fig. 12) with low resistance at 50 Hz and high resistance at 150 and 250 Hz. The mean temperature of the samples was determined from their electrical resistance.

The harmonic components of the temperature oscillations were registered by the supplementary-current method. The supplementary current passing through the sample results in the appearance of a low-frequency voltage across the sample which is proportional to the magnitude of resistance oscillations of the sample and to the value of the supplementary current. This voltage is fed through an *RC*-filter to an amplifier and recorder. Successive tuning of the supplementary-current source to frequencies close to 100 and 200 Hz permits the amplitudes of the harmonic components of the temperature oscillations with corresponding frequencies to be measured. The period of the low-frequency signal was of about 10 s. The method allows us to achieve high selectivity and to measure reliably the second-harmonic signal in the presence of the fundamental-frequency signal stronger by three orders of magnitude.

The calculations and the experimental results obtained show that the direct measurement of the TCSH of metals is, indeed, possible in many cases. The method is suitable for observing the specific-heat anomalies associated with the point-defect formation in equilibrium. It is not quite evident whether this method can be applied for studying specific-heat anomalies accompanying second-order phase transitions. For this purpose temperature oscillations must be sufficiently small to ensure good temperature resolution. It seems that modulated-light heating of the sample and use of thermocouples would be most appropriate.

7. DETERMINATION OF ISOCHORIC SPECIFIC HEAT USING TEMPERATURE FLUCTUATIONS

It is known that the mean square value of temperature fluctuations is $\langle \Delta T^2 \rangle = kT^2/mc_v$ and their spectral density as a function of frequency has the form[39]

$$\langle \Delta T_f^2 \rangle = 4kT^2/P'(1+x^2) \qquad (32)$$

where k is the Boltzmann constant, P' is the heat transfer coefficient, $x = mc_v\omega/P'$, c_v is specific heat at constant volume.

Temperature fluctuations are determined by the isochoric specific heat because the fluctuations of temperature and of volume are statistically independent.

In order to study temperature fluctuations and to determine isochoric specific heat, it is expedient to measure the temperature oscillations in the same samples under conditions when the heating power supplied to the sample is modulated. When such a modulation is performed using a noise generator, the spectral density of the temperature oscillations of the sample is described by the expression

$$\langle \Theta_f^2 \rangle = \langle p_f^2 \rangle / P'^2(1+y^2) \qquad (33)$$

where $\langle p_f^2 \rangle$ is the spectral density of the square of the heating power oscillations, $y = mc_p\omega/P'$, c_p is specific heat at constant pressure.

Comparing the frequency dependence of temperature fluctuations and that of the temperature oscillations caused by the modulation, it is possible to determine the ratio c_p/c_v. At low frequencies ($x^2 \ll 1$)

$$\langle \Delta T_f^2 \rangle / \langle \Theta_f^2 \rangle = 4kT^2P'/\langle p_f^2 \rangle = A \qquad (34)$$

whereas at high frequencies ($x^2 \gg 1$)

$$\langle \Delta T_f^2 \rangle / \langle \Theta_f^2 \rangle = 4kT^2P'y^2/\langle p_f^2 \rangle x^2 = A(c_p/c_v)^2 \qquad (35)$$

Thus, the measurements of the temperature fluctuations, combined with the additional modulation measurements on the same samples, make it possible to determine the c_p/c_v ratio even without knowledge of the amplitude of the temperature oscillations, the mass and the heat transfer coefficient of the sample.

Temperature fluctuations in thin wire samples were registered using photodiodes.[23] The correlation method was employed for suppression of the inherent noise of the photodiodes (Fig. 13). Two identical channels were used, each consisting of a photodiode, a preamplifier, and a power amplifier. An electro-

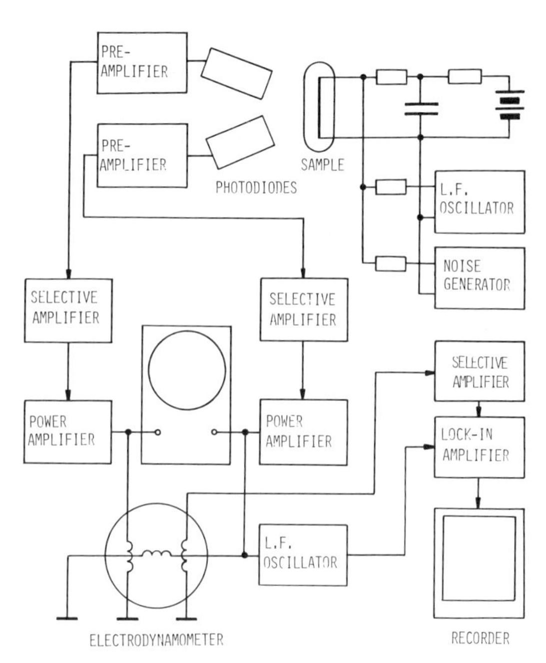

FIGURE 13. Observation of temperature fluctuations in thin wires using photodiodes.[23]

dynamometer was used as a multiplying device. It was capable of an accurate multiplication of the signals which were fed into its moveable and one of the fixed coils. Registration of the displacements of the movable coil was achieved in the following way. An additional low-frequency current was passed through the movable coil from an oscillator. This induced in the second fixed coil a voltage whose amplitude changed with the rotation of the movable coil. The voltage was amplified with a selective amplifier and measured with a lock-in amplifier connected with a recorder.

In the experiments wire samples of tungsten 3.5 μm thick and 1 mm long were used. The measurements were made at 2200 K in the frequency range 5–200 Hz. It was found that the spectrum of the temperature fluctuations is in a good agreement with the theoretical formula (32). At frequencies below 20 Hz the spectral density of the temperature fluctuations is constant and equals approximately $3 \cdot 10^{-11}$ K^2/Hz; at high frequencies it is inversely proportional to the square of frequency. The value of the c_p/c_v ratio obtained in the measurements was found to be 1.4 ± 0.1.[24]

Employing the above system it is possible to detect very small temperature oscillations. For this purpose the modulation of the heating power was performed with a low-frequency oscillator. Only one amplification channel was used, the second signal fed in the multiplying device being the output voltage from the same oscillator. Thus, the measuring system was working as a phase-locked detector. It turned out that in this way it is possible to measure temperature oscillations of the order of 10^{-5} K.[23]

8. MEASUREMENTS AT HIGH MODULATION FREQUENCIES

The frequency of the temperature oscillations can be varied in a wide range. This suggests that the modulation method may be suitable for studying relaxation effects in specific heat. Such effects must be expected, for example, in the ordering of alloys, in solvation of gases in metals, and in formation of equilibrium vacancies.

Relaxation method for studying vacancy formation in solids has been proposed long ago.[40,41] It consists in creating such rapid changes of the sample temperature that the vacancy concentration cannot follow the temperature. Hence, when specific heat is measured at a sufficiently high modulation frequency the result must correspond to specific heat of a vacancy-free crystal (the changes in specific heat due to the presence of static defects are much smaller than the effect due to the changes in their concentration).

The relaxation effect in specific heat manifests itself in the variation of the amplitude and phase of the temperature oscillations caused by the modulation of the heating power. If specific heat of the sample is written in the complex form $C(x) = C + \Delta C/(1 + jx)$ then[42]

$$\begin{aligned} |C(x)|^2 &= (C_0^2 + C^2x^2)/(1 + x^2) \\ \tan \Delta\varphi &= x\Delta C/(C_0 + Cx^2) \end{aligned} \tag{36}$$

Here ΔC and C are relaxing and nonrelaxing parts of specific heat, $x = \omega\tau$ (product of frequency and relaxation time), $C_0 = C + \Delta C$ (specific heat at $x^2 \ll 1$), and $\Delta\varphi$ is the change of the phase of temperature oscillations.

It can be seen from these expressions that at $x < 1$ it is expedient to register the relaxation effect by the phase shift of the temperature oscillations, rather than by their amplitude. In the adiabatic regime of the measurements, which is obtained even at relatively low modulation frequencies, the temperature oscillations are lagged by $90°$ from the power oscillations. The relaxation effect decreases this phase shift.

When observation of the relaxation effects requires very high modulation frequencies, of the order of $10^5 - 10^6$ Hz, serious problems appear associated with diminishing amplitude of the temperature oscillations and with uncontrollable phase shifts in the measuring scheme. The following technique was found to be most suitable for the relaxation measurements at these frequencies. Specific heat is measured simultaneously at low and high frequencies of temperature oscillations (Fig. 14). High-frequency current heating the sample is slightly modulated by a low frequency (modulation coefficient of about 1%). Therefore in the sample simultaneously arise temperature oscillations with low and high frequencies. The relation between the oscillations of power heating the sample at the two frequencies is maintained constant when the heating current is changed. The second method of heating consists in passing through the sample a direct

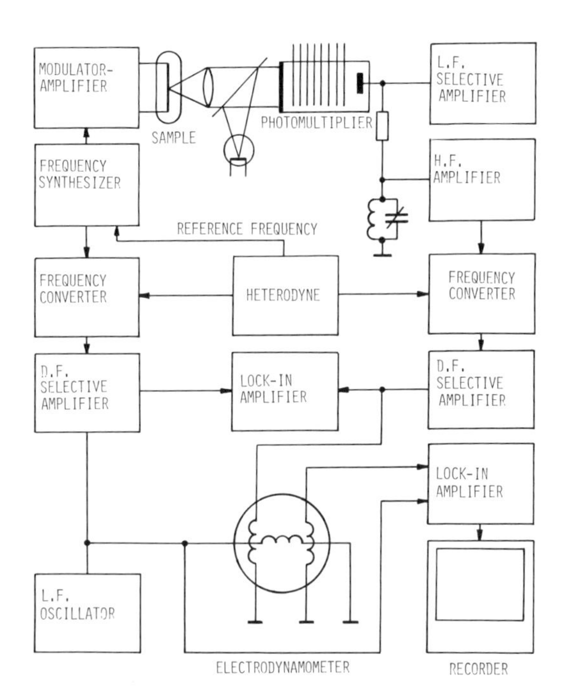

FIGURE 14. Measurement of high-temperature specific heat at high modulation frequencies.[25]

current with two ac components, of high and low frequencies, superimposed on it. In this case the relation between the amplitudes of modulation at the two frequencies must remain constant when the direct current is changed.

The oscillations of the sample temperature are registered with a photomultiplier. The low-frequency component of its output signal is amplified by a selective amplifier. The high-frequency component is selected with a resonant circuit, amplified and fed to a frequency converter. A quartz oscillator is used as a heterodyne. The signal with the difference frequency is amplified with a selective amplifier and fed to the amplitude and phase detectors. Frequency of the current heating the sample is set by the frequency synthesizer, the source of the reference frequency for the synthesizer being the heterodyne. This arrangement makes it possible to set any values of the difference frequency and ensures its stability. Reference voltage for the detectors is taken from an auxiliary frequency converter.

As a phase detector an electrodynamometer was used. A deflection of the movable coil of the meter is proportional to the currents in its coils and cosine of the phase shift between them. For registration of the deflections of the movable coil, an additional low-frequency current is passed through the movable coil. This results in a voltage induced in the second fixed coil being dependent on the orientation of the movable coil. This voltage is amplified with a lock-in amplifier with a sufficient time constant. The output signal from the lock-in amplifier comes to the recorder. The phase shift between the currents in the coils is set close to 90° so that the deflection of the movable coil is minimum while the sensitivity to the variations of the phase shift is maximum. The starting position of the moveable coil is such that the voltage induced in the fixed coil is close to zero. In order to calibrate the scheme, a phase shift of the known magnitude is introduced by means of a phase-shifter either in the signal or in the reference channel. It is possible to use for the calibration the delay of the signal resulting from an increased light-path from the sample to the photomultiplier. Using two mirrors the light-path could be increased to create a phase shift of about 1°.

The observation of the relaxation effects in specific heat can be conveniently carried out on thin foil samples. The sample has the form of a strip narrowing towards one of its ends. When a current is passed through the sample its different areas have different mean temperatures. The temperature of each area was determined by comparing its brightness with the brightness of an auxiliary sample from the same foil whose temperature was measured with a thermocouple. The measurements were reduced to successive projection on the cathode of the photomultiplier of different areas of the sample. By means of a diaphragm the low-frequency component of the photocurrent was maintained constant. The mean current of the photomultiplier was also maintained constant using an additional illumination of the cathode from an incandescent lamp. For each area of the sample output signals were recorded from the amplitude and phase detectors which registered high-frequency temperature oscillation. Thus, the measurements in the whole temperature range were made without changing the heating current, photomultiplier current, and the signal amplitude. This proved to be very useful for excluding undesirable phase shifts.

A similar method is employed for wire samples. Two identical samples are connected in series. One of them is shunted with a resistor so that the samples have different mean temperatures. The heating current is gradually changed and the amplitudes and phases of the temperature oscillations in the two samples are compared.

This technique considerably increases the possibility of observation of the relaxation effects in specific heat. On the samples 0.01 mm thick at temperatures about 2300 K and modulation frequency 3×10^5 Hz the resolution achieved was 1% in the amplitude and 0.2° in the phase of the temperature oscillations. Earlier, similar measurements were made at modulation frequencies not exceeding 10^3 Hz.[43,44]

9. MODULATION METHODS FOR STUDYING THERMAL EXPANSION, ELECTRICAL RESISTIVITY, AND THERMOPOWER

9.1. Thermal Expansion

At the present time the methods for measuring dilatation of solids are known with a sensitivity of the order of 10^{-10} cm and even higher. However, difficulties in studying thermal expansion at high temperatures are associated with the instability of a sample rather than with insufficient sensitivity. Therefore, one has to be content with the determination of average values of the thermal expansion coefficient within wide temperature intervals.

For the measurement of "true" thermal expansion coefficients it is convenient to employ the modulation method which involves oscillating the sample temperature about a mean value and recording corresponding changes in the sample length. Under these conditions the thermal expansion coefficient is directly measured. The use of the periodic temperature oscillations negates irregular changes of the sample length due to external disturbances or creep at high temperatures. In the modulation method only those changes of the sample length are registered which repeat reversibly with periodic changes of the temperature. Measurements can be performed with temperature oscillations of about 0.1 K. At the present time, even measurements with a resolution of about 1 K at high temperatures are of great improvement over traditional methods. Measurements by the modulation method can be performed with different ways of registering sample length oscillations including those which yield the highest sensitivity.

9.1.1. Wire Samples

The sample is heated by ac current or by current with both dc and ac components. One end of the sample is fixed whilst the other is pulled by means of a load or a spring (Fig. 15a). The free end of the sample is projected onto the entrance slit of a photomultiplier with the aid of an optical system. The ac component of the output voltage of the photomultiplier is proportional to the amplitude of the oscillations of the sample length. The measuring system is calibrated under static conditions.[45]

The amplitude of the oscillations of the sample temperature is determined from the oscillations of either the electrical resistance of the sample or its brightness. If the specific heat of the sample is known, then the amplitude of the temperature oscillations can be easily calculated. When the sample is heated by ac, the coefficient of thermal expansion is calculated from the expression

$$\alpha = 2mc\omega V/lPK \tag{37}$$

where l, m, and c are the length, mass, and specific heat of the sample; P and ω are the power and frequency of the heating current; V is the amplitude of the ac component at the output of the photomultiplier; and K is the sensitivity of the photomultiplier to the dilatation of the sample. The frequency of the temperature oscillations is arranged to be sufficiently high that the phase shift between the power and temperature oscillations in close to 90° (adiabatic regime). Otherwise, the temperature oscillations should be calculated from the formulas for nonadiabatic conditions.

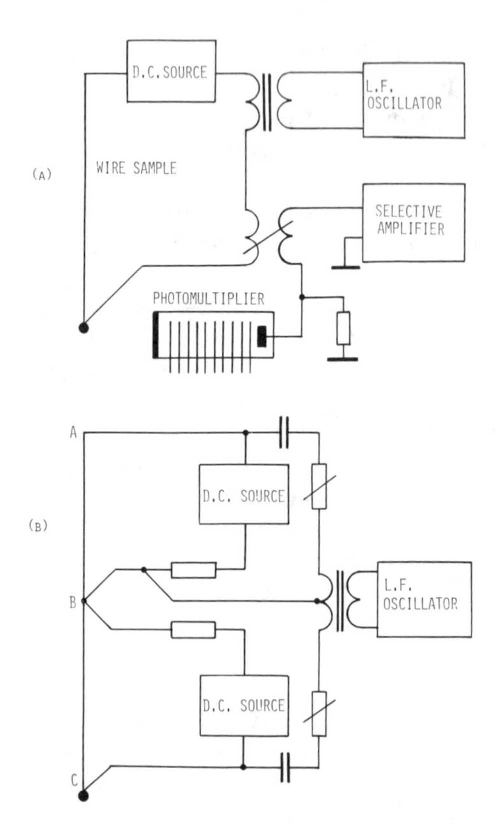

FIGURE 15. (a) Modulation method for measuring the thermal expansion coefficient of wire samples.[45] (b) Compensation method of measurements: oscillations of the length of a sample under study AB are compensated for by antiphase length oscillations of a reference sample BC.[46]

When the sample is heated by a current with both dc and ac components, the coefficient of thermal expansion is calculated from the expression

$$\alpha = mc\omega V/2lKI_0U \tag{38}$$

where U is the amplitude of the ac voltage across the sample. In this case, a system can be employed whose balance is independent of the ac component of the heating current. The ac component at the output of the photomultiplier is compensated for by means of a variable mutual inductance with the heating current being passed through its primary coil. The recording circuit incorporates a narrow-band amplifier which is tuned to the frequency of the sample length oscillations so that the system is insensitive to sample creep and mechanical perturbations.

9.1.2. Compensation Method

The wire sample consists of two portions joined together (Fig. 15b): the sample under study and the reference sample with a known coefficient of thermal expansion. The two portions are heated by dc currents from two separate sources, with an ac current from a common oscillator through both portions. The polarity of the dc currents is chosen such that the temperature oscillations in the two portions of the sample are 180° out of phase. The amplitudes of these oscillations are adjusted so that the oscillations of the length of the portion under study are completely compensated by the oscillations of the reference portion. This is achieved by regulating the ac current in the two parts of the sample. The photomultiplier is here only a zero indicator and the effect of variations in its sensitivity, as well as in the light source intensity, etc., are completely eliminated.[46]

For calculating the coefficient of thermal expansion it is necessary to know the amplitudes of the temperature oscillations in the two parts of the sample. If the specific heat is known then it is easy to obtain the relation which holds when the oscillations in the two portions of the sample balance each other:

$$\alpha_1 I_{10} U_1 l_1 / m_1 c_1 = \alpha_2 I_{20} U_2 l_2 / m_2 c_2 \tag{39}$$

The subscripts 1 and 2 designate the studied and reference portions of the sample, respectively. The reference sample is at a constant mean temperature, and all the quantities except I_{10}, U_1, U_2, c_1, and α_1 are constants. Therefore

$$\alpha_1 = K U_2 c_1 / I_{10} U_1 \tag{40}$$

where K is a proportionality coefficient. Thus the measurements of the coefficient of thermal expansion are reduced to maintaining a zero amplitude of the oscillations of the length of the composite sample and to measuring the dc currents in, and the ac voltages across, the two portions of the sample. The sensitivity of the measuring system is 10^{-7} cm.

9.1.3. *Bulk Samples*

Another version of the modulation method is used for comparatively bulk samples such as rods and foils.[47] The temperature of such samples and the amplitude of its oscillations are measured with a thermocouple while the compensation of the oscillations of the length is achieved by means of an electromechanical transducer attached to the sample and controlled by a device sensing the sample length, such as a small telephone transducer. A blade mounted on the telephone membrane is illuminated by a light source and its image projected at the entrance slit of a photodetector (Fig. 16). The output voltage of the photodetector is partly compensated and then amplified by a dc amplifier. The output of the amplifier is connected to the electromechanical transducer with such a

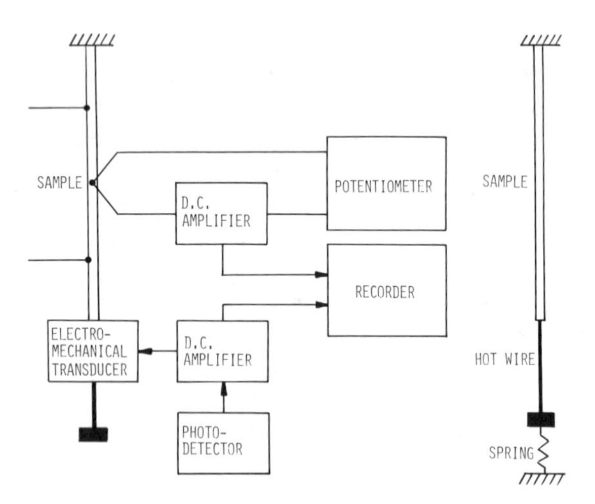

FIGURE 16. Measurement of the thermal expansion coefficient of bulk samples; an electromechanical transducer or a wire heated by current is used for compensating the length oscillations.(47)

polarity that changes of the sample length are compensated by the displacements of the movable part of the transducer. Due to the high gain of the amplifier, almost complete compensation is automatically achieved and the sensitivity of the system is 10^{-6} cm. The changes in the transducer current, which are proportional to the changes of the sample length, are recorded.

For the elimination of cold-end effects, the temperature oscillations are generated only in the central portion of the sample. The sample is heated by ac mains current. A current of the same frequency, but with its phase linearly varying with time, is superimposed on the mains current in the central portion of the sample. As a result of the interaction of the two currents the power dissipated in the central portion of the sample periodically varies and the temperature starts to oscillate around some mean value.

Linearity of the electromechanical transducer used for compensation is very important. For this purpose, one can use the thermal expansion of a thin wire heated by an electric current. One end of the wire is fixed to the sample and the other is pulled by a spring. A blade is mounted on the spring and its image is projected onto the photodetector. Since the temperature oscillations of the sample are small, only small changes of the temperature of the compensating wire are required. With the mean temperature of the wire kept constant, the changes in the temperature and length are proportional to the changes in the current passing through it. A tungsten wire 0.05 mm thick was used for compensation. At the mean temperature of 1500 K the time response of such a transducer is quite sufficient and the linearity is very good. For determining the absolute values of the thermal expansion coefficients, it is necessary to calibrate the electromechanical transducer. For this purpose samples with a known thermal expansion can be used. In some cases (point-defect formation, phase transitions of the second order) it is sufficient to know the character of the temperature dependence of the thermal expansion coefficient, for which purpose only relative values are adequate.

The modulation method for studying thermal expansion proved to be very convenient at high temperatures since it eliminates some difficulties specific to high-temperature measurements. It is possible that this technique will be useful at middle or even at low temperatures. Because of its good temperature resolution, the method seems promising for studying anomalies in thermal expansion near phase transitions.

9.2. *Electrical Resistivity*

The modulation method allows the temperature derivative of electrical resistivity to be measured directly. The method consists in making the sample temperature oscillate about a mean value and measuring the amplitude of resistance oscillations. The sample was heated by a dc current from a stabilized source and by an ac current amplitude-modulated with the period of a few seconds.[48] The amplitude of the oscillations of the sample temperature was measured with a thermocouple; simultaneously measurements were made of voltage oscillations across the sample due to the dc current passing through the sample the resistance of which varies. In order to eliminate the cold-end effects, the measurements were carried out only in the central portion of the sample and thin potential probes were used. The dc component of the thermal emf and of the voltage across the central portion of the sample were measured with potentiometers and time-varying components were aplified and recorded by a two-channel recorder.

The modulation method is particularly suitable for studying effects within narrow temperature intervals, for example for studying the anomalies of electrical resistivity accompanying second-order phase transitions. Direct measurement of the temperature derivative of resistivity is more convenient in these cases since it allows the character of the anomaly to be determined more accurately. This advantage is especially noticeable at high temperatures where accuracy of the conventional methods of studying the temperature dependence of resistivity is substantially lower than at low and medium temperatures.

The temperature derivative of resistivity was measured directly with good temperature resolution by Salamon *et al.*[49] The sample was heated by modulated light. The specific heat and the temperature coefficient of resistivity were measured simultaneously. This method was also used for investigation[50] of the specific heat and resistivity of iron near its Curie point (Fig. 17). The

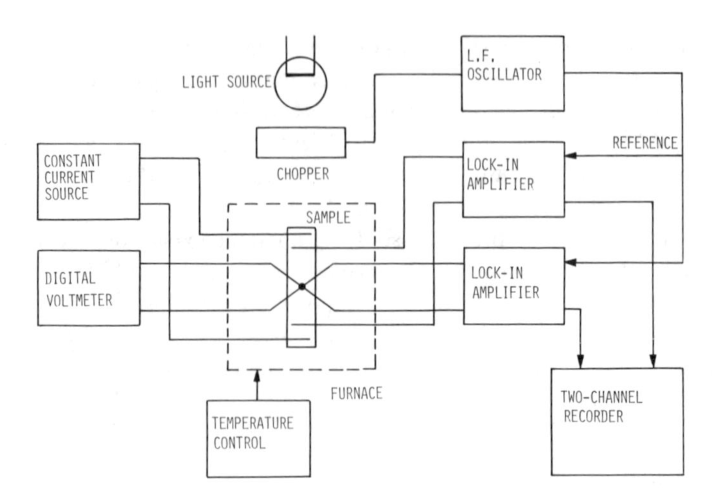

FIGURE 17. Simultaneous measurements of the specific heat and the temperature derivative of resistivity.[50]

ambient temperature within the furnace was linearly swept at a rate of 1 K/min or less. An operating frequency of 8 Hz was chosen to produce temperature oscillations. The resistance oscillations were detected by passing a dc current through the sample. The lock-in amplifiers were employed for measuring the ac voltages corresponding to the temperature and resistance oscillations.

9.3. Direct Measurement of Thermopower

Of all the applications of the modulation method the measurement of the Seebeck coefficient is the simplest and consists in measuring the same periodic oscillations of temperature with the thermocouple which is being studied and with a standard one. The temperature oscillations can be generated in different ways. We used a furnace heated by a modulated current.[51] The modulation period was several seconds. In order to eliminate any magnetic field effects a bifilar or tubular heater was used so that there was no magnetic field in the region where the thermocouple junctions were placed. Signals from the standard and the measured thermocouples, after compensating the dc components, were simultaneously recorded by a two-channel recorder.

Very similar methods of measuring the Seebeck coefficient using modulation were described by Freeman and Bass[52] and Hellenthal and Ostholt.[53] The thermocouple junction is heated by modulated light (Fig. 18). The signal from the thermocouple is amplified and measured by a phase-locked detector. It is possible to use a null method by compensating the signal at the input of the amplifier or to continuously record the measured quantity. High sensitivity of

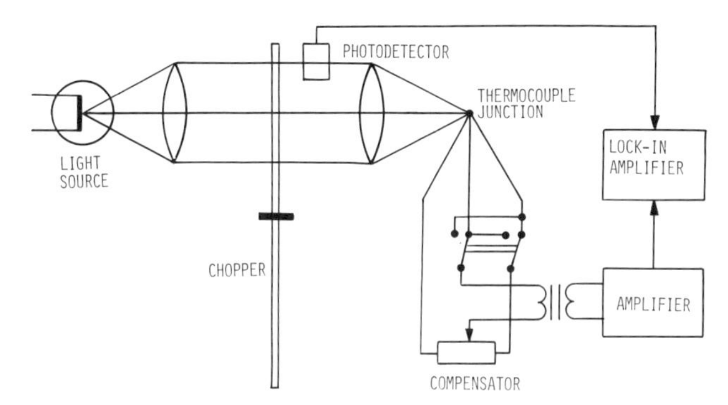

FIGURE 18. Direct measurement of thermopower using modulated-light heating.[53]

the method makes it suitable for studying subtle effects, for example the change in the Seebeck coefficient resulting from deformation.[53]

10. DETERMINATION OF TEMPERATURE OF WIRE SAMPLES BY THERMAL NOISE

The measurements by modulation methods are conveniently performed on thin wire samples heated by current flowing through them. However, accurate determination of the temperature of such samples presents a serious problem. Among other methods, it can be achieved by measuring the thermal electrical noise of the samples. Such noise thermometers have been in use already for a long time.[54] However, when low-resistance samples are used, their thermal noise can be smaller than the noise of the amplifier itself. In such cases a correlation method of amplifying is preferable and it has been used in noise thermometers.[55,56]

In the correlation method of amplification the signal (for example, the thermal noise of a sample) is fed to the inputs of two similar amplifiers (Fig. 19a). The output signals of the amplifiers contain the measured voltage and the inherent noise of the amplifiers. These signals proceed then to a multiplier with an integrating circuit. Since the noise voltages of the two amplifiers are uncorrelated either with each other or with the measured signal, the corresponding terms become zero on averaging and the output signal of the multiplier is proportional to the square of the measured voltage. However, the output signal is sensitive to the gains of the amplifiers and to the stability of the multiplier.

The suggested noise thermometer[57,58] differs from those previously described in that it employs a null method of measurement (Fig. 19b). The noise voltage, U_c, which is being measured (correlated) is fed to the inputs of two channels of a correlation amplifier, and the amplified voltages proceed to the multiplier. In order to compensate the measured correlated voltage, two anti-

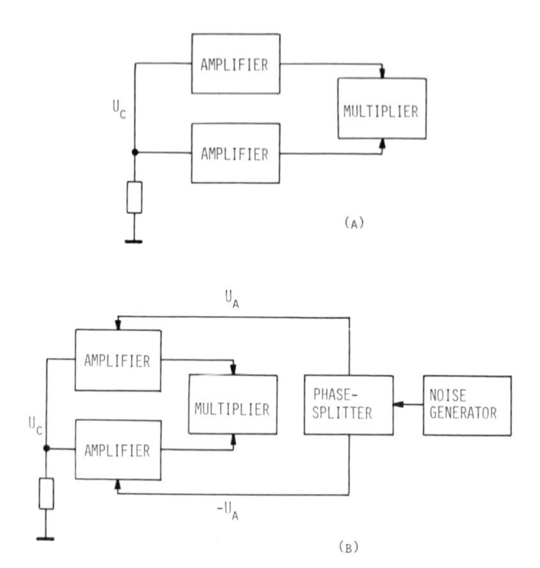

FIGURE 19. Measurement of the temperature of wire samples by their thermal noise: (a) correlation method, (b) compensation of input noise voltage by anticorrelated voltages fed to the inputs.[58]

correlated voltages, U_a and $-U_a$, are additionally fed into the inputs. These voltages are taken from a separate noise generator and a phase-splitter. If the rms voltage to be measured and the rms compensating voltage are equal, then the signal at the output of the multiplier is zero, irrespective of the amplifier gains and of the frequency response of both the amplifier and the multiplier.

The advantages of the noise correlation thermometer based on the compensation principle are especially efficient when the sample temperature is periodically changed. In this case, by measuring the oscillations of the thermal noise and of the electrical resistance of the sample resulting from the changes in the heating current, it is possible to determine the temperature derivative of the electrical resistance and the heat transfer coefficient.[59,60] Two similar wire samples were connected in series with respect to the dc source and in parallel with respect to the input of the correlation amplifier for simplifying the calculations of the noise temperature (Fig. 20). The samples were heated by dc current whose magnitude was periodically changed. The thermal noise of the samples was compensated by the anticorrelated voltage generated by a noise diode and a phase-splitter. The magnitude of the anticorrelated voltage was determined from the anode current of the noise diode measured by a digital voltmeter.

When the sample temperature was periodically changed, the corresponding changes of the multiplier output signal were registered by a recorder. The changes in the heating current and in the thermal noise of the samples were determined. The period of the current changes was sufficiently long to permit the equilibrium temperature to be established. The changes of the electrical resistance of the samples corresponding to definite changes of the heating current

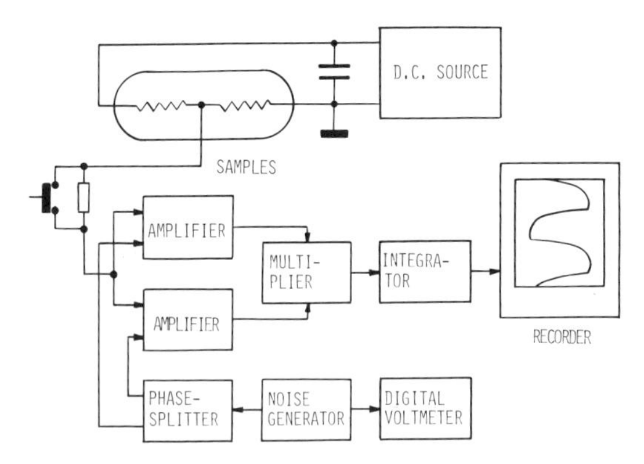

FIGURE 20. Noise correlation thermometer for measuring the oscillations of the wire sample temperature.(59)

were preliminarily determined with a bridge. This allowed one to calculate the changes of the resistance corresponding to the definite changes of the sample temperature. The modulation period was 120 s, the amplitude of the temperature changes was about 50 K. The measuring system was calibrated against a standard resistor at room temperature which was periodically connected to the input of the correlation amplifier in series with the samples. The value of the resistor was chosen such that the magnitude of its thermal noise was close to the changes of the sample thermal noise when the sample temperature was periodically changed.

The expressions for calculating the temperature derivative of resistance R' and the heat transfer coefficient P' are easily obtained:

$$R' = R_1/[\Delta(RT)/\Delta R - T_2] = R_2/[\Delta(RT)/\Delta R - T_1] \tag{41}$$

$$P' = R_1 \Delta P/[\Delta(RT) - T_2 \Delta R] = R_2 \Delta P/[(RT) - T_1 \Delta R] \tag{42}$$

where R_1 and R_2 are the sample resistances at the initial T_1 and final T_2 temperatures, respectively, $\Delta(RT)$ is the quantity determined by the change of the sample thermal noise, ΔR is the corresponding change of the sample resistance, and ΔP represents the change of the heating power when the temperature changes from T_1 to T_2.

The results obtained in this way(59,60) confirm that thermal noise can be used not only for the measurement of the temperature of wire samples heated by an electric current but also for the determination of the temperature derivatives which are required in the modulation measurements.

11. ACCURACY OF MODULATION MEASUREMENTS

As in other cases, the errors of measurements can be grouped into two categories: errors arising from differences between the theoretical model and experimental conditions, and instrumental errors due to the inaccuracy of measuring instruments. The theoretical model is supposed to include the following conditions:

1. The mean temperature of the sample is the same throughout the sample.
2. The amplitude and the phase of temperature oscillations do not vary throughout the sample.
3. The modulation frequency is sufficiently high to meet the criterion of adiabatic conditions: the changes in the heat loss of the sample during oscillations of its temperature are much smaller than the oscillations of the heating power.
4. Heat capacity of the sample is much larger than that of a heater and a thermometer.

Condition (2) supposes that when separate heater and thermometer are used the time for establishing thermal equilibrium between them and the sample is much smaller than the period of modulation. Hence, conditions (2) and (3) pose contradictory requirements for the modulation frequency.

The theoretical model can be implemented the best when the sample, in the form of a wire, foil, or thin rod, is heated by an electric current and the temperature oscillations are registered by the oscillations of sample resistance or brightness. In this case condition (4) is excluded and condition (3) is easily met by increasing the modulation frequency. In order to exclude the cold-end effects, the measurements can be performed only on the central part of the sample where the axial temperature gradients are small. Radial temperature differences are insignificant due to high thermal conductivity of conducting samples and their small thickness. Therefore the conditions (1) and (2) are also easily satisfied with direct electrical heating.

In other methods of heating the conditions (1) and (2) are met by decreasing the sample thickness and the modulated power. When the modulated power is fed into the sample its mean temperature increases. This temperature increase is much larger than the amplitude of the temperature oscillations and can result in nonuniformity of the temperature in the sample. To make this effect negligible, only a small modulated power is supplied to the sample while the mean sample temperature is controlled with a furnace. This method of heating is especially necessary in the phase transition studies where good temperature resolution is one of the important requirements. Thickness of the samples with low thermal conductivity usually does not exceed 0.1–0.2 mm. Decreased sample thickness is favorable for satisfying the conditions (1) and (2) but contradicts requirement (4); this contradiction is removed by making the heater and the thermometer in the form of thin films $10^{-5}-10^{-3}$ mm thick. Therefore, the correction for their heat capacity is usually small.

With the radiative heat exchange when the heat transfer coefficient does not depend on frequency it is possible to decrease the modulation frequency and to measure the specific heat in the nonadiabatic regime or to decrease the radiative heat losses by means of a thermal shield with an oscillating temperature.[36] These methods cannot be used with different mechanisms of the heat exchange. In such cases the modulation frequency should be chosen especially carefully. Usually, one should experimentally verify that there is a frequency range where the product $\omega\Theta_0$ is constant; this means that conditions (2) and (3) are satisfied. It turned out that these conditions cannot be met simultaneosly only in one case, when samples with low thermal diffusivity are studied under high pressures. In this case, in order to obtain the specific heat value, one has to introduce large corrections.

Thus, in almost all cases the conditions for modulation measurements of specific heat are in good agreement with the theoretical model. In modulation measurements of thermal expansion, electrical resistivity, and thermopower such agreement is achieved more easily since it is not necessary to satisfy the adiabatic conditions.

The main sources of errors in modulation measurements are determinations of the amplitude of the heating power and sample temperature oscillations. Modern measuring instruments make it possible to determine very accurately the oscillations of the heating power when direct heating or resistive heaters are used. With other ways of heating the situation is not so good.

Temperature oscillations can be measured with the best accuracy when thermocouples and resistance thermometers are used. With other methods (temperature dependence of resistivity, photoelectric detectors) the errors are determined by the reliability of the data used or by the accuracy of calibration.

Shown in Table 2 are data on the errors involved in different variants of modulation measurements. Ther term "imprecision" refers to the differences of single data points from the smoothed values, and the term "inaccuracy" refers to the estimated total error including random and systematic errors.

12. EMPLOYMENT OF MODULATION METHODS

The modulation method for measuring specific heat has been used in many works; their number is rapidly increasing. Successfully implemented were the main features of the method: wide temperature range, excellent temperature resolution, and high sensitivity to the changes in specific heat. Here are some of the areas of employment of the modulation method:

(1) Specific heat of metals at high temperatures. In this temperature range specific heat was measured, up to recent times, only by the drop method; also the pulse method is successfully used now. Interest in this temperature range was enhanced after the observation of a substantial nonlinear increase of specific

TABLE 2
Estimates of Errors in Modulation Measurements

Source of error	Imprecision (%)	Inaccuracy (%)
Mean temperature of the sample		0.01–1[a]
Mass of the sample		0.1–1[b]
Modulation frequency		$\leqslant$ 0.1[c]
Oscillations of heating power:		
Direct heating		0.2–1[c]
Electron bombardment		1–2[c]
Use of heaters		0.2–1[c]
Modulated-light heating	0.1	2[d]
Induction heating		4–6[e]
Temperature oscillations:		
Supplementary-current method	1	$\geqslant$ 2[f]
Third-harmonic method	1	$\geqslant$ 2[f]
Equivalent-impedance method	0.1	$\geqslant$ 1[f]
Photoelectric detectors	0.5	$\geqslant$ 1[g]
Thermocouples	0.1	0.2–1[a]
Resistance thermometers	0.01–0.1	0.2–1[a]
Oscillations of sample length, resistance, or thermal emf		1
Mean temperature of wire samples determined by their thermal noise		1

[a] Depends on the temperature.
[b] May be greater in microcalorimeters for organic substances.
[c] Depends on the employed equipment.
[d] According to Ref. 22.
[e] According to Ref. 14.
[f] Depends on the accuracy of R' values.
[g] Depends on the accuracy of calibration.

heat.(61) This effect can be interpreted as resulting from the formation of the equilibrium point defects in the crystal lattice. The results of the modulation measurements of specific heat have been used for determination of the formation enthalpy and equilibrium concentration of the vacancies in a number of metals. The equilibrium concentrations of vacancies proved to be very high, of the order of 1% at the melting point. This contradicts the conventional opinion that the vacancy concentrations in metals are small.(62,63) Therefore, many workers believe that the nonlinear increase of the specific heat is of a different nature – for example, caused by the anharmonicity.(64,65) Although there are many arguments favoring the high vacancy concentrations in metals,(66–69) they are all indirect so that the question is still the object of discussion. It can possibly be solved by the use of the relaxation method, i.e., specific heat measurements at such high frequencies of the temperature oscillations that the vacancy concentration could not follow the changes of the temperature.

(2) The anomalies in specific heat of solids at the phase transitions, par-

ticularly of the second order. It was found that the behavior of specific heat near a phase-transition point could not be described by the existing theory. The excellent temperature resolution and high sensitivity peculiar to the modulation method have been used in the numerous measurements of specific heat. In these measurements the exact absolute values of specific heat are not needed, and the modulated-light heating was widely used. The modulation measurements permitted investigating the character of the specific heat anomaly at the second-order phase transitions and in this way contributed to the development of the theory of phase transitions.[70,71] During the recent years the modulation method has been successfully applied for the studies of organic substances. The main objects are liquid crystals. In different temperature regions there are different phases of a liquid crystal. A transition from one phase into another is accompanied by the anomalies in specific heat. With the modulation measurements, the peculiarities of specific heat of liquid crystals were investigated more accurately than by any other method.

(3) Measurements under high pressures present the most serious problem for the modulation method since in this case it is difficult or impossible to satisfy the contradictory requirements to the modulation frequency. However, the modulation method has been successfully applied under high pressures, too.

(4) Low temperatures. In the first low-temperature measurements the modulation method was employed because of its high sensitivity. Specific heat of indium and beryllium was measured as a function of external magnetic field.[11,12,72] Modulation measurements at low temperatures now include studies of phase transitions of different nature,[73-79] properties of thin films,[80-85] measurements under high pressures[86,87] and in strong magnetic fields.[88,89]

In Table 3 are collected modulation measurements of specific heat.[90-146] Not included in the table are measurements below the room temperature and the works in which periodical heating has been used for simultaneous measurements of specific heat and other thermal properties by the temperature wave method.

The use of modulation methods in the studies of thermal expansion, electrical resistivity, and thermopower are presented in Table 4.[147-154]

13. CONCLUSION

During the last 20 years the modulation method has become an important method for measuring specific heat. It is successfully used in the measurements at very high and very low temperatures, under high pressures, for studying metals, semiconductors, dielectrics, and organic substances. Due to the modern measuring techniques, the modulation measurements can be performed with a good temperature resolution unattainable in other methods. Among the variants

TABLE 3
Summary of Modulation Measurements of Specific Heat

Method of measurements	Substance	Temperature range (K)	Ref.
Direct heating			
Thermionic-current oscillations	W	2370–2480	3, 90
	W	1150–2500	91
Supplementary-current method	Ge, Si	300–1000	30
Third-harmonic method	Ti	600–1345	92
	Ge	600–900	93
	Au	500–1200	94
Equivalent-impedance method	W[a]	1500–3600	95
Wire samples,	Ta[a]	1200–2900	96
$f \sim 10–10^2$ Hz	Mo[a]	1300–2500	97
$\Theta_0 \sim 1–10$ K	Nb[a]	1300–2700	98
	Pt[a]	1000–2000	99
	Au[a]	700–1300	100
	W	1300–3000	101
	W–20%Re	1000–2700	60
Photoelectric detectors	W, Ta, Mo, Nb	1200–2400	7
Wire samples,	Mo	1200–2400	102
$f \sim 10–10^2$ Hz,	Nb	1200–2400	103
$\Theta_0 \sim 10$ K	Ir	1500–2500	104
	Pt	1500–2000	105
	Ti[a]	1400–1900	106
	Zr[a]	1300–2000	107
	W–20%Re	1600–2900	108
$\Theta_0 \sim 0.1–1$ K	Ti[b]		109
	Fe[b]		110
	Co[b]		111
Liquid metals	Sn	1300–2000	9
	Cu	1460–1640	35
	Ga	1380–1620	35
	Cs	1200–1700	112
Use of thermocouples	Ni[b]		113
	Cu[a]	500–1250	114
	La[a]	600–1100	115
	Ti	400–1100	116, 117
Pressures to 3 kbar, $\Theta_0 \sim 10^{-4}$ K	Cr[b]	300	18
Electron-bombardment heating			
	Mo	800–1400	118
	Fe	600–1300	29
Use of heaters			
The temperature oscillations are measured with thermocouples or resistance thermometers	Gd	240–330	119
	NH_4Cl, ND_4Cl	200–300	120
Microcalorimeters, $\Theta_0 \sim 10^{-2}$ K	Organic substances	300–400	19–21, 121–127

TABLE 3 (Continued)

Modulated-light heating			
Thin-slab samples $f \sim 1$–10 Hz, $\Theta_0 \sim 10^{-2}$ K	Ni[b]		13, 128–130
	β-brass[b]		131, 132
	Cr[b]		133
	Gd[b]		134, 135
	Fe[b]		50
	Alloys		136, 137
	$LiTaO_3$	300–1000	15
	$NaNO_2$	300–470	138
	$BaTiO_3$	300–450	139, 140
	$SrTiO_3$	80–320	141
	$Gd_2(MoO_4)_3$	300–500	142
	TGSe	100–360	143
	SbSI	80–330	144
	Cu, Ni	300–650	22
Microcalorimeters, $\Theta_0 \sim 10^{-3}$ K	Organic substances	300–400	145, 146

[a]Study of vacancy formation.
[b]Study of phase transitions, measurements only in the vicinity of transition points.

of the modulation method already developed it is possible to find one most suitable for each specific case. At the present time the modulation method for measuring the specific heat has firmly taken its place in the practice of thermophysical experiments.

The principle of temperature modulation has proved to be very useful also in other thermophysical measurements – for studying thermal expansion, electrical resistivity, and thermopower. Here the advantages of the modulation method are also obvious, especially at high temperatures. One may hope that with time the temperature modulation principle will be accepted also in these measurements.

NOTATION

p_0, p	dc and ac components of power fed to the sample
$P(T), P', P''$	Heat loss from the sample and its temperature derivatives (P' is the heat transfer coefficient)
T_0	Mean temperature of the sample
Θ_0, Θ_1	Amplitude of temperature oscillations at the fundamental frequency
Θ_2	Amplitude of the temperature oscillations at the doubled frequency (the second harmonic)
φ, ψ	Phase of the sample temperature oscillations
ω	Frequency ($\omega = 2\pi f$)
C	Heat capacity; capacitance
K	Thermal conductance; proportionality coefficient
R_0, R'	Sample resistance at T_0 and its temperature derivative
β	Temperature coefficient of resistance

TABLE 4
Summary of Modulation Measurements of Thermal Expansion, Electrical Resistivity, and Thermopower

Property	Substance	Temperature range (K)	Ref.
Thermal expansion			
Direct heating, wire samples,	W	1300–2300	45
$f \sim 10\text{–}10^2$ Hz, $\Theta_0 \sim 1$ K	Pt[a]	1100–1900	147
	W[a]	2000–2900	148
Bulk sample, $f = 0.1$ Hz, $\Theta_0 \sim 5$ K	Ni	700–1100	47
Electrical resistivity			
Direct heating, bulk samples,	Ni[b]		48
$f \sim 0.1$ Hz, $\Theta_0 \sim 1$ K	Fe[b]		149
	Sm	650–900	150
	Al[a]	450–900	151
	Pt[a]	1050–1850	152
Modulated-light heating,	Cr[b]		49
$f \sim 10$ Hz, $\Theta_0 \sim 0.01\text{–}0.1$ K	β-brass[b]		132
	Gd[b]		135
	Fe[b]		50
Thermopower			
Furnace, $f \sim 0.1$ Hz,	Fe[b]		51
$\Theta_0 \sim 1$ K	Co[b]		153
	Sm	650–900	150
	Cu[a]	1000–1300	154
Modulated-light heating,	Au-Al	300– 400	52
$f \sim 10$ Hz, $\Theta_0 \sim 0.1$ K	Cu[c]		53

[a]Study of vacancy formation.
[b]Measurements only in the vicinity of transition point.
[c]Changes in thermopower after deformation.

I_0, i	dc and ac components of current
V_0, v	dc and ac components of voltage
Z	Equivalent impedance of the sample
R	Resistance
E'	Thermocouple sensitivity ($E' = dE/dT$)
α	Temperature coefficient of specific heat; thermal expansion coefficient
$\langle \Delta T_f^2 \rangle$	Spectral density of temperature fluctuations
$\langle \Theta_f^2 \rangle$	Spectral density of temperature oscillations caused by noise modulation
k	Boltzmann constant

REFERENCES

1. O.M. Corbino, "Thermische Oszillationen wechselstromdurchflossener Lampen mit dünnem Faden und daraus sich ergebende Gleichrichterwirkung infolge der Anwesenheit geradzahliger Oberschwingungen", *Phys. Z.* **11**, 413–417 (1910).
2. O.M. Corbino, "Periodische Widerstandsänderungen feiner Metallfäden, die durch

Wechselströme zum Glühen gebracht werden, sowie Ableitung ihrer thermischen Eigenschaften bei hoher Temperatur", *Phys. Z.* 12, 292–295 (1911).

3. K.K. Smith and P.W. Bigler, "Oscillations of Temperature of an Incandescent Filaments, and the Specific Heat of Tungsten", *Phys. Rev.* **19**, 268–270 (1922).
4. L.P. Filippov, "Measurement of Thermal Activity of Liquids" (in Russian), *Inzh.-Fiz. Zh.* **3**(7), 121–123 (1960).
5. L.A. Rosenthal, "Thermal Response of Bridge Wires Used in Electroexplosive Devices", *Rev. Sci. Instrum.* **32**, 1033–1036 (1961).
6. Ya.A. Kraftmakher, "Modulation Method for Measuring Specific Heat" (in Russian), *Zh. Prikl. Mekh. Tekhn. Fiz.* (**5**), 176–180 (1962).
7. G.C. Lowenthal, "The Specific Heat of Metals between 1200 K and 2400 K", *Austral. J. Phys.* **16**, 47–67 (1963).
8. L.P. Filippov and R.P. Yurchak, "Measurement of Specific Heat of Solid and Liquid Metals" (in Russian), *Teplofiz. Vys. Temp.* **3**, 901–909 (1965).
9. I.A. Akhmatova, "Measurement of Specific Heat of Liquid Tin at High Temperatures" (in Russian), *Dokl. Akad. Nauk SSSR* **162**, 127–129 (1965).
10. L.R. Holland and R.C. Smith, "Analysis of Temperature Fluctuations in a.c. Heated Filaments", *J. Appl. Phys.* **37**, 4528–4536 (1966).
11. P. Sullivan and G. Seidel, "An a.c. Temperature Technique for Measuring Heat Capacities", *Ann. Acad. Sci. Fennicae, Ser. A, VI. Phys.*, **210**, 58–62 (1966).
12. P.F. Sullivan and G. Seidel, "Steady-State, a.c.-Temperature Calorimetry", *Phys. Rev.* **173**, 679–685 (1968).
13. P. Handler, D.E. Mapother and M. Rayl, "A.c. Measurement of the Heat Capacity of Nickel near Its Critical Point", *Phys. Rev. Lett.* **19**, 356–358 (1967).
14. L.P. Filippov and I.N. Makarenko, "Measurement of Thermal Properties of Metals at High Temperatures" (in Russian), *Teplofiz. Vys. Temp.* **6**, 149–156 (1968).
15. A.M. Glass, "Dielectric, Thermal, and Pyroelectric Properties of Ferroelectric $LiTaO_3$", *Phys. Rev.* **172**, 564–571 (1968).
16. Ya.A. Kraftmakher and V.L. Tonaevskii, "Direct Measurement of the Temperature Coefficient of Specific Heat", *Phys. Status Solidi A* **9**, 573–579 (1972).
17. A.A. Varchenko and Ya.A. Kraftmakher, "Non-Adiabatic Regime in Modulation Calorimetry", *Phys. Status Solidi A* **20**, 387–393 (1973).
18. A. Bonilla and C.W. Garland, "High-Pressure Heat Capacity of Chromium near the Néel Line", *J. Phys. Chem. Solids* **35**, 871–877 (1974).
19. C.A. Schantz and D.L. Johnson, "Specific Heat of the Nematic, Smectic-A and Smectic-C Phases of 4-*n*-pentylphenylthiol-4'-*n*-octyloxybenzoate: Critical Behaviour", *Phys. Rev. A* **17**, 1504–1512 (1978).
20. J.E. Smaardyk and J.M. Mochel, "High Resolution a.c. Calorimeter for Organic Liquids", *Rev. Sci. Instrum.* **49**, 988–993 (1978).
21. O.S. Tanasijczuk and T. Oja, "High Resolution Calorimeter for the Investigation of Melting in Organic and Biological Materials", *Rev. Sci. Instrum.* **49**, 1545–1548 (1978).
22. S. Ikeda and Y. Ishikawa, "Improvement of a.c. Calorimetry", *Jpn. J. Appl. Phys.* **18**, 1367–1372 (1979).
23. Ya. A. Kraftmakher and S.D. Krylov, "Observation of Temperature Fluctuations Using Photoelectric Devices" (in Russian), *Teplofiz. Vys. Temp.* **18**, 317–321 (1980).
24. Ya.A. Kraftmakher and S.D. Krylov, "Temperature Fluctuations and Isochoric Specific Heat of Tungsten" (in Russian), *Fiz. Tverd. Tela* **22**, 3157–3159 (1980).
25. Ya.A. Kraftmakher, "Method for Observing Relaxation Effects in High-Temperature Specific Heat of Metals" (in Russian), *Teplofiz, Vys. Temp.* **19**, 656–658 (1981).

26. Ya.A. Kraftmakher, "The Modulation Method for Measuring Specific Heat", *High Temp. High Pressures* **5**, 433–454 (1973).
27. Ya.A. Kraftmakher, "Modulation Methods for Studying Thermal Expansion, Electrical Resistivity, and the Seebeck Coefficient", *High Temp. High Pressures* **5**, 645–656 (1973).
28. Ya.A. Kraftmakher, in *Thermal Expansion 6* (I.D. Peggs, Ed.), pp. 155–164, Plenum Press, New York (1978).
29. A.A. Varchenko, Ya.A. Kraftmakher and T.Yu. Pinegina, "Use of Electronic Heating in Modulation Measurements of Specific Heat" (in Russian), *Teplofiz. Vys. Temp.* **16**, 844–847 (1978).
30. D. Gerlich, B. Abeles and R.E. Miller, "High Temperature Specific Heat of Ge, Si and Ge–Si Alloys", *J. Appl. Phys.* **36**, 76–79 (1965).
31. R.N. Griesheimer, in *Technique of Microwave Measurements* (C.G. Montgomery, Ed.), pp. 79–220, McGraw-Hill, New York (1947).
32. R.C. Jones, in *Advances in Electronics* (L. Marton, Ed.), Vol. 5, pp. 1–96, Academic Press, New York (1953).
33. A. Van der Ziel, *Solid State Physical Electronics,* Plenum Press, New York (1957).
34. Ya.A. Kraftmakher, "Potentiometer Circuit for Measuring Specific Heat by Modulation Method" (in Russian), *Zh. Prikl. Mekh. Tekhn. Fiz.* **(2)**, 144 (1966).
35. I.A. Akhmatova, "Specific Heat of Melted Gallium and Copper at High Temperatures" (in Russian), *Izmer. Tekh.* **(8)**, 14–17 (1967).
36. Ya.A. Kraftmakher and V.Ya. Cherepanov, "Compensation of Heat Losses in Modulation Measurements of Specific Heat" (in Russian), *Teplofiz. Vys. Temp.* **16**, 647–649 (1978).
37. J.D. Baloga and C.W. Garland, "A.c. Calorimetry at High Pressure", *Rev. Sci. Instrum.* **48**, 105–110 (1977).
38. E.S. Itskevich, V.F. Kraidenov and V.S. Syzranov, "Measurement of Low Temperature Heat Capacity of Metals Under Pressure", *Cryogenics* **18**, 281–284 (1978).
39. J.M.W. Milatz and H.A. Van der Velden, "Natural Limit of Measuring Radiation with a Bolometer", *Physica* **10**, 369–380 (1943).
40. J.J. Jackson and J.S. Koehler, "Production of Vacancies in Pulsed Gold", *Bull. Am. Phys. Soc.* **5**, 154 (1960).
41. E. Korostoff, "Relaxation Method for Studying Vacancies in Solids", *J. Appl. Phys.* **33**, 2078–2079 (1962).
42. J. Van den Sype, "On Temperature Relaxation in Metals", *Phys. Status Solidi* **39**, 659–664 (1970).
43. D.A. Skelskey and J. Van den Sype, "A Relaxation Phenomenon Observed in Fine Gold Wire", *Solid State Commun.* **15**, 1257–1262 (1974).
44. A.H. Seville, "Studies of the Specific Heat of Platinum by Modulation Method", *Phys. Status Solidi A* **21**, 649–658 (1974).
45. Ya.A. Kraftmakher and I.M. Cheremisina, "Modulation Method for Studying Thermal Expansion" (in Russian), *Zh. Prikl. Mekh. Tekhn. Fiz.* **(2)**, 114–115 (1965).
46. Ya.A. Kraftmakher, "Compensation Scheme for Measuring the Thermal Expansion Coefficient" (in Russian), *Zh. Prikl. Mekh. Tekhn. Fiz.* **(4)** 143–144 (1967).
47. Ya.A. Kraftmakher and V.P. Nezhentsev, in *Fizika Tverdogo Tela i Termodinamika,* pp. 233–237, Nauka, Novosibirsk (1971).
48. Ya.A. Kraftmakher, "Electrical Conductivity of Nickel near the Curie Point" (in Russian), *Fiz. Tverd. Tela* **9**, 1529–1530 (1967).
49. M.B. Salamon, D.S. Simons and P.R. Garnier, "Simultaneous Measurement of the Anomalous Heat Capacity and Resistivity of Chromium near T_N", *Solid State Commun.* **7**, 1035–1038 (1969).

50. F.L. Lederman, M.B. Salamon and L.W. Shacklette, "Experimental Verification of Scaling and Test of the Universality Hypothesis from Specific-Heat Data," *Phys. Rev. B* **9**, 2981–2988 (1974).
51. Ya.A. Kraftmakher and T.Yu. Pinegina, "Thermoelectric Power of Iron near the Curie Point", *Phys. Status Solidi* **42**, K151–K152 (1970).
52. R.H. Freeman and J. Bass, "An a.c. System for Measuring Thermopower", *Rev. Sci. Instrum.* **41**, 1171–1174 (1970).
53. W. Hellenthal and H. Ostholt, "Dynamisches Thermokraftmessverfahren und Anwendung für Erholungsuntersuchungen", *Z. angew. Phys.* **28**, 313–316 (1970).
54. J.B. Garrison and A.W. Lawson, "An Absolute Noise Thermometer for High Temperatures and High Pressures", *Rev. Sci. Instrum.* **20**, 785–794 (1949).
55. F.J. Shore and R.S. Williamson, "Suggested Thermometer for Low Temperatures Using Nyquist Noise and Correlator-Amplifier", *Rev. Sci. Instrum.* **37**, 787–788 (1966).
56. L. Storm, "Messung sehr kleiner Rauschspannungen mit einem Korrelator und Rauschthermometrie tiefer Temperaturen", *Z. Angew, Phys.* **28**, 331–333 (1970).
57. Ya.A. Kraftmakher and A.G. Cherevko, "Correlation Amplifier for Studying Fluctuation Phenomena" (in Russian), *Prib. Tekh. Eksp.* (4), 150–151 (1972).
58. Ya.A. Kraftmakher and A.G. Cherevko, "Noise Correlation Thermometer", *Phys. Status Solidi A* **14**, K35–K38 (1972).
59. Ya.A. Kraftmakher and A.G. Cherevko, "Measurement of Temperature Oscillations of Wire Samples Using Their Thermal Noise", *Phys. Status Solidi A* **25**, 691–695 (1974).
60. Ya.A. Kraftmakher and A.G. Cherevko, "Electrical Resistivity and Specific Heat of the W–20%Re Alloy at High Temperatures", *High Temp. High Pressures* 7, 283–286 (1975).
61. N.S. Rasor and J.D. McClelland, "Thermal Properties of Graphite, Molybdenum and Tantalum to Their Destruction Temperatures", *J. Phys. Chem. Solids* **15**, 17–26 (1960).
62. A. Seeger, "The Study of Point Defects in Metals in Thermal Equilibrium. I. The Equilibrium Concentration of Point Defects", *Crystal Lattice Defects* **4**, 221–253 (1973).
63. R.W. Siegel, "Vacancy Concentrations in Metals", *J. Nucl. Mater.* **69–70**, 117–146 (1978).
64. Y. Ida, "Anharmonic Effect on Heat Capacity of Solids up to the Critical Temperature of Lattice Instability", *Phys. Rev. B* **1**, 2488–2496 (1970).
65. R.A. MacDonald, R.D. Mountain and R.C. Shukla, "High-Temperature Specific Heat of Crystals", *Phys. Rev. B* **20**, 4012–4017 (1979).
66. Ya.A. Kraftmakher and P.G. Strelkov, in *Vacancies and Interstitials in Metals* (A. Seeger, D. Schumacher, W. Schilling and J. Diehl, Eds.), pp. 59–78, North-Holland, Amsterdam (1970).
67. Ya.A. Kraftmakher, in *Defect Interactions in Solids* (K.I. Vasu, K.S. Raman, D.H. Sastry and Y.V.R.K. Prasad, Eds.), pp. 64–70, Indian Institute of Science, Bangalore (1974).
68. Ya.A. Kraftmakher, in *Proc. 7th Symposium Thermophys. Properties,* pp. 160–168, ASME, New York (1977).
69. Ya.A. Kraftmakher, "On Equilibrium Vacancy Concentrations in Metals", *Scr. Metall.* **11**, 1033–1038 (1977).
70. M.E. Fisher, *The Nature of Critical Points,* University of Colorado Press, Boulder (1965).

71. H.E. Stanley, *Introduction to Phase Transitions and Critical Phenomena,* Clarendon Press, Oxford (1971).
72. P. Sullivan and G. Seidel, "A.c. Temperature Measurement of Changes in Heat Capacity of Beryllium in a Magnetic Field", *Phys. Lett. A* **25**, 229–230 (1967).
73. M.B. Salamon, "Specific Heat of CoO near T_N: Anisotropy Effects", *Phys. Rev. B* **2**, 214–220 (1970).
74. P. Schwartz, "Order–Disorder Transition in NH_4Cl. III. Specific Heat", *Phys. Rev. B* **4**, 920–928 (1971).
75. M.B. Salamon and H. Ikeda, "Specific Heat of Two-Dimensional Heisenberg Antiferromagnets: K_2MnF_4 and K_2NiF_4", *Phys. Rev. B* **7**, 2017–2024 (1973).
76. F.L. Lederman, M.B. Salamon and H. Peisl, "Evidence for an Order–Disorder Transformation in the Solid Electrolyte $RbAg_4I_5$", *Solid State Commun.* **19**, 147–150 (1976).
77. R. Vargas, M.B. Salamon and C.P. Flynn, "Ionic Conductivity near an Order–Disorder Transition: $RbAg_4I_5$", *Phys. Rev. Lett.* **37**, 1550–1553 (1976).
78. R.A. Vargas, M.B. Salamon and C.P. Flynn, "Ionic Conductivity and Heat Capacity of the Solid Electrolytes MAg_4I_5 near T_c", *Phys. Rev. B* **17**, 269–281 (1978).
79. H. Ikeda, N. Okamura, K. Kato and A. Ikushima, "Experimental Observation of Crossover Phenomena in the Specific Heat of MnF_2", *J. Phys. C* **11**, L231–L235 (1978).
80. G.D. Zally and J.M. Mochel, "Fluctuation Heat Capacity in Superconducting Thin Films of Amorphous BiSb", *Phys. Rev. Lett.* **27**, 1710–1712 (1971).
81. R.L. Greene, C.N. King, R.B. Zubeck and J.J. Hauser, "Specific Heat of Granular Aluminum Films", *Phys. Rev. B* **6**, 3297–3305 (1972).
82. P. Manuel and J.J. Veyssié, "Thin Films Heat Capacity Measurements in the ^3He Temperature Range", *Phys. Lett. A* **41**, 235–236 (1972).
83. G. Kämpf and W. Buckel, "Specific Heat of Amorphous and Polycrystalline Lead–Bismuth Films", *Z. Phys. B* **27**, 315–319 (1977).
84. B.C. Gibson, D.M. Ginsberg and P.C.L. Tai, "Specific Heat of Superconducting Films of Indium and of Indium Alloyed with Magnetic Impurities", *Phys. Rev. B* **19**, 1409–1419 (1979).
85. N.A.H.K. Rao and A.M. Goldman, "Fluctuation Heat Capacity of Amorphous Nb_3Ge Films", *J. Low Temp. Phys.* **42**, 253–276 (1981).
86. A. Eichler and W. Gey, "Method for the Determination of the Specific Heat of Metals at Low Temperatures under High Pressures", *Rev. Sci. Instrum.* **50**, 1445–1452 (1979).
87. A. Eichler, H. Bohn and W. Gey, "Specific Heat of the Superconducting High Pressure Phase Ga II", *Z. Phys. B* **38**, 21–25 (1980).
88. C.C. Huang, A.M. Goldman and L.E. Toth, "Specific Heat of a Transforming V_3Si Crystal", *Solid State Commun.* **33**, 581–584 (1980).
89. P. Garoche and W.L. Johnson, "An Accurate Determination of the Excess Low Temperature Heat Capacity of a Superconducting Metallic Glass", *Solid State Commun.* **39**, 403–406 (1981).
90. L.I. Bockstahler, "The Specific Heat of Incandescent Tungsten by an Improved Method", *Phys. Rev.* **25**, 677–685 (1925).
91. C. Zwikker, "Messungen der spezifischen Wärme von Wolfram zwischen 90 und 2600 K", *Z. Phys.* **52**, 668–677 (1928).
92. L.R. Holland, "Physical Properties of Titanium. III. The Specific Heat", *J. Appl. Phys.* **34**, 2350–2357 (1963).
93. R.C. Smith, "High-Temperature Specific Heat of Germanium", *J. Appl. Phys.* **37**, 4860–4865 (1966).

94. D. Skelskey and J. Van den Sype, "High Temperature Specific Heat of Gold Using the Modulation Method", *J. Appl. Phys.* **41**, 4750–4751 (1970).
95. Ya.A. Kraftmakher and P.G. Strelkov, "Energy of Formation and Concentration of Vacancies in Tungsten" (in Russian), *Fiz. Tverd. Tela* **4**, 2271–2274 (1962).
96. Ya.A. Kraftmakher, "Specific Heat of Tantalum Between 1200 and 2900 K" (in Russian), *Zh. Prikl. Mekh. Tekhn, Fiz.* (**2**), 158–160 (1963).
97. Ya.A. Kraftmakher, "Specific Heat of Tantalum between 1200 and 2900 K" (in *Tela* **6**, 503–505 (1964).
98. Ya.A. Kraftmakher, "Vacancy Formation in Niobium" (in Russian), *Fiz. Tverd. Tela* **5**, 950–951 (1963).
99. Ya.A. Kraftmakher and E.B. Lanina, "Energy of Formation and Concentration of Vacancies in Platinum" (in Russian), *Fiz. Tverd. Tela* **7**, 123–126 (1965).
100. Ya.A. Kraftmakher and P.G. Strelkov, "Energy of Formation and Concentration of Vacancies in Gold" (in Russian), *Fiz. Tverd. Tela* **8**, 580–582 (1966).
101. O.A. Kraev, "Version of Modulation Method for Measuring Specific Heat of Metals" (in Russian), *Teplofiz. Vyz. Temp.* **5**, 817–820 (1967).
102. I.N. Makarenko, L.N. Trukhanova and L.P. Filippov, "Thermal Properties of Molybdenum at High Temperatures" (in Russian), *Teplofiz. Vys. Temp.* **8**, 445–447 (1970).
103. I.N. Makarenko, L.N. Trukhanova and L.P. Filippov, "Thermal Properties of Niobium at High Temperatures" (in Russian), *Teplofiz. Vys. Temp.* **8**, 667–670 (1970).
104. L.N. Trukhanova and L.P. Filippov, "Thermal Properties of Iridium at High Temperatures" (in Russian), *Teplofiz. Vys. Temp.* **8**, 919–920 (1970).
105. L.N. Trukhanova and S.N. Banchila, "Thermal Properties of Platinum at High Temperatures" (in Russian), *Vestn. Mosk. Univ., Ser. Fiz. Astron.* **15**, 599–601 (1974).
106. V.O. Shestopal, "Specific Heat and Vacancy Formation in Titanium at High Temperatures" (in Russian), *Fiz. Tverd. Tela* **7**, 3461–3463 (1965).
107. O.M. Kanel' and Ya.A. Kraftmakher, "Vacancy Formation in Zirconium" (in Russian), *Fiz. Tverd. Tela* **8**, 283–284 (1966).
108. K.S. Sukhovei, "Specific Heat of W-20%Re Alloy between 1600 and 2900 K" (in Russian), *Fiz. Tverd. Tela* **9**, 3660–3663 (1967).
109. G.G. Zaitseva and Ya.A. Kraftmakher, "Specific Heat of Titanium near Its Phase Transition Point" (in Russian), *Zh. Prikl. Mekh. Tekhn. Fiz.* (**3**), 117 (1965).
110. Ya.A. Kraftmakher and T.Yu. Romashina, "Specific Heat of Iron near the Curie Point" (in Russian), *Fiz. Tverd. Tela* **7**, 2532–2533 (1965).
111. Ya.A. Kraftmakher and T.Yu. Romashina, "Specific Heat of Cobalt near the Curie Point" (in Russian), *Fiz. Tverd. Tela* **8**, 1966–1967 (1966).
112. L.P. Filippov, L.A. Blagonravov and V.A. Alekseev, "Specific Heat of Liquid Caesium at Temperatures up to 1700 K Under Pressure", *High Temp. High Pressures* **8**, 658–659 (1976).
113. Ya.A. Kraftmakher, "Specific Heat of Nickel near the Curie Point" (in Russian), *Fiz. Tverd. Tela* **8**, 1306–1308 (1966).
114. Ya.A. Kraftmakher, "Vacancy Formation in Copper" (in Russian), *Fiz. Tverd. Tela* **9**, 1850–1851 (1967).
115. A.I. Akimov and Ya.A. Kraftmakher, "Equilibrium Vacancies in Lanthanum", *Phys. Status Solidi* **42**, K41–K42 (1970).
116. I.I. Novikov, "The Effect of the Degree of Deformation on the Thermophysical Properties of Metals at High Temperatures", *High Temp. High Pressures* **8**, 483–492 (1976).

117. S.V. Boyarskii and I.I. Novikov, "Character of the Temperature Dependence of Specific Heat near the Phase Transition Point" (in Russian), *Teplofiz. Vys. Temp.* **16,** 534–536 (1978).
118. L.A. Pigal'skaya, R.P. Yurchak, I.N. Makarenko and L.P. Filippov, "Thermal Properties of Molybdenum at High Temperatures" (in Russian), *Teplofiz. Vys. Temp.* **4,** 144–147 (1966).
119. G.H.J. Wantenaar, S.J. Campbell, D.H. Chaplin and G.V.H. Wilson, " A simple Application of the a.c. Specific Heat Method to Bulk Samples," *J. Phys. E* **10,** 825–828 (1977).
120. C.W. Garland and J.D. Baloga, "Heat Capacity of NH_4Cl and ND_4Cl Single Crystals at High Pressures", *Phys. Rev. B* **16,** 331–339 (1977).
121. D.L. Johnson, C.F. Hayes, R.J. deHoff and C.A. Schantz, "Specific Heat near the Nematic–Smectic-A Transition of Octyloxycyanobiphenyl", *Phys. Rev. B* **18,** 4902–4912 (1978).
122. C.W. Garland, G.B. Kasting and K.J. Lushington, "High-Resolution Calorimetric Study of the Nematic–Smectic-A Transition in Octyloxycyanobiphenyl (8 OCB)," *Phys. Rev. Lett.* **43,** 1420–1423 (1979).
123. J.D. LeGrange and J.M. Mochel, "High-Resolution Heat-Capacity Studies near the Nematic–Smectic-A Transition in Octyloxycyanobiphenyl (8 OCB)", *Phys. Rev. Lett.* **45,** 35–38 (1980).
124. G.B. Kasting, K.J. Lushington and C.W. Garland, "Critical Heat Capacity near the Nematic–Smectic-A Transition in Octyloxycyanobiphenyl in the Range 1–2000 bar", *Phys. Rev. B* **22,** 321–331 (1980).
125. K.J. Lushington, G.B. Kasting and C.W. Garland, "Calorimetric Investigation of a Reentrant-Nematic Liquid-Crystal Mixture", *Phys. Rev. B* **22,** 2569–2572 (1980).
126. K.J. Lushington, G.B. Kasting and C.W. Garland, "Calorimetric Study of Phase Transitions in the Liquid Crystal Butyloxybenzylidene Octylaniline (40.8)", *J. Phys. (Paris) Lett.* **41,** L419–L422 (1980).
127. E. Bloemen and C.W. Garland, "Calorimetric Investigation of Phase Transitions in Butyloxybenzylidene Heptylaniline (40.7)", *J. Phys. (Paris)* **42,** 1299–1302 (1981).
128. D.L. Connelly, J.S. Loomis and D.E. Mapother, "Specific Heat of Nickel near the Curie Temperature", *Phys. Rev. B* **3,** 924–934 (1971).
129. M. Maszkiewicz, "Specific Heat of High Purity Nickel Single Crystal near the Critical Temperature", *Phys. Status Solidi A* **47,** K77–K80 (1978).
130. M. Maszkiewicz, B. Mrygoń, and K. Wentowska, "Rounding of Specific Heat of Nickel near the Critical Temperature", *Phys. Status Solidi A* **54,** 111–115 (1979).
131. J. Ashman and P. Handler, "Specific Heat of the Order–Disorder Transition in β-Brass", *Phys. Rev. Lett.* **23,** 642–644 (1969).
132. D.S. Simons and M.B. Salamon, "Specific Heat and Resistivity near the Order–Disorder Transition in β-Brass", *Phys. Rev. Lett.* **26,** 750–752 (1971).
133. P.R. Garnier and M.B. Salamon, "First-Order Transition in Chromium at the Néel Temperature", *Phys. Rev. Lett.* **27,** 1523–1526 (1971).
134. E.A.S. Lewis, "Heat Capacity of Gadolinium near the Curie Point", *Phys. Rev. B* **1,** 4368–4377 (1970).
135. D.S. Simons and M.B. Salamon, "Specific Heat and Resistivity of Gadolinium near the Curie Point in External Magnetic Fields", *Phys. Rev. B* **10,** 4680–4686 (1974).
136. L.J. Schowalter, M.B. Salamon, C.C. Tsuei and R.A. Craven, "The Critical Specific Heat of a Glassy Ferromagnet", *Solid State Commun.* **24,** 525–529 (1977).
137. S. Ikeda and Y. Ishikawa, "Critical Phenomena in Amorphous Ferromagnetic Alloys. I. Specific Heat Measurements", *J. Phys. Soc. Jpn.* **49,** 950–956 (1980).

138. I. Hatta and A. Ikushima, "Specific Heat of $NaNO_2$ near Its Transition Points", *J. Phys. Chem. Solids* **34**, 57–66 (1973).
139. I. Hatta and A. Ikushima, "Specific Heat of $BaTiO_3$", *Phys. Lett. A* **40**, 235–236 (1972).
140. I. Hatta and A. Ikushima, "Temperature Dependence of the Heat Capacity in $BaTiO_3$", *J. Phys. Soc. Jpn.* **41**, 558–564 (1976).
141. I. Hatta, Y. Shiroishi, K.A. Müller and W. Berlinger, "Critical Behavior of the Heat Capacity in $SrTiO_3$", *Phys. Rev. B* **16**, 1138–1145 (1977).
142. K.M. Cheung and F.G. Ullman, "Specific Heat of Gadolinium Molybdate at the Ferroelectric Transition", *Phys. Rev. B* **10**, 4760–4764 (1974).
143. K. Ema, K. Hamano, K. Kurihara and I. Hatta, "A.c. Calorimetric Investigations of Specific Heat Anomaly in Ferroelectric TGSe", *J. Phys. Soc. Jpn.* **43**, 1954–1961 (1977).
144. S. Stokka, K. Fossheim and S. Ziolkiewicz, "Specific Heat at a First-Order Phase Transition: SbSI", *Phys. Rev. B* **24**, 2807–2811 (1981).
145. J.M. Viner and C.C. Huang, "A Specific Heat Study of the Nematic–Smectic-A Transition in Octyloxycyanobiphenyl", *Solid State Commun.* **39**, 789–791 (1981).
146. C.C. Huang, J.M. Viner, R. Pindak and J.W. Goodby, "Heat-Capacity Study of the Transition from a Stacked-Hexatic-B Phase to a Smectic-A Phase," *Phys. Rev. Lett.* **46**, 1289–1292 (1981).
147. Ya.A. Kraftmakher, "Vacancy Formation and Thermal Expansion of Platinum" (in Russian), *Fiz. Tverd. Tela* **9**, 1528–1529 (1967).
148. Ya.A. Kraftmakher, "Equilibrium Vacancies and Thermal Expansion of Tungsten at High Temperatures" (in Russian), *Fiz. Tverd. Tela* **14**, 392–394 (1972).
149. Ya.A. Kraftmakher and T.Yu. Pinegina, "On the Anomaly of Electrical Resistivity of Iron near the Curie Point" (in Russian), *Fiz. Tverd. Tela* **16**, 132–137 (1974).
150. Ya.A. Kraftmakher and T.Yu. Pinegina, "Anomalies of Resistivity and Thermopower of Samarium at High Temperatures", *Phys. Status Solidi A* **47**, K81–K83 (1978).
151. Ya.A. Kraftmakher and G.G. Sushakova, "Equilibrium Vacancies and Electrical Conductivity of Aluminium", *Phys. Status Solidi B* **53**, K73–K76 (1972).
152. Ya.A. Kraftmakher and G.G. Sushakova, "Equilibrium Vacancies and Electrical Conductivity of Platinum" (in Russian), *Fiz. Tverd. Tela* **16**, 138–142 (1974).
153. Ya.A. Kraftmakher and T.Yu. Pinegina, "Anomaly of Thermopower of Cobalt near the Curie Point" (in Russian), *Fiz. Tverd. Tela* **13**, 2799–2800 (1971).
154. Ya.A. Kraftmakher, "Vacancy Formation and Thermopower of Copper at High Temperatures" (in Russian), *Fiz. Tverd. Tela* **13**, 3454–3455 (1971).

16

Pulse Calorimetry

ARED CEZAIRLIYAN

1. INTRODUCTION

Conventional steady-state and quasi-steady-state techniques for the accurate measurements of specific heat are generally limited to temperatures below 2000 K. This limitation is the result of severe problems (chemical reactions, heat transfer, evaporation, specimen containment, loss of mechanical strength and electrical insulation, etc.) which are created by the exposure of the specimen and its immediate environment to high temperatures for extended periods of time (minutes to hours). An approach to minimize the effect of these problems and thus to permit the extension of the measurements to higher temperatures is to perform the entire experiment in a very short period of time (less than a second). It is in this context that most of the pulse techniques for the measurement of specific heat at high temperatures were developed.

Although the advantages of the pulse measurement methods were realized for over 60 years, their development had been slow for a number of reasons. The most important of these was the lack of adequate electrical pulse generation, control, measurement, and recording equipment with short response times. Rapid advances in the electronics field during the last two decades, coupled with the increased demand for accurate data on specific heat of substances at high temperatures, revived the interest in the pulse measurement techniques.

The objective of this chapter is to present the description and a general survey of the pulse techniques used for the measurement of specific heat of electrically conducting substances. The presentation is limited to methods in which energy is imparted uniformly to the entire volume of the specimen through resistive self-heating of the specimen. Methods that utilize surface heating (laser pulse or other energy source), cyclic heating (modulation methods), and free cooling are excluded. Techniques which yield specific heat as

ARED CEZAIRLIYAN • Thermophysics Division, National Bureau of Standards, Washington, D.C. 20234.

a by-product, such as in some pulse thermal diffusivity measurements are not included.

In this writing "pulse" refers to experiments which are of subsecond duration. However, for the sake of completeness and continuity between pulse and conventional techniques, quasidynamic experiments with durations greater than one second are also included. Although present primary interest in pulse experiments is for measurements at high temperatures, some earlier investigations, near room or at moderate temperatures, are also discussed.

2. DESCRIPTION OF METHODS

2.1. *General Description*

The principle of the pulse methods for specific heat measurements is based on rapid resistive self-heating of the specimen by the passage of an electrical current pulse through it and on measuring the pertinent experimental quantities with appropriate time resolution. The required quantities are: current through the specimen, voltage across the specimen, and specimen temperature.

Basically the masurement system consists of an electrical power-pulsing circuit and associated high-speed measuring circuits. The power-pulsing circuit includes the specimen in series with an electric pulse power source, an adjustable resistance, a standard resistance, and a fast-acting switch. The specimen is contained in a controlled-environment chamber. The high-speed measuring circuit include detectors and recording systems. A simplified block diagram of a generalized system for specific heat measurements by the pulse method is shown in Figure 1.

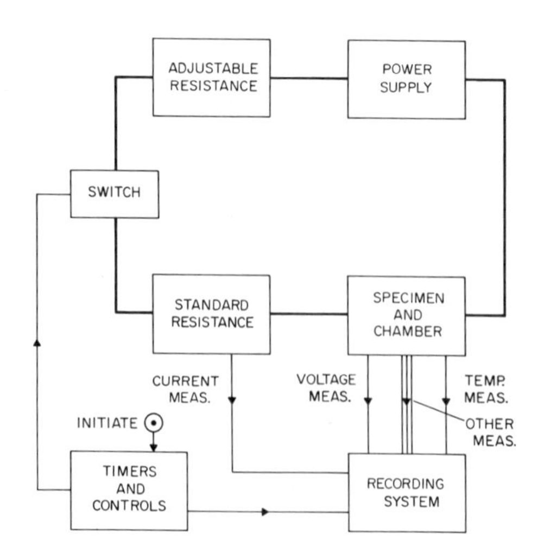

FIGURE 1. Block diagram of a typical system for the measurement of specific heat by the pulse method.

2.2. *Classification of Methods*

Although the general principle of all the pulse methods is the same, there are considerable variations that are used by different investigators. In order to be able to compare the techniques and discuss their advantages as well as disadvantages, it may be appropriate to classify them into the following categories:

I. The specimen is initially at a temperature above room temperature under steady-state conditions. Then, a current pulse is sent through it to raise its temperature in short times. The techniques in this category may be divided into two groups:
   a. The specimen is initially under steady-state conditions at a high temperature as the result of resistive self-heating.
   b. The specimen is initially under steady-state conditions at a high temperature as the result of being in a high-temperature environment (furnace).

II. The specimen is initially at or near room temperature. Then, a current pulse is sent through it to raise its temperature to the desired level in short times.

A graphical representation of the specimen temperature excursion during experiments pertaining to the above categories is shown in Fig. 2. For purposes of clarity and simplicity, the shapes of the curves are exaggerated.

In the following paragraphs, the techniques that belong to the above categories are discussed in detail.

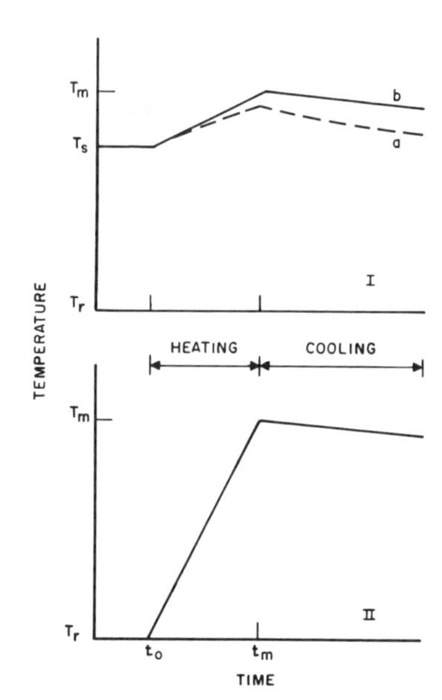

FIGURE 2. Variation of the specimen temperature during different pulse experiments. Explanation of categories Ia, Ib, and II is given in the text (Section 2.2). T_r is room temperature; T_s is specimen initial high steady-state temperature; T_m is specimen maximum temperature after pulse; t_0 is time indicating start of pulse; and t_m is time indicating end of pulse.

The methods under category I are a combination of steady-state and pulse experiments. Desired temperature level is achieved by steady-state power dissipation which allows the measurement of initial temperature and other pertinent quantities with conventional equipment. Then, a current pulse is sent through the specimen to raise its temperature. Quantities required for the determination of power and temperature are measured during this time. Usually the temperature increase is of the order of a few degrees. Because of the small temperature rise, the magnitude of the pulse current is usually small, less than 10 A. Of course, this depends on the specimen geometry. In the case where the specimen is brought to a steady-state temperature by resistive self-heating (category Ia), radiative heat losses during the pulse-heating period become appreciable. This implies that a correction to imparted power has to be made based on either an estimated emittance value for the specimen or on supplementary data which can be obtained in the course of the experiments. For example, from the measurements of voltage and current during the steady-state period, power loss from the specimen can be evaluated. An alternate technique is to determine the power loss from data obtained during the initial cooling period following the heating period. Because of the additional uncertainties resulting from large power loss corrections, experiments in category Ia have to be rapid. However, in the case of those of category Ib, where the specimen is initially in thermal equilibrium with its surroundings, net temperature difference between the specimen and its surroundings during the pulse heating period is small; thus, radiative heat loss is very small. This allows the performance of slower pulse experiments (slower heating rates than those in category Ia), which have certain advantages from instrumentation as well as from theoretical viewpoints.

The methods under category II represent truly fast experiments, in the sense that the entire measurement system is initially at or near room temperature and only the specimen is heated to high temperatures during the very short pulse period. Thus, these techniques do not have the limitations of those in category I, which result from the exposure of the specimen and the surrounding to high temperatures for extended periods. However, they are subject to other difficulties. Since the specimen surrounding is maintained near room temperature, radiation heat loss from the specimen during pulse heating becomes appreciable. In order to reduce this effect, high heating rates are required. In general, the speed of the experiments in category II are much greater than those in category I. This presents difficulties in measuring and recording the pertinent quantities. High pulse currents pose additional problems from both design and operational viewpoints. For example, rapidly varying high currents create varying electromagnetic fields which affect the operation of the transducers and the entire measuring equipment.

In each of the above categories, methods may be further classified according to the speed of the experiments. For subsecond-duration pulse experiments a convenient classification may be as follows: (1) "subsecond" techniques, also referred to as millisecond-resolution techniques, and (2) "submillisecond" techniques, also referred to as microsecond-resolution techniques.

The techniques in all the above categories will be briefly discussed in Section 3 in relation with the experiments reported in the literature.

2.3. Formulation of Equations

In this section, formulation of equations pertaining to the determination of specific heat by the pulse method is given.

In the general case, power balance for the specimen may be expressed as

$$\text{Power imparted} = \text{power stored} + \text{power losses}$$

which becomes

$$ei = c_p m(dT/dt)_h + Q \tag{1}$$

Solving equation (1) for c_p one obtains

$$c_p = \frac{ei - Q}{m(dT/dt)_h} \tag{2}$$

where e is the voltage across the effective specimen in V, i is the current through the specimen in A, c_p is the specific heat of specimen in J mol^{-1} K^{-1} at constant pressure, m is the effective quantity of specimen in mol, Q is the total power loss from effective specimen in W, and $(dT/dt)_h$ is the heating rate of specimen in K s^{-1}. In pulse experiments in which the temperature rise is only a few degrees, the quantity dT/dt in equation (2) may be replaces by $\Delta T/\Delta t$.

The quantity Q may be obtained from data during the initial cooling period. Power balance for this period gives

$$-c_p m(dT/dt)_c = Q \tag{3}$$

where $(dT/dt)_c$ is the cooling rate of the specimen in K s^{-1}. Substituting equation (3) in equation (2) for Q one obtains

$$c_p = \frac{ei}{m(dT/dt)_h\,(1 + 1/M)} \tag{4}$$

where

$$M = -\frac{(dT/dt)_h}{(dT/dt)_c} \tag{5}$$

Specific heat can be obtained from equation (4) provided that experimental data on voltage, current and temperature during the heating period and temperature during the initial cooling period, all as a function of time, are available.

At temperatures above 1500 K, in high-speed experiments of 0.001 to 1 s duration, thermal radiation is the major source of power loss. When data during the initial cooling period are not available, power loss may be estimated using the relation for thermal radiation

$$Q_r = \epsilon \sigma A_s (T^4 - T_0^4) \tag{6}$$

where ϵ is the hemispherical total emittance, σ is the Stephan–Boltzmann constant (5.67032×10^{-8} W m^{-2} K^{-4}), A_s is the effective surface area in m^2, T is the specimen temperature in K, and T_0 is the ambient temperature in K. Equation (6) requires a knowledge of hemispherical total emittance for the specimen. However, if power loss by thermal radiation is only a few percent of imparted power, the effect of even an appreciable error (10%) in estimated emittance on specific heat is much less than one percent.

2.4. Measurement of Experimental Quantities

In this section, a summary is given of the various techniques that were reported in the literature for the measurement, including recording, of the experimental quantities. Because of the recent advances in electronics and optics, most of the measurement and recording techniques used by the investigators before 1970 are obsolete at this time. However, they are included here for historical and completeness reasons as well as for the purpose of demonstrating the magnitude of the advances in the pulse measurement methods.

2.4.1. Measurement of Power

In general, power imparted to the specimens during the heating period was determined from the measurement of current through the specimen and voltage across it as a function of time. During the early developmental years, in some slow experiments, power was measured with specially designed wattmeters; in others, the specimen was placed in one of the arms of a bridge (Wheatstone or Kelvin) and sudden deflection of the galvanometer resulting from the passage of pulse current was used to determine power input to the specimen. The selection of a particular power source was governed by the desired duration of the pulse experiment. Batteries were used for most of the experiments in the 1–1000 ms range. Capacitors were used in experiments shorter than 1 ms. In some of the slow experiments regulated dc power supplies were used.

2.4.2. Measurement of Temperature

Most investigators determined specimen temperature either directly with thermocouples or indirectly by measuring its resistance as a function of time. In general, thermocouples were formed by spot welding the wires individually to the specimen at a plane perpendicular to the current flow. In the resistance method, a separate steady-state experiment was required to obtain the resistance-temperature relation for the specimen, which was used to convert experimentally obtained resistance-time relation to the required temperature-time relation. During the last decade, a few investigators used optical methods to measure specimen temperature.

2.4.3. Recording of Quantities

During the early years of pulse experiments, recording of electrical quantities under transient conditions was achieved by relatively crude methods. These methods were based on the observation or recording of the deflection of a ballistic instrument (galvanometer) placed either in a bridge (Wheatstone, Kelvin) or a potentiometric circuit. Several variations of this method were used by different investigators. In most cases, the magnitude of galvanometer deflection gave the magnitude of the quantity to be measured. In other cases the bridge or the potentiometer circuit was preadjusted, after several trial experiments, to give zero galvanometer deflection under pulse conditions. The preadjusted value gave the magnitude of the quantity to be measured. In two cases, voltage and current were recorded using a fast two-channel recorder. A few investigators have reported the use of oscillographs as a means of recording the electrical quantities. All of the above recording techniques are relatively crude and are limited in speed; thus, they are applicable only to slow experiments in which specimen temperature increases slowly.

The first truly high-speed recording started around 1950 with the use of oscilloscopic techniques. Because of the very fast response characteristics (as low as nanoseconds), oscilloscopes were used for a wide range of experimental conditions, from millisecond to nanosecond resolution. However, they too have limitations which stem primarily from their relatively poor recording resolution. Using oscilloscopes, one may not expect to have better than 1%–2% full-scale resolution per recorded quantity under the most favorable conditions.

During the last decade, major advances in high-speed digital recording techniques and their application to pulse measurements in the millisecond and microsecond time regimes have improved the recording resolution by approximately one to two orders of magnitude (yielding 0.01% in millisecond-resolution experiments and 0.1% in microsecond-resolution experiments).

3. CHRONOLOGY OF DEVELOPMENTS

3.1. Developments Prior to 1970

In the literature, pulse methods of measuring specific heat are referred to as "dynamic", "high-speed", "pulse", "transient", etc. Although the proper usage may sometimes depend on the details of the techniques, the wording has been somewhat arbitrary. In this section, all such methods are referred to as "pulse".

Historically the earliest attempts in using pulse techniques were confined to those in category I (classification is given in Section 2.2). The choice was most likely dictated by the fact that smaller (in amplitude) and longer pulses could have been used for the methods in this category which, in turn, would have allowed the measurement of pertinent quantities with instruments available at those time. During the last decade development of instruments with faster response times and increased accuracy, and also growing interest in the measurement of properties at temperatures above the reliable operation of systems under steady-state conditions, have encouraged the concentration of efforts in methods belonging to category II.

In the following paragraphs a summary of the pulse experiments performed by various investigators for the measurement of specific heat of electrical conductors is presented. In order to avoid repetitions, some of the items relevant to all the techniques and experiments reported in the literature are given in Table 1. The presentation of the experiments is classified under categories discussed in Section 2.2.

3.1.1. Category I(a)

The first successful attempt to use a pulse technique for the measurement of specific heat of electrical conductors may be attributed to Worthing.[1] The technique, reported in 1918, was based upon the measurement of absorbed power by the specimen (tungsten and carbon filaments) while going from one steady-state condition to another as the result of the passage of electric current through it in a short time (1 s). Batteries were used as the pulse source. Electrical measurements were made with a modified potentiometer-galvanometer technique, in which the galvanometer was preadjusted to give zero deflection under pulse conditions. This required a number of trials and was severely limited by the time constant of the galvanometer. Temperature of the specimen was determined from the measurement of specimen resistance during the pulse experiment and from the knowledge of the resistance-temperature relation which was obtained separately under steady-state conditions.

This general concept was used, after considerable modifications and refinements, by Pasternak *et al.*,[17,18] Parker,[20] and Kollie.[21]

Pasternak *et al.*[18] made an exploratory study of a method in which total hemispherical emittance could be obtained in addition to specific heat. This required the performance of two pulse experiments with different heating rates. One experiment was made at such a high rate that radiative heat loss was small relative to absorbed power by the specimen. A second experiment was made at a slower rate in which radiative heat loss was comparable in magnitude to absorbed power. Results of these two experiments gave specific heat and total hemispherical emittance.

Parker[20] used a high-speed capacitive discharge system primarily for the determination of heats of transformation and their rates involved in solid–solid transformations. The same system was used for specific heat measurements.

Kollie[21] described a pulse calorimeter capable of operating up to 1200 K. The novel feature of this was the utilization of a digital voltmeter and a memory which permitted the recording of 400 readings per second with 0.1% full-scale resolution. Relatively low heating rates (10–20 K s^{-1}) combined with the high time resolution of the recordings made this system a very attractive tool for the measurement of specific heat around transition points at moderately high temperatures.

3.1.2. Category I(b)

The difference between categories I(a) and I(b) is that in I(b) the specimen is initially in a furnace at a constant temperature. Apparently, Lapp[2] was the first investigator to use this approach. Measurements were made on nickel at temperatures up to 730 K. A small current pulse was used which increased the specimen temperature by approximately 1–2 K. Duration of the pulse was of the order of 30 s. Power imparted to the specimen was measured with a specially designed wattmeter. Temperature of the specimen was measured with a thermocouple. Since the experiments were slow, a correction for heat losses was made from data obtained during the cooling period following the heating period.

This technique was used after considerable modifications and improvements by several other investigators: Grew,[3] Néel and Persoz,[5] Kurrelmeyer *et al.*,[6] Pallister,[8] Pochapsky,[11,12] Rasor and McClelland,[14,59] Wallace *et al.*,[15] Wallace,[16] and Kollie *et al.*[26]

Grew's[3] approach was almost identical with that of Lapp[2] with the exception that he measured pulse current and specimen resistance separately instead of determining the power using a wattmeter.

Néel and Persoz[5] used a method similar to that of Lapp;[2] however, the specimen temperature was determined from its resistance–temperature relation.

Kurrelmeyer *et al.*,[6] placed the specimen in the arm of a Wheatstone bridge and balanced the bridge with the specimen in a furnace at constant temperature. Then, they discharged a small capacitor through the specimen. Specific heat was determined from the resulting ballistic deflection of the galvanometer. An alternate method was also tried. In this case, the bridge was initially unbalanced in such a manner that the discharge of the capacitor gave zero deflection. The measurements were of a preliminary nature and were conducted near room temperature.

Pallister[8] used a larger specimen than any other used before. An interesting addition, in this case, was a concentric shield placed around the specimen. The shield was electrically in series with the specimen and was designed in such a way that the passage of pulse current heated both the specimen and the shield coincidentally in magnitude and time. This scheme greatly reduced the heat loss from the specimen.

Pochapsky[11,12] conducted experiments with a pulse calorimeter in which the current was supplied by a thyratron-controlled capacitive pulse generator.

Rasor and McClelland[14,59] described a system where temperature rise during the pulse was measured optically. In this system a photomultiplier was used to detect radiation emitted from a slot in the specimen. A typical temperature rise was about 5 K. The steady-state output of the photomultiplier (when specimen was at a constant high temperature in the furnace) was suppressed potentiometrically and only relatively small signals arising from temperature changes during the pulse were recorded.

Wallace *et al.*[15] and Wallace[16] reported the use of a Kelvin bridge as a means of measuring specimen resistance during the pulse. An oscilloscope was connected to the galvanometer terminals of the bridge to record the unbalanced voltage.

Kollie *et al.*,[26] described a pulse calorimeter which essentially is an extension of an earlier version.[21] A furnace was added to the system and the data acquisition system was modified to yield a full-scale resolution of 0.01%, with a time resolution of 2000 readings per second. Measurements up to 1500 K were made. Thermocouples were used for temperature measurements. Heating rate of the specimen was $20\,K\,s^{-1}$. Additional results (iron, nickel, and Ni_3Fe) with this measurement system were reported.

3.1.3. Category II

As mentioned earlier, the technique in this category is a true pulse technique and is free, to a very large extent, from problems that result from subjecting the specimen to high temperatures for prolonged periods, which is the case in techniques belonging to categories Ia and Ib. Apparently, Avramescu[4] was the first investigator to use this technique. The specimen (rod) was placed in series with a battery bank and a switch. The entire system was initially at room temperature. Upon closing the switch, high currents (up to 2000 A) passed through the specimen, heating it to its melting point in a few seconds. Voltage and current were recorded with an oscillograph. Specimen temperature was determined from its resistance–temperature relation. The principle of this technique was used by several other investigators: Baxter,[7] Khotkovich and Bagrov,[10] Nathan,[9] Strittmater *et al.*,[13] Taylor and Finch,[19] Finch and Taylor,[24] Affortit and Lallement,[22] Affortit,[23] and Jura and Stark.[25]

Baxter[7] in a short note described a technique for heating metallic wires to their melting point in about 50 ms.

Khotkovich and Bagrov[10] reported experiments on metallic wires in which pulse lengths as short as 10 ms were used. The general approach of the measurements was the same as that of Avramescu.[4]

Nathan[9] departed from the above by using thermocouples (spot welded to the specimen) for temperature measurements. He studied the behavior of specific heat around the transition points of steels. Recording of electrical quantities was made with a dual-beam oscilloscope. One of the channels was used for the thermocouple signal and the other channel was used to display voltage and current signals chopped with an electronic switch.

Strittmater *et al.*,[13] conducted pulse experiments on platinum and nickel in the range 300 to 720 K. Specimen temperature was determined from resistance measurements. A dual-beam oscilloscope was used for recording voltage and current signals.

Taylor and Finch[19] pulse-heated metallic wires. Heating rates ranged from 1000 to 60 000 K s^{-1}. Resistance of the specimen, obtained from the instantaneous values of voltage and current, served as the means of determining the specimen temperature.

A similar technique with a few modifications was used by Finch and Taylor.[24] A thermocouple, spring-loaded against the specimen, was used to measure final specimen temperature. In addition to this, specimen temperatures during certain times in the heating period were obtained from the resistance–temperature relation. For temperatures above 1400 K, the use of a pyrometer was mentioned but no details were given regarding its construction and operational characteristics.

Affortit and Lallement[22] reported measurements up to 3600 K. Pulse lengths were in the range 0.1–1 s. In the range 300–1500 K, temperature measurements were made with a thermocouple spot welded to the specimen. At higher temperatures, a photomultiplier was used. In the determination of specific heat, a correction for radiative heat loss was made on an assumed value for the hemispherical total emittance of the specimen.

Affortit[23] used the above technique to measure specific heats of some uranium compounds. Since at room temperature these compounds were relatively poor electrical conductors, they were heated to 1300 K, either using a furnace or passing alternating current through the specimen, before applying the pulse current. Because of the initial high steady-state temperature this might have been classified under category I. However, since the only purpose of preheating was to decrease the specimen resistance and since the system was designed primarily for experiments belonging to category II, it was included in this section.

An interesting extension of the pulse methods was given by Jura and Stark.[25] They described a technique for measuring specific heat of metals under high pressure. The technique is basically the same as that presented above with the exception that the specimen was placed in a high-pressure environment (anvil device). The specimen was in the form of a thin wire. Recordings of voltage and current were made with an oscilloscope. Results on iron up to 10 GPa (about 100 kbar) and in the temperature range 77–300 K were given.

3.2. Developments Since 1970

During the decade starting with 1970, major advances in the pulse methods of measuring specific heat took place. For convenience of presentation in this section, they are divided into two groups, namely millisecond and microsecond resolution techniques.

3.2.1. Millisecond-Resolution Techniques

Cezairliyan *et al.*,[27] at the National Bureau of Standards described a technique for measuring specific heat and other properties of electrical conductors above about 1500 K and up to the melting point of the specimen. The technique belongs to category II, in that the specimen was initially at room temperature and then a current pulse (about 2000 A) was sent through the specimen to heat it to any desired high temperature in less than one second. The specimen was a tube of the following nominal dimensions: length, 75–100 mm; outside diameter, 6 mm; wall thickness. 0.5 mm. A small rectangular hole (1 × 0.5 mm) in the wall at the middle of the specimen provided an approximation to blackbody conditions for optical temperature measurements. The experiment chamber was designed for conducting experiments with the specimen in vacuum or in a controlled atmosphere.

Current through the specimen was determined from the measurement of the voltage across the standard resistance placed in series with the specimen. Voltage across the middle two thirds was measured using spring-loaded, knife-edge probes. A high-speed photoelectric pyrometer,[70] designed specifically for this system, permitted 1200 evaluations of the specimen temperature per second. The pyrometer operated at 0.65 μm. The experimental quantities were recorded with a high-speed digital data acquisition system consisting of a multiplexer, an analog-to-digital converter, a core memory, and various control and interfacing equipment. The system was capable of recording sets of data corresponding to temperature, voltage, and current approximately every 0.4 ms with a full-scale signal resolution of about 0.01%. In a subsecond duration experiment, about 2000 data could be digitized and stored in the memory. At the end of the experiment, data were retrieved via a teletypewriter and processed immediately using a time-sharing computer. Details of the design and the operational characteristics of the system are given in the literature.[27,68] The system was used to measure the specific heat of several refractory substances. The main features of the measurements are summarized in Table 1. In addition to specific heat, the system was used for measurements of electrical resistivity, hemispherical total emittance, normal spectral emittance, melting point, thermal expansion, and for studies related to solid–solid phase transformations. A summary of these measurements is given in the literature.[69]

A millisecond-resolution system for the measurement of specific heat and other properties which was developed recently at the Istituto di Metrologia "G. Colonnetti" (IMGC) in Italy is very similar to that at the National Bureau of Standards (NBS). The IMGC system was described in the literature by Righini *et al.*,[65] Some of the modifications and additions to the system may be found in a recent publication.[49] The main difference between the system at IMGC and that at NBS is the pyrometer. The NBS pyrometer[70] uses a photomultiplier tube as a detector and operates in the chopped mode near 0.65 μm. The IMGC pyrometer uses a silicon photodiode as a detector and operates in the dc mode near 1 μm. The design of the IMGC instrument was based on the original work by Ruffino *et al.*,[66] on pyrometry with silicon detectors. The high-speed pyrometer at IMGC was described by Coslovi *et al.*,[67]; a more recent modified version with a shorter time resolution (5 μs) was presented by Coslovi *et al.*,[74] The measurements performed with the IMGC system are summarized in Table 1.

The above described systems belong to category II. Two systems that belong to category Ib have been developed by Petrova and Chekhovskoi[50] and by Naito *et al.*,[52]

In the system described by Petrova and Chekhovskoi,[50] the specimen was

TABLE 1
Summary of Pulse Experiments for the Measurement of Specific Heat of Electrical Conductors[a]

No.	Investigator	Ref.	Year	Category	Power source	Pulse length (s)	Heating rate ($K s^{-1}$)	Substance	Temp. range (K)	Spec. geom.	Power meas.	Temp. meas.	Recording	Inaccuracy
1	Worthing	1	1918	Ia	B	1		W,C	1200–2400	W	EI	R	V	
2	Lapp	2	1929	Ib	B			Ni	100–730	R	W	TC	V	2
3	Grew	3	1934	Ib	B			Ni	90–720	R	EI	TC	V	2
4	Avramescu	4	1939	II	B	1–2		Al,Cu	400–1300	R	EI	R	G	2
5	Néel and Persoz	5	1939	Ib	B	0.1		Cu,Ni,Pt	300–1300	W	W	R	V	2
6	Kurrelmeyer *et al.*	6	1943	Ib	C	0.002–0.05		Pt	300	W	B	R	V	0.5
7	Baxter	7	1944	II		0.05				W	EI	R	G	
8	Pallister	8	1949	Ib	B		1	Fe	273–1500	R	EI	TC	V	2
9	Nathan	9	1951	II	B		15–1000	Steel	770	R	EI	TC	S	
10	Khotkovich and Bagrov	10	1951	II	B	0.01	10^4–5×10^4	Cu,W,Mo,Cd		W	EI	R	G	3
11	Pochapsky	11	1953	Ib	C	0.001		Al,Pb	273–920	W	B	R	V	5
12	Pochapsky	12	1954	Ib	C	0.001		Pt	273–900	W	B	R	V	5
13	Strittmater *et al.*	13	1957	II	B	0.05–0.15	3000–9000	Pt,Ni	300–720	W	EI	R	S	5–10
14	Rasor and McClelland	14	1960	Ib	B		50	Mo,Ta,C	1300–3920	R	EI	0	G	5
15	Wallace *et al.*	15	1960	Ib	B	0.04		Fe	300–1300	W	B	R	S	2
16	Wallace	16	1960	Ib	B	0.03		Th	300–1300	W	B	R	S	2
17	Pasternak *et al.*	17	1962	Ia	B	1–10	100–1000	Pt	300–1100	W	EI	R	R	
18	Pasternak *et al.*	18	1963	Ia	B		35–2000	Pt	400–1300	W	EI	R	R	
19	Taylor and Finch	19	1964	II	B		10^3–6×10^4	Mo,Ta	100–3200	W	EI	R	S	4–7
20	Parker	20	1965	Ia	C	10^{-5}	10^4–10^9	Ti	300–1100	S	C	TC	S	
21	Kollie	21	1967	Ia	R	4–60	10–20	Fe	300–1200	R	EI	TC	D	1
22	Affortit and Lallement	22	1968	II	B	0.1–1		Nb,W	300–3600	W	EI	TC,O	S	3–5
23	Affortit	23	1969	II	B	< 1		UN,UC,UO_2	700–3100	R	EI	TC,O	S	5
24	Finch and Taylor	24	1969	II	B		7×10^3–1.6×10^5	$ZrU_{0.04}H_{0.5}$	300–800	R	EI	R,O	S	5
25	Jura and Stark	25	1969	II	B	0.1		Fe	80–300	W	EI	TC	S	5

26	Kollie *et al.*	26	1969	Ib	R	10–35	20	Fe	300–1500	R	EI	TC	D	1–2
27	Cezairliyan *et al.*	27	1970	II	B	0.3–0.7	4000–8000	Mo	1900–2800	T	EI	O	D	2–3
28	Dikhter and Lebedev	28	1970	II	C	$< 5 \times 10^{-5}$		W	2600–4500	W	EI	O	S	
29	Dikhter and Lebedev	29	1971	II	C			Mo	2200–3700	W	EI	O	S	
30	Cezairliyan *et al.*	30	1971	II	B	0.3–0.5	4000–6000	Ta	1900–3200	T	EI	O	D	2–3
31	Cezairliyan and McClure	31	1971	II	B	0.4–0.6	4000–7000	W	2000–3600	T	EI	O	D	2–3
32	Cezairliyan	32	1971	II	B	0.4–0.5	5000	Nb	1500–2700	T	EI	O	D	2
33	Cezairliyan	33	1972	II	B	0.4–0.5	4000–7000	Ta–10W	1500–3200	T	EI	O	D	3
34	Cezairliyan	34	1973	II	B	0.4	5000–7000	Nb–1Zr	1500–2700	T	EI	O	D	3
35	Cezairliyan and McClure	35	1974	II	B	0.5–0.9	2000–4000	Fe	1500–1800	T	EI	O	D	3
36	Cezairliyan *et al.*	36	1974	II	B	0.4	4000–5000	V	1500–2100	T	EI	O	D	3
37	Cezairliyan and Righini	37	1974	II	B	0.4–0.5	3000–5000	Zr	1500–2100	T	EI	O	D	3
38	Cezairliyan	38	1974	II	B	0.4–0.5	4000–5000	Nb–10Ta–10W	1500–2800	T	EI	O	D	3
39	Cezairliyan and Righini	39	1975	II	B	0.2–0.5	3000–8000	C	1500–3000	T	EI	O	D	3
40	Cezairliyan and McClure	40	1975	II	B	0.3–0.5	4000–8000	Hf–3Zr	1500–2400	T	EI	O	D	3
41	Lebedev *et al.*	41	1976	II	C		10^{8}	Mo,W	1500–3700	W	EI	O	S	
42	Gathers *et al.*	42	1976	II	C	$< 10^{-4}$		Ta	2500–8000	W	EI	O	S	
43	Shaner *et al.*	43	1976	II	C	$< 10^{-4}$		Mo,Nb,Ta,W	2500–5000	W	EI	O	S	
44	Shaner *et al.*	44	1977	II	C	$< 10^{-4}$		Mo,Ta	2000–7500	W	EI	O	S	
45	Shaner *et al.*	45	1977	II	C	$< 10^{-4}$		Nb,Pb	2000–6000	W	EI	O	S	
46	Lebedev and Mozharov	46	1977	II	C		10^{8}–10^{9}	Ta	2300–3900	T	EI	O	S	8
47	Cezairliyan and Miiller	47	1977	II	B	0.5–0.6	3000–4000	Ti	1500–1900	T	EI	O	D	3
48	Cezairliyan *et al.*	48	1977	II	B	0.4–0.5	3000–4000	Ti–6Al–4V	1500–1900	T	EI	O	D	3
49	Righini *et al.*	49	1977	II	B	0.6–1.5	1000–3000	Zircaloy-2	1300–2000	T	EI	O	D	4
50	Petrova and Chekhovskoi	50	1978	Ib	B		300–500	NbC,TaC,ZrC	1600–2300	R	EI	O	S,D	3–4

TABLE 1 (Continued)

No.	Investigator	Ref.	Year	Category	Power source	Pulse length (s)	Heating rate ($K s^{-1}$)	Substance	Temp. range (K)	Spec. geom.	Power meas.	Temp. meas.	Recording	Inaccuracy
51	Taylor	51	1978	Ia	R	6		W	1500–2400	R	EI	R	D	
52	Naito *et al.*	52	1979	Ib				C,SiC	300–1200	R	EI	TC		
53	Gathers *et al.*	53	1979	II	C	$< 5 \times 10^{-5}$		Pt	2000–7500	W	EI	O	S	
54	Seydel *et al.*	54	1979	II	C		10^9	Mo	2900–6000	W	EI	O	S	3–5
55	Cezairliyan and Miiller	55	1980	II	B	0.6	2000	Steel	1400–1700	T	EI	O	D	3
56	Miiller and Cezairliyan	56	1980	II	B	0.5–0.6	3000–4000	Pd	1400–1800	T	EI	O	D	3
57	Cezairliyan and Miiller	57	1980	II	B	0.4–0.8	4000	C-composite	1500–3000	T	EI	O	D	3
58	Righini and Rosso	58	1980	II	B	0.7–1.1	1700–3000	Pt	1000–2000	T	EI	O	D	3

[a] Abbreviations and notes: Category: Designation of categories are described in Section 2.2. Power Source: B, battery; C, capacitor; R, regulated dc power supply. Temp. range: Range between two extreme temperatures covered by the investigator regardless of particular substance. Spec. Geom: R, rod (diam > 1 mm); S, strip; T, tube; W, wire (diam ⩽ 1 mm). Power Meas: B, bridge; C, capacitor energy; EI, voltage–current; W, wattmeter. Temp. Meas: O, optical; R, resistive; TC, thermocouple. Recording: D, digital; G, oscillographic; R, chart recording; S, oscilloscope; V, visual manual. Inaccuracy: Total error in specific heat as reported in the literature by the investigators. (In some instances where the measured quantity was enthalpy and the authors gave only the error in enthalpy the entry in this column is left blank.)

initially in a furnace at a desired high temperature and then it was rapidly heated to a higher temperature. The temperature change in the specimen as the result of pulse heating was of the order of 50 K. The steady-state temperature and the temperature of the specimen during the pulse heating period were measured optically. The signal corresponding to the temperature pulse was recorded with a digital voltmeter. The signals corresponding to pulse current and voltage were recorded with an oscilloscope.

The system described by Naito *et al.*,[52] also incorporated a steady-state furnace to establish the desired high temperature. An added feature was an adiabatic shield that enclosed the specimen in the furnace. During the pulse heating period, current was passed through the shield as well as through the specimen (separate circuits) in such a way that the temperature rise in the shield was approximately the same as that in the specimen. This arrangement minimized the heat loss from the specimen during the pulse period. The temperature rise in the specimen during the pulse was in the range 2 to 4 K, and was measured with a thermocouple.

Measurements of specific heat with a system developed primarily for thermal conductivity measurements was described by Taylor.[51] The technique belongs to category Ia. The specimen was initially under a steady-state condition at a high temperature as the result of resistive self-heating of the specimen. The specific heat measurements were made by subjecting the specimen to a small step-function change in the electrical power. Change in the specimen temperature was determined from the change in its resistivity.

3.2.2. Microsecond-Resolution Techniques

During the last decade, increased interest in properties of high melting substances, especially in their liquid phase, resulted in the exploration of microsecond-resolution techniques. Primarily, the following four laboratories were involved in research related to specific heat measurements: Institute for High Temperatures (USSR), Lawrence Livermore Laboratory (USA), University of Kiel (FRG), and National Bureau of Standards (USA).

In the Institute for High Temperatures, Dikhter and Lebedev[28,64] developed a system for the measurement of specific heat of metals in their solid and liquid phases. The system included a capacitor bank (5.7 μF), a series resistance (48.8 Ω), a standard resistance (0.445 Ω), and a thyratron for switching the current. The specimen was in the form of a wire; 0.08 mm in diameter, and 20 mm in length. Energy imparted to the specimen was obtained from the integral of the product of the current through the specimen and the voltage across it. The specimen temperature was obtained from the measurements of radiation at two wavelengths with the use of photomultipliers. The experimental quantities were recorded with oscilloscopes. Later on, Lebedev and Mozharov[46] introduced refinements in their temperature measurements by utilizing tubular specimens which approximated blackbody conditions. A summary of the measurements of specific heat performed with the microsecond-resolution system in the Institute for High Temperatures is given in Table 1.

The system developed in the Lawrence Livermore Laboratory started with the work of Henry *et al.*,[72] who developed a capacitor discharge system for thermodynamic measurements on liquid metals. The system included a capacitor bank (20 kV, 17 kJ) and a high pressure (200 MPa, about 2000 atm) cell. Modifications and refinements of the system were described by Gathers *et al.*,[63] The changes included a larger capacitor bank (45 kJ), an improved voltage probe assembly and a streaking camera. Other improvements to the above system were made by Gathers *et al.*,[42] The new features were a rapid temperature measurement capability and a new pressure cell (400 MPa) to accommodate an optical window. The temperature measurements were based on three-color pyrometry. To ensure that each channel views the same spot on the specimen, a randomized trifurcated glass fiber-optic bundle was used to divide the signal. The output of each of the three fiber bundles was passed through interference filters (50 nm bandwidth) centered at 450, 600, and 700 nm, respectively. Silicon photodiodes were used as detectors, the signal amplification was achieved with logarithmic amplifiers, and the outputs were recorded with oscilloscopes. In subsequent publications, Shaner *et al.*,[43–45] and Gathers *et al.*,[53] have discussed the operational characteristics of the system and have reported the results of the measurements of the enthalpy of selected metals in their solid and liquid phases. A summary of the work performed in the Lawrence Livermore Laboratory is given in Table 1.

The system developed in the University of Kiel was described by Seydel *et al.*,[60] and by Seydel and Fucke.[61] The system had a 5.1 μF capacitor bank rated at 35 kV. The electrical characteristics of the system were: short-circuit ringing period, 6 μs; circuit resistance, 20 mΩ; circuit inductance, 200 nH. The heating rate of the specimen (thin wire) was typically 10^9 K s^{-1}. The electrical measuring circuits had a rise-time of about 8 ns. In a recent paper, Seydel *et al.*,[62] have described modifications and refinements to their system which included the addition of a fast pyrometer operating at two wavelengths (in the range 450 to 950 nm), and an experiment chamber for pressures up to about 400 MPa (about 4000 atm). Measurements of the enthalpy of liquid molybdenum with this system were discussed by Seydel *et al.*,[54]

At the National Bureau of Standards, a system is under development for the measurement of specific heat of electrically conducting substances in their solid and liquid phases up to about 10 000 K. The capacitor bank has a capacity of 24 kJ (120 μF, 20 kV). A combination of series resistors and inductors are used to approximate the desired pulse shape during the discharge. The capacitor discharge system is enclosed in a shielded room to minimize its interference with the measurement and the recording equipment. A special pyrometer has been developed which will be used to measure the specimen temperature.[75] The pyrometer operates at two wavelengths (0.65 and 0.9 μm) and has a circular target of 0.5 mm in diameter. The optical components of the pyrometer are located near the experiment chamber; however, all the amplifier electronics including the radiation detectors (silicon photodiodes) are placed outside of the shielded room. Fiber-optic cables are used to link the pyrometer optics to the detector-electronic circuits. Recording of the experimental quantities is performed with a digital data acquisition system which is capable of recording five quantities (current, voltage, two temperatures, and one spare) simultaneously approximately every one microsecond with a full-scale signal resolution of about 0.1%. The memory has a capacity of 5000 "data points". Oscilloscopes are used only to monitor the general pattern of the experimental results and to detect the presence of anomalies. At present, the construction of the system is completed and tests are underway to check the operation of the various components and the characteristics of the overall system. The description of the system and preliminary results on its operational characteristics are given in Ref. 73. Once fully operational, this system will complement the millisecond-resolution system described in Section 3.2.1. The two systems have a considerable overlap in temperature, making the combination attractive from the viewpoint of cross-checking the measurements.

An extention of the pulse technique for measuring specific heat under very high pressures, originally described by Jura and Stark,[25] was given by Loriers-Susse *et al.*,[71] They made significant improvements in the technique and performed measurements on copper up to 10 GPa (about 100 kbar) in experiments of 100–200 μs duration.

4. EXPERIMENTAL DIFFICULTIES AND SOURCES OF ERRORS

Although the basic principle of the pulse method of measuring specific heat is simple, the development of systems and performance of accurate experiments present difficulties and require the solution of serious problems. In this section, some of the experimental difficulties which give rise to errors in the measurements are discussed briefly.

4.1. Errors in Directly Measured Quantities

4.1.1. Errors in Electrical Measurements

Identifiable sources of errors that affect the measurement of electrical quantities such as current and voltage are: skin effect, inductive effects, thermoelectric effects, and errors due to calibrations. The contributions of the above depend on the specific systems and on the speed of the experiments, in particular.

The contribution of the skin effect in millisecond-resolution experiments is generally negligible. However, it should be taken into consideration in the design of the microsecond-resolution systems and especially in the selection of the specimen geometry.

The inductive effects should be considered in all pulse experiments. In millisecond-resolution experiments their contribution is generally small and can be eliminated by proper shielding of pertinent cables and components. In microsecond-resolution experiments, both self and mutual inductive effects can cause serious measurement problems by creating large induced voltages in the electrical circuits. In these cases, conventional shielding is usually inadequate and other techniques, such as compensation, have to be employed.

The thermoelectric effects can be neglected in almost all pulse experiments, since either they take place outside the "effective" specimen or their contribution is small compared to the resolution of the measurements.

Normally, calibrations of the electrical measuring circuits, which include amplifiers, resistors, etc., do not present any serious problems. If they are performed under steady-state conditions, it is essential to establish the adequacy of the time-response of the circuits.

At present, techniques have been developed to permit the millisecond-resolution measurements of electrical quantities (current, voltage) with an uncertainty of about 0.1%. Measurements of the same quantities with microsecond-resolutions are considerably less accurate: about 1% under the most favorable conditions.

4.1.2. Errors in Temperature Measurements

Temperature, is by far, the most critical experimental quantity, as it appears in the form dT/dt in the relation for specific heat [equation (4)]. Each type of temperature measurement (resistive, thermoelectric, optical) has its own unique sources of error. Since most of the modern pulse techniques utilize optical techniques for temperature measurements, some of the problems associated with such techniques are discussed in the following paragraphs.

Optical pyrometry with appropriate time resolution (millisecond and microsecond) requires sophistication in both optics and electronics. Because of the short times and the small target sizes involved, the pyrometer should have a large aperture and minimum internal attenuation as well as high quality optical elements to minimize scattering of light.

A source of temperature error is the temperature nonuniformities along the specimen during its rapid heating. This may be due mainly to nonuniformities in the specimen's cross-sectional area. Temperature gradients, both radial and axial, may also contribute to the temperature error if the dimensions of the specimen and/or the speed of the experiments are not carefully selected.

Ability to measure the true temperature of the specimen is probably the most important issue in optical pyrometry. When the pyrometer is sighted at the surface of the specimen, the resultant temperature is the surface radiance temperature. Conversion to true temperature requires a knowledge of the normal spectral emittance of the specimen at the wavelength of the measurements. Usually, normal spectral emittance data are not readily or easily available and one cannot rely on literature values (even if available) since this property depends very strongly on the surface conditions of the specimen. The most direct approach to measuring the true temperature is to have a specimen that approximates blackbody conditions. This can be achieved if the specimen is in the form of a tube, with a small hole fabricated in the wall of the middle of the length. The blackbody quality for a realistic size specimen (length, 100 mm; outside diameter, 6 mm; wall thickness, 0.5 mm) may be as high as 99% when the sighting hole is 0.5 x 1 mm. Such a specimen can be used in millisecond-resolution experiments. In cases where specimens in the form of a solid rod are to be used (in most of the microsecond-resolution experiments), an alternate approach for obtaining true temperature, based on some assumptions regarding the dependence of normal spectral emittance on wavelength, is to measure radiance temperature at more than one wavelength. Simultaneous measurements at two and three wavelengths have been tried by a few investigators. However, the reliability of true temperature determinations with the multiwavelength measurements approach is not as good as that obtained from one wavelength measurements on a blackbody.

In addition to the above items, accurate pyrometry requires careful calibration and data reduction procedures. Calibration procedures are generally based on the use of calibrated tungsten filament lamps. However, such lamp calibrations are usually performed at 0.65 μm, the wavelength at which conventional slow pyrometers operate. Calibration of lamps at wavelengths other than 0.65 μm presents difficulties, and is not readily available except at some standards laboratories. In some cases, it may be desirable to perform pyrometer calibrations directly with blackbody radiation sources.

At present, pyrometry techniques have been developed to permit the millisecond-resolution measurements of temperature with an uncertainty of about 5 K at 2000 K, and 10 K at about 3500 K. The existing microsecond-resolution pyrometry techniques are considerably less accurate and are likely to yield results with an uncertainty of at least 50 K at temperatures above 3000 K. Extensive research is underway to improve the accuracy of microsecond-resolution pyrometry.

4.1.3. Errors in Data Acquisition

In most of the modern pulse experiments, recordings of the experimental quantities are done either with oscilloscopes or digital data acquisition systems. In millisecond-resolution experiments, one may use digital data acquisition systems which have a full-scale signal resolution of about 0.01%. In microsecond-resolution experiments, researchers generally use oscilloscopes, which under the best conditions may have a resolution of about 1%. Recent advances in the electronics field have brought about the capability of developing microsecond-resolution digital data acquisition systems with a full-scale signal resolution of about 0.1%. While the above figures refer to measurement resolution, accuracy of the data recording systems may be lower depending on factors such as calibration procedures and the effect of external electromagnetic fields, etc.

4.2. Errors Due to Departure from Assumed Conditions

The accuracy with which specific heat can be determined depends also on the degree of experimental realization of the conditions imposed in the formulation of the equation (equation (4)] for specific heat. This formulation assumes accurate accounting of heat loss, no specimen evaporation, no thermionic emission from or near the specimen, and that specimen is in thermodynamic equilibrium at all times during its rapid heating period.

Because of the speed of the pulse experiments and the geometry of the specimen, the only significant heat loss from the specimen is due to thermal radiation. In millisecond-resolution experiments, this can be a significant portion (about 10%) of the imparted power at temperatures near 3000 K. However, a correction for radiation heat loss can be made, based on data obtained either from the free radiative cooling period, or from a literature value for the hemispherical total emittance. Even a significant uncertainty (10%) in the correction does not introduce more than 1% error in the specific heat results.

Other errors may result from specimen evaporation. However, because of the short times involved, this can be neglected for metals up to their melting points. For experiments on liquid metals considerably above their melting points, evaporation can introduce significant errors. In order to avoid this, experiments should be made with the specimen in a pressurized environment such as argon at pressures 200 MPa (about 2000 atm) and higher, and in some cases in water or some other transparent liquid.

Thermionic emission from the specimen usually does not introduce any serious problems, since either the temperature is low when the specimen is in a vacuum environment or the specimen is surrounded with a pressurized fluid when the temperature is high.

Whether a specimen is under thermodynamic equilibrium while measurements are performed during a rapid heating period depends upon several factors, such as crystalline structure, magnitude of impurities and imperfections, temperature, etc. In pure electrical conductors at high temperatures, relaxation time of electron–phonon interactions is of the order of 10^{-10} s. However, other processes, such as formation and migration of vacancies, may have relaxation times of the order of 10^{-6} to 10^{-3} s.

4.3. *Errors in Specific Heat*

Errors in measured specific heat reported in the literature by most investigators (as listed in Table 1) are based on crude estimates and most probably are somewhat optimistic. An analysis of sources and magnitudes of errors in the best pulse experiments indicate that in the range 1000–2000 K, specific heat can be measured as accurately as with conventional methods; above 2000 K the pulse methods may yield an accuracy that surpasses the accuracy of any conventional method.

With the best millisecond-resolution pulse method available at present, it is possible to measure specific heat with an uncertainty of about 2% at 2000 K and 3% at 3000 K. Reproducibility of measurements may be approximately 0.5% over this temperature range.

Uncertainties in the specific heat obtained from microsecond-resolution experiments are considerably greater than those given above. The estimates indicate that, in the best experiments reported in the literature, the uncertainty in specific heat is not better than 10%, and is likely that it is considerably worse than 10%. However, new developments indicate that it may be possible to measure specific heat with an uncertainty of about 5% up to about 5000 K.

5. SUMMARY AND CONCLUSIONS

In the previous sections, the description, classification, and a summary of the experiments for the measurement of specific heat of electrically conducting substances with the use of pulse heating techniques were presented. The experimental difficulties and the sources of errors associated with such experiments were also discussed. It was pointed out that the main advantage of the pulse techniques is in measurements at high temperatures (above 2000 K), above the limits of reliable operation of accurate steady-state systems.

During the last decade, major advances were made in developing pulse techniques for the measurements of specific heat. These advances may be attributed primarily to two factors: (1) the need for high-temperature properties of materials in various applications, such as in aerospace, nuclear energy, etc., and (2) the emergence of highly sophisticated electronics, particularly in the area of fast digital data acquisition systems.

The millisecond-resolution techniques have been perfected to a level of competitiveness with the most accurate conventional steady-state techniques in the overlapping pyrometric temperature ranges (1000 to 2000 K) for the two types of techniques. In this temperature range, the pulse techniques have the added advantage of covering a wide temperature range in a single experiment and of yielding high resolution in temperature measurements in comparison with most of the conventional techniques, such as drop calorimetry. This is an important feature especially in measurements near and at phase transitions. While the low-temperature limit of millisecond-resolution pulse techniques is governed by the operation of the millisecond-resolution pyrometers, the upper temperature limit is governed by the melting temperature of the specimen.

For measurements above the melting temperatures of the specimens, microsecond-resolution techniques show great promise. Because of the severity of the measurement conditions, these techniques have not yet been perfected to the level of millisecond-resolution techniques. However, the results of the preliminary research indicate the potential of the microsecond-resolution techniques as a means of obtaining accurate data at temperatures approaching 10 000 K. Their lower limit is governed by the operation of the microsecond-resolution pyrometers, which may be about 2000 K. This indicates that for a great number of refractory materials there is a considerable overlapping temperature range between the millisecond and the microsecond resolution techniques, which is important in establishing the reliability of the measurements.

Methods belonging to category I, where the specimen is initially at a high steady-state temperature, are primarily used for measurements below 2500 K. In general, small electrical pulses are used and the resultant temperature rise is only a few degrees. These are suited for investigations that require high resolution in temperature measurements.

Methods belonging to category II, where the specimen is initially at room temperature, are useful for measurements above 2000 K. Rapid measurements over a wide temperature range makes this very attractive for research at very high temperatures.

It may be concluded that the pulse techniques, both millisecond and microsecond resolution, have a great potential in extending the accurate measurements of specific heat to temperatures beyond the limit of steady-state techniques. In addition to measuring specific heat, the pulse techniques may be used to measure other thermal, electrical and related properties in the solid and liquid phases. Also, in addition to simple metallic specimens, the pulse techniques may be used for measurements on other more complex electrically conducting substances, such as certain carbides, oxides, borides, and nitrides.

REFERENCES

1. A.G. Worthing, *Phys. Rev.* **12**, 199–225 (1918).
2. E. Lapp, *Ann. Phys.* **12**, 442–521 (1929).
3. K.E. Grew, *Proc. R. Soc. (London)*, **145**, 509–522 (1934).
4. A. Avramescu, *Z. Tech. Phys.* **20**, 213–217 (1939).
5. L. Néel and B. Persoz, *Compt. Rend.* **208**, 642–643 (1939).
6. B. Kurrelmeyer, W.H. Mais and E.H. Green, *Rev. Sci. Instr.* **14**, 349–355 (1943).
7. H.W. Baxter, *Nature* **153**, 316 (1944).
8. P.R. Pallister, *J. Iron Steel Inst.* **161**, 87–90 (1949).
9. A.M. Nathan, *J. Appl. Phys.* **22**, 234–235 (1951).
10. V.I. Khotkovich and N.N. Bagrov, *Dokl. Akad. Nauk* **81**, 1055–1057 (1951); English Trans., U.S. AEC (Oak Ridge), Tr. No. 1817.
11. T.E. Pochapsky, *Acta Met.* **1**, 747–751 (1953).
12. T.E. Pochapsky, *Rev. Sci. Instr.* **25**, 238–242 (1954).
13. R.C. Strittmater, G.J. Pearson and G.C. Danielson, *Proc. Iowa Acad. Sci.* **64**, 466–470 (1957).
14. N.S. Rasor and J.D. McClelland, *J. Phys. Chem. Solids* **15**, 17–26 (1960).
15. D.C. Wallace, P.H. Sidles and G.C. Danielson, *J. Appl. Phys.* **31**, 168–176 (1960).
16. D.C. Wallace, *Phys. Rev.* **120**, 84–88 (1960).
17. R.A. Pasternak, E.C. Fraser, B.B. Hansen and H.U.D. Wiesendanger, *Rev. Sci. Instr.* **33**, 1320–1323 (1962).
18. R.A. Pasternak, H.U.D. Wiesendanger and B.B. Hansen, *J. Appl. Phys.* **34**, 3416–3417 (1963).
19. R.E. Taylor and R.A. Finch, *J. Less-Common Metals* **6**, 283–294 (1964).
20. R. Parker, *Trans. Met. Soc. AIME* **233**, 1545–1549 (1965).
21. T.G. Kollie, *Rev. Sci. Instr.* **38**, 1452–1463 (1967).
22. C. Affortit and R. Lallement, *Rev. Int. Hautes Tempér. Réfract.* **5**, 19–26 (1968).
23. C. Affortit, *High Temp. High Pressures*, **1**, 27–33 (1969).
24. R.A. Finch and R.E. Taylor, *Rev. Sci. Instr.*, **40**, 1195–1199 (1969).
25. G. Jura and W.A. Stark, *Rev. Sci. Instr.* **40**, 656–660 (1969).
26. T.G. Kollie, M. Barisoni, D.L. McElroy and C.R. Brooks, *High Temp. High Pressures*, **1**, 167–184 (1969).
27. A. Cezairliyan, M.S. Morse, H.A. Berman and C.W. Beckett, *J. Res. Natl. Bur. Stand. (U.S.)* **74A**, 65–92 (1970).
28. I.Ya. Dikhter and S.V. Lebedev, *High Temp. (USSR)* **8**, 51–54 (1970).
29. I.Ya. Dikhter and S.V. Lebedev, *High Temp. (USSR)* **9**, 845–849 (1971).
30. A. Cezairliyan, J.L. McClure and C.W. Beckett, *J. Res. Nat. Bur. Stand. (U.S.)* 75A, 1–13 (1971).
31. A. Cezairliyan and J.L. McClure, *J. Res. Natl. Bur. Stand. (U.S.)* **75A**, 283–290 (1971).
32. A. Cezairliyan, *J. Res. Natl. Bur. Stand. (U.S.)* **75A**, 565–571 (1971).
33. A. Cezairliyan, *High Temp. High Pressures* **4**, 541–550 (1972).
34. A. Cezairliyan, *J. Res. Natl. Bur. Stand. (U.S.)* **77A**, 45–48 (1973).
35. A. Cezairliyan and J.L. McClure, *J. Res. Natl. Bur. Stand. (U.S.)* **78A**, 1–4 (1974).
36. A. Cezairliyan, F. Righini and J.L. McClure, *J. Res. Nat. Bur. Stand. (U.S.)* **78A**, 143–147 (1974).
37. A. Cezairliyan and F. Righini, *J. Res. Nat. Bur. Stand. (U.S.)* **78A**, 509–514 (1974).
38. A. Cezairliyan, *J. Chem. Thermodynamics* **6**, 735–742 (1974).
39. A. Cezairliyan and F. Righini, *Rev. Int. Hautes Tempér. Réfract.* **12**, 124–131 (1975).
40. A. Cezairliyan and J.L. McClure, *J. Res. Nat. Bur. Stand. (U.S.)* **79A**, 431–436 (1975).
41. S.V. Lebedev, A.I. Savvatimskii and M.A. Sheindlin, *High Temp. (USSR)* **14**, 259–263 (1976).

42. G.R. Gathers, J.W. Shaner and R.L. Brier, *Rev. Sci. Instr.* **47**, 471–479 (1976).
43. J.W. Shaner, G.R. Gathers and C. Minichino, *High Temp. High Pressures* **8**, 425–429 (1976).
44. J.W. Shaner, G.R. Gathers and C. Minichino, *High Temp. High Pressures* **9**, 331–343 (1977).
45. J.W. Shaner, G.R. Gathers and W.M. Hogson, in the *Proceedings of the Seventh Symposium on Thermophysical Properties* (A. Cezairliyan, Ed.), Am. Soc. Mech. Eng. New York (1977), pp. 896–903.
46. S.V. Lebedev and G.I. Mozharov, *High Temp. (USSR)* **15**, 45–48 (1977).
47. A. Cezairliyan and A.P. Miiller, *High Temp. High Pressures,* **9**, 319–324 (1977).
48. A. Cezairliyan, J.L. McClure and R. Taylor, *J. Res. Natl. Bur. Stand. (U.S.)* **81A**, 251–256 (1977).
49. F. Righini, A. Rosso and L. Coslovi in the *Proceedings of the Seventh Symposium on Thermophysical Properties,* (A. Cezairliyan, Ed.), Am. Soc. Mech. Eng., New York (1977), pp. 358–368.
50. I.I. Petrova dn V.Ya. Chekhovskoi, *High Temp. (USSR),* **16**, 1045–1050 (1978).
51. R.E. Taylor, *J. Heat Transfer* **100**, 330–333 (1978).
52. K. Naito, H. Inaba, M. Ishida and K. Seta, *J. Phys. (E): Sci. Instrum.* **12**, 712–718 (1979).
53. G.R. Gathers, J.W. Shaner and W.M. Hodgson, *High Temp. High Pressures* **11**, 529–538 (1979).
54. U. Seydel, H. Bauhof, W. Fucke and H. Wadle, *High Temp. High Pressures* **11**, 635–642 (1979).
55. A. Cezairliyan and A.P. Miiller, *Int. J. Thermophys.* **1**, 83–95 (1980).
56. A.P. Miiller and A. Cezairliyan, *Int. J. Thermophys.* **1**, 217–224 (1980).
57. A. Cezairliyan and A.P. Miiller, *Int. J. Thermophys.* **1**, 317–331 (1980).
58. F. Righini and A. Rosso, *High Temp. High Pressures,* **12**, 335–349 (1980).
59. N.S. Rasor and J.D. McClelland, *Rev. Sci. Instrum.* **31**, 595–604 (1960).
60. U. Seydel, W. Fucke and B. Moller, *Z. Naturforsch* **32a**, 147–151 (1977).
61. U. Seydel and W. Fucke, *Z. Naturforsch* **32a**, 994–1002 (1977).
62. U. Seydel, H. Bauhof, W. Fucke and H. Wadle, *High Temp. High Pressures,* **11**, 35–42 (1979).
63. G.R. Gathers, J.W. Shaner and D.A. Young, *Phys. Rev. Lett.* **33**, 70–72 (1974).
64. I.Ya. Dikhter and S.V. Lebedev, *High Temp. High Pressures* **2**, 55–58 (1970).
65. F. Righini, A. Rosso and G. Ruffino, *High Temp. High Pressures* **4**, 597–603 (1972).
66. G. Ruffino, F. Righini and A. Rosso, in *Temperature: Its Measurement and Control in Science and Industry* (H.H. Plumb, Ed.), Vol. 4, Part 1, ISA, Pittsburgh (1972), pp. 531–537.
67. L. Coslovi, F. Righini and A. Rosso, *Alta Frequenza* **44**, 592–596 (1975).
68. A. Cezairliyan, *J. Res. Natl. Bur. Stand. (U.S.)* **75C**, 7–18 (1971).
69. A. Cezairliyan, *High Temp. Sci.* **13**, 117–133 (1980).
70. G.M. Foley, *Rev. Sci. Instrum* **41**, 827–834 (1970).
71. C. Loriers-Susse, J.P. Bastide and G. Bäckström, *Rev. Sci. Instrum.* **44**, 1344–1349 (1973).
72. K.W. Henry, D.R. Stephens, D.J. Steinberg and E.B. Royce, *Rev. Sci. Instrum.* **43**, 1777–1784 (1972).
73. A. Cezairliyan, M.S. Morse, G.M. Foley and N.E. Erickson, in *Proceedings of the Eighth Symposium on Thermophysical Properties* (J.V. Sengers, Ed.), Am. Soc. Mech. Eng., New York (1982), pp. 45–50.
74. L. Coslovi, F. Righini and A. Rosso, *J. Phys. (E): Sci. Instrum.* **12**, 216–224 (1979).
75. G.M. Foley, M.S. Morse and A. Cezairliyan, in *Temperature: Its Measurement and Control in Science and Industry* (J.F. Schooley, Ed.), Vol. 5, ISA, Pittsburgh (1982), pp. 447–452.

17

Application of Differential Scanning Calorimetry to the Measurement of Specific Heat

M. J. RICHARDSON

1. INTRODUCTION

The current widespread application of specific-heat–temperature (c_p–T) curves to the characterization of materials would have been inconceivable just a few years ago when c_p could only be obtained via adiabatic or drop calorimetry. For either of these, large samples (several grams) are needed and it takes, literally, days to cover adequately any appreciable range of temperature. In addition each instrument represents a major financial investment, not only in the initial construction of the calorimeter, but also in the labor required for day-to-day operation. The transformation of calorimetry from a cumbersome research technique to a tool with an ease of use and a range of applications comparable with that of infrared spectroscopy, for example, was brought about by another transformation, that of differential thermal analysis (DTA) from a qualitative to a quantitative technique. Although quantitative DTA is nowadays synonymous with differential scanning calorimetry (DSC; this abbreviation will also be used for the instrument) this latter expression was first used with respect to a particular instrument of novel design (see below) which may be considered to have initiated the recent revolution in thermal analysis. The net result is a simple technique whereby c_p–T curves can be rapidly determined using only milligram quantities of solids or liquids over a temperature range of 100–1000 K. The form (disk, film, powder, irregular granules) of the solid is, at least in principle, immaterial and vapor pressures up to about two bars can be routinely accommodated.

M.J. RICHARDSON • National Physical Laboratory, Teddington, Middlesex TW11 0LW, England.

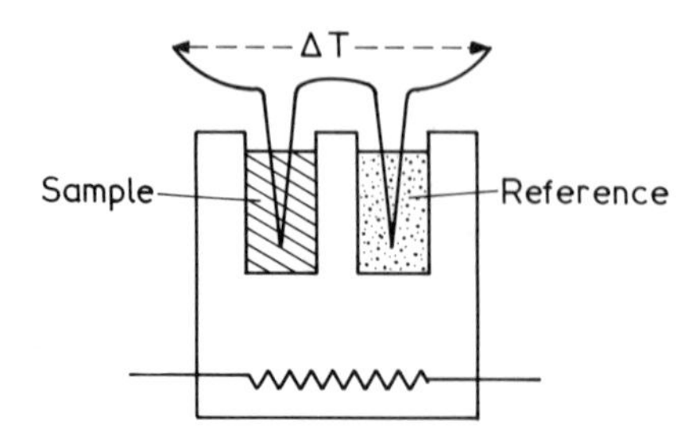

FIGURE 1. Conventional DTA.

If a particular thermal event (phase change, annealing, chemical reaction, etc.) is encountered, the calorimeter can be thought of as treating it as an unusual c_p requirement so that by integration overall enthalpy changes can be found and these in turn yield heats of fusion, transition, reaction etc. Thus calorimetry, as exemplified by DSC, has many valuable applications beyond the simple generation of c_p data.

2. DIFFERENTIAL THERMAL ANALYSIS AND DIFFERENTIAL SCANNING CALORIMETRY

An understanding of the similarities and differences between the variety of instruments now available as "differential scanning calorimeters" is most readily gained by considering how conventional qualitative DTA has developed from the pioneering ideas of Le Chatelier and Roberts-Austen towards the end of the last century. In DTA the temperature difference ΔT between the sample and an inert reference material is measured as the two are uniformly heated (Fig. 1). When a constant heating rate has been established ΔT is a slowly varying function of temperature (Fig. 2) and reflects differences in packing, thermal conductivity, and heat capacity between the two materials. If the sample now undergoes a phase change, for example, it requires a certain transition enthalpy and while this is being absorbed its temperature remains constant. However, the reference material temperature is still rising steadily and so ΔT changes rapidly, as shown in Fig. 2, before returning once again to constant heating rate conditions when

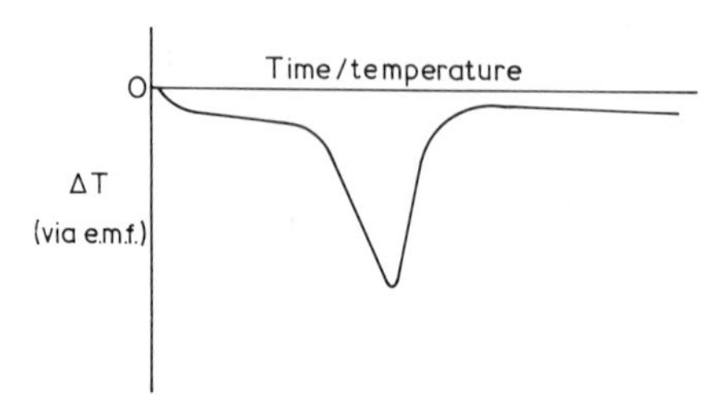

FIGURE 2. DTA trace; the abscissa may be time only, or the temperature of the sample, reference, or some point in the heating block.

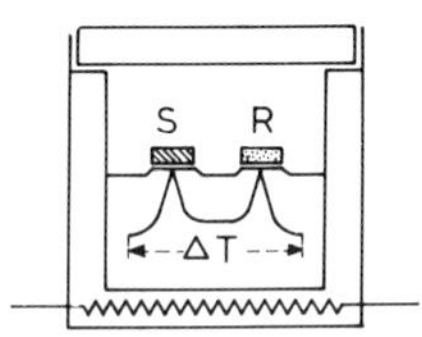

FIGURE 3. Boersma–DTA; the thermal path from heat source to thermocouples no longer includes the sample or reference.

the "thermal event" is complete. If the sample under investigation shows several transitions and/or reacts on heating (e.g., dehydration, decomposition) the resultant curve is a useful fingerprint unique to that material and DTA was used thus for over half a century for the characterization of, mainly, rocks and minerals.

It is clear that the peak area of Fig. 2 will depend on the sample size but it is almost impossible to use this type of DTA apparatus quantitatively because ΔT is influenced by many other parameters all of which basically affect the rate at which heat arrives at the sample thermocouple. To take an extreme example: at its melting point a lightly packed material may collapse completely away from the thermocouple giving a totally meaningless signal. In 1955 Boersma[1] showed how many of these disadvantages could be overcome by allowing a controlled heat leak between sample and reference containers and by removing the thermocouple from direct contact with the specimen (Fig. 3). This advance was later allied with concurrent progress in the techniques of signal amplification and linear temperature programing and quantitative DTA apparatus became commercially available.

DSC was first described in the literature in 1964.[2,3] Instead of measuring ΔT this is reduced to zero and the experimental quantity is the *differential power* needed to achieve the $\Delta T = 0$ condition. Obviously individual heaters are required for both sample and reference holders (Fig. 4) and this is perhaps the major distinction between differential power and differential temperature calorimeters.

A good impression of the thermal analysis scene at the time when DSC was introduced can be gained from the "Handbook" of Smothers and Chiang[4] which was published in 1966. The Perkin–Elmer DSC rates a brief mention in the section on Commercial Equipment but the extraordinary advances in quantitative work that it would initiate were not obvious at the time – indeed the

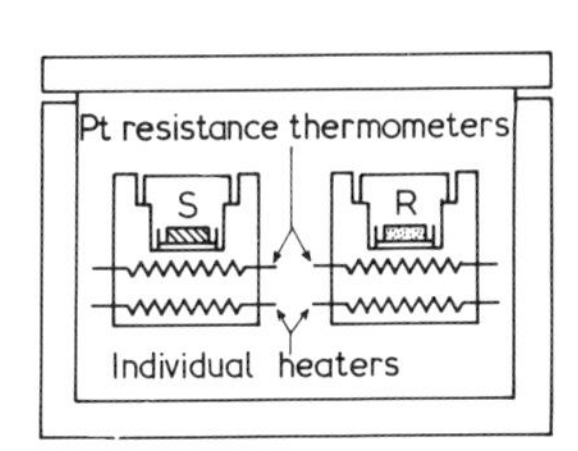

FIGURE 4. Perkin–Elmer DSC; ΔT is reduced to zero by varying the heat supply to the sample and reference containers.

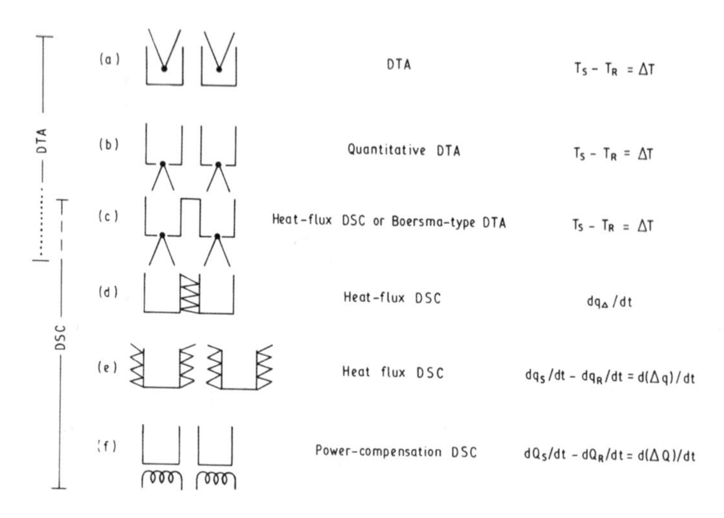

FIGURE 5. The change from DTA to DSC. S, sample; R, reference; Δ, difference; dq/dt, spontaneous thermal flux; dQ/dt, compensatory thermal flux; ∧, thermocouple; ⦚, thermopile; ෴, heater. Reproduced from Ref. 6 with the permission of the author and the Chemical Society, London.

Preface expressed a doubt, widely held at that time, regarding the quantitative capabilities of DTA. Although complex heat transport theories existed, all were reduced to common uncertainty by the simple practical problem of sample packing. Papers frequently discussed particle size effects (as they affected packing and heat transfer, as opposed to surface energy considerations) but these did not engender confidence in those whose interests were in materials, rather than the technique of DTA. Significantly, however, the final reference (number 4248) in Smothers and Chiang's bibliography of DTA (to 1965) was to a paper by Wunderlich,[5] on the heat capacity of polyethylene, which was the first to fully exploit the quantitative potential of DSC. Power compensation calorimetry rapidly became so successful that most manufactuters of thermal analytical instruments soon offered "DSC" facilities. These are generally thermocouple systems with a common source of heat for both sample and reference cells. It would have prevented much later confusion had the Perkin–Elmer Company also been able to protect the expression DSC in addition to patenting the power compensation principle. As used nowadays "DSC" implies quantitative operation but gives no indication as to which of the several methods of measurement an instrument might use. An excellent summary of the many sensor/heater configurations now available has recently been given by Mackenzie[6] and this clearly shows the transition from traditional DTA to Perkin–Elmer DSC (Fig. 5). In addition to conventional DTA equipment (ideal if temperatures alone are needed) which is available from several manufacturers the more specialized configurations of Fig. 5 are represented by the DuPont DSC cell (Fig. 5c), the Mettler 2000 and 3000 systems (Fig. 5d) and the Calvet-type calorimeters marketed by Setaram (Fig. 5e).

2.1. *Theory and Implications*

Although a rigorous theory of DSC or DTA is mainly of interest to instrument designers and manufacturers, even a simple treatment clearly demonstrates the similarities and differences. The following presentation is based on the thermal analysis cell of Fig. 6 and was originally given by Gray.[7] As we shall be mainly concerned with heat capacity measurements alone, only this form of energy uptake is considered and the resulting equations are very easily derived. The sample and sample holder, of total heat capacity C_s (Fig. 6) receive heat at a rate dq_s/dt from a source at a temperature T_p which rises at a programmed rate dT_p/dt, the thermal resistance between source and sample assembly is R. Similar conditions apply to the reference cell (subscript r) which for simplicity is assumed to have the same R value. The rate of rise of the sample cell temperature is given by

$$C_s dT_s/dt = dq_s/dt \tag{1}$$

subtracting the analogous equation for the reference cell gives

$$\text{DSC:} \quad dq/dt = dq_s/dt - dq_r/dt = (C_s - C_r)dT_p/dt \tag{2}$$

showing that the experimental quantity dq/dt, the differential power, is very simply related to heat capacities through a well-defined parameter, the heating rate, dT_p/dt ($= dT_s/dt$). Newton's law of cooling (often described in this context as the thermal analog of Ohm's law) gives a transformation to DTA conditions

$$dq_s/dt = (T_p - T_s)/R \tag{3}$$

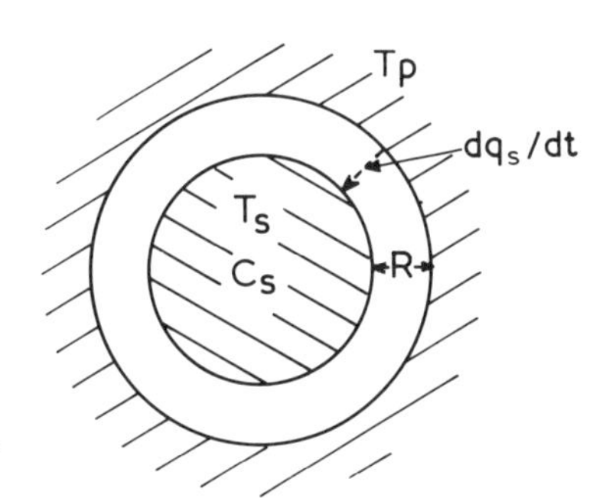

FIGURE 6. Heat flow in a thermal analysis cell (see Section 2.1 for definitions).

so that, again subtracting the reference material equations, (1) and (3) give

$$\text{DTA:} \quad \Delta T = T_s - T_r = -R(C_s - C_r)dT_p/dt \tag{4}$$

It is clear from (2) and (4) that the response of DSC and DTA instruments, although in some respects similar, differs in one important aspect – inclusion of the thermal resistance term, R, in (4). R is a function of the properties of the materials used in the construction of the calorimeter and, most important, is a function of temperature. By contrast the calibration constant of equation (2) is independent of temperature. Although it will be seen later that it is not difficult to calibrate any type of thermal analysis instrument over a range of temperature there is always an advantage for those of the differential power type in that the observed constancy, or otherwise, of the calibration factor gives an immediate indication of instrumental performance. Another result of the presence of R in equation (4) should be noted; the larger the value of R the larger the signal ΔT will be – but thermal lag will also increase. DTA design must always be a compromise between these conflicting effects although the problem has become less acute with modern advances in signal amplification. It must still be remembered, however, that in any series of multiple experiments (as are required in heat capacity work) it is considerably more critical in Boersma-DTA, as opposed to power compensation DSC, to reposition pans carefully so that R is exactly reproduced.

2.2. Heat Capacities

The application of DSC to the determination of heat capacities is based upon equations (2) or (4). As these are formally very similar it is useful to recast them in the common form $S = K\Delta C$ where S is the signal, K is a constant and ΔC is the heat capacity difference between the sample and reference cells. Use of DSC for the determination of heat capacity calls for three "runs" with the sample cell (i) empty (subscript e) (ii) + sample (s) (iii) + calibrant (c) so that the above equation becomes

$$S_e = K\Delta C \tag{5}$$

$$S_s = K(\Delta C + m_s c_{ps}) \tag{6}$$

$$S_c = K(\Delta C + m_c c_{pc}) \tag{7}$$

where m and c_p refer to mass and heat capacity, respectively. Eliminating $K\Delta C$ and dividing to remove K gives

$$c_{ps} = (S_s - S_e)m_c c_{pc}/(S_c - S_e)m_s \tag{8}$$

This is the basic DSC heat capacity equation, many of the difficulties in its application are centered around the meaning of the several quantities S. The direct use of instrumental readings demands absolute baseline reproducibility and to varying extents this is never achieved in practice so that methods must be developed which will allow for experimental changes of this type. These are considered in the following section after a general survey of operating variables.

3. OPERATING VARIABLES

The maximum temperature range of commercially available DSC instruments is 100–1000 K and in the most recent models the desired limits can be preselected so that unattended operation is feasible. Scan rates of from a fraction of a degree to upwards of a hundred degrees per minute in both heating *and* cooling are routine. The slowest rates are generally used for annealing or for purity determination where the geometry of the melting peak is strongly influenced by impurities, slow heating rates ensure that samples are never far from thermal equilibrium. Straightforward heat capacity work normally uses scan rates of 5–20 K min^{-1}. The signal S of equations (5)–(7) is, by (2) or (4), directly proportional to dT_p/dt, but it is not good practice to increase S indefinitely by use of too high a scan rate for there is a corresponding increase in thermal lag which, although derivable (see Section 3.2.2), must always be a source of uncertainty in absolute work.

Formally, any thermal "event," such as annealing, a phase change, or a chemical reaction may be considered to be an unusual heat capacity requirement and can be measured in a DSC. There may be problems in the proper characterization of conditions in melting, for example (where the theory becomes more complex), but the overall enthalpy change is directly accessible and problems concerning the subsequent decomposition into latent and specific heat contributions are the same as those encountered using adiabatic calorimetry. Samples are normally contained in aluminum pans ca. 1.5 x 6 mm diameter. It is rare to need anything approaching the full pan volume (40–50 mm^3) and the ability to make quantitative measurements on small samples is one of the great advantages of DSC. Sample geometry is immaterial (but see the later discussion concerning thermal lag) although it is clearly more convenient to handle disks (especially) or powders as opposed to irregular chunks. Liquids are equally amenable to measurement in a DSC unless they have an appreciable vapor pressure and even then sealable pans with a small vapor space can extend the range until the vapor pressure is perhaps 2 bar (sealed pans are also needed for solids that show any tendency to sublime, for example, benzoic acid, a popular calibrant, near its melting point). Aluminum pans are widely used up to 700 K because they are cheap and readily available but a wide range of special materials (gold, platinum,

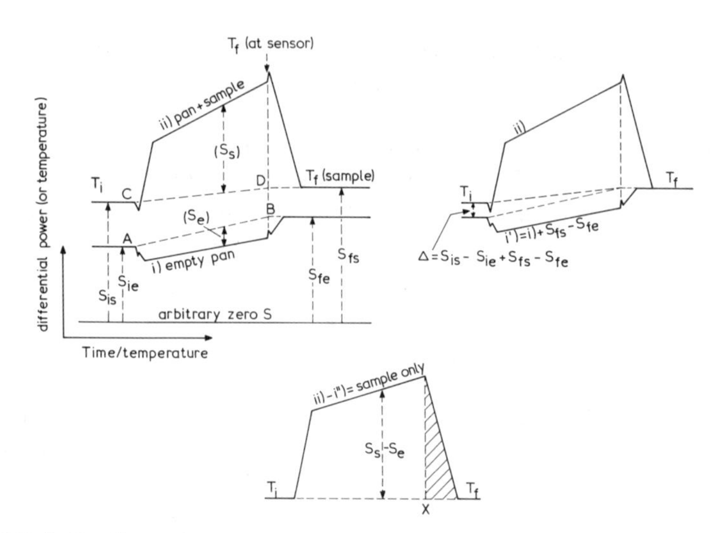

FIGURE 7. Baseline balancing in the determination of heat capacity (calibrant curve omitted). See Section 3.2 for a discussion of the conditions when time and temperature are equivalent. (a) Original curves for (i) empty pan and (ii) sample. (b) Initial correction, $S_{fs}-S_{fe}$ has been added to all points on curve (i) to give (i′). (c) Base lines balanced: the curve shows S_s-S_e after the correction $\alpha\Delta$ (see text) to curve (i′) of Fig. 6b.

graphite, silica) can be substituted as the occasion demands. For example, care is even needed at 300 K, the melting point of another useful calibrant, gallium, which readily alloys with aluminum. Within these non-too-severe limitations a DSC will accept an enormous range of materials from disciplines as adverse as biology, chemistry, metallurgy, and engineering.

Work is normally carried out at ambient pressure but a DuPont accessory is available that can be operated from vacuum conditions to 60 bar and a research instrument reaching 2.5 K bars has been described.[8]

3.1. Baseline Problems

Schematic diagrams showing the use of a DSC to measure heat capacities are shown in Fig. 7. The curves have been drawn to emphasize the problems that will inevitably be found. Initially, the calorimeter is at a steady temperature T_i and the output, which is usually the ordinate on a chart recorder, is some constant value S_i. The programmed temperature rise (commonly at 10–20 K min^{-1}) is then started and after perhaps a minute "steady rise" conditions are attained and the signal gradually changes until the selected final temperature (T_f) is reached when it decays to the constant final value S_f. In the initial run (i) with only an empty pan, the dynamic signal may be greater or less than S_i – for clarity Fig. 7 shows a lower value. It is customary to "back off" the empty pan signal with a similar pan in the reference side, this then remains untouched throughout the experiment (Figs. 3 and 4 show added inert material in the "reference" pans, this is often useful to back off a large signal). At the end of the "empty" run the temperature is returned to T_i and the whole procedure repeated for the pan + sample (or calibrant). Here the first problems appear. In general $S_{ie} \neq S_{is}$, $S_{fe} \neq S_{fs}$ but this would be only a minor inconvenience if the inequalities merely reflected a uniform vertical displacement of, say, curve (ii) with respect to (i) because absolute values of S are in themselves of little significance. The problem is normally aggravated because $S_{is} - S_{ie} \neq S_{fs} - S_{fe}$ so that even if the final isothermal regions are forced to coincide ($S_{fe} = S_{fs}$) the initial regions do not (Fig. 7b) and it is then not clear to what $S_s - S_e$ [equation (8)] refers for some intermediate temperature T. The difficulty is overcome by drawing baselines AB and CD (Fig. 7a) between the two steady program temperatures T_i and T_f; S_e and S_s may then be taken to be as shown (bracketed) in Fig. 7a where S_e would be a negative value. Later on it will be shown that a variety of corrections complicate the simple application of equation (8) and to make full use of the potential of DSC instruments some form of computerized data treatment is essential. One method is to incorporate the constructions of Fig. 7a into a program which gives $S_s - S_e$ of Fig. 7c directly. Alternatively, the baselines AB and CD may be dispensed with by forcing the steady "initial" or "final" values into coincidence, e.g., for the latter by adding $S_{fs} - S_{fe}$ to *all* readings for the empty pan to give curve (i′), Fig. 7b. The new difference at T_i, $\Delta [= S_{is} - (S_{ie} + S_{fs} - S_{fe})]$ is assumed to steadily decrease to zero at T_f so that the final correction is to add scaled [by a factor $\alpha = (T_f - T)/(T_f - T_i)$] values of Δ to the readings for the empty pan. The resultant curve (i″) (not shown) closes the "full minus empty" loop so that (ii)–(i″) gives the final baseline-corrected quantities of Fig. 7c. The last correction $\alpha\Delta$ ($0 < \alpha < 1$) is a commonsense operation but it can only be justified by appeal to experiment; a comparison of results for a given sample shows no trends when Δ is varied within wide limits. Although computing times are comparable for either route to Fig. 7c it is helpful to have Δ as an indication of baseline reproducibility.

Errors caused by the combination of an incorrectly *shaped* curve (i) with curve (ii) (even if $\Delta = 0$) cannot be treated analytically but can be recognized by careful attention to the design of experiments (see Section 4). They must be regarded as defining the limits of quantitative operation if they persist after eliminating obvious causes such as incorrect pan placement or sublimation from a volatile or contaminated sample.

3.2. *Temperatures*

Even when all curves have been normalized to consistent baselines, equation (8) still cannot be immediately applied because conditions along the abscissa remain undefined. In Figs. 2 and 7 the abscissa is given as "time/temperature" and the significance of this must now be considered.

First a trivial but important point must be made. Time and temperature are only synonymous under steady heating (or cooling) conditions. In an isothermal region (e.g., T_i, T_f Fig. 7) time alone is the variable, again in the buildup to, or decline from, steady temperature programming conditions (rate $\dot{T}$) the rate will vary from $0 \rightarrow \dot{T}$. Once steady programming conditions have been attained heating rates in most modern calorimeters are truly linear functions of time so that "temperature" and "time" are synonymous. This has not always been the case and even now is only true when instrumental settings are correct. The temperature-sensing devices in Perkin–Elmer DSC are platinum resistance thermometers and in quantitative DTA they are thermocouples but both types of sensor show nonlinear variations with temperature. Linearizing circuitry overcomes the resultant problems but to allow for instrument-to-instrument variations optimum linearization is only achieved by individual "tuning" and it is therefore essential to know how to calibrate the temperature scale to ensure that the instrument is correctly set up – or, for older models, to know what correction must be applied.

3.2.1. Isothermal Conditions

Calibration applies to both isothermal and scanning conditions. The former gives a direct indication of the validity, or otherwise, of the instrumental temperature scale. It is very easily carried out by using pure materials with known melting points (T_m). The temperature is raised stepwise through the melting region, as shown in Fig. 8a. Both before and after melting the response to a given step is a relatively minor fluctuation, but in the melting region, when the sample is taking in heat of fusion (ΔH) much larger effects are found and the magnitude of these increases to T_m after which "no ΔH" conditions are abruptly restored. Sample purity determines the temperature range over which the "pre-melting" buildup (A, B, C, Fig. 8a) to T_m occurs. Very pure metals are readily available and are recommended temperature calibrants – the premelting

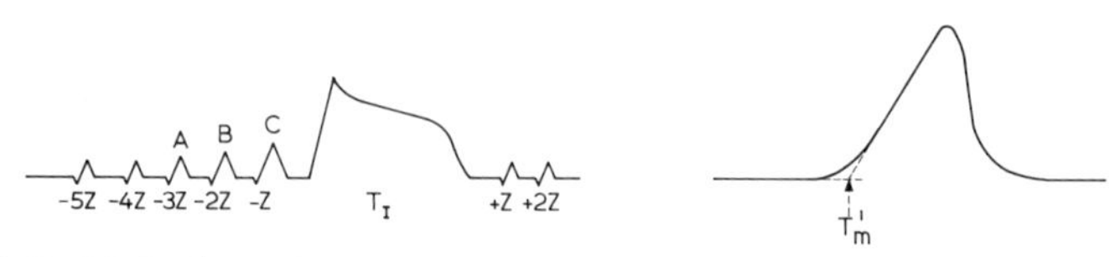

FIGURE 8. (a) Isothermal temperature calibration; premelting shown at A, B, C; when correctly calibrated $T_I = T_m$; temperature increments Z can vary from 5 K (rough survey) to 0.1 K (final definition). (b) Dynamic (scanning) temperature calibration showing the definition of T'_m.

range for indium, for example, may be only 0.1 K whereas for an organic material it can be many times greater. Intermediate temperatures (400–700 K) are conveniently covered by such metals as indium, tin, lead, and zinc but other materials are needed outside this range because of either the lack of suitable metals (at low temperatures) or alloy formation with the sample pan (or worse, the pan holder itself). Transition temperatures may also be used as calibration points and indeed solid–solid phase changes form a majority of the samples available from the National Bureau of Standards as International Confederation of Thermal Analysis (ICTA)-approved calibrants. This is because these are also intended for use with conventional DTA instruments (Fig. 1) when melting as such might both break thermal contact and give problems in cell cleaning. The net result of this type of calibration will be a graph of temperature correction δT_0 $(= T_m - T_I)$ versus indicated instrumental temperature T_I and, as noted in the previous paragraph, instrumental settings can be adjusted to make the correction very small. Even if this cannot be achieved, as with the older models of either class of instrument, corrected temperatures can be readily deduced for subsequent calculations.

3.2.2. Scanning Conditions

As heat arrives at the sample via the sensor there is, under scanning conditions, a thermal gradient (lag) between the two. Clearly the lag (δT) will be greater for a large sample of irregular geometry compared with a thin film of the same material. The implication is that the dynamic temperature calibration must be individual to each sample and must therefore be reflected in some way in an individual DSC curve.[9] This can be done by using the tail area (shaded, Fig. 7c) to define δT. In the absence of thermal lag there would be an immediate return to the base line at X, Fig. 7c, when the final program temperature T_f is reached. In reality there is a gradual return to isothermal conditions and the shaded area is proportional to $m\bar{c}_p\delta T$, where m is the sample mass and $\bar{c}_p$ is, strictly speaking, the average heat capacity over the range $T_f - \delta T \rightarrow T_f$, this refinement is, however, quite unnecessary and an adequate estimate can be made by using equation (8) and the $S_s - S_e$, etc, values at T_f. δT can then be calculated and the final corrected temperature T is given by

$$T = T_I + \delta T_0 + \delta T \tag{9}$$

Here δT_0 may be of either sign but δT is always negative in heating and positive in cooling.

Manufacturers' manuals normally give a different method of temperature calibration. Isothermal conditions are not considered and a universal calibration is obtained via the melting curves of pure materials. These show a linear approach

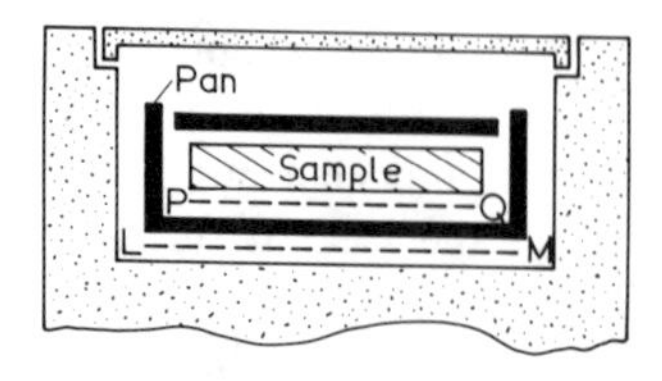

FIGURE 9. DSC (Perkin–Elmer DSC-2) showing major sources LM, PQ of thermal lag.

to the peak (Fig. 8b) followed by an exponential decay. The melting point (T_m') is obtained as the point of intersection of the extrapolated leading edge with the low-temperature base line. There are sound theoretical reasons for this construction but the calibration curve of $\delta T'$ ($= T - T_I$, at a calibration point $T = T_m$ and $T_I = T_m'$) versus T_I is unique so that the relationship $\delta T' = \delta T_0 + \delta T$ must only be valid under very restricted circumstances. These are best understood by a more detailed consideration of conditions within an individual cell. An expanded view a Perkin–Elmer DSC-2 cell is shown in Fig. 9. The two interfaces LM and PQ are serious barriers to efficient heat flow and are thus major sources of thermal lag. At T_m' (Fig. 8b) the sample starts to melt – obviously at the point nearest the heat source. The dynamic calibration $\delta T'$ therefore gives the thermal lag at the face of the sample in contact with the pan (PQ) and $\delta T' = \delta T_0 + \delta T$ holds only in the limit when sample sizes are very small; as sample sizes increase so do the errors due to the size of the "universal" correction $\delta T'$ as opposed to the correct quantity $\delta T_0 + \delta T$. When experiments are being carried out in cooling a δT type of calibration is the only one possible because every material supercools to some extent and it is impossible to supply freezing standards of comparable precision to those used in melting. The lowest degrees of supercooling are probably found in liquid crystal transitions, but readily available materials cover only a relatively narrow temperature range.

Some emphasis has been placed on a clear definition of temperature conditions in a DSC because, although T does not appear as such in equation (8), c_p is a function of temperature and it is important to know to what value S of Fig. 7 refers.

4. PRACTICAL ASPECTS

The application of equation (8) implies that a DSC is being used in a relative, rather than absolute, mode. This is an obvious step for Boersma-type instruments in that it eliminates problems due to the temperature variation of R [equation (4)] but it might be thought unnecessary for power compensation DSC [equation (2)] and in principle this is true. However any electrical instrument requires an initial calibration which should be periodically checked and in practice both types of DSC are conveniently operated in the same way. The calibrant chosen must have been thermodynamically well characterized and for temperatures above ambient α-alumina is a near-universal choice. This material has been the subject of several investigations using adiabatic calorimetry and these generally agree to within a few tenths of a percent. Below room temperature dc_p/dT becomes large for alumina; accurate abscissa calibration is therefore essential and a less demanding standard is benzoic acid. Tabulated data for both materials are available.[10, 11]

The use of a DSC calls for no special experimental skill. In fact the simplicity of operation has in the past been a definite disadvantage in that it has tended to encourage a similar simplistic philosophy in the subsequent interpretation of results which can at times be rather complex as, for example, in the vicinity of a glass transition. Details of the temperature calibration and baseline balancing needed have already been given and with computer assistance it is a routine operation to obtain a *continuous* heat capacity versus temperature curve covering perhaps 200 K or more.[12] If calculations must be done by hand it is better to use several short temperature ranges rather than one long one – a useful range is $1.5(dT_p/dt)$ K, which ensures that constant heating rate conditions have been attained at T_f; with short runs uncertainties due to non-computed baseline balancing are minimized. Even with full data treatment it is sensible to check the consistency of results in as many ways as possible, at least in the initial proving of any apparatus. The main uncertainty must concern the correctness of any given pair of empty and full curves [(i) and (ii), Fig. 7a]. Either of the treatments mentioned in Section 3.1 is appropriate when the only problem is $\Delta \neq 0$ but, even if $\Delta = 0$, there will be errors if the shape of (i) is incorrect with respect to (ii). In this case the errors will be greatest at temperatures between T_i and T_f and an independent determination should reveal any problem. If, for example, a range of 200 K has been covered in one run it is useful to see how the midpoint value compares with the final value in a run that covers only the first 100 K; again, if results are thermodynamically significant, values obtained in cooling should reproduce those obtained in heating ("shape" effects are probably the cause of the loops in Fig. 10, Section 4.1).

4.1. *Performance*

The following discussion is based on the performances of a Perkin–Elmer DSC-1B or DSC-2 which have successively been in use in this laboratory for more than 10 years. Data have always been handled using computer techniques so that a large amount of standardized information is available. Where the comparison is possible (mainly using calibrant figures) the reproducibility of all data is generally within the range ± 1% except at the highest temperatures where it falls (in the range 800–1000 K) to ± 2.5%. Deviations from these figures can almost invariably be traced to incorrectly positioned lids which, if not actually observed when removing the sample, show up as gross base-line mismatches. The

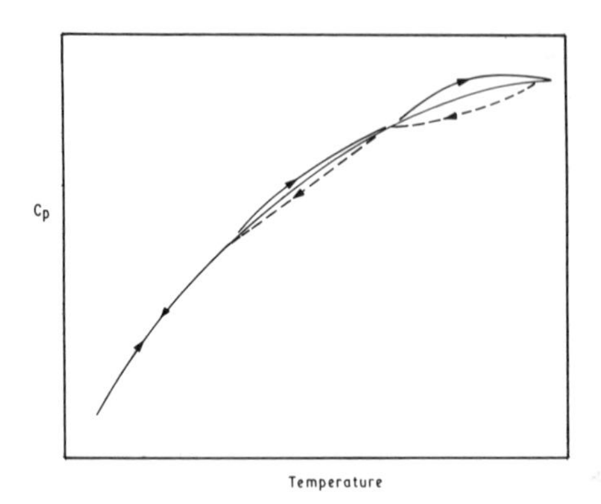

FIGURE 10. Schematic representation (differences grossly exaggerated) of data obtained in heating (full lines) and cooling (broken lines) modes over a wide range of temperature that has been covered in three stages.

larger errors at high temperatures are probably due to difficulties in reproducing identical radiation conditions for empty and full runs. The problems can be seen in a comparison of heating and cooling experiments when results of the type shown in Fig. 10 can be found. There is good agreement at lower temperatures but as radiation effects become important the loops of Fig. 10 often appear. (Here it should be emphasized that when operating at high temperatures it is important to ensure that samples are fully equilibrated with respect to both phase behavior and also loss of volatile components or contaminants). The overall accuracy has been assessed using a variety of materials that have been characterized in these laboratories by adiabatic calorimetry. Again a figure of $\pm 1\%$ covers nearly all results, including measurements made in cooling – a mode of operation not available to adiabatic calorimetry. Errors of $\pm 1\%$ for both unknown and calibrant would naturally be expected to give larger uncertainties in the final answer and the eventual figure of $\pm 1\%$ shows how errors tend to cancel when using equation (8): calibrant readings that approach the high limit $+ 1\%$, for example, are invariably associated with sample readings that also tend to larger values. For a homogeneous sample these figures are independent of sample size, geometry, and heating (or cooling) rate provided the appropriate correction is made for thermal lag.[9] If they were not, no reliance could be placed on DSC as a quantitative technique.

The error limits are based on routine measurements with no special precautions in sample preparation. (In Boersma-DTA more care is needed to ensure exact reproducibility of conditions in the empty–calibrant–sample cycle and this may require a special positioning jig[13] before a comparable overall accuracy is achieved.) In one important respect, however, the measurements are atypical – a digital voltmeter was used rather than attempting to read from a chart recorder, a procedure that *in itself* can lead to overall uncertainties greater than anything considered above. Inadequacies external to the calorimeter (another, more obvious, source of error is the use of a low-precision balance) should not be allowed to obscure the full potential of the technique.

4.2. Advantages and Disadvantages

Some of the benefits of DSC have already been noted but a compilation of the advantages will emphasize the versatility of this relative newcomer to the field of calorimetry. Three aspects are dominant: simplicity, rapidity, and economy in its sample requirements.

Complete newcomers can make meaningful measurements after, literally, a few minutes' acquaintance – at times this may be counted a disadvantage for it often encourages a somewhat cursory interpretation of what may be a very complex curve.

Heating rates in adiabatic calorimetry are of the order of degrees per hour whereas common rates in DSC are 5–40 K min^{-1}. Not only does this give a simple time advantage in that a given temperature range is more rapidly traversed, but low-energy phenomena are more clearly revealed. Consider a material that sinters or anneals on heating: a metastable form *may* transform rapidly, but often the process is sluggish and encompasses a wide range of temperatures. Under these circumstances the apparent heat capacity is only slightly depressed (for an exothermic change), but, because a DSC provides a continuous c_p versus T record, any anomalous behavior is immediately apparent and the sample, now in more stable form, can be rerun so that the sintering or transition energy is obtained with little more effort than subtraction of the "before" from the "after" curve. By contrast an adiabatic calorimeter works in a discontinuous mode and great care is needed to ensure that a full energy balance is maintained as the temperature drifts (due to changes in the sample) in the normally isothermal regions. Because rapid measurements can also be made in *cooling* metastable phases are often available to direct experimentation. Such phases are not merely of academic interest for there are direct applications to the determination of such basic quantities as heats of fusion or transition. Thus if a material (e.g., urea) is thermally unstable only a few degrees above its melting point, only measurements on the supercooled liquid will give an adequate description of the liquid heat capacity.

Sample sizes in a DSC are usually measured in milligrams, as opposed to the grams of an adiabatic calorimeter. Simple heat capacity determinations require more material than when large energies, as with phase changes or chemical reactions, are involved. Typical organic samples may range from 1–20 mg, metals will perhaps be a factor of 10 heavier. The small sample sizes called for in DSC have an additional advantage or disadvantage, depending on the final requirement. Any material that is at all heterogeneous will need careful treatment to ensure that a representative sample is used, this may never be possible because homogenization may so change the material that it is no longer representative of the original. By contrast the procedure can be inverted and the technique deliberately used to monitor materials for uniformity.

The main disadvantage of DSC is that it is relative and not absolute but even this is fairly unimportant given the simplicity of calibration against a well-defined material such as alumina.

5. CONCLUDING REMARKS

The whole field of calorimetry has changed dramatically over the last 15 years from a few laboratories making precision measurements (on materials that were often of secondary importance to the technique) to one where calorimetry is now a routine tool of materials science. The transformation is due to the commercial development of DSC and its ready acceptance in both academic and industrial laboratories. Unfortunately in some early work a too-ready, uncritical, acceptance of the technique led to near-absurd results which, rightly, led those unfamiliar with DSC to view the whole trend to differential methods with grave misgivings. Hopefully usage of DSC has now matured sufficiently to provide substantial evidence of a quantitative ability that can, under ideal conditions, approach that of adiabatic calorimetry.

The response of modern calorimeters is such that it is already rare to need to use the most sensitive ranges. Obviously the future *will* see even further improvements in sensitivity which will have the concomitant benefit of reducing the noise level at the present limit. Although a reduction of sample size brings the equality $\delta T'(m \to 0) = \delta T_0 + \delta T$ closer, a quite unrelated complication should be noted: unusual metastable phases can persist for increasing periods so that if a sample is melted and subsequently recrystallized in the calorimeter the behavior may be a strong function of sample size.

Many manufacturers now offer microprocessor-based data handling facilities which help considerably in the treatment of results. Usage will undoubtedly increase and should be encouraged with the caution that even the most sophisticated system cannot overcome instrumental or programming inadequacies. One very beneficial aspect of the move to microprocessors is the ability to preprogram the instrument for automatic operation including, if necessary, complex thermal histories. Productivity will be further increased by facilities for the automatic changing of samples.

The current status of DSC is such that true heat capacity curves are produced almost in parallel with the actual measurement and users are free to concentrate on the interpretation of results without major concern for the technique itself. At this stage of maturity DSC will presumably be married with other techniques because the potential for *in situ* experiments is most attractive, magnetic field effects, for example, could already be studied with relatively minor modifications to existing apparatus.

REFERENCES

1. S.L. Boersma, *J. Am. Ceram. Soc.* **38**, 281–284 (1955).
2. E.S. Watson, M.J. O'Neill, J. Justin, and N. Brenner, *Anal. Chem.* **36**, 1233–1238 (1964).
3. M.J. O'Neill, *Anal. Chem.* **36**, 1238–1245 (1964).
4. W.J. Smothers and Y. Chiang, *Handbook of Differential Thermal Analysis,* Chemical Publishing Company, New York (1966).
5. B. Wunderlich, *J. Phys. Chem.* **69**, 2078–2081 (1965).
6. R.C. Mackenzie, *Proc. Anal. Div. Chem. Soc.* **17**, 217–220 (1980).
7. A.P. Gray, in *Analytical Calorimetry* (R.S. Porter and J.F. Johnson, Eds.), Vol. 1, pp. 209–218, Plenum Press, New York (1968).
8. M. Kamphausen and G.M. Schneider, *Thermochim. Acta* **22**, 371–378 (1978).
9. M.J. Richardson and N.G. Savill, *Thermochim. Acta* **12**, 213–220 (1975).
10. D.A. Ditmars and T.B. Douglas, *J. Res. Natl. Bur. Stand.* **75A**, 401–420 (1971).
11. G.T. Furukawa, R.E. McKoskey, and G.J. King, *J. Res. Natl. Bur. Stand.* **47**, 256–261 (1951).
12. M.J. Richardson, *J. Polymer Sci., Part C* **38**, 251–259 (1972).
13. H.K. Yuen and C.J. Yosel, *Thermochim. Acta* **33**, 281–291 (1979).

V

THERMAL EXPANSION MEASUREMENT METHODS

18

Thermal Expansion Measurement by Interferometry

G. RUFFINO

1. INTRODUCTION

The basic principle of heat expansion measurement by interferometry consists in converting sample length variations into variations of optical path difference of two interfering monochromatic light beams. The interference produces a pattern of fringes that shift on their localization plane as the sample length varies. In this way expansion is measured by counting the fringe number passing across a reference.

This operation is performed by means of an interferometer entirely contained in an experimental cell, or including this cell as an integral part. The temperature of the experimental cell can be varied, controlled, measured, and recorded.

In the historical development of the interferometric technique of heat expansion measurement we may distinguish two periods, before and after the mid-sixties.

The first period is characterized by the variety of configurations of interferometers, experimental cells, detection systems developed in order to cope with the severe limits imposed by light levels (and fringe contrast) obtained from the monochromatic sources available up to that time, namely, discharge tubes with low-pressure vapors.[1, 5–7, 10, 11, 14–16] A review of these instruments was made by E. Mezzetti.[8]

Around the mid-sixties the experimental approach has been substantially changed by the availability of the laser. The high degree of coherence and substantial power level of this light source allows a drastic simplification of the

---

G. RUFFINO • Microtecnica, Via Madama Cristina 147, 10126 Turin, Italy. The author's present address is: Department of Mechanical Engineering, University of Rome, Via Orazio Raimondi, 00173 Rome, Italy.

optical system and detection, as well as the general use of the absolute expansion measurement on relatively long samples. The last fact reduces the relative error, thus shifting the emphasis of the measurement technique from length to temperature, which must be kept uniform over large samples.

It is the scope of this chapter to present a modern approach to measurements of thermal expansion by optical interferometry. Therefore it only considers those features of the past which are compatible and advantageous with the use of laser beams, leaving aside all details of merely historical significance.

After a brief outline on (a) the fringe patterns and (b) the interferometer types relevant to thermal expansion, we will focus our attention on (c) the experimental cell, (d) the system layout, and (e) the data acquisition and treatment.

2. INTERFERENCE PATTERNS

The laser yields a highly coherent collimated light beam. Coherence allows high fringe contrast for long optical path difference. A collimated beam is a *point source* at infinity, which can be brought to a finite distance with a converging lens, in other words, a plane wave easily convertible into a spherical wave.

Interference occurs when a light wave is reflected by two plane surfaces and successively recombined on another surface. The latter is usually a plane in which a fringe pattern appears; this plane is called the localization plane. If the position of this plane is arbitrary, the interference fringes are *unlocalized*: this is always the case when the light source is a point source.

The reflecting surfaces belong to two plates which are separated by the specimen; therefore they delimit a (vacuum or air) gap whose thickness equals the sample length.

Thus, the use of the laser limits the number of interference cases to the ones which result from the combination of finite of infinite source distance with parallel or wedged reflecting planes. Among these four cases two have practical relevance:

a. point source and parallel planes: *circular fringes of equal inclination*;
b. parallel beam and wedged planes: *straight fringes of equal thickness.*

2.1. Fringes of Equal Inclination

The light beam emitted by a point source on two transparent plates (Fig. 1) are partially refracted and reflected on each face. The medium is vacuum or air. To avoid confusion among the different interference patterns the

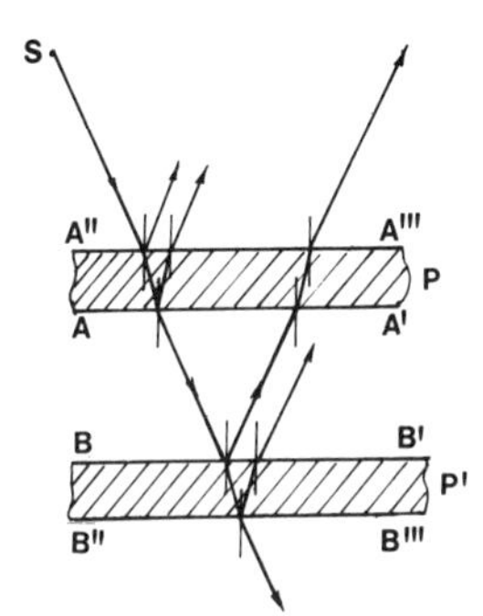

FIGURE 1. Interference by two plates.

reflections on faces A″A‴ and B″B‴ are excluded by making the first anti-reflecting and the second absorbing. More conveniently, face A″A‴ is ground at a small angle (about 1°) with respect to AA′, so that the beam reflected by the former is deflected out of the observation field.

In the following analysis we will consider only the beam splitting face AA′ and the reflecting plane BB′. We will speak of rays only in a geometrical sense as the perpendiculars to the wavefronts, which, in the present instance, are spherical.

The point source is produced by focusing the parallel beam generated by a laser with a converging lens.

Interference takes place wherever two rays reflected by faces AA′ and BB′ (Figs. 2a, 2b, 2c) intersect with each other. Interference fringes are located on one plate (Fig. 2a), at infinity (Fig. 2b), or in any parallel plane in between (Fig. 2c).

Reflection on the vacuum-to-plate face (BB′) causes a path jump $\lambda/2$ to be added to the optical path difference (opd). The two interfering waves are virtually emitted by the two mirror images S′ and S″ given by the reflecting

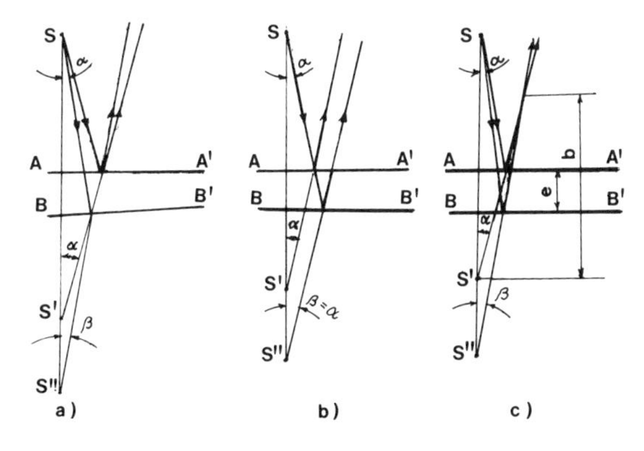

FIGURE 2. Localization of fringes: (a) on the plate; (b) at infinity; (c) in space in between.

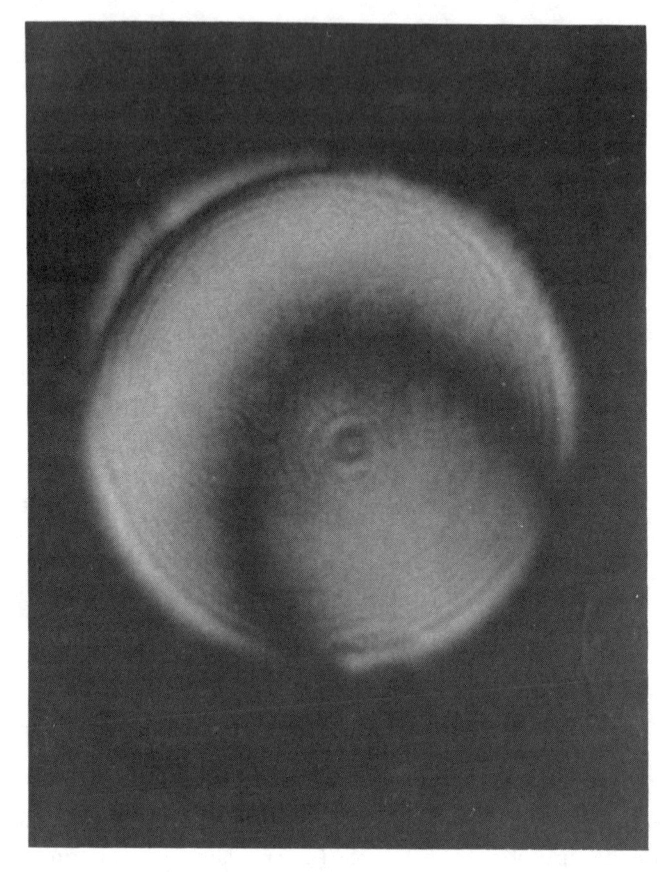

FIGURE 3. Circular fringes of equal inclination (Newton rings).

faces. The opd of the ray perpendicular to the plates is $2e + \lambda/2$, e being the plate spacing. Any opd is a function of the angle α of the ray inclination to the plate normal and is radially symmetrical with respect to the perpendicular to the plates passing through the source (optical axis). Thus, we get circular fringes of equal inclination (Fig. 3).

Any *parallel shift* of a plate with respect to the other produces a *variation of the opd.* A shift of $n\ \lambda/2$ causes n fringes to emerge on a point of the optical axis. Their count measures the plane displacement in term of the wavelengths of a monochromatic radiation.

Since the fringes are counted by a light detector placed in the center of the fringe pattern, a certain diameter ratio between detector and first fringe must exist to achieve sufficient modulation as the fringe moves on the localization plate.

Let the direction of the first luminance maximum to form an angle α with the optical axis (Fig. 2c). For a luminance maximum on the optical axis it is

$$2e = \frac{2n+1}{2}\lambda \tag{1}$$

where n is an integer and λ is the laser wavelength. The difference of the opd between the central and the first maximum is λ. Therefore, we have

$$2e - \left(\frac{b+2e}{\cos\beta} - \frac{b}{\cos\alpha}\right) = \lambda \tag{2}$$

Moreover, α and β satisfy the condition

$$b \tan\alpha = (b+2e)\tan\beta \tag{3}$$

Angles α and β are so small that the following approximations are valid:

$$\tan\alpha \cong \alpha, \qquad \tan\beta \cong \beta, \qquad \frac{1}{\cos\alpha} = 1 + \frac{\alpha^2}{2}, \qquad \frac{1}{\cos\beta} = 1 + \frac{\beta^2}{2}$$

By introducing these approximations into equations (2) and (3) and solving the system, we get

$$\alpha = (1 + 2e/b)^{1/2}(\lambda/e)^{1/2} \tag{4}$$

For fringes at infinity (Fig. 2b), it is

$$\alpha_\infty = (\lambda/e)^{1/2} \tag{5}$$

Fringes are located on the first reflecting plane when b is equal to the distance of the latter from the source. In this case we have typically $2e/d = 0.01$ and therefore $(1 + 2e/b)^{1/2} \approx 1.005$. The variability limits of α are very small indeed and therefore we conclude that the fringe maxima are aligned with the mirror image of the source.

The spacing e is equal to the sample length, which in practical cases is typically 20 mm. Since $\lambda \approx 0.6\,\mu$m, we get $\alpha = (0.6 \times 10^{-3}/20)^{1/2}$ rad $= 5.5 \times 10^{-3}$ rad. At a distance of 1 m from the virtual source the diameter of the first maximum would barely be 5.5 mm. This size is inadequate to accommodate a light detector whose collecting surface has a diamater equal to 1/8 of the first interference ring. We therefore need to magnify the interference figure in order to compress the overall length of the optical system.

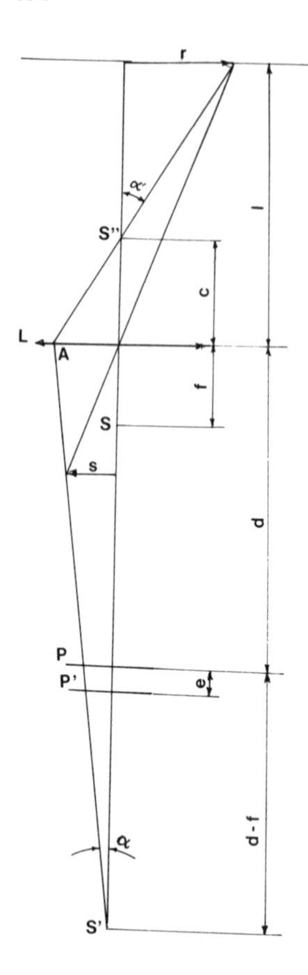

FIGURE 4. Optical diagram of the fringe magnifier.

The magnification is achieved by placing a lens L (Fig. 4) centered on the optical axis behind the point source S. As fringes are unlocalized, we get an interference pattern on any plane beyond the conjugate of the reflecting surface P. The illuminance maxima lay on the conjugate to the direction S′A, which passes through S″, the conjugated of the mirror image of the source. It is easy to see that in Gauss' approximation, with symbols of Fig. 4, the radius of the first fringe at distance l from the lens is

$$r = 2\,[(d-f)(l/f-1)-f\,]\,\alpha \tag{6}$$

As the focal length f is small with respect to distances d and l, we can write with enough accuracy

$$r \approx 2dl(\lambda/e)^{1/2}/f \tag{7}$$

For $\lambda = 0.6\,\mu\text{m}$, $e = 20$ mm, $f = 30$ mm, $d = l = 200$ mm, it would be $r = 15$ mm,

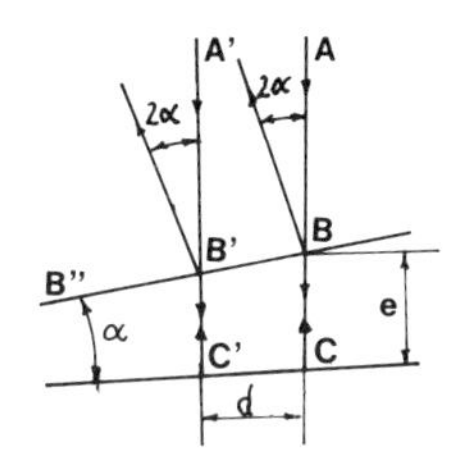

FIGURE 5. Interference produced by a wedge.

A detector with 3 mm diameter of collecting surface would occupy 1/10 of the fringe diameter.

Lens L can perform the double function of laser beam focalization and fringe magnification (S.D. Preston, Ref. 12). Beam obscuration is avoided by placing the detector, a compact silicon photodiode, out of axis. An alternative configuration could be to couple the detection spot with any sensitive area (including photocathodes) by means of a fiber optics.

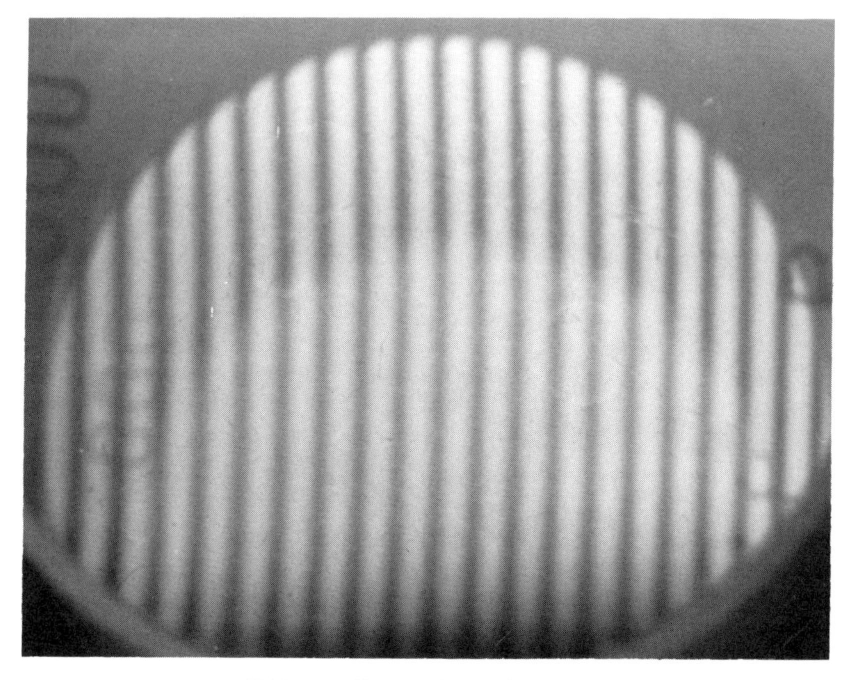

FIGURE 6. Fringes of equal inclination.

2.2. *Fringes of Equal Thickness*

When a parallel beam is reflected by a wedged pair of surfaces (Fig. 5) straight fringes of equal thickness are generated. The fringes are localized on face BB″. On point B we have maximum luminance, if it is (*in vacuo*)

$$e + \lambda/2 = n\lambda \tag{8}$$

or

$$e = (2n + 1)\lambda/2 \tag{9}$$

The next maximum is located on B′ at a distance d such that:

$$\tan \alpha = \lambda/2d \tag{10}$$

The fringes are straight and equidistant (Fig. 6): they shift on the localization plane while one face moves along the beam direction.

3. DILATOMETRIC INTERFEROMETERS

Interferometric measurement of thermal expansion has traditionally used the Fizeau interferometer.[5] However, the laser light opens the possibility of utilizing the Michelson interferometer.

3.1. *Fizeau Interferometer*

With this interferometer (Fig. 7), the specimen is kept within two parallel plates illuminated by a point source. Sample expansion causes spatial variation between the plates and radial motion of the circular fringes with the emergence of new fringes on the optical axis. A light detector and a counter measure dimensional variations, while sample temperature is measured with some kind of thermometer.

Plate P_1 is wedge-shaped with angle of about 1°, so that light reflected by its upper face is diverted from the field of view. The lower face of plate P_2

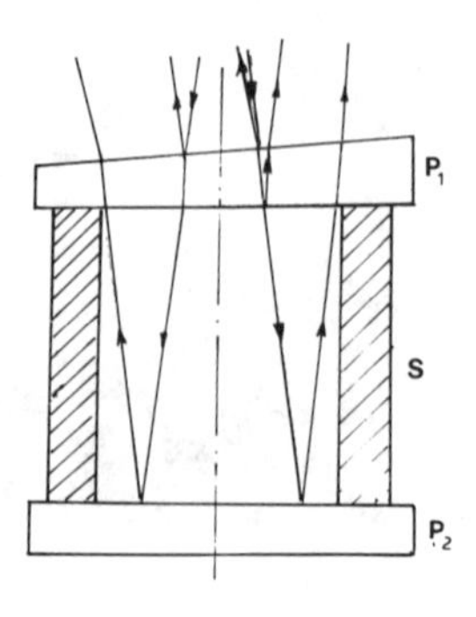

FIGURE 7. Fizeau interferometer.

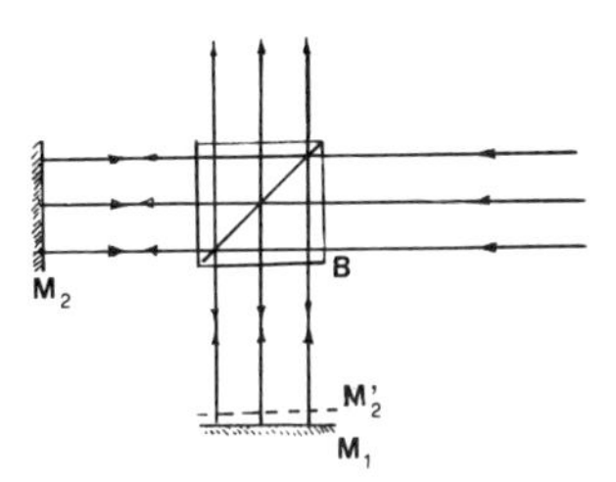

FIGURE 8. Michelson interferometer.

is made to absorb the incident light. The reflecting surfaces should be flat within $\lambda/10$ up to the limit of the working temperature. Suitable plate materials are fused silica up to 900°C and sapphire, presumably up to 1500°C (upper limit to be investigated).

3.2. *Michelson Interferometer*

In the Michelson interferometer a parallel light beam (Fig. 8) generated by a laser through a beam expander is split by beamsplitter B. The resulting beams are reflected by mirrors M_1 and M_2 and recombined on plate B. If M_2 is slightly inclined over the light beam, its mirror image M_2' forms a small angle with M_1 : fringes of equal thickness are located on the virtual face M_2'.

The dilatometer uses Michelson interferometer to measure the relative shift of the plane parallel faces of the sample. Two types of instrument are used: the single and the double Michelson interferometer.

3.2.1. *Single Michelson Interferometer*

In the *contact interferometric dilatometer* (Fig. 9), the sample is a prism or cylinder placed on mirror M, so that a polished face is opposed to the incident

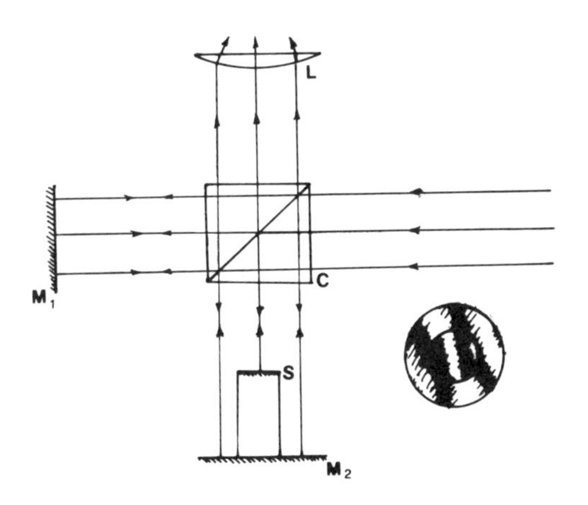

FIGURE 9. Contact Michelson interferometric dilatometer (symbols as in Fig. 8).

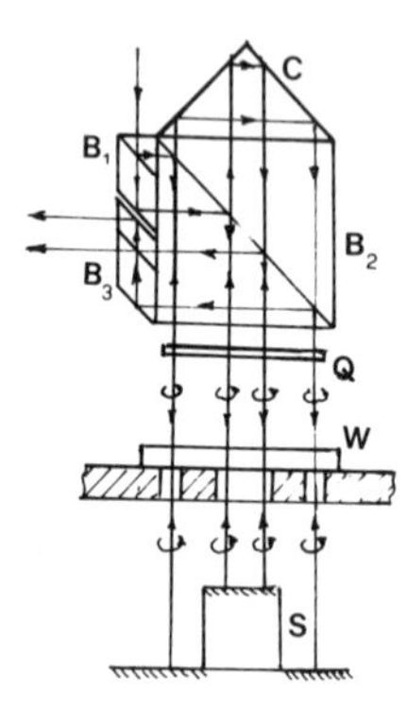

FIGURE 10. Optical diagram of the N.P.L. Michelson interferometric dilatometer. B_1, B_2, B_3: beamsplitters; C: corner cube; Q: quarter-wavelength plate; W: window in front of the vacuum chamber; S: sample. The laser beam enters into beamsplitter B_1.

beam. In this way the interference pattern is divided into two fields, which correspond to both ends of the sample (Fig. 9a). Lens L projects the image of the fringes on a plane where two detectors are placed, one on the sample and the other on the baseplate fields. On heating, both the sample and the sample support expand causing surfaces S and M_2 to move relative to surface M_1 at different rates. The difference of the fringe counts yields the net absolute expansion of the sample.

A more sophisticated setup is implemented by the N.P.L. interferometric dilatometer described by S.J. Bennett.[2,3] Two main features (Fig. 10) characterize this instrument: (a) it is a double-pass interferometer, in which the interfering beams are reflected two times from each of the sample end faces, this interferometer therefore yields double sensitivity as it measures twice the expansion; (b) no reference arm is needed since the doubly reflected beams are recombined at B_3.

The fast pulse method to measure thermal expansion led Miiller and Cezairliyan to resort to a *contactless interferometer.*[9] In this dilatometer (Fig. 11), the beamsplitter B_1 generates the reference and the measuring beams. The former, B_1–P_1 is made to bypass the specimen by means of pentaprism P_1, lenses L_1 and L_2 and mirror M_1, and is recombined, on beamsplitter B_2, with the measuring beam. The latter hits both faces of the sample and, between reflections, bypasses the specimen through the combined action of pentaprism P_2, mirror M_2, and beamsplitter B_2. The use of light circularly polarized by quarter wavelength plates Q_1 and Q_2 prevents reflected parasitic light from reaching the detector.

3.2.2. *Double Michelson Interferometer*

A typical example of double interferometer is described by Wolff and Eselun.[18] This setup also achieves contactless measurement of thermal expansion.

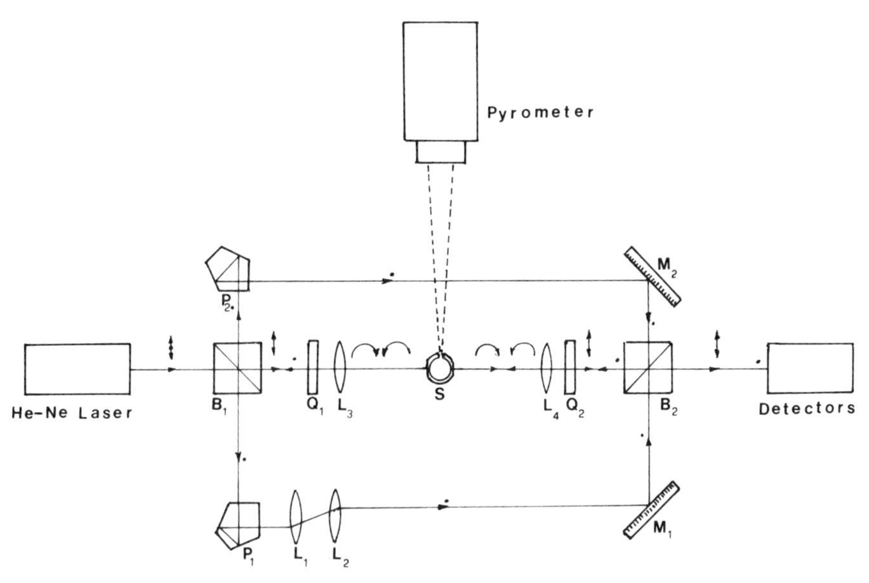

FIGURE 11. Müller and Cezairliyan pulse dilatometer. The laser light is polarized: double arrows mean that the polarization plane is parallel to the drawing sheet; dot is for the perpendicular direction; circular arrow means circular polarization.

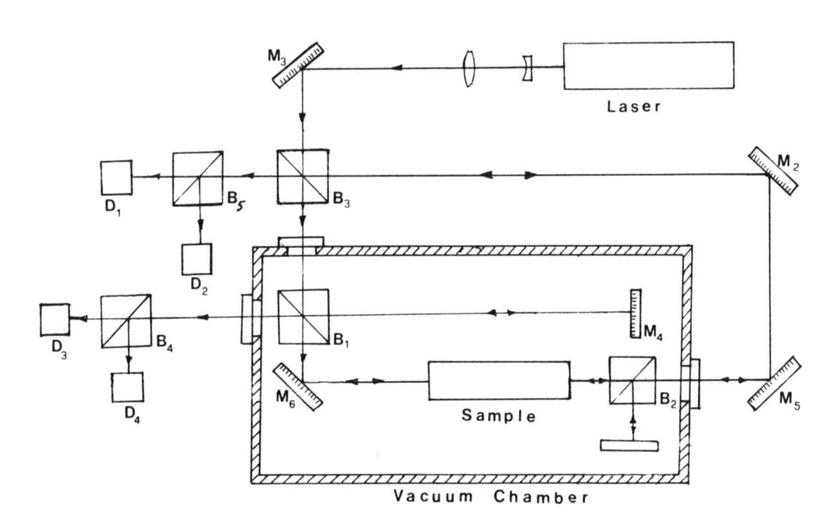

FIGURE 12. Wolff and Eselun contactless dilatometer. Beamsplitters B_3 and B_4 are polarizing and send beams in quadrature to pairs of detectors D_1 and D_2, D_3 and D_4.

Sample and interferometers are contained in a vacuum chamber (Fig. 12) to off-set optical drifts caused by instability of the air refractive index. Beamsplitter B_3 generates light beams for each interferometer. One of them consists of beamsplitter B_1, mirror M_6 and an end face of the specimen. The other is made with beamsplitter B_2, mirror M_4 and the other face of the sample.

To discriminate the motion sense and for fractional fringe measurement, each beam issued by an interferometer is split by polarizing beamsplitters into two beams polarized in quadrature, each sent to a detector. Counting, subtracting, and interpolation are performed electronically.

Another polarizing double interferometer of remarkable sensitivity has been described by Roblin and Souche.[13]

3.3. Polarization and Beam Modulation

It is common practice with dilatometric interferometers to use polarized laser light and quarterwave plates to generate circularly polarized light. In this way detectors in combination with appropriate analyzers generate, for each light beam, two signals conveying information on fringe number, fraction and motion sense. Appropriate electronics processes these signals.

Fractional fringe measurement can be made by light modulation resulting in a subnanometer sensitivity (1/1000 of a wavelength). Devices and analysis are presented by Eselun and Wolff[4] and Souche *et al.*[17]

4. THE EXPERIMENTAL CELL

The sample may be a hollow or a solid cylinder (Figs. 13a, 13b). When the material is rare, samples can be smaller, namely, three cylinders of equal height (within a fraction of a wavelength), located at the vertexes of an equilateral triangle, (c). Approximate dimensions are given in the figures. Configurations (a) and (c) are the most stable, the latter being the most difficult to machine. With configuration (b), which is less stable mechanically, but the easiest to build, the laser beam must be sent laterally to the specimen, which is easy if the zone is of the order of 10 mm. The contact Michelson interferometer requires the setup of Fig. 14.

The Fizeau interferometer (Fig. 13a) stands on a baseplate supported by a tube: both elements are made with low thermal conducting and very stable ceramic material. As stated before, the sample baseplate in the Michelson interferometer (Fig. 9) is specular, but not necessarily of the same material as the sample. It stands on a ceramic tube with the same characteristics as above.

Heat inputs into the supporting structure must be as symmetrical as possible so as to avoid the slightest tilting of the reflecting surfaces.

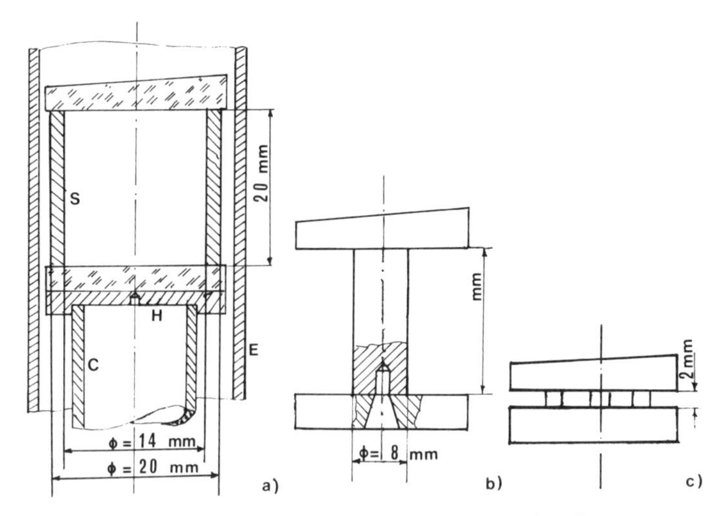

FIGURE 13. Types of experimental cells for Fizeau interferometer.

For high-temperature systems, where radiation pyrometers are used, blackbody conditions are created by drilling a hole (a few millimeters deep, with about 2 mm diameter) into the specimen (Fig. 13b; Fig. 14) or at least in the baseplate. The optical path for the pyrometer runs in the opposite direction as the interfering beams.

5. EXPERIMENTAL SYSTEM LAYOUT

The experimental setup consists of three parts: the interferometer, the thermostat which totally or partially surrounds the former and, most frequently, a high-vacuum system.

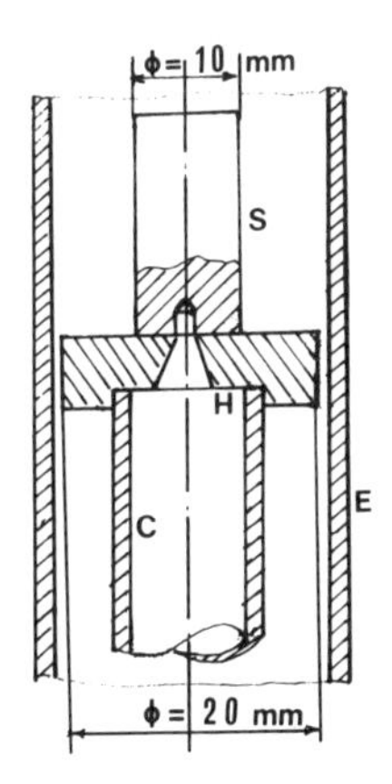

FIGURE 14. Blackbody simulator for Michelson interferometer.

The type of thermostat depends upon the temperature range in which experiments are carried out. Basically, they are cryostats, liquid baths, and furnaces. A common feature of this piece of equipment is that it provides a vertical optical passage a couple of centimeters wide to admit the laser beam into the experimental cell, where all elements are kept in place by gravity alone.

The vacuum system is a need in cryogenic measurements, or with materials which react with air, or with Michelson interferometers where turbulence disturbs the fringe counting. For these reasons, at least at high temperature, vacuum is recommended also in Fizeau interferometers. The evacuated space may only be a tube introduced into the thermostat. But this setup is viable only when the tube material is stable at an interferometric level in the thermostated environment. This is not the case at high temperature. Thus, the furnace must operate *in vacuo,* which requires compactness, clean materials, that do not release gases, provision for heat transmission by radiation, and nonetheless high temperature uniformity over the specimen. This condition may be favored by an equalizing screen (item E in Figs. 13 and 14).

A few details are given here for high-temperature measurements with Fizeau and Michelson interferometers.

5.1. High-Temperature Fizeau Interferometer

A schematic diagram of this interferometer is represented in Fig. 15. The optical system is basically the one of S.D. Preston.[12] A laser (A) beam is focused by lens B and produces fringes of equal inclination by the experimental cell E as described in Sections 2.1, 3.1. The latter can be any described in Section 4. The temperature is measured by radiation pyrometer N focused on a hole drilled on the baseplate, as in Fig. 13a or 13b. The pyrometer must be calibrated jointly with window L and mirror M, namely, in operating conditions.

The furnace heater is a tungsten helical ribbon suspended at its end on solid tungsten rods which also act as current leads (F). Heat is transmitted by radiation to the specimen through a tungsten equalizing screen. Thermal insulation is obtained by two tantalum screens and a roll of graphite paper H. The whole furnace is enclosed in two coaxial cylindrical shells. The inner one, a cylindrical cavity made with copper plate, absorbs heat by water flowing through a copper tube brazed on its surfaces. The outer shell performs a mechanical function. It is a stable stainless steel cylinder flanged, at its upper end, to the baseplate and, at its base, to the sample-holder. In this way the mechanical path of the sample cell attachment to the rest of the optical system runs at room temperature except for the tubular ceramic support which is located within the furnace. The optical paths into the furnace pass through sealed windows K and L and the outer shell is connected to a high vacuum system. The complete

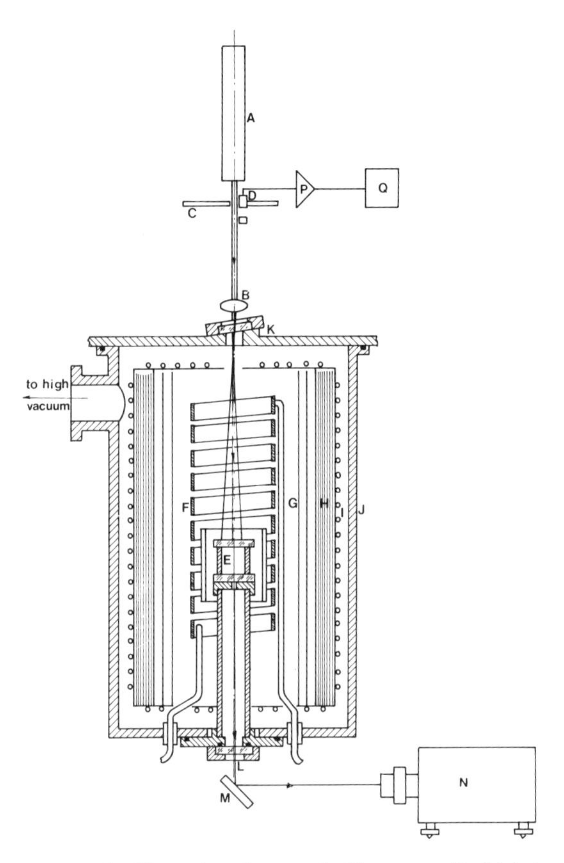

FIGURE 15. Layout of a Fizeau interferometric dilatometer for high temperature.

apparatus must rest on vibration-free stands. It is advisable to install it on a concrete block embedded in a sand bath.

5.2. High-Temperature Michelson Interferometer

Figure 16 presents a proposal for a Michelson interferometer. The furnace and the temperature measuring system are the same as in the previous apparatus. The only difference is the experimental cell, which is the one of Fig. 14.

The parallel light beam is generated by laser A and beam expander B. The interferometer is enclosed in an extension of the evacuated shell which contains the furnace. It consists of the beamsplitter D, the fixed mirror E, and the double mirror F created by the upper face of the sample and the polished face of the supporting plate. Light paths into and from the interferometer shell cross

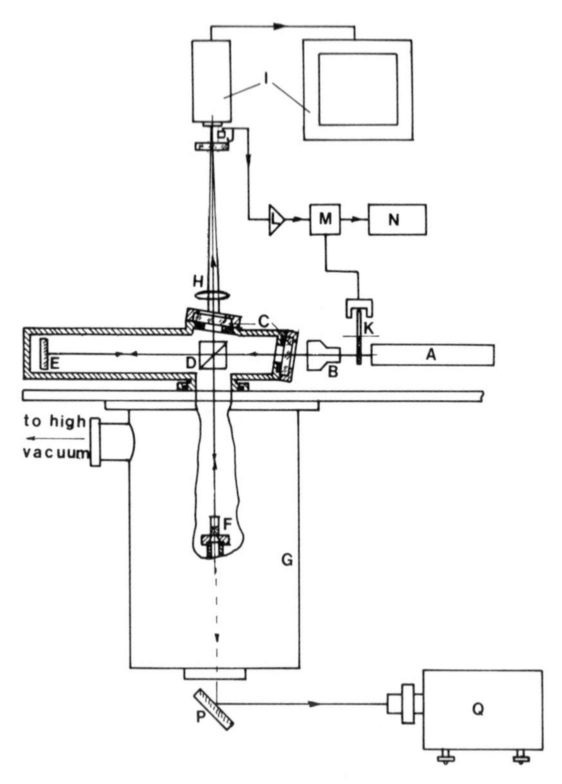

FIGURE 16. Layout of a Michelson interferometric dilatometer for high temperature.

window C obliquely to the optical axis. Lens H projects the fringe image through filter J. As orientation and spacing of the fringes may change during the experiment, it is advisable to monitor the fringe pattern by means of TV camera and screen I. Obviously, lens H, thin and nonachromatic, acts as the TV camera objective. Fringe number can be sensed either by the vidicon or by a small detector located close to the tube photocathode where the fringe image is collected. In this way, the detector position appears on the TV monitor so that it can be easily adjusted.

A common feature of the interferometers here described (only represented in Fig. 16) is the modulation of the laser beam by means of chopper K, which also generates, through a LED–photodiode pair, a square wave. The latter is used for synchronous demodulation (M) of the amplified (L) signal generated by the detector. Thus the furnace background radiation, which is drastically attenuated by a narrow-band interference filter placed in front of the detector, is completely eliminated. The filter prevents amplifier saturation at any furnace temperature, so that demodulation is feasible. Item N is a counter.

Temperature measurement is made by any means appropriate to the temperature range. A peculiar one (F.C. Nix, D. MacNair, Refs. 10 and 11) is integrated into the interferometers: by knowing the temperature dependence of the refractive index and the thermal expansion of the upper Fizeau plate, temperature is calculated from the number of interference fringes on both faces of that plate.

6. DATA ACQUISITION AND TREATMENT

Linear thermal expansion is the ratio between length variation from a reference length l_0 and the latter quantity. Reference length is the sample length at a reference temperature t_0 (0°C or room temperature: 20°C or 25°C). Thermal expansion is related to temperature variation with respect to reference temperature.

An interference dilatometer yields two sets of data: fringe counts and temperature.

The length variation is

$$\Delta l = 2n\lambda \tag{11}$$

where n is the fringe count, and λ is the laser wavelength *in vacuo*. As this wavelength must be stable and accurately known; lasers used in interferometry must be stabilized both in power and wavelength.

Thermal expansion is plotted as a function of temperature:

$$\Delta l/l_0 = f(t) \tag{12}$$

It is straightforward to fit the two sets of data, namely, expansion and temperature variation $\Delta t = t - t_0$, by the least squares method, to a polynomial. Usually the second degree is adequate except when a discontinuity of the function or its derivative is present due to a phase transformation. In this case, the transformation is detected either on the graph of $\Delta l/l_0$ vs. Δt, or on the attempts of data fitting.

Thus, we get

$$\Delta l/l_0 = a + b\Delta t + c(\Delta t)^2 \tag{13}$$

From this equation it is easy to get the coefficient of linear thermal expansion by differentiation

$$\alpha = b + 2c\Delta t \tag{14}$$

Automatic data acquisition and control systems are available to run thermal expansion measurements. They program the temperature steps, control the furnace, take fringe counts and temperature readings, all on-line, and at the end, treat data and print results.

REFERENCES

1. F.G. Barron, *Phys. Rev.* **105,** 418 (1957).
2. S.J. Bennett, *J. Phys. E* **10,** 525 (1976).
3. S.J. Bennett, in press.
4. S.A. Eselun and E.G. Wolff, *24th Int. Instr. Symp.,* ISA, Albuquerque, New Mexico (1978).
5. A.H.L. Fizeau, *Ann. Chem. Phys.* **2,** 143 (1864); **8,** 335 (1866).
6. D.F. Gibbons, *Phys. Rev.* **112,** (1958).
7. R.W. Meyerhoff and J.F. Smith, *J. Appl. Phys.* **33,** 219 (1962).
8. E. Mezzetti, *Ric. Sci.* **34,** 473 (1974).
9. A.P. Miiller and A. Cezairliyan, *Int. J. Thermophys.* **3,** 259 (1982).
10. F.C. Nix and D. McNair, *Phys. Rev.* **60,** 597 (1941).
11. F.C. Nix and D. McNair, *Rev. Sci. Instrum.* **12,** 66 (1941).
12. S.D. Preston, *High Temp. High Pressures* **12,** 441 (1980).
13. G. Roblin and Y. Souche, *Nov. Rev. Opt.* **5,** 287 (1974).
14. T. Rubin *et al., J. Am. Chem. Soc.* **76,** 5289 (1954).
15. G. Ruffino, *Acta IMEKO (Budapest)* **21-II,** 169 (1961).
16. G. Ruffino, *ASME Prog. Intrn. Res. on Thermod. and Transp. Prop.,* p. 185, Academic Press, New York (1962).
17. Y. Souche, R. Vergne, J.C. Cotillard, J.L. Porteseil, G. Roblin, and G. Nomarski, in press.
18. E.G. Wolff and S.A. Eselun, *Proc. SPIE* 192, *Interferometry,* p. 204 (1979).

VI

THERMAL RADIATIVE PROPERTY MEASUREMENT

19

Measurement of Thermal Radiation Properties of Materials

J.C. RICHMOND

1. INTRODUCTION

The thermal radiation properties of materials are reflectance, absorptance, transmittance, and emittance. They are called thermal radiation properties because they control the rate of heat transfer by radiation between noncontacting bodies at different temperatures, and between a body and its surroundings.

In radiant heat transfer energy is transferred in the form of photons, or electromagnetic waves (the two terms are different ways of describing the same form of energy). No heat transfer medium, solid, liquid or gas, is required for radiant heat transfer, hence it is the only mode of heat transfer to or from a satellite or space vehicle in the vacuum of space, or a thermally isolated sample in an evacuated chamber in the laboratory, that does not also involve the simultaneous transfer of mass in the form of matter.

Photons do not interact with any matter through which they pass, from the point of generation to the point of absorption, and hence generate no thermal gradient between these points. The generation and absorption of photons are statistical processes, and it is not possible to predict in advance where or when a photon will be generated, or in what direction it will be propagated from the point of generation, or where along its path it will be absorbed. The statistical parameters governing the generation and absorption of photons are well known, and the rate of emission of photons from a given volume of material, and the geometric and energy distribution of the photons, can be computed from the thermal radiation properties and temperature of the material. Also the fraction of the photons absorbed per unit path length in the material

---

J.C. RICHMOND • National Bureau of Standards, Washington, D.C. 20234.

can also be computed. It is thus possible to compute a mean path length for the photons, and a distribution curve for the lengths of the paths for all photons absorbed. No single photon generates a thermal gradient, but since photons are being emitted and absorbed throughout the material, a portion of the thermal gradient in the material is due to radiant heat transfer, but it is always smaller – and usually much smaller – than the portion due to conductive heat transfer.

The rate of radiant heat transfer between two objects in space is proportional to the difference in the fourth power of the temperature of the two bodies. Within a body, the rate of radiant heat transfer is an exponential function of the average temperature over the conduction path, with an exponent varying from about 3 for materials with an absorption coefficient that is constant with wavelength, to a much higher value, of as much as 10 in some cases, for some materials that are transparent at short wavelengths and strongly absorbing at longer wavelengths.

Radiant heat transfer is the only mode of heat transfer in a vacuum and becomes the dominant mode of heat transfer between noncontacting solids at high temperatures, such as exist in many industrial furnaces.

The nomenclature used in this chapter is given in the Appendix, and follows in general that in the International Lighting Vocabulary.[1]

2. THERMAL RADIATION PROPERTIES

2.1. Interrelation of Thermal Radiation Properties

All matter is continually emitting energy in the form of electromagnetic waves or photons, as a result of the thermal vibration of the particles of which it is composed – electrons, ions, atoms, and molecules. This process is called thermal radiation, and the energy so emitted is called thermal radiant energy. The rate of emission and the spectral and geometric distribution of the emitted energy are determined by the temperature of the material and its thermal radiation properties.

The terms "photon" and "electromagnetic wave" describe aspects of radiant energy on a micro and macro scale, respectively.

Each solid body is not only constantly emitting radiant energy, but is also constantly irradiated by radiant energy from its surroundings, some of which is absorbed. The difference between the rates of emission and absorption is the net rate of heat transfer to or from the body by radiation. The properties that control these rates are called thermal radiation properties.

When radiant flux is incident on a body part of it will be reflected, part will be absorbed, and the remainder will be transmitted. Nothing else can happen to it. Hence we can write

$$\Phi_i = \Phi_r + \Phi_a + \Phi_t \tag{1}$$

where Φ_i is the incident flux, Φ_r is the reflected flux, Φ_a is the absorbed flux, and Φ_t is the transmitted flux. If we now divide both sides of equation (1) by Φ_i, we get

$$1 = \Phi_r/\Phi_i + \Phi_a/\Phi_i + \Phi_t/\Phi_i \tag{2}$$

Φ_r/Φ_i is reflectance, ρ, Φ_a/Φ_i is absorptance, α, and Φ_t/Φ_i is transmittance, τ. Equation (2) thus becomes

$$1 = \rho + \alpha + \tau \tag{3}$$

For an opaque material, $\tau = 0$, and equation (3) becomes

$$\rho + \alpha = 1 \qquad (\tau = 0) \tag{4}$$

Kirchhoff's law* states that absorptance equals emittance, and equation (3) can be written as

$$\rho + \epsilon + \tau = 1 \tag{5}$$

and equation (4) as

$$\rho + \epsilon = 1 \qquad (\tau = 0) \tag{6}$$

For an opaque body the thermal radiation properties are thus completely described by either the reflectance or the emittance.

There are restrictions that limit the application of equations (1)–(6). For equation (2) to be valid the Φ_i's must be identical quantitatively, geometrically, and spectrally. For equations (3) and (4) to be valid the requirement that the incident flux for measurement of ρ, α, and τ be quantitatively identical can be relaxed, since thermal radiation properties of almost all materials are independent of the irradiance at all normal levels of irradiance. Nonlinear optical effects are usually restricted to the extremely high irradiances produced by high-powered Q-switched lasers. Photochromic glass used in sunglasses is an exception. Its transmittance varies with the irradiance at the levels normally encountered in an outdoor environment. For equation (5) and (6) to be valid, the incident irradiance for the reflectance and transmittance must have the spectral distribution of blackbody radiation at the temperature of the sample, and the solid angle of the incident beam for reflectance and transmittance must be identical to the solid angle of collection for emittance, and all of the reflected and transmitted flux must be collected for measurement. For spectral properties, the spectral bandwidth of the incident flux for the reflectance and transmittance measurements must be identical to the spectral bandwidth of the monochromator

*A derivation of Kirchhoff's law is given in Ref. 2.

used for measuring the emitted flux for the emittance determination. Within the narrow wavelength range involved in spectral measurements, the spectral distribution of almost any thermal source is essentially equivalent to a blackbody distribution at almost any temperature.

2.2. Blackbody Radiation

A blackbody radiator at a known temperature is the primary standard for all radiometric quantities. A blackbody absorbs all radiant energy incident upon it, and emits the maximum spectral concentration of radiant flux at all wavelengths for any body at its temperature. The amount and the spectral and geometric distribution of the flux emitted by a blackbody radiator obey well-known laws of physics. These laws are expressed as the Planck radiation equation, the Stefan–Boltzmann equation, and Lambert's cosine law.

The Planck equation relates the spectral radiance or exitance of a blackbody radiator to its temperature. For spectral radiance it is

$$L_{b,\lambda}(\lambda, T) = c_{1L} \cdot \lambda^{-5} \cdot [\exp(c_2/\lambda T) - 1]^{-1} \tag{7}$$

where $L_{b,\lambda}(\lambda, T)$ is the spectral radiance at wavelength λ in m of a blackbody radiator at temperature T in K, c_{1L} is the first radiation constant for radiance, 1.191062×10^{-16} W m^2 sr^{-1}, and c_2 is the second radiation constant, 1.438786×10^{-2} m K. For wavelength in m and temperature in K, the units of $L_{b,\lambda}(\lambda, T)$ are W m^{-2} sr^{-1} m^{-1}. The m^{-2} is per area of square meter, and the m^{-1} is per meter of wavelength. For spectral exitance it is

$$M_{b,\lambda}(\lambda, T) = c_1 \cdot \lambda^{-5} \cdot [\exp(c_2/\lambda T) - 1]^{-1} \tag{8}$$

where $M_{b,\lambda}$ is the spectral exitance at wavelength λ in m of a blackbody at temperature T in K, and c_1 is the first radiation constant for exitance. $c_1 = 3.741832 \times 10^{-16}$ W m^2. The units of $M_{\lambda,b}(\lambda, T)$ are W m^{-2} m^{-1}.

The Stefan–Boltzmann equation relates the total radiance or exitance of a blackbody radiator to its temperature. For total radiance it is

$$L_{t,b}(T) = \sigma_L T^4 \tag{9}$$

where $L_{t,b}$ is total radiance of a blackbody radiator at temperature T in K, and σ_L is the Stefan–Boltzmann constant for radiance, 1.80149×10^{-8} W m^{-2} sr^{-1} K^{-4}. For total exitance, $M_{t,b}(T)$, it becomes

$$M_{t,b}(T) = \sigma T^4 \tag{10}$$

where σ is 5.67032×10^{-8} W m^{-2} K^{-4}, and $M_{t,b}$ is in W m^{-2}.

Lambert's cosine law states that the intensity of an element of area of a blackbody radiator is maximum in a direction normal to the surface, and at other directions is reduced by the cosine of the angle between the direction and the normal to the surface. Expressed mathematically

$$I_\lambda(\theta, \phi; T) = I_\lambda(0; T) \cos\theta \tag{11}$$

where $I_\lambda(\theta, \phi; T)$ is the spectral intensity, in direction θ, ϕ to the surface, of an element of the area of a blackbody radiator at temperature T in K, and $I_\lambda(0; T)$ is its spectral intensity in a direction normal to the surface. The same equation applies for total intensity, $I_t(\theta, \phi; T)$, where the total intensity normal to the surface is $I_t(0; T)$. The law can be restated as "the radiance of a blackbody radiator is isotropic over a hemisphere."

From Lambert's cosine law it is evident that for a blackbody radiator $M = \pi L$, since

$$M = \int_0^{2\pi} \int_0^{\pi/2} L \cos\theta \cdot \sin\theta \cdot d\theta \cdot d\phi \tag{12}$$

No real material is a true blackbody radiator, but a completely enclosed cavity with opaque walls at a constant uniform temperature behaves like a true blackbody radiator, regardless of the emittance of its walls. Any photon of radiant energy emitted into the cavity will be multiply reflected until it is absorbed, since it cannot escape through the opaque walls, and the cavity has an absorptance of one, and hence by Kirchhoff's law, the emittance must also be one. Since, by definition, the walls are at a constant uniform temperature, they must be in thermal equilibrium, and the rates of absorption and emission must be equal for all areas of the wall. The exitent radiance, the reflected plus emitted, will be the same at every point on the wall and in every direction from each point, even if the emittance of the cavity varies over its surface. The radiance will also be the maximum possible emitted radiance for any body at that temperature, at all wavelengths, hence it is true blackbody radiance.

The radiance of a small portion of the opaque wall of the isothermal cavity can be observed through a small hole in that wall without significantly disturbing the equilibrium that determines its spectral distribution and magnitude. Ideally the magnitude and spectral distribution of the radiance of each exitent ray from the internal wall of the cavity is a function only of the equilibrium temperature.

Most high-temperature laboratory blackbody furnaces are designed to have outer surfaces that are not significantly above room temperature. Such blackbodies must of necessity have thick walls, with a large thermal gradient from the hot interior wall of the cavity to the cool outer wall of the furnace. If the outer end of the hole, through which the interior wall of the cavity is viewed, is considered to be the aperture of the blackbody radiator, it will be filled with rays originating from the interior wall of the cavity, all of the same radiance (or appear to be a Lambertian radiator) over a range of directions that comprise a relatively small solid angle, the magnitude of which can be computed from the geometry of the hole. Over this restricted solid angle the intensity of the aperture, the plane of which is normal to the axis of the hole, will vary as the cosine of the angle between the direction of viewing and the normal to the plane of the aperture. Because of thermal gradients along the walls of the hole through which the internal wall of the cavity is viewed, rays outside the solid angle of isoradiance rays will decrease in radiance as the angle between the direction of viewing and the normal to the plane of the aperture increases. Thus practical high-temperature laboratory sources can be good blackbody simulators only over limited solid angles. However, it is relatively easy to build a blackbody cavity with effective directional emittance normal to the plane of the aperture of 0.995, and with careful design emittances in excess of 0.999 have been attained.

The spectral distribution of blackbody flux may be computed from equation (7) or (8) above. The relative spectral distribution is a function of λT, as can be shown by multiplying both sides of the equation by T^{-5}, which gives, for equation (7),

$$L_{b,\lambda}(\lambda, T)T^{-5} = c_{1L}(\lambda T)^{-5}\,[\exp(c_2/\lambda T) - 1]^{-1} \tag{13}$$

Equation (8) or (13) can be differentiated with respect to T to give the following expression for the fractional increase in flux (radiance or exitance) produced by a small fractional increase in temperature. For equation (13)

$$\frac{dL_{b,\lambda}}{L_{b,\lambda}(\lambda, T)} \cdot \frac{T}{dT} = \frac{c_2}{\lambda T} \cdot \frac{\exp(c_2/\lambda T)}{[\exp(c_2/\lambda T)]^{-1}} \tag{14}$$

The symbol R is used for the term on the left of equation (14). The same symbol is also used for reflectance factor. Any confusion as to which meaning is meant should be easily resolved from the context in which it is used. At λT of 2898 μm K, R has the value of 5. For values of λT less than 3000 μm K, R can be approximated by the following equations:

$$R = 5 \times 2898(\lambda T)^{-1} \tag{15}$$

$$R = c_2/\lambda T \tag{16}$$

with an error of less than 1%. Equation (15) is more accurate at values of λT near 3000, and equation (16) is more accurate as λT approaches zero.

Figure 1 is a plot of the relative blackbody spectral distribution as a

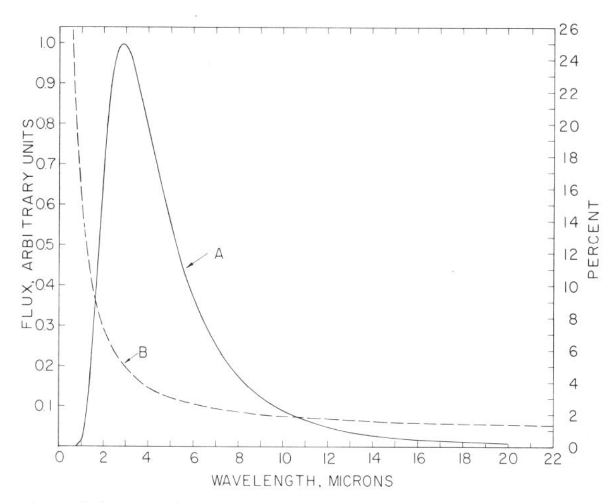

FIGURE 1. A plot of the relative spectral distribution of blackbody flux, solid line A, and the ratio R, $c_2/\lambda T\,(\exp c_2/\lambda t) \cdot [(\exp c_2/\lambda T) - 1]^{-1}$, dashed line B, plotted as a function of λT in cm K.

TABLE 1
Allowable Temperature Difference between Sample and Blackbody Reference for Uncertainty of 0.5% in Emittance

λ	Temperature (K)								
(μm)	300	400	500	750	1000	1250	1500	2000	3000
0.5	0.016	0.028	0.043	0.098	0.174	0.271	0.391	0.695	1.564
1.0	0.031	0.056	0.087	0.195	0.348	0.543	0.782	1.389	3.102
2.0	0.063	0.111	0.174	0.391	0.695	1.083	1.551	2.704	5.687
3.0	0.094	0.167	0.261	0.585	1.034	1.594	2.249	3.791	7.485
4.0	0.125	0.222	0.347	0.775	1.352	2.050	2.843	4.640	8.739
5.0	0.156	0.278	0.433	0.956	1.640	2.443	3.335	5.302	9.645
6.0	0.188	0.333	0.517	1.125	1.896	2.780	3.743	5.826	10.33
7.0	0.219	0.387	0.598	1.280	2.121	3.066	4.082	6.249	10.86
8.0	0.250	0.440	0.676	1.422	2.320	3.135	4.369	6.596	11.28
9.0	0.280	0.491	0.750	1.551	2.495	3.527	4.613	6.885	11.63
10.0	0.310	0.541	0.820	1.668	2.651	3.712	4.823	7.130	11.91
12.0	0.368	0.634	0.948	1.871	2.913	4.019	5.164	7.522	12.36
14.0	0.424	0.719	1.061	2.041	3.124	4.261	5.429	7.820	12.70
16.0	0.475	0.796	1.150	2.185	3.297	4.456	5.641	8.054	12.96
18.0	0.524	0.865	1.248	2.307	3.443	4.617	5.814	8.243	13.17
20.0	0.569	0.928	1.325	2.411	3.565	4.752	5.958	8.399	13.34
22.5	0.620	0.998	1.411	2.523	3.694	4.892	6.106	8.559	13.51
25.0	0.667	1.060	1.485	2.618	3.801	5.009	6.229	8.690	13.65
27.5	0.710	1.116	1.550	2.700	3.893	5.107	6.332	8.799	13.76
30.0	0.749	1.165	1.608	2.770	3.972	5.191	6.419	8.891	13.86
∞	1.500	2.000	2.500	3.750	5.000	6.250	7.500	10.00	15.00

function of λT, normalized to a value of 1.0 at $\lambda T = 2898\,\mu\text{m K}$, solid line, and the ratio R, as a function of λT, dashed line.

The ratio R can also be used as an exponent, to show how the exitance or radiance changes with temperature, by use of the equation

$$L(\lambda, T) \propto T^R, \qquad M(\lambda, T) \propto T^R \tag{17}$$

The equations hold only for values $\Delta T/T$ small enough that the value λT does not change significantly as T changes.

Table 1 shows the temperature tolerance for $\pm 0.5\%$ uncertainty in radiometric spectral emittance measurements. The values in the table are the temperature change, in K, required to change the spectral radiance of a blackbody radiator by 0.5% at the indicated combination of temperature and wavelength. Note that extremely close control of the temperature difference between sample and blackbody reference is required for any combination of wavelength and temperature that gives a low value of λT, as can be seen in Fig. 1, where the curve for R rises steeply as λT decreases below $2898\,\mu\text{m K}$.

It should be emphasized that for radiometric evaluation of emittance, either spectral or total, the critical temperature control problem is to keep the sample and reference blackbody as nearly as possible at the same temperature. Emittance, and particularly spectral emittance, the ratio of the spectral radiance of the sample to that of a blackbody radiator at the same, temperature, is not a strong function of temperature, and errors of up to at least 25 K or 5%, whichever is smaller, in the absolute temperature measurement usually do not result in a detectable change in the measured spectral emittance unless such temperature differences result in a phase change in the sample, or a shift in an absorption edge in the sample. The spectral radiance of both the sample and blackbody radiator changes rapidly with temperature, as shown in Fig. 1, especially at low values of λT, but the ratio remains nearly constant.

The spectral range and distribution of blackbody flux (exitance or radiance) is shown in Table 2, which gives the wavelengths below which the indicated fraction of the total exitent flux is emitted, first in terms of λT, and then for selected temperatures from 300 to 4000 K. Wavelengths for any temperature not included in the table may be obtained by dividing each λT value by the desired temperature. The fractions vary by increments of 0.01 from 0.005 to 0.995. These wavelengths are specifically selected for use in the 100 selected ordinate method of computing total emittance from spectral emittance, as described below.

Total (spectrally total) emittance is the weighted average of spectral emittance, with the relative spectral distribution of blackbody flux at the temperature of the sample as the weighting function. Expressed mathematically,

TABLE 2

Fraction of Total Flux Emitted at Wavelengths Below the Indicated Wavelengths, for Selected Temperatures

No.	λT $\mu m \cdot K$	$\frac{M(0-\lambda)}{M(0-\infty)}$	K 300 °F 80	350 170	400 260	450 350	500 440	550 530	600 620	650 710	700 800	750 890	800 980	850 1070	900 1160
			μm												
1	1323.36	.005	4.411	3.781	3.308	2.941	2.647	2.406	2.206	2.036	1.891	1.764	1.654	1.557	1.470
2	1534.44	.015	5.115	4.384	3.836	3.410	3.069	2.790	2.557	2.361	2.192	2.046	1.918	1.805	1.705
3	1662.56	.025	5.542	4.750	4.156	3.695	3.325	3.023	2.771	2.558	2.375	2.217	2.078	1.956	1.847
4	1762.01	.035	5.873	5.034	4.405	3.916	3.524	3.204	2.937	2.711	2.517	2.349	2.203	2.073	1.958
5	1846.20	.045	6.154	5.275	4.616	4.103	3.692	3.357	3.077	2.840	2.637	2.462	2.308	2.172	2.051
6	1920.81	.055	6.403	5.488	4.802	4.268	3.842	3.492	3.201	2.955	2.744	2.561	2.401	2.260	2.134
7	1988.79	.065	6.629	5.682	4.972	4.420	3.978	3.616	3.315	3.060	2.841	2.652	2.486	2.340	2.210
8	2051.91	.075	6.840	5.863	5.130	4.560	4.104	3.731	3.420	3.157	2.931	2.736	2.565	2.414	2.280
9	2111.33	.085	7.038	6.032	5.278	4.692	4.223	3.839	3.519	3.248	3.016	2.815	2.639	2.484	2.346
10	2167.85	.095	7.226	6.194	5.420	4.817	4.336	3.942	3.613	3.335	3.097	2.890	2.710	2.550	2.409
11	2222.05	.105	7.407	6.349	5.555	4.938	4.444	4.040	3.703	3.419	3.174	2.963	2.778	2.614	2.469
12	2274.32	.115	7.581	6.498	5.684	5.056	4.549	4.135	3.791	3.499	3.249	3.032	2.843	2.676	2.527
13	2325.05	.125	7.750	6.643	5.813	5.167	4.650	4.227	3.875	3.577	3.322	3.100	2.906	2.735	2.583
14	2374.48	.135	7.915	6.784	5.936	5.277	4.749	4.317	3.957	3.653	3.392	3.166	2.968	2.794	2.638
15	2422.82	.145	8.076	6.922	6.057	5.384	4.846	4.405	4.038	3.727	3.461	3.230	3.029	2.850	2.692
16	2470.26	.155	8.234	7.058	6.176	5.489	4.941	4.491	4.117	3.800	3.529	3.294	3.088	2.906	2.745
17	2516.95	.165	8.390	7.191	6.292	5.593	5.034	4.576	4.195	3.872	3.596	3.356	3.146	2.961	2.979
18	2563.00	.175	8.543	7.323	6.408	5.696	5.126	4.660	4.272	3.943	3.661	3.417	3.204	3.015	2.848
19	2608.54	.185	8.695	7.453	6.521	5.797	5.217	4.743	4.348	4.013	3.726	3.478	3.261	3.069	2.898
20	2653.65	.195	8.846	7.582	6.634	5.897	5.307	4.825	4.423	4.083	3.791	3.538	3.317	3.122	2.948
21	2698.42	.205	8.995	7.710	6.746	5.996	5.397	4.906	4.497	4.151	3.855	3.598	3.373	3.175	2.998
22	2742.93	.215	9.143	7.837	6.857	6.095	5.486	4.987	4.572	4.220	3.918	3.657	3.429	3.227	3.048
23	2787.24	.225	9.291	7.964	6.968	6.194	5.574	5.068	4.645	4.288	3.982	3.716	3.484	3.279	3.097
24	2831.42	.235	9.438	8.090	7.079	6.292	5.663	5.148	4.719	4.356	4.045	3.775	3.539	3.331	3.146
25	2875.52	.245	9.585	8.216	7.189	6.390	5.751	5.228	4.793	4.424	4.108	3.834	3.594	3.383	3.195

TABLE 2 (Continued)

No.	λT $\mu m \cdot K$	$\frac{M(0-\lambda)}{M(0-\infty)}$	μm K 300 °F 80	350 170	400 260	450 350	500 440	550 530	600 620	650 710	700 800	750 890	800 980	850 1070	900 1160
26	2919.59	.255	9.732	8.342	7.299	6.488	5.839	5.308	4.866	4.492	4.171	3.893	3.649	3.435	3.244
27	2963.68	.265	9.879	8.468	7.409	6.586	5.927	5.389	4.939	4.560	4.234	3.952	3.705	3.487	3.293
28	3007.85	.275	10.026	8.594	7.520	6.684	6.016	5.469	5.013	4.627	4.297	4.010	3.760	3.539	3.342
29	3052.13	.285	10.174	8.720	7.630	6.783	6.104	5.549	5.087	4.696	4.360	4.070	3.815	3.591	3.391
30	3096.56	.295	10.322	8.847	7.741	6.881	6.193	5.630	5.161	4.764	4.424	4.129	3.871	3.643	3.441
31	3141.20	.305	10.471	8.975	7.853	6.980	6.282	5.711	5.235	4.833	4.487	4.188	3.926	3.696	3.490
32	3186.07	.315	10.620	9.103	7.965	7.080	6.372	5.793	5.310	4.902	4.552	4.248	3.983	3.748	3.540
33	3231.22	.325	10.771	9.232	8.078	7.180	6.462	5.875	5.385	4.971	4.616	4.308	4.039	3.801	3.590
34	3276.68	.335	10.922	9.362	8.192	7.282	6.553	5.958	5.461	5.041	4.681	4.369	4.096	3.855	3.641
35	3322.49	.345	11.075	9.493	8.306	7.383	6.645	6.041	5.537	5.112	4.746	4.430	4.153	3.909	3.692
36	3368.69	.355	11.229	9.625	8.422	7.486	6.737	6.125	5.614	5.183	4.812	4.492	4.211	3.963	3.743
37	3415.33	.365	11.384	9.758	8.538	7.590	6.831	6.210	5.692	5.254	4.879	4.554	4.269	4.018	3.795
38	3462.42	.375	11.541	9.893	8.656	7.694	6.925	6.295	5.771	5.327	4.946	4.617	4.328	4.073	3.847
39	3510.02	.385	11.700	10.029	8.775	7.800	7.020	6.382	5.850	5.400	5.014	4.680	4.388	4.129	3.900
40	3558.17	.395	11.861	10.166	8.895	7.907	7.116	6.469	5.930	5.474	5.083	4.744	4.448	4.186	3.954
41	3606.89	.405	12.023	10.305	9.017	8.015	7.214	6.558	6.011	5.549	5.153	4.809	4.509	4.243	4.008
42	3656.24	.415	12.187	10.446	9.141	8.125	7.312	6.648	6.094	5.625	5.223	4.875	4.570	4.301	4.062
43	3706.26	.425	12.354	10.589	9.266	8.236	7.413	6.739	6.177	5.702	5.295	4.942	4.633	4.360	4.118
44	3756.98	.435	12.523	10.734	9.392	8.349	7.514	6.831	6.262	5.780	5.367	5.009	4.696	4.420	4.174
45	3808.46	.445	12.695	10.881	9.521	8.463	7.617	6.924	6.347	5.859	5.441	5.078	4.761	4.481	4.232
46	3860.74	.455	12.869	11.031	9.652	8.579	7.721	7.020	6.435	5.940	5.515	5.148	4.826	4.542	4.290
47	3913.86	.465	13.046	11.182	9.785	8.697	7.828	7.116	6.523	6.021	5.591	5.218	4.892	4.605	4.349
48	3967.90	.475	13.226	11.337	9.920	8.818	7.936	7.214	6.613	6.104	5.668	5.291	4.960	4.668	4.409
49	4022.88	.485	13.410	11.494	10.057	8.940	8.046	7.314	6.705	6.189	5.747	5.364	5.029	4.733	4.470
50	4078.88	.495	13.596	11.654	10.197	9.064	8.158	7.416	6.798	6.275	5.827	5.439	5.099	4.799	4.532
51	4135.95	.505	13.786	11.817	10.340	9.191	8.272	7.520	6.893	6.363	5.908	5.515	5.170	4.866	4.596
52	4194.15	.515	13.980	11.983	10.485	9.320	8.388	7.626	6.990	6.453	5.992	5.592	5.243	4.934	4.660
53	4253.54	.525	14.178	12.153	10.634	9.452	8.507	7.734	7.089	6.544	6.076	5.671	5.317	5.004	4.726
54	4314.21	.535	14.381	12.326	10.786	9.587	8.628	7.844	7.190	6.637	6.163	5.752	5.393	5.076	4.794
55	4376.24	.545	14.587	12.504	10.941	9.725	8.752	7.957	7.294	6.733	6.252	5.835	5.470	5.149	4.862

56	4439.66	.555	14.799	12.685	11.099	9.866	8.879	8.072	7.399	6.830	6.342	5.920	5.550	5.223	4.933
57	4504.61	.565	15.015	12.870	11.262	10.010	9.009	8.190	7.508	6.930	6.435	6.006	5.631	5.300	5.005
58	4571.15	.575	15.237	13.060	11.428	10.158	9.142	8.311	7.619	7.033	6.530	6.095	5.714	5.378	5.079
59	4639.40	.585	15.465	13.255	11.598	10.310	8.279	8.435	7.732	7.138	6.628	6.186	5.799	5.458	5.155
60	4709.46	.595	15.698	13.456	11.774	10.465	9.419	8.563	7.849	7.245	6.728	6.279	5.887	5.541	5.233

No.	λT $\mu m \cdot K$	$\frac{M(0-\lambda)}{M(0-\infty)}$	K 300 °F 80	350 170	400 260	450 350	500 440	550 530	600 620	650 710	700 800	750 890	800 980	850 1160	900 1250
			μm												
61	4781.43	0.605	15.938	13.661	11.954	10.625	9.563	8.694	7.969	7.356	6.831	6.375	5.977	5.625	5.313
62	4855.45	.615	16.185	13.873	12.139	10.790	9.711	8.828	8.092	7.470	6.936	6.474	6.069	5.712	5.394
63	4931.65	.625	16.439	14.090	12.329	10.959	9.863	8.967	8.219	7.587	7.045	6.576	6.165	5.802	5.480
64	5010.17	.635	16.701	14.315	12.525	11.134	10.020	9.109	8.350	7.708	7.157	6.680	6.263	5.894	5.567
65	5091.18	.645	16.971	14.546	12.728	11.314	10.182	9.257	8.485	7.833	7.273	6.788	6.364	5.990	5.657
66	5174.85	.655	17.250	14.785	12.937	11.500	10.350	9.409	8.625	7.961	7.393	6.900	6.469	6.088	5.750
67	5261.35	.665	17.538	15.032	13.153	11.692	10.523	9.566	8.769	8.094	7.516	7.015	6.577	6.190	5.846
68	5350.19	.675	17.834	15.286	13.375	11.889	10.700	9.728	8.917	8.231	7.643	7.134	6.688	6.294	5.945
69	5443.77	.685	18.146	15.554	13.609	12.097	10.888	9.898	9.073	8.375	7.777	7.258	6.805	6.404	6.049
70	5540.18	.695	18.467	15.829	13.850	12.312	11.080	10.073	9.234	8.523	7.915	7.387	6.925	6.518	6.156
71	5640.40	.705	18.801	16.115	14.101	12.534	11.281	10.255	9.401	8.678	8.058	7.521	7.050	6.636	6.267
72	5744.77	.715	19.149	16.414	14.362	12.766	11.490	10.445	9.575	8.838	8.207	7.660	7.181	6.759	6.383
73	5853.62	.725	19.512	16.725	14.634	13.008	11.707	10.643	9.756	9.006	8.362	7.805	7.317	6.887	6.504
74	5967.38	.735	19.891	17.050	14.918	13.261	11.935	10.850	9.946	9.181	8.525	7.957	7.459	7.020	6.630
75	6086.48	.745	20.288	17.390	15.216	13.526	12.173	11.066	10.144	9.364	8.695	8.115	7.608	7.161	6.763
76	6211.41	.755	20.705	17.747	15.529	13.803	12.423	11.293	10.352	9.556	8.873	8.282	7.764	7.308	6.902
77	6342.76	.765	21.143	18.122	15.857	14.095	12.686	11.532	10.571	9.758	9.061	8.457	7.928	7.462	7.048
78	6481.20	.775	21.604	18.518	16.203	14.403	12.962	11.784	10.802	9.971	9.259	8.642	8.102	7.625	7.201
79	6627.48	.785	22.092	18.936	16.569	14.728	13.255	12.050	11.046	10.196	9.468	8.837	8.284	7.797	7.364
80	6782.48	.795	22.608	19.379	16.956	15.072	13.565	12.332	11.304	10.435	9.689	9.043	8.478	7.979	7.536
81	6947.24	.805	23.157	19.849	17.368	15.438	13.894	12.631	11.579	10.688	9.925	9.263	8.684	8.173	7.719
82	7122.96	.815	23.743	20.351	17.807	15.829	14.246	12.951	11.872	10.958	10.176	9.497	8.904	8.380	7.914
83	7311.08	.825	24.370	20.889	18.278	16.247	14.622	13.293	12.185	11.248	10.444	9.748	9.139	8.601	8.123
84	7513.32	.835	25.044	21.467	18.783	16.696	15.027	13.661	12.522	11.559	10.733	10.018	9.392	8.839	8.348
85	7731.75	.845	25.772	22.091	19.329	17.182	15.464	14.058	12.886	11.895	11.045	10.309	9.665	9.096	8.591

TABLE 2 (Continued)

No.	λT μm·K	$\frac{M(0-\lambda)}{M(0-\infty)}$	K 300 °F 80	350 170	400 260	450 350	500 440	550 530	600 620	650 710	700 800	750 890	800 980	850 1160	900 1250
			μm												
86	7968.92	.855	26.563	22.768	19.922	17.709	15.938	14.489	13.282	12.260	11.384	10.625	9.961	9.375	8.854
87	8228.00	.865	27.427	23.509	20.570	18.284	16.456	14.960	13.713	12.658	11.754	10.971	10.285	9.680	9.142
88	8512.94	.875	28.376	24.323	21.282	18.918	17.026	15.478	14.188	13.097	12.161	11.351	10.641	10.015	9.459
89	8828.87	.885	29.430	25.225	22.072	19.620	17.658	16.052	14.715	13.583	12.613	11.772	11.036	10.387	9.810
90	9182.41	.895	30.608	26.235	22.956	20.405	18.365	16.695	15.304	14.127	13.118	12.243	11.478	10.803	10.203
91	9582.50	.905	31.942	27.379	23.956	21.294	19.165	17.423	15.971	14.742	13.689	12.777	11.978	11.273	10.647
92	10041.40	.915	33.471	28.690	25.104	22.314	20.083	18.257	16.736	15.448	14.345	13.389	12.552	11.813	11.157
93	10576.70	.925	35.256	30.219	26.442	23.504	21.153	19.230	17.628	16.272	15.110	14.102	13.221	12.443	11.752
94	11214.36	.935	37.381	32.041	28.036	24.921	22.429	20.390	18.691	17.253	16.021	14.952	14.018	13.193	12.460
95	11995.43	.945	39.985	34.273	29.989	26.657	23.991	21.810	19.992	18.455	17.136	15.994	14.994	14.112	13.328
96	12989.52	.955	43.298	37.113	32.474	28.866	25.979	23.617	21.649	19.984	18.556	17.319	16.237	15.282	14.433
97	14326.94	.965	47.756	40.934	35.817	31.838	28.654	26.049	23.878	22.041	20.467	19.103	17.909	16.855	15.919
98	16294.90	.975	54.316	46.557	40.737	36.211	32.590	29.627	27.158	25.069	23.278	21.727	20.369	19.170	18.105
99	19724.33	.985	65.748	56.355	49.311	43.832	39.449	35.862	32.874	30.345	28.178	26.299	24.655	23.205	21.916
100	29367.00	.995	97.890	83.906	73.418	65.260	58.734	53.395	48.945	45.180	41.953	39.156	36.709	34.509	32.630

No.	K 950 °F 1250	1000 1340	1100 1520	1200 1700	1300 1880	1400 2060	1500 2240	1600 2420	1700 2600	1800 2780	1900 2960	2000 3140	2500 4040	3000 4940	4000 6740
	μm														
1	1.393	1.323	1.203	1.103	1.018	0.945	0.882	0.827	0.778	0.735	0.697	0.662	0.529	0.441	0.331
2	1.615	1.534	1.395	1.279	1.180	1.096	1.023	.959	.903	.852	.808	.767	.614	.511	.384
3	1.750	1.663	1.511	1.385	1.279	1.188	1.108	1.039	.978	.924	.875	.831	.665	.554	.416
4	1.855	1.762	1.602	1.468	1.355	1.259	1.175	1.101	1.036	.979	.927	.881	.705	.589	.441
5	1.943	1.846	1.678	1.538	1.420	1.319	1.231	1.154	1.086	1.026	.972	.923	.738	.615	.462
6	2.022	1.921	1.746	1.601	1.478	1.372	1.281	1.201	1.130	1.067	1.011	.960	.768	.640	.480
7	2.093	1.989	1.808	1.657	1.530	1.421	1.326	1.243	1.170	1.105	1.047	.994	.796	.663	.497
8	2.160	2.052	1.865	1.710	1.578	1.466	1.368	1.282	1.207	1.140	1.080	1.026	.821	.684	.513
9	2.222	2.111	1.919	1.759	1.624	1.508	1.408	1.320	1.242	1.173	1.111	1.056	.845	.704	.528
10	2.282	2.168	1.971	1.807	1.668	1.548	1.445	1.355	1.275	1.204	1.141	1.084	.867	.723	.542

11	2.339	2.222	2.020	1.852	1.709	1.587	1.481	1.389	1.307	1.234	1.170	1.111	.889	.741	.556
12	2.394	2.274	2.068	1.895	1.749	1.625	1.516	1.421	1.338	1.264	1.197	1.137	.910	.758	.569
13	2.447	2.325	2.114	1.938	1.788	1.661	1.550	1.453	1.368	1.292	1.224	1.163	.930	.775	.581
14	2.499	2.374	2.159	1.979	1.827	1.696	1.583	1.484	1.397	1.319	1.250	1.187	.950	.791	.594
15	2.550	2.423	2.203	2.019	1.864	1.731	1.615	1.514	1.425	1.346	1.275	1.211	.969	.808	.606
16	2.600	2.470	2.246	2.059	1.900	1.764	1.647	1.544	1.453	1.372	1.300	1.235	.988	.823	.618
17	2.649	2.517	2.288	2.097	1.936	1.798	1.678	1.573	1.481	1.398	1.325	1.258	1.007	.839	.629
18	2.698	2.563	2.330	2.136	1.972	1.831	1.709	1.602	1.508	1.424	1.349	1.282	1.025	.854	.641
19	2.746	2.609	2.371	2.174	2.007	1.863	1.739	1.630	1.534	1.449	1.373	1.304	1.043	.870	.652
20	2.793	2.654	2.412	2.211	2.041	1.895	1.769	1.659	1.561	1.474	1.397	1.327	1.061	.885	.663
21	2.840	2.698	2.453	2.249	2.076	1.927	1.799	1.687	1.587	1.499	1.420	1.349	1.079	.899	.675
22	2.887	2.743	2.494	2.286	2.110	1.959	1.829	1.714	1.613	1.524	1.444	1.371	1.097	.914	.686
23	2.934	2.787	2.534	2.323	2.144	1.991	1.858	1.742	1.640	1.548	1.467	1.394	1.115	.929	.697
24	2.980	2.831	2.574	2.360	2.178	2.022	1.888	1.770	1.666	1.573	1.490	1.416	1.133	.944	.708
25	3.027	2.876	2.614	2.396	2.212	2.054	1.917	1.797	1.691	1.598	1.513	1.438	1.150	.959	.719
26	3.073	2.920	2.654	2.433	2.246	2.085	1.946	1.825	1.717	1.622	1.537	1.460	1.168	.973	.730
27	3.120	2.964	2.694	2.280	2.117	2.117	1.976	1.852	1.743	1.646	1.560	1.481	1.185	.988	.741
28	3.166	3.008	2.734	2.507	2.314	2.148	2.005	1.880	1.769	1.671	1.583	1.504	1.203	1.003	.752
29	3.213	3.052	2.775	2.543	2.348	2.180	2.035	1.908	1.795	1.696	1.606	1.526	1.221	1.017	.763
30	3.260	3.097	2.815	2.580	2.382	2.212	2.064	1.935	1.822	1.720	1.630	1.548	1.239	1.032	.774
31	3.307	3.141	2.856	2.618	2.416	2.244	2.094	1.963	1.848	1.745	1.653	1.571	1.256	1.047	.785
32	3.354	3.186	2.896	2.655	2.451	2.276	2.124	1.991	1.874	1.770	1.677	1.593	1.274	1.062	.797
33	3.401	3.231	2.937	2.693	2.486	2.308	2.154	2.020	1.901	1.795	1.101	1.616	1.292	1.077	.808
34	3.449	3.277	2.979	2.731	2.521	2.340	2.184	2.048	1.927	1.821	1.725	1.638	1.311	1.092	.819
35	3.497	3.322	3.020	2.769	2.556	2.373	2.215	2.077	1.954	1.846	1.749	1.661	1.329	1.107	.831
36	3.546	3.369	3.062	2.807	2.591	2.406	2.246	2.105	1.982	1.871	1.773	1.684	1.347	1.123	.842
37	3.595	3.415	3.105	2.846	2.627	2.440	2.277	2.135	2.009	1.897	1.798	1.708	1.366	1.138	.854
38	3.645	3.462	3.148	2.885	2.663	2.473	2.308	2.164	2.037	1.924	1.822	1.731	1.385	1.154	.866
39	3.695	3.510	3.191	2.925	2.700	2.507	2.340	2.194	2.065	1.950	1.847	1.755	1.404	1.170	.878
40	3.745	3.558	3.235	2.965	2.737	2.542	2.372	2.224	2.093	1.977	1.873	1.779	1.423	1.186	.890

TABLE 2 (Continued)

	μm														
No.	K 950 °F 1250	1000 1340	1100 1520	1200 1700	1300 1880	1400 2060	1500 2240	1600 2420	1700 2600	1800 2780	1900 2960	2000 3140	2500 4040	3000 4940	4000 6740
41	3.797	3.607	3.279	3.006	2.775	2.576	2.405	2.254	2.122	2.004	1.898	1.803	1.443	1.202	.902
42	3.849	3.656	3.324	3.047	2.812	2.612	2.437	2.285	2.151	2.031	1.924	1.828	1.462	1.219	.914
43	3.901	3.706	3.369	3.089	2.851	2.647	2.421	2.316	2.180	2.059	1.951	1.853	1.483	1.235	.927
44	3.955	3.757	3.415	3.131	2.890	2.684	2.505	2.348	2.210	2.087	1.977	1.878	1.503	1.252	.939
45	4.009	3.808	3.462	3.174	2.930	2.720	2.539	2.380	2.240	2.116	2.004	1.904	1.523	1.269	.952
46	4.064	3.861	3.510	3.217	2.970	2.758	2.574	2.413	2.271	2.145	2.032	1.930	1.544	1.287	.965
47	4.120	3.914	3.558	3.262	3.011	2.796	2.609	2.446	2.302	2.174	2.060	1.957	1.566	1.305	.978
48	4.177	3.968	3.607	3.307	3.052	2.834	2.645	2.480	2.334	2.204	2.088	1.984	1.587	1.323	.992
49	4.235	4.023	3.657	3.352	3.095	2.873	2.682	2.514	2.366	2.235	2.117	2.011	1.609	1.341	1.006
50	4.294	4.079	3.708	3.399	3.138	2.913	2.719	2.549	2.399	2.266	2.147	2.039	1.632	1.360	1.020
51	4.354	4.136	3.760	3.447	3.182	2.954	2.757	2.585	2.433	2.298	2.177	2.068	1.654	1.379	1.034
52	4.415	4.194	3.813	3.495	3.226	2.996	2.796	2.621	2.467	2.330	2.207	2.097	1.678	1.398	1.049
53	4.477	4.254	3.867	3.545	3.272	3.038	2.836	2.658	2.502	2.363	2.239	2.127	1.701	1.418	1.063
54	4.541	4.314	3.922	3.595	3.319	3.082	2.876	2.696	2.538	2.397	2.271	2.157	1.726	1.438	1.079
55	4.607	4.376	3.978	3.647	3.366	3.126	2.917	2.735	2.574	2.431	2.303	2.188	1.750	1.459	1.094
56	4.673	4.440	4.036	3.700	3.415	3.171	2.960	2.775	2.612	2.466	2.337	2.220	1.776	1.480	1.110
57	4.742	4.505	4.095	3.754	3.465	3.218	3.003	2.815	2.650	2.503	2.371	2.252	1.802	1.502	1.126
58	4.812	4.571	4.156	3.809	3.516	3.265	3.047	2.857	2.689	2.540	2.406	2.286	1.828	1.524	1.143
59	4.884	4.639	4.218	3.866	3.569	3.314	3.093	2.900	2.729	2.577	2.442	2.320	1.856	1.546	1.160
60	4.957	4.709	4.281	3.925	3.623	3.364	3.140	2.943	2.770	2.616	2.479	2.355	1.884	1.570	1.177
61	5.033	4.781	4.347	3.985	3.678	3.415	3.188	2.988	2.813	2.656	2.517	2.391	1.913	1.594	1.195
62	5.111	4.855	4.414	4.046	3.735	3.468	3.237	3.035	2.856	2.697	2.556	2.427	1.942	1.618	1.214
63	5.191	4.932	4.483	4.110	3.793	3.523	3.288	3.082	2.901	2.740	2.596	2.466	1.973	1.644	1.233
64	5.274	5.010	4.555	4.175	3.854	3.579	3.340	3.131	2.947	2.783	2.637	2.505	2.004	1.670	1.253
65	5.359	5.091	4.628	4.243	3.916	3.637	3.394	3.182	2.995	2.828	2.680	2.546	2.036	1.697	1.273
66	5.447	5.175	4.704	4.312	3.981	3.696	3.450	3.234	3.044	2.875	2.724	2.587	2.070	1.725	1.294
67	5.538	5.261	4.783	4.384	4.047	3.758	3.508	3.288	3.095	2.923	2.769	2.631	2.105	1.754	1.315
68	5.631	5.350	4.864	4.458	4.116	3.822	3.567	3.344	3.147	2.972	2.816	2.675	2.140	1.783	1.338
69	5.730	5.444	4.949	4.536	4.188	3.884	3.629	3.402	3.202	3.024	2.865	2.722	2.178	1.815	1.361
70	5.831	5.540	5.037	4.617	4.262	3.957	3.693	3.463	3.259	3.078	2.916	2.770	2.216	1.847	1.385

71	5.937	5.640	5.128	4.700	4.339	4.029	3.760	3.525	3.318	3.134	2.969	2.820	2.256	1.880	1.410
72	6.047	5.745	5.223	4.787	4.419	4.103	3.830	3.590	3.379	3.192	3.024	2.872	2.298	1.915	1.436
73	6.162	5.853	5.321	4.878	4.503	4.181	3.902	3.659	3.443	3.252	3.081	2.927	2.341	1.951	1.463
74	6.281	5.967	5.425	4.973	4.590	4.262	3.978	3.730	3.510	3.315	3.141	2.983	2.387	1.989	1.492
75	6.407	6.086	5.533	5.072	4.682	4.347	4.058	3.804	3.580	3.381	3.203	3.043	2.435	2.029	1.522
76	6.538	6.211	5.647	5.176	4.778	4.437	4.141	3.882	3.654	3.451	3.269	3.106	2.485	2.070	1.553
77	6.677	6.343	5.766	5.286	4.879	4.531	4.229	3.964	3.731	3.524	3.338	3.171	2.537	2.114	1.586
78	6.822	6.481	5.892	5.401	4.986	4.629	4.321	4.051	3.812	3.601	3.411	3.241	2.592	2.160	1.620
79	6.976	6.627	6.025	5.523	5.098	4.734	4.418	4.142	3.899	3.682	3.488	3.314	2.651	2.209	1.657
80	7.139	6.782	6.166	5.652	5.217	4.845	4.522	4.239	3.990	3.768	3.570	3.391	2.713	2.261	1.696
81	7.313	6.947	6.316	5.789	5.344	4.962	4.631	4.342	4.087	3.860	3.656	3.474	2.779	2.316	1.737
82	7.498	7.123	6.475	5.936	5.479	5.088	4.749	4.452	4.190	3.957	3.749	3.561	2.849	2.374	1.781
83	7.696	7.311	6.646	6.093	5.624	5.222	4.874	4.569	4.301	4.062	3.848	3.656	2.924	2.437	1.828
84	7.909	7.513	6.830	6.261	5.779	5.367	5.009	4.696	4.420	4.174	3.954	3.757	3.005	2.504	1.878
85	8.139	7.732	7.029	6.443	5.948	5.523	5.154	4.832	4.548	4.295	4.069	3.866	3.093	2.577	1.933
86	8.388	7.969	7.244	6.641	6.130	5.692	5.313	4.981	4.688	4.427	4.194	3.984	3.188	2.656	1.992
87	8.661	8.228	7.480	6.857	6.329	5.877	5.485	5.142	4.840	4.571	4.331	4.114	3.291	2.743	2.057
88	8.961	8.513	7.739	7.094	6.548	6.081	5.675	5.321	5.008	4.729	4.480	4.256	3.405	2.838	2.128
89	9.294	8.829	8.026	7.357	6.791	6.306	5.886	5.518	5.193	4.905	4.647	4.414	3.522	2.943	2.207
90	9.666	9.182	8.348	7.652	7.063	6.559	6.122	5.739	5.401	5.101	4.833	4.591	3.673	3.061	2.296
91	10.087	9.582	8.711	7.985	7.371	6.845	6.388	5.989	5.637	5.324	5.043	4.791	3.833	3.194	2.396
92	10.570	10.041	9.129	8.368	7.724	7.172	6.694	6.276	5.907	5.579	5.285	5.021	4.017	3.347	2.510
93	11.133	10.577	9.615	8.814	8.136	7.555	7.051	6.610	6.222	5.876	5.567	5.288	4.231	3.526	2.644
94	11.805	11.214	10.195	9.345	8.626	8.010	7.476	7.009	6.597	6.230	5.902	5.607	4.486	3.738	2.804
95	12.627	11.995	10.905	9.996	9.227	8.568	7.997	7.497	7.056	6.664	6.313	5.998	4.798	3.998	2.999
96	13.673	12.990	11.809	10.825	9.992	9.278	8.660	8.118	7.641	7.216	6.837	6.495	5.196	4.330	3.247
97	15.081	14.327	13.024	11.939	11.021	10.234	9.551	8.954	8.428	7.959	7.540	7.163	5.731	4.776	3.582
98	17.153	16.295	14.814	13.579	12.535	11.639	10.863	10.184	9.585	9.053	8.576	8.147	6.518	5.432	4.074
99	20.762	19.724	17.931	16.437	15.173	14.089	13.150	12.328	11.603	10.958	10.381	9.862	7.890	6.575	4.931
100	30.913	29.367	26.697	24.472	22.590	20.976	19.578	18.354	17.275	16.315	15.456	14.684	11.747	9.789	7.342

[a] These are the wavelengths for use in the 100 selected ordinate method of computing total thermal radiation properties from measured spectral data.

the ideal equation is

$$\epsilon_t(T) = \frac{\int_0^\infty \epsilon(\lambda, T) \cdot L_{b,\lambda}(T) \cdot d\lambda}{\int_0^\infty L_{b,\lambda}(T) \cdot d\lambda} \tag{18}$$

Unfortunately, the expression in the numerator of equation (18) cannot be integrated analytically, because $\epsilon(\lambda, T)$ is never known as an algebraic expression. The integral in the denominator is given as the Stefan–Boltzmann equation (10). The numerator must be evaluated by summation, and the ideal equation becomes

$$\epsilon_t(T) = \frac{\Sigma_{\lambda_1}^{\lambda_2} \epsilon(\lambda, T) \cdot L_{b,\lambda}(T) \cdot \Delta\lambda}{\sigma L T^4} \tag{19}$$

The product $L_{b,\lambda}(T) \cdot \Delta\lambda$ can be evaluated in two ways, first with $\Delta\lambda$ remaining constant and $L_{b,\lambda}(T)$ varying with wavelength, and second with the product $L_{b,\lambda}(T) \cdot \Delta\lambda$ made constant by varying $\Delta\lambda$. The first is called the weighted ordinate method, and the second is called the selected ordinate method.

In the weighted ordinate method, equation (19), the limits of summation, λ_1 and λ_2, are selected to include most of the total blackbody radiance (usually 99%, but sometimes as little as 95%). The computation of $L_{b,\lambda}(T)$, the multiplication $\epsilon(\lambda, T) \cdot L_{b,\lambda}(T) \cdot \Delta\lambda$ and the summation is usually accomplished with the aid of an electronic digital computer.

In the selected ordinate method equation (18) becomes

$$\epsilon_t(T) = 1/N \sum_{i=1}^{i=N} \epsilon(\lambda_i, T) \tag{20}$$

The N values of λ_i are selected by dividing the wavelength range 0 to ∞ into N segments, $\Delta\lambda$, within each of which $1/N$ of the total blackbody radiance is emitted, and then taking the centroid wavelength, λ_i, of each segment $(\Delta\lambda)_i$. Table 2 gives the λ_i values for the 100 selected ordinate method at values of $M(0-\lambda)/M(0-\infty)$ at increments of 0.01 from 0.005 to 0.995. For the 20 selected ordinate method, values of λ_i are taken from the table at values of $M(0-\lambda)/M(0-\infty)$ at increments of 0.05 from 0.025 to 0.975.

The 100 selected ordinate method is recommended for computing total emittance of materials for which spectral emittance varies significantly with wavelength. The 20 selected ordinate method is satisfactory for computing total emittance of materials for which the spectral emittance is essentially constant (within 10%–15%) with wavelength.

Hemispherical emittance is also a weighted average of the directional emittance over a complete hemisphere, with $\cos\theta$ as the weighting function. The angle θ is measured from the direction of emission to the surface normal. The

TABLE 3
Approximate Ratio for Hemispherical to Normal Emittance for Polished Samples[a]

Metals		Dielectrics	
$\epsilon(0)$	$\epsilon(2\pi)/\epsilon(0)$	$\epsilon(0)$	$\epsilon(2\pi)/\epsilon(0)$
0.0	1.33	0.35	1.085
0.05	1.27	0.40	1.06
0.10	1.225	0.45	1.04
0.15	1.185	0.50	1.025
0.20	1.145	0.55	1.005
0.25	1.105	0.60	0.99
0.30	1.075	0.70	0.97
0.35	1.055	0.75	0.96
		0.80	0.95
		0.85	0.94
		0.90	0.935
		0.95	0.935
		1.0	1.0

[a] From Jakob.[3]

mathematical equation is

$$\epsilon(2\pi; T) = 1/\pi \int_0^{2\pi} \int_0^{\pi/2} \epsilon(\theta, \phi; T) \sin\theta \cos\theta \, d\theta \, d\phi \tag{21}$$

The directional emittance over a complete hemisphere is rarely measured. The hemispherical emittance can be approximated by multiplying the normal emittance by an approximate conversion factor. The conversion factors from Jakob[3] are given in Table 3. The values for metals apply only to polished or vacuum-deposited metals. For rough metallic surfaces the conversion factors will be lower. Nonmetallic materials in general are less sensitive to surface topography. The values computed by use of the conversion factors for metals will have an uncertainty on the order of 0.05 in emittance. The uncertainties are somewhat lower for nonmetallic materials. The values apply to optically homogeneous materials with clean surfaces, of thickness sufficient to be opaque at all wavelengths. Optical inhomogeneity (internal scattering), surface films (such as oxide layers on metals), and nonopaque coatings on an opaque substrate (such as some paints on metal) can cause significant deviations from the quoted factors.

2.3. *Radiation of Real Materials*

Equations (7)–(11) apply only to true blackbody radiators, but they can be modified to apply to any sample by using the proper emittance of the sample as a correction factor. Equation (7) then becomes

$$L_{\lambda,s}(\theta,\phi;\lambda;T) = \epsilon_s(\theta,\phi;\lambda;T)c_{1L}\lambda^{-5}[\exp(c_2/\lambda_T)-1]^{-1} \tag{22}$$

where $L_{\lambda,s}(\theta,\phi;\lambda;T)$ is the spectral radiance of a sample at temperature T, wavelength λ, and direction θ, ϕ, and $\epsilon_s(\theta,\phi;\lambda;T)$ is the directional spectral emittance of the sample under the same conditions. In like manner equation (8) becomes

$$M_{\lambda,s}(\lambda;T) = \epsilon_s(\lambda;T)c_1\lambda^{-5}[\exp(c_2/\lambda T)-1]^{-1} \tag{23}$$

where $\epsilon_s(\lambda;T)$ is the spectral hemispherical emittance of the sample at temperature T.

Equation (9) for total radiance becomes

$$L_{t,s}(\theta,\phi;T) = \epsilon_s(t;\theta,\phi;T)\sigma_L T^4 \tag{24}$$

where $L_{t,s}(\theta,\phi;T)$ is total radiance of the sample at temperature T in the direction θ,ϕ, and $\epsilon_s(t;\theta,\phi;T)$ is the total directional emittance of the sample at temperature T in the direction θ,ϕ.

For total exitance equation (10) becomes

$$M_{t,s}(T) = \epsilon_s(t;2\pi;T)\sigma T^4 \tag{25}$$

where $M_{t,s}(T)$ is the total exitance of a sample at temperature T, and $\epsilon_s(t;2\pi;T)$ is the total hemispherical emittance of the sample at the same temperature.

Conversion of equation (11) is more complex because two emittances are required. It becomes

$$I_{s,\lambda}(\theta,\phi;T) = I_{s,\lambda}(0;T)[\epsilon_s(0;\lambda;T)]^{-1}\epsilon_s(\theta,\phi;\lambda T)\cos\theta \tag{26}$$

where $I_{s,\lambda}(\theta,\phi;T)$ is the spectral intensity of an element of area of a sample at temperature T in the direction θ,ϕ and $I_{s,\lambda}(0;T)$ is its spectral intensity normal to its surface, $\epsilon_s(0;\lambda;T)$ is the normal spectral emittance, and $\epsilon_s(\theta,\phi;\lambda;T)$ is the directional emittance in direction θ,ϕ, both for the sample at temperature T. The same equation can be used for total intensity by changing the subscript λ to the subscript t and λ to t wherever it appears, and each spectral quantity or property thus becomes a total quantity or property.

Equations (22)–(26) can be used to compute the flux emitted by a sample at any desired temperature into any desired solid angle if the required values of emittance of the sample are known.

3. MEASUREMENT OF THERMAL RADIATION PROPERTIES

3.1. *Measurement of Reflectance*

The most commonly used reflectance measuring techniques are (1) the heated-cavity reflectometer, (2) the integrating sphere reflectometer, (3) the integrating mirror reflectometer, (4) the specular reflectometer, and (5) the gonioreflectometer.

Before discussing these procedures the differences between reflectance, ρ, and reflectance factor, R, should be explained. As mentioned previously, if absorptance or emittance is evaluated by measuring reflectance, all of the reflected flux must be collected for measurement. This is physically impossible, because the solid angles of the incident beam and the reflected beam collected for measurement are mutually exclusive. In a true reflectance measurement, the incident flux and reflected flux are measured separately, which may be difficult to do. Reflectance factor is defined as the ratio of the flux reflected by a sample under specified conditions of irradiation and collection to the flux reflected, under identical conditions of irradiation and collection, by the ideal completely reflecting perfectly diffusing surface. While such a surface does not exist, there are several materials that have high reflectance ($>97\%$), over restricted spectral bands, that are near perfect diffusers. In measuring reflectance factor the flux reflected by the sample is measured and recorded. The sample is replaced by a diffusely reflecting standard of known reflectance, and the flux it reflects is measured and recorded. The reflectance factor is then computed as the ratio of the flux reflected by the sample to that reflected by the standard times the known reflectance of the standard. For diffusely reflecting samples the flux losses for sample and standard tend to compensate, so that the uncertainty due to such losses is much less than in a true reflectance measurement, where incident and reflected flux are measured separately. Reflectance factor measurements are usually made only for directional–hemispherical, or 45°-normal geometrical conditions, where the reflectance factor and reflectance are numerically equal. For specularly reflecting samples, the bidirectional reflectance factor in the specular direction may have very large values, approaching infinity in the limiting case.[4]

3.1.1. *Heated Cavity Reflectometers*

The heated cavity reflectometer consists of a cylindrical cavity with a shallow conical end. The walls are usually of a high-emittance material such as oxidized Inconel or other high-temperature alloy, and are maintained at a uniform temperature in the range of 800 to 1100 K. The water-cooled sample is introduced into the cavity, where it is irradiated over nearly a complete

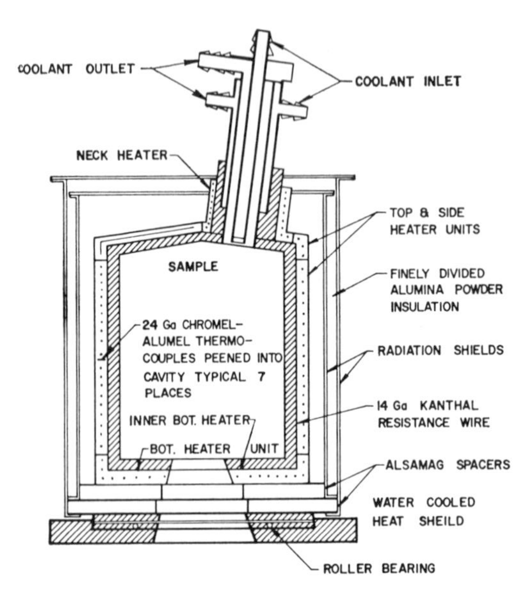

FIGURE 2. Cross-sectional view of a heated-cavity reflectometer in which the water-cooled sample is positioned to form a part of the wall of the cavity.

hemisphere by blackbody radiation from the cavity walls. The sample and an area on the cavity wall are viewed through an aperture in the cavity, either alternately by a single-beam monochromator, or simultaneously by the two beams of a double-beam monochromator. The ratio of the flux reflected by the sample to that emitted and reflected by the cavity wall is the hemispherical–directional reflectance factor of the sample. The flux emitted and reflected by the spot viewed on the cavity wall gives it all of the characteristics of the ideal perfectly diffusing, completely reflecting surface irradiated by the blackbody radiation of the cavity. The normal range of spectral reflectance factor measurement is 1 to 15 μm, and in a few cases 1 to 35 μm.

In the original instrument[5-8] the water-cooled sample formed a portion of the wall of the cavity, as shown in Fig. 2. There were errors in this version,[7] due to thermal gradients in the cavity walls caused by the cooling water for the sample. In an improved version,[9] the sample, supported by the tubes carrying the cooling water, is introduced into the center of the cavity through an aperture in the bottom, thus eliminating the cooling of the walls by the cooling water for the sample. This arrangement has the added advantage that the sample can be viewed at angles from near normal to about 80° from normal.

There are many sources of error in heated-cavity reflectometers, which have been well documented.[5-7] The principal sources of error are nonuniform irradiation of the sample due to the apertures and thermal gradients in the cavity walls, emission from the sample due to heating by radiation, and failure of the portion of the cavity wall viewed during a measurement to represent the average radiance of the cavity walls. Accurate measurements depend upon the sample temperature being low enough that the radiant energy emitted by the sample is negligibly small compared to that reflected by the sample. The method is obviously not suitable for use on materials of low thermal conductivity, and particularly not for organic materials that may be damaged by heating due to radiation, conduction, and convection in the cavity.

The hohlraum irradiates the sample over nearly a complete hemisphere and nearly uniformly, and the sample is viewed directionally over a small solid angle. The measured property is thus the hemispherical–directional reflectance factor. which, by Helmholtz reciprocity, is equivalent to the directional–hemispherical reflectance factor, and is numerically equal to the directional–hemispherical reflectance.

3.1.2. Integrating Sphere Reflectometers

The integrating sphere reflectometer is probably the most common type of reflectometer in use today. An integrating sphere is used to collect the flux reflected by the sample over nearly a complete hemisphere.

The theory of the integrating sphere is based on two fundamental laws: (1) The flux incident on a surface from a point source varies inversely as the square of the distance from the source, and directly as the cosine of the angle between the normal to the surface and the direction of incidence, and (2) a perfectly diffuse reflector of a given size reflects the incidence flux isotropically in all directions, so that the flux reflected within a given solid angle in a given direction is proportional to the cosine of the angle between the given direction and the normal to the reflecting surface.

As a consequence of these laws, any flux incident on the diffusely reflecting internal wall of an integrating sphere is uniformly distributed over the surfaces of the sphere after the first reflection. A detailed analysis of an integrating sphere is given in Ref. 2, and general integrating sphere theory in Refs. 10 and 11.

The sample in an integrating sphere can be irradiated either directly or indirectly, and these types of irradiation are sometimes referred to as the direct and inverse modes of operation. In direct mode operation the sample is irradiated directly, and the detector views a portion of the sphere wall. In inverse mode operation, the irradiating beam is incident on the sphere wall, and the sample is irradiated diffusely by flux reflected by the sphere wall, and viewed directly by the detector. Most integrating sphere reflectometers incorporate a monochromator to provide spectral data. In the direct mode the monochromator is usually in the irradiating beam optics, and the sample is irradiated with a monochromatic beam. In inverse mode operation, the irradiating beam is usually heterochromatic, and the monochromator is incorporated in the reflected beam optics.

The integrating sphere reflectometer measures reflectance factor, because the flux reflected by the sample is compared to that reflected by a standard of known reflectance. In direct mode operation the measured property is the directional–hemispherical reflectance factor. In the inverse mode, it is the hemispherical–directional reflectance factor. These two quantities are equivalent through Helmholz reciprocity, and are each numerically equal to the directional–hemispherical reflectance, and are frequently reported as such.

In the operation of an integrating sphere reflectometer the sample may be removed and replaced by the standard, which is called the substitution method. In other types of integrating spheres there may be separate apertures for sample and standard, which are alternately irradiated and viewed, which is called the comparison method. Almost all double-beam integrating sphere reflectometers use the comparison method. The sample and standard are irradiated alternately, and the complex signal produced by the detector is separated electronically into two separate signals. The signal is electronically amplified, either before or after separation, and the two signals are then compared. In the comparison method the sample and standard apertures should be symmetrically located relative to (1) the entrance aperture for the irradiating beam, (2) the sphere wall area that is irradiated in the inverse mode, and (3) the detector aperture or area on the sphere wall viewed by the detector.

Errors that may be significant for some samples can be avoided if there is an internal baffle in the sphere that prevents flux reflected by the sample in direct mode operation from falling on the area of the sphere wall viewed by the detector, or in inverse mode operation that prevents the flux reflected by the sphere wall on the first reflection from falling on the sample. All baffles should be coated with the sphere paint.

The spectral range of most integrating sphere reflectometers is limited to about 0.25 to 2.5 μm by the reflection properties of the sphere coating. Smoked magnesium oxide[12, 13] was the standard sphere coating for many years. Barium sulfate[14,17] is now commonly used, and Halon† [15,16] is coming into use.

In some integrating sphere instruments an auxiliary port is located at the point where the specularly reflected beam hits the sphere wall. This aperture is closed with a plug whose inner surface conforms to the contour of the sphere wall, and the sphere coating is applied to the plug. The plug remains in place for measurement of the directional–hemispherical reflectance factor. With the plug removed and replaced by a blackbody absorber, the directional–diffuse

†Halon is registered trademark of the Allied Chemical Company. It is a fluorinated polycarbon. It is available in small lots from C.G. Leete, Manufacturers Council on Color and Appearance, 9416 Gama Ct., Vienna, Virginia 22180, U.S.A.

reflectance factor is measured, and the specularly reflected flux passes out through the aperture and is absorbed by the blackbody absorber, which may be in the form of an exponential horn. The same effect can be obtained by removing the plug and making the measurement in a darkroom with black walls.

Most integrating sphere reflectometers measure directional–hemispherical or hemispherical–directional reflectance factor only for a fixed direction of irradiation or viewing. In these instruments the sample is clamped over an aperture in the wall of the sphere, in a position where the sample essentially forms a part of the sphere wall. The direction of incidence or viewing is near normal, usually in the range of 6° to 10° from normal, and the measured quantity may be referred to as the normal–hemispherical or hemispherical–normal reflectance factor, since the difference between values obtained with incidence or viewing at 0° and 10° from normal is negligibly small for diffuse reflectors. Incidence or viewing normal to the surface automatically eliminates the specularly reflected flux, which for many common white materials may be on the order of 0.05 in reflectance. The specularly reflected beam is lost out the entrance port for the condition of normal incidence, and the image of the dark detector port is seen for the condition of normal viewing.

The integrating sphere developed by Edwards *et al.*[18] is unusual in that it is designed to measure directional–hemispherical reflectance at directions of incidence from 0° to 80° from normal. This instrument, sketched in Figs. 3 and 4, has a sphere in which the sample is located in the center of the sphere, and can be rotated about its support rod to provide the different directions of

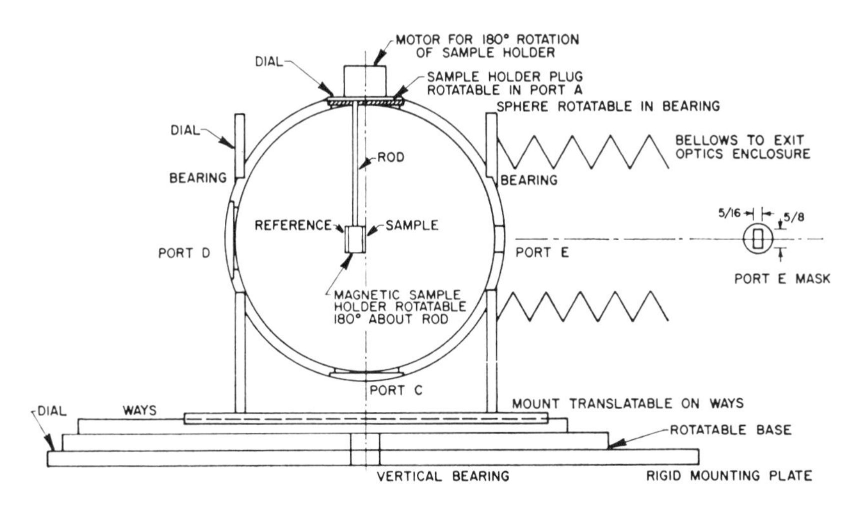

FIGURE 3. The integrating sphere developed by Edwards *et al.*,[18] in which the sample, at the center of the sphere, can be rotated to vary the direction of incidence.

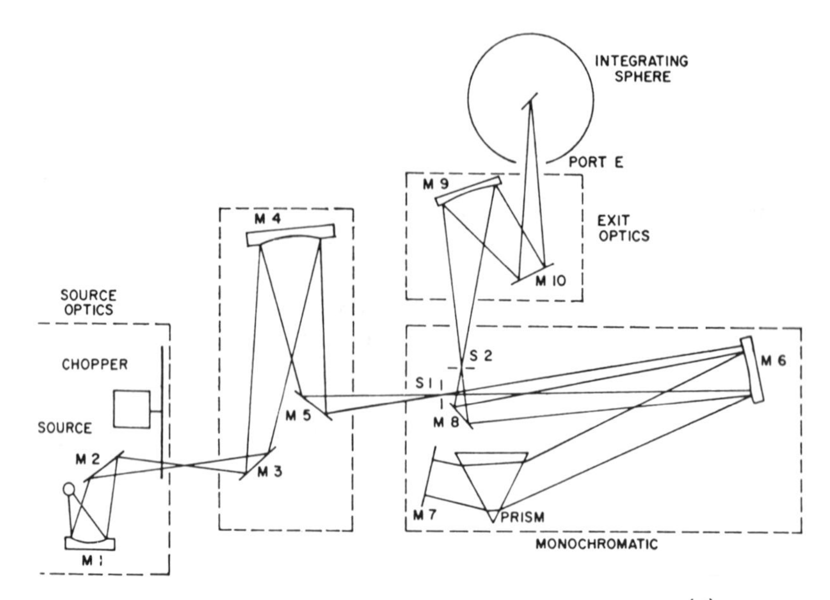

FIGURE 4. The optical path in the Edwards *et al.* reflectometer.[18]

incidence. For absolute reflectance measurements the wall of the sphere can be irradiated by moving and rotating the sphere, as indicated in Fig. 5.

An integrating sphere instrument[24] for measurment of reflectance at temperatures up to 2500 K uses a helium–neon laser as a source. The high-intensity narrow-bandwidth chopped source, a spike filter, transmitting at the laser wavelength, in front of the detector and synchronous amplification of the signal from the detector combine to produce a very high signal-to-noise ratio in the presence

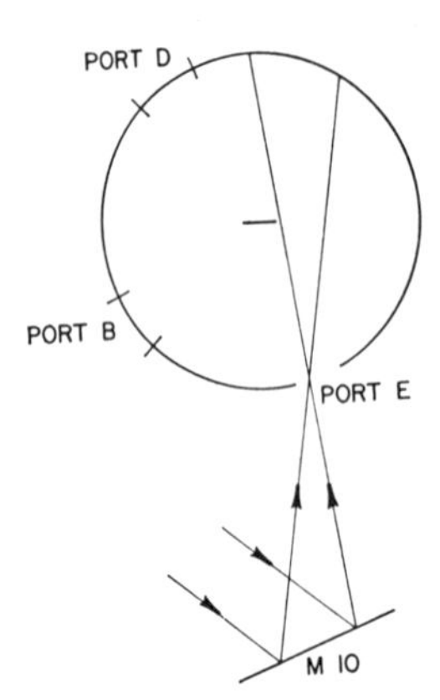

FIGURE 5. The integrating sphere of the Edwards *et al.* reflectometer after rotation so that the incident beam misses the sample and hits the sphere wall.

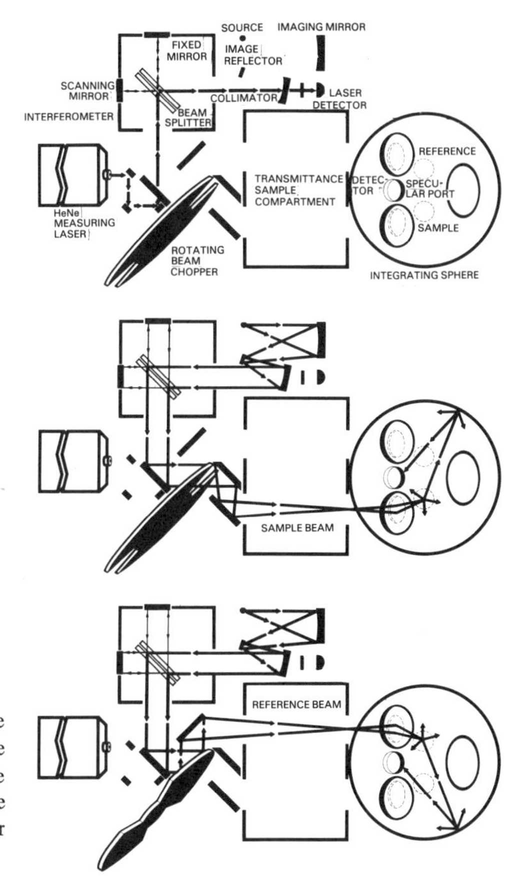

FIGURE 6. Optical paths in the Willey[19,20] integrating sphere reflectometer—top, path for the laser beam; center, path for the sample beam; bottom, path for the reference beam.

of the large amount of flux emitted by the hot sample, at the laser wavelengths of 632.8 and 1150 nm.

A recent development[19,20] is an integrating sphere reflectometer for use in the 2–20 μm wavelength range. The instrument uses a roughened, gold-coated sphere, with a wall reflectance of about 0.95, and a Fourier transform spectrometer as the dispersing element. Sketches of the three optical paths in the instrument are shown in Fig. 6. Motion of the movable mirror in the Michelson interferometer is sensed by the beam from a helium–neon laser, as shown in the top sketch. The beam passes through small holes in the plane mirror directly in front of the chopper, and the collimator in front of the laser detector, and passes through the interferometer in a direction opposite that of the infrared measurement beam. As the interferometer mirror moves at a uniform velocity, the laser detector produces a sine-wave signal, in which one full cycle is produced for each change of one wavelength of light in the path length difference for the two beams, or 1/2 wavelength motion of the mirror. The zero points of the signal, 1/2 wavelength apart, thus correspond to motion of the mirror through 1/4 wavelength, or 0.1582 μm. The zero points, or polarity reversals, are counted electronically, and the infrared interferograms are sampled at intervals of any preset number of counts from 1 to 15. The usual sampling interval is 3 or 4 counts, or 0.4746 or 0.6328 μm.

The source for the reflectance beam is a Nernst glower, about 3 mm diameter and 3 mm long. The infrared beam from the source is collimated and passes through the interferometer, as shown in the bottom two sketches, and is reflected to the chopper. The chopper has 8 segments, each 45° wide, two openings and two mirrors, separated by four nonreflecting, nontransmitting segments. The mirror segments reflect the beam into the reference beam path, as shown in the bottom sketch, and the two open segments permit the beam to pass through the chopper into the sample beam path, as shown in the center sketch. The four nontransmitting, nonreflecting segments block the beam completely, and provide a reference zero signal from the detector. The signal from the detector is thus a repeating four-segment signal: zero, reference signal, zero, sample signal. In essence, the chopper multiplexes the zero, reference, and sample signals. The chopper frequency is 26 Hz.

The reflecting surface of the integrating sphere is a textured gold surface, with directional–hemispherical reflectance of about 0.95 throughout the wavelength range of interest, and good diffusion. There is provision for measurement of both transmittance and reflectance factor. An aperture with a replaceable cover is located opposite the sample port. When the cover is removed, the specularly reflected component passes out of the sphere, and is not measured.

In the electronics, the signal is amplified and demultiplexed to produce two interferograms, one for the sample and one for the reference. These interferograms are sampled simultaneously at fixed intervals of mirror movement, as mentioned previously. The signals are amplified, digitized, and stored in the memory of an integral minicomputer. The computer later performs the Fourier transforms to produce the usual spectra, which are then processed to produce the true spectrum of the sample, which is stored in the memory. This spectrum can be displayed on a CRT display, or recorded on a chart recorder, with either wave number or wavelength as the ordinate.

Other materials for use as integrating sphere coatings for use at wavelengths beyond about 2.5 μm include mu sulfur[20, 21, 22] and sodium chloride.[2, 23] Gold-coated roughened surfaces have also been used by other investigators.[19, 22]

Although not employing an integrating sphere, the technique of McNicholas[25] and Martin[26] gives data equivalent to those obtained with an integrating sphere. The sample is located at the center of a hemispherical source and is viewed directionally through a small hole in the source. Hemispherical irradiation is achieved by distributed light sources about the exterior of a translucent diffusing hemisphere.

3.1.3. Integrating Mirror Reflectometers

An alternative to the integrating sphere is to replace the sphere with an integrating mirror. There are both advantages and disadvantages to this approach. Mirrors have high reflectance in the infrared, and are much more efficient than integrating spheres, hence they are most advantageous in the infrared, where energy is low. On the other hand, it is physically impossible to convert the radiant flux collected over a hemisphere into a parallel beam of small cross-sectional area. Hence an integrating mirror reflectometer requires either a large-area detector that is equally responsive to flux incident upon it from different directions within a large solid angle, and that is equally sensitive at all points over its sensitive area, or a flux-averaging device[23] in front of the detector.

Arc-image reflectometers[27, 28] will be mentioned only briefly. These devices employ two focusing mirrors to image the flux from a carbon arc onto a sample, usually on a nonmetallic material. The sample is heated by the incident flux to a temperature in the 2500–3700 K range, and viewed directionally by a detector through a small hole in the concentrating mirror. Elaborate synchronized shutters between the arc and the sample and between the sample and the detector permit the separate measurement of emitted and emitted plus reflected flux. The reflectance is taken as the difference in the two measured quantities.

The hemispherical mirror reflectometer developed by Coblentz[29] was one of the earliest reflectance techniques. A sketch of the instrument is shown in Fig. 7. The basic principle is that the flux leaving a point source in the plane of the edge of a hemisphere will be focused into a small image in the same plane,

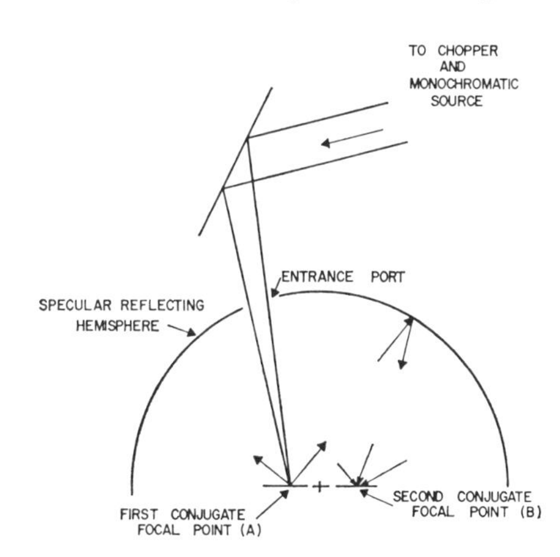

FIGURE 7. Schematic diagram of the Coblentz[29, 30] hemispherical mirror reflectometer. The sample is at A and the detector at B.

along the diameter containing the source and on the opposite side from the source and equally distant from the center of the hemisphere.[30] These two points are called conjugate focal points. The size of the image of the point source is determined by spherical aberrations which increase with the distance between the points.

The sample is centered on one conjugate focal point and the detector is centered on the other. The two points are close together to reduce aberrations. The sample is irradiated directionally through a small hole in the hemisphere, and the reflected flux is focused into an enlarged image on the detector. The detector can be moved to the sample position to measure the incident flux directly, or a diffusely reflecting sample of known reflectance can be substituted for the sample in order to evaluate the incident flux. Major losses in the system are due to absorption by the mirror and the atmosphere, flux lost out the entrance aperture, and flux that misses the detector due to aberrations. It is important that both the sample and detector surfaces be accurately located in the plane of the center of curvature of the mirror, or the error due to flux missing the detector can be large. Other errors are due to variations in the sensitivity of the detector for flux incident at different angles, or on different areas of the detector.

Janssen and Torborg[31] modified the original Coblentz instrument as

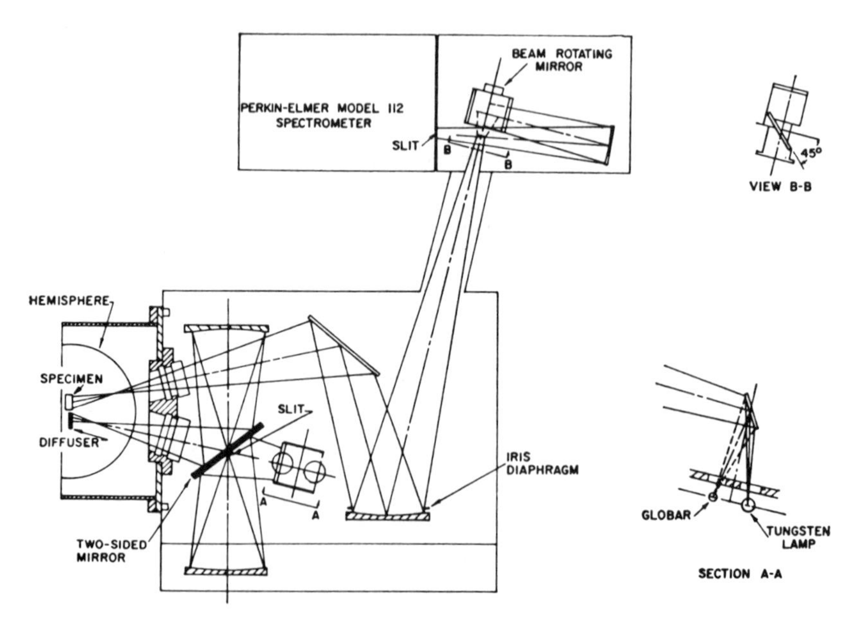

FIGURE 8. The Janssen–Torborg[31] modification to the Coblentz hemispherical reflectometer.

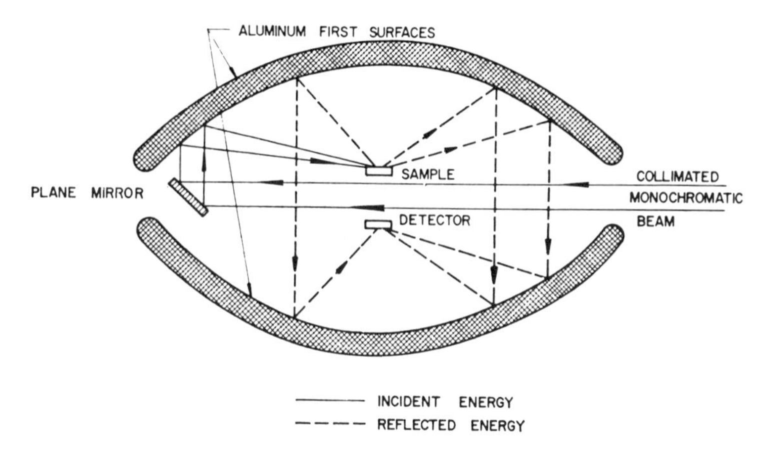

FIGURE 9. The paraboloidal mirror reflectometer developed by Dunkle.[5]

shown in Fig. 8. A diffuser at one conjugate focal point is irradiated through a hole in the mirror, and the sample, at the second conjugate focal point, is irradiated over a hemisphere by the reflected flux, and viewed directionally through a second hole in the mirror. A comparison standard is required to evaluate the incident flux. The instrument operates in the 0.4 to 20 μm range, and at sample temperatures up to about 575 K.

White[32] designed an automatic-recording double-beam instrument with a hemispherical reflectometer. A Nichrome wire source at the first focal point irradiates the sample at the second. Source and sample are viewed through a single hole in the mirror, by optical systems that focus beams on the entrance slits of the two beams of a double-beam spectroradiometer. The wavelength range is 2.5 to 22.5 μm. A unique internal chopper permits measurement of samples of temperatures up to about 1250 K. Measurements have also been made at temperatures as low as 77 K.[33].

Birkebak and Hartnett[34] extended the hemisphere to about 3/4 of a sphere, which permitted the sample and detector to be tilted, to vary the direction of incidence. Kozyrev and Vershinin[35] used baffles within the hemisphere to measure flux reflected into preselected solid angles. Other variations of hemispherical-mirror reflectometers have been described.[36–38]

A paraboloidal mirror reflectometer[5] is shown in Fig. 9. The basic principle is the same as for a hemispherical mirror reflectometer. The movable mirror between the two parabolic mirror permits the sample to be irradiated in any desired direction. Neher and Edwards[39] used off-axis paraboloids, as shown in Fig. 10. The sample is irradiated hemispherically by a large-area source and is viewed directionally. The mirrors marked FM in the figure reflect only in

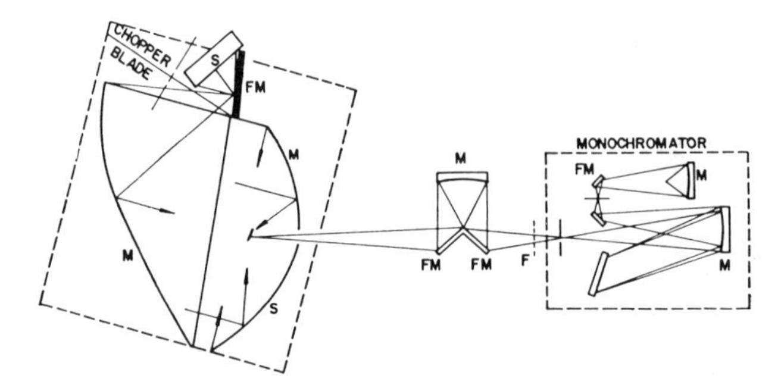

FIGURE 10. The off-axis paraboloid-mirror reflectometer developed by Neher and Edwards.[39]

the infrared, and absorb at wavelengths shorter than 2 μm, thus reducing errors due to scattered flux. The wavelength is 2 to 100 μm.

Aberrations are reduced when an ellipsoidal mirror, which has true focal points, replaces a hemispherical mirror with conjugate focal points. Such mirrors are prolate ellipsoids, which are symmetrical about the major axis. Such an ellipsoidal mirror will form an undistorted image, located at the second focal point, of a point source at the first focal point. A plane area normal or parallel to the major axis and surrounding one focal point will be imaged on a plane parallel to the first and will be centered on the second focal point, but the edges of the image will be blurred because of the variation in magnification for rays reflected from different areas of the ellipsoid. The amount of such blurring will increase with the maximum magnification, which is a function of the eccentricity of the ellipsoid. For evaluation of hemispherical reflectance, all of the flux reflected into the hemisphere above the sample must be collected for measurement, which requires only a portion of an ellipsoid.

If a prolate ellipsoid is cut along a plane through its major axis, one half may be used to replace the hemisphere in a hemispherical mirror reflectometer, with considerable reduction in the blurring at the edges of the image, but the other disadvantages of the hemispherical mirror remain – the necessity for a detector that senses equally from all directions over a complete hemisphere, and the necessity for very accurate location of both sample and detector in the plane of the edge of the mirror through the focal points. Neu[40] and Heinesch[41] have designed instruments of this type. If a prolate ellipsoid of appreciable eccentricity is cut in a plane that passes through a focal point and is normal to the major axis, the small portion can be used in the configuration shown in Fig. 11.[42] The sample is located at the first focal point and is irradiated by a small-diameter chopped monochromatic beam through a hole in the mirror. An enlarged image of the irradiated area on the sample is formed on the detector at

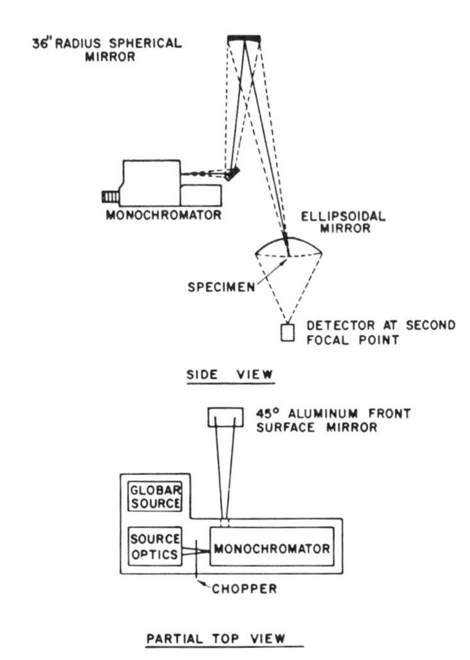

FIGURE 11. The ellipsoidal mirror reflectometer developed by Dunn, Richmond, and Wiebelt.[42]

the second focal point. The image is blurred, hence a flux-averaging device,[23] such as an averaging sphere, is required if a detector of area smaller than the image, or that varies in sensitivity over its surface, is used. As an alternative, the detector can be scanned over the image area[43] and the signal integrated during the scanning period. A second alternative is to use a large area detector, such as a pyroelectric, that has uniform response over its entire area. The reflectance of the sulfur coating on the averaging sphere used in Ref. 43 limited the spectral range to about 1 to 8 μm. With a scanning or large-area detector the range is limited only by the spectral sensitivity of the detector. In the inverse mode a large-area source with a chopper just above the source is located at the second focal point. The sample is irradiated uniformly over a hemisphere, and viewed directionally by the monochromator.[43] In any of these configurations shields can be placed in the first focal plane to confine the reflected (or incident) flux to any desired solid angle. Corrections can then be made for all flux losses on the basis of the actual geometrical distribution of reflected flux from the sample, and the final uncertainty in the measured reflectance can be reduced to 1% or less. A high-quality first-surface mirror of known reflectance was used as the reference standard. The spectral range of the device in the inverse mode is limited to that of the monochromator.

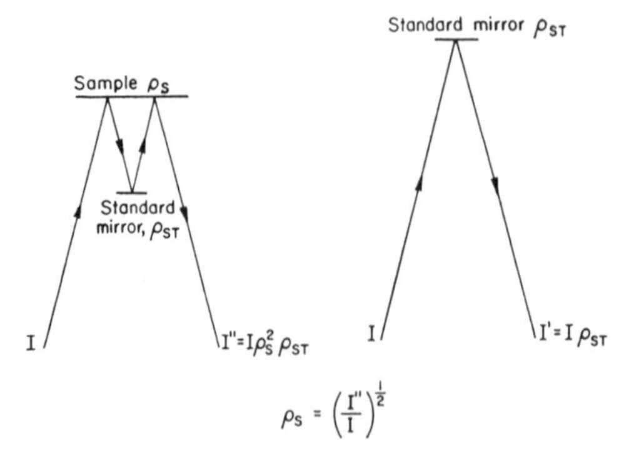

FIGURE 12. The Strong spectral reflectometer.[44-46]

3.1.4. Specular Reflectometers

For high-quality mirrors the multiple-reflection technique[44-46] is simple and accurate. Figure 12 shows a schematic of the optical path. The specular reflectance is computed as the ratio of the two measured fluxes, hence the uncertainty in the computed reflectance is about half that of the measured ratio. Very accurate alignment of the mirrors, and optically flat mirrors, are required to maintain the same beam geometry for the two measurements. The major source of error is slight displacement of the reflected beam on the detector, since for most detectors the output varies with the position of the incident beam on the detector. Flux averaging devices in front of the detector can help to minimize the error.[2, 23] Bennett and Koehler[47] have improved the optical system, as

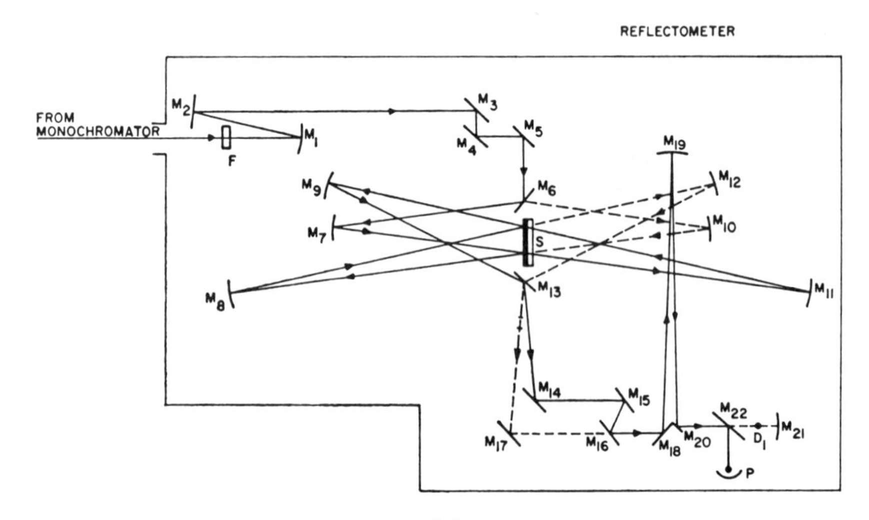

FIGURE 13. The Bennett and Koehler[47] high-precision specular reflectometer.

shown in Fig. 13, to reduce errors due to slight shifts in the alignment of the sample mirror or slight deviations of the sample mirror from flatness. They report uncertainties of 0.001 reflectance units over the wavelength range 0.45 to 22.5 μm.

3.1.5. Gonioreflectometers

A gonioreflectometer measures the bidirectional reflectance as a function of angle of reflection for any given angle of incidence. Several instruments[2, 48, 49] have been developed for measuring bidirectional reflectance in the plane of incidence. More elaborate instruments[50–53] have been developed to measure bidirectional reflectance over nearly an entire hemisphere for any given direction of incidence. In addition, the measurements can be made with monochromatic flux from about 0.4 μm to 2.5 μm.

3.2. Emittance Evaluation

Spectral directional emittance can be computed from the measured directional–hemispherical reflectance by use of equation (6). The direction for the computed emittance is the same as the direction of incidence for the reflectance measurement, and for the temperature of the sample at the time of the reflectance measurement.

Direct measurements of emittance are made calorimetrically, or radiometrically. In radiometric measurements the radiances of a sample and of a blackbody radiator at the same temperature are measured, and the emittance is computed as the ratio of the two radiances. Both total directional and spectral directional emittance can be measured radiometrically.

In calorimetric measurements the sample is placed in an environment where essentially all heat loss is by radiation, and where the heat loss from, or heat input to, the sample can be measured. The total hemispherical emittance is computed from the measured temperature or rate of temperature change of the sample and the rate of heat transfer to or from the sample (and the heat capacity of the sample for dynamic measurements) by use of an equation derived from the Stefan–Boltzmann equation (9).

A major source of uncertainty in the measured emittance is the uncertainty in the measured temperature of the sample. In steady-state calorimetric measurements, the percentage uncertainty in the measured emittance is four times the percentage uncertainty in the absolute temperature measurement. In radiometric measurements of emittance, the ratio R of the fractional uncertainty in the measured emittance to the fractional uncertainty in the measured temperature is a function of λT, the product of wavelength and temperature, as discussed previously, in Section 2.2.

Two developments over the last decade have grearly increased the attainable signal-to-noise ratio in thermal radiation measurements, and particularly in thermal emittance measurements. The first is the development of the pyroelectric detector.[54–58] The pyroelectric detector has advantages over both thermal and photon detectors for use in the far infrared, beyond about 4 μm. It has very rapid response as compared to other thermal detectors so that chopping frequencies in the 1000 Hz range are commonly used, compared to the chopping frequencies of 13 or 33 Hz commonly used with thermocouple or thermopile detectors. Pyroelectric detectors can be made in almost any desired size, with very uniform response over the entire area, while large-area thermopile detectors can show very large differences in response with position over the sensitive area, even exhibiting negative response in some areas in certain conditions.[43] Pyroelectric detectors operate at room temperature and have flat spectral response over a wide spectral band, compared to photon detectors which usually require cryogenic cooling for operation at wavelengths beyond 4 μm, and display marked changes in responsivity with wavelength.

The second recent development is the use of Fourier transform spectroscopy.[58,59] The Michelson interferometer has been known for many years, but its practical use did not come about until the development of digital data processing equipment to record interferograms and computer programs for fast Fourier transform of the interferogram into normal spectra. Richmond and Geist[58] give a good explanation of the operation of the Michelson interferometer. Kneubuhl[60] gives an extensive bibliography on spectroscopy in general including 77 references on Fourier transform spectroscopy.

For stable sources of radiant flux the statistical uncertainty in measured radiometric quantities is inversely related to the signal-to-noise ratio of the measuring instrument and the square root of the time interval over which the signal is averaged. Fourier transform spectrometers have the advantage of greatly increased signal-to-noise ratio as compared to dispersion (prism or grating) spectrometers. This increased signal-to-noise ratio can be exploited in three ways; separately or in combination: (1) to increase spectral resolution or (2) decrease time required for measurement at flux levels normally adequate for dispersion spectrometers, or (3) to make measurements with acceptable uncertainties at flux levels very much lower than those required for dispersion spectrometers. As examples, multiple scans per second can be made in studying the spectra of rapidly varying sources, or good spectra can be obtained for very faint sources, such as small stars, or low-temperature objects, such as the human body.

3.2.1. Radiometric Emittance Measurements

Direct radiometric emittance evaluations are made by measuring the radiance of a heated sample and of a blackbody radiator at the same temperature and under the same spectral and geometric conditions, and computing the emittance as the ratio of the two radiances. The recorded signals need not be reduced to radiance, total, in $W m^{-2} sr^{-1}$, or spectral, in $W m^{-2} sr^{-1} \mu m^{-1}$, if the geometric and spectral conditions are identical for the two measurements made with the same detector assembly. The geometric conditions are the angular field of view, which controls the area on the sample or aperture of the blackbody that emits the measured flux, and the direction of the axial ray of the received beam relative to the surface of the sample. The spectral conditions are the central wavelenths, bandpass, and slit function of the spectrometer. The measured property is the directional emittance of the sample, total if a spectrally nonselective detector is used to measure all wavelengths simultaneously, or spectral if the measurement is confined to a narrow spectral band at one time. Spectral measurements are frequently made with a double-beam ratio-recording spectrophotometer, that scans over an extended wavelength range.

The comparison blackbody may be an integral blackbody cavity whose walls are formed by the sample[61–65] or a separate blackbody controlled at the temperature of the sample.[66,67]

The integral blackbody is preferred at very high temperatures, where temperature measurement and control may be difficult, or at very short wavelengths, where close temperature control is required for accurate measurements. The R in equations (15)–(17) is based on the temperature difference between sample and blackbody, and not on the uncertainty in either temperature. The integral blackbody technique is particularly suitable for measurements where the sample is heated in vacuum or controlled atmosphere and viewed through a window for measurement. The transmittance of the window may change due to condensation of materials volatilized from the hot sample. The effect of the window transmittance, and changes in the transmittance, are essentially compensated for when sample and reference are viewed through closely adjacent areas of the window. Figure 14[61] is a drawing of a completely instrumented sample with a blackbody cavity. A thin-walled tubular sample is heated by passing a current through it. The small hole in the wall of the sample is a good approximation of a blackbody at the temperature of the wall, if the tube walls are uniform in thickness and thin, the hole diameter is small compared to the diameter of the sample, and the hole is viewed off-normal by 5 to 10°. Triangular samples formed from flat sheet have been used as well as cylinders. Larrabee[68,69] has made very precise measurements on tungsten by the integral blackbody technique.

The separate blackbody technique is most accurate at temperatures where conventional thermocouples can be used for temperature measurement and control. Such thermocouples may be unstable if used at temperatures above about 1500 K. This technique is particularly suitable for use with a double-beam spectrophotometer, with the sample and blackbody serving as sources for the

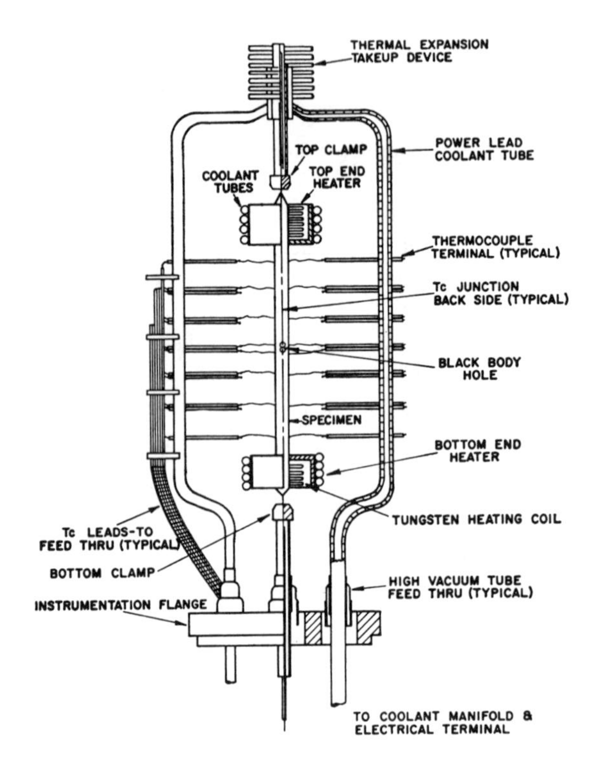

FIGURE 14. An instrumented sample for integral-blackbody emittance measurement.[61]

two beams. Particular care must be taken to see that the geometry (area viewed and solid angle of collection) is identical for the two beams. The radiance is rarely evaluated as such, since these geometrical parameters are not easily evaluated, but if the two beams are identical, the geometrical factors cancel out, and the radiance need not be evaluated.

Figure 15 is a drawing of the optical system used by Richmond *et al.*[67] The sample, in the form of a flat metal strip, about 6.3 mm wide and 20 cm long was heated by passing a current through it, and was enclosed in a water-cooled shield of an iron–nickel–chromium alloy, with an inside surface that had been roughened and oxidized, to reduce thermal gradients due to heat loss by convection from the hot sample, and errors in measurement of the emitted flux due to reflection from the enclosure walls. The sample temperature was measured by means of a thin-wire noble-metal thermocouple formed by separately spot welding the two wires to the sample on the side opposite and near the center of the area viewed. Two identical blackbody furnaces were used, which were designed to have effective emittances of better than 0.998. The

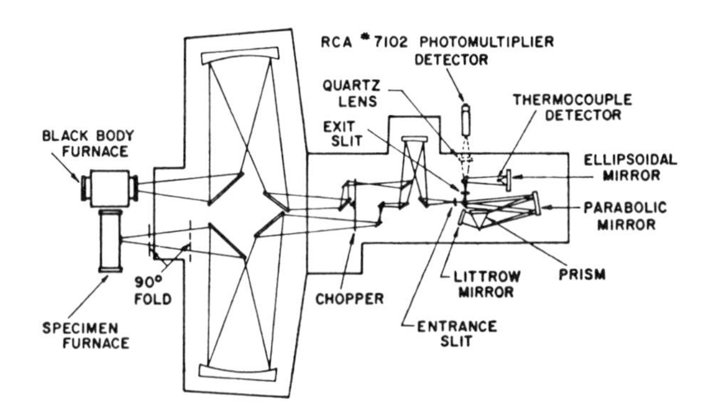

FIGURE 15. The optical system used by Richmond *et al.*[65] for separate-blackbody emittance measurements.

temperature of the blackbody furnaces was measured by means of a noble metal thermocouple whose bead was exposed inside and near the back of the cavity.

One blackbody, called the comparison blackbody, was rigidly mounted in position to serve as the source for the comparison beam of the spectrophotometer. The sample, in its housing, and the reference blackbody, were rigidly mounted side by side on a slide that moved on a lathe bed so that either could be locked in position to serve as the source for the sample beam. Rigidly positioned stops on the lathe bed permitted accurate positioning of the sample and blackbody. The complete measurement technique is described in detail by Harrison *et al.*[70]

In brief, the comparison blackbody was brought to the desired temperature by manual adjustment of the power input. Its temperature fluctuated with time over a few degrees. The reference blackbody was positioned to serve as the source for the sample beam, and its platinum thermocouple lead was connected to the platinum lead of the comparison blackbody thermocouple, and the two remaining leads were connected to the temperature control equipment. In essence, the two thermocouples became a differential thermocouple measuring the temperature difference between the two blackbodies. The differential thermocouple was connected to a center-zero recorder-controller that produced an output voltage that was proportional to the temperature difference. This voltage was fed to a current actuating type (CAT) controller, whose output was made up of the sum of three voltages, proportional to the input voltage, the first time derivative of the input voltage, and the second time derivative of the input voltage. The proportions of the three voltages could be independently varied, so that the response time of the controller could be matched to that of the heating load. The inputs were adjusted until the maximum temperature between the two blackbodies was less than 0.5 K.

The output voltage of the CAT control was fed to a thyristor-controlled saturable core reactor which supplied the power to the reference blackbody. The spectrophotometer was then scanned over the wavelength range of 1 to 15 μm, and the recorded ratio of the two signals was used as the 100% line for correction of spectrometer errors. The sample was then moved into position to serve as the source for the sample beam, and its temperature was controlled to be the same as that of the comparison blackbody by means of the same control system, adjusted to match the time constant of the sample. The spectrometer was again scanned over the wavelength range of 1 to 15 μm, and the ratio of the two signals was again recorded. Finally the sample beam was blocked near the sample by a water-cooled shutter, and the ratio of the two signals was recorded as the zero signal. The emittance $\epsilon(\lambda)$ at wavelength λ was computed as

$$\epsilon(\lambda) = \frac{S(\lambda) - Z(\lambda)}{H(\lambda) - Z(\lambda)} \tag{27}$$

where $S(\lambda)$ is the ratio for the sample, $Z(\lambda)$ is the ratio when the sample beam was blocked, and $H(\lambda)$ is the ratio when the reference blackbody was in the sample beam, all at wavelength λ. $\epsilon(\lambda)$ was evaluated at 100 wavelengths between 1 and 15 μm, and plotted as a function of wavelength to obtain the spectral emittance curve of the sample.

Metal samples are easy to heat by passing a current through them and usually have high thermal conductivity so that thermal gradients are small. Most metals are highly opaque, so that the emitted flux originates in a very thin layer below the surface, only a few or a few hundred atomic diameters thick. Nonmetallic samples, such as ceramics, present problems in several respects. Their electrical conductivity is usually so low that they cannot be heated by passing a current through them. In general, they tend to be somewhat translucent, at least at some wavelengths, so that the emitted flux originates in a layer of thickness measured in millimeters to meters. They tend to have low thermal conductivity and high thermal emittance, as compared to metals. These properties combine to produce thermal gradients in a heated sample unless precautions are taken to prevent them, and the gradients tend to be normal to an emitting or absorbing surface.

When a sample is emitting from a surface layer of appreciable thickness with a large thermal gradient normal to the surface, it has no unique temperature, and it is difficult to define an effective temperature for the emitting layer. Emittance is defined as the ratio of the radiance (or exitance) of a sample to that of a blackbody radiator at the same temperature, and under the same geometrical and spectral conditions. An effective temperature for a nonisothermal sample must be defined before its emittance can be evaluated. If the defined temperature is that of the surface, a sample with a positive thermal gradient (surface cooler than interior) will emit more flux than an isothermal sample at the defined temperature. In an extreme case the measured emittance could exceed 1.0. If the gradient is negative (surface hotter than interior) the sample will emit less flux than a sample at the defined temperature. If the effective temperature is defined as the temperature of an isothermal sample that emits the same flux, the effective temperature is difficult to evaluate, and if evaluated spectrally, will vary with wavelength, because the extinction coefficient, which determines the thickness of the emitting layer, varies with wavelength. The Planck equation does not apply to nonisothermal samples, and any attempt to so apply it will result in errors. For these reasons all measurements of the emittance of nonmetallic samples should be made under conditions where the thermal gradients normal to the surface are insignificant.

There are essentially three methods of reducing the thermal gradients to a negligible value: (1) alternate heating and viewing of a moving sample; (2) alternative heating by radiation and viewing of a stationary sample, and (3) heating a sample with monochromatic flux, as with a high-powered CW laser, and evaluating the emittance at wavelengths other than that used for heating. Moving samples, alternately heated and viewed, have been tested in the form of a rotating disk,[71–74] a rotating cylinder,[75,76] or as disks on an oscillating beam.[77] Stationary samples, alternately heated and viewed by use of rotating shutters, have been heated in an arc-image furnace[78,79] and in a solar furnace.[80]

In the case of a moving sample that is alternately heated and cooled the sample will reach a steady-state temperature, and if the motion is rapid enough that the viewing time is short, the temperature fluctuation of the viewed surface will be so small that is can be neglected.

Photon conduction does not generate a thermal gradient in a material. The photons are transmitted from the point of emission to the point of absorption without affecting the material in any way. As a result, the thermal gradient of a sample heated by radiation tends to be small near the surface. The radiant flux is absorbed within the layer from which the emitted flux originates, and when the sample has reached a steady-state condition there is little net conduction in the surface layer, either photon or phonon. If the time periods during which the sample is alternately heated and viewed are brief, the temperature fluctuation in the surface layer will be so small it can be neglected.

Although method (3) above – heating a sample with monochromatic flux and evaluating emittance at other wavelengths – does not seem to have been used, the very great advances made in laser technology in recent years suggest that this procedure may be the best of the three. A true steady state can be established, in which the gradient normal to the surface is essentially zero for the full thickness of the emitting layer.

In the rotating cylinder method developed at NBS[76] the sample was a

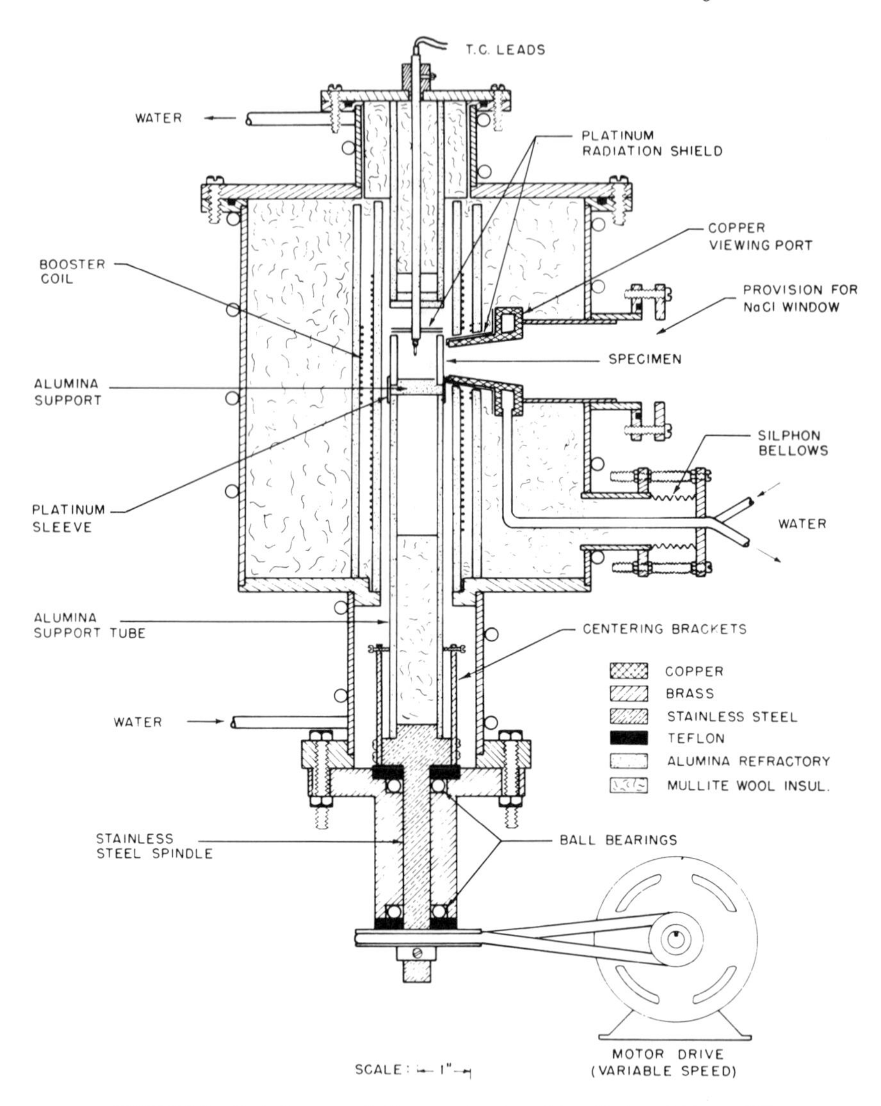

FIGURE 16. Rotating sample furnace developed by Clark and Moore.[76]

hollow cylinder 25 mm high and 25 mm in diameter, with walls ground to be round to within about 0.05 mm. A cross-sectional view of the furnace is shown in Fig. 16. The sample was mounted on top of an alumina pedestal which revolved inside a platinum-wound furnace. The winding of the furnace was designed so that there was no axial thermal gradient along the sample. The sample revolved in front of a water-cooled viewing port. A theoretical analysis[81] confirmed by temperature measurements at a point near the surface of the rotating sample, showed (1) that the temperature drop on the surface of the rotating sample as it passed the viewing port was inversely proportional to the speed of rotation of the sample, and was only about 2 K at 50 rpm, and (2) that the temperature measured by a stationary radiation-shielded thermocouple at the center of the rotating sample was within about 1 K of that measured by the thermocouple embedded near the surface of the rotating sample.

The flux from the hot sample was focused on the entrance slit of the sample beam of a double-beam infrared spectrophotometer, and the flux from a laboratory blackbody furnace, whose temperature was controlled to be the same as that of the sample, was focused onto the enhance slit of the reference beam of the spectrophotometer. The spectrophotometer automatically scanned over the spectral range of 1 to 15 μm, and plotted the ratio of the signals from the two beams as a function of wavelength, which, after correction for 100% line and zero line errors as previously discussed, was the normal spectral emittance of the sample.

Total normal emittance can be measured by the techniques described above if a detector with a flat spectral response over the wavelength range of the emitted radiant flux is used, or if the emittance of the sample is independent of wavelength (a graybody emitter).[82] This technique has also been used to study the variation of total directional emittance with angle of viewing.[83] Total normal emittance can also be computed from measured normal spectral emittance, as has been discussed previously.

The pulse heating technique has been used to measure total hemispherical emittance calorimetrically, and normal spectral emittance at the optical pyrometer wavelength radiometrically, of a wide variety of materials. The equipment is described in detail by Cezairliyan.[84] A sample in the form of a tube 102 mm long, 6.3 mm o.d., and a wall thickness of 0.5 mm was heated very rapidly (in a fraction of a second) to a very high temperature by a high-amperage pulse of dc electrical power. The data were recorded in the core memory of a high-speed data acquisition system in the form of a 13-bit-plus-sign-bit binary signal. A signal for each of three parameters was recorded each 0.4 ms. The data included (1) the potential across probes on the sample, (2) the current in the pulse, and (3) the temperature of the sample, and (4) the radiance temperature of the sample surface. The temperatures were measured by a high speed photoelectric pyrometer[85] sighted on a rectangular aperture in the wall of the sample, 0.5 $\times$ 1.0 mm in size, for the sample temperature, or on the exterior of the sample for the radiance temperature. The power input was cut off on reaching a preselected temperature below the melting point of the sample, and the rate of temperature change during cooling of the sample was recorded. Total hemispherical emittance was computed from the power input to, the temperature, and rates of heating and cooling of the sample. Normal spectral emittance at the optical pyrometer wavelength was evaluated by measuring the radiance temperatures of the aperture and surface of the sample. The measured temperatures were converted to radiances, and the normal spectral emittance was computed as the ratio of the radiances. Data have been published on molybdenum,[86] tantalum,[87] tungsten,[88] niobium,[89] iron,[90] vanadium,[91] zirconium,[92] graphite,[93] titanium,[94] tantalum–10% tungsten,[95] niobium–1% zirconium,[96] niobium–tantalum–tungsten,[97] hafnium–3% zirconium,[98] 90 Ti–6 Al–4 V,[99] titanium,[100] and some refractory alloys.[101]

3.2.2. *Calorimetric Emittance and Absorptance Measurements*

In calorimetric emittance and absorptance measurements the rate of radiant heat transfer to or from a sample is measured in terms of the heat lost or gained by the sample. The ratio of the measured rate of heat transfer to that of a blackbody ratiator or absorber under the same conditions is the emittance or absorptance of the sample. The evaluated emittance is usually the total hemispherical emittance, but absorptance may be for flux of any spectral distribution, including monochromatic (narrow bandwidth), if the irradiance is high enough to permit the rate of heat gain to be measured.

The sample must be thermally insulated from its surroundings to prevent unwanted heat leaks to or from the sample. The rate of heat change may be evaluated in terms of the rate of temperature change of a sample of known heat capacity in dynamic measurements, or in terms of the measured heat transfer by other means required to maintain the sample in thermal equilibrium in steady-state measurements.

It is convenient to classify calorimetric emittance measurements into three temperature ranges – above 500 K, 270–500 K, and below 270 K.

The simplest calorimetric technique, for measuring total hemispherical emittance above about 500 K, is the hot-filament method described by Worthing.[102] The sample, in the form of a long filament, is heated in vacuum by passing a current through it. If the wire is of uniform cross-sectional area and resistance, a section at the center will quickly come to a uniform equilibrium temperature. If a small section at the center has been instrumented with fine-wire thermocouples (with very small diameters compared to the filament) the temperature can be measured directly and the potential across the center portion can be measured by using the thermocouples as potential leads. The length and diameter of the central section is also measured. The total hemispherical emittance $\epsilon(2\pi; t; T)$ at the sample temperature T is computed by use of the equation

$$\epsilon(2\pi; t; T) = \frac{V \cdot I}{A \cdot \sigma \cdot (T_1^4 - T_2^4)} \tag{28}$$

where V is the potential, in volts, across the central section, I is the current in amperes flowing in the wire, A is the surface area of the central section, T_1 is the temperature of the central section, and T_2 is the temperature of the walls of the vacuum chamber. There are several assumptions made in the derivation of equation (28), which are discussed in the literature. Care must be taken to see that the thermocouples do not pick up a potential due to the current flowing in the sample, particularly if dc power is used to heat the sample, and that true rms voltage is measured and that correction is made for any power factor loss if ac power is used to heat the sample.

The same general method has been used by Richmond and Harrison[103] and Cezairliyan *et al.*[86] for metal strip samples with the addition of (1) water cooling of the walls of the vacuum chamber, (2) provision for expansion of the heated sample, and (3) guard heaters for the ends of the strip. Figure 17 is a drawing of one such apparatus.

The overall accuracy depends upon the magnitude of $(T_2 - T_1)$ and the magnitude, relative to the heat loss by radiation, of the heat loss by conduction

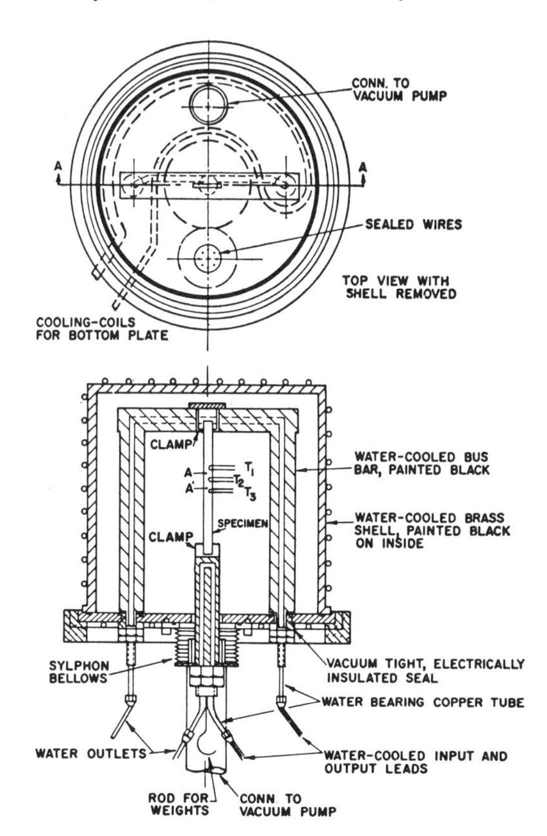

FIGURE 17. An instrumented sample for calorimetric measurement of total hemispherical emittance, Richmond and Harrison.[103]

and convection through the atmosphere of the vacuum chamber and by conduction through the thermocouple leads, and the assumptions made in deriving equation (28). With careful work, uncertainties on the order of 2% are easily attained for sample temperatures above 500 K. Error analyses of the method have been published by McElroy and Kollie[104] and Nelson and Bevans.[105] Measurements to 1800 K are relatively simple and reliable with conventional noble-metal thermocouples. Measurements have been made to 3000 K with the use of refractory metal thermocouples, with a significant increase in uncertainty. The use of a hollow cylindrical[86] or triangular[68] sample, with a small aperture on which an electronic optical pyrometer is sighted extends the range to near the melting point of the most refractory materials. In this case potential taps of a refractory metal are required.

The hot filament method can be used at lower temperatures by decreasing T_2. This is commonly done by cooling the vacuum chamber with liquid nitrogen, which extends the lower temperature limit down from about 500 K to about 270 K.[105] Heat losses by conduction become a much larger fraction of the total heat loss at these temperatures, and more stringent precautions must be taken to reduce them. Also longer times are required to reach thermal equilibrium in steady-state measurements.

In the 270 to 500 K temperature range the heat input is frequently by radiation.[106–108] The sample may be heated by a beam from a solar simulator, and $\alpha(0;S)/\epsilon(2\pi;t)$, the ratio of normal solar absorptance (for extraterrestrial solar flux) to total hemispherical emittance is computed from T_1, the equilibrium temperature of the sample, and T_2, the temperature of the chamber walls, by use of the equation

$$\frac{\alpha(0;S)}{\epsilon(2\pi;t)} = \frac{A_2\sigma(T_1^4 - T_2^4)}{EA_1} \tag{29}$$

where A_2 is the total surface area of the sample, σ is the Stefan–Boltzmann constant, E is the irradiance on the sample, and A_1 is the irradiated area of the sample; or the absorptance, $\alpha(0;S)$ during heating by use of the equation

$$\alpha(0;S) = \frac{\epsilon(2\pi;t)A_2\sigma(T_1^4 - T_2^4) + Mc_p(dT_1/dt)}{EA_1} \tag{30}$$

where M is the mass of the sample, c_p is the specific heat of the sample, and dT_1/dt is the rate of temperature rise of the sample. The source is then blocked or turned off, and the sample is allowed to cool. The total hemispherical emittance is then computed by use of the equation

$$\epsilon(2\pi;t) = \frac{Mc_p}{\sigma A_2(T_1^4 - T_2^4)}\,\frac{dT_1}{dt} \tag{31}$$

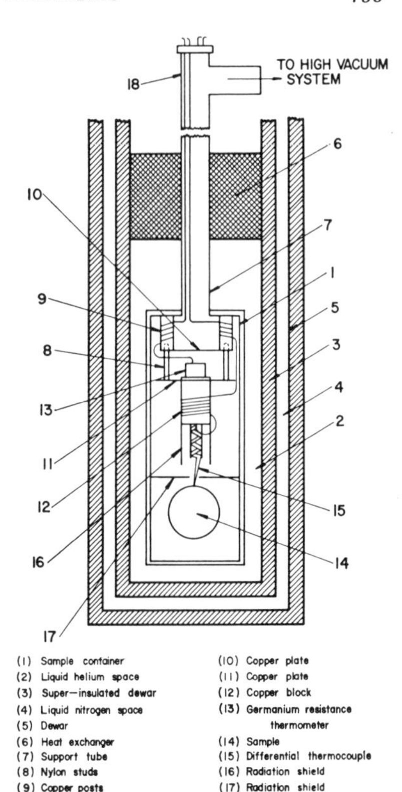

FIGURE 18. A helium-cooled chamber, with electrically heated spherical sample, for calorimetric total hemispherical emittance measurements in the 20–300 K temperature range (Caren, Ref. 113).

where dT_1/dt in this case is the cooling rate of the sample.

For measurements below 270 K the sample can be surrounded with a shroud cooled to liquid-helium temperature,[109,110] and $\alpha(0; S)/\epsilon(2\pi; t)$, $\alpha(0; S)$, and $\epsilon(2\pi; t)$ can be measured as outlined above. For steady state measurements of $\epsilon(2\pi; t)$ the heat input may be electrical.[111,112] Figure 18 shows a helium-cooled chamber used for measurements in the 20–300 K range,[111] in which electrical heating is used. The total hemispherical emittance can be computed from the net heat transfer rate between two parallel plates at different temperatures. Figure 19 shows one such apparatus,[113,114] in which two samples of the same material are used. The lower sample, 20 cm in diameter, is electrically heated and maintained at about 300 K. The upper sample, 10 cm in diameter, is cooled with liquid nitrogen to about 80 K. The net heat transfer rate was measured by the rate of boiloff of the liquid nitrogen from the inner Dewar. The emittance ϵ was computed by use of the equation

$$\Phi = \sigma \frac{\epsilon}{2-\epsilon} A(T_1^4 - T_2^4) \tag{32}$$

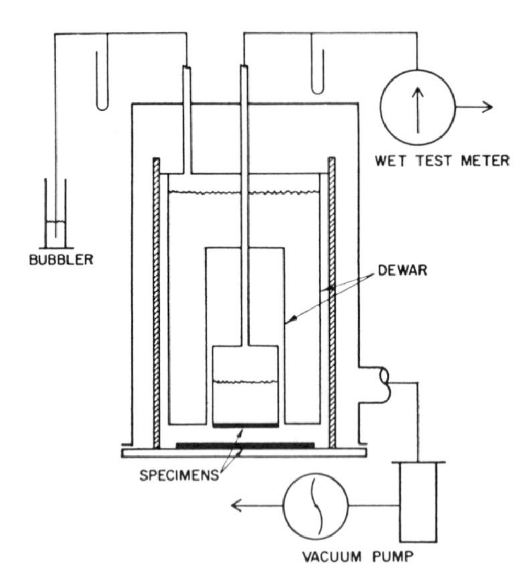

FIGURE 19. A parallel-plate total hemispherical emittance apparatus (Haury, Refs. 115, 116).

where Φ is the measured heat transfer rate, A is the area of the upper cold sample, T_1 is the temperature of the hot sample, and T_2 is the temperature of the cold sample. The emittance ϵ is an average total hemispherical emittance of the material at the two temperatures. The ratio $\epsilon/(2-\epsilon)$ in the equation is the expression for the exchange factor between infinite parallel plates. The small separation between the plates, about 6 mm, compared to the diameter of the smaller (upper) plate of 10 cm approximates this condition. Caren[(116)] describes a similar instrument for use in the range of 10 to 300 K.

Biondi[(116)] measured the spectral absorptance and reflectance of specularly reflecting samples of copper and silver at 4.2 K. The sample and a black absorber were thermally bonded to copper stages of appreciable thermal mass, which were in turn thermally connected to a liquid-helium heat sink through leaks of the proper thermal conductivity to give the stage a time constant of about 10 s. The black absorber was positioned to receive the flux specularly reflected by the sample. The stages could also be electrically heated for calibration. The sample was irradiated by a beam of monochromatic flux producing a known irradiance, and allowed to come to thermal equilibrium. The temperatures of the copper stages and of the liquid helium sink were then measured. The incident flux was then turned off, and the copper blocks were heated electrically to produce the same equilibrium temperatures. The power required to maintain the stages at the equilibrium temperature was used as a measure of the flux absorbed by the sample and black absorber, and the absorptance and reflectance of the sample were computed.

3.3. Measurement of Transmittance

Transmitting materials are of two types, transparent and translucent. A transparent material is optically homogeneous, so that each ray maintains its identity in passing through the material, and if the sample has optically smooth parallel surfaces, objects viewed through the sample will appear undistorted and sharply defined. A translucent material is optically inhomogeneous, and any incident ray is scattered and loses its identity in traversing a small thickness, so that objects cannot be seen through it. Essentially all of the flux transmitted by a translucent sample is scattered over a complete hemisphere.

Transmittance varies with the path length in the material. For this reason, samples for transmittance measurements should be of uniform thickness, with flat, optically smooth, faces. For translucent samples the thickness should be small compared to the minimum width dimension of the sample.

The transmittance of a transparent sample is easily measured. A simple optical system of a source, collimator, and detector-amplifier with linear response to incident flux is all that is required. The signal is recorded with and without the sample in the beam, and the ratio of the two signals is the transmittance of the sample. The measured transmittance is a spectrally total transmittance, weighted by the combination of the spectral distribution of the source and the spectral response of the detector. For spectral transmittance a monochromator is introduced into the system, between the source and sample or the sample and detector.

Commercial double-beam spectrophotometers suitable for measuring spectral transmittance of transparent samples are widely available.

Specular reflectance occurs at both surfaces of a sample. This can cause errors if multiple reflections occur between the sample and optical elements in the measuring system. Such multiple reflections can be avoided by slight tilting of the sample, so that the incident beam strikes it off normal. A tilt of about one degree is usually sufficient.

Measurement of transmittance of translucent samples, which requires the use of an integrating sphere or integrating mirror, is much more difficult than for transparent samples. Unfortunately, there does not appear to be any readily available commercial instrument specifically designed for directional–hemispherical transmittance measurements. A single-beam directional–hemispherical reflectometer can be altered to make directional–hemispherical transmittance measurements if the optical path can be altered to be incident on the back, rather than the front, of the sample. With an integrating sphere reflectomer, such measurements can be made by placing the sample over the entrance port to the sphere, and closing the sample port with a plug whose inner surface conforms to the shape of the sphere and is coated with the sphere coating. Double-beam reflectometers are much more difficult to adapt to such measurements. Modification of the entrance port will probably be required, so that the sample is at least tangent to the inner surface of the integrating sphere, and preferably is in a position where the rim of the port is at the inner surface of the sphere. The sample and port must be large compared to the area of the incident beam, to avoid edge losses of transmitted flux, because the beam will spread widely in passing through the sample, particularly if it is thick. The exitent area through which the flux leaves the sample is much larger than the incident beam. For integrating mirror reflectometers, the sample remains in the same position as for reflectance measurements, but the optical path must be altered to irradiate the sample on the rear surface, rather than from the front. Such measurements are rather rare.

4. DISCUSSION AND CONCLUSIONS

The thermal radiation properties of materials are reflectance, absorptance, transmittance, and emittance. These properties vary with the temperature and surface topography of the sample, the spectral distribution, geometry, amount and direction of polarization of the incident beam for reflectance, absorptance and transmittance, and the geometry of the flux collected for measurement and the spectral, polarization, and directional response of the detector, for reflectance, transmittance, and emittance measurements. Fortunately properties involving near-normal incident or exitent flux are unaffected by polarization, and the variation of the thermal radiation properties of optically homogeneous samples with an optically smooth surface can be computed if the (complex) index of refraction is known. For opaque samples at any one temperature, a single evaluation of spectral near-normal reflectance, or spectral normal emittance, supplies data from which the other properties can be computed, if the spectral range is wide enough. Reflectance measurements are most widely used at room temperatures, and emittance measurements are most commonly used at high temperatures.

The integrating sphere reflectometer is the most widely used instrument for measuring directional–hemispherical reflectance at room temperature and at wavelengths from 0.25 to 2.5 μm. At wavelengths longer than 2.5 μm the heated-cavity or integrating mirror reflectometer is most commonly used, but the Fourier transform reflectometer with gold-coated integrating sphere is coming into increasing use in the 2–20 μm range. The measurement of spectral emittance of metallic samples at temperatures above 500 K presents no problem, but the thermal gradient effects present in most nonmetallic materials makes the evaluation of emittance of such materials much more difficult. Transmission measurements of transparent materials present no serious problem, but transmittance of translucent materials presents serious measurement problems.

Calorimetric procedures are most commonly used to evaluate absorptance, particularly extraterrestrial solar absorptance, and total hemispherical emittance. The extraterrestrial solar absorptance is usually evaluated at relatively low temperatures, from about 140 to 500 K. Total hemispherical emittance can be evaluated at temperatures from about 20 K to near the melting point of any metallic sample. Similar measurements on dielectric materials are more difficult to make.

Three developments over the last decade show promise of greatly facilitating the radiometric evaluation of thermal radiation properties – the laser, the pyroelectric detector, and the use of Fourier transform spectrometers.

APPENDIX: NOMENCLATURE

In discussing radiant heat transfer and thermal radiation properties of materials in detail, it is necessary to define the terms that are used. Because processes and measurements usually involve continuous beams of radiant flux, the time rate of flow of radiant energy, most of the quantities will be defined in terms of flux rather than in terms of energy. Quantities will be defined with the symbol for the quantity following the term, then a mathematical equation for the quantity in terms of flux or energy, in parentheses, then the unit dimensions of the quantity in square brackets and finally the verbal definition. The definitions are consistent with those in the International Lighting Vocabulary.(1)

A.1. Processes

Absorption. Process by which radiant flux (energy) is converted into another form of energy, usually heat, by interaction with matter.

Detection. Process by which radiant flux (energy) is absorbed by a device which produces a signal (physical change) that is related to the flux (energy) absorbed.

Propagation. Process by which radiant flux is transferred from one point to another in space. It may include reflection and refraction in addition to transmission.

Radiation. (1) Process by which radiant flux is emitted by a body. (2) Process by which energy is transferred in the form of electromagnetic waves or photons. (3) Sometimes used as a synonym for the more precise terms *radiant flux* or *radiant energy*.

Reflection. Process by which radiant flux incident on a stationary surface or medium leaves that surface or medium from the incident side without change in frequency.

Refraction. Process by which a beam of radiant flux, obliquely incident on an interface between two media of different indices of refraction, is deviated so that the angle between the ray and the normal to the interface is greater in the medium of lower index of refraction. See *Snell's law.*

Transmission. Process by which radiant flux incident on a stationary surface or body leaves that surface or body through a surface other than the incident surface, without change in frequency.

Note. The frequency of flux incident on a moving surface will be changed by a doppler shift on reflection or transmission.

A.2. Objects

Absorber. A body that absorbs incident radiant flux.

Blackbody. An ideal body that absorbs all radiant flux incident upon it, and emits the maximum possible spectral concentration of flux at all wavelengths for any body at its temperature.

Detector. A device that absorbs incident radiant flux and produces a signal (physical change) related to the amount of incident flux.

Radiator. A source of radiant flux.

Reflector. A body that reflects incident radiant flux.

Retroreflector. A body that reflects incident radiant flux predominantly in directions close to the direction of incidence.

Thermal radiator. A source that emits radiant flux as a consequence of its temperature only.

Transmitter. A body that transmits incident radiant flux.

A.3. Quantities

Exitance, M. ($M = d\Phi/dA$) [$\mathrm{W\,m^{-2}}$] (At a point on a surface.) Flux per unit area leaving the surface. Exitance may include emitted, reflected, and/or transmitted flux.

Fluence, F. ($F = \int_t F_t dt$) [$\mathrm{J\,m^{-2}}$] (At a point in space.) Quotient of the radiant energy incident on an elementary sphere with the point at its center by the cross sectional area of the sphere. Time integral of fluence rate.

Fluence rate, F_t. ($F_t = \int_{4\pi} L d\omega$) [$\mathrm{W\,m^{-2}}$] (At a point in space.) Quotient of the flux incident on an elementary sphere with the point at its center by the cross-sectional area of the sphere.

Irradiance, E. ($E = d\Phi/dA$) [$\mathrm{W\,m^{-2}}$] (At a point on a surface.) Quotient of the flux incidence on an element of area surrounding the point by the element of area.

Radiance, L. [$L(\theta,\phi) = d^2\Phi(d\omega\, dA \cos\theta)^{-1}$] [$\mathrm{W\,m^{-2}\,sr^{-1}}$] (At a point on a real or imaginary surface, and in a given direction.) Quotient of the flux incident on, passing through, or emitted by an element of area surrounding the point and contained within an element of solid angle surrounding the given direction, by the product of the element of solid angle and the orthogonal projection of the element of area normal to the given direction.

Radiant density, w. ($w = dW/dV$) [J m^{-3}] (At a point in space.) Quotient of the radiant energy contained in an element of volume surrounding the point by the element of volume.

Radiant energy, Q. [J]. Energy in the form of electromagnetic waves or photons.

NOTE. The terms "photon" and "electromagnetic wave" are different terms for the same form of energy. The emission and absorption of radiant energy by atoms and molecules is more easily understood in terms of quantum mechanics, based on its photon properties. Propagation and interactions with matter on a macroscopic scale are more easily understood on the basis of physical optics, based on the wave properties of radiant energy.

Radiant flux, Φ. ($\Phi = dQ/dt$) [W]. Time rate of flow of radiant energy.

Radiant intensity, I. ($I = d\Phi/d\omega$) [W sr^{-1}] (Of a point source in a given direction.) Quotient of the flux emitted by a point source† and contained in an element of solid angle surrounding the direction, by the element of solid angle.

Refractive index, n. ($n = V_s \cdot c^{-1}$) [dimensionless]. Ratio of the velocity of light in a medium, V_s, to the velocity of light in vacuum, c.

Snell's law. Law of refraction.

$$n_i \cdot \sin\theta_i = n_r \cdot \sin\theta_r \tag{34}$$

where n_i is index of refraction of medium on incident side, θ_i is the angle of incidence, n_r is the index of refraction on the refracted side, and θ_r is the angle of the refracted ray. Angles θ are measured from surface normal.

Thermal radiant flux, Φ. ($\Phi = dQ/dt$) [W]. Radiant flux emitted by a thermal radiator. Sometimes called thermal radiation.

A.4. Properties

Absorptance, α. ($\alpha = \Phi_a/\Phi_i$) [dimensionless]. Ratio of absorbed flux, Φ_a, to incident flux, Φ_i.

†Intensity is based on the inverse square law, which is exact for a point source, and states that the irradiance from a point source varies inversely as the square of the distance from the source to the plane of measurement. Mathematically

$$E = I \cdot D^{-2} \tag{33}$$

where E is the irradiance, I is the intensity, and D is the distance from the source to a plane of measurement normal to the ray from the source. The deviation from the inverse square law is less than 1% if the maximum dimension of the source projected normal to the direction is less than $0.1\,D$.

Absorption coefficient, a. $(a = d\Phi \cdot (\Phi \cdot dx)^{-1})$ [m^{-1}]. A property of a transmitting material. The fraction of internally transmitted flux absorbed per unit path length. The coefficient a is part of the exponent in the Bouguer's law equation

$$\Phi_x = \Phi_0 \cdot e^{-ax} \tag{35}$$

where Φ_x is the flux in a beam at a point at a distance x from point 0 where the flux in the beam is Φ_0.

Attenuation coefficient, μ. $(\mu = d\Phi \cdot (\Phi \cdot dx)^{-1})$ [m^{-1}]. A property of a translucent material. The fraction of internally transmitted flux lost by absorption and/or scattering per unit path length. Equation (35) also applies to attenuation coefficient, with μ replacing a.

Emittance, ϵ. $(\epsilon = \Phi_s/\Phi_{bb})$ [dimensionless]. Ratio of flux (radiance or exitance) emitted by a sample to that emitted by a blackbody radiator at the same temperature and under the same spectral and geometric (including polarization) conditions of measurement.

Reflectance, ρ. $(\rho = \Phi_r/\Phi_i)$ [dimensionless]. Ratio of reflected flux to incident flux.

Reflectance factor, R. $(R = \Phi_{r,s}/\Phi_{r,d})$ [dimensionless]. Ratio of flux $\Phi_{r,s}$ reflected by a sample under specified conditions of irradiation and collection, to $\Phi_{r,d}$, that reflected by the ideal completely reflecting, perfectly diffusing surface under identical conditions of irradiation and collection.

Transmittance, τ. $(\tau = \Phi_t/\Phi_i)$ [dimensionless]. Ratio of transmitted flux to incident flux.

NOTE. The thermal radiation properties of materials, especially those of metals, are strongly influenced by the surface topography, surface films, and, for partially translucent materials such as many nonmetallics, by the sample thickness and internal structure of the sample. The terms above ending in "ance" are the properties of a sample of a material. Similar properties ending in "ivity" – reflectivity, absorptivity, and emissivity – are properties of the material under ideal conditions. They can be approximated by measuring the properties of clean, optically smooth samples of thickness sufficient to be completely opaque. Transmissivity is also a property of a material, and is defined as the internal transmittance per unit path length in the material. A more useful material property for transmittance is the attenuation coefficient, which is equal to the absorption coefficient for nonscattering materials.

The thermal radiation properties are strongly influenced by the conditions of measurement, including (1) the geometric and spectral distribution and the polarization (amount and direction relative to the plane of incidence), of incident flux for reflectance, absorptance, and transmittance, and (2) the size and direction of the solid angle over which flux is collected for measurement, and the spectral, directional, and polarization sensitivity of the detector used for the measurement, for emittance, reflectance, and transmittance.

The modifiers defined in the next section may be used to specify the geometrical and spectral conditions under which properties are evaluated. Unless otherwise indicated, the incident flux is assumed to be isotropically diffuse over the solid angle of irradiance and unpolarized, the detector is assumed to be insensitive to differences in polarization and the geometric distribution of the collected flux.

A.5. Modifiers

Conical. Over a finite solid angle smaller than a hemisphere and larger than an element of solid angle. Both the size the the direction of the solid angle must be specified. For right circular cones the symbol is (θ, ϕ, α),† where θ, ϕ is the axial direction, and α is the plane half-angle of the cone. For other geometries the symbol is (g), and the solid angle must be described in detail in the text.

Diffuse. (1) For a reflector or transmitter, reflecting or transmitting in all possible directions over a hemisphere, (2) see *hemispherical.* Symbol: subscript d.

Directional. In a given direction relative to a surface. The direction is specified by two angles, the angle θ between the direction and the normal to the surface, and the angle ϕ to the plane of incidence, measured counterclockwise from a reference direction on the surface. Symbol: (θ, ϕ) in the general case. The size of the angles, in radians or degrees, is given in a specific case. The azimuth angle ϕ may be omitted for azimuthally isotropic samples.

Exitent. From a surface, as exitent flux or exitent radiance.

Hemispherical. Over all directions contained in a complete hemisphere. Symbol (2π).

NOTE. Both conical and hemispherical incidence imply isotropically diffuse incident flux over the indicated solid angle.

Isotropically diffuse. With the same radiance in all directions.

Linearly polarized. (For a beam of radiant flux.) With the electric vectors of all photons parallel to each other.

Normal. (1) In a direction normal to the surface. Symbol (0). (2) (For a beam of plane polarized flux.) With the electric vector normal to the plane of incidence. Symbol, subscript $\perp$.

Parallel. (For a beam of plane polarized flux.) With the electric vector parallel to the plane of incidence. Symbol, subscript $\|$.

Perfectly diffuse. See Isotropically diffuse.

Perpendicular. See Normal.

†The symbol α is also used for absorptance. This should not lead to confusion, since the α used here is always given at (θ, ϕ, α).

Plane of polarization. (For a beam of linearly polarized flux.) The plane containing the electric vector.

Solar. (1) (For a quantity.) Having the relative spectral distribution of solar energy. Symbol: subscript S. (2) (For a property.) Weighted average of the spectral property, with the relative spectral distribution of solar energy as the weighting function. Symbol (S).

Spectral. (1) (For a quantity) Spectral concentration (a) as a function of wavelength; symbol: subscript λ, $(L_\lambda = dL/d\lambda)$ $[\mathrm{W\,m^{-2}\,sr^{-1}\,nm^{-1}}]$; (b) at a specific wavelength; symbol: wavelength in nm or μm in parentheses as L_λ (600 nm). (2) (For a property.) (a) As a function of wavelength; symbol (λ); (b) at a specific wavelength; symbol, wavelength in nm or μm in parentheses, as α (600 nm).

REFERENCES

1. *International Lighting Vocabulary,* 3rd Edition. Publication CIE No. 17 (E1.1), Bureau Central de la CIE, Paris, France (1970).
2. G.J. Kneissl and J.C. Richmond, "A Laser-Source Integrating Sphere Reflectometer," NBS Tech. Note 439 (February 1968) Appendix B gives a derivation of Kirchhoff's Law.
3. M. Jakob, *Heat Transfer,* Vol. 1, John Wiley and Sons, New York (1949), pp. 43 and 51.
4. F.E. Nicodemus, J.C. Richmond, J.J. Hsia, I.W. Ginsberg, and T. Limperis, "Geometrical Considerations and Nomenclature for Reflectance," NBS Monograph 160 (October 1977).
5. R.V. Dunkle, "Spectral Reflectance Measurements," in *Surface Effects on Spacecraft Materials* (F.J. Clauss, Ed.), Wiley, New York (1960), pp. 117–137.
6. R.S. Hembach, L. Hemmerdinger, and A.J. Katz, "Heated Cavity Reflectometer Modifications," in *Measurement of Thermal Radiation Properties of Solids* (J.C. Richmond, Ed.), NASA SP-31. U.S. Government Printing Office, Washington, D.C. (1963), pp. 153–167.
7. E.R. Streed, L.A. McKellar, R. Rolling, Jr., and C.A. Smith, "Errors Associated with Hohlraum Radiation Characteristics Determinations," in *Measurement of Thermal Radiation Properties of Solids* (J.C. Richmond, Ed.), NASA SP-31, U.S. Government Printing Office, Washington, D.C. (1963), pp. 237–252.
8. J.T. Gier, R.V. Dunkle, and J.T. Bevans, "Measurement of Absolute Reflectivity from 1.0 to 15 Microns," *J. Opt. Soc. Am.* **44,** 558 (1954).
9. R.V. Dunkle, D.K. Edwards, J.T. Gier, K.E. Nelson, and R.D. Roddick, "Heated Cavity Reflectometer for Angular Reflectance Measurement," in *Progress in International Research on Thermodynamics and Transport Properties,* ASME, New York (1962), pp. 100–106.
10. D.G. Goebel, "Generalized Integrating Sphere Theory," *Appl. Opt.* **6,** 125 (1967).
11. J.A. Jacquez and H.F. Kuppenheim, "Theory of the Integrating Sphere," *J. Opt. Soc. Am.* **45,** 460 (1954).
12. "Preparation of Reference White Reflectance Standards," ASTM Designation E 259–66. 1973 Annual Book of ASTM Standards, Pt. 30, 775–777 (1973).

13. W.E.K. Middleton and C.L. Sanders, "The Absolute Spectral Diffuse Reflectance of Magnesium Oxide," *J. Opt. Soc. Am.* **41,** 419 (1951).
14. W.E.K. Middleton and C.L. Sanders, "An Improved Sphere Paint," *Illum. Eng.* **48,** 254 (1953).
15. K.E. Eckerle, W.H. Venable, Jr., and V.R. Weidner, "Averaging Sphere for Ultraviolet, Visible and Near Infrared Wavelengths: A Highly Effective Design," *Appl. Opt.* **15,** 703–707 (March, 1976).
16. F. Grum and M. Saltzman, "New White Standard of Reflectance," P-75-77, Compte Rendue, 18° Session, pp. 91–98, Publication CIE No. 36 (1976). Bureau Central de la CIE, Paris, France.
17. F. Grum and G.W. Luckey, "Optical Sphere Paint and a Working Standard of Reflectance," *Appl. Opt.* **7,** 2289 (1968).
18. D.K. Edwards, J.T. Gier, E.K. Nelson, and R.D. Roddick, "Integrating Sphere for Imperfectly Diffuse Samples," *Appl. Opt.* **51,** 1279 (1961).
19. R.R. Willey, "Fourier Transform Infrared Spectrophotometer for Transmittance and Diffuse Reflectance Measurements," *Appl. Spectrosc.* **30,** 593–601.
20. W.R. Willey, "Infrared Reflectance. Independent Measurements Yield Good Agreement," *Appl. Opt.* **15,** 1124 (1976).
21. M. Kronstein, R.J. Krauschaar, and R.E. Deacle, "Sulphur as a Standard of Reflectance in the Infrared," *J. Opt. Soc. Am.* **53,** 458 (1963).
22. S.T. Dunn, "Application of Sulfur Coatings to Integrating Spheres," *Appl. Opt.* **4,** 377 (1965).
23. S.T. Dunn, "Flux Averaging Devices for the Infrared," NBS Technical Note 279. National Bureau of Standards, Washington, D.C. (1965).
24. G.J. Kneissl, J.C. Richmond, and J.A. Wiebelt, "A Laser Source Integrating Sphere for Measurement of Directional Hemispherical Reflectance at High Temperatures," in *Progress in Aeronautics and Astronautics,* Vol. 20 (G.B. Heller, Ed.), Academic Press, New York, pp. 177–202.
25. H.J. McNicholas, "Absolute Methods in Reflectometry," *J. Res. Natl. Bur. Stand.* **1,** 29 (1928).
26. W.E. Martin, "Hemispherical Spectral Reflectance of Solids," in *Measurement of Thermal Radiation Properties of Solids* (J.C. Richmond, Ed.), NASA SP-31, U.S. Government Printing Office, Washington, D.C. (1963), pp. 183–192.
27. M.R. Null and W.W. Lozier, "Measurement of Reflectance and Emittance at High Temperatures with a Carbon Arc Image Furnace," in *Measurement of Thermal Radiation Properties of Solids* (J.C. Richmond, Ed.), NASA SP-31, U.S. Government Printing Office, Washington, D.C. (1963), pp. 535–551.
28. R.G. Wilson, "Hemispherical Spectral Emittance of Ablation Chars, Carbon and Zirconia (to 3700 K)," in *Symposium on Thermal Radiation of Solids* (S. Katzoff, Ed.), NASA SP-55, U.S. Government Printing Office, Washington, D.C. (1965), pp. 259–275.
29. W.W. Coblentz, "The Diffuse Reflecting Power of Various Substances," *Bull. Bur. Stand.* **9,** 283 (1913).
30. W.M. Brandenberg, "Focusing Properties of Hemispherical and Ellipsoidal Mirror Reflectometers," *J. Opt. Soc. Am.* **54,** 1235 (1964).
31. J.E. Janssen and R.H. Torborg, "Measurement of Spectral Reflectance Using an Integrating Hemisphere," in *Measurement of Thermal Radiation Properties of Solids* (J.C. Richmond, Ed.), NASA SP-31, U.S. Government Printing Office, Washington, D.C. (1963), pp. 169–182.
32. J.U. White, "New Method for Measuring Diffuse Reflectance in the Infrared," *J. Opt. Soc. Am.* **54,** 1332 (1964).

33. H.J. Keegan and V.R. Weidner, "Infrared Spectral Reflectance of Frost," *J. Opt. Soc. Am.* **56,** 523 (1966).
34. R.C. Birkebak and J.P. Hartnett, "Measurements of Total Absorptivity for Solar Radiation of Several Engineering Materials," *Trans. ASME* **80,** 373 (1958).
35. R.P. Kozyrev and O.E. Vershinin, "Determination of Spectral Coefficients of Diffuse Reflection of Infrared Radiation from Blackened Surfaces," *Opt. Spectrosc.* **6,** 345 (1959).
36. W.L. Derksen and T.I. Monahan, "Automatic Recording Reflectometer for Measuring Diffuse Reflectance in the Visible and Infrared Regions," *J. Opt. Soc. Am.* **42,** 263 (1952).
37. W.L. Derksen, T.I. Monahan, and A.J. Lawes, "Automatic Recording Reflectometer for Measuring Diffuse Reflectance in the Visible and Infrared Regions," *J. Opt. Soc. Am.* **47,** 995 (1957).
38. J.A. Sanderson, "The Diffuse Spectral Reflectance of Paints in the Near Infrared," *J. Opt. Soc. Am.* **37,** 771 (1947).
39. R.T. Neher and D.K. Edwards, "Far Infrared Reflectometer for Imperfectly Diffuse Specimens," *Appl. Opt.* **4,** 775 (1965).
40. J.T. Neu, "Design, Fabrication and Performance of an Ellipsoidal Spectroreflectometer," NASA-CR-73193 (1968).
41. R.P. Heinisch, F.J. Bradac, and D.B. Perlick, "On the Fabrication and Evaluation of an Integrating Hemiellipsoid," *Appl. Opt.* **9,** 483–489 (1970).
42. S.T. Dunn, J.C. Richmond, and J.A. Wiebelt, "Ellipsoidal Mirror Reflectometer," *J. Res. Natl. Bur. Stand.* **C70,** 75 (1966).
43. J.C. Richmond and J.C. Geist, "Intrared Reflectance Measurements," National Bureau of Standards Report 10,071 (1970).
44. P. Fowler, "Far Infrared Absorptance of Gold," National Technical Information Service, U.S. Government AD418 456 (1960).
45. D.M. Gates, C.C. Shaw, and D. Beaumont, "Infrared Reflectance of Evaporated Metal Films," *J. Opt. Soc. Am.* **48,** 88 (1958).
46. J. Strong, *Procedures in Experimental Physics,* Prentice-Hall, Englewood Cliffs, New Jersey (1938), p. 376.
47. H.E. Bennett and W.F. Koehler, "Precision Measurement of Absolute Specular Reflectance with Minimized Systematic Errors," *J. Opt. Soc. Am.* **50,** 1 (1960).
48. R.K. Brookshier, Paper presented at the Optical Society of America meeting, 1966.
49. D.G. Goebel, personal communication, to be published in *Appl. Opt.*
50. E.R. Miller and R.S. VunKannon, "Development and Use of Bidirectional Spectroreflectometer," in *Thermophysics of Spacecraft and Planetary Bodies, Progress in Aeronautics and Astronautics,* Vol. 20 (G.B. Heller, Ed.), Academic Press, New York (1967), pp. 219–223.
51. W.M. Bradenberg and J.T. Neu, "Unidirectional Reflectance of Imperfectly Diffuse Surfaces," *J. Opt. Soc. Am.* **56,** 97 (1966).
52. D.F. Comstock, Jr., "A.D. Little Report to Jet Propulsion Laboratory," Contract No 950867, Subcontract Under NAS7-100, March 1966; Presented at Optical Society of America Meeting, 1967.
53. J.J. Hsia and J.C. Richmond, "Bidirectional Reflectometry, Part I. A High Resolution Laser Bidirectional Reflectometer with Results on Several Optical Coatings," *NBS J. Res.* **80A** 189–206 (1976).
54. W.R. Blevin and J.C. Geist, "Influence of Black Coatings on Pyroelectric Detectors," *Appl. Opt.* **13,** 1171 (1974).
55. E.H. Putley, "The Pyroelectric Detector," in *Semiconductors and Semimetals* (R.K.

Willardson and H.C. Beer, Eds.), Academic Press, New York (1970), Vol. 5, pp. 259–285.
56. B.R. Holeman, "Sinusoidally Modulated Heat Flow and the Pyroelectric Effect," *Infrared Phys.* **12,** 125 (1972).
57. R.J. Phelan, R.J. Mahler, and A.R. Cook, "High D^* Pyroelectric Polyvinyefluoride Detectors," *Appl. Phys. Lett.* **19,** 337 (1971).
58. F.J. Phelan and A.R. Cook, "Electrically Calibrated Pyroelectric Optical-Radiation Detector," *Appl. Opt.* **12,** 2494 (1973).
59. L. Mertz, *Transformations in Optics,* J. Wiley and Sons, Co. New York (1965).
60. F. Kneubuhl, "Diffraction Grating Spectroscopy," *Appl. Opt.* **8,** 505–519 (1969).
61. J.C. DeVos, "A New Determination of the Emissivity of Tungsten Ribbon," *Physica* **20,** 690 (1954).
62. W.H. Askwyth, R.J. Yahes, R.D. House, and G. Mikk, *Determination of Emissivity of Materials,* Vol. I, NASA-CR-56496 (1962), Vol. II NASA-CR-56497 (1963), Vol. III NASA-CR-56498 (1964), U.S. Government Printing Office, Washington, D.C.
63. R.D. House, G.J. Lyons, and W.H. Askwyth, "Measurement of Spectral Normal Emittance of Materials Under Simulated Spacecraft Powerplant Operating Conditions," in *Measurement of Thermal Radiation Properties of Solids* (J.C. Richmond, Ed.), NASA SP-31, U.S. Government Printing Office, Washington, D.C. (1963), pp. 343–355.
64. D.G. Moore, "Investigation of Shallow Reference Cavities for High Temperature Emittance Measurements," in *Measurement of Thermal Radiation Properties of Solids,* (J.C. Richmond, Ed.), NASA SP-31, U.S. Government Printing Office, Washington, D.C. (1963), pp. 515–525.
65. T.R. Riethof and V.J. DeSantis, "Techniques of Measuring Normal Spectral Emissivity of Conductive Refractory Compounds at High Temperatures," in *Measurement of Thermal Radiation Properties of Solids* (J.C. Richmond, Ed.), NASA SP-31, U.S. Government Printing Office, Washington, D.C. (1963), pp. 565–584.
66. A. Gravina, R. Bastian, and J. Dyer, "Instrumentation for Emittance Measurements in the 400 to 1800 F Temperature Range," in *Measurement of Thermal Radiation Properties of Solids* (J.C. Richmond, Ed.), U.S. Government Printing Office, Washington, D.C. (1963), pp. 329–336.
67. J.C. Richmond, W.N. Harrison, and F.J. Shorten, "An Approach to Thermal Emittance Standards," in *Measurement of Thermal Radiation Properties of Solids* (J.C. Richmond, Ed.), NASA SP-31, U.S. Government Printing Office, Washington, D.C. (1963), pp. 403–423.
68. R.D. Larrabee, "The Spectral Emissitivy and Optical Properties of Tungsten," Technical Report 328, AD156602, Research Laboratory of Electronics, MIT, April (1957).
69. R.D. Larrabee, "Spectral Emissivity of Tungsten," *J. Opt. Soc. Am.* **49,** 619 (1959).
70. W.N. Harrison, J.C. Richmond, F.J. Shorten, and H.M. Joseph, "Standardization of Thermal Emittance Measurements," WADC-TR-59-510 pt IV. Air Force Materials Laboratory, Wright-Patterson AFB, Ohio (1963).
71. H.O. McMahon, "Thermal Radiation Characteristics of Some Glasses," *J. Am. Ceram. Soc.* **34** (3), 91 (1951).
72. W.A. Clayton, "A 500° to 4500°F Thermal Radiation Test Facility for Transparent Materials," in *Thermal Radiation Properties of Solids* (J.C. Richmond, Ed.), NASA SP-31, U.S. Government Printing Office, Washington, D.C. (1963), pp. 445–460.
73. A.S. Kjelby, "Emittance Measurement Capability for Temperatures up to 3000°F," in *Measurement of Thermal Radiation Properties of Solids* (J.C. Richmond, Ed.), NASA SP-31, U.S. Government Printing Office, Washington, D.C. (1963), pp. 499–503.

74. W.S. Slemp and W.R. Wade, "A Method for Measuring the Spectral Normal Emittance in Air of a Variety of Materials Having Stable Emittance Characteristics," in *Measurement of Thermal Radiation Properties of Solids* (J.C. Richmond, Ed.), NASA SP-31, U.S. Government Printing Office, Washington, D.C. (1963), pp. 433–439.
75. R.C. Folweiler, "Thermal Radiation Characteristics of Transparent, Semi-Transparent and Translucent Materials Under Non-Equilibrium Conditions," ASD-TDR-62-719, Air Force Materials Laboratory, Wright-Patterson AFB, Ohio (1964).
76. H.F. Clark and D.G. Moore, "A Rotating Cylinder Method for Measuring Normal Spectral Emittance of Ceramic Oxide Specimens from 1200 to 1600°K," *J. Res. NBS* **70A**(5), 394–415 (1966).
77. H.T. Betz, O.H. Olson, B.D. Schurin, and J.C. Morris, "Determination of Emissivity and Reflectivity Data on Aircraft Structural Materials," WADC-TR-56-222, Pt III, Wright Air Development Center, Wright-Patterson AFB, Ohio (1958).
78. D.F. Comstock, "A Radiation Technique for Determining the Emittance of Refractory Oxides," in *Measurement of Thermal Radiation Properties of Solids* (J.C. Richmond, Ed.), NASA SP-31, U.S. Government Printing Office, Washington, D.C. (1963), pp. 461–468.
79. R.J. Evans, W.A. Clayton, and M. Fries, "A Very Rapid 3000°F Technique for Measuring Emittance of Opaque Solid Materials," in *Measurement of Thermal Radiation Properties of Solids* (J.C. Richmond, Ed.), NASA SP-31, U.S. Government Printing Office, Washington, D.C. (1963), pp. 483–488.
80. T.S. Laszlo, R.E. Gannon, and P.C. Sheehan, "Emittance Measurements of Solids Above 2000°C," in *Symposium on Thermal Radiation of Solids* (S. Katzoff, Ed.), NASA SP-55, U.S. Government Printing Office, Washington, D.C. (1965), pp. 277–286.
81. B.A. Peavey and A.G. Eubanks, "Periodic Heat Flow in a Hollow Cylinder Rotating in a Furnace with a Viewing Port," in *Measurement of Thermal Radiation Properties of Solids* (J.C. Richmond, Ed.), NASA SP-31, U.S. Government Printing Office, Washington, D.C. (1963), pp. 553–563.
82. T. Limperis, D.M. Szeles, and W.L. Wolfe, "The Measurement of Total Normal Emittance of Three Nuclear Reactor Materials," in *Measurement of Thermal Radiation Properties of Solids* (J.C. Richmond, Ed.), NASA SP-31, U.S. Government Printing Office, Washington, D.C. (1963), pp. 357–364.
83. G.L. Abbott, "Total Normal and Total Hemispherical Emittance of Polished Metals," in *Measurement of Thermal Radiation Properties of Solids* (J.C. Richmond, Ed.), NASA SP-31, U.S. Government Printing Office, Washington, D.C. (1963), pp. 293–306.
84. A. Cezairliyan, "Design and Operational Characteristics of a High-Speed (Millisecond) System for the Measurement of Thermophysical Properties of High Temperatures," *NBS J. Res.* **75C**(1), 7–18 (1971).
85. G.M. Foley, "High Speed Optical Pyrometer," *Rev. Sci. Instr.* **41**, 827 (1970).
86. A. Cezairliyan, M.S. Morse, H.A. Berman, and C.W. Beckett, "High-Speed (Subsecond) Measurement of Heat Capacity, Electrical Resistivity, and Thermal Radiation Properties of Molybedenum in the Range 1900 to 2800°K," *J. Res. NBS* **74A**(1), 65–92 (1970).
87. A. Cezairliyan, J.L. McClure, and C.W. Beckett, "High-Speed (Subsecond) Measurement of Heat Capacity, Electrical Resistivity, and Thermal Radiation Properties of Tantalum in the Range 1900 to 3200°K," *J. Res. NBS* **75A**(1), 1–13 (1971).
88. A. Cezairliyan and J.L. McClure, "High-Speed (Subsecond) Measurement of Heat Capacity, Electrical Resistivity, and Thermal Radiation Properties of Tungsten in the Range 2000 to 3600°K," *J. Res. NBS* **75A**(4), 283–290 (1971).

89. A. Cezairliyan, "High-Speed (Subsecond) Measurement of Heat Capacity, Electrical Resistivity, and Thermal Radiation Properties of Niobium in the Range 1500 to 2700°K," *J. Res. NBS* **75A** (6), 565–571 (1971).
90. A. Cezairliyan and J.L. McClure, "Thermophysical Measurements on Iron Above 1500 K, Using a Transient (Subsecond) Technique," *J. Res. NBS* **78A** (1), 1–4 (1974).
91. A. Cezairliyan, F. Righini, and J.L. McClure, "Simultaneous Measurements of Heat Capacity, Electrical Resistivity, and Hemispherical Total Emittance by a Pulse Heating Technique: Vanadium 1500 to 2100°K," *J. Res. NBS* **78A**(2), 143–146 (1974).
92. A. Cezairliyan and F. Righini, "Simultaneous Measurements of Heat Capacity, Electrical Resistivity and Hemispherical Total Emittance by a Pulse Heating Technique: Zirconium, 1500 to 2100°K," *J. Res. NBS* **78A**(4), 509–514 (1974).
93. A. Cezairliyan and F. Righini, "Measurements of Heat Capacity, Electrical Resistivity and Hemispherical Total Emittance of Two Grades of Graphite in the Range 1500 to 3000°K by a Pulse Heating Technique, *Rev. Int. Hautes Temp. Refract.* **12,** 124–131 (1975).
94. A. Cezairliyan and A.P. Miller, "Heat Capacity and Electrical Resistivity of Titanium in the Range 1500 to 1900°K by a Pulse Heating Method," *High Temp. High Pressures* **9,** 319–324 (1977).
95. A. Cezairliyan, "High-Speed (Subsecond) Simultaneous Measurement of Specific Heat, Electrical Resistivity, and Hemispherical Total Emittance of Ta–10 (Wt.%) W Alloy in the Range 1500 to 3200°K," *High Temp. High Pressures* **4,** 541–550 (1972).
96. A. Cezairliyan, "Simultaneous Measurement of Specific Heat, Electrical Resistivity, and Hemispherical Total Emittance of Niobium 1 (Wt. %) Zirconium Alloy in the Range 1500 to 2700°K by a Transient (Subsecond) Technique," *J. Res. NBS* **77A** (1), 45–48 (1973).
97. A. Cezairliyan, "Simultaneous Measurement of Heat Capacity, Electrical Resistivity, and Hemispherical Total Emittance of an Alloy of Niobium, Tantalum and Tungsten in the Range 1500 to 2800°K," *J. Chem. Thermodynamics* **6,** 735–742 (1974).
98. A. Cezairliyan and J.L. McClure, "Simultaneous Measurements of Specific Heat Electrical Resistivity, and Hemispherical Total Emittance by a Pulse Heating Technique: Hafnium-3 (Wt. %) Zirconium, 1500 to 2400°K," *J. Res. NBS* **79A**(2), 431–436 (1975).
99. A. Cezairliyan, J.L. McClure, and R. Taylor, "Thermophysical Measurements on 90 Ti-6 Ae-4V Alloy Above 1400°K Using a Transient (Subsecond) Technique," *NBS J. Res.* **81A,** 251–256 (1977).
100. A. Cezairliyan and A.P. Miiller, "Melting Point, Normal Spectral Emittance (at the Melting Point) and Electrical Resistivity (Above 1900K) of Titanium by a Pulse Heating Method," *J. Res. NBS* **82,** 119–122 (1977).
101. A. Cezairliyan, "Measurement of Melting Point, Normal Spectral Emittance (at Melting Point) and Electrical Resistivity (Near Melting Point) of Some Refractory Alloys," *J. Res. NBS* **78A**(1), 5–8 (1974).
102. A.G. Worthing, "Temperature Radiation Emissivities and Emittances," in *Temperature and its Control in Science and Industry,* Am. Inst. of Physics–Reinhold Publishing Corp., New York (1941), pp. 1164–1187.
103. J.C. Richmond and W.N. Harrison, "Equipment and Procedures for Evaluation of Total Hemispherical Emittance," *Bull. Am. Ceram. Soc.* **39**(11), 668–673 (1960).
104. D.L. McElroy and T.G. Kollie, "The Total Hemispherical Emittance of Platinum, Colombium–1% Zirconium, and Polished and Oxidized INOR-8 in the Range 100° to 1200°C," in *Measurement of Thermal Radiation Properties of Solids* (J.C. Richmond, Ed.), NASA SP-31, U.S. Government Printing Office, Washington, D.C. (1963), pp. 365–379.

105. K.E. Nelson and J.T. Bevans, "Errors of the Calorimetric Method of Total Emittance Measurement," in *Measurement of Thermal Radiation Properties of Solids* (J.C. Richmond, Ed.), NASA SP-31, U.S. Government Printing Office, Washington, D.C. (1963), pp. 55–65.
106. T.W. Nyland, "Apparatus for the Measurement of Hemispherical Emittance and Solar Absorptance from 270° to 650°K," in *Measurement of Thermal Radiation Properties of Solids* (J.C. Richmond, Ed.), NASA SP-31, U.S. Government Printing Office, Washington, D.C. (1963), pp. 393–401.
107. W.B. Fussell, J.J. Triolo, and J.H. Henninger, "A Dynamic Thermal Vacuum Technique for Measuring the Solar Absorption and Thermal Emittance of Spacecraft Coatings," in *Measurement of Thermal Radiation Properties of Solids* (J.C. Richmond, Ed.), NASA SP-31, U.S. Government Printing Office, Washington, D.C. (1963), pp. 83–101.
108. R.F. Gaumer and J.V. Stewart, "Calorimetric Determination of Infrared Emittance and the α_s/ϵ Ratio," in *Measurement of Thermal Radiation Properties of Solids* (J.C. Richmond, Ed.), NASA SP-31, U.S. Government Printing Office, Washington, D.C. (1963), pp. 127–133.
109. C.P. Butler and R.J. Jenkins, "Space Chamber Emittance Measurements," in *Measurement of Thermal Radiation Properties of Solids* (J.C. Richmond, Ed.), NASA SP-31, U.S. Government Printing Office, Washington, D.C. (1963), pp. 39–43.
110. R.J. Jenkins, C.P. Butler, and W.J. Parker, "Total Hemispherical Emittance Measurements Over the Temperature Range 77° to 300°K," USNRDL-TR-663, AD-419067 National Technical Information Service, Springfield, Virginia (1963).
111. R.P. Caren, "Cryogenic Emittance Measurements," in *Measurement of Thermal Radiation Properties of Solids* (J.C. Richmond, Ed.), NASA SP-31, U.S. Government Printing Office, Washington, D.C. (1963), pp. 45–49.
112. M.M. Fulk, M.M. Reynolds, and O.E. Park, "Thermal Radiation Absorption by Metals," in *Advances in Cryogenic Engineering* (K.O. Timmerhaus, Ed.), Plenum Press, New York (1960), pp. 224–229.
113. G.L. Haury, "An Apparatus for the Measurement of Total Hemispherical Emissivity and Thermal Conductivity Between Ambient and Liquid Nitrogen Temperatures," ASD-TDR-63-146, 16 pp. (1960), AD411140, National Technical Information Service, Springfield, Virginia.
114. G.L. Haury, "An Apparatus for Measuring Total Hemispherical Emittance Between Ambient and Liquid Nitrogen Temperatures," in *Measurement of Thermal Radiation Properties of Solids* (J.C. Richmond, Ed.), NASA SP-31, U.S. Government Printing Office, Washington, D.C. (1963), pp. 51–54.
115. R.P. Caren, "Low Temperature Emittance Determinations," in *Progress in Astronautics and Aeronautics,* Vol. 18 (G.B. Heller, Ed.), Academic Press, New York (1966), pp. 61–73.
116. A. Biondi, "Optical Absorption of Copper and Silver at 4.2°K," *Phys. Rev.* **102,** 964–967 (1956).

VII

THERMOPHYSICAL PROPERTY STANDARD REFERENCE MATERIALS

20

Certified Reference Materials for Thermophysical Properties

R.K. KIRBY

Reference materials for use in calibrating either the temperature scale of equipment or a physical property measured by the equipment as a function of temperature are available from certifying agencies in at least five countries. These reference materials are certified for properties that include thermal conductivity, electrical resistivity, heat capacity, thermal expansion, and freezing and melting points. Certified reference materials (CRMs) are also useful in developing new test methods and ensuring the compatibility of measurements on a worldwide basis.

A certified reference material is a homogeneous lot of material that has been certified by an authorized agency to have a stated composition, structure, and/or property. The value(s) certified is generally the best estimate of the "true" value for that lot of material as determined by careful measurements and is as free of random and systematic errors as is practical for its intended use. Certification generally includes the estimated uncertainty of the certified value.

CRMs for selected thermophysical properties are available from the following sources:

Bündesanstalt für Materialprüfung Unter den Eichen 87 D-1000 Berlin 45 Federal Republic of Germany	BAM
Gosstandart of the USSR 9, Leninsky Prospekt 117049, Moscow USSR	GOST

R.K. KIRBY • National Bureau of Standards, Washington, D.C. 20234.

National Bureau of Standards NBS
Office of Standard Reference Materials
Washington, DC 20234
USA

National Physical Laboratory NPL
Office of Reference Materials
Teddington, Middlesex TW11 0LW
UK

Service des Matériaux de Référence SMR
1, rue Gaston Boissier
75015 Paris
FRANCE

The following material provides short descriptions of the CRMs that are available according to the ISO Directory of Certified Reference Materials,[1] the 1976 Catalogue of Reference Materials by IUPAC,[2] as well as recent catalogs that were available from some of the above sources.

ELECTRICAL RESISTIVITY

Source	Material	Temperature range (K)	Value* at 293 K ($\mu\Omega$ m)	Uncertainty (%)
NBS	Iron	4–1000	0.10	2
NBS	Stainless steel	5–1200	0.81	2
NBS	Tungsten	4–3000	0.05	2

HEAT CAPACITY

Source	Material	Temperature range (K)	Value* at 293 K (J mol^{-1} K^{-1})	Uncertainty (%)
GOST	4 CRMs including potassium nitrate	173–700	Not known	Not known
GOST	Quartz	293–900	45	Not known
GOST	Molybdenum	293–2600	24	Not known
GOST	Sapphire	293–2700	73	Not known
NBS	Copper	1–300	24	1
NBS	Polystyrene	10–350	124	0.2
NBS	Polyethylene	5–360	22	1
NBS	Sapphire	5–2250	73	0.3
NBS	Molybdenum	293–2800	24	0.5
SMR	Copper	350–1300	24	3
SMR	Platinum	298–2000	26	Not known

*Nominal value.

THERMAL CONDUCTIVITY

Source	Material	Temperature range (K)	Value* at 293 K ($W m^{-1} K^{-1}$)	Uncertainty (%)
BAM	Platinum	273–373	73	Not known
GOST	Low-carbon steel	293–900	65	Not known
GOST	Cr–V–Mg alloy	293–900	9	Not known
NBS	Fiberglass board	255–330	0.03	2
NBS	Iron	4–1000	78	3
NBS	Stainless steel	5–1200	14	5
NBS	Tungsten	4–3000	172	5
SMR	Fiberglass board	273–350	0.03	Not known
SMR	Copper	230–850	375	3
SMR	Platinum	293–1273	73	Not known

THERMAL EXPANSION

Source	Material	Temperature range (K)	Value* at 293 K ($10^{-6} \cdot K^{-1}$)	Uncertainty (%)
GOST	Molybdenum	293–2800	5	Not known
NBS	Borosilicate glass	80–680	5	3
NBS	Vitreous quartz	80–1000	0.5	12
NBS	Tungsten	80–1800	4	1
SMR	Invar	273–350	0.4	Not known
SMR	Copper	323–873	18	3
SMR	Vitreous quartz	293–1300	0.5	Not known
SMR	Pyros	293–1400	Not known	Not known
SMR	Zircon	193–1600	3	Not known
SMR	Magnesia	193–1800	10	Not known

THERMOMETRIC FIXED POINTS

Source	Material	Temperature* (K)	Uncertainty (%)
BAM	Tin	505	Not known
BAM	Zinc	693	Not known
BAM	Ag–Cu eutectic	1053	Not known
BAM	Cadmium	594	Not known
BAM	Aluminum	933	Not known
BAM	Sulfur	718	Not known
BAM	Silver	1235	Not known
BAM	Gold	1338	Not known
GOST	Copper	1328	Not known

*Nominal value.

Source	Material	Temperature* (K)	Uncertainty (%)
GOST	Corundum	2326	Not known
GOST	Molybdenum	2896	Not known
GOST	W–Re	Thermoelectric force	Not known
GOST	Cu–Ni	Thermoelectric force	Not known
NBS	Mercury	234	0.0001
NBS	Gallium	303	0.0002
NBS	Tin	505	0.0001
NBS	Lead	600	0.0008
NBS	Zinc	693	0.0001
NBS	Aluminum	933	0.021
NBS	Copper	1358	0.037
NBS	Alumina	2326	0.22
NBS	Superconducting device (high range)		0.02
	Lead	7.2	
	Indium	3.4	
	Aluminum	1.2	
	Zinc	0.8	
	Cadmium	0.5	
NBS	Superconducting device (low range)		0.2
	Gold–indium	0.205	
	Gold–aluminum	0.157	
	Iridium	0.098	
	Beryllium	0.024	
	Tungsten	0.015	
NPL	4-Nitrotoluene	325	0.1
NPL	Napthalene	353	0.1
NPL	Benzil	368	0.1
NPL	Acetanilide	388	0.1
NPL	Benzoic acid	396	0.1
NPL	Diphenylacetic acid	420	0.1
NPL	Anisic acid	456	0.1
NPL	2-Chloroanthraquinone	483	0.1
NPL	Carbazole	519	0.1
NPL	Anthraquinone	558	0.1
NPL	Indium	430	Not known
NPL	Water triple point cell	273	Not known
SMR	Helium	2	Not known
SMR	Oxygen	54	Not known
SMR	Argon	84	Not known
SMR	Silver	1235	Not known
SMR	Gold	1338	Not known
SMR	Palladium	1827	Not known

*Nominal value.

It should be obvious that the above list is not complete and it is hoped that more information will eventually be available in the REMCO Directory.[1]

Information about unit size, availability, and price should be obtained directly from the source of each CRM. Information can also be obtained from each source about activities on the certification of new CRMs.

REFERENCES

1. *Directory of Certified Reference Materials* (CRM): Sources of Supply and Suggested Uses, ISO/REMCO Directory, 1982, Prepared and Disseminated by International Organization for Standardization, Case Postale 56, CH-1211, Geneve 20, Switzerland.
2. "Physicochemical Measurements: Catalogue of Reference Materials from National Laboratories," International Union of Pure and Applied Chemistry, *Pure Appl. Chem.* **48**, 503–515 (1976).

Index